# THE SEED IS MINE

Kas Maine

*Charles van Onselen*

CHARLES VAN ONSELEN

# THE SEED IS MINE

*

The Life of Kas Maine, a South African Sharecropper

1894–1985

HILL AND WANG

A DIVISION OF FARRAR, STRAUS AND GIROUX

NEW YORK

Printed in the United States of America
Published simultaneously in Canada by HarperCollins*CanadaLtd*
First edition, 1996
First Hill and Wang paperback edition, 1997

LIBRARY OF CONGRESS CATALOG CARD NUMBER
95-079980

# PREFACE AND ACKNOWLEDGMENTS

My father, the oldest son in an eastern Cape family of seven children, joined the South African Police during the Great Depression. In the early 1950s, after two decades of highly successful service, the incoming Afrikaner Nationalist government decided that there was no room in the force for career policemen who did not subscribe to their political vision. His career came to an abrupt end and he left the police. He then found employment in the estates division of one of the smaller Rand mining houses, and spent the next twenty years of his life as stand-in landlord for the company, overseeing several large properties in the southwestern Transvaal and northern Orange Free State. Much of my childhood and all of my adolescence was spent in small mining towns—mere industrial islands set amidst seas of green maize that were ruled by bands of rugged Afrikaner farmers. To the extent that I have roots anywhere, I am a son of the South African highveld.

I know that there are other, more attractive, verdant, and densely settled parts of the country. South Africa has a narrow, fairly well-watered east coast littoral where thick bush and occasional forest is the historic home of indigenous Nguni-speakers such as the Xhosa and the Zulu and many nineteenth-century European settlers. But an escarpment, which curls from north to south divides this 'wet' eastern third of the country from the dry western two-thirds, and behind this lies the highveld—the elevated plateau that dominates the interior. Between five and six thousand feet above sea level and tilting gently towards the distant Atlantic, the highveld's scattered acacia bush and thin grasslands eventually concede defeat to a demanding climate and make way for the Kalahari Desert. These hot, dry and expansive plains of the subcontinent provide the country with its most characteristic landscape, and a distinctive terminology which singles out for special attention any geomorphological feature which so much as hints at either height or water. *Koppie, krans, pan, platteland, rand, sloot, spruit, vlakte,* and *vlei*—are all Afrikaans words that defy easy translation and which have been incorporated holus-bolus into the vocabulary of all urban South Africans. They are quintessentially highveld

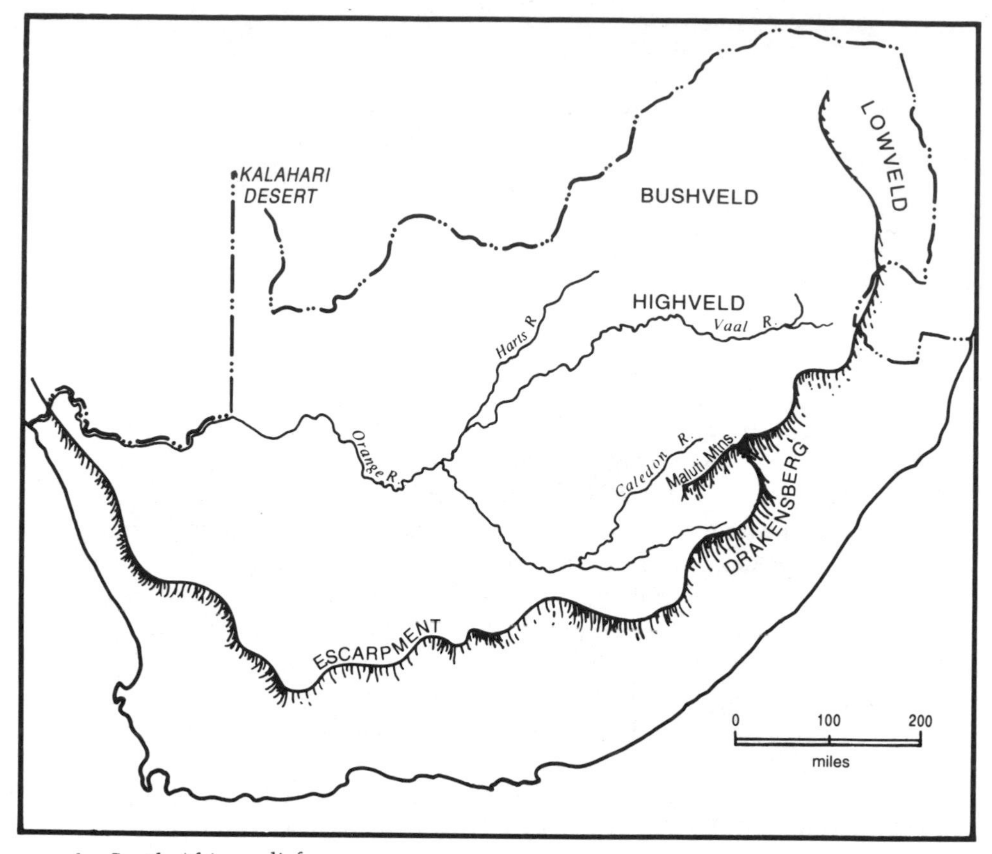


1 South Africa, relief

words; words that give life to the ways in which one senses and experiences much of what it is to be South African.

The highveld has, for more than a century, also been the site of some of the most intense, intimate and searing interactions between Afrikaner landlords and their predominantly Sotho-speaking labour tenants and farm labourers. Bitter-sweet relationships born of paternalism and unending rural hardship have seen the emergence of peculiar quasi-kinship terms such as *outa* (venerable father) and *ousie* (older sister)—nouns that are as embedded in the modern SeTswana lexicon as they are in Afrikaans. When an authentic South African identity eventually emerges from this troubled country it will, in large part, have come from painful shared experiences on the highveld.

Only twenty-five years ago most South Africans still lived and laboured in the countryside. Now noisy, young, mass-schooled, semi-literate urban insiders dominate the political order and pointedly ignore our rural origins, the lingering ideologies of the highveld, and experiences that moulded an often silent, bemused and largely illiterate older generation. Those intent on building our future around industry may do well to pause and reflect on the fact that we live in times when the field has barely given way to the factory, the peasant to the proletarian, and the patriarch to his family.

This book—a work of biography—seeks to establish some of the deep-seated personal, psychological, social and structural reasons that underlay one family's gradual move away from the highveld into the twilight world of labour migrancy, peri-urban space and industry. It does so knowing, as the German historian Meinecke warned us more than half a century ago, that 'behind the search for causalities there always lies, directly or indirectly, the search for values.' Contemporary South African values evoke hope and despair in equal measure. Perhaps there could be no other way for we are in the adolescence of our nationhood.

I have explored some of the factors that govern the attitudes, beliefs, cultural practices and values of South Africa's highveld inhabitants elsewhere—notably in essays which appeared in *The American Historical Review* (1990) and in the *Journal of Historical Sociology* (1992). Readers interested in the theoretical issues that underpin this narrative of Kas Maine's life, such as the ideology of paternalism, are invited to consult these articles. Likewise, those wishing to engage with the hidden methodological assumptions that inform this study as well as the real and imagined limitations of oral history as a research technique might turn to *The Journal of Peasant Studies* (1993). But writing articles and writing a book—especially a rather lengthy book—are two very different exercises.

*

We live in an era dominated by the state, big business and the mass media. They are, we are told, great supporters of the arts and social sciences. Maybe. In my experience, however, statutory funding agencies serve the government of the day far too zealously, and large corporations often hire narrow eyes and silky tongues to protect their 'social responsibility' budgets from academic projects with unpredictable outcomes. Historians could do with a few of the monarchs, monks and madmen of yesteryear to patronise the pursuit of ideas. Progress has its down side. But I have been lucky. This book has been generously supported by many foundations, institutions and their intermediaries.

Back in 1979 Michael O'Dowd and the Anglo American & De Beers' Chairman's Fund provided seed-money for a research programme in oral history based at what was then the African Studies Institute at the University of the Witwatersrand. Jean Copans and the Centre D'Etudes Africaines at the Ecole des Hautes Etudes en Sciences Sociales gave me time and space to work through some ideas with colleagues in Paris in 1991. When it wasn't always easy, Bill Carmichael and the Ford Foundation were willing to be associated with a white man in a city that abounded with more politically correct causes. The Social Science Research Council and the American Council of Learned Societies' Joint Committee on African Studies came to my rescue when all work on this project had ground to a halt. And the managers of the Smuts Memorial Fund in Cambridge, England, helped in the most generous way possible. They underwrote an Overseas Visiting Fellowship for me during 1989–90; Michael Allen at Churchill College and John Lonsdale of Trinity made my stay in Cambridge as pleasant as it was productive.

But if half of this book originated in a room overlooking the quad at Churchill College, the other half emerged around Sutton Close. My greatest debt is to the University of the Witwatersrand. Not once did my research efforts over a decade and a half receive less than wholehearted support from it and its most senior office-bearers. Friedel Sellschop, in particular, was exemplary in his patience.

Institutions pay the bills; people do the work. At Wits's Institute for Advanced Social Research I have been spurred on by a group of remarkably warm and supportive colleagues. Russell Ally, Celeste Emmanuel, Arlene Harris, Paul la Hausse, Jeanette Kruger, Steve Lebelo, Nita Lelyveld, Hosea Mahlobane, Ephraim Msimango, Theti Matsetela and Karin Shapiro all helped in ways too numerous to mention. David Goldblatt and Santu Mofokeng do not need me to sing their praises, for their photographs speak for themselves. I value the friendship, professionalism and support of all these people more than they can imagine. My only regret is that two of our number—the late Thomas Matsetela and Moss Molepo—brother historians lost amidst the mindless battles of a bloody society, did not live to see the conclusion of this project. I like to think that they would have approved of a work they helped to shape and they would have been first to understand why I want to single out Thomas Nkadimeng for special praise. Historical research is for stayers rather than sprinters. For fourteen years Thomas Nkadimeng lent me his eyes and his ears. He never hesitated to crisscross the Transvaal in search of data that might have struck him as boring, crass, irrelevant, insensitive, repetitive, vague or plain stupid. On one occasion during South Africa's long undeclared civil war, he was woken by a nervous landlord prodding him in the head with a double-barrelled shotgun. It is my hope that, through our efforts in bad times, we may just have created something of value that our grandchildren can share in better times.

Family-centred research is both intrusive and interactive. At a turbulent time in their own lives and in South African history, the Maines answered a knock, opened their doors, and allowed strangers to wander round their homes at will. Those who possess least in material terms often give most by way of friendship, hospitality and kindness; but the Maine miracle surpasses what one expects of good hosts. Their courage, honesty and commitment to the pursuit of historical knowledge through all its painful twists and turns are all extraordinary. Sons and daughters of a formidable father, they acted in the tradition of Kas, who once observed, 'I have always differed in spirit from others insofar as I wanted to preserve knowledge, values and wisdom.'

I would also like to thank the many doctors, farmers, labour tenants, landlords, lawyers, magistrates, politicians, sharecroppers, shopkeepers and traders from Bloemhof, Schweizer-Reneke and Wolmaransstad who assisted me in the research for this book. The Triangle is part of South Africa's Deep South. Its Asians, blacks, coloureds and whites all had complex reasons for fearing the changes in South African society that were only just becoming apparent when this project was coming to fruition. They, too, took me in and willingly shared insights that sharpened my understanding of their lives between the two world wars. If any of the characters in this work shake themselves free of the word-shackles that bind them to the page and walk three-dimensionally into the reader's mind, it is a tribute to the story-telling abilities of these informants. Oom Schalk Lourens and his friends remain amongst the best social historians in South Africa.

The task of writing biography—history without boundaries—is a daunting one. In consequence my debts to friends and family have grown out of all proportion during the composition of this book. Stanley Trapido's ability to read societies comparatively, stand back, and then ask wonderful questions about South African history is, in my experience, without parallel. His answers are models of clarity and professionalism, while his ear for accents of person, place and moment makes him the most persuasive historian I know. This book is dedicated to him. Without his encouragement, friendship, support and vision this project would simply never have come to fruition. He and the ever-generous Barbara—along with Anna and Joe—have made Southmoor Road and Oxford a refuge. This book, like its precursors, was talked through in the Trapido kitchen long before it was written.

I am also indebted to another friend and historian. Shula Marks, whose vision pierces the mists of historical obscurity better than most of ours, has always actively encouraged an interest in topics that lie on the margins of the discipline. She, along with Stanley Trapido, set the standards for the history writing I aspire to.

Stanley Greenberg is the only one in my circle who had the wit to abandon universities completely, set up a business, and further the interests of his family. He retains a residual sympathy for those of us remaining in academe, and has always had the enviable knack of knowing what would appeal to me even before I realised it. He eased me into the world of Nate Shaw and, by so doing, unwittingly set me on a road that led to Kas Maine. American historians know more about sharecropping than I ever will, and I am grateful for the advice, guidance and encouragement of

Paul Gaston, Eugene D. Genovese, George Frederickson and Theodore Rosengarten.

Several people were good enough to read all or parts of the manuscript prior to publication or to help in other important ways. Josie Adler, Keith Beavon, Jim Campbell, Anthony Costa, Peter Delius, Hermann Giliomee, Georgina Hamilton, Phyllis Lewsen, Bruce Murray, Graham Neame, Richard Rathbone, Phillip Stickler, Tony Traill and Gavin Williams all made valiant efforts to save me from myself. Ivan Vladislavić was smarter and tried to save the readers; giving incisive written comment early on, and suggesting dozens of ways in which I might ease the readers' burden. I am also indebted to Eve Horwitz and—especially—Pat Tucker.

My friend Joseph Lelyveld did a wonderful thing by introducing me to Elisabeth Sifton. She just has to be the best editor in the business. Her understanding of what I was trying to achieve was total, her ability to bring out clearly the cadences buried deep beneath my stilted prose, unerring.

I made extraordinary demands on three other friends whom I asked to comment on early drafts, a task that tests friendship to breaking point. They are still my friends and I want to emphasise how much I value their support. Helen Bradford asks unnerving questions about interpretation I cannot answer. I have agonised over her criticisms and observations and tried to incorporate my muttered 'replies' in the text.

Ian Phimister understands my terror of the black dog and has often given me a safe haven in Cape Town. He is also as fine a historian and as honest a reader as an author is ever likely to encounter. His forthright answers to embarrassing questions and his unwavering attention to detail helped me to avoid many self-indulgent interpretations or unnecessary errors.

Johannesburg is an awful setting for anyone in search of academic peace. Visions of ivory towers are, for the most part, the by-products of executive business lunches or of the smoking habits of students. In order to survive intellectually one has to find a star to steer by, and follow the direction of one's choice. The social injustices of apartheid have left their mark on South African university life, and there are always more 'relevant' things to do than write.

Tim Couzens is a star. His dedication to creative research and writing of the highest standard is absolute. Whenever my own commitment wavered, he reminded me of my initial choice. Offering me more than good advice, he has taken on scores of tasks that should have fallen to me so that I might have more time to write—and this at a time when he had work of his own to do.

If research and writing is hard on friends, it can be positively destructive when it comes to families. Belinda Bozzoli holds our family together and somehow combines that with a demanding career of her own. As an historian I recommend marriage to a sceptical sociologist with a formidable intellect. 'The more sociological history becomes, and the more historical sociology becomes,' E. H. Carr once observed, 'the better for both.' I don't know that this is what he had in mind, but I could not have asked for a better, more loyal, or tolerant tutor.

Gareth, Jessica and Matthew—children of the late twentieth century who will live out most of their lives in the twenty-first—have long puzzled over my obsession with Kas Maine. Let me make one last attempt at providing them with an answer by quoting Tacitus. 'It seems to me,' he argued, that 'a historian's foremost duty is to ensure that merit is recorded, and to confront evil deeds and words with the fear of posterity's denunciations.' It is my hope that in the story of the life and times of Kas, they will see an attempt by one historian to accommodate that dictum.

C. v. O.
*Johannesburg*
*July 1995*

*For Stan*

# CONTENTS

MAPS

*The seed is mine. The ploughshares are mine.*
*The span of oxen is mine. Everything is mine.*
*Only the land is theirs.*

—KAS MAINE

# THE SEED IS MINE

# Introduction

This is a biography of a man who, if one went by the official record alone, never was. It is the story of a family who have no documentary existence, of farming folk who lived out their lives in a part of South Africa that few people loved, in a century that the country will always want to forget. The State Archives, supposedly the mainspring of the nation's memory, has but one line referring to Kas Maine. The Register of the Periodic Criminal Court at Makwassie records that on 8 September 1931, a thirty-four-year-old 'labourer' from Kareepoort named 'Kas Tau' appeared before the magistrate for contravening Section Two, Paragraph One of Act 23 of 1907. A heavy bound volume reveals that 'Tau,' resident of Police District No. 41, was fined five shillings for being unable to produce a dog licence. Other than that, we know nothing of the man.

Life transcends bureaucratic notation and legal formulations, however. Words—no matter how precisely chosen—mislead, phrases obscure, and sentences deceive. The man's name was Ramabonela Maine. But depending on when and where you met him in a life that spanned ninety-one years, he was—without ever wishing to deceive—also Kasianyane Maine, Phillip Maine, Kas Deeu, Kas Teeu, Kas Teu or just 'Old Kas.' Born in 1894 into a BaSotho family that had emigrated to the Transvaal a decade earlier, he had been raised amongst SeTswana-speakers on farmland belonging to white landlords who, depending on the protagonist's agenda, were described variously as 'Afrikaners,' 'Boers' or 'Dutchmen.' By culture part-MoSotho, part-MoTswana, part-Afrikaner, the one thing that Maine most certainly was not was a 'labourer.' With the exception of some casual work undertaken as a young man he never 'laboured' for anyone. In an industrialising state founded on mineral wealth, which for the first fifty years of its existence devoted most of its effort to rendering African labour cheap, docile and plentiful, Kas Maine retained his economic independence. Kas worked for no man—black or white.

It was never easy. At the time of his arrest at Kareepoort in 1931 he seemed to be a relatively well-off tenant farmer. A fine of five shillings did not tax him unduly. Twenty-four months later, the Great Depression had reduced him to the point where 'I was starving, my cattle were starving,' and 'the horse and the children ate from the same last half-bag of grain.' From this low point in his career he used his skills as blacksmith, cobbler, grain farmer, herbalist, leather-worker, tailor, thatcher and livestock speculator to lift him to the position where in 1948—the year that the Afrikaners' National Party gained political ascendancy in South Africa—he, a landless black sharecropper in the country of his birth, reaped a thousand bags of sorghum and five hundred and seventy bags of maize. A harvest on this scale conjured up from poor soil in an area of low rainfall was a spectacular achievement.

But in racist societies, where dominant groups greeted failure with silent satisfaction because they fitted stereotypes, success, too, could be fatal for a black man. In South Africa a misplaced emphasis on the supposed 'failures' of blacks perpetuated a vision of Africans as victims. But while the tribulations of the vulnerable in the face of oppression cry out for sympathy—a response of the heart—the state's capacity to propagate evil can be better understood by demonstrating how it destroyed the strong and the resilient. A family like the Maines—Kas, two wives, three sons, and six daughters—was condemned by a racist state for its economic achievements rather than its supposed social shortcomings. Along with thousands of other sharecropping families on the Transvaal highveld in the years between the world wars, the Maines had the misfortune of reaching the zenith of their productive powers at a critical historical juncture, namely at that point where old-world paternalism and the social proximity of master and servant on white farms were giving way to virulent forms of racism associated with pseudo-science, physical distance and technological progress.

Old and new social forces, contested in surprising ways in the day-to-day interactions between black and white folk who lived in the Transvaal, lie at the core of this book. Currents of anger, betrayal, hatred and humiliation surge through many accounts of modern South Africa's race relations, but what analysts sometimes fail to understand is that without prior compassion, dignity, love or a feeling of trust—no matter how small, poorly, or unevenly developed—there could have been no anger, betrayal, hatred or humiliation. The troubled relationship of black and white South Africans cannot be fully understood by focusing on what tore them apart and ignoring what held them together. The history of a marriage, even an unhappy one, is inscribed in the wedding banns as well as the divorce notice.

*

For most South Africans, let alone for people drawn from farther afield, the Triangle of land between the small southwestern Transvaal towns of

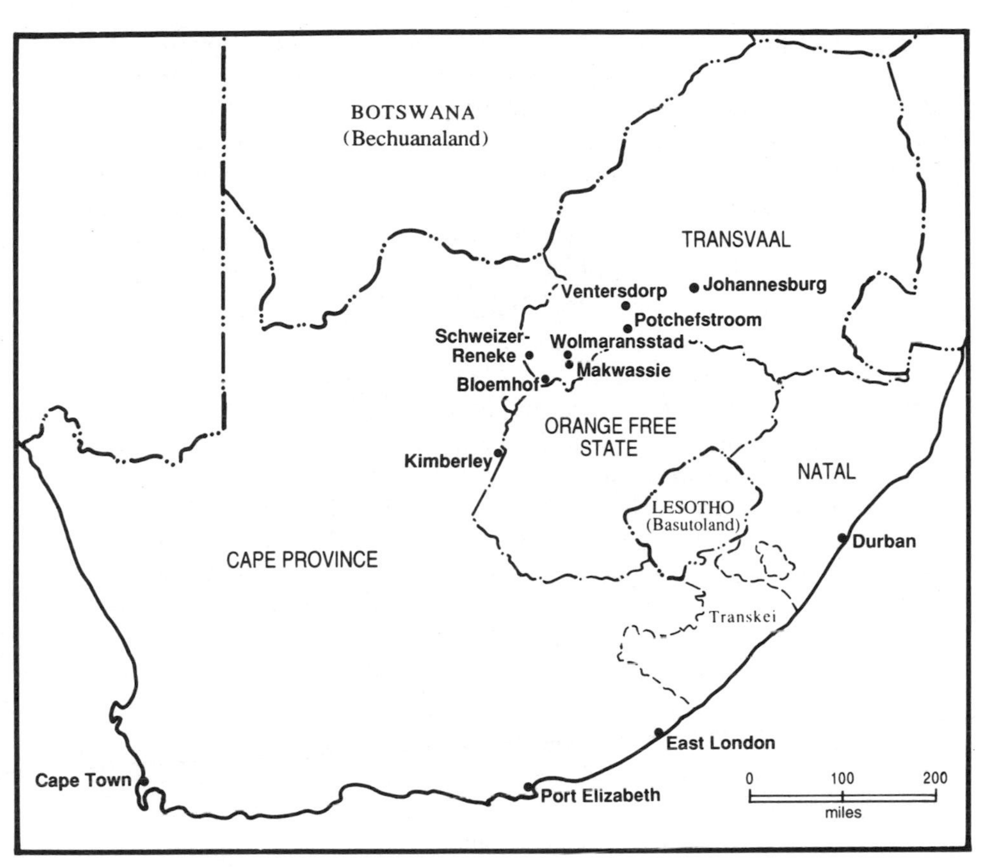


2 South Africa, political

Bloemhof, Schweizer-Reneke and Wolmaransstad forms part of a nondescript expanse reaching westward to the Kalahari Desert. As a resentful reporter sent there by an editor of a national daily noted recently, 'towns don't come more unremarkable than Bloemhof.'[1] Situated half-way between the gold mines of Johannesburg to the north and the diamond fields of Kimberley to the south, these country towns—*platteland dorps*—represent little more than names on a map fleetingly encountered out of the corner of an eye in the course of an uncomfortable car journey. As small farming centres on the periphery of the bigger maize-producing country to the south and east, the three towns figure almost as poorly in South African history books as they do in travel guides. Yet for hundreds of black sharecropping families, the hot dusty plains of the Triangle formed the very centre of their universe between the two world wars.

Sharecropping had its origins in the industrial revolution triggered by diamond discoveries in the northwestern Cape Colony in the late 1860s. White landlords and black tenants entered into verbal agreements to share harvests in proportion to the economic inputs they made to the farms. The emergence of Kimberley as a mining centre meant that for the first time black peasants and white farmers on the highveld had easy access to a rapidly expanding market for their products. Among those quick to respond to new opportunities and expand their production were the SeSotho-speakers of the Maluti Mountains and the better-watered plains a long way east of Kimberley. By the 1870s, Basutoland—now Lesotho—was said to be the 'granary of the Free State and parts of the Cape Colony.'[2] For various reasons—including land dispossession in the upper reaches of the Caledon River, the arrival of the railroad, the importation of cheaper grain from the United States of America and an economic recession—BaSotho agricultural production declined in the mid 1880s. Then, just as grain prices reached their lowest point, the discovery of gold on the far off Witwatersrand in 1886 ensured the development of an even larger urban market to the north and west.[3] With interest in the established but sluggish Cape markets waning, the new opportunities north of the Vaal River attracted first the white farmers in the areas of higher rainfall of the eastern Orange Free State and then the Boer newcomers in the drier northwestern regions of thc highveld plateau. But while conquest made the white farmers rich in land, they were less well endowed with the capital, draught oxen, harrows, labour, and ploughs necessary to expand grain production. At about the same time, land-hungry black peasants from western Basutoland, the northern Cape Colony and the northwestern Transkei sought new pastures for their livestock and were looking for places to put their farming skills and equipment to work. This great historic meeting between landless blacks and property-owning whites led to the first significant inter-racial sharecropping contracts on the platteland.

As elsewhere in the world, details of these agreements varied over time and from place to place. Stated crudely, the chances of the crop being shared equally between landlord and tenant were greatest where the resources and needs of the two parties were most evenly matched—that is, where the white landlord's need for labour was balanced by the black peasant's need for access to land. But as agricultural economists and development experts have often noted, the risks in sharecropping contracts do not devolve equally on the partners. In the long run insecurities weigh more heavily on landless tenants than they do on property-owning farmers, and this increases rather than decreases inequalities; in time the landlord gains the upper hand and, when it comes to renegotiating agreements, he tends to press more heavily on the tenants, either demanding a larger share of the crop or insisting on ever-greater labour inputs from the peasant family.

In South Africa the economic distance between landlord and tenant was widened by the racial and political inequities that came first with conquest, then with segregation, later still with the policy of apartheid. Unequal access to state resources such as credit from the Land Bank, deepened the divide between landlord and tenant and hastened the decline of sharecropping as an institution. In the countryside, the interaction of these class forces made for a sharecropping frontier that shifted steadily westward. As white landlords in the 'wet' east slowly accumulated capital and put on the economic muscle that enabled them to mechanise production and expand the areas under cultivation, so black tenants were pressured into accepting wage labour; when they refused they were evicted, and in a renewed search for land-rich but labour-poor white landlords, they were driven farther north and west into drier areas where grain farming was less dependable.

By the late nineteenth century this market-driven northwesterly movement of black tenants was already well advanced and, by the first decade of the twentieth, the institution of sharecropping itself had outlived its usefulness in economically progressive grain-producing regions. Indeed, South Africa's infamous Natives Land Act of 1913 not only restricted the black peasantry's access to land by preventing African acquisition of property outside designated areas, but expressly forbade sharecropping ('farming-on-the-halves') in the agricultural heartland of the highveld.

Relatively well-off BaSotho, BaTswana or Xhosa sharecroppers could hardly be legislated out of existence, however. Unwilling to abandon their role as agricultural producers and step down the economic ladder to become labour tenants or wage labourers, they continued to move westward, crossed the Vaal River and moved into the sparsely populated areas of the southwestern Transvaal. There they sought out white landlords even less well capitalised than those in the Orange Free State and, entering into a new round of sharecropping agreements, tried to earn a living on some of the driest, most extensive maize- and sorghum-producing properties in South Africa.

Verbal sharecropping contracts binding these reasonably affluent black tenants to rather poor white landowners in geographically isolated regions of the highveld were widespread for nearly half a century after they were officially outlawed by the Natives Land Act. Their stubborn persistence gave rise to some unforeseen consequences. Farming-on-the-halves facilitated inter-racial social practices that transcended the clauses of the economic contract binding the parties together. With no great economic distance separating them, and with both parties far from their more densely populated cultural hinterlands, white landlords and black sharecroppers developed a *modus vivendi* embracing shared ideas about dress, health, justice, language, production, recreation and religious life. Poor Afrikaners emerged from the comparative seclusion of this shared experience far more 'Africanised' than their protestations would lead one to believe, while better-off Africans were far more 'Afrikanerised' than cultural purists are willing to concede. Class stroked away at the fibres of the emerging culture until it, like the fur on the cat, was best reconciled with the underlying shape. In the Triangle the passion of inter-war nationalist political rhetoric of economic advancement—black and white alike—was, in part, a public disavowal of private practices. The fur occasionally got ruffled, but the underlying profile was unmistakably paternalistic and rural.

In the twentieth-century western Transvaal, no less than in the nineteenth-century Orange Free State, such emerging social practices were always vulnerable to erosion by the encroaching tide of capitalism. In the Triangle a swelling wave of mechanisation between the mid-1930s and the mid-1950s undercut the security and well-being of the black tenantry. The introduction of petrol-driven tractors, along with the Marketing Act of 1937, which guaranteed white farmers a minimum price for their grain, made landlords less dependent on tenants' draught oxen at precisely the moment they were seeking to extend the proportion of land devoted to maize and sorghum production. Black sharecroppers, confronted by the growing obsolescence of their production techniques and by declining access to land on which to raise crops or graze cattle, were being forced to choose between labour tenancy, wage-labour agreements and abandoning white-owned properties altogether. With the sharecropping frontier pushing up against the outer limits of the Kalahari Desert, black families with ageing agricultural equipment had few new vistas to conquer. Sharecroppers were diverted into pools of cheap labour on white farms or made to undertake the last and most painful journey of all—the move into the crowded 'native reserves.' Racist legislation in 1913 and 1936 had designated these areas for the sole occupation of blacks, thereby relegating 80 percent of South Africa's population to 13 percent of the land.[4]

*

With the benefit of hindsight, then, we can see that, between the mid-nineteenth and mid-twentieth centuries, the emerging South African state engaged in a hundred year war to seal off the sharecropping frontier so as to deliver to politically privileged white landlords a black labour force that capitalist agriculture demanded. It is within the context of this long march north and west that we must situate our understanding of those black farmers and white landlords whose whispered verbal agreements remain muffled to this day by the sigh of the highveld breeze. Only when we place the strivings of the Maines in this broad context can we make sense of the limited options they could exercise when forced to choose between 'right' and 'wrong,' 'good' and 'evil.' Kas Maine's odyssey was but a moment in a tiny corner of a wider world that thousands of black South African sharecropping families came to know on a journey to nowhere.

The 'family,' let alone the 'sharecropping family,' is a problematic category. Indeed, as we know, the black family constantly changes ideological shape and size to occupy the spaces opened by economic, political or social opportunity—now huddling in near-nuclear profile at the 'western industrial' end of the spectrum, the next spreading itself at the end marked 'traditional farming,' and more often than not, simply straddling the nondescript socioeconomic terrain in the middle.

As should be apparent from the travails of the Maines, the black family was a dynamic, mutating, tempestuous and vibrant social form forever in the process of adaptation—both to the economic demands imposed upon it by the institution of sharecropping itself, and to the more elusive inner processes of social transformation that came through the subterranean structures of kinship. These latter changes, which accompany the establishment, growth, maturation, rise and eventual decline of the family as a social entity and which anthropologists refer to as the 'developmental cycle,' pose yet another challenge for those who are eager to penetrate the hidden universe of a patriarch like Kas Maine.

As with all families, sharecropping families move through both historical and cyclical time. What makes the sharecropping family unlike others, however, is that it is, at one and the same time, a unit of economic production and a social entity. With the Maines, we must be sensitive to the way in which the logic of their 'cyclical time' was in or out of phase with 'historical time' and the processes of agricultural production. Seen through social lenses, the sharecropping family profile constantly modifies by age, gender, and size so as to bear on production processes that are themselves responsive to externally induced changes. This changing profile is a weakness as well as a strength. On the one hand, it makes for economic flexibility, because labour power drawn from within the family can be adjusted and readjusted to meet different combinations of mechanised technology. On the other, as the family grows, matures and then declines, any number of cultural arguments—predicated on age, gender or sibling order—can be advanced by family members to contest or reject the economic role allocated them.

The latter tendency—the willingness of daughters, sons and wives to question the places assigned to them in the production process—lies at the centre of the storms that rage around the sharecropping family as it moves through 'cyclical' and 'historical' time. And at the very eye of this storm stands the head of the household. Janus-faced, the black patriarch is called upon to look inward—in on his family, to mobilise indigenous customs and practices for productive purposes in accordance with African 'tradition'—and to look outward to negotiate the demands of a changing economic dispensation in the 'modern' idiom of his European landlord's alien culture. As Kas Maine once put it, successful negotiation of the economic and social intricacies of Triangle life demanded that he sometimes play at being 'a chameleon amongst the Boers.'

A bifurcated existence in a colonial order calls for great social dexterity, formidable inter-personal skills and a positive mastery of the codes and practices of patriarchy and paternalism. Between the orbits of two planets, the one 'black,' the other 'white,' lies the greatest challenge for those wishing to understand the inner complexities of a black sharecropper's life. Like other African tenants, Kas Maine oscillated between two fields of gravity that were constantly shifting, now closer, then farther apart, now attracting, then repelling, at one moment overlapping to produce a challenging new cultural synthesis, at another separating out so as to be light years apart.

If the lives of South Africa's sharecroppers are forever lost to official memory in the shape of documentary archives, then the inner sanctuaries of black family life and the secret anguish of members trying to cope with a modernising society are even more difficult to retrieve. It takes men and women of rare ability, courage, dedication and vision to dredge through their public and private memories and then lay them out for the critical scrutiny of outsiders. The well-known inadequacies of the spoken word make the production of a body of historically verifiable facts based on oral evidence an exercise fraught with difficulty.

Kas Maine, more than any other member of his family, was equal to that task. From that moment in 1979 when a fieldworker from the Oral History Project at the University of the Witwatersrand first met him in one of South Africa's notorious resettlement camps, he astounded with his ability to recall, in sequence, the names of more than a dozen of his former landlords as well as the nature and size of each of the harvests they had shared. Over the six years that followed until his death in 1985, he never once ceased to amaze with the accuracy, depth and extent of his insights into the social, political and economic structures that dominated the southwestern Transvaal. He proved that in an industrialising society characterised by a high level of illiteracy, history lives on in the minds of its people far more powerfully than the cracked parchment of its officialdom might know.

Using Kas's life as the inner core of a carefully structured programme of interviewing, fieldworkers took more than a decade to add layers of evidence gathered from his wives, children and grandchildren to supplement and verify his testimony. The material was strengthened further by inserting it into the broader context of evidence provided by scores of other black sharecroppers, labour tenants, local traders, white landlords and lawyers. Finally, wherever possible, the oral evidence was checked against documentary evidence held by agricultural co-operatives, the Land Bank, the office of the Registrar of Deeds, official meteorological records, state crop reports and last, but by no means least, the five hundred or more scraps of paper which Kas Maine himself had been careful to preserve.[5]

From these resources—oral and written—it has been possible to set right a historic wrong, to recreate the life of a man who deserves to be remembered for far more than his failure to produce a dog licence in 1931. Kas Maine, the members of his family, and thousands like them were central to the building, feeding and shaping of this tortured country as it struggled to brush aside the racial goblins that guarded entry to the modern world.

The Maines' story is one of great complexity and infinite subtlety, filled with those ambiguities, complexities, ironies and paradoxes that always chase but never quite catch the fleeing spectre of wisdom. Kas Maine was simultaneously a very ordinary man and an extraordinary countryman. Those who are small of stature will seek to appropriate his searing experiences for political ends. Others, who pay obeisance to Queen History, will stand back, admire his achievements, share in his pain, and reflect on a life that often tells us as much about ourselves as it does about Kas.

PART ONE

# CHILDHOOD

*

*'The South African history which is really significant is that which tells us about the every day life of the people, how they lived, what they thought, and what they worked at, when they did think and work, what they produced and what and where they marketed, and the whole of their social organisation. Such a history of South Africa remains to be written.'*

W. H. MACMILLAN,
*The South African Problem and its Historical Development*, 1919

CHAPTER ONE

# Origins

## c. 1780–1902

Lethebe Maine, so it was said, was the oldest of the ancestors. It was he who at some point during the eighteenth century established the family at Sekameng, on the eastern extremity of the highveld, where only the distant foothills of the Maluti Mountains suggested the great southern African escarpment beyond. There, on the open plains that rose gently from the upper reaches of the Caledon River, much game was still to be had. Indeed, Makaoteng, the name of the area in which Chief Theko of the BaSotho granted Lethebe a place of refuge, meant 'the place of the people who live by hunting.' Lethebe was a skilled hunter and a man known for his ability to prepare the leather hides that shielded his family from the unforgiving wintry blasts of the snow-clad Malutis.[1]

Sekameng, which lies in the Mafeteng district of modern-day Lesotho, had other claims to fame. As a 'place of ilmenite,' it was popular for its deposits of shiny black crystalline *sekama*, a mineral that was crushed into powder and mixed with fat for cosmetic purposes. Makaoteng itself was even more famous for the sandstone mountain named Kolo that peered down over the village. There, under the shadow of the mountain, at a time when there was still sufficient game to feed a growing family, Lethebe fathered a son, Seonya.[2]

Young Seonya walked in the footsteps of his MoSotho father, but by the time he came to take a wife, the economy of the open plains (Mabalane, as the BaSotho called it) had already changed. It seems likely that when he married the Maines were already firmly wedded to the soil, because his son was named Hwai—'one who cultivates much and well.' But the boy was born in troubled times, and it was difficult to translate the promise of his name into practice. By the 1820s, disturbances that emanated from deep within the troubled Nguni-speaking kingdoms, which lay beyond the escarpment and nearer the coast, were starting to decant bands of marauding refugees into the region of the Caledon River headwaters. The plains farther south where BaSotho groupings were settled were particularly vulnerable, and the resulting armed incursions by Zulu-speakers gave rise to what became known widely as The Scattering, or *difaqane*.[3]

Whether because of the difaqane or because of a famine that it helped to bring down on the people of the plains is unclear, but at some point before 1825, Seonya and his family abandoned Sekameng and made their way to the distant safety of kinsmen who lived on the northwestern perimeter of the highveld. Seonya led his family north for almost three hundred miles, along the cultural trails left by his forebears, across the Vaal River and then west toward Molote, near the present-day town of Rustenburg. There, Chief Mathope of the BaKubung—an offshoot of the culturally diverse Hoja grouping that clustered towards the Tswana end of the great Sotho-Tswana continuum spanning southern Africa—took in the Maines.[4]

The family's new-found security was soon threatened by ominous political rumblings. When Chief Mathope died the Maines sided with his pregnant wife, Madubane, endorsing the claim that any male child of hers would have on the throne. But when the claim of the child, Lesele, was effectively denied by the late chief's brothers, she and a group of her supporters abandoned the BaKubung for her ancestral home amongst yet other Tswana-speakers, the BaHurutshe. This exodus left the late Chief Mathope's remaining clients, including the Maines, without any obvious source of political patronage or protection.[5]

The sense of unease that this squabble engendered in the family gave way to outright terror when the outer waves of the difaqane threatened BaKubung society as a whole. With the family now vulnerable from both within and without, Seonya decided to uncouple what remaining links they had to his hosts and to return to the Malutis, where the greatest Sotho-speaking chief of all, Moshweshwe, was drawing together the strands of the emerging BaSotho nation. For the newly ascendant faction of the BaKubung, however, Seonya's decision to leave amounted only to treachery compounded; it took a hundred and twenty years for them to forgive the Maines.

With the outlines of what were eventually to become the Boer Republics just beginning to emerge, the Maines went back across a thinly populated central highveld that seemed to belong to all who were willing to consider southern Africa their home, crossed the Caledon River, and reestablished themselves at Sekameng. The worst of the violence unleashed by the difaqane had abated, and, under the umbrella of political protection afforded them by Moshweshwe's astute diplomacy, the plains once again seemed to offer opportunities to settled agriculturalists. This time the young Hwai could live up to his name.

Scurrying around the hem of Kolo's skirts, Hwai acquired the basic knowledge that rural society demanded of a boy working with his father's livestock and plough. The great sandstone mountain that loomed above the village and its even more impressive kinsmen—the mighty Malutis that rose to meet the sky in the east—had many skills to bestow on favoured children. Shepherding, the command and love of mountain ponies, and an enviable ability to dress stone and build with it all fell to any young man willing to learn. Grafted on to this was undoubtedly the greatest gift of all—the knowledge of herbal medicine that made Hwai a traditional doctor, a *ngaka*. By the time his father died Hwai was a fully fledged adult, a MoSotho, and the Maines' earlier experience amongst the far-off BaKubung was no more than a distant memory.[6]

Hwai's talents blossomed at a propitious moment. As early as 1863 the Lesotho lowlands were described in official documents as constituting 'the granary of the Free State and parts of the Cape Colony.' The prosperity of the plains was further enhanced when, four years later, diamonds were discovered at Kimberley and yet more new markets opened to the grain farmers of the interior. By 1873 the BaSotho were considered to be a 'thriving and well-ordered people'; that year they exported 100,000 bags of grain—maize, sorghum and wheat—as well as 2,000 bags of wool to adjacent territories which were themselves starting to attract a growing number of commercial farmers of European descent. The highveld proper, which had seemed so open to all in the immediate wake of the difaqane, was becoming a patchwork quilt of white-owned farms where boundary pegs divided off the promise of the earth as a whole into that legal thing referred to as 'property.'[7]

The Maines shared in much of the early BaSotho prosperity. In the two decades 1850–1870, Hwai contracted four marriages which between them yielded at least a dozen children. The first of these, to Modiehi, was an especially felicitous union from which issued three sons, Tshilo, Mpoko and Sekwala. Like their father and grandfather before them, each of the boys acquired the traditional craft skills of a MoSotho male. Perhaps at least as importantly, they were all socialised in an era when the power of an expanding market was opening unprecedented horizons for southern Africa's black peasants.[8]

But in an economic cycle, as in life, the apex of a trajectory is by definition also the start of a downturn. In the mid-1860s the BaSotho became embroiled in a long and debilitating war with the burghers of the neighbouring Orange Free State, and shortly after it ended, the founder of the nation, Moshweshwe, died. By 1875, 15,000 of the younger men in a population of 130,000 had been drawn into the cash nexus of migrant labour on the highveld, and by 1884 the number was said to have doubled. Peasant societies that bleed manpower haemorrhage away their life blood; just as Hwai and his family had profited from the upturn in the market, so they were now called upon to share in the decline of the plains economy.[9]

In the late 1880s Kolo, who had for so long looked down benignly on the Maines, seemed to lose her smile. Whether it was the death of Modiehi, the strain of feeding a growing family or simply the capriciousness of a climate that produced one of its periodic famines is unknown but, at some point after 1889, Hwai, too, decided to abandon Makaoteng. Unlike his father before him, however, Hwai did not want to risk severing all links with Sekameng and therefore sought out a location that would ensure the family continued access to the cultural headwaters of Sotho society. Making his way up the valley of the Caledon River and then beyond it into the broken country that shied away from the Malutis, he eventually found a place on a farm at Mequatling, in the eastern Orange Free State. The choice of Mequatling, meaning 'I am at home in the hills,' proved to be a wise one.

Formerly an out-station of the Paris Evangelical Mission Society, a French Protestant grouping that had been active in Basutoland since the 1830s, and proclaimed as a farm only in 1874, Mequatling occupied a spectacular setting in the hills not far from the village named Clocolan. Here Hwai negotiated favourable terms as a tenant with a young white man named Max Woldmann, then in his late twenties. Woldmann, who had made a fortune during the previous three years by buying horses cheaply in the Free State and then selling them at a profit on the newly opened Witwatersrand goldfields, had acquired the property in March 1889. Having secured the farm of his choice, Woldmann required the services of several BaSotho men who had a knowledge of horses in order to allow him to undertake a journey home, where he hoped to re-establish contact with his family in Germany.[10]

Hwai and his sons seemed like especially attractive labour tenants, and their stay at Mequatling was most profitable. Family tradition has it that the Maines arrived at the farm with only a few pack-oxen but left with a large number of horses. Hwai certainly garnered sufficient resources during their stay for him to send at least one of his unmarried sons, Sekwala, back to Sekameng to secure a bride of the family's choice. Unfortunately this young woman died during childbirth soon after her arrival at the farm; almost immediately thereafter, Hwai decided to cut short the family's stay at Mequatling. This time he moved them about fifty miles northwest, onto a grain farm owned by a certain Dolf Brits in the district of Winburg.[11]

But in the early 1890s the commercial farmers around Winburg were riding the crest of a wave, and white landlords pressed hard on their black tenants. The Maines spent even less time on Brits's farm than they had at Mequatling. Away from the mountains and back on the highveld proper, Hwai now heard a tale which, if true, would help the family in its search for a permanent home and free them from some of the uncertainties associated with working for European landlords. Local blacks spoke of how a number of BaKubung families who had been dispersed by the

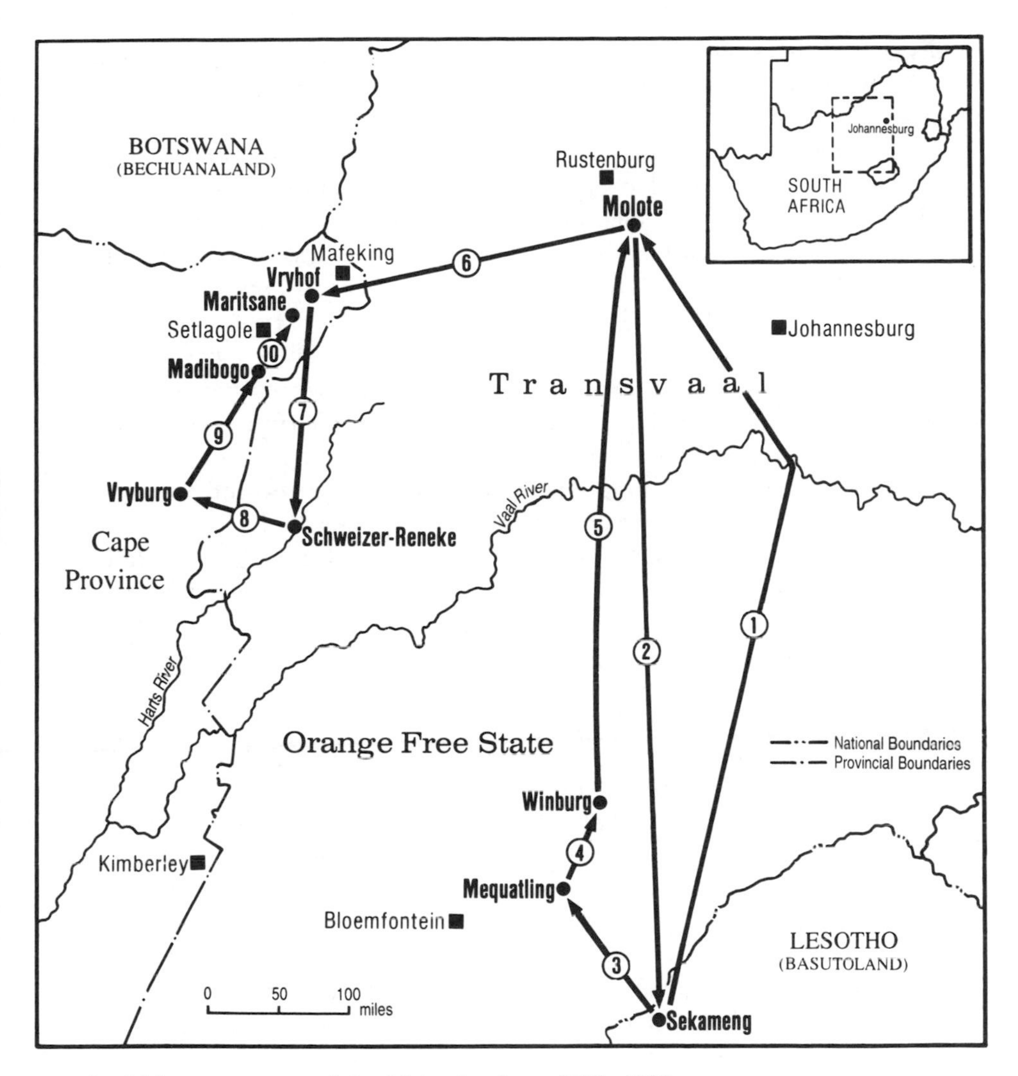


3 Major movements of the Maine family, c. 1780–1902

difaqane and who had until fairly recently been living in the Heilbron district of the northern Free State, had departed for Molote, in the western Transvaal, where they had purchased farms of their own with the assistance of a well-disposed white missionary. The possibilities that this held for the Maines must have seemed all the more exciting when Hwai learned that Lesele, the very boy whose cause Seonya had unsuccessfully championed in the chieftaincy struggle more than fifty years earlier, was said to preside over the BaKubung and their farm.[12]

The long and arduous trek north to the Rustenburg district was fuelled by optimism, but it ended in disappointment. On reaching Molote Hwai discovered that Lesele had indeed governed the BaKubung but only until 1884 when, in a recurrence of the old succession dispute, he had been ousted by a commoner drawn from amongst Free State emigrants. Denied his role for a second time, Lesele had left with a small group of loyal followers to seek out sharecropping possibilities amongst surrounding white farmers. The new chief, in essence a descendant of the same group which Seonya had run foul of a half-century before, was quick to remind Hwai of his father's desertion of the BaKubung in their hour of need and refused his request for land.

Stranded in the Transvaal and responsible for the well-being of a family comprising at least thirty members as well as their horses and livestock, Hwai was confronted with a difficult choice: either to return to the mountains, defeated like Seonya before him, or to make one last effort at conquering the highveld. Hoping somehow to weave weakness into strength, he opted for the latter. Drawing on the precedent set by Lesele's mother at the time of her expulsion from the BaKubung and seeking to capitalise on his father's erstwhile loyalty to her, Hwai and his family set off in a westerly direction, toward the desert, hoping to find refuge amongst the people of Lehurutshe, whose totem, like that of the Maines, derived from the delicate moisture of *phoka*, or dew.[13]

As the family soon discovered, however, such limited prospects as there were for settlement in the early 1890s lay farther south and west, around Mafeking, where elements of the southern Tswana had recently been dispossessed of land between the upper reaches of the Setlagole and Maritsane Rivers. There, amidst the political uncertainties of the Setlagole reserve, Hwai and his entourage were taken in at the small village of Vryhof.[14]

Soon the Maines started to put down roots—roots that at least partially helped consolidate the family's tenuous hold on the highveld's thin unyielding soil. While looking round for a spouse to replace the wife whom Sekwala had lost, Hwai was steered in the direction of the Mokawane family, who formed part of a small community of immigrant BaSotho at a nearby village. These Mokawanes, who had fallen under the sway of the Methodism that dominated the region, boasted twin daughters of marriageable age. Despite his misgivings about Christians, Hwai was of the opinion that either of these women would make Sekwala an admirable wife, help to restore his flagging spirits, and assist the family in forging new links in what, for them, was still an unfamiliar society.[15]

In the end Hwai decided on the one named Motheba, and after the customary negotiations between the families had been satisfactorily concluded, the marriage was celebrated. When almost immediately thereafter the couple produced a son whom they named Mphaka, Hwai considered his choice of daughter-in-law vindicated. But though Sekwala was now at one with himself and the rest of the family was more settled than it had been for some time, the Maines continued to feel economically vulnerable because the area around the reserve was relatively densely inhabited and characterised by much livestock theft and fierce competition for limited grazing.[16]

Unease gave way to unhappiness when, shortly after Sekwala's marriage, the family lost its horses to a group of white rustlers who took the animals across the border to Bechuanaland and sold them. This loss, which stripped the family of the principal asset that it had acquired during the course of the long trek across the subcontinent, prompted Hwai to scan the horizon yet again. He had to restock, for this was essentially cattle country, and he had to keep his family intact. Once again he was forced to consider becoming a labour tenant for a white farmer. The nearest opportunities lay south, around Mmamusa, the former stronghold of a powerful Koranna chief which the Boers had over-run and renamed Schweizer-Reneke when they established a permanent presence there in 1888.[17]

Back in the early 1880s the dry, dusty and demanding region between the Vaal and Harts Rivers had been the site of considerable competition for grazing and firewood between the indigenous Korannas of Mossweu, based at Mmamusa, and the BaThlaping of Mankurwane, centred on Taung to the southwest. In the ensuing war both parties enlisted the help of white mercenaries, an arrangement that on the Koranna side necessitated payments in land and culminated in the establishment of the short-lived Boer Republic of Stellaland. Following this, and the subsequent defeat of Mossweu by the forces of the South African Republic under President S. J. P. Kruger in 1885, a number of white farmers entrenched themselves in the area around Schweizer-Reneke but they lacked access to a supply of cheap black labour. It was against this broad background that Hwai moved his family into the district in late 1892 or early 1893.[18]

At the farm Rietput, on the banks of the Harts River and within sight of the ruins of Mossweu's former stronghold on a nearby flat-topped hillock, or *koppie*, Hwai negotiated terms with an elderly white farmer named Reyneke. Each of Hwai's three sons would pledge their labour for a year and, in return, be rewarded with a heifer and the right to graze their livestock on the property. In addition, the Maines would provide the services of a young girl to assist with the customary tasks in and around the farmhouse. This contract, typical of those entered into by black labour tenants at the time, suited the needs of both parties, albeit unequally.[19]

Jacobus Cornelius Reyneke, who was by then almost sixty years old, had been born in the Cradock district of the Cape Colony in May 1834. As a young man he, like so many others in his cohort, had moved north into the Rouxville district of the Orange Free State, where he had married and subsequently become a reasonably successful sheep farmer. In 1888, shortly after the discovery of gold in the Transvaal had shifted the country's centre of economic gravity even farther north, he and his family crossed the Vaal and purchased Rietput.[20]

On taking possession of his new property, however, Jacobus Reyneke was distressed to find that the village of Schweizer-Reneke lacked a church where he and other followers of the Nederduitse Gereformeerde Kerk could worship. A devout Christian, he set about organising a committee to erect a suitable building and, with the help of two Italian stonemasons by the name of Scribante, arranged for sandstone to be brought in from across the Vaal River by ox wagon. Not long thereafter he was instrumental in arranging for Reverend George Faustman to take up the first ministry in Schweizer-Reneke.[21]

Reyneke's enthusiasm for religion was not confined to a weekly journey into Schweizer-Reneke, however, and his tenant's children later recalled how he often arranged for ministers drawn from denominations other than his own to hold services on the property, where the Maines and another family of BaSotho immigrants, the Mashodis, were joined by several of the local Korannas. For Motheba Maine these gatherings merely served to supplement her more regular attendance at the Methodist Church in Schweizer-Reneke where, so it was said, the congregation was even more racially integrated than that at farmhouse gatherings.[22]

At first the Maines enjoyed steady economic progress under 'Ou Koos' Reyneke. They acquired cattle, just as grandfather Hwai had hoped for, and subsisted by making use of hand-hoes to plant maize and sorghum. These modest gains were further consolidated when the patriarch acquired a cheap imported American plough with a single share and a wooden beam. Hwai's acquisition of the '75' plough enabled the family to increase the area under cultivation and thus to accumulate surplus grain, which they stored in specially woven grass baskets, or *disiu*.[23]

In a setting where family size and the ability to mobilise labour were at least as important as agricultural equipment, Hwai detected other signs of progress. All his older children saw significant additions to their families during their stay at Rietput, and Sekwala and his new wife were especially well blessed when it came to sons. In addition to Mphaka, the boy born soon after the marriage in Setlagole, Motheba had produced another son, Phitise, shortly before the family's arrival at Mmamusa. In late 1894 the couple had yet another son. In accordance with custom, the boy was given a BaSotho name, Ramabonela, but in deference to the social dominance of the Koranna around Schweizer-Reneke in general and the redoubtable Mossweu in particular, the boy's praise poem came to reflect not only his immediate ancestry but the importance of those who had once been the overlords in the area around his place of birth.

> I am Ramabonela of the place of Kasianyane, perfectly black, Son of Shield and Plough, from the wife of Phokane of the Ndebele.

MmaPhokane, return the knobkerrie [stick] and continue the fight on your knees. Return the Red Bull to your home, return it to your father, Son of the Maines.[24]

As might have been expected, praise poetry that referred to Kasianyane, Mossweu's redoubt on the koppie, went down well amongst Sekwala's Koranna friends, and it was not long before the baby was so often being called Kasianyane that his real name fell into disuse even with his parents. Perhaps even more significantly, the word 'Kasianyane' was so eponymous with the Dutch 'Casper'—especially its abbreviated form, 'Cas'—that the landlord and other Afrikaans-speakers slid naturally into the practice of referring to the child as 'Kas.' Thus it was that the very name accorded Sekwala's third son was a form of adaptation to the wider BaSotho, Koranna and Boer society in which he found himself.

Despite his daughter-in-law's admirable ability to produce a string of male grandchildren, Hwai cherished the hope that one day Sekwala would have a wife in keeping with tradition, drawn from BaSotho kin rather than from amongst the BaKwena of the highveld. To this end he sent an emissary to instruct a younger brother living in the eastern Free State to find a cousin for Sekwala to marry. This search culminated in Sekwala's marriage to Maleshwane, his third wife, who, besides being a MoFokeng traditionalist, had the added advantage of being about seven years younger than Motheba.[25]

The Maine family continued to make steady social and economic progress and, not long after the birth of Kas, Hwai was invited to move to a neighbouring property by a local notable who was on the lookout for labour tenants. A land and store owner who hailed from the Christiana district, Gerrit van Niekerk had played a leading role in the war of dispossession between Mossweu and Mankurwane and had subsequently become the president of the Stellaland Republic. In compensation for his loss of political standing when the Republic collapsed, the Land Commission of 1886 had awarded him 'four and a half farms.' Two of these, Niekerksrust and Zorgvliet, lay on the Harts River some ten miles southwest of Schweizer-Reneke, but the grant was so tied up in red tape that it was not until the mid-1890s that he was able to take possession of them. The prospect of gaining access to virtually unrestricted grazing appealed to Hwai, and at some point in 1895 the Maine family moved across to Zorgvliet.[26]

But Zorgvliet was never to fulfill its promise. In 1896 it and neighbouring farms were ravaged by the rinderpest epidemic which carried before it 90 percent of the western Transvaal's cattle. In a matter of days the disease consumed the labour of years and the hard-won gains of Rietput lay rotting on the banks of the Harts River. Yet, despite their losses, the Maines left van Niekerk's farm relieved that the disease—which they knew as *bolawane*—had taken only their cattle and not one of their children. In the midst of the rinderpest, fever-ridden cattle had bellowed and stampeded their way through the homestead threatening to trample underfoot a stranded toddler. While the men stood transfixed, Motheba, showing the same sort of courage that marked her spiritual commitment, darted through the oncoming animals to sweep little Kas to safety.[27]

The family's retreat to Rietput was, as before, both congenial and profitable. Kas, as a small boy, spent much of his time playing with the half dozen cousins who were closest to him in age. His most constant companion was his Uncle Mpoko's son whose name was Sempane but whom the Boers had been quick to rename Champagne. They and the other boys on the farm spent their time herding kids and playing along the banks of the Harts which, according to the local Koranna children, harboured *kganyapa*—fearsome water-snakes that emitted a strange glowing light once darkness settled over the river.[28]

As Kas made the transition from looking after the kids to being responsible for the goats, his father continued to labour for Reyneke and restock his herd so as to make good the losses sustained at Zorgvliet. Although trek oxen remained in short supply after the rinderpest epidemic, Sekwala did manage to acquire a few other animals: one he was particularly proud of, a black cow called Flower, Blom. He also raised the money for a wagon which, he hoped, would enable him to undertake the transport-riding that could, in turn, generate a modest cash income.[29]

As the men in the Maine family understood only too well, even this modest aspiration, if it were to be pursued seriously, would require more freedom from ploughing and harvesting than their landlord would grant them. As had happened once before, Hwai and his sons used Rietput as a base to find their feet and then, having made the necessary gains, debated the need to move on in search of yet more economic space. Like any other white farmer, Old Koos Reyneke, in the final analysis gave priority to his own needs over those of his tenants. This time a solution was closer at hand on the neighbouring property of Holpan.

Holpan, like several farms in the district, was an established property that had fallen into the hands of an 'English' land and mineral speculator after the discovery of gold on the Witwatersrand in 1886. By the mid-1890s its owner, disillusioned with its mineral-bearing potential, sold it to a company which he controlled, Henderson Consolidated. But the corporation, although better placed than an individual owner, was no more capable than he of digesting this tract of dry southwestern Transvaal land and in the end did as most such companies did and leased the property out to an undercapitalised Afrikaner farmer. The Boer, in turn, raised the rent as best he could and turned to what was commonly known as 'kaffir farming.' Thus it was that in 1898 Hwai and his family 'jumped the fence' and joined Sarel Niewoudt as rent-paying tenants at Holpan.[30]

Sekwala and Motheba celebrated their arrival on the property by producing their first daughter, Moselantja, who joined her brothers Mphaka, Phitise and Kas in a rapidly expanding family. Another daughter, Motlakamang, died a year or so later. Under the circumstances Kas's mother was grateful for the assistance that not only her kin but the many other black families living at Holpan could give her.

Motheba was not alone in benefiting from the warmth and generosity of the inhabitants of the village, or *stadt*, that developed on the Henderson property. Nor was such help confined to the ranks of blacks for, at some point just prior to the arrival of the Maines, Sarel Niewoudt had given permission to a destitute widow, Mrs. Swanepoel and her children, to live on the property. Mrs. Swanepoel, like the Reynekes, had originally come from the Rouxville district of the Orange Free State but, having lost her tenant-farmer husband nearly twenty years earlier, had been forced to move from farm to farm as a 'poor white' in an attempt to find a place for her family.[31]

The drought, disease and locust invasions that were such a marked feature of the 1890s pressed especially hard on the widow, and during this acute rural distress she and her oldest son had become more and more dependent on the assistance of well-disposed black neighbours. By the time Sekwala got to Holpan, Mrs. Swanepoel and her youngsters had long been in the care of one of the most prosperous black tenants on the property.[32]

A man of substance, Thinyane Nthimule had the distinction of owning a full span of draught oxen in the post-rinderpest period, and it was to him that Sekwala turned for assistance when the time came to plough. While hiring Nthimule's oxen he was introduced to Hendrik Swanepoel, the oldest of the widow's sons. Sekwala, a staunch traditionalist, took an immediate liking to this thirty-year-old SeSotho-speaking white, who had been strongly influenced by black culture during the years that he and his family had spent wandering across the highveld. The two became firm friends and, shortly thereafter, started a transport-riding business that specialised in ferrying grain to the small railhead established at Vryburg in 1890.[33]

The close business ties and village intimacy that linked his MoSotho father to his partly deracinated Afrikaner partner meant that for the young Kas white men held few terrors. As a boy much of his time was spent clambering about his father's wagon going from farm to farm, watching the men labouring together and then interacting as social equals during the long journey to Vryburg. Besides priming the boy's entrepreneurial skills, these excursions gave Kas a self-confidence when dealing with Afrikaans and Afrikaners that never quite left him.[34]

If in one small and isolated corner of the highveld, the late 1890s were marked by this slightly unusual example of inter-racial co-operation between an Afrikaner, a black man and his child, the same was hardly true of the broader politics emerging in southern Africa. Indeed, ever since gold had been discovered on the Witwatersrand in the mid-1880s the British government had watched the development of the Boer Republics with a mixture of awe and apprehension; it nurtured a deep suspicion that a self-sufficient Afrikaner state beyond the Vaal River would pose a long-term challenge to the existing pattern of international trade as well as to British hegemony in the region.

It was precisely these fears which prompted official British connivance in the ill-fated Jameson Raid of 1895. A small group of mine owners on the Witwatersrand, with the tacit approval of Chamberlain, Secretary of State for the Colonies, triggered an armed insurrection which they hoped would lead to the overthrow of President Kruger's government. Although the raid failed, it did succeed in causing such a serious deterioration in Anglo-Boer relations that the entire subcontinent was set on a course for war. Somewhat mistakenly, or so it later transpired, this was not a prospect the British government or its principal agent in the Cape Colony, Lord Alfred Milner, baulked at.[35]

Out at Holpan, where they were busy bringing in the harvest in the autumn of 1899, the political tremors emanating from the distant Witwatersrand goldfields were only faintly detectable. During the transport-riding season that followed, however, it became clear to Sekwala that his family's future was being threatened by forces over which not even the neighbouring landlords had control. By mid-1899 anxious Boers were talking openly about the coming war with the British. When these disturbing prophecies had still not materialised by the end of winter, Sekwala ploughed the soil as usual in the mistaken belief that he was a servant of the seasons rather than of distant white politicians. But no sooner had the October rains arrived than war was declared, and within days one of Ou Koos Reyneke's grandsons appeared on the farm brandishing a piece of paper. Both Sarel Niewoudt and Hendrik Swanepoel were told to join the commando, and the reluctant Sekwala was instructed to accompany them as one of many black *agterryers* (auxiliary troops) who would assist the Boer forces with the supply of ammunition, food and transport. This was hardly the 'white man's war' that the Boer and British propagandists proudly proclaimed it to be.[36]

Men drawn from the Schweizer-Reneke district formed part of the Bloemhof Commando under Commandant Tollie de Beer, and in the weeks that followed moved steadily southwestward, to beyond Kimberley, in order to assist General Piet Cronje's forces in intercepting British troops moving north under Lord Methuen. They first saw action at the Modder and Riet Rivers in late November 1899, but it was the events in the fortnight that followed at Magersfontein that imprinted themselves on Sekwala's mind most vividly. At Scholtznek he and thousands of other commandeered blacks dug the trenches that featured so prominently in the much-vaunted Boer resistance, while British officers hovered overhead in three enormous balloons designed to facilitate aerial reconnaissance. When, nearly two months later, Piet Cronje and his forces eventually abandoned the earthworks at Magersfontein and moved east, there were yet more trenches to be dug, this time at Paardeberg Drift. On this occasion Boer resistance was short-lived and, after the first major British victory in the war, General Cronje's forces surrendered to Lord Roberts on 19 February 1900.[37]

Stranded behind the British lines and cut off from his kin at Holpan, Sekwala decided to seek refuge at the home of his Uncle Mokentsa, in the Orange Free State. A few months later he was joined by his youngest wife, Maleshwane, who had fled from the conflict in the western Transvaal and made her way back to her parents' home. The couple set up a temporary home and, not long thereafter, Maleshwane was pregnant—a welcome development since back at Schweizer-Reneke she had already lost two baby daughters, each dying shortly after birth. In 1901 she gave birth to a boy, and Sekwala, in a conscious effort to shake off the bad luck that had been dogging their footsteps, broke with tradition and gave the child the distinctive name of 'Balloon,' a tribute befitting a veteran who, along with the Boers, had survived the great Battle of Magersfontein.[38]

Once the road to the north was effectively open, British forces methodically extended their occupation of the Boer Republics and, early in 1900, swung through the western Transvaal, pursuing a scorched-earth policy designed to dry up Boer supplies in the rapidly developing guerilla war. When they eventually reached Holpan, the 'Tommies' put the grain stores and huts to the torch and then offered to escort the destitute black tenants and their livestock to a distant place of safety; Hwai and the rest of the Maines were marched off in a westerly direction. These events left an understandably vivid impression on the five-year-old Kas.[39]

Sekwala's youngsters were put in charge of their father's cattle, but as the column wound its way through the streets of Schweizer-Reneke, Kas and the others were distressed to see Blom break ranks and head back in the direction of Holpan, where, in the hasty retreat from the farm, her calf had been left tethered to a stake. The panic this caused was soon forgotten, though Motheba, struggling to keep her family together amidst the constantly swelling ranks of black refugees, became separated from her oldest boy. By the time the troops and their unhappy charges entered the village of Vryburg, several days later, nine-year-old Mphaka as well as several other members of the family were well and truly lost in the clouds of dust and confusion enveloping everything.[40]

Within hours of their arrival at Vryburg, Hwai, Motheba and the children were joined by yet more refugees before they were confined to a makeshift camp in a 'location' situated on the outskirts of the white village. Reluctant to abandon the small measure of safety offered by this

British umbrella, and unwilling to risk allowing the boys to take the remaining cattle out to graze beyond the confines of the common, Motheba and Hwai soon came up against the realities of a dwindling food supply. Fortunately the camp was not far from Motheba's home, and she arranged for a message to be sent to her family at Setlagole. The people in and around Mafeking were experiencing their own distinctive taste of a supposedly 'white man's war,' however, and, within a matter of weeks she was having to feed Kas and Phitise as well as the infant Moselantja porridge made from such ground barley as the troops' horses could be deprived of. Starvation was a real possibility. But one of Motheba's brothers, who happened to be serving with the British, put in an unexpected appearance.[41]

Mofubathi Mokawane's job as scout for the British was one of which his family's highveld hosts at Vryhof wholeheartedly approved since the rival Rapulana BaRolong had used the opportunity of the war to settle an old score with their neighbours by siding with the Boers. These wartime complications in Tswana society greatly increased the dangers of any journey between Vryburg and Mafeking. But Motheba, still tortured by the disappearance of her son Mphaka and the thought of what awaited his siblings if they were forced to stay in the camp, was determined to reach the sanctuary of her parents' home at Setlagole. After bidding grandfather Hwai and the remaining Maines farewell, she and her three youngest children set off for Mafeking with her brother.[42]

The long move to the Setlagole reserve was successfully accomplished only by avoiding all the main roads and moving at night. At Madibogo, where the hungry small boys were struck by the relative abundance of cattle and milk, Mofubathi placed his sister and her family in the care of their brother, Ngwanapudi, and the little group could spend time with Motheba's parents and recover from the ordeal of the Vryburg camp, before moving even farther north to Maritsane, where yet another brother, Musi Mokawane, took them in for the remainder of the war. Young Kas never forgot the Maines' stay at his Uncle Musi's home because, on at least one memorable occasion, he was soundly thrashed for having neglected his herding duties.[43]

The war ground slowly to a halt. When the Peace of Vereeniging was eventually signed in May 1902, both the Boers and the British could look back on one of the bloodiest and most costly wars of all time. But they could also look forward with a measure of guarded optimism, for in differing degrees they remained masters of their own destinies in a country of infinite promise. For Hwai, however, the war meant more long journeys during which his family was dispersed and impoverished by a conflict not of their making; a son and a grandson were reported missing, and now they all faced the prospect of homelessness in what the authorities promised would be a new South Africa.

CHAPTER TWO

# Foundations

## 1902–13

Early in 1902 Sekwala left the greenery of the eastern Orange Free State and returned to that dry and dusty part of the southwestern Transvaal which, before the war, had looked as if it might one day offer the Maines a permanent home. If truth be told, this ugly extended plain was not the most promising part of the country that the family had encountered in its decade-long withdrawal from Makaoteng. The rainfall was infrequent, only about twenty inches a year, and extremely unreliable: when the rains did come, in torrential thunderstorms between October and March, the water rushed away along dry watercourses, percolated down into the belly of the Kalahari sands or stood about in pans of shallow brackish water that taunted man and beast. The thin topsoil, often broken by unsightly veins of protruding white limestone, seldom sustained more than a modest covering of summer grass and struggled to host the tough little acacias that clustered together in occasional pockets of clay. The challenge that this and the accompanying heat presented was enough to make any man yearn for the long cool shadows of the distant Maluti Mountains.

But there was more to making a living from the soil. Viewed by a landless black family several rungs down the social ladder, the barren triangle that lay between the towns of Schweizer-Reneke, Bloemhof and Wolmaransstad held some promise. On the periphery of the great Kimberley diamond fields, which lay some way off to the southwest, and even more distant from the large agricultural markets of the Witwatersrand goldfields to the northeast, the region was something of an economic backwater. True, several large land companies had bought up farms in anticipation of eventual speculative gains, but the area had shown limited potential for mining and now it was dominated by vulnerable Boer tenants unable to pursue agricultural profit with any sustained vigour or absentee landlords who 'farmed kaffirs' for rent. Where white landlords were weakest, black tenants stood to gain most.

Sekwala hoped to pick up the reins of his life at the place where he had been forced to put them down, in the small community at Holpan. But when he got there he found the *stadt* in ruins and he searched in vain for any member of his family. A few old men of Hwai's age told him that most of Sarel Niewoudt's tenants had moved off with the British as refugees during the turbulent months of early 1900. Many of these exiled families had since returned to regroup on various farms in the Triangle, in some cases on properties that appeared to have been abandoned by their Boer landlords. A search of the district uncovered a marvellous settlement prospect on one such abandoned farm, Mahemspanne, but failed to reveal any trace of his wife or children. The only thing left to Sekwala was to go back to Setlagole.[1]

The long, painful trudge north was worth it when he was reunited with Motheba, Phitise, Kas and the smallest child, Moselantja, whom he had never seen before. The joy of this reunion was complete when, not long thereafter, they learned that Mphaka had been found unharmed and living at Taung, where Motheba's twin sister had taken him in, having spotted the child at a nearby camp.

Still, they were disheartened whenever they thought of Hwai and those whom they had left behind at Vryburg two years earlier. Motheba's younger brother came to their assistance once again. Making use of contacts within the British army, Mofubathi Mokawane established that Hwai and others had been moved to a concentration camp at Thaba Nchu, a small town hundreds of miles away in the Orange River Colony, and then gave Sekwala a horse to fetch them. Weeks later they were all reunited in the Setlagole reserve. Sekwala started singing the praises of the deserted farm deep in the heart of Boer country.[2]

Mahemspanne, which means 'the pans of the crowned cranes,' was one of two properties at the very heart of the Bloemhof–Schweizer-Reneke–Wolmaransstad Triangle that had caught the eye of the descendants of the legendary Voortrekker leader Andries Pretorius. While an older son, the distinguished state president of the South African Republic, Marthinus Wessel Pretorius, had acquired another farm, Soutpan, in 1870, a less distinguished kinsman, Matthys Wynand Pretorius, had been granted the much smaller property at Mahemspanne, which he quickly sold after being given title to it in 1871. Mahemspanne had never qualified as a much-loved 'family farm.'[3]

In the next two decades, the property passed through the hands of several land speculators and companies until William Pope acquired it in 1890. He, in turn, let the farm to a Boer tenant who was said to have died during the war. Pope, a man of greater means than his predecessors and blessed with that tolerance which comes more readily to the wealthy than to the poor, took little interest in the farm thereafter, beyond asking W. E. Wood, a Bloemhof agent who had acted for several of the larger companies, to keep an eye on it. Purely by chance the Maines and a half dozen other families, including the Ramotswaholes and the Kutoanes, had chosen a place where neither the owner nor his agent at first showed any inclination to collect rent from a self-selected tenantry.[4]

Pope's lack of economic appetite was especially appreciated by the Maines, when food was still at a premium. Without grain reserves to fall back on or small livestock to tide them over, Sekwala and his family were forced back into the most basic of survival strategies—hunting and food gathering. Their meagre new diet came as a shock to Motheba's youngsters, who had grown accustomed to the more traditional and generous fare offered them by their mother's people at Madibogo. Resident BaTswana, drawing on a seemingly inexhaustible fund of knowledge about indigenous plants, taught the women and the children which roots and berries were edible while Sekwala and his brothers used hunting dogs to flush meercats, springhares and antbears. Likewise, the old Koranna technique—digging large pits into which springbok were driven before being speared to death—provided them with meat.[5]

But right from their very first day on the property, Sekwala knew he would have to acquire livestock if the Maines were to free themselves from the hazardous existence offered by hunting and the gathering of *veldkos*. But to acquire cattle or goats, it would be necessary first to accept a lengthy period of employment with one of the local farmers, leaving his family while it was still economically vulnerable. This idea was especially unattractive since Maleshwane was pregnant and Motheba had just given birth to a fourth son, Sebubudi.[6] Under the circumstances he had little choice but to send the older boys out to work as cattle herders amongst the Boers.

Initially, most of this burden was placed on the shoulders of Mphaka, then aged eleven. Sekwala found him a position as herdboy on a neighbouring farm, Hartsfontein, where, in return for a year's labour, he was promised a splendid brown and white calf named Skilder.[7] This agreement with the proprietor, Gert Meyer, appealed to the boy's mother because it meant that although Mphaka left the house at sunrise, he could return at sunset and spend his nights at home. This first rather tentative venture was followed by a less successful experience the following year when the boy was sent to work for old Sam Elliot on yet another neighbouring property, Langverwacht. There, Mphaka, already endowed with a fiery temper, got involved in a fist fight with some of the young Elliot boys who, blacks and whites alike agreed, were known to be 'very wild.'[8] After this unfortunate incident he was sent back to work at Hartsfontein where he was less likely to be called upon to interact with whites because Gert Meyer tended to spend most of his time on a more favoured family farm, Rietfontein, some twenty miles away on the banks of the Vaal.

In due course, when they were old enough to herd cattle, Phitise and Kas followed in the footsteps of their older brother, and by 1908 three of the four Maine boys were at work on the Meyer property, each earning either a goat a month or a calf a year for their labours. All these animals found their way into their father's kraal.[9] Stock thus acquired, and the fact that he still had not been called upon to pay rent, meant that Sekwala could expand production. Skilder, the cow that Mphaka had brought back from Hartsfontein, was soon supplemented by a second which he managed to acquire by trading goats with a Boer. When these two animals calved shortly thereafter, their offspring were promptly traded for two burly oxen named Goudman and Swartman. This small herd not only provided the children with much-needed milk, but allowed for a ploughing team; soon the harvests far transcended those from fields worked only with hand-hoes. Sekwala made a forty-mile round trip to Schweizer-Reneke, where the Stirling Brothers, remembering him from Rietput days, extended him credit for a small 75 plough; he used the four beasts as collateral. The spirit of enterprise, born on the plains of Makaoteng at the height of BaSotho prosperity in the late nineteenth century was being transplanted into a new and even more challenging environment.[10]

With its single share and wonderfully light wooden beam, the *Nkakatlele*, or 'Balance Me' plough, was ideally suited to the needs of the Maines at a time when they owned only a few livestock and could not call on the labour of the two oldest boys, who were at Hartsfontein. With Motheba at the helm and either Kas or Moselantja leading the team, Sekwala was free to use the whip on the hard-worked oxen. Indeed, the 75 was so light that even little Moselantja could cope with it if Motheba were called away to attend to the infant Sebubudi. The plough appears to have paid for itself fairly rapidly, because not very long after its acquisition Sekwala set his sights even higher.[11]

During a visit to Bloemhof in the winter of 1907 he called in at a new general dealers store, Gabbe's, where a plough much larger than the 75 caught his eye. The Star was a heavy two-share metal plough that needed at least twelve oxen to draw it. It was also an unwieldy thing that would readily topple over unless held firmly, and the blades had to be set precisely if it was to work at anything approaching maximum efficiency. It left him in a quandary: on the one hand it would require a much greater investment in animal and human muscle power and would stretch the family's newly acquired resources to their utmost; on the other, if the rains did come, the plough could help produce a truly bumper crop. So much had been won during the preceding seasons that it seemed to be a chance worth taking.

A few days later Daniel Gabbe, a tall bespectacled Jew from Riga, in Latvia, arrived at the farm. Sekwala and the boys helped him unload the plough, then stack the twenty bags of grain that Sekwala had pledged to him onto the back of his wagon. Gabbe left, but when the plough was removed from its crate, the boys watched as their father's brow slowly furrowed. Unlike the model in the store, the one before them came in bits and pieces and needed skilful assembly. As he stood contemplating the bewildering array of parts before him, all the old reservations about the cost and size of the Star again surged through Sekwala's mind until he consciously steadied himself and resolved to look around for help. He knew to whom to turn.

In the spring of 1906 Sekwala had resumed his friendship with Hendrik Swanepoel when his former transport-riding partner had bought a portion of the farm Koppie Alleen, which lay immediately south of Mahemspanne, from one of the most notable landowners in the district, Leopold Stern.[12] Swanepoel's return to the district (after a lengthy spell in a British prisoner of war camp in Bermuda) was welcomed by the Maine menfolk, who were hampered by the absence of a sympathetic white who could act as broker in a colonial society, especially when they needed written passes to move about the district. Swanepoel was bound to know how to assemble the plough.

Indeed, Swanepoel not only taught Sekwala how to use the plough but willingly lent him the additional animals needed to draw it. The Star, true to name, helped to produce bigger and better crops. Turning two furrows rather than one, Sekwala managed to increase the area under cultivation with only marginally more effort. Being considerably heavier than the old 75, the new plough cut deeper into the soil and helped to preserve more of the life-giving spring moisture that made for higher yields. Perhaps it was the Star's efficiency at soil-breaking that gave Mahemspanne its SeSotho praise name of Thubang, meaning 'to break.' The resulting increase in acreage planted with grain meant mobilising more family labour at harvest time, but it brought its own reward once the crop had been threshed, which was done by having it trampled underfoot by the cattle or horses. The surplus of sorghum that the new plough helped produce was loaded onto Hendrik Swanepoel's wagon by the boys and taken into Bloemhof, where it was traded at Gabbe's store.[13]

Prosperity based on grain farming, supplemented by four or five seasons of reasonable rainfall, was expressed in the Maines' rapidly expanding herd of cattle. In name at least, this was presided over by Hwai, patriarch of all the Maines. By the end of the decade it was said that the old man, his twelve sons, and their numerous children controlled no less than two hundred cattle—and that at a time when the total livestock holdings of blacks throughout the district numbered little more than two thousand. Far from being a collection of mere scrub cattle, this herd was serviced by four quality bulls named Hartman, Kruisman, Lapa and Rooilan. Rooilan, an enormous Afrikander bull, was especially admired by the local Boers and so assiduously pursued by Gert Meyer that the Maines eventually accepted one of his best oxen in exchange for it. From its new home at Hartsfontein the bull then found its way to the herds of many other prominent white farmers in the Bloemhof district.[14]

The task of looking after these cattle—a demanding assignment, since most properties were unfenced and animals were likely to invade neighbouring fields or get lost—fell to Hwai's grandsons. A generation of ten- to fifteen-year-old cousins, with the inseparable Kas and Sempane at its core, spent most of their early years at Mahemspanne undergoing an apprenticeship as herdboys. Some of this instruction occurred around the fire at night, when the elders taught the boys how to count or recall small differences among similar objects by playing traditional games such as *molatadiane* ('the follower') or *phupwe-ka-lefeng* ('where is the stone?'). Most of the instruction, however, was more practical, involving everything from milking goats and cows to the skills associated with stick and stone fighting. And, as in all schools, at least part of supposedly productive time was spent simply in testing the limits of authority by indulging in activities expressly forbidden by the elders, such as mounting and racing cattle, or arranging for the most formidable bulls to fight each other.[15]

Sekwala's boys were also offered more personalised tuition in specialist skills. A mild-mannered man with great reservoirs of patience, Sekwala spent most of the off-season working cattle and game hides into yokes, harnesses, bellows, shoes or karosses, activities that attracted the attention of Phitise and young Kas. While virtually all the children were fascinated by the forging and sharpening of the short handheld hoes, called *petlwane*, Mphaka and Phitise were keen to help with the arduous tasks of dressing stone, building and thatching—operations in which the hard-working Sekwala became increasingly involved as the family settled in at Mahemspanne.[16]

Ironically, at just this point new and challenging developments that might jeopardise their future were being set in motion elsewhere in the district, unleashing shockwaves that could undermine the emerging social and economic structures at Mahemspanne and, in the long term, do much to destroy the quality of life for rural blacks throughout the southwestern Transvaal. The unlikely epicentre of this upheaval lay some twenty miles south, in the unassuming little village where Sekwala did most of his grain trading, Bloemhof.

After the monopolisation of the Kimberley diamond fields by the giant De Beers' Corporation in the 1880s, undercapitalised independent diggers had worked their way along the course of the Vaal River and thicker deposits of gravel close to its banks. By 1906, the diggers had reached as far north as Christiana, where they experienced some success with the thinner deposits of gravel lying a little way from the river. These 'dry diggings,' sometimes more extensive than their riverine counterparts, were often less well endowed and tended to enjoy a limited economic life.

When the Christiana diggings faltered after a year or two, the diggers, many of them so-called poor whites, continued to go north, and by 1908 members of the advance guard were already at work in the vicinity of Bloemhof.[17]

In 1909, the Bloemhof diamond fields yielded a mere 783 carats valued at £2,886. From this very modest base, production doubled over the following twelve months. Such a gentle tap on the shoulder produced but few stirrings in an economic body that had previously slumbered. A positively rude awakening followed, however, when a far more extensive field was discovered at Mooifontein, about fifteen miles northwest of Bloemhof, in 1911. The proclamation of this field attracted close on 5,000 people, while a further 1,200 fortune-seekers made their way to the diggings on the Bloemhof 'townlands.' By the end of 1911, the new fields had yielded more than 37,000 carats valued at close on £200,000, and this development was sustained over the following two years.[18]

Wrinkled and weather-beaten old Bloemhof, which had languished in the enervating western Transvaal sun ever since 1866, when it had come into being as an outgrowth of a convenient drift over the Vaal, was suddenly attracting younger friends from near and far. An influx of attorneys, traders and storekeepers soon found their commercial activities underwritten by branches of Rand banks, while at least a dozen diamond buyers representing the largest houses in London made a weekly visit to the new industrial centre. A brewery, several liquor stores, two hotels, the erection of the Palace Theatre, regular visits from Pagel's Circus and the establishment of a newspaper, *The Diggers' Friend*, all testified to this newfound vitality. The town was accorded the right to run its affairs through a Health Committee, and by 1912 many of its leading inhabitants were of the opinion that it should be accorded full municipal status.[19]

All these developments were carefully observed by Bloemhof's Resident Justice of the Peace, Matthew Wood. Wood, a tall man and one of the many postwar British 'settlers' in the district, lived on the outskirts of the town on a farm, Kareefontein, which he hired from different land companies before eventually purchasing it in 1922. From his vantage point he was well placed to assess the effect which the new industry was having on both the town and the surrounding countryside, and not all of it met with his approval. Shortly after the first major discovery of diamonds in the district, he became involved in an acrimonious dispute that developed between the townsfolk and the diggers regarding grazing rights on the Bloemhof common.[20] Likewise, by 1911, his own personal experience made it superfluous for the local magistrate to inform him that 'native labour was badly wanted by the farmers' and that 'the diamond diggings were paying such high wages that the farmers were deprived of all their farm hands.'[21] A shortage of labour was something Wood had had to struggle with for years before it became a widespread complaint.

Kareefontein, bordering the Vaal and straddling the road from Bloemhof to Christiana, was poorly situated for any struggle revolving around black labour since it lay directly in the path of diggers advancing up the river. Certainly by 1907–8 Wood felt the effects of this growing competition for unskilled labour; he suddenly did not have enough labour to run his farm. When he happened to mention this problem to his brother, W. E. Wood the land agent, he first heard of the community of independent blacks at Mahemspanne.

Matthew Wood arrived at Mahemspanne on horseback, introduced himself to Sekwala as the 'magistrate' from Bloemhof who was empowered to inspect abandoned properties, asked how long the Maines had been living on the farm and inquired how much rent they paid. When he was told that to date no landlord had ever appeared to ask for rent, Wood suggested that they best find work on neighbouring farms if they wished to continue living on the property and remarked that he needed two workers to assist him at Kareefontein. Sekwala reluctantly agreed to the proposition that his two oldest sons would each undertake a spell of labour on the 'magistrate's' farm in return for the Maines' right to stay on the property.[22]

This arrangement held little appeal for the family, which in 1907 had acquired an additional mouth to feed when Motheba had been delivered of a third daughter, Sellwane. Sekwala lost the services of two of his most valued helpers at the very moment that he was poised to make most use of the Star, and Motheba, given her wartime trauma, was reluctant to lose sight of the children under any circumstances. In addition, Mphaka, now sixteen and newly graduated from a traditional circumcision school overseen by his grandfather Hwai, was wanting to acquire stock of his own and did not like to work in exchange for something as intangible as rent. Yet, despite their misgivings, Mphaka and Phitise dutifully laboured at Wood's farm for six months, during which time they were taken under the wing of yet another family of BaSotho labour tenants, the Masihus, whose origins lay in far-off Taung.[23] Afterward, the boys assisted Sekwala with the harvest at Mahemspanne; although not as large as hoped for, it nevertheless yielded enough resources to rekindle Sekwala's enthusiasm for a venture he had been planning for some months.

The time had come for the ageing and sickly Hwai to go on a visit to Makaoteng, so that he could take final leave of the kin whom they had left behind at Sekameng some twenty years earlier. In the spring of 1908, after the first rains and the planting of the crop, Sekwala helped Hwai mount the grey mare and, with him relying on shanks's pony, they set out for the distant Malutis.[24]

Throughout the long hot months of 1909 Mphaka and his brothers attended to the fields at Mahemspanne on their own. Summer crawled slowly by and the days were already noticeably shorter when, late one afternoon, they heard the familiar strains of Maine praise poetry being recited in the distance:

> The moon rose very early in the morning,
> Raphoolo was born very early in the morning,
> Rasehai scattered the dawn and the feet,
> The round one of fame, a MoFokeng.
>
> This one belongs to the gorges, he is a
> black wildebees of Majara's.
> The round beast of fame, a MoFokeng.
> He is an insect which continues to buzz.
> They say he is 'Mmamohohojoe.'
>
> The moon came very early in the morning at Raseaheng's,
> It came with the dawn and bent the darkness.
> The round beast of fame, a MoFokeng.
> He belongs to Legotho-wa-Dipudungwana-Majara.[25]

It was Sekwala, seated on a new grey named Bloutjie, accompanied by Hwai and two distant kinsmen from Makaoteng, who had come to see where these wandering Maines were attempting to take root.[26]

In the cold winter months that followed, the family neither saw nor heard of the 'magistrate' from Bloemhof. Rather optimistically, Sekwala hoped that they had perhaps seen the last of the tall Englishman on horseback and that the boys could spend the coming season at Mahemspanne. But when spring came and Wood reappeared with his familiar 'suggestion,' Sekwala realised that the family was probably coming to the end of an era. Some months earlier Hwai had already taken part of the herd and re-established himself on the farm Soutpan, and, if Sekwala and his sons were ever to free themselves of their newly acquired master, then they too would probably have to find a new home. The 1909–10 season's crop was again planted without the assistance of the two older boys, as the services of Mphaka and Phitise were commandeered by Matthew Wood.[27]

By the time Halley's comet appeared the following year, Wood was no longer alone scouring the district for independent labour bottled up on the properties of absentee landlords. Just as Mphaka and Phitise were completing what was their final spell of duty at Kareefontein in 1910, Wood's counterpart at Schweizer-Reneke, Thomas McLetchie, Resident Justice of the Peace, telegraphed the Secretary of Native Affairs in Pretoria urgently requesting the ' "application of the Squatters Law to this Ward to distribute Native Labour. Companies (sic) farms are full, some farmers have more than they need, others have none." '[28]

This appeal to officialdom produced a survey of all properties in the

Bloemhof, Christiana and Schweizer-Reneke wards occupied by five or more black families, the number allowed for by the provisions of the Squatters' Law of 1887, 'whose explicit purpose was to redistribute African households.' Thus, by June 1910 the authorities knew that McLetchie had eight extended families living on his farm at Zandfontein, that Norman Anderson had seven residing at Mooilaagte, that W. K. Jelliman had six at Niekerk's Rust, and that at least two large established black households were still farming at Mahemspanne.[29]

Despite the prevailing drought, The Year of the Star, *Naledi*, saw reasonable rain at Mahemspanne, and by the time that Sekwala and his family were told to relocate themselves on a farm under a white landlord, the crops were already stretching out toward the sun. While Sekwala did the rounds looking for a new home, his wives and children brought in a bumper harvest which, after provision of twenty bags had been made for the insecure year in the offing, left them with a handsome surplus. The Maines had exceeded themselves, and when he returned from his tour Sekwala used his old friend Hendrik Swanepoel's wagon to take thirty bags of sorghum to Bloemhof.[30]

He returned home from Gabbe's store that year laden with gifts. As an expression of pleasure at the size of the harvest and in recognition of what had been achieved during their eight-year stay on the property, Sekwala gave distinctively marked blankets to his wives and their older sons, each one in keeping with tradition, and appropriate to the recipient's status. The Maines left Mahemspanne at the end of winter in 1910 with the dignity and style that befitted a BaSotho family of substance, Sekwala mounted on Bloutjie, Motheba and Maleshwane each wearing the red, yellow and black *Rope Sa Motswetse* blankets portraying their standing as married women with children, Mphaka and Phitise each clad in the black and white stripes of *Konyana,* which told the world that they were unmarried men, and fifteen-year-old Kas—a mere uncircumcised boy—wearing his red blanket, *Qidi*.[31]

*

Although their destination, the farm Vaalboschfontein, lay only seven miles north, Sekwala had chosen the new location with more than mere convenience in mind. Makgalo, as Vaalboschfontein was more commonly known amongst SeTswana-speakers, was an enormous and largely undeveloped property of more than 5,700 morgen—at least three and a half times the size of Mahemspanne. Land on this scale held out the promise of unrestricted grazing as well as more than enough space in which to do maize or sorghum farming. Moreover, while it was true that the farm was overseen by a 'Boer' supervisor, it was said that the new owner was an 'Englishman' from Johannesburg who was seldom seen on the property.

Like most tenants, Sekwala was of the opinion that the best landlord was the one least encountered.

In fact the new owner was not an 'Englishman' at all but the thirty-eight-year-old son of a Swedish sea captain, Olaf Lindbergh, who had settled in Wales during the late nineteenth century. In 1892, Albert Victor Lindbergh had given up being a school teacher and gone to South Africa, where he worked in the publishing department of *The Star* newspaper before moving on to establish the Central News Agency. This modest enterprise made spectacular progress in the immediate postwar period, when both the Argus Printing and Publishing Company and the *Cape Times* decided to place all their publishing contracts with the C.N.A. By 1904, A. V. Lindbergh was well enough established to be involved with the mining magnate Abe Bailey and others in taking over the *Rand Daily Mail*, and two years later he became a part owner of the even more successful *Sunday Times*.[32]

This meteoric rise thrust Johannesburg's new press baron into elevated social circles; by 1908, Victor Lindbergh, like several other Randlords, was looking to underline his arrival amongst the gentry by acquiring the accoutrements necessary for country life. At this Edwardian juncture—only a matter of months before the diamond discoveries triggered an increase in land prices and the arrival of less elevated elements into the south-western Transvaal—Lindbergh acquired Vaalboschfontein in the knowledge that it could house a game lodge for shooting parties, and be a stud farm for his race horses and pedigree cattle.[33]

But in Wolmaransstad no less than in Wales, a gentleman's farm required steady hands at reasonable rates, and by the time that Lindbergh was recruiting a workforce, the diggings were siphoning off local labour at an alarming rate. These underlying forces, at least as much as the landlord's undoubted generosity, enabled Sekwala to extract favourable terms from Lindbergh's intermediary on the estate, Hendrik van der Walt.[34]

Unlike the local Afrikaner farmers, who were mostly hopelessly under-capitalised and reliant on labour tenants, the foreman at Vaalboschfontein was in the fortunate position of being able to pay monthly wages in hard cash. In return for grazing rights and access to fields, Van der Walt wanted a guarantee that at least two of Sekwala's older sons would work on the estate at any one time; in addition to ten shillings a month for herding or stable duties the Maine boys were to get a weekly ration of half a bag of mealie meal. In accordance with custom they, like all the other workers on the estate, would also be eligible to remove the meat of any animal that died on the farm—an important perquisite on a property where sheep and cattle were plentiful.[35]

This accord suited Sekwala well since it left Kas and Sebubudi free to tend the fifty cattle they had brought to Makgalo (most of the family herd remained with Hwai and the others at Soutpan). But for all that the future did not necessarily belong to cattle farming: in the next few months, Sekwala noticed that farmers were expanding their sheep holdings, for the price of wool was rising steadily. With his access to virtually unlimited grazing at Makgalo, he followed the market trend and bought sheep to supplement the few Angora goats they had brought from Mahemspanne.[36] (It so happened that this diversification came at a good time. Gal-lamziekte, a form of virulent animal paralysis, took a very heavy toll amongst the cattle of the district in 1911.)

In spring, once he had seen the last of the frost, Sekwala set about teaching Motheba's boys how to shear sheep, a skill they soon mastered, earning an enviable reputation amongst white farmers for miles around. But the reward for all this effort—which came only after most of the wool had been baled, transported to Vryburg by cart, and forwarded by rail to a Port Elizabeth merchant—proved to be rather small, since the price for both wool and mohair slumped badly in late 1911. But not even this setback could dampen the business spirits of young Kas who, showing the same sharp entrepreneurial eye that characterised his father and grandfather, used the remaining mohair to knit skull-caps, which he sold to workers on the estate.[37]

This new-found enthusiasm for sheep was not without cost, for as Sekwala's interest in sheep farming grew, so the family's involvement in grain production decreased. An adequate supply of fresh meat and milk as well as an assured ration of mealie meal from van der Walt meant that Sekwala could afford, for the first time in years, to avert his eyes from the merchants in Bloemhof and Schweizer-Reneke. He continued to use the Star for planting sorghum and supplemented it with a new Chain Plough, but the Maines both sowed and reaped less at Vaalboschfontein than they had at Mahemspanne.[38]

Somewhat paradoxically, this move away from arable farming occurred when they were becoming more involved in the traditional social rituals normally associated with grain planting. There was a sizeable village at Makgalo with many strangers in it, and this necessitated taking precautions against witchcraft which, if left unchecked, could lead to disturbing variations in the size of the harvest. Prior to the spring planting each year, Sekwala and all the other heads of household on the property appeared before the village headman and presented him with a selection of the seeds intended for use in the forthcoming season. These samples were pooled and the resulting mixture, called *mosuelo*, treated with a traditional potion such as *phetola* which was said to hasten the ripening of the grain. All the participants in the ceremony would then be given a small quantity of the communally generated mosuelo which, on their return home, would be re-integrated into the bulk of their supplies, thereby minimising the risk of antisocial behaviour.

The passing of the seasons also nudged other rituals to the fore that helped strengthen the bonds among the black families on the estate.[39] Once he was through with ploughing and planting and the footsore oxen had been released into the fields to recover their strength, Sekwala and the others would call upon their wives to brew beer from the excess sorghum so that everyone in the village could join in 'drinking the cattle's legs,' *ho noa maoto a dikgomo*. Likewise, if the rains held off for too long, the men and boys would assemble for a traditional communal hunt, or *molutsoane*, which, so the elders said, had never yet failed to bring on a storm. More predictably perhaps, the agricultural year would end with a round of 'work parties' where a liberal provision of beer ensured that bringing in the harvest remained the most memorable of all occasions.

Social activities such as these undoubtedly helped to integrate extended families into the wider community at Vaalboschfontein, but they were less successful at preventing a small but worrying crack from developing within Sekwala's own household. A short, mild-mannered man with a marked dislike for any form of open confrontation, Sekwala had for some time shown a preference for the company of his second wife, Maleshwane. The precise reason for this was difficult to determine—perhaps the more pressing demands of the junior house, a greater social compatibility between an avowed traditionalist and a pagan wife, or simply the sexual advantage that age bestowed on the younger woman.[40]

This development did not escape the notice of the tall woman of the senior house, who could, as her children well knew, manifest a truly explosive temper when provoked. Kas, for example, recalled an occasion at Mahemspanne when Motheba had snatched a stick from the cooking pot and struck out at him so powerfully that he had fallen to the ground unconscious. Motheba, who for a horrible moment believed she had killed her child, was instantly chastened. Although fiercely independent when challenged, she normally carried herself with a quiet dignity that could be traced to her deeply held religious beliefs. These same convictions, strengthened by her new allegiance to the then rapidly expanding African Methodist Episcopal Church, gave her the strength to tolerate Sekwala's declining interest in their marriage, and she steadfastly refused to make an issue of it. Instead, she devoted herself to her children and, in so doing, drew them closer not only to herself but to her church.[41]

Although it took four or five years to become fully manifest, the effects of this gradual realignment within the extended family had first become apparent at Mahemspanne when Sekwala's absence had caused Motheba's oldest son to take on a more paternal and responsible role within the senior house; Mphaka became something of a father to his sisters Moselantja and Sellwane and wielded more power over his three younger brothers than might otherwise have been the case. A short-tempered disciplinarian who set demanding standards for any form of agricultural labour, he made certain that the gentle psychological heat which Motheba kindled within the household was not allowed to dissipate out in the fields. But by the time the Maines got to Vaalboschfontein, Mphaka was finding it hard to play this supportive role to the full. As a man of twenty he knew he would have to get out and earn more cash than he was getting from van der Walt if he were to raise the bridewealth that would enable him to start a family of his own. He had managed to acquire a few beasts after a spell of labour on railway construction sites a few years earlier, animals which Sekwala held in trust for him, but he knew that there was still insufficient *bohadi* for a wife. Eager for wage labour, but at the same time bound by his sense of duty to his mother, Mphaka compromised and sought work close to home.[42]

In 1912, the small general dealer's store situated on the western border of the Lindbergh estate at Hessie had been taken over by Norman Anderson and William Hambly, two settlers who, having completed their service with the British Imperial forces in the South African War of 1899–1902, had chosen to stay on and establish themselves in the Schweizer-Reneke district. Anderson, a New Zealander, had for some years been a farmer at Mooilaagte while Hambly, an Australian, had been Secretary to the Wolmaransstad School Board before he, too, had turned to farming at Vaalboschbult. Earlier that year Hambly had met and taken a liking to Sekwala, so when he and Anderson started looking around for an intelligent young man whom they could put in charge of a horse and cart to make deliveries to the surrounding farms, they knew whom to approach. Sekwala steered them toward Mphaka.[43]

Although a full-time commitment at Hambly's store met Mphaka's needs admirably, it gave rise to a problem back at the estate: van der Walt was quick to point out to Sekwala that the terms of their contract specified that he had to ensure the services of two of his sons, not just one. Sekwala had to rearrange the duties of the three remaining boys so that Kas could join Phitise on the estate, and Sebubudi could join his half-brothers in looking after the family's livestock. For the seventeen-year-old Kas, it was his first exposure to the rigours of wage labour under direct 'European' or white supervision, and, from the accounts which his father received, the young man appeared to revel in the experience.

Initially, Kas assisted the other unskilled labourers in fencing the farm's perimeter and dividing it into camps. Although largely a response to the requirements of the Fencing Act of 1912, this effort would also have helped Lindbergh in the management of his pedigreed livestock and in warding off the unwanted attention of poachers. When this work ended the foreman called Kas aside and outlined his new duties. Van der Walt had noticed that the young MoSotho had a special talent with horses and he assigned him to the stables, where the man in charge, 'Maarman' Mokomela, helped the lad to extend and refine the skills that came with the cleaning, grooming and exercising of a dozen or more highly temperamental race horses.[44]

Kas soon established himself as the best stable hand on the property, and, not long thereafter, when Albert Lindbergh acquired Sam—the first race horse to be imported from Australia—he was entrusted with sole care of the new stallion. This exposure to pedigreed animals increased his love of horses, and this enthusiasm, in turn, reawakened his father's interest in the animals that had helped the Maines to get their first position under Woldmann back at Mequatling thirty years earlier. With one eye on his son's developing skills and the other on the market, Sekwala invested in several additional horses.[45]

The father's delight at his son's achievements was matched by his pleasure at other signs indicating the emergence of a well-rounded young man. Shortly after Mphaka had started working at Hambly's, Sekwala and his brothers had requested the foreman to be allowed to convene a circumcision school for the boys on the farm who were coming of age. Van der Walt, eager to accommodate the wishes of the community and quick to appreciate that graduates from an initiation lodge would help swell the pool of labour available on the farm, had readily granted them permission to hold what his BaSotho tenants somewhat misleadingly persisted in referring to as a 'Mountain School.' For seven months centring on the winter of 1912, Hwai and his older sons instructed twelve young Maine cousins and fifteen of their BaKwena and BaRolong peers in the mysteries and responsibilities of manhood. Kas emerged from this exacting exercise with great honour and a significantly enhanced respect for the richness and complexity of his native BaSotho culture.[46]

Motheba, in keeping with her pragmatic philosophy, raised no objection to her sons having this traditional initiation. Instead, as an avid Bible reader, she endorsed her children's search for literacy by encouraging them to attend the conventional classes that were, for a short period, a feature of life at Vaalboschfontein. In the evenings Motshwarateu Pholwana, the oldest son of one of the more prosperous BaRolong tenants on the property, would, for a shilling a month and the provision of the paraffin for the lamps, teach young and old alike to read and write. Although not excelling, in this too Kas made enough progress to enable him to cope with the normal exigencies of farm life, while his rapidly growing proficiency in the Afrikaans language benefited from his casual interaction with van der Walt's three children.[47]

The work in the stables did not always leave him with as much time for recreation as he would have liked, but Kas did find moments to take advantage of his recent change in social status. Some of these precious hours were squandered on adolescent peers, as when he and the herdboys experimented by smoking *dagga* obtained from the marijuana plants which some of the more adventurous tenants at Vaalboschfontein cultivated between their rows of sorghum. But most of this valuable free time was jealously guarded and spent in the company of his cousin Sempane. The two of them often disappeared on short hunting expeditions to the far corners of the estate, or else mounted specially trained cattle for more leisurely visits to the girls in some of the nearby villages.[48]

*

By 1913, Sekwala found himself under increasing pressure, both as patriarch and as labour tenant. Having found a cousin for Mphaka to marry, he made the painful discovery that, if the price in bridewealth cattle was to be met in full, the balance between milch and draught animals in the Maine herd would be seriously disturbed. He could extricate himself from this potentially embarrassing situation only by persuading Kas to allow a few of his cows, rather than all of Mphaka's trained oxen, to be used as part of the bohadi settlement. But no sooner had this difficulty been overcome than he was confronted by a new and even more serious problem.[49]

In 1912 the spring rains had failed to materialise and, by the summer of 1913, the whole country was in the grip of one of the most serious droughts in decades. Even Vaalboschfontein's extensive fields were not enough to provide all the animals on the estate with grazing, and it soon became apparent that, if the landlord's game and pedigreed stock were to survive, the tenants' livestock would have to be culled. Sekwala was in no hurry to respond when Hendrik van der Walt requested that the tenants reduce their cattle, and van der Walt, valuing the Maines's labour, was reluctant to press for an answer. But the resulting silence placed Sekwala on the defensive, and when the normally mild-mannered Kas became involved in a clash with the foreman, he found his space to manoeuvre restricted.[50]

At dawn one morning, Kas and Sempane started the day as usual by separating the calves from the cows, isolating the two sections of the herd in different camps and then reporting back to the stable for duty. In midmorning they paused for a break and were busy having their breakfast when van der Walt burst in on them to tell them that the calves had somehow managed to get out of their camp and to suckle from the cows before the milking had commenced. As he spoke, the foreman took out a leather strap from his pocket and lashed out at them. The cousins leapt to their feet and fled the room, but when van der Walt ran after them, still wielding the belt, Kas stopped, picked up a stone and threatened to retaliate. This unexpected challenge stopped the foreman in his tracks; apparently regaining his composure, he told them to finish their meal before returning to their chores.

But van der Walt had no intention of allowing a *klein kaffer* to question the authority of a white man, and he waited for a suitable moment to complete the business he had started. That evening, when the oxen were being returned to the kraal next to the farmhouse, the foreman lured Kas within his reach, caught hold of him by the arm and then—while he and the stable-hand danced the disciplinary jig—called for his wife to remove the leather strap that he had hidden in his back pocket. Before the lady could oblige, Kas, inspired by the need for improvisation in a novel social setting, sank his teeth into a conveniently situated white finger, which promptly spurted enough blood to elicit a chorus of shouting and swearing from a clearly impressed Mrs. van der Walt. A half-dozen blacks, suspecting that oaths and cries on such a scale could only be a summons for their services, suddenly appeared from nowhere to witness some deeply concerned Maine kinsmen persuading Kas to release the hapless foreman's finger.[51]

This potentially serious altercation had no immediate sequel. In the days that followed van der Walt made no further overt attempt to discipline the 'little kaffir.' He stalked his quarry so surreptitiously that it took several weeks before it became apparent from which direction he was approaching. The young Maine's duties around the farm gradually, at first almost imperceptibly, increased; as the weeks and months passed Kas became aware of the progressive erosion of his leisure time. But the son of a labour tenant, caught in the vice of a quasi-feudal relationship where he was accountable to father and foreman alike, found it almost impossible to mount a counterchallenge that did not jeopardise his family's position on the estate. Instead, he spoke to his father, who on more than one occasion remonstrated with the foreman about what he considered to be his son's unrealistic work load. Despite this attempt at paternal mediation, the relationship between foreman and worker continued to deteriorate.[52]

Matters came to a head at an awkward moment which—like the ragged line between loyalty to the family and wage labour—was poised uncertainly between the hours of work and leisure. One Sunday an ox in a herd grazing on a distant part of the farm sickened and died. When the foreman was told of this, he accused Kas, on duty around the farmhouse that afternoon, of being so obsessed with doing the social rounds that he had failed to keep track of the livestock. Kas pointed out that others were supposedly responsible for animals kept at a distance from the homestead and asked for a clearer definition of his own duties; at that point van der Walt moved to occupy the space he had opened up during the preceding months. He insisted on holding Kas responsible for the death of the ox, the exchange became heated, and it ended only when Kas demanded to be released from all his duties. The foreman, seizing the chance to ease his cattle-culling problems and simultaneously end the dispute with his independent-minded stable-hand, readily agreed, pausing only to remind him that this meant all the Maines must leave Vaalboschfontein.[53]

Sekwala was now trapped between an unstated appeal for loyalty from his son on the one hand, and the need to ensure access to grazing for their cattle on the other, and he realised that the problem had marched on beyond the reach of mere diplomacy. A devastating drought, the danger of having the herd culled and a son who had become a man—all became concatenated in such a way as to leave him without choice. How was he to find a place large enough to accommodate all his family and all their cattle at a time when grazing was at a premium? He discussed the matter with Mphaka, and between them they struck on the idea of temporarily splitting the family along the cleavage line that had been developing between the patriarch and his two wives and seeking a home on not one, but two farms.

Late in the summer of 1913 Sekwala, Maleshwane and their children rounded up most of the livestock and went to Hessie, a farm that bordered Vaalboschfontein, to become sharecroppers under William Hambly; it was a property which ancient BaThlaping calling it *Mphaku*, had singled out for its fertility. Motheba and her children, along with Mphaka and his young wife, Elisa, collected the remaining cattle and the Angora goats and took them about twelve miles farther north, to a property on the main road between Schweizer-Reneke to the west and Wolmaransstad to the east. There, still under Mphaka's tutelage, they became labour tenants of Hendrik Koekemoer, who had hired the farm Doornhoek to supplement his grazing during the great drought.[54]

PART TWO

# YOUTH

*

*'After all that has been said on these more tangible matters of human contact, there still remains a part essential to a proper description of the South which it is difficult to describe or fix in terms easily understood by strangers. It is, in fine, the atmosphere of the land, the thought and feeling, the thousand and one little actions which go to make up life. In any community or nation it is these little things which are most elusive to the grasp and yet most essential to any clear conception of the group life taken as a whole. What is thus true of all communities is peculiarly true of the South, where, outside of written history and outside of printed law, there has been going on for a generation as deep a storm and stress of human souls, as intense a ferment of feeling, as intricate a writhing of spirit, as ever a people experienced. Within and without the sombre veil of color vast social forces have been at work—efforts for human betterment, movements toward disintegration and despair, tragedies and comedies in social and economic life, and a swaying and lifting and sinking of human hearts which have made this land a land of mingled sorrow and joy, of change and excitement and unrest.'*

W. E. B. DU BOIS,
1896

CHAPTER THREE

# Preparation

## 1913–21

Doornhoek was an exceptionally large property traversed not only by the Schweizer-Reneke–Wolmaransstad road but by several other minor roads, notably the one linking an adjacent farm, Katbosfontein, with the nearby hamlet of Glaudina and then, farther north, the village of Migdol. The self-confident manner in which these upstart country tracks strode across the property testified to the fact that the farm did not lie close to anybody's heart. In an area notorious for its vulnerability to drought, Doornhoek had a distinctive reputation as a place without water.

Through most of late summer and early winter 1913, the younger Maine brothers tended Hendrik Koekemoer's livestock and their own under the watchful eye of the formidable Mphaka—a demanding task, for the stock had lost condition and had to be driven over long distances to get water on neighbouring properties. Kas, accustomed to the varied duties and richer social life of Vaalboschfontein, found life at Doornhoek tedious and, after several months of unrelieved herding, was waiting for a new challenge to present itself when the rural tracks that crisscrossed the property suddenly dumped an unexpected opportunity at the family doorstep.

On a midwinter afternoon, a small gang of BaHurutshe roadworkers in search of off-duty company and sorghum beer were directed to Mphaka's homestead. During the conversation that followed, it emerged that they were BaKwena who not only shared Motheba's crocodile totem but, like her immediate family at Setlagole, had placed themselves under BaRolong jurisdiction and therefore qualified as distant kinsmen. 'Charlie,' the oldest of the men and leader of the party, sensing Kas's interest in the group's wanderings, suggested that he join it as a labourer on road construction works that would take them on a sweep through the southwestern Transvaal.[1]

The idea appealed to Kas: he could shake off the dust of Doornhoek and earn cash for a few more cattle and sheep of his own. Mphaka, who had met the family's immediate obligations to Koekemoer and already deputed Phitise to assist the landlord during the coming season, foresaw no problem, and Motheba, mollified by the presence of BaKwena guardians, raised no objection. It was to be expected that a young man newly graduated from the Mountain School would want to go out into the world to gain experience.

Kas joined the road gang at Katbosfontein as an unskilled labourer loading gravel into Scotch carts for ten shillings a month. For a while the contractor, an Englishman named Ellis who hailed from the Orange Free State, deployed the team near Doornhoek, and Kas managed to see something of the family over weekends. But a few weeks later the gang moved out toward Wolmaransstad, where they embarked on a longer-term project to improve the road linking the rail siding at Makwassie with Bothaville, just across the Vaal River.[2]

Bothaville was on the fringes of the most capital-intensive maize producing region in the country, and it was while Kas was at work on the roads around the district that the journalist Solomon Plaatje passed through the area and recorded the effects of one of the great rural spasms associated with South Africa's uneven transformation into an industrial society. Designed to eventually confine 80 percent of the country's population to the ownership or occupancy of about 13 percent of the land, the racially discriminatory clauses of the Natives Land Act of 1913 made provision for the state to prohibit white landlords and black tenants from farming-on-the-halves or sharecropping in regions where agricultural producers were said to be in the greatest need of wage labour.[3]

When these repressive measures were enforced in the northern Orange Free State the same year, they resulted in the eviction of hundreds of black families who then either had desperately to search the province for new homes, or had to cross the river into the drier southwestern Transvaal, where they sought out land-rich but undercapitalised white farmers who might need labour tenants. Neither the passing of the law nor its sequel escaped Kas's attention, and nearly seven decades later, he could clearly recall 'the law passed by the Boers which stipulated that we should no longer farm-on-the-halves but work on the farms as wage labourers.' 'But,' he went on to note, 'sharecropping continued because many white farmers could not afford to pay their labourers cash wages.'[4]

By December 1913, the road gang was at work on the banks of the Vaal in the vicinity of Kommandodrif when Kas returned briefly to Doornhoek to re-establish contact with his family. But by then water had become so scarce at Hendrik Koekemoer's cattle outpost that Mphaka and the others had decided to link up with Hwai at Soutpan, the farm to which he had moved in 1909 when he left Mahemspanne. Around Bloemhof there were yet more signs of displaced tenants moving into the district from south of the river, and Kas found the Maines, including members of his father's family regrouping on a large property which, it was said, belonged to an Englishman whose name, they thought, was 'Ou Stories.' As had happened with the first wages he had earned in the stables at Makgalo, Kas handed his earnings to his father, secure in the knowledge that Sekwala would invest it in livestock which he would find waiting when he returned from his travels.[5]

On rejoining Ellis and his team, Kas was assigned the role of cook, a task less strenuous than the heavy manual labour to which he had become accustomed and, at thirty shillings a month, better paid. For several months he rose before dawn to light the fires, prepare the relish and cook the drums full of mealie meal that the workers were served in midmorning and late afternoon. His culinary efforts were clearly appreciated by his BaKwena kinsmen since there was an outcry when for a brief period he was seconded to look after the oxen and a replacement was taken on as cook.[6]

Kas continued to remit money to Sekwala via trustworthy intermediaries whenever he could, but he did not get the chance to visit the farm in person for more than eight months. His younger brother, Sebubudi, teasingly suggested he had joined the ranks of the *lekgolwa*—migrant labourers so satisfied with their lot that they saw no reason to communicate with their families. But Kas was not in danger of becoming permanently reconciled to life on the road, and nor were others in the construction team. In winter 1914, as they approached Bothaville, he noticed signs of growing unease amongst Afrikaners in the party, who talked about a gathering conflict between the British and the Germans and the need to return to their families.[7]

When war between Britain and Germany was declared in August 1914, Ellis dismissed his remaining employees, suggesting that they, too, go home. But neither Kas nor Motsoko, a young MoHurutshe, could see the need for heeding the advice, though it held obvious appeal for the married men in the party. Instead the two bachelors went back across the Vaal and set out for the small town of Leeudoringstad, where they hoped to get work on the railway line linking the two great mining centres of Kimberley and Johannesburg.

The thirty-mile hike back to the heart of the western Transvaal was rewarded with one shilling and sixpence worth of wage labour that lasted for exactly a day. On the second morning the resident subcontractor appeared on site and, eager to avoid the practical problems and instability that came with employing young casual labourers, promptly dismissed them when he discovered that they had nowhere to live and no wives to cook for them. Caught in a world which for a brief moment seemed obsessed with a man's marital status, Kas remembered that there was one place where the appetite for the labour of single young black men seemed insatiable, the Bloemhof diamond diggings.[8]

But Kas had been gone for a long time, and the local demand for black labour had changed radically. The most promising of the diggings, at the farm Mooifontein, had been virtually worked out during the previous year, and he and Motsoko were forced to go farther north—to the new discoveries at London, not far from Vaalboschfontein. Even there conditions were not as buoyant as they had been when he had left Doornhoek. With the advent of war the demand for diamonds had dropped, and as a consequence the always moody economy of the southwestern Transvaal, already stressed by several seasons of drought, was plunged into a full-scale recession that lasted for most of 1915. The resulting reduction in wages made hard manual work on the diggings even less attractive than usual, and after only a few weeks Kas left his long-time MoHurutshe companion and went back to Soutpan.[9]

Soutpan, it will be recalled, had belonged to the Voortrekker leader and state president of the South African Republic, M. W. Pretorius, in the 1870s. The property had then passed through the hands of two other owners before Hendrik Pistorius acquired it in 1898. But the new landlord found a 5,500-morgen property to be much larger than he needed and had therefore let out sizeable portions to less well-endowed Boers established in the district.

When he rejoined his extended family in October 1914, Kas found that his father had entered into a labour tenancy agreement with one of Pistorius's Afrikaans-speaking tenants named 'Leeu' ('Lion') Badenhorst. This verbal contract specified that, in return for access to grazing for his animals and the right to plough a field or two, Sekwala would ensure that Badenhorst got year-round access to the labour of at least one of his sons. In practice, this obligation had to be met by his unmarried sons, who were expected to relieve each other at six-month intervals. Kas, having been exempted from family labour for virtually a year, was now called on to take over Phitise's duties.[10]

Despite his forbidding name and towering frame, 'Lion' Badenhorst was a phlegmatic man who had married Christina, the daughter of 'Ou Koos' Barnard, from Rietpan, the neighbouring property. She, by contrast, was a highly irascible woman with little respect for or patience with her husband's black tenants, and she spent most of her day concerned with a simple-minded son, Hendrik, and a daughter, Miemie. The gentle 'Leeu' was usually called on to restore the peace whenever Christina had cause to interact with the farm labourers. This was a largely thankless task which earned him his wife's swift rebuke: *'Ja, jou lang ding. Jy is mos al agter die kaffers aan, die kaffers maak soos hulle wil met jou!'* 'Yes, you misery. You curry favour with the kaffirs, and they do as they please with you!'[11] With the events in the stable at Vaalboschfontein still very much in his mind, Kas was pleased that his new duties centred on the livestock and kept him well away from the farmhouse.

There are things in the world more formidable than a woman's temper, and by late 1914 Kas could detect the distant thunder of a storm that threatened to engulf all of the southwestern Transvaal. The initial rumblings of discontent echoed sentiments he had first heard expressed on the road to Bothaville, and quite unexpectedly, Soutpan itself now seemed to lie directly in the path of the approaching tempest.

*

Following the declaration of war in August 1914, Britain had requested that South African government forces invade the neighbouring German colony of South West Africa. Whilst this request commanded a sympathetic hearing from those in high office, notably Prime Minister Louis Botha and his minister of defence, J. C. Smuts, it evoked far more equivocal responses from Afrikaners who did not have direct access to the levers of power. Veterans of the South African War, amongst them Generals J. H. de la Rey, C. R. de Wet and S. G. Maritz, either hankered for a restoration of the Boer Republics or felt unable to assist Britain in a struggle against their former political ally. Many of these reservations were shared by the Commander of the newly formed Active Citizen Force, General C. F. Beyers, and yet others by another high-ranking officer, General J. C. G. Kemp. These custodians of Afrikaner political independence tried to profit from the British plight by instigating an armed rebellion.

In the Orange Free State and southwestern Transvaal the fires of revolt ignited by the generals were soon stoked by mystics and desperate Afrikaner farmers who, in addition to simply drawing comfort from the heat, wished to burn off the miseries of drought and debt. Within days of the announcement of war, there was an excited republican demonstration in Lichtenburg, while in neighbouring Wolmaransstad, a famous local 'prophet,' Nicholas van Rensburg, or 'Siener,' as he was more popularly known, had a vision that gave succour to rural radicals accustomed to drawing on the idiom of the Old Testament for their inspiration. Within three months several thousand armed and mounted rebels were being pursued by government forces who, in addition to horses, had the advantage of access to motorised transport and the expanding South African railway system.[12]

At Soutpan, where the owner of the property was an English-speaker, the first signs of conflict became apparent when two of the tenants, Leeu Badenhorst and Andries Holloway, left to join the Citizen Force based at Makwassie. Most of the Afrikaners in the district, on the other hand, chose to join the rebels under P. J. K. van Vuuren, a *Veld-Kornet* or district officer from the farm Vaalbank in the Lichtenburg district who had pledged his allegiance to General Kemp. This sudden exodus of Boers occasioned a strange lull in activity at a time when preparations for the new season were usually in full swing. The departure of Badenhorst caused the Maines no immediate hardship, although a prohibition on movement by night made it hard for them to arrange Phitise's marriage.[13]

The unnatural spring calm was broken on 1 November when van Vuuren and three rebels swept into Soutpan, rode up to the farmhouse and called out the waspish Christina Badenhorst. They issued her with a receipt for a half-dozen sheep which they intended commandeering as part of the supplies needed for General Kemp's men, who were on their way to Schweizer-Reneke. There, they would rendezvous with General Beyers and his followers before pushing on to South West Africa, where they hoped to link up with the forces under the command of General Manie Maritz. A campaign on this scale, over what was some of the most testing terrain in the country, required dozens of mounts, and having determined where Kas had taken the sheep to graze, the Veld-Kornet started looking around for any horses they could use.[14]

But the sturdiest horses had long since been pressed into service by Badenhorst and Holloway, and the remaining animals belonged to the tenants. In what was held to be yet another 'white man's war,' this disposed of the need to issue any further receipts, and van Vuuren and his men simply 'exchanged' their injured white horse for two other animals, a brown one belonging to RaMofasiwa and a white-faced one owned by RaKolobe, before moving out into the fields at the far end of the farm to collect sheep. The four armed men had already isolated part of the flock when they were challenged by a surprised Kas who, when he dared question them, was dismissed with a perfunctory, '*Kaffer, hou jou bek!*' 'Kaffir, shut your mouth!'[15]

He and another tenant ran to the farmhouse where an angry Christina Badenhorst shouted at them to follow van Vuuren and his men even if it meant coming under fire. Reluctantly, they traced the rebels to an isolated part of the farm where, from a hiding place, they witnessed the arrival of yet more armed men. The Veld-Kornet issued an order that six rams be slaughtered and roasted, and while this meal was still in progress scouts arrived to report that a detachment of the Citizen Force had been seen in the vicinity. Before van Vuuren could arrange for his troops' withdrawal, the government forces—which included Leeu Badenhorst, Holloway and several friends—arrived on the scene and engaged the rebels, who, forced to take shelter behind a kraal wall, returned fire before disappearing in a northwesterly direction, hotly pursued by the Citizen Force.[16]

Meanwhile Motheba and Sellwane were on their way to Kommissierust, where they hoped to collect Phitise's bride-to-be and escort her home. On their trip they were overtaken by the thunder of horses carrying men in the direction of Schweizer-Reneke. But the rebels did the women no harm, and when Kas appeared with his companion some time later they could only point them in the right direction. By late that afternoon, they had lost both contact with the rebels and any remaining enthusiasm for a rather pointless chase. They left the rebels near the farm Makouskop and returned to Soutpan, where they recovered the skins of the slaughtered rams before turning to face the almost inevitable wrath of Christina Badenhorst.[17]

Van Vuuren and his supporters spent the night in the bush near Mooistroom before going to Vleeskraal, where, on 2 November, more than two thousand rebels were addressed by Generals Beyers and Kemp, the Reverend H. D. van Broekhuizen and the visionary 'Siener' van Rensburg.[18] After the meeting General Kemp and his followers occupied Schweizer-Reneke and wounded two policemen before abandoning the town and setting out to join Maritz's forces in South West Africa. Van Vuuren and his party linked up with the men under the command of General Beyers and rode south to the farm Kareepan, adjoining Soutpan, before moving on to the rail siding at Kingswood where they were involved in a skirmish with government forces.

This flurry of military activity around Soutpan was followed by weeks of relative calm until another small band of men, this time under the command of General de Wet, passed through the district and went to the farm Vleeskraal, where, on 24 November, they too clashed with units of the Citizen Force. Two weeks later General Beyers, hotly pursued by government forces, was drowned while attempting to cross the Vaal on horseback near Makwassie: gradually the government forces acquired the upper hand. By January 1915, the rebellion was over, and the southwestern Transvaal slowly eased its way back into the more sedate rhythms of the countryside.[19]

The Maines had watched the unfolding spectacle of Boer against Boer, where one was 'a government supporter and his brother a rebel,' with mixed feelings. Kas's thoughts mirrored some of the tenants' confused and wishful thinking about the rebels' failure to overthrow the government. On the one hand he, along with the rest of the peasantry, had long since lost faith in the capacity of the British Empire to force the Boers to improve the lot of blacks on white farms, and, although he knew the rebels' German allies had an unenviable reputation for cruelty, he nevertheless hoped that as a new imperial authority they would somehow preside over a more enlightened regime and usher in an era of material prosperity for rural folk. On the other hand, like the vast majority of blacks in the district, he found it difficult to see white rebels in the role of liberators, and joined in reciting some of the mock 'praise poetry' that accompanied the death of the hapless General Beyers:

> Beyers, you were a real man! You were clumsy and fell into the water with your trousers and jacket on, and a black horse jumped over your head. Your darkened blood mingled with the water and flowed to the ships until it reached a white woman—an Englishwoman—because it was a black horse that overturned Beyers the man![20]

Black society, although lacking a 'Siener' van Rensburg, was not without resources when it came to drawing out a deeper, more symbolic explanation for the unfortunate events that had played themselves out in the river at Makwassie.[21]

Whatever the cause of Beyers's death, the rebellion severely disrupted production in the 1914-15 season in a district where one in three farms was already heavily bonded to the Land Bank. This, coupled with a slump in the price of diamonds, released rank vapours of pessimism that soon enveloped town and countryside alike, and in professional banking circles it was officially reported that 'An air of poverty and little business prevails in Bloemhof.' At Soutpan, where Sekwala correctly interpreted the signals of a recession, the family turned its back on the market and contracted their agricultural efforts to mere subsistence levels. Indeed, they produced so little maize that they decided to sell off the flock of Angora goats which, besides requiring a great deal of mollycoddling in the cold highveld winters, needed a grain supplement to keep them in peak condition.[22]

With the markets for grain and stock equally depressed, the Maines unpacked some of their other skills and set to work generating a flow of cash that would see them through the hard winter of 1915. Sekwala and Hendrik Swanepoel, as they had done so often before, got together to do some transport-riding, but the poor harvests made it one of their less successful seasons. When Swanepoel returned to his home at Koppie Alleen, Sekwala and Mphaka moved across to the de Lange family at Kareebosfontein, who ran an off-season contracting business, and helped erect *rondavels*, wagon sheds or kraal walls, for the wealthier whites in the district. And when Phitise returned from his six-month spell on the diamond diggings to meet the family's obligations to Leeu Badenhorst, Kas engaged in some casual labour on neighbouring properties at one shilling and sixpence a day before moving back to the diggings for a few months.[23]

Despite this great effort, one member of the family did not see out the winter. Hwai, by then well into his eighties—repository of the BaSotho folklore, convenor of the initiation schools, and skilled ngaka—sensed the moment of his departure and called Sekwala, to hand over his divining bones to him. Shortly thereafter the freezing east wind that swept down from the Malutis arrived to remind him for the last time of the call of Kolo and the mountains of Sekameng. The old man was laid to rest by Sekwala and his brothers in the characteristically aligned east-west grave reserved for a true MoFokeng—the first of three generations of Maines to be buried at Soutpan. Kas and his siblings mourned the departure of their grandfather who, despite his advanced years, had been a familiar figure as he moved about the property on Skammoi, the horse which he praised to his last days as the animal that had carried him back to the Makaoteng of his youth.[24]

Hwai's death left more than an emotional imprint on the family. The departure of the patriarch uncoupled what had undoubtedly been the family's strongest personal link to the owner of Soutpan. Perhaps it was because of this, or the presence of the old man's grave, or simply because Sekwala had always seen the move to Doornhoek and Soutpan as an expedient response to the demands of the drought, that shortly after the burial they decided to abandon the Pistorius property and seek alternative pastures. Spring saw most of the Maines back at Vaalboschfontein.

*

The return to the Lindbergh estate at what was supposed to be the advent of the new rainy season was not auspicious. Nor was the rest of that summer. Indeed, the whole of 1916 saw only fifteen inches of rain recorded at Bloemhof, and a few miles west of the town the Vaal River, showing the cumulative effect of three years of cloudless skies, stopped flowing for several months. Withering beneath the hostile glare of the sun, land in the district lost between 12 and 15 percent of its value, and farmers scurried to replace cattle with sheep in the hope that the wartime rise in the price of wool would compensate for the ravages of *gal-lamziekte* and drought.[25]

Sekwala, who had eyes only for the extensive grazing that Makgalo could offer the family herd, either did not see this switch to sheep farming or concluded that 'blue-tongue' and 'wireworm' posed at least as many problems for wool farmers as did gal-lamziekte for cattle keepers. Enthusiastically, he set out to renew the family's labour tenancy agreement with Hendrik van der Walt—the same man who had been in charge of the property two years earlier. The foreman was still appreciative of the Maines' skills, and seemed willing to re-enter into a contract with them despite his earlier collision with Kas.

The resulting accord gave Sekwala access to a limited amount of arable land but, more importantly, once again guaranteed unlimited grazing for the cattle in return for the labour of two of his sons, who would also be eligible for the generous cash wages and rations that distinguished Vaalboschfontein from most of the surrounding properties. Any remaining tension between the parties was dissipated when Kas, in keeping with the niceties of the paternalistic ideology that dominated rural race relations, issued van der Walt a timely, albeit highly political apology for his earlier behaviour. What was interesting about this almost ritualised form of submission was the confidence of the nominally subordinate partner that his 'apology' would be accepted.[26]

Van der Walt lost no time in assigning his former adversary to the stables. Kas and Maarman Mokomelo once again did the routine chores, mucking out and grooming, in addition to occasionally helping the farrier, a local Boer called 'Ou Mapogo,' to shoe the horses. Given his skill at handling animals, it was not long before Kas was put in charge of breaking in and exercising the growing number of race horses at Vaalboschfontein, and when toward the end of the war Lindbergh imported yet another stallion, this time from Europe, he was entrusted with taking delivery of the animal at the Kingswood siding and escorting it back to the estate. He managed this admirably, despite having picked up a very severe bout of 'flu' somewhere along the way.[27]

Like most highly trained animals, the race horses on the Lindbergh estate were often nervous and temperamental, but any demands they made were well within the reach of the young Kas Maine, who had matured into a truly excellent all-round horseman. In addition to his skill as a groom and largely as a result of his exposure to his father and Hendrik Swanepoel's off-season transport-riding ventures, he had also developed the knack of working with horse and cart. This too was a useful qualification, and the foreman soon put it to the test.

In 1914, the forty-two-year-old A. V. Lindbergh, happy in the knowledge of having established himself as 'the W. H. Smith of South Africa,' had taken Gladys St. Leger, the daughter of a prominent Capetonian, as his wife. Gladys Lindbergh's arrival at his residence at Parktown, in Johannesburg, led to certain changes in his lifestyle, some of which filtered down to his estate in the southwestern Transvaal. Whereas Vaalboschfontein had once served largely as a game farm to provide sport for the owner and friends like Abe Bailey, it now became more of a family farm, and the Lindberghs used it as a seasonal retreat. Their year-end excursions to the countryside soon came to be marked by an annual Christmas party at which they generously hosted some of the less privileged children from surrounding properties, including the numerous offspring of the Dauth and Swanepoel families. Although primarily exercises in good neighbourliness, these gestures were open to other interpretations: for politically sensitive Afrikaners still smarting in the aftermath of their failed rebellion, such as the Niemans of Hartsfontein, the Lindberghs' Christmas parties came to epitomise the patronage of local 'poor whites' by a class-bound, city-based English-speaking gentry whose liberalism was anathema to their own nationalist project. Further evidence of the outsiders' capacity to undermine the rural social order and breach its unwritten codes of racial etiquette arose when Hendrik van der Walt asked Kas to take Mrs. Lindbergh on her shopping expeditions to Paradise Bros. Store in Wolmaransstad.[28]

These twenty-mile round trips into town by horse and cart were a revelation to Kas, whose previous experience of interacting with white women had been confined to chance meetings around the farmhouse with the shrewish Christina Badenhorst. Unlike Boer women, who were reluctant to travel with a black man unless escorted by their husbands, Mrs. Lindbergh did not baulk at the prospect of being taken into town by a black servant. Not only did she prattle away in Afrikaans during these long journeys, but she openly flaunted convention by joining him on the front of the cart, a practice that prompted comment in town and countryside alike. In addition, on their return home Mrs. Lindbergh, on more than one occasion, showed herself willing to assist by helping him return the horses to the stables. Kas attributed this social flexibility to cultural considerations rather than class, and thought that this derived from her being 'English' and the fact that 'she did not care' rather than from any confidence she might have acquired from lifelong exposure to black servants.[29]

But so marked was the hardship in the countryside after the drought and the rebellion that not everyone was willing to wait for Gladys Lindbergh's Christmas shopping expeditions to gain access to the fruits of Vaalboschfontein. As 1915 drew to a close it was noticed that there was a small but steady loss of sheep from the estate, and since Kas was responsible both for driving the sheep out into the fields at first light and for bringing them home at dusk, van der Walt's suspicions began to centre on him. Kas thought these thinly veiled accusations were inspired largely by racial antipathy and resolved to be extra vigilant by checking on the flock during the day.

One morning shortly before Christmas he took Davis, one of the work horses, and set out to check on sheep he had left grazing in a remote and wooded part of the estate. As he approached the copse, he heard what he took to be the crack of an enormous stock whip. On drawing closer, he was surprised to see the son of a neighbouring farmer running off into the bush. He dismounted and ran off into the thicket after him, only to stumble across the boy's father who—having fired the shot Kas had heard, which had killed one of the sheep—grabbed his rifle and also bolted. Reluctant to follow an armed white man, Kas picked up the carcass, slung it across the front of his saddle and took it back to the farmhouse, where van der Walt scribbled him a note and dispatched him to Makwassie to summon the police.

As Kas and a policeman returned to Vaalboschfontein they turned off the road before reaching the estate boundary, and went down a smaller and even dustier track that led to a tumbledown brick structure belonging to Klaas Verster. Verster, a *bywoner* whom the drought and the collapse of the diamond market had reduced to repairing chicken runs and kraal walls, was clearly in desperate straits. Unable to feed his family, and on the brink of starvation, he had taken to poaching sheep from Lindbergh; confronted with the evidence, he made a ready confession. The constable arrested him and took him to Wolmaransstad where, so Kas was later told, the magistrate sentenced him to a month's imprisonment.[30]

Vindicated by these unfortunate events, twenty-year-old Kas nevertheless saw in Verster's plight a timely reminder of the need to increase his own security by acquiring livestock in addition to that which his father already held in trust for him. His older brothers Mphaka and Phitise were both married men, and if he too were to take on the responsibility of raising a family, he would have to earn money for *bohadi* by doing more wage labour. When autumn set in and the pace around the farm slackened, he asked his father and the foreman to release him from his duties: armed with a pass signed by Hendrik van der Walt, he set out for the nearest diamond diggings, which lay twenty miles northwest of Vaalboschfontein.[31]

Although it lacked a Piccadilly Circus and was a trifle more modest in size than its famous namesake, London, even in 1916, was not without landmarks of its own. Dominated by the offices of the London Diamond & Estate Company, registered in 1911, this stretch of veld astride the main Bloemhof–Schweizer-Reneke road had blossomed in the period before World War I when the price of diamonds escalated. The boom had attracted hundreds of diggers, given shape to a small but tidy core of white settlement and then spewed out the surrounding sprawl of corrugated-iron shacks that housed the mass of black workers. But in the desert even the plainest flowers somehow succeed in getting pollinated, and the London diggings already displayed, albeit in outline form only, the typical profile of a southwestern Transvaal hamlet. The Half-way Hotel and Bottle Store, reserved for the exclusive patronage of whites, helped slake the European diggers' unquenchable thirst, four or five Asian-owned stores catered for the day-to-day needs of poor folk of all colours, while a Kaffir Eating House disgorged the coarser fare that thousands of black migrant labourers were forced to rely on for sustenance.[32]

But much of this precocious economic development had been checked by a collapse in the price of diamonds in 1914–15. In February 1914, it had been proposed that the holding company be sold and, although this proved unnecessary, the directors of London Diamond & Estate were forced to find new ways of supplementing the company's income. One consequence of this was that L.D. & E., which already laid claim to 10 percent of the value of all the diggers' finds, leased the sole water rights on the property to a concessionaire named Nicholas in return for a third of his gross profits. The monopolist levied a charge of thruppence a barrel of water, an unavoidable cost for the diggers who required large quantities of water for gravel washing. Not content with this, the company tried to increase its income even further by imposing a grazing fee on the diggers' livestock.[33]

The diggers, for the most part already badly undercapitalised, buckled beneath the weight of these additional costs. Those who did not collapse outright and end up as unskilled wage labourers working for their better-off white brothers tried to steady themselves by rearranging the cost structure of their already modest enterprises. For many, this meant cutting the wages paid to their black workers and, for a time during the war, wages fell to the previously unheard-of level of five shillings a week.[34] Others, but one shaky step away from economic ruin, dispensed with paying cash wages altogether and entered into illegal partnerships with their black labourers by promising to share with them the value of any diamonds they recovered. This arrangement met with opposition from better-off diggers, who believed that it encouraged theft and the flow of diamonds from their own underpaid wage labourers to more independent black 'partners' on adjacent claims.[35]

By the time Kas reached the diggings in the winter of 1916, the price of diamonds had started to stabilise, although most of the diggers were still very hard pressed and complaints about the number of inter-racial partnerships between poor whites and blacks were increasing. This diamond-sharing, or *mohlathe* system, at first glance the industrial analogue of equally illegitimate rural sharecropping agreements that bridged the racial divide on white-owned properties, must have held a certain attraction for the sons of black peasants who were accustomed to living out the economic lies of the southwestern Transvaal.

But Kas, along with the sons of most local sharecroppers, understood the important differences between the two systems. Thus, while the agents of the state made no effort to penetrate or police the domestic arrangements of tempestuous Afrikaner Nationalists—who after all had only recently been deeply implicated in an armed rebellion against the government in an area of marginal agricultural importance—active and intrusive government inspectors did make it their business to invade the crowded, more cosmopolitan diamond diggings, where they could actively help keep up the price of an industrial commodity that significantly contributed to the national economy. Moreover, as always in South Africa, a racial double standard was applied to any infringement of the law; in the case of the mohlathe system, Kas heard that 'whites would just receive light sentences but blacks were given jail sentences ranging from seven to ten years.' Under these circumstances he chose to restrict his own involvement to orthodox wage labour and was merely an interested onlooker when white diggers coached their black partners in the lines they would have to recite before the government inspectors.[36]

Given the shortage of cheap labour, it did not take long for Kas to find work with a digger by the name of Davids, a Johannesburg Jew who, it is said, dragged around a paunch of memorable proportions. With the assistance of Maphale, an older MoSotho worker who took a paternalistic interest in his well-being, Kas laboured for some weeks at digging, washing and sorting. Then early one morning the police suddenly arrived at the claim, turned the workers out of their tents and corrugated-iron shacks, ordered them to form a queue and demanded to see their passes. Maphale, sensing a tax raid and anticipating the sequel, drew his country cousin aside and advised him to flee the diggings until he had obtained the necessary documents. Kas, who like his brothers had never before been requested to produce anything other than a tatty bit of paper signed by a local farmer in order to move about a farming district, gingerly returned to Vaalboschfontein, where he called on the assistance of Hendrik van der Walt.[37]

A week later, after the foreman had paid his poll tax and got him an official pass in Wolmaransstad, Kas returned to London, with greater confidence. He found the reassuring figure of Maphale at Davids's claim and soon settled into the monotonous rhythms of uninterrupted heavy manual labour. Work commenced at seven in the morning and continued without pause until midmorning; then, after a break for a meal, resumed until late afternoon. Despite the quasi-urban setting, this was a rural work pattern; diggers—some of whom were more comfortable with an industrial day built around three meals—had to accept it if they wished to retain their black workers. Cultural realities meant that in diamond mining no less than in sharecropping, white employers sometimes had to meet their black employees half-way along the road that led migrants out of the peasantry and into the modern world.

For eight to nine hours a day, Davids would sit atop a strategically sited gravel dump and keep watch over the workers, never once taking his eyes off them, and rewarding any unnatural break in their rhythm with an accurately thrown stone. At midday on Saturday this energy sapping effort would be rewarded with a payment of five shillings.[38]

Their meagre income was, of necessity, mostly spent on food. Each evening after work Kas, along with other workers on the diggings, would go to the Kaffir Eating House. There, the proprietor, a small Jew whom the whites with customary flair had nicknamed 'Shorty' but whom the blacks saw as RreWabanana, since he seemed capable of fathering only daughters, served up plates full of mealie meal costing a 'tickey' (thruppence) or sixpence. Sometimes these generous portions of dry crumbed maize porridge were made more palatable by a tin of fish which, for the outlay of a further sixpence, gave the staple some badly needed relish.[39]

But since it required only two or three meals on this lavish scale to deplete his five shillings' wage, Kas was forced into looking around for even cheaper food. Following the example of hard-pressed mohlathe workers who were expected to live on only a shilling a week until their ship came in, he developed a nose for the bargains available at the local butcher shop. A sheep's head, if properly prepared and shared amongst several workers, could go a long way and cost only eighteen pence. Smaller bits of offal were even cheaper, and if one was lucky enough to be around when Willem Greef slaughtered animals, sheep legs could be picked up at no cost at all. Most of these offcuts were cooked on a fire that was kept burning outside the butchery; with a plate of RreWabanana's mealie meal, they made for a tasty and reasonably nutritious repast.[40]

After six months of this frugal existence Kas was delighted to return to the bounty of Vaalboschfontein, where he handed Sekwala his savings to purchase additional bohadi livestock. Then, without pausing for rest, he went back into harness, in the belief that the advent of the new season would give him the chance to meet his share of the family's commitment to Hendrik van der Walt as well as allow his tired body a chance to recover from the deprivations of the diggings. It did neither.

In 1917 Victor Lindbergh, for reasons that can only be speculated on, decided that the foreman no longer met the needs of the estate. Perhaps van der Walt, an untutored son of the western Transvaal, lacked the social graces necessary to survive on a property that was changing from male game lodge to family retreat, or, just as likely, he had not mastered the modern farming practices that would have appealed to a businessman like Lindbergh, who was by then pursuing a passion for breeding Herefords. Whatever the reason, that summer saw the departure of the foreman and, with him, the demise of the old order with which most of the black tenants were familiar and the arrival of a new English-speaking 'manager' from Johannesburg.[41]

The changing of the guard at Vaalboschfontein was greeted with deep dismay amongst the tenantry. Abel Bezant embodied the energised spirit of capitalism and its dedicated pursuit of profit just as surely as Hendrik van der Walt had exuded the ethos of a precapitalist order characterised by paternalism and tenancy. The English-speaking 'manager' and the Afrikaans-speaking *voorman* whom he replaced came from, and related to, different social and economic universes. Bezant, deceptively soft-spoken, lost no time at all in launching an attack on what he considered to be the antiquated, wasteful, or downright dangerous practices of the *ancien régime*. His drive for efficiency earned the resentment of black families who, unaccustomed to the logic of more profit-orientated enterprises, cast around for a cultural explanation of what was happening.

> That Englishman was full of shit. He was the one who destroyed the villages at Vaalboschfontein. You see, before that we used to get rations from Hendrik every Saturday and we were quite happy about that. When Bezant came he introduced his English system. He started measuring out our rations with a mug although we had previously got half a bag of mealie meal from van der Walt. Bezant decreed a mug-full for the morning, a second for the day and a third for the evening. It was so little that it fell far short of the half a bag of mealie meal that we had got accustomed to from van der Walt.[42]

Not content with cutting back on the staple in this way, Bezant also stopped the practice of slaughtering animals for the labour tenants; perhaps most puzzling of all, 'when cattle died he said that they should not be eaten, but buried in a hole.'[43]

These measures, which amounted to an arbitrary reduction in the income of the tenants and almost certainly constituted a breach of the contract they had had with Hendrik van der Walt, came at a bad moment for the Maines. Sekwala, who had consciously turned his back on arable farming during the long drought and concentrated on the family's livestock holdings, was in no position to make good the sudden shortfall in grain. And for Kas, who had just worked his way through a hard winter at reduced wages on the diggings the new dispensation was equally disappointing. In the end, the family decided to cut its losses: Kas spent the rest of summer working in the stables, where again he excelled. Once it became known that the Maines were bent on leaving, Bezant pleaded with him to stay on and offered him a substantial wage increase. When this financial ploy failed, Bezant refused to issue him with a pass and withheld more than three pounds in wages that were still due to him.[44]

But Kas was undeterred. He forfeited the three pounds and, as soon as autumn set in, absconded and made his way back to the diggings at London. This time his progress was less ambiguous. He was put in charge of the 'baby,' the gently rocking gravel sieve mounted above the sorting table. The market for gems had revived during the year (the price of diamonds rose to over six pounds per carat in 1917) and the upward momentum was maintained for some time. The renewed vitality of the diamond industry was also reflected in rising wages which, much to the chagrin of the white farmers, reached fifteen shillings a week. As a result Kas not only lived more comfortably than he had the year before, but could also save more. In spring he bought a bicycle for two pounds ten shillings from a migrant worker who was leaving for Taung, and set off to the southwest on a twenty-five-mile ride to the farm where Maine family members had regrouped during his absence.[45]

*

Kareepan, one of the oldest farms in the region, had a family history with which the Maines, given their own movement over the face of the subcontinent for the better part of a century, could readily identify. As early as 1851, C. J. S. Coetzee had left the area around Burghersdorp, in the Cape Colony, to establish himself as a farmer in the vicinity of what was later to become Bloemhof. In 1859, he was given the title to Soutpan—the same property which Pretorius and eventually Pistorius acquired—and in 1872 he was granted a second farm, Kareepan, by the Republican government. But he had no real need of a second farm and so soon sold it; it passed into the hands of strangers for more than a decade. Several members of the Coetzee family moved to the Orange Free State, but in 1886, one of Coetzee's sons, left the Bultfontein area where he had been farming and came back across the Vaal River to repurchase the original 'family farm' at Kareepan. In the decade that followed J. S. Coetzee and his offspring entrenched themselves at Kareepan, each farming a portion of the 4,600 morgen and then spreading their operations to neighbouring properties.[46]

It was the third and youngest of J. S. Coetzee's sons, the twenty-six-year-old 'Hans'—like his father before him he bore the traditional family name Johannes Stephanus—with whom Sekwala, his wives and their fourteen children chose to ally themselves when they left Vaalboschfontein. In addition, several other members of Sekwala's extended family as well as their children placed themselves under the protection of Hans Coetzee's brothers.[47]

At Kareepan, or Ralekala, as the property was better known amongst the local Tswana-speakers, Kas found that he figured prominently in the agreement that his father had entered into with the young white farmer. Predictably, Sekwala had believed that priority should be given to protecting his family's livestock holdings, while the landlord, already heavily involved in sheep farming and horse breeding, was more interested in extending the portion of the farm devoted to crops and therefore keen to use the black family's draught animals and labour. The resulting compromise allowed the Maines grazing for their cattle and the right to cultivate a plot of their own, in return for the use of their oxen and the year-round labour of one of the sons, which could come in the customary six-month spells.

One additional clause to this otherwise orthodox agreement appealed greatly to Sekwala and his two oldest unmarried sons. Coetzee, who had himself made an early start in life by taking a wife at the age of nineteen and setting up on his own, made special provision for young Kas and Sebubudi by giving them each access to a two-acre plot which they were free to cultivate as they saw fit—an unusual and generous concession to young men who did not own spans of oxen of their own.[48]

This interest in grain farming by the landlord and therefore indirectly by his tenants came at a fortunate moment, since spring 1917 brought with it the first sign that the great drought which had gripped South Africa for the better part of a decade was, at last, starting to loosen its hold. Indeed, 1918 saw the best rains in the district for nearly twelve years, and north of Kareepan, optimists, sensing the pulse of new life in the diamond diggings, promptly increased the area they put under the plough by more than 10 percent. Although the ever-cautious Maine family did not concentrate on large-scale production for the market during their first year on Coetzee's property, the subsequent rise in maize prices was not lost on Kas. Newly returned from a spell of migrant labour, he was impatient to establish his independence as a farmer. 'I prayed to God that if he spared me I would work for myself so that I should make progress. I had no wish to starve, to steal or be destructive. I wanted to live by the sweat of my brow.'[49]

But, while Hans Coetzee was cheered by the upturn in the market for maize and sorghum which was such a feature of the years 1917–20, he and his brothers had a greater interest in sheep farming and reserved most of their enthusiasm for horse and mule breeding. The pure white horses that Richard Coetzee and 'Ou Ernst the Bushman' bred on a neighbouring property were, it was said, in great demand throughout the southwestern Transvaal: Hans himself earned a reputation for being amongst the first men in the district to use horses for ploughing, an innovation necessitated by the destruction of oxen wrought by gal-lamziekte. This passion for horses, which Hans shared with tenants who had been influenced by their experiences at Mequatling and Vaalboschfontein, soon gave rise to a round of bartering: he swopped a strong but very nervous and untamed reddish mule named Fillet for a docile draught-ox belonging to Sekwala. Not surprisingly, the task of breaking in this spirited animal fell to Kas, who, conjuring up just the correct mixture of concession and coercion, succeeded in mastering its difficult temperament.[50]

The Maines' love of horses, their limited involvement in the rigours of maize and sorghum farming, and the improvement in the condition of their livestock, which now had access to unusually abundant grazing, appeared to bring a new sense of contentment to most of the family during the wet summer of 1918. Some of the tensions which had crept into Sekwala's relationship with his senior wife during the years at Vaalboschfontein appeared to ease, and the patriarch became more reconciled to Motheba and her children's involvement with the African Methodist Episcopal Church, believing that the Church respected the 'customs and traditions' of black folk. The only slightly jarring note came from Kas who, although always a dutiful son, thought that his father's reluctance to be separated from yet more livestock after the heavy outlay on the older brothers' bohadi stood between him and marriage.[51]

Kas realised that the only way around this problem was to go out to work again and earn the money. He therefore applied himself to his duties at Kareepan and, instead of waiting for winter to arrive, asked the landlord for a pass to seek work as soon as the planting and ploughing were over. Coetzee readily agreed to his request, and with Sebubudi once again deputising for him on the farm, Kas made his way back to the diamond diggings by bicycle—this time to Kameelkuil, a few miles northeast of London, where he found work at the rate of fifteen shillings a week.[52]

By 1918 the diggings were beginning to assume a new aspect. There were still many undercapitalised diggers, as was evident from the persistence of the mohlathe system and the number of poor whites who begged for food from black workers, but new and younger employers had appeared, usually the sons of more prosperous white farmers in the district. Yet neither these better-off employers nor steadier wage rates seem to have made life on the diggings easier for the black workers and their families. On the contrary, a dramatic increase in the cost of living during the last months of the war forced black women on the diggings into brewing ever larger quantities of beer in an attempt to make ends meet. This, and increasing theft of diamonds, attracted greater police attention, which in turn saw an increase in harassment, bribery and ill-treatment. Escalating tensions showed in strikes, riots and disturbances amongst the more proletarianised workers over the next two years. But none of these found an echo in Kas's consciousness. Single, frugal and a mere seasonal labourer, he managed to save much of his wages and apparently avoided the worst of the hardship that most of the resident black families suffered. This relative security, not to mention caution born of an upbringing in an essentially conservative environment, made him wary of alcohol and prostitution.[53]

Yet, for all this care and caution, one menace seemed unavoidable. Early in October 1918, scores of diggers and workers started going down with influenza. The virus, known as *'Drie-dag Siekte'* because the unfortunate victim was said either to have died or to be on his way to recovery by the third day, sent waves of fear and apprehension through the congested hovels on the diggings. Hundreds died. At Kameelkuil, as at London and elsewhere, dozens of black bodies were dumped into mass graves, 'like cattle,' Kas thought. Burials on this scale, conducted with none of the customary attention to ritual and tradition, induced terror amongst those remaining, and thousands of people fled the diggings, only to die along the country roads or spread the disease farther afield. In the surrounding reserves, such as Taung, unhappy villagers sang: *'Kgolopane wee boela hae'*—'Influenza please go back to your home'—but to no avail, since by then the virus was everywhere.[54]

Kas was amongst the fortunate few at Kameelkuil who did not get sick—which he attributed to having been immunised by the flu he had contracted the previous winter. But the immunity had its price, for, as he later explained, 'you had to carry a corpse all by yourself and bury it.' Nevertheless, his primary concern was for the members of his family, and once he had done his duty, he too left for home, stopping only to do the decent thing and bury an unknown corpse found on the roadside.[55]

Kas reached Kareepan shortly before the pandemic, which the Tswanas termed *boklhoko ba febere*, and found his family as yet unaffected. Hans Coetzee took the precaution of sealing off what he considered to be the disease's most obvious route into the farmhouse by instructing the younger Maines not to play with his children, Dolly and John. But the foreign virus, unfamiliar with South Africa's complex racial codes, chose to enter the property via the front door, and on the very day his ban was imposed, the landlord himself was struck down.[56] Mrs. Coetzee at once summoned Kas and told him to dash into Bloemhof and obtain the help of the district surgeon. But the fifteen-mile ride into town on the landlord's prized brown stallion was as pointless as it was exhilarating. Dr. Hutchinson himself had been laid low, and so Kas simply remounted the exhausted horse for a more measured canter home, where he told the deeply distressed Maria Coetzee that almost everyone in Bloemhof seemed to have influenza.

The second day saw a marked deterioration in the landlord's condition, and Coetzee, sensing his end, called for 'Andries'—the name Sekwala had adopted for himself in dealing with whites who were more comfortable speaking Afrikaans than SeSotho. In the dying man's bedroom and in the presence of his wife, the landlord instructed 'Old Andries' to take good care of the property, which, he said, would not be sold after his death; he expressed the wish that his tenants remain at Kareepan and then, singling Kas out by name, 'gave explicit instructions that he should be allowed to continue living on the farm.' Coetzee then instructed him to go to Abraham Pretorius, his fifteen-year-old brother-in-law, on the farm Leeubos and make one more effort to secure the services of the doctor. Hutchinson and Pretorius arrived at Kareepan late that afternoon, but, despite the doctor's efforts, Hans Coetzee died that evening, 26 October 1918.[57]

Not content with a casual visit to the farmhouse, death moved out into the tenants' quarters, where the Maines, without the services of a resident herbal doctor since the loss of Hwai at Soutpan three years before, were especially vulnerable. Despite the prevailing belief that drinking sorghum beer could stay a fever, the virus soon claimed Uncle Makale, Sekwala's brother-in-law. Then it invaded Phitise and Nneso's home where the couple, married during the rebellion, were forced to hand over their young son, Dikomo. What had started out as a winter of unusual warmth and contentment had given way to a chilly spring, and the seasons seemed strangely out of character.[58]

Dikomo's was the last of the deaths at Kareepan itself, but Kas remained exceptionally busy throughout the crisis. His horsemanship and impressive command of Afrikaans made him an obvious choice for being unofficial district messenger, and he spent several days in the saddle gathering information from neighbouring properties, calling in at the rooms of the hard-pressed district surgeon and then hurrying back to the farms to deliver whatever medicine Hutchinson had dispensed. It was during these visits that he got to know and admire the doctor's professionalism.

Born in former Boer 'New Republic' territory around Utrecht, in Natal, the young Henry Hutchinson had grown up in northern Zululand and was a fluent Zulu-speaker. On completing his schooling he had proceeded to Edinburgh for a medical education, but at the outbreak of the South African War he abandoned his studies and returned to the country to enlist with the Republican forces. His active commitment to the Boer war effort ended when he was made to serve a lengthy spell in a British prisoner of war camp in Ceylon. It was only in 1911, at the age of thirty-six, that he eventually qualified as a doctor in Dublin. His loyalty and dedication to the Boer cause were not forgotten, however, and four years after the formation of the Union of South Africa, he was rewarded with the offer of the position of district surgeon in Bloemhof.[59]

Kas, although always inclined to place greater faith in the herbal medicines recommended by his late grandfather than in modern concoctions, was impressed by Hutchinson's dedication to his patients and enough of a pragmatist for the entire experience to leave a deep imprint on his mind at an important moment in his own career. The status and importance attached to the healing professions, especially during an acute crisis, were not lost on him, and he resolved that, should he or his family ever require the services of a white doctor, he would not hesitate to consult Henry Hutchinson. It was the start of a warm relationship that was to span more than three decades.[60]

As the epidemic retreated, life at Kareepan slowly returned to normal and the logic of spring gradually began to reimpose itself on rural life. Sekwala, whose day-to-day supervision of the property was monitored by the widow Coetzee's father on the adjacent farm, Leeubos, was also called upon to decide on the Maines' own plans for the new season. Mindful of the pressure coming from his oldest unmarried son and still ambitious to rebuild the family's livestock holdings, he ignored the continued rise in the price of summer grains. Instead, he took the eight pounds Kas had given him and bought six sheep from a distant kinsman at a cost of ten shillings a head.[61]

Sekwala's decision not to devote too many resources to maize and sorghum farming was vindicated by the elements, because 1919 saw a brief return to the drier conditions of earlier seasons; an average of only eleven inches of rain was recorded in the Bloemhof district that year. Fortunately, farms like Kareepan that stretched out towards Wolmaransstad, were not as badly affected as those in other parts of the highveld, and the Maines, profiting from the plight of Free State farmers, acquired a further ten sheep for only ten shillings each. Kas found it a frustrating season. Most of the summer was spent out in the veld preventing cattle from grazing too much *mohau*, a fibrous shrub which, if consumed in large quantities and followed by a large intake of water, was said to cause possibly fatal obstruction in the animal's stomach. The rather tedious task of checking on the daily diet of the livestock was made tolerable only by the acquisition of a small black terrier named 'Bull' which, once properly trained, was an able assistant out in the fields.[62]

Summer went by so slowly that it seemed in danger of missing its annual appointment with autumn. The two eventually met for a typically brief highveld encounter, before autumn, always the more purposeful in demeanour, scurried off to pass on its message to winter. By then the Maines were ready to bring in the harvest which, as expected, was modest, in sharp contrast with the yield elsewhere in the district. At the start of the season better-off Boers and their bywoners, believing that the price of grain would continue to rise, had put an unprecedented acreage under the plough and reaped a record harvest. This helped to raise land values, and prosperous buyers moved in from the more commercialised maize farming areas across the Vaal. It also brought the first signs of trouble for the Maines, who up to that point believed, rather naively, that Hans Coetzee's dying wishes would give them all the long-term security and protection they needed.[63]

In July 1919, the widow Coetzee decided to let the greater part of her property to Cornelius Badenhorst, a man whose eyes were firmly fixed on the rising price of maize. Badenhorst immediately let it be known that there was no longer any room for labour tenants on Kareepan and proposed instead that black families on the farm enter into sharecropping agreements with him for the forthcoming season. The Maines, more especially Kas, considered this a gross betrayal of the former landlord's wishes. They had no desire to work on large-scale maize production, which would divert them from their stock farming enterprise and diminish the amount of grazing. Hemmed in between Badenhorst on the one hand and the new obsession with mealies on the other, the family saw no alternative but to look around for a new home if they were to protect their investment. A white landlord's expanding maize fields, it seemed, could eat a black tenant's cattle just as certainly as the cattle ate maize.[64]

It was a difficult moment for Sekwala to move his family. With the price of both maize and diamonds rising, most farmers, like the newcomer at Kareepan, were expanding grain production, and it was hard to obtain reasonable terms as a labour tenant. To complicate matters further, he was still under pressure from Kas who would, as had been the case under the late Hans Coetzee, require special consideration in any new agreement. What Sekwala needed was a farmer who, like himself, was committed to stock farming and who also had enough grazing to be able to offer reasonable terms. Largely by chance he bumped into an old acquaintance before winter had passed and the spring rains could trap them at Kareepan.

Gert Meyer (whom the local Boers irreverently called *'Ou Koeël-boud'* Meyer because he had been rewarded with a shot in the buttocks by a British patrol whilst out on a scouting expedition during the South African War) had known Sekwala and his four sons for more than a decade, since the period when the Maines were based at Mahemspanne. Indeed, it was Meyer who had taken Mphaka and his younger brothers into service as herdboys shortly after the Boer War, an arrangement that had lasted until the Maines moved on to Vaalboschfontein; not long thereafter, the Meyers had retreated to their property at Rietfontein, on the banks of the Vaal.[65]

At Rietfontein, Meyer, who had lost two children to disease in the crowded British concentration camps during the South African War, tried to rebuild his life and apply himself to farming with renewed vigour. By 1918 he had achieved some success as a stock farmer and was looking to expand wool production. But his sheep needed more space than his riverside property could provide and he looked around for additional grazing. Not surprisingly, this search took him back to familiar haunts, and a little way to the southwest of his family farm at Hartsfontein, he found what he was looking for.

Kommissierust, like several neighbouring properties, had been declared a farm in the early 1870s and then passed through the hands of various nondescript speculators until, right after the discovery of the Witwatersrand goldfields, it had attracted the attention of a syndicate led by the undisputed king of all Transvaal speculators, A. H. Nellmapius. Nellmapius, a Hungarian Jewish immigrant and a close confidant of the late President Kruger, had bought the 3,700-morgen property and then sold it to one of the many land companies operating in the Republic, which, in turn, sold it to Northern Minerals Limited during the 1912 boom in alluvial diamonds. But Kommissierust again failed to live up to expectations and Northern Minerals, like previous owners, found that all it could do was to rent out portions of the farm to stock farmers until it could sell at a reasonable price.[66]

When Sekwala asked Meyer about a place to live, the farmer saw a chance to recover a portion of the rent he was paying to Northern Minerals and to get some of the ablest shepherds and wool shearers in the district to take care of his flocks. He offered to accommodate Sekwala and his three oldest sons as well as his brother-in-law for four pounds a year each; Kas and young Sebubudi would also look after his sheep. In return, the Maines would have as much grazing as their stock needed and be allowed to plough and plant as they saw fit. Although the rental cost was perhaps a bit high, this contract would enable the Maine family to safeguard its livestock holdings at a time when Kas was likely to be drawing on it for bridewealth, while unrestricted access to arable land meant they might recover some of their cash outlay at a moment when grain prices were holding up well. For Sekwala, any arrangement that gave Kas access to a field of his own had the added attraction of coping with his son's desire to farm for himself. By spring 1919, most of the Maines were at Kommissierust.[67]

❦

As had happened a decade earlier at Mahemspanne, when Matthew Wood had inflicted his labour demands on blacks who were farming in the absence of a visible white landlord, so the small cluster of Maines at Kommissierust in 1919 soon drew envious glances from labour-hungry Boers.

And as in 1908–10, this interest was prompted by a familiar configuration. Whenever the prices of maize and diamonds rose simultaneously, there was a marked escalation in the competition between white farmers and diggers for cheap black labour. It was then usually only a matter of time before one or other state agency investigated any group of blacks who appeared to be operating without the benefit of direct 'European' supervision. On this occasion it was days rather than weeks before the new residents at Kommissierust were summoned to present their poll-tax receipts at the offices of the local Native Commissioner.[68]

In Bloemhof, where white diggers had been reduced to paying the fines of black prisoners at the magistrate's court in order to mobilise labour, the Maines succeeded in persuading state officials that they were *bona fide* employees of Gert Meyer, of the farm Rietfontein, and they paid their outstanding taxes. When Sekwala asked for receipts, however, the frustrated Commissioner raised another long-standing grievance. For some years officials in the post office had complained about the growing number of 'Maines,' he said, all bearing remarkably similar names. Inadequately addressed letters, written in the poor cursive that often characterised self-instruction, bedevilled the delivery of mail. Now the Native Commissioner proposed giving all Maines around Kingswood new names based on their clan totem. Sekwala was asked what his family totem was and he replied '*phoka,*' or 'dew,' so the Native Commissioner's clerk issued his tax receipt in its Afrikaans equivalent, '*dou.*' As far as the South African state was concerned, from that moment on the members of his family were to be known as Dou. In the course of time Dou, pronounced with a Tswana accent, was transformed first into 'Deeu' or 'Deu' and, even later still, into 'Teeu' or 'Teu.' This arbitrarily enforced change in legal identity hardly succeeded in deracinating the members of a family who could easily trace their origins back to the eighteenth century, and the more pragmatically inclined amongst them did thereafter tend to use the phonetic variations of 'Dou' when dealing with whites or state officials.[69]

Back at Kommissierust Sekwala had time to reflect on this development and the family traditions. Perhaps it was the spur of their being issued new names or, more simply, the sense of vulnerability they had experienced during the flu epidemic, but whatever the reason, he now found his mind harking back to the death of his father at Soutpan four years earlier. Hwai, a ngaka as well as the man then most responsible for maintaining the family traditions, had handed him his divining bones shortly before his death in the hope that the skill would be passed down to the next generation. But Sekwala, who had no desire to become either a diviner or a herbalist himself, had done nothing to respect his father's wishes, and this worried him. The ancestors were not to be trifled with, and the very least he could do was to ensure that one of his own sons took up the challenge. When this proposition was put to the adult males of the senior house, however, only the third son, the one who had become involved with the white doctor during the flu crisis, displayed any interest. Confident that his father would have approved, Sekwala set about finding a practitioner who could train Kas.[70]

Among the more famous traditional doctors then at work in the south-western Transvaal were two brothers distantly related to Motheba. The older of the two, Mohane Mosia, was an obvious choice to train someone who was, after all, a kinsman. Using this convenient conduit into Ba-Thlaping society, Sekwala sent a message to the reserve at Taung inviting the great man to visit the farm and train Kas and a brother-in-law of Phitise's named 'Duister' who had also expressed an interest in becoming a herbalist.[71]

When Mosia arrived at Kommissierust some weeks later, it was agreed in discussion amongst the elders that he would be given four sheep for each kinsman he trained. He then led his novices off deep into the veld where, by day, they were taught the mysteries of the craft: how to 'read' the bones that pointed you in the direction of the herbs you needed and what the healing properties of the various plants were. At night he and his charges returned to the small cluster of huts that the Maines, Thamagas and other black families occupied, and rounded up the younger boys. Under the watchful eye of the master, they would then be examined by the apprentices and offered diagnoses of their various ailments. After three weeks of this intensive instruction he pronounced the novices fully qualified and returned to his home in Taung.[72]

In many societies, perhaps more markedly so in Africa, the opaque crafts of an outsider with the gift of healing often take precedence over the more familiar skills of an insider. Just as an exotic edge had once helped to enhance Hwai's reputation, so Kas now found himself benefiting as the numerically preponderant BaTswana in the district came to welcome the presence of a MoSotho herbalist amongst them. The news that the Maines had a ngaka to succeed Hwai did the rounds amongst the local farm workers and labour tenants, and soon the newly qualified young doctor was called out on his first case. He could not have imagined a sterner challenge.

At Leeubos, Kas was introduced to a man named Tjhorilwe who told him that his son was experiencing deeply disturbing bouts of 'madness.' In accordance with his master's teachings, Kas wandered off to collect *mokumetso* and other herbs, which he later ground into a fine powder. A large fire was lit and then allowed to burn down, Kas approached its still searing heat and sprinkled his mixture over the embers. The fire spluttered back to life for a second or two before releasing clouds of highly aromatic smoke which he told the afflicted youngster to inhale in deep breaths.

Although the treatment was distinctly unpleasant Kas assured the worried parents that it would clear their son's muddled head, and that his condition would undergo more lasting improvement if they took the added precaution of slaughtering a sheep to propitiate the ancestors. As was customary, he levied no immediate charge for his services but left the family to await the outcome of the treatment. A few days later he was relieved to see Tjhorilwe approaching the farm, driving before him six sheep which he offered in payment for the improvement in his son's condition. Kas, in accordance with a prior agreement, passed the sheep on to Sekwala thereby repaying his father for livestock that had been advanced to Mohane Mosia for his training as a ngaka.[73]

With the first spring rains now only days away, Kas recalled the late Hans Coetzee's prowess with horse and plough, and told his father he planned to use the family's donkeys to cultivate a field of his own. Sekwala, who had been living with these pressures for full economic independence for several seasons, asked him how he hoped to cope without a wife who could bring him food when he was out in the fields. But this objection received short shrift: Kas told him Motheba could perfectly well do the job.[74]

If the father had residual reservations, he kept them to himself. This was probably just as well because Kas's next step put the issue almost beyond debate. Like most small-scale producers in the district, black or white, Kas did not own any fixed property and was therefore without recourse to the state-owned Land Bank which could give him credit through an entire agricultural cycle. In the southwestern Transvaal, as elsewhere, it was often left to the local grain merchant to act as the poor man's banker by making resources available in spring that could be repaid once the harvest had been disposed of in the autumn. During more buoyant economic periods, such as 1915–19, most merchants were only too pleased to be able to give new producers credit, and Kas, exploiting the happy coincidence of seniority and prevailing market forces, went to seek an advance in his own right.

At Hessie, ten miles north of Kommissierust, he found William Hambly, still running the general dealer's business that he and his onetime partner, Norman Anderson, had acquired in 1912. Hambly was quick to welcome him into the shop not only as a distant relative of his messenger, Sisindane Maine, but as part of the extended family with whom he had had extensive dealings during their stay at Vaalboschfontein. The young Maine was obviously attracted to a single-share plough, the Canadian Chief, which, having the advantage of being light enough to be drawn by a team of donkeys, had the disadvantage of costing twelve pounds. Kas resolved to buy the plough despite its price and, because he had left it so late in the season and would not have time to make its leather accoutrements himself, he also had to get a harness and supplementary *trekgoed*, additional purchases that amounted to more than the value of the plough itself. By the time that he left the store, he owed William Hambly the sum of twenty-five pounds. Many a successful career had started out on an even less secure footing and, in any case, so the storekeeper assured him, the debt could be settled over more than one season.[75]

Even before the Canadian Chief could be put to work, Sekwala was talking to Motheba about the need to find their son a wife. The herd would not yet readily sustain another large withdrawal of cattle so soon after Mphaka's and Phitise's marriages, but there could be no further delay in finding a suitable companion for a twenty-five-year-old man who had contracted a large debt and was set on ploughing for himself. Moreover, someone who wanted to follow the calling of a ngaka would, in accordance with BaSotho custom, have to have a marriage partner selected from amongst his cousins. Motheba thought of her nieces—she already had someone in mind.

Shortly after her own marriage to Sekwala in the early 1890s, Motheba's twin sister, Tshaletseng, had married Tshounyane Lepholletse, a member of yet another immigrant BaSotho family on the highveld. After their marriage the Lepholletses had left Setlagole, like the Maines, but, whereas Sekwala and his wife had chosen to establish themselves amongst the Boers, Tshounyane and his wife had moved even farther south and west and placed themselves under a minor BaThlaping chief at Mathanthabe only a few miles from Taung. There, Tshaletseng had produced no less than nine children while Tshounyane's second wife, a MoTswana woman, raised two more.

Tshaletseng's male offspring seemed to be dogged by illness and misfortune, with the result that the Lepholletses were left with a disproportionate number of girls, amongst them three sisters named Leetwane, Lebitsa and Tseleng. Because their father had had to farm without the assistance of boys, these girls, especially Lebitsa, the oldest, had helped their father with ploughing and planting until the family moved farther east to the village of Matsheng, on the eastern border of the reserve. Motheba, deeply impressed by the skills and virtues of her nieces, thought that Leetwane, although not a Christian and a year or two older than Kas, would make an ideal partner for a son who wanted to be a grain farmer.[76]

When Sekwala and Motheba told Kas they had come to an agreement, he accepted it with equanimity. A dutiful son did not question the wisdom of his parents, and the choice of a cousin whom he had seen once or twice during family visits seemed acceptable and in line with the teachings of the elders. Sekwala got the local scribe to draft a note to the Lepholletses broaching the question of marriage and probing the delicate issue of what his family would be expected to offer as bridewealth. While waiting for this to be composed, Kas managed to track down Gert Meyer, who moved constantly between Kommissierust and his river property, and got him to give him a 'pass' to negotiate the countryside—this at a time when diggers, farmers and policemen alike were all on the lookout for black labour. Armed with a letter to the Lepholletses, with the written authority of a white man to move about the country of his birth, and with enough provisions for a fifty-mile cycle ride, Kas set out on a rainy day in September 1919 to contract the marriage which, until then, had always threatened to be just one more season away.[77]

But on his arrival at Matsheng Kas was told that his aunt was away on a visit to one of her younger daughters. Undeterred, he entrusted Sekwala's note to the oldest married daughter who in turn passed it on to her father, Tshounyane. She and Leetwane gave him food and drink while the old man disappeared into the village to find someone who could read the letter to him. Kas, driven by the moisture in the spring air and the desire to get on with his work, was becoming impatient: 'I told them that I would only stay one night, and that I would be returning home on the following day because I had left my family ploughing.' This announcement spurred Tshounyane to uncharacteristic action: that very afternoon the mother suddenly appeared as if from nowhere, and was handed Sekwala's letter.

His aunt's arrival precipitated a flurry of activity in the usually slow-moving Lepholletse household. She and her husband, who appeared to be in agreement about the matter, called in Leetwane and told her what she had probably already guessed, namely, that her cousin was there to ask about her eligibility as a wife. Like her future husband, she accepted that fate ran along the tracks of tradition and gave no thought to questioning their decision. 'Your parents decided, you had no choice. You had to do as you were told.' Tshounyane then went to find her paternal uncles; in keeping with the custom, the men settled on an appropriate price for the bride, terms conveyed to a scribe who hastily committed them to paper in a letter handed to Kas but addressed to Sekwala.[78]

When Kas got home, it was Sekwala's turn to find a literate friend, who told him that the Lepholletses wanted twenty head of cattle or their equivalent to bind the families together, ten sheep as *bodumo*, and that their daughter would be released for the wedding only when ten beasts had 'broken' the bohadi. Sekwala, whose heart must have sunk, did not flinch as he relayed these terms to his son. For the third time in seven years the patriarch, whose own daughters were not yet of marriageable age and therefore had not yet brought any cattle into the family herd, would have to dig into the Maines' reserves.[79]

Before these customary demands could be addressed the ploughing had to be completed—a task which, thanks to Kas, the Canadian Chief and the more junior members of the family, was accomplished with relative ease. In October, once the early spring rains had coaxed the first reluctant maize plants into sticking out their heads and taking a look at their unpromising surroundings, Kas addressed himself more squarely to raising bohadi. He had given the matter careful thought and decided to pay Jan Greef a visit.[80]

Greef, a member of one of the many closely related Afrikaner families struggling to eke out livings on the smaller sections of Kommissierust, owned a brown cow and a red one, each with a calf: four beasts that Kas had eyed with interest for some time, an ideal nucleus around which to build a herd that could eventually make its way to Matsheng. But the Boer, hoping to profit from rising livestock prices, quoted Kas an unrealistic price. Thwarted, Kas decided to dispose of a few of the wethers he had bought at Kareepan and use the proceeds to buy other cattle at better prices.

*

The monthly stock fairs held at Bloemhof, like most of those in the western Transvaal, were organised by the Klerksdorp-based firm of Neser & Campbell and usually presided over by one of the local agents such as the attorney A. J. Marais. These noisy outdoor gatherings tended to attract a few dedicated 'foreign' professional livestock speculators who followed the circuit, some of the district's more progressive Afrikaner stock farmers and a fair number of butchers and 'Native Eating House Keepers' drawn mainly from the diamond diggings. But when Greef learned that Kas intended taking his wethers to the market, he too suddenly decided to participate and asked Kas to help him by driving his four cattle to the market. On the appointed day Kas left the farm driving before him the very animals he had originally wanted to buy, as well as the six sheep he had decided to sell.[81]

Kas's debut at the fair at this crucial juncture in his life resulted in a minor coup and it left him with a lifelong appetite for speculation in livestock. His half-dozen wethers soon fetched the handsome price of fifteen shillings each. With these ninety shillings and a little additional cash in his pocket he settled down to await the sale of Greef's cattle; to his great delight, he eventually acquired his neighbour's animals at a price below that which Greef had quoted to him. That a disinterested intermediary in the shape of an auctioneer and a market had delivered better prices than he could obtain from face-to-face negotiations was a lesson that embedded itself deeply in his consciousness. He drove the cows and their calves home in triumph—never had the twelve-mile walk seemed shorter.[82]

In the weeks that followed Kas kept his eyes peeled, scrutinising the herds of neighbouring farmers in the hope of uncovering yet another bargain, but nothing presented itself. Then, as he mounted his horse one morning, he suddenly thought that the solution lay not so much in looking longingly at other people's livestock as in re-examining his own holdings. Clearly, the moment had arrived to check whether the Griesels who lived down at the river were still interested in buying his short-tailed grey roan.

De Beersrust was a farm well known to the Maines. Hwai had once spent a rather unhappy time there under 'Ou Willem' Griesel, one of three brothers from Heilbron, in the Orange Free State, who had established themselves on the banks of the Vaal in the 1890s. The older Griesel had brought with him a son who shared both his father's name and some of his less appealing features. As a thirteen-year-old, the boy had joined the Boer forces in the fight against the British and then rapidly matured into a cantankerous hard-drinking man with no great love of blacks. 'Willem wore his hat in such a way that it hid his eyes and would say: "Come here, you shit." He never addressed one politely, he was always rude,' Kas remembered. But, despite these misgivings, Kas exchanged his roan for two of Griesel's cattle.[83]

As the year end approached, Kas bought two more cows at a very reasonable price from another white neighbour—an athletic Englishman by the name of Forsyth. With eight of the ten animals needed to 'break the bohadi' already safely on the farm, he now looked to his father. Sekwala, drawing on the remittances which Kas had sent him during spells of labour on the roads or at the diamond diggings, came forward with a 'gift' of twenty head of sheep—ten sheep being the equivalent of the two outstanding cattle required to make up the bohadi, and the remaining ten to cover the cost of bodumo.

*

One morning in January 1920, Sekwala, Mphaka, Phitise and their cousin Kgofu assembled at Kommissierust and then set off for Taung, driving Kas's animals before them. When the party reached the outskirts of Matsheng a day later, everything went according to custom: one of Kas's brothers was sent ahead to warn the Lepholletse family that the Maines were approaching with the bohadi and to request that a member of the bride's family accompany them on the last stage of the trip. One of Leetwane's married sisters came to join her cousins who, again in keeping with BaSotho custom, cut her a walking stick, or *lere*, to help her steer the party and its cattle and sheep into her parents' village the next day.[84]

Early in the morning, with the prospective bride and her closest friends in an appropriately secluded hut so that the incoming bohadi was safely out of their sight, the Maine men and their guide, brandishing her lere, entered Matsheng. The Lepholletses inspected and took charge of the livestock, and word was put out that the bohadi had been accepted. Leetwane and her friends were told to proceed to the river to draw water, then return to her parents' house, where the visiting kinsmen slaughtered an animal for a feast. The remainder of a memorable day was devoted to eating, drinking, singing and dancing, and Tshounyane recited his family's praise poetry:

> I am of the cat family, Mosia,
> I am a motobatsi,
> I am the person who takes out the shields.
> And then gets them dry, before the war.
>
> They asked, "Where does the trouble come from?"
> It comes from the cracks in stones.
> They said it was from Konkong.
> They said that we, of the cat family,
> are really birds.
> We die easily.
> I did not see them with my eyes.
>
> They say there is no smoke from my home.
> How can there be smoke when there is no one
> to make a fire?
> It is because they have all died.
>
> They call me Tshele.
> They asked me these questions at the time
> that two people gave me the name Tshele.
> It was my parents who named me Tshele.[85]

The feasting over, Sekwala and his sons and nephew made their way home, where they gave the family, and most importantly Kas himself, an account of what had happened at Matsheng. Kas had thus far seen precious little of his prospective bride or her family. Nor did he see much more of them in the months that followed. As the summer wore on and the maize slowly responded to the summer rain, he devoted himself to the field he had opened up, confident that the end of the season would see his wife drawing on the fruits of his labour to help to start a new line of Maines.

In April, when the days were becoming shorter, Leetwane appeared at Kommissierust, with a small entourage of sisters and well-wishers from her native village. The following morning her husband slaughtered a wether, and the rest of the day was devoted to celebrating the arrival of his bride in a new round of feasting. On the third day the women of Matsheng, having been given a pound for escorting the bride to her new home, withdrew, singing the poignant BaSotho refrain:

> We are leaving you, you will manage on your own.
> They are your family.

You will manage on your own, we are leaving you.
We brought you, and leave with the cake.
We are leaving you.[86]

For Kas, too, this marked a point in his life that had been long in the making: 'By the time that I got married most of my cohort were married with children of their own.'[87]

But a summer heralded by hope suddenly wheeled round to renege on its promise. Kas had bought the plough, waited for the rain, turned the soil and watched patiently as the maize plants drew on their reserves of hidden moisture to produce the first cobs. But one morning he found the field invaded by locusts that swarmed as far as the eye could see. Within hours these ugly visitors, like unloved guests discovered in the pantry prying off the lids of the best preserves, devoured a season's labour at a single go. The last of the long summer greenery was reduced to the stubbled brown and grey of winter, and with it Kas's dream of providing for himself and his wife from his own granary.[88]

In midyear, when those young men fortunate enough to have escaped the devastation caused by the locusts were starting to leave for work on the diggings, Kas had little to do apart from taking care of his donkeys and a few sheep. As someone whose natural inclination had always been to challenge the dormancy of the off-season, the enforced idleness of this first full winter spent at home since the transition to manhood was especially hard to bear. Leetwane, without maize of her own to harvest or grind, busied herself by applying a layer of cow dung to the floor of the hut which the Maine men had built for her. Kommissierust and its surroundings were dry and dead.

In August, once the harvest had been delivered and the farmers paid, the pace of rural life gradually picked up in anticipation of the new season. As the icy breath of the Malutis was sucked back into Basutoland a few Indian and Jewish hawkers ventured out of town and along the country roads in the hope that their trinkets could siphon off some of the financial liquidity that characterised the preseason's more substantial purchases. The occasional mounted black messenger, who in the midwinter months confined his outdoor activities to dashing between the trading store and the Kingswood post office, now negotiated the obscure farm tracks at a more sedate pace, reminding rural folk of their debts. Kas needed no reminding.[89]

Ever since the locust visitation Kas had watched nervously as Sisindane and 'Norman'—a particularly fine horse that William Hambly had named in honour of his former partner—slowly closed the circle around Kommissierust. But struggle as he might, he could see no way to pay for his plough; even his father owed the storekeeper money for a saddle. When Sisindane arrived to issue them with a summons to attend a meeting of all of those who had not yet met their debts, he still had no solution. The Canadian Chief was an imposter—an implement that was meant to have helped usher in a new era of freedom and independence, but proved no more than a shackle to the general dealer's store.

On the day of the meeting he released his ten donkeys from their kraal, mounted the sturdiest of them and, with his father, set off on the uncomfortable ride to Hessie. They tethered the animals outside Hambly's store, where they joined more than a dozen other debtors of all colours, reduced to the same lowly status by nature's little levellers. This unhappy fraternity sought shelter from a merciless sun beneath the corrugated-iron overhang at the front of the store, and farmers, bywoners or sharecroppers waited their turn for the painful discussion that was bound to follow. Not all the negotiations went smoothly, and as the day wore on, Hambly had several heated exchanges with his debtors. What happened next is best told by Kas himself:

> When my turn came I told him to take my ten donkeys in part payment of my debt since I had no money with which to pay him. He then told me to wait and, after he had completed his discussions with the others, he reappeared on the *stoep* of the store and announced to the crowd: "Gather round. I want you to take a look at these donkeys and tell me what you think they are worth." Some called out "a pound" while others suggested "three pounds." Hambly then announced that he would take my donkeys and that he valued them at a half-a-crown each. I thanked him and was about to leave when he called me back and once again addressed the crowd.
>
> "I want you all to know that there is not a better person amongst you than this man. He has carried the burden of his debt on his heart and came here today driving his team of donkeys to pay me. In desperation he has offered to pay me by disposing of his draught animals because the locusts have destroyed his crop and, when I offered him only a half-a-crown an animal, he readily agreed to the proposition."
>
> Hambly then turned to me, put his arm around my shoulders and said: "Go, take your donkeys and use them to plough and you will settle your debt next year." I then left.[90]

Deeply moved by the warmth and generosity of the 'Englishman,' Kas nudged the donkeys back to Kommissierust with the same bewildering mixture of joy and uncertainty he had experienced when driving home the animals he had bought at the stock fair the year before. Then he had been a young man in search of bohadi, but now he was a married man faced with a more perplexing difficulty. Although he had been with his wife for five months, she had failed to become pregnant. This was disturbing for a young farmer bent on developing a career of his own, 'because everybody wishes to have a baby boy. A boy is a foundation.'[91]

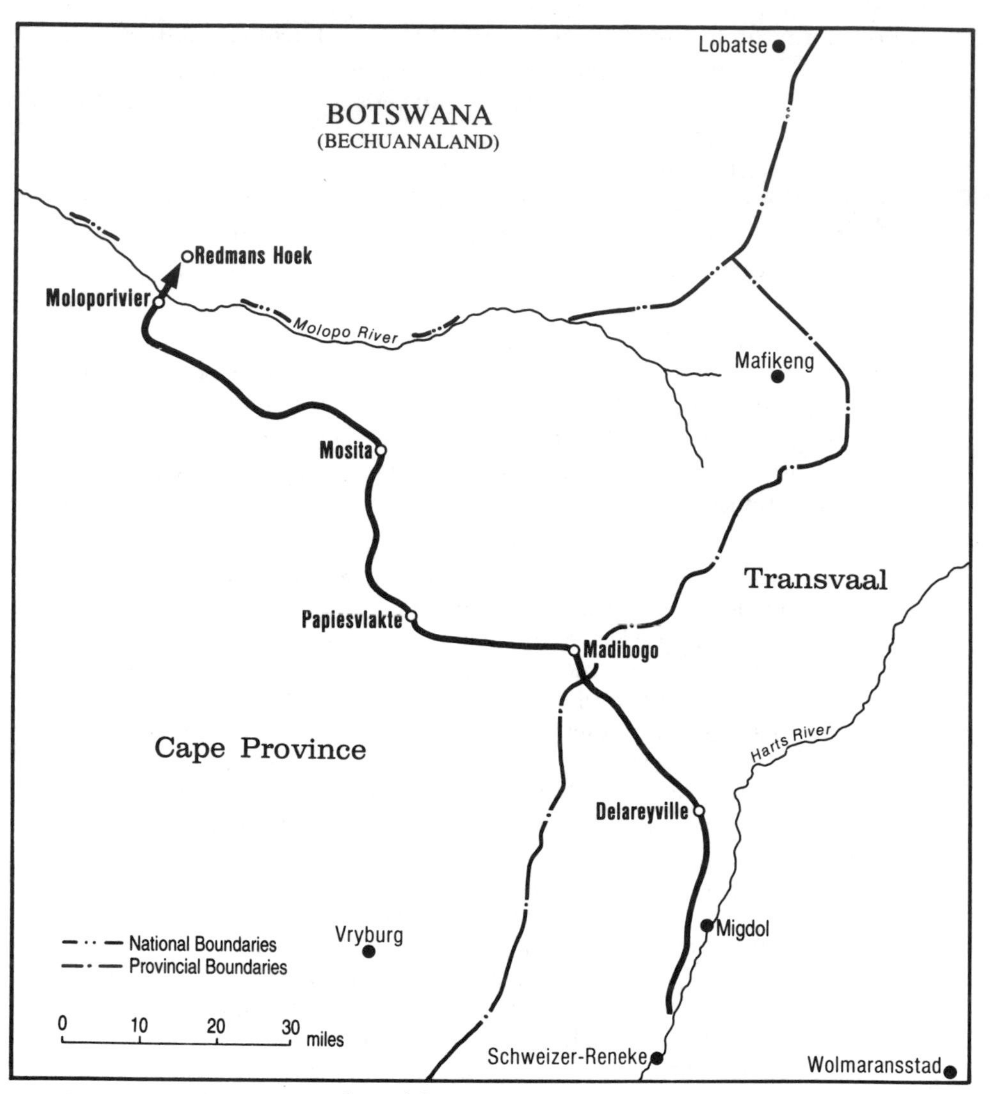


4 Kas Maine's journey to the Kalahari Bushmen, 1920

This puzzle, coupled with the unexpected arrival of the locusts in his field, echoed discussions he had once had with his mentor Mosia Mohane, and forced him to consider alternative explanations. It seemed as though there were ill-disposed people abroad who 'were bewitching me so that I should not have children and be made to suffer.' Before the new season's ploughing could start, such evil forces would have to be countered and, when he sought his father's advice, Sekwala told him of the wondrous powers enjoyed by an old MoSarwa woman who lived in a distant San village in a far-off country where the BaKgalakgadi held sway. The craft of the outsider took precedence over that of the insider, and since Sekwala knew of a guide, and only a fool would leave such malevolence unchecked, Kas resolved to make the long journey into the depths of the Kalahari Desert and visit the mysterious 'Bush' woman before the rains came.[92]

*

Modise Tsubane, who lived on a neighbouring portion of Kommissierust and who had himself visited the land of the Bushmen, agreed to escort Kas on a journey that would take them to the very margins of Tswana society. According to Tsubane, the woman did possess powerful herbs for treating infertility, but would demand to be paid in dagga. This did not pose much of a problem in an area where many people cultivated the drug either for domestic consumption or to sell to migrant workers on the diamond diggings. Kas asked around and soon succeeded in obtaining a matchbox-full from Mmapata, a neighbour's son. Armed with a little of the currency of the countryside as well as a few more orthodox provisions, he selected one of Sekwala's best horses for the two-hundred-mile-long journey into Bechuanaland.

Kas and Tsubane left Kommissierust, headed out north along the familiar road to Doornhoek, and then swung west through Boer farming country towards Migdol before making their way to Delareyville. From there they pushed on for several hours before eventually stopping at a spruit where, after a hard day's riding that had covered sixty miles, they spent the night. Long before daybreak they again set off in a northwesterly direction, this time toward Madibogo, where Kas remembered how, as a very small boy, he and his mother had sought refuge amongst her kin during the Boer War. From there they took a more westerly course, traversed the 'native reserve,' crossed the border into the northern Cape Province and made for the hamlet of Papiesvlakte. After a brief rest, they continued moving north towards the safety of yet another reserve for a night stop at the village of Mosita.

Two-thirds of the outward leg completed, the trek was resumed for a third day, again in a northwesterly direction but this time toward the Cape–Bechuanaland border. With the greens and browns of Afrikaner farming country and BaTswana 'native reserves' well behind them, isolated cattle outposts that threatened to sink beneath the yellow sands of the Kalahari now started to spew out bizarre names that betrayed the contested terrain of frontier society—Tshokong, Donkerhoek, Chiselhurst, Vermont, Vooruitsig, Wexford, Toledo and Exeter. Tsubane claimed to recognise some of these weird names from his earlier journey and, at Redmond's Hoek, turned them toward the BaKgalakgadi village of Molopo, situated on the banks of a seasonal river with the same name.[93]

At that point Tsubane suggested that Kas abandon his horse, consult the villagers and then proceed on foot. The people of Molopo gave him a guide who spoke the strange clicking tongue of the BaSarwa and who led him across the dry river bed into the sandy distance of a Bushman village, where he was taken to a hut and introduced to a small, grey-haired old woman. Wrinkled and dust-cured, she slowly sifted through his problems until she had established that he was looking for herbs that could help make him a father. Satisfied that she knew what it was that *he* wanted, she then got down to the equally important business of making sure that he knew what it was that *she* wanted. Kas fumbled about himself for the matchbox and handed it to her. She stretched out, lit up and then settled back waiting for the fumes to take effect.

When she had finished, the old woman told him to remove his clothing, slowly got to her feet, spread her legs and told him to get down on his knees and crawl through the arch she had formed. This unusual act of deference to the powers of the opposite sex completed, he was subjected to physical examination and then told to get dressed. Outside the hut she looked around and then shuffled off into the scrub to locate and identify the plant that would protect him against the malevolent forces unleashed by his neighbours. She gave him instructions on how best to prepare the herb and then told him that, if he took care never to kill a woman, he would live to be as old as she was. Delighted by this and by the general prognosis, he took his leave and, once again accompanied by his guide, made his way back across the dry Molopo River to the village of the BaKgalakgadi.[94]

The return journey to the western Transvaal farming country was safely accomplished and, after an absence of ten days, the two men were relieved to rediscover the comparatively verdant terrain of Kommissierust. Armed with his new potion and satisfied that Leetwane would soon be pregnant, Kas started to take a more charitable view of the Canadian Chief and the challenges of the new season. But the last of winter's emaciated cirrus clouds had not yet disappeared over the horizon when, rather unusually, the family was visited by the landlord.

Gert Meyer, who had known the Maines for the better part of two decades, now told them that resentful Boers on adjacent parts of the property had complained about his having 'hired a farm for kaffirs' and about blacks living on a section of Kommissierust without paying rent. Although this was not strictly true, Meyer was not in a position to disclose the nature of his contract with the Maine family, and their arrangement would therefore have to be terminated. In a harsh political climate, where poor whites were constantly looking for ways to increase the social and economic distance separating them from some of the better-off sharecroppers and labour tenants, blacks not only had to have a master, but had to be *seen* to have a master. Where nature failed to sustain the desired social order, man was forced to step in. The Maines would have to relocate and show the world that if they did not work *for* a 'baas,' then at the very least they worked *with* a baas.[95]

This eviction, coming as it did at the very end of the agricultural year, made the family particularly vulnerable, for it gave them very little time in which to make alternative arrangements. Without the usual postharvest respite to find a new landlord, they had to make temporary arrangements. These unfortunately not only split the family as a social entity but limited its productive capacity. As Kas put it years later: 'It was very useful living together because we would, for example, agree to combine our livestock in such a way as allowed us to put them to work. If I owned only one animal and my brother had two more, we could combine them to form a span.'[96]

Sekwala took Maleshwane and the younger children attached to the junior house and went to the diggings at London, where he put his horse and cart to work transporting water for diggers who wanted to circumvent the Company's water concession. Mphaka and Phitise each found places for their families and livestock on adjacent properties where the farmers who owned the land were not as vulnerable to complaints from poor whites as those who merely rented property. But Kas was stranded at Kommissierust, together with his wife, his mother, and his youngest brother, Sebubudi.

With the spring rains days rather than weeks away, Kas had no option but to sit tight, start ploughing and wait to see what Northern Minerals delivered by way of a new landlord. As Motheba had anticipated, Leetwane more than lived up to her expectations. The couple would rise long before sunrise and Leetwane would assist her husband by inspanning the donkeys. Then, with her leading the animals, they would make their way to the field that Kas had opened up the season before, link the animals to the Canadian Chief, and work until the sun's rays bit deeply into their necks; then she would retreat to the shade of their shack to prepare the midmorning meal.[97]

Any pleasure they derived from their progress in the field was soon eclipsed, for Leetwane became pregnant. This happy revelation, coming so soon after Kas's expedition to the Molopo, showed that the old woman's herbs were working; it was therefore most unlikely, they thought, that there would be any repetition of earlier misfortunes. They redoubled their efforts in the field, waiting for the rains. Ploughing a field for their child, Kas was confident that it would be a son and that 'a boy was a foundation.'

Just as lightning precedes thunder, so the landlord comes before the storm: even before the first rains had fallen, they were visited by Jaap van Deventer. A reasonably amiable man who farmed on an adjacent portion of Kommissierust, van Deventer told Kas and Leetwane that he had hired the land they were working and had been granted an option to purchase it by Northern Minerals. Rising grain prices continued to make arable farming attractive, and he was willing to make Meyer's former tenants his partners in a new venture to expand the area planted in maize.[98]

The older members of the Maine family had always steered clear of sharecropping contracts in the belief that livestock holdings were best protected by labour tenancy agreements that guaranteed access to grazing on Boer farms. Kas, however, had less room for manoeuvring than either his father or his older brothers. Stranded by the seasons, without cattle of his own and still indebted to William Hambly, he could see no other way out of his predicament. In the spring of 1920, he, Leetwane, Motheba and Sebubudi reluctantly entered into partnership with Jaap van Deventer.[99]

Within days of this agreement being reached, Kas got his first insight into the hazards of inter-racial sharecropping. One morning van Deventer simply announced that he was taking over the field that Kas and Leetwane had laboured on for weeks. 'I told him that he could not do that since I had already ploughed the land. He replied that I had not done so under him, and that he had hired both the farm and its fields. I could break new ground.'[100] Despite Leetwane's condition they were forced into the fields a second time, which left them exhausted and van Deventer with double the amount of cultivated land.

Although Kas refused to allow this selfish act to destroy his relationship with a man whom he believed to have some redeeming qualities, it did make him acutely aware of the need for long-term planning. With the first rains behind them and the Molopo medicine obviously protecting the crop, he embarked on a scheme to do transport-riding once winter set in. One of his brothers gave him a cow which he promptly sold in order to buy a small wagon from a recently widowed black woman. With this wagon and a team of donkeys of his own, he could go into partnership with Hendrik Swanepoel, who was living close by at Koppie Alleen.[101]

Looking even further ahead, Kas saw the need to find a landlord with a property larger than the 400 morgen that van Deventer hoped one day to purchase from Northern Minerals. But before he could get around to this, Sekwala reintegrated the extended family. Early in 1921 he told Kas that he and the other Maines had found a place on the neighbouring property of Hartsfontein and that Kas's family should join them as soon as was practicable.

Kas acted swiftly. Ignoring the usual practice of waiting for winter and the harvest before telling the landlord of his wish to leave, he informed van Deventer virtually at once. The risk was worth it. Van Deventer not only accepted his notice with good grace, but asked him to help find a family of labour tenants better suited to his small property. Kas discussed this with Leetwane, and they saw an opportunity to rescue her parents and unmarried siblings from the increasingly barren 'native reserve' at

Taung. Kas borrowed the landlord's large wagon and, leaving behind his crop, set out early one morning to fetch his in-laws from Matsheng.[102]

In autumn 1921, with the Lepholletses safely installed at Kommissierust and Tshounyane keeping an eye on his maize, Kas and Leetwane loaded a few things into his new wagon for the move across the road to Hartsfontein. Their material possessions were few, but Kas was confident about the immediate welfare of what would soon be his family. When the time came to return to harvest his share of van Deventer's crop, he would be the father of a son.

# PART THREE

# MANHOOD

*

*'Tiennon had to be very subservient to his landowner, who could never remember his name, who called him "Thing," and who said to him "Obey and work: I ask nothing else of you. And never bother me with requests for repairs: I do none, on principle." In compensation, the landlord sometimes summoned him to the château and gave him a vast chunk of pork to eat in the kitchen, as a special treat. But Tiennon also had to be subservient to the landlord's manager and the latter's mistress, who pestered his wife with requests for menial services. Pathetic scenes took place which made him exclaim: "We are still slaves." He had to work very hard to get a living, sleeping only five or six hours a day, rising at four in the morning. Nevertheless, as "head of a farm, I felt myself to be something of a king."'*

T. ZELDIN,
*France 1848–1945: Ambition and Love* (Oxford 1973),
drawing on Emile Guillaumin, *Life of a Simple Man*, 1904

CHAPTER FOUR

# Independence

## 1921–24

Willem Adriaan Nieman had an unhappy childhood, and the early years of his adulthood were at least as miserable. Born into a poor Afrikaner family in the Bloemfontein district in 1867, the youngster had lost a much-loved older brother as well as his father by the time he was six. His widowed mother, raising six children on her own on a large farm, soon remarried, choosing as her new partner a choleric Dutch immigrant by the name of Scheepers. Scheepers had problems of his own, including the need to reconcile his foul temper with his profession as a schoolmaster, and young Willem did not prosper under his regime. Denied access to formal schooling in town, the boy was made to do much of the hard manual labour around the farm, where he was regularly assaulted by the querulous Dutchman.

The emotional desert that encroached on the farmhouse produced a plant well adapted to a harsh environment—tough, prickly, uncompromising, bad-tempered and occasionally violent. In 1885, at the age of eighteen, Willem Nieman broke free from the schoolmaster's tyrannical grip and made his way to a farm in the Leeudoringstad district of the western Transvaal, where he was 'adopted' by a wealthy family named Jonker. There, he found some of the warmth that had eluded him in his mother's home and, in the mid-1890s, he married one of his mentor's daughters and shortly thereafter fathered a son of his own, Pieter Jacobus. In the brief but fairly happy period that followed, Willem Nieman put down a few tentative economic roots, showing particular promise as a speculator, horse breeder and sheep farmer.[1]

But at the outbreak of war in 1899, Nieman's republican sympathies drew him back to the Orange Free State, where he became a policeman in Bloemfontein. As an untutored twenty-two-year-old with an almost unquenchable thirst for action, he soon resigned from the police and enlisted with the State Artillery. Under the command of the redoubtable General de la Rey, he was amongst those who engaged the British at the battle of Magersfontein and fought later in the Boer struggle for freedom under

General Potgieter. Armed conflict deepened his commitment to the Afrikaner Nationalist cause, but this loyalty became overlaid with anger when, on returning home after the Peace of Vereeniging, he discovered that the war had claimed the life of his wife. His young son, Pieter Jacobus, was without a mother.[2]

It took time for him to come to terms with this new emotional wound, but, like his mother before him, Nieman eventually succeeded in finding another spouse, and in 1904 he and Jacoba Magdalena Greyling married. The marriage helped bring tranquillity to a troubled man's life, and Nieman laid the economic foundations for his new family. But fortune had never looked him squarely in the eyes, and he sometimes found it difficult to understand her messages. In the very year that he took Jacoba Greyling as his wife, the death of a close friend, D. S. van der Merwe, led to the chance inheritance of 1,400 morgen of land on the farm Hartsfontein, in the Schweizer-Reneke district, as part of an unexpected return on an earlier speculative investment in a wagon and a team of oxen. Mindful of his own vulnerability as a child, Nieman generously put this portion of land into the name of the six-year-old Pieter Jacobus.[3]

For nearly a decade thereafter Nieman, spurred on by steady growth in his new family, battled to regain some of the economic momentum he had lost when the British made their grab for the Transvaal. By 1910 he possessed more than a thousand sheep as well as many other animals, and between them these herds threatened to exhaust the limited grazing at Blinkklip. Like several other wool farmers in the same predicament, Nieman looked west for a solution, where the land was both drier and cheaper, and set his heart on building around the nucleus of property that fortune had so unexpectedly pushed his way at Hartsfontein.

When Gert Meyer's mother died in early 1912 and 'Ou Koeël-boud' decided to concentrate his efforts on the family's riverside property, Nieman was quick to step in and buy about 2,000 morgen of the farm at little more than four shillings a morgen. This, together with the property still in young Pieter's name, gave him a far more substantial domain, and he and two of his *bywoners* left Blinkklip and re-established themselves at Hartsfontein. But then fortune again made one of its unheralded appearances and, as before, handed him a gift wrapped in a shroud: in their first winter on the new farm, young Pieter Jacobus contracted pneumonia and died. For the second time in a decade Nieman found himself inheriting property that had once belonged to D. S. van der Merwe, this time from his fourteen-year-old son.

Putting the pain of his most recent bereavement behind him as best he could, Nieman went about the business of slowly integrating his family into the social life of the small and introverted Afrikaner farming community centred on Wolmaransstad. As a product of the bitterly divided political environment that had sundered in the rebellion of 1914, he had no difficulty in identifying with the populist causes that scarred the southwestern Transvaal countryside. An intense dislike of middlemen such as 'Jews' and 'Englishmen,' as well as of any institutions associated with them, such as trading stores, banks and land companies, helped to fuel the anger of militant republicans, who believed that God had destined for their exclusive use the land from which they were determined to wrest a living. A loyal member of the Nederduitse Gereformeerde Kerk at Wolmaransstad, Nieman particularly enjoyed the company of its fiery minister, Dominee E. J. J. van der Horst, an ardent Afrikaner patriot who had been elected to parliament in the 1915 election; in 1920 he was forced to relinquish his position in the church so as to continue his political career. Nieman, whose political passions easily rivalled those of his clerical counsellor, became one of the moving forces behind the establishment of the first National Party branch at Kleindorings in 1914 and served as its chairman for many years. His very public and uncompromising Nationalist sentiments made him an obvious suspect when rebellion broke out later that year, and the state, fearing his influence, took the precaution of placing him in preventive detention at Bloemhof for several days.[4]

If the populist politics of the Triangle agreed with Willem Nieman, then so too did the slowly changing economic environment, and he made steady financial progress in the two decades after his move to Hartsfontein. Like many others he profited from the sustained rise in the price of wool during World War I, and, by the mid-1920s, he owned a flock of more than two thousand Red Persians. Most of the wool from these sheep, as well as their hides and skins, were sold to Hersch Gabbe, the *smous*, or Jewish hawker, better known among the locals as *'Ou Vel'* Gabbe. Hersch was the poorest, least educated, and most obviously 'foreign' of the three Gabbe brothers who had established themselves as traders in the southwestern Transvaal between 1900 and 1910. He, in turn, usually sold his wares to one of his more established and acculturated brothers —Wolf, who owned a store in Makwassie, or the highly respected and notably anglicised Daniel, whose business was in Bloemhof. From there the hides, skins and wool went to larger coastal merchants, such as Hoffenberg's, in Port Elizabeth.[5]

Willem Nieman's greatest enthusiasm was reserved for horse and mule breeding, a branch of the agricultural industry for which the Bloemhof district enjoyed a good reputation, despite the animals being prone to contract horse sickness during the wettest months of the year. Besides his own prized stallion, Express, Nieman ran scores of other horses and mules which, before the country roads were conquered by the noisier motorised transport of the mid-1920s, found a ready market as draught animals. Some of these were sold directly to transport riders, but much larger numbers, sometimes as many as fifty at a time, were sold to dealers in Wolmaransstad like Max Bach, or to the firm of Kaplan & Paradise.[6]

Wool production and horse breeding held other attractions for the xenophobic Nieman, notably that unlike arable farming they did not require large numbers of unskilled labourers and thus minimised his need to interact with black workers. Partly because of this, his farm was served by few permanent 'non-white' labourers and of these surprisingly few were African.

Much of Nieman's working day was spent in the company of his Afrikaans-speaking bywoner, Jan Coetzee, who lived in a small red-brick dwelling situated just behind the main farmhouse. In an era when a pact between the Nationalist and Labour parties saw the government urging that preference be given to the employment of 'civilised,' or white, labour, it fell to Coetzee as the farm voorman, to pass on most of the landlord's instructions to 'coloured' or mixed-race tenants like 'Ou Gool' (Goliath), who had been extended the privilege of being invited into the farmhouse for the evening prayer meetings, *huisgodsdiens*. Most of the hard manual labour on the property was performed by one of two black stalwarts, Pitso Sebothe, a MoTswana who hailed from Konopo, near Lichtenburg, and a deracinated man remembered as 'Piet,' whose wife 'Mina' also served in the Nieman family's kitchen. The only other person of note amongst the labourers who lived on a distant part of the farm was a 'Baster' or 'bastard' by the name of 'Simon' who appears to have been responsible for some of the more general duties in and around the farmhouse.[7]

This small group of socially and economically marginalised men met Nieman's needs admirably by keeping him at a distance from the district's numerically preponderant blacks. But the psychological luxury exacted a price of its own, and as grain prices continued to rise after 1915, so too did the cost of his self-indulgence. When maize prices reached record levels in 1920, his speculative instincts triumphed over his deep-seated racism, and he decided to look around for African tenants who had the necessary equipment and skills to help him devote more of his property to arable farming. This departure thrust him into contact with black sharecroppers who had a reputation for being 'cheekier' than the more culturally assimilated and dependent 'bastards' and 'coloureds' of the region between the Vaal and the Harts rivers: by early 1921, he had at least three such tenant families resident at Hartsfontein—the Ntshohlos, the Tjalempes and the Maines. Between them, these families illustrated the diverse trajectories that led into sharecropping in the southwestern Transvaal during the interwar years.

The Ntshohlos, distant relatives of the Maines on Motheba's side, were a small family that had moved out of the overcrowded reserve at Taung and were still struggling to garner the two major requisites for successful sharecropping: labour and equipment. With only four very young children and half a span of oxen to draw on, they were looking to the dynamic of sharecropping to lift them above the station of some of the poorer labour tenants they encountered on the property. In this respect they formed the strongest possible contrast with an established family like the Tjalempes.[8]

Born into a Xhosa-speaking family, Nini Tjalempe had spent most of his boyhood in the Herschel district of the Cape Colony during the closing years of the nineteenth century before moving on to neighbouring Basutoland as a young man. From there he went to the Orange Free State, where he was involved in sharecropping until the passage of the Natives Land Act in 1913 drove him and his family across the Vaal and into the Ermelo district. This terrain, in the heartland of maize farming in the Transvaal and therefore more prone to the mechanisation of production, was in turn abandoned, and the family eventually went to the drier southwestern Transvaal.[9] There the father's farming successes became legendary amongst black sharecroppers, for Nini Tjalempe was an exceptionally talented livestock farmer and, by 1921, was among the wealthiest of the black men on white farms in the Bloemhof–Schweizer-Reneke–Wolmaransstad Triangle. In addition to an impressive array of agricultural equipment that enabled him to put no less than four spans of animals into the field at any one time, he possessed scores of donkeys, horses, sheep, goats and a herd of nearly two hundred cattle. Substantial assets and the full-time assistance of a younger 'coloured' brother-in-law 'Geelbooi' (Yellow Boy) meant that Nini Tjalempe had no immediate reason to become involved in grain farming in an area of marginal rainfall. But like other blacks prevented from owning property outside the areas designated by the Natives Land Act, he was pushed to make sharecropping agreements in order to ensure that his livestock got access to grazing on underutilised 'European' or company-owned farms.[10]

Together with their wives and children, the Maine men—Sekwala, Mphaka, Phitise and Kas—had more in common with the immigrant Tjalempes than with their poorer relatives, the Ntshohlos, who were drawn from the nearby reserve. While a spate of *bohadi* payments had sapped Sekwala's herd of much of its former strength, between them the members of the extended Maine family still commanded a full span of fourteen trained oxen and the necessary accoutrements, as well as a prodigious amount of skilled muscle power and a large herd of supplementary livestock. This underlying economic affinity was further strengthened by certain cultural similarities between the two families. The stay in Basutoland had ensured that most of the Tjalempes were accomplished SeSotho-speakers, while Nini, a prominent lay preacher in the African Methodist Episcopal Church, was blessed with the gift of faith healing. This unexpected social congruence partially explains why a family of immigrant 'Xhosas' came to be a model for all the Maines, and Kas in particular.[11]

There was one dimension of the Tjalempe family that Kas and Leetwane had no desire to replicate when they joined the small community of sharecroppers on Willem Nieman's farm in the autumn of 1921. For all his ability, Nini Tjalempe had somehow managed to produce six daughters and but a solitary son, an unpromising configuration for a farming family. The prospect of such a lopsided unit held no appeal for the Maines: Kas yearned for a son who could help them coax a living out of the dry western Transvaal soils. With the birth of the child imminent, he and Leetwane went daily to Kommissierust, where her parents helped them to reap the sorghum they had been made to abandon.

Despite reasonable summer rains, the harvest from the field that Jaap van Deventer had forced upon them at such short notice was disappointingly small. Still, they received generous enough compensation for the otherwise meagre return on the summer's labour. Just as the last of the grain was being brought in, Leetwane, assisted by Motheba and willing kinswomen, gave birth to a boy, whom the delighted parents named Mosebi.

A day or two later Sekwala and the older men, each armed with a sharp-pointed stick, were dispatched to dig for the roots which tradition demanded should be added to large containers of fermenting sorghum. Liberal quantities of this fertilised beer or *mmutedi* (which literally means horse manure), were then offered to family and neighbours in a celebration that saw even married women drinking to celebrate Leetwane's achievement. The Maines' small house was filled with laughter as the sharecroppers gave voice to Sotho tradition in uninhibited revelry. But Kas never forgot the demands of the dominant order and the customs of the big house. With the noise of the festivities still ringing in his ears, he picked his way across the stubbled fields to Nieman's farmhouse for an altogether more ambiguous and complex exchange. Sixty years after the birth of the son he had longed for, he remembered it thus:

> That was the rule on the farms. When a child was born you went to the landlord and said, "We have had a baby boy." The landlord would be pleased and say, "Oh, you have had a little monkey, have you cut off its tail?" Then we would say, "*Nee baas, ek het die stertjie afgesny, is nou 'n mens, is nie meer 'n bobbejaan nie.*" ['Yes master, I have cut off the tail, it's a person now, no longer a baboon.'] That was how the white farmers used to put it to us.[12]

Normally, a price was attached to such joking relationships, since the white landlord, having 'won' a verbal joust that affirmed the Darwinian-cum-social order in his own mind, was expected to give his 'defeated' black vassal a sheep to celebrate the arrival of another potential male labourer on the property. But Nieman, who could be forgiven for his lack of formal schooling, was also a tough-minded social illiterate who prided himself on his inability to read the finer print in the countryside's code of race relations. This embarrassing exchange was therefore left without its customary conclusion, and Kas departed feeling humiliated. 'He was bloody rotten, a big rebel, who was unwilling to take a "kaffir" into consideration.' Kas decided that as soon as Mosebi was old enough, he would slaughter a ram of his own to celebrate the arrival of his son.[13]

With the preliminary rituals behind him, Kas faced the perennial predicament of how to raise additional cash during what was left of winter and the coming spring. By midyear the problem seemed worse than ever. The inflationary whirlwind had blown itself out and 1921 saw an alarming slump in the price of agricultural commodities. With wool and mutton fetching only half of what they had the previous season, it was pointless trying to sell sheep. Nor was there much sense in going to the market with the few bags of grain they had harvested, since the drop in maize and sorghum prices was at least as pronounced. And, as in 1920, when his position had been undercut by the locusts, Kas was haunted by his debt of honour at Hessie.[14]

Ironically, it was at precisely this low point that his earlier planning started to show some return, for he suddenly saw a chance to put to use the wagon he had bought from Gert Meyer. Given the drop in grain prices, hard-pressed white farmers who were indebted to the Land Bank were being forced to sell more and more maize and sorghum. Like his father before him, Kas recognised this as an opportunity and pooled his resources with those of Hendrik Swanepoel in a transport-riding venture. While Sekwala, Mphaka and Phitise made money from several building contracts, he and Swanepoel undertook dozens of journeys to the store of Stirling Brothers at Schweizer-Reneke, or to the railhead at Kingswood, earning ninepence for each bag of grain delivered.[15]

On his return from one of these trips, about six weeks after the birth of Mosebi, Kas ran into Sekwala and Maleshwane entertaining one of the younger wives attached to the house of Mmusi Mokawane, the uncle who had made such generous provision for Motheba and her children during the closing stages of the South African War. Kas, eager to repay his share of the historical debt to the Mokawanes, invited the aunt to spend the night at his home before she moved on to visit other kin in the district. She graciously accepted his invitation, apparently unaware of just how recently Leetwane had become a mother.

The following morning Kas awoke to find Leetwane cradling the baby who, crying incessantly, was suffering from acute diarrhoea. The sight and sound of the distressed infant alarmed more than the parents. The aunt, who had given birth to a stillborn child shortly before she had left Setlagole, shamefacedly came forward to admit that she had neglected to go through the necessary cleansing ritual before setting out on her journey. Since it was well known that a woman who had had an abortion or given birth to a stillborn child without subsequently undergoing ritual purification was capable of passing on a condition known as *phuana* to other children, everybody suddenly feared the worst. The baby's condition worsened, and within a day or two the child was dead. For Leetwane, the silence and loneliness were eased only by the echoes of the refrain her friends had sung to her on her wedding day. 'We are leaving you, you will manage on your own, they are your family.'[16]

In the empty days and weeks that followed friends and neighbours rallied to her side but, above all others, Motheba came to her rescue. Drawing on the great reservoir of faith embodied in her affiliation to the A.M.E.C., Motheba lavished care and attention on her daughter-in-law and much-loved niece. The Maines were indeed Leetwane's family and gave her emotional refuge. But in the dark of the night she could still hear the baby crying.

The funeral took place in a small, partially overgrown roadside cemetery for blacks on the surrounding farms. It drew between twenty and thirty adults, members of the extended family and their children predominating. The somewhat eclectic service, held under a few tall bluegum trees that were struggling to maintain their composure in the teeth of a swirling winter wind, drew the chilled huddle of Christians and traditionalists closer together. In keeping with custom, the corpse was wrapped in a soft sheepskin hide, lowered into the parched earth, and then, in a symbolic plea for the continued fertility of the mother, sprinkled with the froth drawn off the beer, *lekweba*. The A.M.E.C. contingent, led by Nini Tjalempe, took leave of the little one with the much-loved Sotho hymn 'If you ask where my faith lies, it lies in Jesus.'[17]

The funeral over, Kas threw himself into transport-riding with an almost vengeful passion. Yet not even the distraction of exceptionally hard work allowed him to put the loss of the child behind him. Somehow that year all the roads that he and Hendrik Swanepoel used to ferry grain across the stony plain seemed to lead past the tree-marked intersection at Hartsfontein, and the bluegums made sure the men never forgot the baby. Endless hours huddled on the wagon, interrupted only by short energy sapping spells of off-loading grain, left Kas physically exhausted but unable to achieve inner rest.

In August, having met the last of their commitments to the white farmers, he and Swanepoel parted company, each going his way to tally the season's takings and plan his spring campaign. For Kas this brought some consolation, once it was clear that he had earned enough to meet the family's immediate requirements. Indeed, he had enough money to go to Hessie, pay for the Canadian Chief and keep faith with Hambly.

However, a friend told him that the small corrugated-iron building at Hessie no longer housed a business. After the locust invasion, which had left so much devastation and debt in its wake, the market for agricultural products had collapsed and William Hambly reluctantly conceded that he had absorbed enough punishment. With several of the larger land and mineral companies selling off their less promising properties at sharply reduced prices, he had decided to try his hand at farming, and, in August 1921, had bought a farm from Henderson Consolidated and retreated to the far side of the district.[18]

Kas knew exactly where to look for him, and early one morning, before the rains came, he harnessed two donkeys to the cart with the smart iron-rimmed wheels. The trap, which had been sadly neglected when he had acquired it from a neighbour in exchange for five bags of grain earlier in the year, had been carefully restored. A good cart was a prerequisite for any respectable couple going about their business, and with great pride he asked his wife to accompany him on the sixty-mile round trip to Holpan.[19]

The bumpy ride across country roads shook loose old memories. He remembered how, as a small boy, he had watched in horror as the invading British troops had burned down the village grain store at Holpan, and how he and his brothers had been put in charge of Blom only to witness the cow bolting from the column of refugees in an effort to find her calf.

On the outskirts of Schweizer-Reneke he found the old gate that led to the farm. Hambly, who had not told all his former clients of his new address, was surprised to find the young couple when he came out to meet the cart and was struck anew by the young sharecropper's determination to settle a debt in troubled times. His faith in the Maine family vindicated, Hambly, in a tribute that had a biblical ring to it, pressed thirty shillings on Kas, urging him to treat his wife to a meal before setting out on their return journey.[20]

The road home took them past the bluegums. But on this occasion they appeared less sombre, and with the touch of spring already evident in their brighter green apparel, they hinted at the release that would come with the cleansing ceremony of '*ho tlhatswa*.'

One morning at sunrise, about a month after the journey to Hambly's, Kas slaughtered a sheep before a small gathering of his immediate kin. He and Leetwane served them soft porridge. As soon as they heard the scratching of spoons on metal plates, they removed themselves from the company, reappearing a few minutes later stripped to the waist. Then, in keeping with the practice of their forefathers, they were approached by family members who took great delight in pouring a watery mixture of chyme, drawn from the sheep's intestine, over their heads and rubbing it into their hair and on their torsos. Soon they were covered with tiny green flecks of partially digested grass. When the supply of chyme was exhausted the family rinsed them with clear water, after which they retreated to the privacy of the shack for a thorough, less public washing. By the time they reappeared, Mosebi's clothing had been spread out on a mat before them, lightly sprinkled with what remained of the chyme; it was handed out either to those kin who had some immediate use for it, or to those who wanted a small article that could serve to remind them of the child.

The rest of the day they spent in a relaxed mood, eating roast mutton washed down with liberal quantities of sorghum beer. If it were not for the slightly subdued atmosphere, they could have been 'drinking the cattle's legs' but, with the first rains some weeks away, there was much work still to be done before the oxen could be led out into the fields.

Fired by the optimism that marks the arrival of spring in the countryside, the Maine family started ploughing in earnest in September. Despite the earlier startling decline in prices they, like most of those around them, hoped that 1922 would see a return to the better levels they had grown accustomed to. And if higher prices did not prevail, there was all the more reason to try to keep income constant by planting even more grain.

As Sekwala's oldest son and the man with the most trained oxen at his disposal, it was Mphaka who managed to plough most extensively. Using twelve or fourteen animals at a time to draw the heavy two-share Dutchman, he succeeded in cultivating a memorably expansive tract of land. Kas started out by harnessing his donkeys to the lighter Canadian Chief but then decided that he, too, would benefit from the action of the deeper-cutting Dutchman, and waited his turn to use the heavier plough.[21]

Nieman watched the alien ploughs devour his land with decidedly mixed feelings. The entrepreneurial spirit within him made him eager to have a share in the Maine, Tjalempe and Ntshohlo crops, but the instinct to give priority to the sheep-, horse- and mule-breeding business that lay closest to his heart made him wince at the loss of good grassland. This inner struggle, which would have been closely fought in any normal year, was intensified by the fact that while the market for grain was uncertain, the price of mules and other draught animals seemed to hold up surprisingly well.[22]

The way round this dilemma was to buy more land while the recession continued, and in February 1922, Nieman bought a parcel of 380 morgen on the neighbouring property, Kommissierust, when the directors of Northern Minerals decided to divest themselves of their interest in that venture.

This acquisition eased the pressure on Hartsfontein, but Nieman's tenants still had the impression that they were not really welcome on the farm. According to Kas this was not merely because the landlord lacked experience when it came to dealing with sharecroppers, but because Nieman found it difficult to accommodate any attempt by a black man to improve himself. Kas had first encountered this churlishness about his desire to 'pick himself up' at the time of Mosebi's birth, but the tension became even more evident as the ploughing season continued and then drew to a close.[23]

With the locust invasions at the back of his mind, Kas took the precaution of planting a good number of household vegetables and watermelons between the rows of maize and sorghum plants, which began to wilt quite early in the summer heat. Traditionally, such supplementary production was considered beyond the scope of the sharecropping agreement with the fruit of this marginal effort belonging to the tenant alone. In a year that saw only ten inches of rain fall in the Bloemhof district, half the norm, this insurance took on added significance. The melons somehow succeeded in extracting sufficient moisture from the unyielding Kalahari sands and, where God's lesser orders failed, they managed to survive.[24] As the sand storms raged across the plains in yet another drought season, the Maines drew on the succulent fruit in Kas's field.

This pronounced domestic demand gave Kas the idea of loading watermelons into his wagon and taking them to the diggings, where they found a ready market amongst the army of parched diggers around London. But, each time he loaded the wagon and trundled out of the farm gate, Kas felt his landlord's envious and reproachful stare. A man with an eye of his own for the market, it irked Nieman that someone with an entrepreneurial sense at least as well developed as his own had found an opportunity when all that he and his peers could detect were the retreating dust storms. What hurt even more was that a black man was earning between one and three shillings each for watermelons grown on his farm and there was no way of his sharing in the proceeds.[25]

Kas's heart sank when, one morning not long after the sale of the watermelons, he discovered that somehow or other his donkeys had managed to get into Nieman's field and had eaten some of the maize. After removing the animals from the field, he went to the farmhouse and invited the landlord to come with him to the site and pronounce on the extent of the damage. Much to his surprise and relief, Nieman did not seem unduly perturbed and merely remarked that not too much harm had been done.

Only with the benefit of hindsight did the landlord choose to revise this casual assessment. A few weeks later swarms of brown locusts again invaded the western Transvaal, and such was their greed that even Hartsfontein's frazzled fields took on a favourable appearance. Their onslaught was as deadly as ever. Kas lost all his sorghum and, at season's end, had only fifty bags of maize. But the largest locust of all continued to lurk in the farmhouse: when the moment arrived to share the harvest, Nieman chose to recall the donkey invasion and inflict such a severe and belated fine on young Kas that he was left with just fifteen bags to show for the season's labour, 'just about enough to live on.'[26]

Along with the Maines, several other tenants found themselves thrown back on their livestock resources to survive, while Nieman contrived to eat most of what the locusts had left. Whereas Sekwala and his traditionalist sons muttered about their lot, the Tjalempes, Ntshohlos and others drew strength from their faith: whenever this small band of Christians set off for the nearest formal gathering of the African Methodist Episcopal Church they were joined by Motheba, Phitise, Sellwane and, especially after the baby's death, by Leetwane.

The African Methodist Episcopal Church was gaining strength throughout the district. After the passage of the Natives Land Act and the spurt of economic development triggered by World War I, it had made its presence in the southwestern Transvaal more widely felt, and by 1919, the clergyman presiding over the circuit, the Reverend J. T. Gabashane, was meeting the spiritual needs of more than three hundred souls in congregations spread throughout the Bloemhof, Schweizer-Reneke, Makwassie and Wolmaransstad districts. Many, if not most, of these adherents came from among sharecroppers on white farms or from among better-paid workers on the diamond diggings.

In late 1921, this quite prosperous circuit was entrusted to a new minister, the Reverend A. M. Phosa. Like several of the A.M.E.C. pioneers, Ananias Phosa hailed from Ramokhopa, in the northern Transvaal, where he had been one of the first pupils at the original Wilberforce Institute, situated on the Dwars River. After a spell in the South African Police as a constable, he had joined the South African Native Labour Corps during the Great War, where 'he was very highly thought of by his Commanding Officer and was mentioned by him in dispatches.' After the war he worked as a teacher for two years before being accepted into the ministry.[27]

A gifted linguist who commanded six African languages, Ananias Phosa was well suited to dealing with the heterogeneous black communities that had sprung up in the area between the Vaal and Harts Rivers after the discovery of alluvial diamonds, and he applied himself vigorously to his work. In addition to leading the Sunday service, held in one of the larger towns on the circuit each week, he made a point of visiting the more far-flung parishioners at least once every three months. On these occasions he was drawn into closer contact with communities of sharecroppers such as the one on Willem Nieman's farm at Hartsfontein.

Motheba and the others welcomed these quarterly visits by their minister as a valuable supplement to their regular weekly prayer meetings. It meant not only that they had a formal service in their own homes, but a pleasant social gathering thereafter: after collecting membership fees, which the sharecroppers could pay in cash or in kind, the minister presided over communion for those who found it hard to get to church more regularly, and then, after the service, would relax over a meal and a little beer. Leetwane, who had been sickly after Mosebi's death, became increasingly reliant on this circle of friends in which Motheba played such a leading role and, on 18 March 1922, Phosa accepted her as a full member of the African Methodist Episcopal Church.[28]

Kas, who had been exposed to the church throughout his childhood and adolescence, accepted this development with his customary pragmatism. Conscious of how his own parents' marriage had withstood most of the tensions that came with reconciling Christian and Sotho traditions, he tried to build on the experience. He took out the cart and willingly accompanied his wife on monthly trips into Makwassie, where he listened attentively to the Reverend Phosa's sermons. While his calling as diviner excluded him from full membership in the A.M.E.C., he nevertheless thought that there was little to fear from a church which, although much taken with its links to black Americans and the modern world, somehow still managed to tolerate herbalists and men who practised polygamy.[29]

With Leetwane showing new interest in the life around her and becoming well integrated into the family, Kas embarked on the winter campaign in a more settled frame of mind. Although the harvest had effectively failed, he managed to retrieve some of the lost ground when the ever-reliable Hendrik Swanepoel invited him to join forces for yet another round of transport-riding. The season drew to a close on a positively optimistic note when Leetwane informed him that she was once again pregnant. Cheered by this news, Kas looked forward to the coming spring fray with renewed optimism.

But, well before he or his brothers could commence ploughing, they were summoned to the house and informed about a change in plans. Willem Nieman was disappointed by the poor returns from his first serious venture into sharecropping and anxious to avoid his previous mistakes since the stock and wool markets were showing signs of recovery. Hartsfontein was therefore to revert to its more familiar function as a farm devoted to sheep, horses and mules, he said, and the black tenants would have to move onto the smaller parcel of land he had acquired earlier that year. In the spring of 1922 the Maines and other sharecroppers moved over the road to Kommissierust, where they were later joined by two newcomers, Katse and Kwape Moshodi.[30]

As before, Kas worked closely with Mphaka and once again chose to wait for the use of the family's span of oxen and the Dutchman rather than set his donkeys to work with the Canadian Chief. In the end, and contrary to all expectations, the choice of plough made little difference to the eventual size of the harvest. As the first clouds of summer crept over the horizon toward the already simmering highveld, reports were received of enormous swarms of locusts, rumoured amongst the sharecroppers to have originated in India, massing on the distant fringes of the Kalahari Desert and preparing for their programmed flights to the east. In a district where much of the grain trade was in the hands of Asian storekeepers, rumours about the Indian origins of the locusts found a ready audience amongst the predominantly Christian landlords and sharecroppers.[31]

Having long since abandoned the hope that these four-winged enemies could be deflected from their path of devastation by the use of traditional medicines, Kas resigned himself to waiting and seeing whether the devil's dice would come to rest amongst his own crops. One afternoon in November with both the maize and sorghum seedlings already making headway against their other adversary, drought, the first swarms were observed approaching Kommissierust, but on this occasion they chose to overfly the farm and continue eastward. However, they were followed by even larger swarms. In January locusts came in numbers that beggared description and settled down to an orgy of destruction that lasted until they had devoured every last morsel of plant life. With locusts stretching almost as far as the eye could see, small clusters of farm folk drawn from every walk of rural life battled to lift and shake out the brown blanket suffocating everything beneath it. At dusk, when the swarms settled for the night, kraals would be emptied of sheep, cattle, donkeys, mules and horses and every available animal would be taken out and driven over seemingly endless layers of crunchy locust bodies. Others used sledges or heavy bluegum branches to the same end, until the cool night air was heavy with the stench of crushed and rotting insects.

A man might as well have attempted to deflect a dust storm with a tea strainer. As Kas recalled it many years later,

> . . . the locusts ate all of my beautiful red sorghum and maize crop. They caused great trouble for all the people, because everyone had been advanced credit through the trading stores. Man, there was confusion! We lost badly!

For the third consecutive season Kas saw the fruit of his labour eaten in the fields, while all around the veld was stripped of grass. With the Triangle in the midst of what seemed like seven lean years, the price of land fell and economic recovery in the district once again faltered.[32]

But sharecroppers could not afford to wait for the estate agents in town to announce the return of better times and, with a cunning born of hardship, they found ways of feeding the very agents of death back into the rural life cycle. Like the BaTswana on the neighbouring properties, the Maines collected baskets-full of locusts which they boiled and dried to serve as protein-rich relish with mealie meal. Prepared imaginatively, the insects were tasty enough to draw the younger Nieman children into the tenants' homes where, against their mother's wishes, they partook of locust meals as readily as any of their black peers.[33]

Most of the insects were dried and stored in maize bags and used as a supplementary livestock feed during the winter that followed. Cattle took readily to being fed on a mixture of dried locusts and coarse salt—indeed, so readily that they had to be restricted to small measures followed by immediate watering if they were not to become constipated. The livestock's condition picked up so rapidly on this unusual diet that Kas could make an early return to Neser & Campbell's monthly stock fair at Bloemhof. Although short of cash, he succeeded in buying several animals which, fattened on the mixture, he later sold at a handsome profit at a time when the price of beef was rising faster than that of other agricultural commodities.[34]

Livestock speculation and the absence of midyear harvesting duties left Kas with more time on his hands than he had had in other winters, and he watched with relief as Leetwane's pregnancy drew to a safe conclusion. When the magical day dawned in late May 1923, his mother was once again on hand to help deliver a boy whom the overjoyed parents named Mmusetsi, 'the replacement,' the child who 'wiped away our tears.' This time Kas didn't bother to inform Nieman.[35]

When Leetwane had recovered, Kas slaughtered a ram for a gathering attended by the family and Nieman's established tenants, along with the newly arrived Moshodis. A pleasant day was made positively memorable when Kwape Moshodi, who in addition to his reputation as a good farmer was also known to be a snappy dresser, staged frequent disappearances only to reappear on each occasion dressed in a more fetching outfit. This self-conscious display of sartorial elegance earned him the mocking title of 'Mister Kwape' amongst the amazed sharecroppers; Kas was uncharacteristically contemptuous of such self-indulgent behaviour. Sensible men worked for cattle, not clothing.[36] Kas's reservations about Kwape Moshodi notwithstanding, the community of tenants at Kommissierust was socially cohesive. The underlying affinities of Sotho-Tswana culture and the common experience of hard times did much to offset the disparities of wealth and to blur the distinction between Christians and non-Christians. The real potential for social disruption, it seemed, lay elsewhere.

One afternoon, not long after the birth of Mmusetsi, Kas heard a huge commotion outside his shack. He emerged to find nine-year-old Hansie Nieman angrily disputing the ownership of a penknife with one of Nini Tjalempe's youngsters of about the same age. Their shouting match soon gave way to punching, with the young Nieman managing to deliver one or two telling blows before the black boy effectively immobilised him by grabbing hold of his testicles. This unorthodox, highly effective countermanoeuvre unleashed a round of even more anguished howling, which in turn saw the arrival of several small Tjalempe girls armed with sticks and more than willing to go into battle on behalf of their brother.[37]

At this interesting juncture one of the older Nieman boys arrived to put a stop to the fray. But his intervention did nothing to determine ownership of the knife, nor did it console the injured parties, each of whom rushed off to tell his father what had happened. Young Hansie managed to get to his father first, and Willem Nieman, by temperament always more inclined to action than thought, jumped on his horse and dashed off after his son's adversary. He chased the terrified black child all the way back to his parents' shack and then, his anger spent, went home. In the meantime Nini Tjalempe and his wife listened in distress to their son's account of the clash and went with the boy, still clutching the penknife, back to the farmhouse. The issue of ownership was now conceded by a somewhat calmer Willem Nieman, who had questioned young Hansie and reflected on his own response. It was largely as a face-saving exercise that he insisted that the tenant's son was nevertheless wrong for having resorted to violence to protect his property. Mrs. Tjalempe would have nothing of it, and she continued to defend her son. Her uncompromising stand reignited the embers of Nieman's smouldering anger, so her husband, fearing a truly fiery end to the confrontation, suggested that it would be in everyone's interests, including the children's, if the Tjalempes were to leave the property at the end of the season.

Caught between his desire to retain the services of the black farmer on the one hand, and his pride as a white man on the other, Nieman was stranded with only the weakest of replies.

> It is up to you really. I am not expelling you. We have been involved in transport-riding to Kuruman for a long time, and you have been a sharecropper with me.[38]

But there was no way out. As was so often the case, a dispute over a tenant's son did most to undo the agreement between black sharecropper and white landlord. By winter's end Nini Tjalempe had abandoned Hartsfontein and linked up with another prominent black sharecropper, Mdeboniso Tabu, and an ambitious young white farmer named Koos Meyer, to hire the property Pienaarsfontein.[39]

Another factor might account for Tjalempe's willingness to retreat before the storm. Willem Nieman's monopolisation of the resources at Hartsfontein for his horse- and mule-breeding business had restricted his tenants' cattle to the smaller property at Kommissierust. This might have been a viable strategy in normal seasons, but the combination of drought and locusts in successive years had severely eroded the land's carrying capacity. By late 1923, many of Nieman's tenants, including the Maines, were looking around for more grazing and farming land.

When this problem had arisen in the past, either Sekwala's entire entourage had moved to a larger property, such as Soutpan, or a smaller and quasi-autonomous section of the family had split off and gone to a temporary redoubt, such as Doornhoek, until it was safe for the extended family to regroup. On this occasion they chose the latter option and, in the course of their search, Kas and Mphaka renewed the Maines' relationship with the Reyneke family, a connection that could be traced back to the 1890s when Sekwala had first become a tenant under 'Ou Koos' Reyneke on the farm Rietput, in the Schweizer-Reneke district.[40]

Petrus Albertus Reyneke, better known simply as 'Piet' Reyneke, was the oldest son of 'Ou Koos.' After having served in the South African War, young Piet had returned to Schweizer-Reneke, where he had worked as a telegraph clerk in the local post office before marrying sixteen-year-old Anna Griesel. But, like one of his younger brothers who had taken over operations at Rietput, he had his heart set on farming and, shortly after World War I, had hired a property to try his hand at wool production. The price of wool held up fairly well, and by the time Northern Minerals were forced to dispose of some of their holdings during the slump of 1922, he had enough capital to buy a section of Kommissierust.[41]

Within months of acquiring the property, which he named Vlakfontein, Piet Reyneke also started hiring land on the adjacent farm, Koppie Alleen, a measure forced on him by locust visitations and the need to find grazing for a flock of sheep that numbered close on a thousand. When the Maine brothers asked him about the possibility of taking them in as labour tenants, he jumped at the chance. Not only were the two well known to his family, but they were also widely respected as stock farmers. If Kas and Mphaka assumed responsibility for the sheep grazing on Koppie Alleen, then he would leave them free to devote what was left of their time to the pursuit of their own interests. The brothers readily accepted his generous offer.[42]

The move across the road from Willem Nieman's farm to Piet Reyneke's hired property posed few logistical problems, especially since it had been decided that the Maine cattle would remain in Sekwala's care at Kommissierust. A week or so later they loaded their heavier equipment onto the wagons and trundled across the way to Koppie Alleen where they set about erecting suitable shelter for their families. With the help of his older and more experienced brother, Kas soon got the mud walls of what would be his and Leetwane's new home to roof height.

Mphaka, a skilled builder, then erected a pitched timber frame for the roof of his own house, which he proceeded to thatch in traditional Sotho fashion. But Kas hesitated. Ever since the locust invasion termites had taken to attacking grass roofing with unusual vigour. Hendrik Swanepoel, by then a concerned neighbour as well as an old friend and business partner, suggested that he make the roof of corrugated-iron. Kas was not opposed, but he lacked the resources. Swanepoel offered to purchase the material against his own name and let Kas settle the account after the next season's transport-riding. Kas agreed, and soon Swanepoel helped unload ten sheets of flashing metal which the Stirling Brothers had sold to him for ten shillings each. A debt of five pounds was a small price to pay for a roof that would protect both his wife and his son.[43]

Once their families had been safely installed in the new houses, Kas and Mphaka turned to ploughing. Since the oxen had been left at Kommissierust with Sekwala, it was no longer practical for them to use the heavy Dutchman plough, so they reverted to Kas's eight donkeys and the lighter Canadian Chief, which, given the assistance of Mphaka's thirteen-year-old son, Baefesi, was able to till fields to their requirements.[44]

Kas's daily routine revolved around the needs of Piet Reyneke's flock of Merinos. It was not the most demanding task he had ever undertaken, and it seemed it would become easier still when one of the landlord's brothers presented him with a small bitch terrier to replace Bull whom he had lost to a snake bite at Hartsfontein. But the puppy developed into a foul-tempered and unloved dog, and when he had to pay ten shillings in tax for it to a visiting inspector, he passed it on to a friend who, tiring of it equally quickly, disposed of the unfortunate creature by feeding it particles of ground glass.[45]

Even without a dog, Kas found that he and Mphaka could get through most of the day's chores fairly easily because they could always call on young Baefesi or, on more demanding occasions, such as when shearing sheep, Phitise and Sebubudi. This meant that for most of the year they spent on Reyneke's hired property, Kas had better access to uncommitted labour time than when he had been bound by sharecropping contracts. Labour tenancy as opposed to sharecropping enabled him to engage in a rare spell of wage labour of his own.

Once the sun reached its highest point in the seasonal arc, Kas made his way down the road to a place which the BaTswana, since time immemorial, had known as Letswaeng but which the first Boers who settled there permanently named Sewefontein. Much of the Triangle's geography was mapped in such alternative ethnic codes, and, as a fluent linguist, Kas slipped in and out of them as the occasion demanded. Depending on your origins, whom you were visiting or the nature of your business, the Edwards Brothers' salt works was located at Letswaeng or Sewefontein. He was on his way to Sewefontein.

When the sun was hottest and the evaporation rate above the salt pans at its highest, it was possible for several dozen black men to find unskilled work at the Sewefontein works. Although the wages were slightly lower than those paid on diamond diggings, the work was marginally more congenial, and since the salt works were only eight miles away, they had the great advantage of allowing Kas to go home over weekends to help with the Merinos and keep an eye on his family.[46]

Along with the larger Vaal River Salt Works Company at Zoutpan, in the neighbouring district of Wolmaransstad, Sewefontein had been known as a source of fine quality table salt since the property was first proclaimed as a farm in the early 1860s. But, as with most of South Africa's slowly developing secondary industries, it was only during the First World War that the Edwards brothers found their product to be in greater demand amongst undersupplied domestic users, livestock farmers and commercial dairies. Supplying this growing market with sodium chloride was easy, since the farm was blessed with a seemingly inexhaustible supply of brack water which required only a little help from nature to be converted into excellent table salt.[47]

Brine was extracted from two pits, 'Coleman' and 'Stone,' and pumped into a large storage dam built on the floor of a natural depression more than a square mile in extent.[48] Once much of the water had evaporated, the thickened brine was led off into a series of shallow evaporating pans. When the solution reached jelly-like consistency men would agitate the mixture with long sticks. This labour-intensive technique separated the sulphates from the chlorides, and allowed the salt to dry in a thick white powdered crust. The layered salt was then scraped into piles, poured into two-hundred-pound bags, loaded onto the waiting wagons of Boer transport riders, and sent to the railhead at Kingswood.[49]

As someone who had spent virtually all of his life out of doors, Kas readily adapted his farming skills to a production process driven largely by nature. His ability to read the vagaries of the western Transvaal summer weather helped him anticipate winds and squalls whose untimely appearance gave rise to turbulence capable of remixing the salts at exactly the moment they most needed to be kept separate. But he also learned to develop a greater respect for the destructive properties of salt, which, once extracted from its relatively benign liquid base, tended to counterattack both men and machines with a vengeance.

Like most of the workers, he appreciated the need to apply a thick layer of axle grease to his hands and feet if his skin was to be kept from drying out, cracking and stinging continuously. For the same reason he and the other labourers usually donned rough home-crafted wooden sandals whenever they ventured onto or near the pan surface. Unlike the others, however, Kas considered it unnecessary to put on dark glasses to protect his eyes from the harsh white reflection off the salt beneath a clear blue sky. And, with or without glasses, anyone could see that only a few months' exposure to the salt necessitated the replacement of corroded pumps and shovels.[50]

With two months of hot and sticky labour at Sewefontein behind him and a few notes carefully folded into a small leather purse, Kas returned to Koppie Alleen, where he became a junior partner in a more lucrative venture. Reyneke, having taken note of the almost inexhaustible demand for water on the diggings, had invested in the sinking of a borehole at Vlakfontein, erected a windmill, and built himself a large reservoir. But, lacking the labour and transport necessary to bring the venture to fruition, he had to find a subcontractor with a cart and donkeys who could deliver a minimum of five barrels of water at a time to his customers. Kas was willing to provide this service for thruppence a barrel. This arrangement, which lasted throughout the late summer months of 1924, met with Reyneke's ready approval since Kas somehow still managed, between all the comings and goings, to find the time to keep one eye on his landlord's flock and the other on his fields.[51]

The Maine brothers' fields had been ignored by several swarms of locusts, which had flown over them and gone on to the outer fringes of the district, where huge unsightly scars hacked out of the wilting autumn colours reminded the less fortunate of their summer visitation. With Koppie Alleen fortunately unscathed, the women and children were sent into the fields and, for the first time in several seasons, emerged with a reasonable harvest. Kas was more than satisfied when they recouped fifty bags of sorghum; Mphaka, too, did reasonably well on the patch he had planted.[52]

Playing the roles of wage labourer, water carrier, shepherd and grain farmer hardly exhausted Kas's repertoire, for he was brimming with the energy of manhood. His reputation as a herbalist was slowly percolating through the district, and within the seclusion of a hired property where there was no resident landlord to monitor entry and egress, he now found a few bashful Afrikaner males appearing amongst his more regular clients. His introduction to this less privileged and educated element of Boer society came via the white man whom he knew to have the greatest possible respect for the mysteries of black culture, Hendrik Swanepoel.

One afternoon, Kas was asked to call on Swanepoel, who was at his home at the far end of Koppie Alleen suffering from backache. Backache was a familiar complaint amongst transport riders who prided themselves on their strength and, in Swanepoel's case, Kas attributed it to the awkward way in which his friend tended to twist his spine when transferring the weight of a two-hundred-pound grain bag from his shoulders to a wagon. The back could be permanently strengthened through the application of a herb known as *kgomodikae*, but if the treatment was to be effective, the herb would have to be freshly gathered from the veld. Kas advised his partner to rest and told him he would pay him another visit when he was better.

Not long thereafter Swanepoel appeared at the house and together they combed the surrounding veld until they found the plant Kas was looking for. Kas dug out several roots with a stick and then disappeared to prepare the herb in a way known only to a ngaka. On his return, he sought out a patch of warm soft sand, laid out the flattened roots in the shape of a cross, and ordered Swanepoel to remove his shirt. Next, he instructed him to lie down, place the affected part of his body over the cross and wait for several minutes as the powers of the herbal medicine penetrated his body. He knew this would relieve his friend of his backache, and it did.[53]

Swanepoel, delighted, recommended Kas's services to several of his male acquaintances. Some of them, who suffered from ailments more complicated than backache, were shy of treatment by general practitioners in small platteland towns, where doctors and ministers sometimes shared a good deal more than an elevated social standing. Piet van Schalkwyk, an eccentric *bywoner* from the farm Vaalboschbult, was one of those who had reservations about the wisdom of consulting white doctors about potentially embarrassing complaints. Known to local SeTswana-speakers as Ramatekwane, 'the dagga smoker,' not so much because he smoked cannabis but because he behaved as if he had, van Schalkwyk, like his friend, enjoyed a reputation for being well disposed to blacks.[54]

Van Schalkwyk, whom the Maines had known from their time at Vaalboschfontein, used a simple rural idiom to explain his problem: ' "*een van my osse werk nie*," ' 'one of my oxen is not working.' In addition to the problem with his testicle, he complained of having a scrotum distended by accumulated fluid. The latter complaint, which Kas had first heard of while working on the diamond diggings, was one that was incorrectly attributed to having come into contact with polluted women suffering from venereal diseases whereas, in fact, it was part of an inherited condition passed from father to son. Fortunately, in Kas's experience, it responded readily to herbal treatment; it was possible to drain off the excessive liquid. He prepared the *doepa*, handed the potion to van Schalkwyk, instructed him on how to use it and, when Swanepoel informed him a few weeks later that his client had been cured, he levied a charge of a pound to supplement the shilling consultancy fee he had obtained for their first meeting. Just as surely as the social dynamic of the middle class drew together orthodox medical practitioners and ministers of religion in the small towns, so too a lack of formal education, landlessness, and a need for economic cooperation in rural production drew together the bywoners, sharecroppers and transport riders of the countryside.[55]

The ubiquitous Hendrik Swanepoel, a central figure in Kas's social network, alerted him to yet another end-of-season opportunity. To the west of Koppie Alleen, Koos Meyer and his army of sharecroppers (among them the Tjalempes and the Tabus) had enjoyed an enormously successful season and needed hired help to bring in a massive sorghum harvest. With one eye already firmly fixed on the road leading to the railhead at Kingswood, Swanepoel argued that if Kas were to help with the harvesting at Pienaarsfontein, they might well win the transport-riding contract that was bound to follow.[56]

The peasant wisdom in this suggestion appealed to Kas and, having no landlord of his own to work for at Koppie Alleen, he promptly went to Pienaarsfontein, where he joined the harvesting teams who flooded into the district from Taung each autumn in the hope of finding seasonal employment at two shillings and sixpence a day. In earlier years many of the same men had used their wagons and labour to sell firewood on the diggings, but the combined impact of drought and their sustained assault on the struggling acacia bush had begun to take a toll. This, along with a decline in the amount of arable land available in the reserve, had pushed more and more former entrepreneurs into wage labour. Others, much like Leetwane's parents, had abandoned the reserves altogether and taken up positions as full-time workers on the comparatively underdeveloped white farms of the southwestern Transvaal.[57]

Casual labour on Koos Meyer's hired farm not only brought in yet more money in what was turning out to be a truly exceptional season, but prompted Kas to think of other ways of benefiting from the larger harvests. With more cash than he had had for some time, he hurried home and persuaded Piet Reyneke's young son, Willem, to let him use one of the three large underutilised wagons at Vlakfontein at a rental of five pounds a year.[58] With no less than sixteen donkeys harnessed to the hired wagon, he increased its carrying capacity by thirty to forty bags a trip, and, with the bearded Hendrik Swanepoel ostensibly in control of the partnership, they succeeded in winning most of Koos Meyer's business with a tender to deliver sorghum from Pienaarsfontein to the traders at Schweizer-Reneke for ninepence a bag. A coup on this scale was potentially threatening to the rural social order, and resentful landlords put pressure on Swanepoel to abandon his partner and enter into a contract that would ensure that Meyer's business flowed through the conduits of colour rather than class.[59]

But Hendrik Swanepoel was no more willing to betray a family friend than was Koos Meyer willing to see ethnic preference triumph over sound business practice. Although it was technically illegal, Meyer expanded his inter-racial sharecropping practices to embrace nearly half a dozen farms between 1924 and 1929, and given the state's reluctance to prosecute a white landlord in an area renowned for its Afrikaner populism, he went on to establish a reputation as the local Kaffir Corn King. While other Triangle farmers contemplated such conspicuous success with alternating awe and aversion, Hendrik Swanepoel and his black partner won the lion's share of the new transport contracts spawned by the expansion of his business. There was much to support Kas's contention that 'Koos Meyer was the man who made me rich.'[60]

Maine and Swanepoel valued their partnership well beyond the extent to which it enabled the latter to pay off his farm or the former to dispense with Willem Reyneke's hired wagon and buy a much larger vehicle of his own, a new plough and yet more livestock. The long hours spent on the road between farms like Diamantfontein, Hartsfontein, London and Pienaarsfontein and the small railheads or towns like Bloemhof, Kingswood, Kuruman, Makwassie, Schweizer-Reneke, Taung or Wolmaransstad drew them close together. For Kas, the memory of time spent sharing a meal cooked over an open fire on a cold clear winter's night before retiring to sleep under the protection of the wagon retained its special quality more than fifty years later.[61]

In part the contentment experienced on the road reflected the domestic tranquillity Kas was also enjoying. Only a contented man could afford the psychological luxury of being away from his family for days at a time. Kas, satisfied with the progress shown by their infant son, Mmusetsi, was eager to accommodate his wife who, in turn, continued to flourish under the tutelage of his mother and the A.M.E. Church. For example, when Leetwane once expressed a dislike of the smoke from his odd pipe-full of Buffalo tobacco, an occasional indulgence that he had started on the diamond diggings years earlier, he promptly gave up the practice and turned instead to what became a lifelong habit of taking snuff in prodigious quantities.[62]

Small items, such as his snuff, were usually procured while out on his ever more frequent visits to the monthly stock fairs at venues round and about the Triangle. On other occasions, especially once the harvest was in and a settlement had been reached with the grain merchants, the sharecropping homesteads would be visited by Indian hawkers such as the brothers Abraham and Ishmael Asvat or Essop Suliman. In addition to selling clothing, materials and inexpensive trinkets to women trapped on the farms, the hawkers also stocked a range of items that appealed to the menfolk. Abraham Asvat in particular became a regular caller after Kas had given him a herbal remedy for an uncharacteristic bout of sterility. This cure ensured Kas a warm and generous reception whenever he visited the Asvats at the Asian bazaar in Bloemhof.[63]

Because he had personal experience of the vagaries of the road, Kas was also ready to extend the hospitality of his home to Hersch Gabbe, the Jewish hawker, whenever the *smous* found himself in the vicinity of Vlakfontein around nightfall. Seated on a horse-drawn cart dripping with hides, skins and wool, 'Ou Vel' was a familiar figure around the sheep farms, for the market for these commodities had regained some of its wartime vigour. Gabbe, who readily accepted invitations to spend an evening with black peasants, was clearly more at home with his BaSotho hosts than across the way at Hartsfontein, where Willem Nieman's thinly veiled anti-Semitism relegated him to the barn for the night. He showed his appreciation for these kindnesses by never failing to present Leetwane with tea, sugar, eggs or a chicken before leaving her home.[64]

Leetwane, too, found the years after the locust visitations easier. With more money in the kitty, the Maines could help her parents and unmarried sisters in small but welcome ways, and on little outings to Kommissierust she and Mmusetsi were often accompanied by Kas. With the passage of time his interest in these excursions became less and less altruistic. He developed an interest in Lebitsa, the oldest of the Lepholletse daughters, and as a man of thirty making a success of his career, gave serious thought to taking her as his second wife. Marriage to a cousin would enjoy the sanction of tradition, assist his hard-pressed parents-in-law by taking another adult off their hands, and be unlikely to run into any serious opposition from Leetwane, who seemed to get on well with her sister.[65]

As a young man who was economically independent and freed from the need to arrange a second or subsequent marriage through the agency of his parents, Kas put his proposal of marriage directly to Lebitsa. When she agreed, he took their plans directly to her parents. Tshounyane and Tshaletseng believed that an exchange of sixteen cattle by way of *bohadi* would consolidate the already close ties that existed between the Maine and Lepholletse families. The wedding took place shortly thereafter, and Lebitsa moved across to Koppie Alleen, where, within weeks of taking occupation of her new accommodation, she became pregnant. Leetwane, not to be outdone by an older sister, also became pregnant and, by winter of 1924, Kas was contemplating the pleasing prospect of having to feed three children.[66]

But fortune winds its way down many a twisted lane, and before he could detect the silhouette of the coming spring, Kas was dealt a blow from behind. Just as winter dipped into its coldest months, the lease on Koppie Alleen expired and the easygoing Piet Reyneke suddenly found himself unable to match the rental offered by the land-hungry, still expanding business of Koos Meyer. Caught in the middle of the off-season with a considerable investment in a new house and a need to have shelter for his pregnant wives as well as a small child, Kas was transformed from a labour tenant with shepherding duties who enjoyed independent access to the land into a sharecropper-cum-labour tenant under the Kaffir Corn King. It was not a fate he would have chosen.[67]

*

At the age of twenty-seven, J. J. 'Koos' Meyer was, by 1924, already something of a phenomenon in the southwestern Transvaal. One of ten sons of R. S. C. L. Meyer, a poverty-stricken bywoner who had moved from the Cape Colony town of Oudtshoorn into the Bloemhof district during the late nineteenth century, the boy manifested extraordinary ability, energy and drive from an early age. At sixteen, while his father was based at Hartsfontein, Koos Meyer had set out to conquer the local countryside by the simple expedient of inviting black sharecroppers to plant maize and sorghum on every square inch of neglected or underutilised Triangle land which he, as a white man, could arrange to hire.

Written contracts that portrayed his black partners as labour tenants rather than sharecroppers offered the only legal record of productive arrangements which yielded excellent returns during the wartime commodity boom. Freed from the mortgage repayments that crippled Triangle farmers trying to buy property of their own, Meyer managed to protect his modest financial resources and scale down his operations during the recessionary interlude of 1921–22 and, when conditions improved thereafter, manoeuvred himself into a position from which to benefit from the upturn in the grain market. In 1924, he bought himself an Indian Scout motor cycle and this, together with seemingly limitless energy, enabled him to visit more of the bought and hired properties that he had manned by scores of black sharecroppers. Indeed, it was claimed by the end of the 1920s that he enjoyed indirect control of sixty spans of oxen ploughing on seven different farms spread through the district.[68]

But single-minded ambition is seldom achieved without hidden cost, and even in the Triangle, where men prided themselves on their masculine subculture almost as much as on their radical populist politics, Koos Meyer was known and singled out as an awkward, difficult and fiercely independent man. He was frequently in court, with adversaries ranging from neighbouring farmers who had failed to respect rights of access through to the state, which tried to place telegraph poles on his property without paying for the privilege. Blessed with the ability to get through life on very little sleep, Meyer performed most of his own chores on the trot and had nothing but contempt for those who did not share his passion for hard work. Features that singled him out amongst Afrikaner landlords lost nothing in translation when they were taken out into the open veld, where he spent most of his time conversing and interacting with his sharecroppers. In those circles, too, Koos Meyer had a reputation for being an uncompromisingly hard man who, although never guilty of assaulting a tenant, had a way of making a black man discover new dimensions to the dour world of manual labour.[69]

So demanding a landlord at Koppie Alleen would have required a considerable adjustment at the best of times. As it was, after a period when the Maines had devoted a great deal of time to the pursuit of their own interests under a sheep farmer, the transition was even trickier. Kas Maine and Koos Meyer were both highly motivated grain farmers eager to extract maximum economic advantage from the meagre natural resources of the western Transvaal, and thus it was preordained that they would clash over the allocation of time and labour.[70]

Soon after getting access to his new property Meyer announced that some of the land previously devoted to grazing would be put under the plough, in order to ease the burden of work that would come once the spring rains set in. Breaking virgin land in midwinter necessitated the use of a full span of oxen and a heavy plough, which he and Kas took turns at handling. He appeared on the farm day after day, and worked for so long without pausing to rest that Kas too could believe the story that, on more than one occasion, Koos Meyer had worked his oxen to the point where they had literally collapsed.[71]

Koos Meyer and his sharp-edged demands fell upon Koppie Alleen's black residents with all the focused energy of a shrike descending upon a cricket. Yet, the cricket, even as it is being aligned for swallowing, continues to struggle and resist. Stride for stride, hour after hour, Kas reflected on how it was that fate had suddenly inflicted a landlord on him, when normally the system left space for the sharecropper to find a partner of his own choosing. Acute personal discomfort unleashed some uncharacteristically radical thinking about social justice and the rural order, and at last, when his arms felt as though they would be wrenched from their sockets by the very effort of guiding the plough, he turned on the intruder, saying:

> You know, one day God will allow us to purchase property—just like you—and I will hire you, and overwork you just as you are doing to me.

There was no immediate response. But a dart thrown with such accuracy at the son of a bywoner who was still trying to turn his back on rural poverty could not fail to find its mark. It lodged in a psychic nerve, which became inflamed.[72]

One bitterly cold morning a few weeks later, Meyer arrived at the sharecropper's shack to tell Kas that he wanted him to drive some oxen to the auction at Bloemhof, where he hoped to sell the animals at a certain price. True to character, he also left his tenant with instructions for the following day's work, in case he himself should be detained at one of his more distant properties and inadvertently miss out on a day's labour.

Sensing the approach of a winter storm, Kas wrapped himself in a coat, got out his horse and, with a shrill wind from the Malutis steadily slicing away at his cheeks, slowly nudged the animals to town. But the twenty-mile trek into Bloemhof was pointless: by late afternoon it was clear that the oxen were incapable of attracting buyers at the price Meyer wanted, and Kas regrouped the tired animals for the journey home.

He was hardly clear of Bloemhof when the clinical cut of the Maluti wind gave way to the cloying dampness of sleet and then, most unusually, to snow. With cunning perversity, snowflakes and drops penetrated each layer of his clothing, finding their way into every crease and fold, until even his shirt surrendered its last remaining patch of warmth. By the time Kas reached home, close on midnight, he was wet, cold and hungry. The uncomfortable long night was spent trying to stoke up bodily heat in a frame that seemed unwilling to relinquish the posture of a man frozen on horseback.

For once, the weak winter sun succeeded in slipping through the doorway before Kas could ease his way out of the shack to survey the frost-encrusted veld. He was still rubbing his hands when Meyer was upon him, berating him for the late start to the day. The labour in the fields, the futility of the trek into Bloemhof with the oxen, and the pains of the night made for an explosive combination, and Kas rounded on Meyer, telling him he would leave the property before the day was through. Then, knowing exactly where the white man's most exposed nerve lay, he twisted the dart by repeating his threat to buy a farm of his own and make Meyer work for him.

The landlord turned on his heel and left. But the threat of an inversion in the rural social order must have conjured up a forewarning at the very heart of many a highveld landlord's recurring nightmare. Coming as it did from a black sharecropper and known 'witchdoctor,' and directed against someone in a family rumoured to have psychological instability, it was doubly potent. For years thereafter, Koos Meyer was heard to enquire from other blacks in half-jesting tones as to whether or not Kas Maine had somehow managed to buy a property.[73]

For a man with no access to land of his own, the immediate problems that arose from this blunt exchange on a cold morning at Koppie Alleen lay in the practical rather than psychological realm. There were mouths to feed, and spring was coming. By late that afternoon Kas had persuaded Piet Reyneke to take him and his two older brothers in as sharecroppers at Vlakfontein. It was a decision he never regretted.

CHAPTER FIVE

# Consolidation

## 1924–29

By the mid-1920s Piet Reyneke, then in his early forties and an increasingly successful wool farmer, presided over a sizeable family dominated by his sons. He and his wife Anna's six surviving children were all boys, their only daughter having died at a very early age. The oldest, Petrus, in a practice reminiscent more of African than of Afrikaner families, had been 'given' to his maternal grandfather when, shortly after the South African War, Oupa Griesel had decided that his seventeen-year-old daughter and her young husband were not coping with the demands of parenthood. Fortunately the couple found themselves with enough other children to care for not to resent this unsolicited help. Young Petrus grew up under the guidance of his grandparents on the banks of the Vaal River and eventually launched out on a career of his own.

Two younger boys, seventeen-year-old Willem and fifteen-year-old Koos, although still linked to the main family via a *rondavel* located not far from the farmhouse, were by the mid-1920s also virtually independent. They were responsible for farming a portion of a hired property at Zorgvliet which their father hoped they would eventually acquire in their own right. In addition to this loosely linked trio of older children, three younger boys also clustered together: seven-year-old Frikkie, two-year-old Fransie and the baby of the family, Benjamin, who, having been born only in 1926, was separated from the oldest brother, Petrus, by more than twenty years.[1]

The patriarch, a generous, soft-spoken, and deeply religious man who for many years served as an elder in the Nederduitse Gereformeerde Kerk at Bloemhof, devoted most of his efforts to providing his wife and sons with some of the material comfort and support to which a middle-class Afrikaner family in the countryside could then aspire. The task of dealing with the three smallest children, and with the two older rondavel inhabitants, who sometimes lapsed into the minor indiscretions of late adolescence, he left largely to his wife. Although capable of great kindness, she was, like most members of the Griesel family, always ready to dish up a helping of cold tongue or hot temper when she thought the occasion demanded it.[2]

The contrasting personalities of the two older Reynekes made for a surprisingly strong marriage, and between them they managed to keep the family ship on an even keel. The older boys were afforded a basic education at nearby schools, supplied with motor cycles when they moved out of the house, and handed one of the keys to life's door when they were given access to the hired property at Zorgvliet. None of this drew too heavily on their father's resources. In addition to Vlakfontein, which was by then virtually paid for, Piet Reyneke had inherited property at Klipdrift, on the Orange Free State side of the Vaal, and bought himself a black Chevrolet sedan. By local standards, the Reynekes were a comfortably placed family with little to complain about.

But, as was the case with most successful agriculturalists in the southwestern Transvaal, the productive base to this success did not rest squarely or exclusively on farming. In addition to the return from his large flock of Merino sheep, Piet Reyneke benefited from the occasional windfall that came his way via diamond diggings that he had developed on the lower section of the property. Drawing on the labour of some of the poorer whites from neighbouring Koppie Alleen—such as the three Swanepoel brothers and members of the Dauth family—he made certain that winter and the Triangle's alluvial gravel yielded what summer and poor soils tended to deny him: namely, a cash income. Indeed, it was partly because his bright, bespectacled eyes focused more readily on the price of wool and diamonds than that of grain that he was so willing to make provision for a few black tenants on his farm. We can see from this perspective that Piet Reyneke's interest in sharecropping was simply one more way of spreading risk in commodity markets that had, at times, fluctuated wildly.[3]

When Kas Maine and his brothers went to work under the affable Piet Reyneke at Vlakfontein in 1924, they were settling on the property of a relatively successful capitalising farmer whose diversified interests made it unnecessary for him to push his black tenants to the limits of their ability to produce grain. Moreover, Reyneke's willingness to hire additional grazing when circumstances demanded it reduced the pressure on the Maines' cattle; and this at a time when many tenants elsewhere were being told to reduce their livestock holdings.[4]

Nor did their good fortune end there. While commercial agriculture continued to experience short-term vicissitudes, the prices of grain, diamonds and wool all remained buoyant during the first three seasons the Maines spent on Reyneke's property. These propitious circumstances made for cohesion in the normally fissile world of sharecropping, and helped to make the Maines' stay at Vlakfontein long and relatively happy.

Paradoxically, this stability occurred when many blacks in the South African countryside were enduring considerable social, political and economic turbulence.[5]

The turbulence that most interested Kas in 1924 came from a plough drawn by eight hard-working donkeys. Given the late start to the season after the precipitate move from Koppie Alleen, it was urgent to plough quickly; using the lighter Canadian Chief, Kas succeeded in turning over the largest patch of land he had yet tried to cultivate. With Leetwane and Lebitsa both pregnant, there would soon be more mouths to feed and, since Reyneke had imposed no limit on the size of the plots they could cultivate, it made sense to make the most of the opportunity. Kas was not immune to the competitive ethos that many landlords in the district consciously instilled amongst their more ambitious sharecroppers, and after his unfortunate encounter with the Kaffir Corn King, he was keen to prove to black and white alike that, in the right circumstances, he could produce along with the best of them.[6]

The good rains that fell over much of the southwestern Transvaal that spring soon had Kas and his brothers scurrying for their stock of seed and pondering anew the virtues of the most popular varieties of sorghum. Given the early rains some opted for *Kobo Kgolo*, the 'Big Blanket,' a pale-coloured slow-ripening sorghum that got its name from the way in which the grain was tucked beneath a protective sheath. Other adventurous souls, eager to capitalise on the early start and longer growing period which the higher-yielding varieties needed, chose either the white *Tshwidi* or grey *Lesehla*. But Kas (and those who like him had mistrusted Old Man Climate's willingness to be generous for long), stayed with the small drought-resistant red sorghum which, because of its very short ripening period, had earned itself the accolade of *Tshabatsie*, 'Elude the Locust.' In the southwestern Transvaal, which never knew whether it owed its allegiance to the dry Kalahari in the west or to the marginally better-watered highveld in the east, the codes of survival were most explicitly stated in the languages of conservatism and pragmatism.

Likewise, it was Kas's scepticism about the chances of adequate mid-summer rain that did most to influence his choice of maize seed. Whereas a few gambled on the slower growing *Soetman* or *Antvel* varieties, believing that they were in for a long rainy season, he opted for smaller but safer returns to be reaped from the more familiar *Boesman* ('Bushman') or *Botman* mealie, a quick-growing yellow maize that dominated local production.[7]

Speculation about the weather aside, it was already clear that the wet start to the season increased the risk of worms getting into the seedlings, which necessitated special precautions. Kas combed the surrounding veld for the characteristically grey *Vaalbos* plants, added to them a few stubby leaves from the aloes strewn round the yard and then left them to the mercy of the Triangle sun. This dried vegetation was scraped together, ignited and reduced to a pile of acrid-smelling ash which he mixed with the maize until each individual seed was coated and almost looked burnt. This laborious procedure yielded an effective organic pesticide quite unlike the modern chemicals that some white farmers had used during the locust invasions and of which Kas strongly disapproved.[8]

But his caution was misplaced. It was not that the worms did not arrive, but that the rains did. In nine out of ten seasons the smile of spring would indeed give way to the grimace of summer, but 1924–25 was destined to be a tenth season. A generous start to the season was followed by such copious rains that the sorghum soon buckled beneath its unaccustomedly heavy burden. Those slow to sharpen their sickles found that overloaded plants toppled over long before the crop was ready to be brought in. When the drooping sorghum plants came into contact with damp soil, the grain turned black and started sprouting before it could be reaped.[9]

*'Lemo sa Mabele a Matsho,'* 'The Year of the Black Sorghum,' was memorable. Lebitsa, with some help from her delighted spouse, brought in well over a hundred bags of Tshabatsie, which meant that even after the landlord had been given his share, they were left with more than fifty bags. A harvest on this scale, supplemented by dozens of bags of maize, could feed a family well beyond the winter: Kas mixed the grain with wood ash and packed it into a hastily constructed store next to his hut.[10]

Across the way, at Hartsfontein, his younger brother Sebubudi had an even more pressing problem with his four hundred bags of sorghum. According to his agreement with Nieman, two-thirds of this went into the landlord's open-mouthed barn, while several other bags were swallowed as fines. But in 1925 not even Willem Nieman could do much to deny a black man a bumper harvest. The problem was that a harvest on this scale delivered all farmers, black and white alike, into the hands of bloated grain merchants, and the price of sorghum dropped dramatically. When the market plummeted, large producers like Koos Meyer responded by releasing their cattle into their unharvested fields.[11]

Not everybody could afford to shun the market so dramatically. Mphaka, in debt to Jaap van Deventer for a wagon that he had bought earlier in the year, and already giving serious thought to taking a second wife from a family on a neighbouring property, was hardly in a position to turn his back on the grain merchants. Forced to market more of his harvest than he would have liked, he asked Kas to help him transport thirty bags of sorghum to Schweizer-Reneke. But by the time they reached Stirling's Store the price had fallen to five shillings and sixpence a bag, a price that not even Mphaka was willing to entertain. Hearing that there were better prices to be obtained farther down the railway line that linked the Triangle to the Witwatersrand goldfields, they wheeled round and set off to explore the possibilities at the agricultural co-operative in Makwassie.

The branch of the South-Western Transvaal Agricultural Co-operative at Makwassie was an offshoot of the original Wolmaransstad Koöperatiewe Landbouw Vereeniging, established in 1909. During the postwar agricultural boom the village had made solid but not spectacular progress, as cash-starved farmers in the district continued to market most of their grain through traders rather than through the co-operative, a practice that was estimated by the mid-1920s to leave hated 'foreign' middlemen with almost a third of the profit.[12]

The solution to Mphaka's problem did not lie in Makwassie. Indeed, officials at the S.W.T.L. offered him sixpence a bag less than he had been offered in Schweizer-Reneke. With the retreat to Stirling Brothers cut off by time as well as pride, Kas advised him to sell his grain at the offered price anyway and then, as somebody who had himself experienced the weight of a debt of honour, offered to sell an additional wagon-load of his own grain at the same price to help raise the cash needed to pay van Deventer. Mphaka accepted his offer, and a few days later they went back to Makwassie with Kas's surplus grain. The two brothers, already close, had been brought still closer by Lemo sa Mabele a Matsho.[13]

Sharecropping families who only a few years earlier had been eating locusts had good reason to celebrate a season when affluent white farmers could afford to let their cattle feast on sorghum and grain prices plunged to five shillings a bag. But Kas had at least two other reasons for remembering the 'Year of the Black Sorghum.' Toward the end of that year Lebitsa, as her pregnancy drew to a close, left Vlakfontein and went to her parents' home at Kommissierust. Unable or unwilling to rely on the support that Leetwane got from Motheba and her circle of committed A.M.E.C. supporters, Lebitsa felt more comfortable with her immediate family, and there, on 5 February 1925, she gave birth to a daughter who was named Thakane. Three months later, while the new harvest was still being brought in, Leetwane, who had stayed with Kas at Vlakfontein, presented him with a second daughter, to whom the couple gave an old family name, Morwesi, 'The Burden.'[14]

Despite unspoken differences that informed the Lepholletse sisters' choice of venue for the birth of their children, the two babies were soon drawn into a ritual that helped reinforce their identity as Maines and as BaSotho, and integrate them into the family. Unlike Willem Nieman, Piet Reyneke readily agreed to give his tenants a sheep for a feast and, as soon as the weather obliged, Motheba stripped Thakane and Morwesi of their clothing and placed the infants out in the open to be washed by the falling rain. Symbolically 'baptised,' the protesting babies were then snatched back from the elements, taken into the protection of the huts and formally accorded their names.[15]

With a veritable flood of sorghum beer available to lubricate social occasions, this and numerous other rural gatherings drew the Maines into especially close contact with their black neighbours. Predictably, both the Tjalempe and Tabu families had benefited greatly from the truly remarkable harvests. Indeed, Mdeboniso Tabu—a Xhosa-speaker, who had moved into the district from the Orange Free State after the passage of the Natives Land Act, and an extremely talented farmer—had used his oxen to produce 1,600 bags of sorghum on Pienaarsfontein which he and Nini Tjalempe had covertly hired through Koos Meyer for twenty-five pounds a year.[16]

The benefits of the Year of the Black Sorghum were not confined to the ranks of exceptional tenants; large harvests were enjoyed by other black farmers too. Throughout the district, BaSotho and non-BaSotho sharecroppers—including the Moshodi brothers of Hartsfontein, the Molohlanyis of Katbosfontein, the Masihus of Leeubos and the Kadis of Prairieflower—all did well even though they had far fewer resources. (Mmereki Molohlanyi was nominally a 'coloured,' being the son of a white farmer at Rietfontein named King and a BaTswana woman named Motlalepule.) Not surprisingly, it was the core of the middle-ranking Sotho-speaking tenantry that formed Kas's most immediate point of reference for his farming operations and whose company he enjoyed most.[17]

Most members of this group were drawn together not only by their interest in sharecropping and differing degrees of BaSotho ancestry, but by their exposure to, and interest in, the more formal elements of education. For example, as a young man in his twenties, Mmereki Molohlanyi had often cycled into town from the farm in order to become literate at the Reverend Thompson's Lutheran night school in Wolmaransstad. Mosothonyanye Masihu, following the lead of Nini Tjalempe, who had arranged for a daughter to be sent home to the Transkei to be educated, sent his own son, Lerata, back to Morija in Basutoland for three years of basic education before recalling the lad to the more modest farm school run at the salt works at Sewefontein. In 1925, Lerata, by then twenty-one, badgered his father to support his application to the A.M.E.C.'s Wilberforce Institute at Evaton, where he hoped to qualify either as a carpenter or a motor mechanic. It was not only the flashily dressed 'Mr. Kwape' Moshodi of Hartsfontein who looked forward to the day when he could shake off the dust of farming. Others, too, had their dreams.[18]

But Kas, who had received his own rudimentary education by the light of a paraffin lamp at Vaalboschfontein, was less persuaded about the supposed virtues of book learning. He was willing to concede that a man needed to be numerate in order to account for his family's cattle or to determine his share of a crop. Likewise, a little literacy was undoubtedly a good thing because it enabled one to record transactions with traders. But beyond that it really suited only priests and teachers, men who believed that the pen held some sort of elevated status over the plough. At Vlakfontein he had no call for a pen beyond forging an occasional pass when he wanted to move about the district without waiting for his landlord's written permission. Longer letters could be written by Pata, a literate Koranna who resided on the property, while incoming correspondence in SeTswana was read to the family by Reyneke, who was a fluent SeTswana-speaker. With each season bringing new chores, there was seldom time for the downright irksome business of reading or writing. Even when the harvest was good, as was the case in the winter of 1925, there were always off-season tasks, such as making yokes or repairing ploughs, that took precedence over written words. All the books in the world could not tell a man what rain the new season would bring. A sharecropper must prepare himself as best he could and wait. And wait Kas did.[19]

September was unusually dry. Occasionally an energetic gust of wind managed to sneak in from across the plain and for a few wonderful minutes would lift the blanket of thick hot air hovering over the Triangle, only to retreat as a whisper to a distant place. By mid-October, when herds of fat white cumulus clouds usually moved in to reclaim the blue highveld sky from the last of winter's thin cirrus stragglers, there was still no sign of rain. Then the west wind, revitalised, brought with it the first traces of the red desert dust. With only a few remaining tufts of scraggly natural vegetation to impede its progress, it sped across the bald pate of the Vaal and Harts plain, picking up dry earth and Kalahari sand, which it directed at passers-by in short blasts or else tossed into the air in great orange-tinted plumes.

The Maines and other sharecropping families watched the dance of these dust devils with strange fascination. Early-season turbulence had become common in recent years, as farmers placed ever-greater stretches of marginal agricultural land under the plough. The wind, freed of resistance from the stocky indigenous bush, rushed in, collected the sandy topsoil, and hurled it at the nearby diamond diggings with such ferocity that work was often suspended for days on end. But that year the 'Red Dust,' *Lerole le Lefubedu*, took on such proportions that both man and beast became fearful.[20]

Out on Jaap van Deventer's portion of Kommissierust, Motheba kept a wary eye on the storms. Her child-minding duties had expanded well beyond the normal grandmotherly bounds, for in addition to taking care of Mphaka and Phitise's two boys, she had the responsibility of looking after three of the children's cousins. She was fortunate to have her daughter, Sellwane, to help look after the five youngsters. Together, they listened to the wind.

But older eyes no longer saw everything. One afternoon, long before sunset, Phitise and Sebubudi suddenly appeared at the house without the usual accompanying sounds of the cattle and pointed to a gigantic red fire-ball hurtling toward them from the western horizon. Sensing the scale of the impending onslaught, Motheba told Sellwane to find the children and to bring inside the utensils that lay scattered about the yard after a late lunch. By the time that Sellwane had rounded up the children, the sky had changed from red to black and the air was thick with dust. All thought of pots and pans was abandoned; people buried their noses in bits of cloth and darted for cover. With visibility down to a few feet, Motheba and Sellwane managed to herd the frightened children into the refuge of her small mud-walled hut while her sons bolted for the protection of their own homes. Inside, she remembered being warned of the danger of suffocation, so she opened the windows to allow a little air to circulate. But opening these small glass eyes to the full horror of the storm terrified the children. When a truly monstrous wind began to shake the house, she too became deeply apprehensive and told the children to close their eyes as she began to pray. Before she had uttered three words, the tornado gave a mighty roar and ripped off the thatched roof. At the same instant the back wall of the hut gave way and fell in on the screaming children. A moment's silence followed.

When Sellwane opened her eyes, most of the house was gone. She heard Motheba asking her to run to Phitise's home and tell him that the children were dying. Choking, covered in bits of thatch, and shaking with panic, she called out that it was too dark to find her way. One of the older boys ventured out into the teeth of the gale. A few yards down the road he staggered into Phitise and Sebubudi and told them what had happened. Phitise himself had narrowly escaped injury when his roof had blown off and the heavy boulders used to weigh down the corrugated-iron sheets had crashed into the living room. He and Sebubudi rushed to what was left of their mother's home.

The first sound Phitise heard was the cry of his ten-year-old son, half-buried beneath the rubble of the collapsed wall. With Sebubudi's help he tore at the chunks of hardened mud and, within minutes, managed to free Malefane, who was in great distress. But before he could attend to the boy he had to pull out Mokowane, one of the cousins, who was trapped beneath a fallen rafter.

The men carried Malefane and Mokowane to a small corrugated-iron shanty which had miraculously managed to withstand the blast. There they were joined by the other badly shaken occupants of what had been Motheba's home. As the storm raged on in a lower key, it was Malefane who gave them most cause for concern: the area around one of his hips and kidneys had been badly bruised and he seemed to be suffering from an internal injury that made it painful and difficult to urinate.

An hour later, when the wind had somewhat abated, the heavens assumed a more familiar steely-grey hue and then, with loud thunder and streaks of forked lightning, proceeded to pelt hail at the countryside. The hail beat out its distinctive tattoo on the tin roof until it too moved off in the direction of Bloemhof. The sky eventually cleared by late afternoon and the battered veld was bathed in apologetic sunshine. Phitise and Sebubudi's thoughts turned for the first time to the livestock they had abandoned several hours earlier. Leaving Malefane in care of Motheba and their sister, they hurried out to recover the animals and return the herd to the safety of the kraal.

Within minutes they stumbled on yet more evidence of the Red Dust's appetite for destruction. The carcasses of two cows marked the spot where the storm had revelled in the absence of the herdboys. The magnetic attraction that wet nostrils have for dry sand had ensured that the Lerole le Lefubedu had choked the animals on a mixture of mud and mucus. They found the remaining cattle, cleaned them up as best they could, and escorted the still bewildered beasts back to the homestead and the shelter of the stockade.[21]

The next morning Phitise and Willem Coetzee, the son of the bywoner at Hartsfontein, lifted Malefane into a cart and took him to Bloemhof, where they consulted Henry Hutchinson. The visit to Kas's old friend the district surgeon took up most of the day and it was dark by the time they returned. In their absence Sebubudi had carved up the carcasses of the two animals, and that night, over a meal, the family was told that Malefane was already well on the road to recovery. This and their own miraculous escape from the tornado convinced Motheba and Sellwane of God's mercy, and together they drew renewed strength from their faith.

Across the way, at Vlakfontein, Kas and Mphaka were equally relieved to hear of the outcome of the family's encounter with the storm. True to character, Kas chose to ignore its effects and focused instead on the little moisture that had followed. Within hours he was out of the shack and ploughing with all his customary energy. But scant rainfall and exceptional heat made the arduous task even more demanding than usual. The day's work commenced before dawn with the inspanning of the donkeys and, after only an hour or two behind the plough, Kas found himself covered in sweat and with a throat as parched as a Schweizer-Reneke *sloot*.[22] He could think only of the moment in midmorning when Leetwane would arrive, baby on back and Mmusetsi toddling behind her, with thirst-quenching beer and a generous helping of sorghum porridge.

Despite his marriage to her sister during the previous season, Kas remained especially close to Leetwane. The death of Mosebi, the appearance of 'the Replacement,' Mmusetsi, and the arrival of the baby Morwesi had created powerful emotional bonds that tied him to her in a way that was not possible with Lebitsa. Besides according Leetwane the proprieties that tradition demanded for a senior spouse, he devised other ways of letting her know that she commanded a unique position in his extended household. She alone knew where he hid his money, and he made time and a little cash available to support her interest in the church. She, in turn, applied herself with great diligence to the work expected of her.[23]

But perhaps Leetwane had been more concerned about the arrival of the second wife than she cared to admit, for her health now became cause for growing concern amongst those who surrounded her. Shortly after the birth of Morwesi she had started experiencing problems with her legs, and by the time the ploughing was in full swing she was walking with great difficulty. Motheba, despite her endless fund of sage advice and Christian counsel when it came to matters of childbirth and its after-effects, could not account for her condition; Kas, reluctant to offer herbal treatment when he was uncertain about the exact cause of the affliction, cast around for a practical, professional solution.

He found a famous MoPedi herbalist who hailed from the northern Transvaal and brought him in to see Leetwane, but despite an outlay of three pounds in cash, the treatment did nothing to halt the slide in his wife's condition. The man from BoPedi was followed by two other 'outside,' or non-Sotho, practitioners: a Zulu herbalist or *isangoma* from Natal who, although unsuccessful, was given a donkey for his trouble; and a well-meaning MoThlaping *ngaka*, who was equally unsuccessful. With Leetwane finding it harder and harder to cope with her household chores as well as the two small children, it was decided to bring in one of her young unmarried sisters from Kommissierust to help manage the domestic routine at Vlakfontein.

Tseleng's arrival eased the burden of housework but did nothing to improve her sister's condition. By mid-December Leetwane had lost nearly all use of her legs and, to everyone's distress, was reduced to crawling about the house on all fours. Alarmed, Kas for the first time gave serious thought to an alternative strategy. If the solution to Leetwane's problem did not lie in traditional herbal medicine, then perhaps it was indeed to be found in the belief systems of his wife and mother. This sort of pragmatic thinking had yielded dividends in the past, and he resolved to explore seriously the possibilities held out by faith healing and the A.M.E. Church. He found Tjalempe at work at Pienaarsfontein, and was only mildly surprised when the tall sharecropper told him that he already knew about Leetwane's illness. The church, like the ngakas, had a network of its own. Kas explained the seriousness of the position and their need for outside help; Tjalempe, sensing his desperation, agreed to accompany him home.

Although unable to come up with a definitive diagnosis, Tjalempe thought that there were grounds for optimism. A long time before, back in the Cape Colony, he had encountered a similar ailment among Xhosa-speakers. Known as *Umshosha Phantsi*, 'that which crawls underneath,' it had been treated successfully by binding strips of white cloth soaked in fresh milk around the affected limbs. If Leetwane was suffering from the same disease, and if Kas followed his instructions to the letter, he felt confident her condition would improve.[24]

Reyneke, told of Leetwane's illness, readily agreed to give the family all the milk they needed, and Kas visited Asvat's store in Bloemhof and brought back several yards of broadcloth. Just before Christmas the treatment commenced in earnest: each afternoon, just before dusk, one of the children would arrive at the house bearing a canister of fresh white foamy milk in which to soak the cloth, Kas would bind his wife's legs from ankle to thigh, while Nini Tjalempe led the gathering in prayer, an aspect of the proceedings that had a special appeal for Leetwane and Motheba. The faith healer then withdrew.

In two weeks there had been no discernible improvement in Leetwane's condition. Tseleng, who was as attractive as she was industrious, still had to do most of the work around the house. Then, just as suddenly as his wife's legs had been shackled by some malign influence, so some new unknown force liberated her limbs. By the time the new year came around Leetwane was walking without the aid of a stick, and a fortnight later she was fully cured.

Delighted that his hunch was correct that 'God worked through his own people,' Kas hurried across to Pienaarsfontein to share the news with Tjalempe and to ask what should be done to avoid any relapse. But this time the faith healer was less forthcoming and merely told him to go home, prepare two pots of sorghum beer and await his arrival on a day he nominated. The day came, and Kas and Leetwane, together with their parents, assembled at the house to await their guest of honour and two of his friends. (Significantly, this tightly defined gathering left no space for the junior wife, and Lebitsa took no part in the ritual that followed.) Tjalempe and his assistants arrived, examined the brew, and the faith healer instructed one of his friends, a man named Matjolo, to ease a little of the beer over Leetwane's head, which he rubbed vigorously into her hair. He then shaved Leetwane's head, and gave her husband the hair to bury in a place known only to him. Matjolo took a pot of beer and placed it before the BaSotho family, while Tjalempe handed the second to the Xhosa-speakers, and they settled down to celebrate in earnest.

Kas continued to fret that they might have incurred a greater debt, and that while it remained unpaid his wife would be vulnerable to another bout of illness. He again raised the topic with the faith healer but was no more successful than the first time. This time, Tjalempe, as if at some pains to make explicit the underlying dynamic at work, emphasised that the cure had been effected by Kas himself and that his own role had been marginal. Leetwane, publicly and privately reassured about the place that she occupied in Kas's affections at a crucial juncture in the development of his extended household, suffered no further bout of paralysis.[25]

These pressing domestic issues had diverted his attention from the fields for some weeks, but Kas was keenly aware that nature was busy turning her back on grain farmers. A merciless sun had climbed to the highest point in its arc and roasted the southwestern Transvaal plain until the normally clear line that demarcated the distant horizon lay blurred beneath the shimmering veils of a thousand dancing mirages beckoning the credulous toward water.[26]

By the time that Tseleng had left and Leetwane was once again coping with domestic chores on her own, it was clear that the drought was not restricted to the Triangle. A local minister, the Reverend Mkhwanazi, encouraged the farmworkers and sharecroppers to join him and thousands of others throughout South Africa in a 'day of prayer.' At Vlakfontein a modest outdoor service conducted in Afrikaans and SeSotho was joined by the landlord's family, and 'Ou Rosinah,' as Motheba was known among the whites, was called on to lead the congregation in prayer. God and the drought drew together those whom men and money contrived to keep apart.[27]

But not even a united appeal altered the course of a season designed in hell. In a district where even at the best of times it was hard to eke out a living from the soil, there was a frightening reduction in rainfall. By the time summer gave way to the short highveld autumn, it was clear the harvest would be very small. The Maine brothers had to find off-season economic opportunities.[28]

The winter of 1926 was unusual in that it was the first time in years that Kas and Hendrik Swanepoel failed to team up for transport-riding. Not only was the harvest smaller, but the move from Koppie Alleen to Vlakfontein had put distance between them. More importantly, Koos Meyer had invested in the first of several Chevrolet trucks and had less need of outside contractors to move his harvests to the grain merchants and rail sidings. This loss of the Kaffir Corn King's business was only partly compensated for by the transport-riding Kas managed to coax out of the Reynekes, for they were never really extensive grain farmers.[29]

Mphaka, who had in the interim married the woman he had selected as his second wife, was equally hard pressed for cash. So too was Phitise. It was therefore a relief when Piet Reyneke mentioned that his brother-in-law was looking around for some builders. The boorish Willem Griesel (whom Kas had last encountered while disposing of his short-tailed grey roan as part of the effort to raise Leetwane's bohadi), had moved from De Beersrust to the farm Leeubos, a few miles north of the rail siding at Kingswood. The house had been too small to accommodate his generously proportioned wife, Jacomina, let alone their five children, and, in February 1926, he had been forced to sell part of the property to his brother-in-law in order to raise the capital to build another, larger house at Leeubos. This was exactly the sort of off-season opportunity the three cash-starved Maine brothers had been looking for.[30]

While the building was going on for several months, the Griesels moved out of Leeubos and into the farmhouse at Vlakfontein. Although convenient enough for them, this arrangement was a mixed blessing for the Maine brothers, who had to contend with the cantankerous Willem not only while away at work, but at home in the evenings and over the weekends. Still, it all went well enough. Mphaka and Phitise, skilled bricklayers, erected the walls while Kas was responsible for the roofing. Inside plastering and finishing touches were left to du Plessis, one of Griesel's bywoners. The completed structure was said to have had a pleasing aspect to it and did much to enhance Mphaka's reputation as a builder. Ironically, this came at precisely the moment when the Afrikaner Nationalists and their Labour Party allies in the 'Pact government' were most anxious to restrict any such opportunities to white artisans.[31]

During the building operations, Kas found himself more exposed to the Griesel family than he had bargained for. Since he was a principal beneficiary of the construction work at Leeubos, both Reyneke and Griesel felt free to ask him to take on any number of small additional chores. Not all of these were unpleasant and some of them were decidedly amusing. Thus, as the driver who had once escorted no less a person than Mrs. Victor Lindbergh on her shopping rounds, it fell to him to escort Jacomina Griesel into town. But, whereas the former Gladys St. Leger of Cape Town had been a petite and sociable person, Ou Leeu van Zyl's daughter was a large, laconic woman cast in a formidable Boer mould. This took its toll on man and beast alike. On their weekly journeys into Bloemhof it was not long before the horse on Mrs. Griesel's side of the cart was so exhausted that Kas was forced to stop, outspan, and then swop the animals around. 'Jesus, that woman was fat!'[32]

Willem Griesel, on the other hand, gave less cause for mirth. A notoriously hard-drinking man on a very short fuse, he would often return home on Saturday evenings with his shirt torn or his nose bloodied, after a thirsty foray in town where the local diggers obviously gave as good as they got. Not content with weekend away-games, he also brought some of this love of conflict home and, on at least one occasion, traded punches with his mild-mannered brother-in-law.

Kas observed these incidents with disbelief and disapproval. He was determined to give Reyneke's coarse guest as wide a berth as possible. Despite this, it was not long before they clashed and exchanged hard words. The trivial cause of their argument was soon lost amidst other issues, but it deepened their developing dislike of one another. Unlike Griesel, however, Kas was unwilling to ascribe the friction solely to racial differences. For him an individual's behaviour was a more reliable guide to the success of a relationship than skin colour, and he was reluctant to draw any broad conclusion.[33] How else could he account for yet another Griesel's behaviour to him that same winter?

One evening he and the landlord's son had returned to the farmhouse having spent the better part of a bitterly cold, wet afternoon trying to persuade several hundred sheep to seek shelter in the kraal. Unlike Willem Reyneke, who wore protective clothing, Kas was well and truly drenched, and he stood huddled outside the back door, teeth chattering, waiting for the landlord to appear. When Anna Reyneke found him there, she took one look and said, 'God, man, you will die of cold.' She scrabbled around and found a tot of brandy which she pressed on him, and then, when he had recovered his composure, marched him right through the house and into her son's bedroom, where she offered him a change of clothes, presented him with a raincoat, and sent him home with the instruction that he should not report for work in the morning. 'That,' he was quick to remind sceptics in later years, 'was done by a white woman.'[34]

There were other instances, including several that revolved around nineteen-year-old Willem Reyneke, which showed how the human spirit could rise above the restrictions that the dominant order sought to impose on the countryside. As an old man, Kas recalled with a smile that it was Piet Reyneke who had asked him to accompany his son when the young man first set out to woo the young ladies of the district. In theory, all that was expected of him was to look after the horse and cart while Willem regaled the local beauties with the treacly tales that hopeful young men tell cautious women. In practice, the God-fearing landlord hoped that Kas would have a restraining influence on a son with a talent for mischief.

Willem, whose plausibility was exceeded only by his persuasiveness, put these journeys to and from the neighbouring farms to good effect and soon had not only the pride of Schweizer-Reneke under his spell, but his unofficial guardian as well. Confidences shared on the cart developed into a special friendship and, before long, reached the point where they were meeting at one of the more secluded spots on the farm to share the occasional tot of brandy stolen from Piet Reyneke's cupboard. Willem's thirst and love of adventure eventually outgrew both this modest supply and their irregular meetings, and he looked to Kas to help him lay in a more substantial quantity of brandy.

They borrowed the landlord's black Chev for the short ride into London. Willem went into the off-licence and emerged with several bottles wrapped in the tell-tale brown paper that was meant to keep the contents hidden from blacks, minors and the innocent folk who might be found sitting around a hotel frequented by diamond diggers. The journey home took longer. At several points along the way they stopped to test the quality of their purchase. By the time they reached the farm gate the Chev was moving slowly enough for the sound of the engine to attract the attention of its short-sighted owner, who followed its progress from the partial seclusion of the stoep.

The car halted opposite a clump of bluegums about half-way down the track to the farmhouse. Two figures clutching elongated parcels got out, disappeared into the trees and re-emerged empty-handed. Reyneke heard the car start up and continue down the road with exaggerated care. It drew up at the house and the two passengers climbed out, Willem effervescent even by his own standards and Kas, normally laconic, strangely talkative. It took only a moment for Ouderling Petrus Reyneke to assimilate the attraction of the bluegums, but being a judicious as well as a religious man, he said nothing. A few days later son and sharecropper alike were puzzled to find that the cache concealed in the copse had been discovered and removed. Only weeks later did the discreet landlord have a quiet word with the wolf in shepherd's garb and let him know that he sometimes saw more than he was given credit for, poor eyesight notwithstanding.[35]

If Kas had disappointed Reyneke by his performance as Willem's guardian, no one knew. His reputation as a hard-working tenant remained untarnished and, as the season changed, he eased the draught animals out into the fields to help rouse the soil from its winter slumber. The donkeys, gentle to a fault, moved slowly up and down the fields, drawing the plough at a pace that might have been set by nature itself—thus it had been since his childhood. There was something deeply reassuring about the swish of the sandy earth as it peeled open beneath the share and then fell quietly apart. But that year, if one stopped and listened carefully enough, a new sound could be detected in the spring air. An insistent drone, muffled only by distance, filtered in across the fields as a few of the more successful farmers conducted their first experiments with paraffin-driven monsters.[36]

Like most of the bywoners and sharecroppers he spoke to, Kas saw nothing ominous or menacing in the arrival of the tractor. 'Some laughed and dismissed it as a lot of shit, others said that it was a good thing.' While 90 percent of all arable farming in the western Transvaal rested on a crop-sharing basis, the tractor posed no immediate threat to black tenants. Or: 'Boers wanted anyone who had a span of donkeys or oxen to plough on the half.' So the arrival of the tractor passed largely unnoticed and Kas, ever ready to improvise, positively welcomed the planter that Reyneke introduced to Vlakfontein as part of his concession to the new drive toward mechanisation.[37]

Reyneke, who knew more about wool farming and diamond digging than he did about growing grain, saw the move away from sowing mealies broadcast to the use of the planter as a huge step forward. But he had no particular mechanical aptitude or patience, and it was soon clear that the assembly and workings of the machine were beyond him. In the end he was forced to call in Kas: between them, they established how the machine operated and used it to plant his fields the modern way. As soon as the seeds germinated, though, the planter's drawbacks became clear. Enormous gaps and unexpected clusters of seedlings left the field looking more like the stubble on a digger's chin than a maize field. Reyneke blamed Kas for not having put enough seed into the machine, but Kas was far from convinced by this explanation. He ferreted out one of the new breed of progressive young Boers who told him that, unless one first sifted the maize kernels down to a uniform size, the holes in the bucket would become blocked, causing the erratic pattern he had seen at Vlakfontein. Reyneke was sceptical, but Kas did some experimentation that proved the theory. He went out and replanted the field with gratifying results: 'Later he admitted that I was right. The seed needed to be sifted. From then on we sifted the grain before planting. After that, he left everything in my hands as he did not really know how to use the planter.' Still, Reyneke was never fully reconciled to the machine and a year later sold it to Kas for five pounds.[38]

Buoyed by his initial success, Kas abandoned the practice of sowing broadcast and soon became an ardent champion of mechanised planting. Yet he was firmly convinced that other problems of a different order were impervious to orthodox scientific logic. It was one thing to use a machine to plant seed, another to protect a promising crop from an invasion of birds wished upon one by a jealous neighbour.

Long before the grain was in the ground Kas had taken the usual precaution of preparing a traditional potion which, sprinkled round the perimeter of the field, helped to ward off the attention of long-tailed widows, red bishops and weaver birds. In the past this had been so successful that several Boers had asked him in private to perform the same ritual on their fields. For this he was usually paid five bags of maize or sorghum at the end of a successful season, a valuable supplement to his own efforts on the land. In a harsh environment where bywoner and sharecropper were equally involved in the struggle for survival, the membranes of class often permitted a cultural osmosis that was otherwise impossible in a colour-coded state.[39]

In the same pragmatic vein Kas asked Reyneke for permission to organise a *molutsoane* or communal hunt that was believed to bring on rain, when it became clear that they were in for an abnormally dry season. (Their last had taken place fifteen years earlier during the great drought of 1913, when Sekwala had presided over the family at Vaalboschfontein.) The landlord gave his approval for the gathering, and the unusual goings-on at Vlakfontein attracted a goodly number of black participants as well as a clutch of interested white onlookers—the usual mix of sceptics, the genuinely curious, and those who secretly called on ngakas to protect their fields from witchcraft. This time the osmotic process was reversed: at the end of a long day in the fields, the BaSotho hunters joined the white farmers in a prayer for rain. In the countryside at moments of crisis it was sometimes as difficult to separate 'pagan' from 'Christian' as to know where 'tradition' ended and 'modernity' began.[40]

When three inches of rain bucketed down on the northern part of the district in one day in early December, at least some of those who had been at the meeting must have believed that a potent mixture of BaSotho custom and Afrikaner Calvinism had done the trick. But in the Triangle, where the new year and high summer creep over the horizon at the same time, 1927 arrived looking more like an old man weighed down with familiar terrors than a youth dispensing hope. Between 1925 and 1928 less rain fell than in any previously recorded four-year period. By the end of the 1927 season, the Bloemhof, Schweizer-Reneke and Wolmaransstad magisterial districts had all been listed as 'proclaimed areas' in terms of the Drought Distress Relief Act.[41]

But state-sponsored relief was confined to whites. Even then, it was so modest that it could not rescue the Triangle's least capitalised farmers or their bywoners. As the drought cycle deepened, hundreds of whites and thousands of blacks forsook the land and flocked to the diamond diggings in a desperate attempt to make a living. This region-wide trend was intensified in the Triangle when large new public diggings were proclaimed at Lichtenburg, fifty miles northeast of Schweizer-Reneke, in 1927.[42]

The rise of Lichtenburg saw the decline of Bloemhof, where the Great Depression arrived almost two years earlier than elsewhere in South Africa. Whereas in 1925 and 1926 Bloemhof had been the centre for five thousand diggers producing diamonds worth more than one million pounds annually, by 1928 only five hundred diggers were left, hawking gems worth a paltry £5,000 a year. This, the drought and a steady decline in the price of most farm products exacerbated a population exodus. An official visiting the local branch of the Standard Bank in 1929 was moved to observe that 'agriculture [there] is not a quantity to be reckoned with,' and that 'Bloemhof today is a decaying town.'[43]

Piet Reyneke, protected by the relative diversity of his operations, was more secure than many of his neighbours. The price of wool increased steadily over the three years between 1926 and 1929. This and the return from his diamond diggings (although declining in value because of a rampant overproduction that threatened the overall stability of the market) left him reasonably safe. And when he asked Kas to transport some of his digging equipment to Lichtenburg in 1927, he, unlike many others, made the move from a position of strength rather than weakness.[44] But his tenants were exposed, not least of all because there was not enough grazing at Vlakfontein for them to run sheep of their own. Without income from wool or diamonds, with falling grain prices and a harvest reduced by drought, the sharecroppers were far more vulnerable than their landlord. The Maines experienced the first tremors of the gathering depression with far less equanimity than did Piet Reyneke.

As in 1926, the brothers looked to the winter's economic activities to make good the losses caused by shrivelled summer grain. Mphaka and his thirteen-year-old son Baefesi did not even have to venture off the premises to get work. Reyneke, pumping out ever-increasing quantities of underground water to survive the drought, had installed another windmill and wanted a large reservoir built to help water the sheep. His wife, prompted by the experience of having had the Griesels as guests for several months, wanted changes made to the house. These operations kept Mphaka and some of the older boys in the extended household fully occupied until well into the summer.[15]

But while Mphaka was busy doing construction work, Kas found that the type of tasks that he and old Hendrik Swanepoel had undertaken in the past were not in great demand. A smaller harvest and competition from more and more lorries meant that the chances of getting a contract to deliver grain to the railhead at Kingswood were not good. Instead, Kas spent hours conveying water to diggers who had been unable or unwilling to move to Lichtenburg with the rest of their fraternity. Tedious as this was, it gave him steady income and meant he could always avail himself of small cash loans at the farmhouse at short notice.[46]

Of course, other springs in the rural economy never quite dried up and, even the driest of times, such as 1927, could be relied upon to produce a trickle of cash. Leetwane and Lebitsa collected, dried and stacked cow dung, which found a ready market amongst the men on the diggings, who had discovered that wood for cooking purposes was hard to come by now the tractors had eaten the indigenous bush and spat out the remains as so many grain fields or pasturage. Also, Kas often earned extra shillings or half-crowns by dispensing *morokori* or *monnamotsho* to patients who consulted him about herbal remedies for headaches, stomach cramps or other minor ailments.[47]

Stranded without a winter building project of his own, Phitise dusted down an old family craft and turned his hand to shoe repairing. As times got harder, folks were making their shoes walk farther. Kas, who fancied himself as a better and more precise worker than his brother, found that this rekindled his own interest in cobbling, and he, too, was soon busy cutting soles from leather bought from 'Ou Vel' Gabbe. All this activity attracted Reyneke's attention and, with a more ambitious plan in mind, he seized on the moment to show an enthusiastic Kas how to cut, cure and stretch a set of yokes.[48]

Piet Reyneke's new-found enthusiasm for yokes, like the alterations that his wife Anna wanted made to the farmhouse, had its origin in the changing composition of his household. Willem, recently married, had moved across to the part of Zorgvliet that had at long last been acquired in the Reyneke family's name. His brother Koos had his heart set on a lass by the name of Anna Kleynhans, and it was only a matter of time before he, too, would abandon the parental home and try to find a patch of his own to work. Piet thought the ideal solution would be for Kas to work with Willem on the more extensive property at Zorgvliet where Koos and Anna could join them. This would help ease the pressure on the grazing at Vlakfontein, and Kas and Willem's sharecropping arrangement would benefit everybody: Willem would have access to his mentor's experience as a grain farmer and Kas could use Willem's oxen. Between them, they would all help to pay for Zorgvliet.[49]

Kas liked the idea. Although Zorgvliet lay a full twenty-five miles west of Vlakfontein, the property was well known to him. It was one of the first farms that grandfather Hwai had worked on when the Maines came to the Schweizer-Reneke district nearly thirty years earlier. Moreover, his cousin and closest boyhood friend, Sempane, was a sharecropper there. Piet's plan was an attractive proposition.[50]

But sentiment was the sauce rather than the substance of good farming practice. There were other, more deep-seated reasons to shift to Zorgvliet, and, as in the case of the Reynekes, several of these could be traced back to changes in the Maines' household structure and domestic strategy. Ever since Sekwala had first shown a preference for his second wife, the family had expected Kas's younger brother, Sebubudi, to look after Motheba. This burden had become more onerous once Sekwala had abandoned Hartsfontein and moved to the farm Houtvolop, near the London diggings. Not only was Sebubudi being called upon to shoulder a disproportionately heavy load at a crucial juncture in his own development, but the family was in danger of being too widely dispersed to benefit from the economies of scale in an integrated effort.

Kas thought that the moment might have arrived to get Sebubudi a span and a plough of his own, and to set him on course for a career at Vlakfontein. If this could be accomplished it would free him from the clutches of Adriaan Nieman, give his mother access to a broader support structure, and facilitate the physical reintegration of the family. But all of this hinged on getting the implements Sebubudi would need for independent sharecropping long before the start of the season.

Lacking capital, Kas knew there was no alternative but to sit, watch and wait for an opportunity. Like the black-shouldered kite that sat perched on the telephone line for hours on end, he eschewed unnecessary movement and hoped that the power of the senses alone would suffice to penetrate the camouflage of chance. But nothing stirred. The silence seemed impenetrable. If he turned his head, he could hear the ticking of nature's clock and see the last of winter's cloudless days disappearing over the distant horizon. Then, out of the very corner of an unblinkered eye, he suddenly caught a glimpse of a movement that was as slight as it was awkward. Talons extended, he swooped down to claim his prize.

Governed from afar by the dictates of the Land Bank, the Reynekes were eager, nay, anxious, to get on with the business of ploughing at Zorgvliet. One afternoon, frustrated at the thought of having to train a team of oxen before he could get moving, Willem Reyneke idly let slip that he would consider swopping his span of uninitiated oxen for Kas's team of well-trained donkeys. Training oxen was difficult and expensive. (Less scrupulous landlords, such as van der Byl of Boschplaas, passed the cost of doing it on to their sharecroppers who, each year, would have to train a new span that would then sell for a profit at the end of the season.)[51] Kas closed the deal almost before the young man's words were out. True, donkeys were easier to handle, but oxen could draw a heavier plough, and in the Triangle it was the man who cut the soil deepest who preserved most of what little moisture the dry Kalahari sands retained.[52]

The closing weeks of winter were devoted to the time-consuming business of training the oxen. The two most promising animals were approached on foot with a leather thong lasso dangling from the end of a long stick. When the moment came, the lasso would be hooked over the head of each unhappy ox. Once secured and the animal restored to relative calm, the yoke would be placed on the necks of both animals. With Kas making weight at the rear end, one of his nephews would half-pull, half-lead the pair until they became used to taking the strain and to having someone walk directly in front of them. This procedure would then be repeated, each time linking two additional beasts to the team until the optimum combination of draught power was reached and the animals could take the strain without having to be led.[53]

With a full span of oxen at his disposal for the first time, Kas could take his place beside the better established sharecroppers in the district. The Tabus, Tjalempes, Moshodis and Kadis almost all used oxen rather than donkeys for ploughing. But they also owned the heavier ploughs that went along with such draught power, and that was something he still lacked.[54] Fortunately, he soon heard of equipment on sale at Maokanashomo (the name by which the farm Schoonsig, ten miles west of Vlakfontein, was known to the locals).

On Tuesday morning, 9 August 1927, Kas took out his bicycle and, with the half-dozen or so sovereigns that he had tucked away in his trouser pocket calling out the pace each time his feet completed a circle on the pedals, set off down the London road. At Pienaarsfontein, a spot near where the Tjalempes lived, someone told him how to find a Boer named de Beer who was selling ploughs and things. When he saw what was on offer, he knew he wanted to own it. The two-share Canadian Wonder was the logical successor to the single-share Canadian Chief acquired from William Hambly that had served him with such distinction for eight years. He handed de Beer the money for the plough but then decided he had

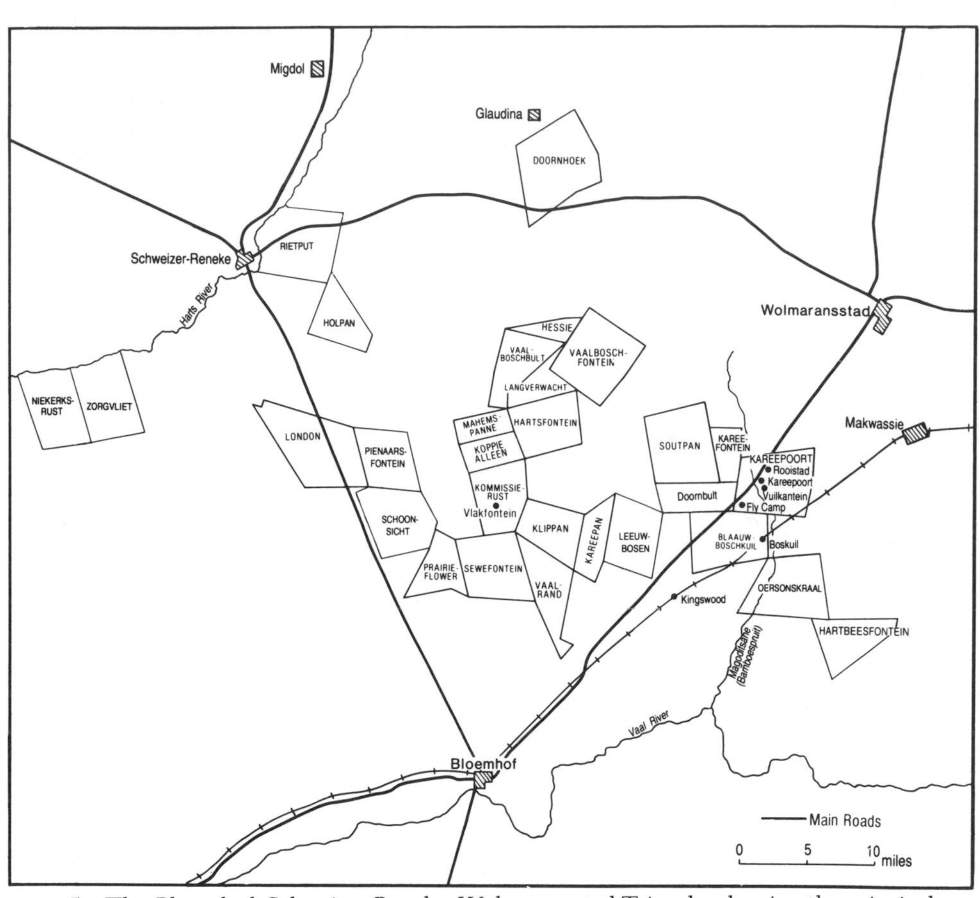


5 The Bloemhof–Schweizer-Reneke–Wolmaransstad Triangle, showing the principal farms on which the Maine family were economically active, 1902–49

better ask him for a receipt. Buying equipment from a well-established trader in the district who also happened to be a friend of the family was one thing; buying a plough from an unknown farmer was another.[55]

But writing did not come any more easily to the Boers than it did to most of the sharecropping folk Kas knew, and only after a long hunt for a scrap of suitably lined paper did De Beer sit down, compose himself, and then with great deliberation write:

> Resiet ver een Ploeg
> ik de ondergetekinde het een Ploeg
> ver Koop aan die Jong Kas vir £ 6.10 Kontant
> ondertekinde
> G. W. de Beer
> de 9 Agustus 1927

Kas folded the document and placed it in a pocket for safekeeping before turning to take a more relaxed look at what else was on offer. Fourteen donkeys with harnesses and yokes had an appeal of their own. Donkeys, no less than his old plough, had served him well. With access to a second span he and his younger brother would be set for several seasons to come. By the time he left Schoonsig that day, he had bought animals and equipment worth no less than sixty pounds.[56]

The thought of owing so large a sum to someone whom he did not really know exercised his mind, much as the property at Zorgvliet and the Land Bank apparently worried his landlord. On the way home Kas decided that if he was going to be in debt, he would rather be indebted to the Reynekes than some faceless man at Maokanashomo. Easy access to personalised credit was one of the most appealing features of life on Boer farms, and 'the Reynekes were reasonable whites who knew that a "boy" could borrow money.' But, with Willem away at Zorgvliet, he had to turn to his new and younger charge for a loan. Koos, who like his brother before him had become an occasional drinking partner and a confidant during trips around the district, was ready to oblige; Kas took the fifty pounds he gave him to settle his debt with de Beer.[57]

The ripple effect of these interconnected preseason manoeuvres soon worked across the extended family. Kas had the delicate task of determining exactly how much of his new load he could pass on to his youngest brother. In the end he and Phitise sold Sebubudi enough donkeys at good prices for the latter to have a full span of draught animals for himself and Motheba. In addition, Kas let him have an old chain plough in exchange for eight bags of grain to be paid at the end of the season.[58]

The way this web of deals hung together was a source of satisfaction: as Kas bade his family farewell for the first of several short journeys between Vlakfontein and Zorgvliet in 1927–28, he did so content in the knowledge that he had taken significant steps to enhance his own productivity as a grain farmer. At the same moment that the half-dozen most successful Boers in the district were beginning to experiment with mechanised planters and harrows, or were cautiously introducing their first paraffin-, petrol- or steam-driven tractors, trucks and threshers, Kas had acquired a planter, invested in a heavy plough and upgraded his draught power from a team of donkeys to a span of oxen. Admittedly this still left him way behind innovative white farmers in the Triangle, but it helped to close the gap between himself and the most progressive black sharecroppers on the surrounding farms.[59]

At Zorgvliet Kas harnessed his new span and attacked the season's ploughing schedule with a ferocity that startled even his brothers and nephews. But the Bloemhof sun too had seen many an ambitious campaign go awry, and before long it counter-attacked so strongly that he was forced to retreat. Working the familiar midmorning shift sapped younger oxen of their energy, while the combination of heat and sweat at that time of day made the animals' skin break out and tear at the point where the yoke bit into the neck. Kas took to inspanning his animals long before dawn or in very late afternoon. From there it was only a matter of time before he started ploughing at night: whenever there was enough moonlight, he would work from dusk to dawn.[60]

At first it seemed that even this Herculean effort would come to nothing. Only a few scattered showers fell during the earliest part of the season, reminding anyone who dared to forget that, although there might be temporary relief in any one year, they were still locked into what seemed like a long-term drought cycle. Indeed, when Kas returned to Vlakfontein in December for the cash that came with the midsummer shearing of the Merinos and a short Christmas break, the fields looked unpromising. Back at Zorgvliet early in the new year things took on a different colour. Leetwane was once again pregnant, and out in the fields the sorghum and maize too were swelling visibly, nourished by welcome summer rain. For once January's promise was not reneged on in March, and an otherwise uneventful summer saw enough rain for landlords and sharecroppers alike to look forward to autumn.[61]

Unfortunately, not everyone else was equally confident of a winter food supply. Forty miles away, in the increasingly dry and crowded 'native reserves,' the shorter days and colder nights in April and May heralded the marshalling of the hungry into seasonal harvesting teams. Platoons of men, women and children armed with sickles marched east and invaded the less densely populated white farmlands in the hope of finding short-term employment that would be paid in kind. As Kas recalled years later, 'That is how the seasonal labourers from Taung and elsewhere were paid because there was no food where they came from. They came to look for grain at our place and then took it back to Taung.'[62]

At Zorgvliet the strongest of the insurgents were made to fall upon the sorghum with sharpened sickles while the weaker were dispatched to hand pick the maize, remove the husks, stuff the cobs into bags, and stack the bags on the wooden sledges left between the fields. From there the oxen dragged the bags to the threshing point, the contents were emptied on the floor and the donkeys driven over the crop until the grain was readily separable from the chaff.[63]

The seasonal raiders were seen off at the standard rate of five bags for every hundred reaped; they slowly and reluctantly staged their retreat into the Taung reserve to wait until communal hunger drove them out again in even greater numbers. Once they had withdrawn into the seclusion of the reserves, the landlords and sharecroppers were sheltered from the gaze of their resentful eyes and, in the sacrosanct confines of private property, shared the bulk of the harvest amongst themselves.[64]

Kas had reason to be pleased. Even after the Reynekes had taken their

Max Woldmann, the landlord at Mequatling, c. 1890
*Dr. M.A.A.M. Geertshen and family*

J. C. (Ou Piet) Reyneke, the landlord at Rietput in the mid-1890s
*Mrs. H. D. Jonker*

The Reyneke family: (left to right) Frikkie, Koos, Willem, Petrus, with Piet and his wife, Anna. Kas worked with the Reynekes for many years
*J. C. Reyneke, Jr.*

A. V. Lindbergh, the landlord at Vaalboschfontein in 1910-13
*F. M. Lindbergh*

J. S. (Hans) Coetzee, the landlord at Kareepan in 1917-19
*J. S. Coetzee, Jr.*

Gert Meyer, the landlord at Kommissierust in 1919, and his wife
*J.J.P. Meyer and Mrs. H. H. de Waal*

Willem Adriaan Nieman, the landlord at Hartsfontein in 1921-22 and at Kommissierust in 1922-23
*R. J. Nieman*

P. A. (Piet) Reyneke, the landlord at Vlakfontein and Zorgvliet
*J. C. Reyneke, Jr.*

J. J. (Koos) Meyer, the landlord at Koppie Alleen in 1924
*R. S. Meyer*

S. E. Seedat, a trader of Kingswood and Rietpan
*Mrs. M. Hafejee*

William Hambly, a trader at Hessie
*Mrs. F. M. Hambly*

Wolfe, Daniel, and Arthur Gabbe, traders at Bloemhof and Makwassie
*Gabbe family*

Jan Coetzee's farm at Kareepan; photograph from 1991
*S. Mofokeng*

Bloemhof, showing a diamond-trader's shop, in the mid-1920s
*Glenn family*

The transport rider Hendrik Swanepoel
*H.F.B. Swanepoel*

G.M.H. Patel: in the early 1920s when he helped set up a cricket league in the southwestern Transvaal; and with his cousin Essop Moola, outside the Makwassie store in 1927
*Mrs. Z. Isane*

The store at Hesse, near Schweizer-Reneke
*Mrs. M. M. Kathrada*

The Empire Hotel in Schweizer-Reneke, c. 1925
*Africana Museum*

Sekwala Maine: in a retouched image probably from the 1920s; and with his sons Mpholletse and Phitise
*Maine family*

Vaal River diggings, c. 1895

Diamond diggings at Bloemhof in 1914

Farmers on the threshing floor, 1931
*Pienaar family*

A group of diamond buyers with Mr. J. Zinger, a diamond trader in Bloemhof, 1925
*Glenn family*

Grasfontein diggings, Lichtenburg, c. 1929

Jason Jingoes
*Oxford University Press*

Jacob Lebone
*Mrs. S. Kadi*

RaKapari Kadi
*Padimole Kadi*

Padimole Kadi, his son, at a heap of cow dung, Klippan, 1983
*M. Nkadimeng*

share, he was left with more than two hundred bags of maize and sorghum. Between them, man and nature had once again conspired to make certain that the social order remained intact. In Taung, harvesters sat down to eat the usual stiff porridge of maize meal; at Zorgvliet the tenants used sorghum beer to 'drink the cattle's legs,' while at Vlakfontein the landlord and his family gathered round to roast a sheep for their end-of-season *braaivleis* or barbecue.[65]

Piet Reyneke's share of the 1928 harvest was big, and he had to give careful thought as to where it should be marketed. With Bloemhof in decline and the district as a whole losing white diggers and black labour to the Lichtenburg diamond fields, there was little point in going to the local grain merchants like Gabbe or Stirling Brothers. Nor was there much point in venturing to the catchment area of Wolmaransstad and Makwassie, where after nearly two decades of effort, the South-Western Transvaal Agricultural Co-operative was beginning to make headway in its struggle against what it saw as too many 'foreign' middlemen.[66] Beyond that, most of the local white farmers would be looking to sell their grain directly on the new northern diamond fields.

They had to look elsewhere. All that was left was the drier west, where virtually the only crop was sorghum, and where the grain merchants could still try to make speculative gains at the expense of black families locked into the reserves. Diverted through the gloriously impersonal channels of the market, grain produced on white farms thus eventually found its way back into the hands of the very teams that earlier in the season had helped to harvest it, but now at enhanced off-season cash prices.

The decline of the native reserves—from the relatively self-sustaining timber and grain-producing enclaves they had been in the late nineteenth century to labour-exporting economic backwaters—meant that they were not well served by a railway network. Piet Reyneke suggested that Kas take the maize to Taung and he, Reyneke, would negotiate with the merchants there. Kas, although eager not to lose a potentially lucrative off-season contract, was worried about being away from home during the final month of Leetwane's pregnancy, but the landlord promised that he would see to it that Motheba and Tshaletseng were brought across from Jaap van Deventer's place to attend to his wife during his absence. Reassured, Kas enlisted the help of RaMputi, a cousin of Leetwane's, and, with two wagons, prepared for the long-distance haulage.[67]

Kas knew that oxen were not the best draught animals for long journeys. Sekwala had once taken a family of white diggers and their equipment from London to Barkly West in the northwestern Cape Province, and the hard roads had harmed the oxen's hooves so badly that he had been forced to sell them along the way and replace them with donkeys. Fortunately, Kas had the donkeys he had bought from de Beer at Schoonsig that spring. Long hauls also required plenty of light but nutritious food. There again, his experiences during his long apprenticeship under his father and Hendrik Swanepoel helped him out: he roasted and coarsely ground up half a bag of mealies and then sprinkled salt over it. When added to water and eaten in small quantities, this unappetising mixture generated a surprising amount of energy and helped to stave off a man's hunger pangs for hours on end.[68]

With the preparations behind them and the two wagons piled high with grain, Kas and RaMputi set off for Matsheng, making their way to the section of the reserve whence the Lepholletses hailed. The next morning they were overtaken by a cloud of fine dust that eventually settled to reveal the bespectacled face of Piet Reyneke in his black Chev. Reyneke told them he was driving on ahead to see what the merchants were offering. But, whatever it was, it was not enough. When they caught up with him at Taung, he told them to push on to Kuruman, a further long haul to a point a hundred and twenty miles west of Vlakfontein.

Kas—who had made this journey once before, in the company of Hendrik Swanepoel while delivering grain for the Kaffir Corn King—knew that ferrying maize to Kuruman would produce a return of at least two shillings and sixpence a bag. If he had reservations, and he did not raise them with Reyneke, they were only about being away from home for longer than he had anticipated. But the Kuruman dealers gave the men no greater cause for joy than those at Taung. In the end Reyneke had to trade the grain for some sheep and goats, and when Kas and RaMputi eventually drove through the gates at Vlakfontein several days later, they had earned three pounds and fifteen shillings each for a round trip of two hundred and fifty miles that had kept them away for close on two weeks.

As soon as he saw Reyneke, however, Kas knew there would always be other reasons for remembering the trip. The landlord greeted him with news couched in the Darwinian phrases that seemed to lie so close to Boer hearts. In his absence, Leetwane had given birth to a healthy little monkey, he said: it had since had its tail cut off and now stood revealed as a baby boy. But, unlike Willem Nieman, Reyneke accompanied this potentially explosive formulation with an outpouring of such genuine warmth that the words somehow slipped by without giving offence. When he asked what the child's name was to be, Kas heard himself say 'Isaac,' a name that had come to him somewhere along the long and dusty road from Kuruman. Reyneke, deeply approving of any biblical inspiration, offered a sheep, which was slaughtered at the earliest opportunity. But names had to ring in ears other than those of landlords, and the little fellow whom whites were free to think of as Isaac came to be known within the Maine family as Bodule.[69]

The celebrations marking Bodule's birth were still in progress when Reyneke again asked Kas to help transport grain but this time only as far as Bloemhof. The new contract, with more bags but a shorter distance, further increased his winter income. By the end of that season he could again look back on twelve highly satisfying months: he had generated a substantial cash flow from an impressive range of off-season activities, ploughed for the first time with his own oxen, and fathered a second son.[70]

These successes had been achieved at a time when the shadows of the Depression were already lengthening, and many around him were finding it hard to make a living in the unsympathetic environment. On two occasions in four seasons Kas had felt he had to help out one or other of his brothers: Mphaka had benefited from the gift of grain that had allowed him to settle his debt with Jaap van Deventer, and Sebubudi had virtually been given the plough and donkeys which enabled him to sharecrop independently. Although freely given in a context of familial responsibility, these 'gifts' nonetheless conjured up ambivalent feelings within the hearts of the recipients: on the one hand genuine gratitude for help given at a moment of economic vulnerability, but on the other a tinge of resentment about Kas's somewhat elevated status that allowed him to dispense such patronage. Sebubudi, for example, once acknowledged, 'Kas usually reaped more than any of us' but asked, rather ambiguously, 'How could we compete with him? He ploughed at night whilst we were asleep!' Kas found it disconcerting that his two older brothers seemed to draw into an alliance that kept him at arm's length. 'Mphaka and Phitise were always in agreement, they were never really willing to understand me. They knew that I was a traditional doctor and that I was respected by many people. That made them jealous.'[71]

Fortunately these strains developed no further, and certainly never approached breaking point. But their very existence distressed Kas, especially so in the case of Mphaka. They had worked the land together ever since the Maines had moved to Hartsfontein, and he considered himself indebted to Mphaka for having helped to develop his farming abilities and having assisted him just as he and Phitise had helped Sebubudi. In retrospect, perhaps the kernel of the problem was that while Kas was making steady progress in his career as a farmer and ngaka, Mphaka was more and more reliant on his skills as an artisan and was finding it hard to make headway.

In part Mphaka's difficulties could be traced back to 1926, when he had taken as a second wife Motlagomang Mothibi, a farm labourer's daughter from Mahemspanne, even though he lacked the livestock to meet his bohadi commitments in full. This, coupled with a growing commitment to and aptitude for construction work, saw him hire labour drawn from outside the family for the first time and his building operations soon occupied not only the off-season but a good part of the summer as well. Before long he was hiring a cousin from Madibogo to do his ploughing.[72]

Although in bumper seasons Mphaka continued to make a decent income from the land, the size of the crop and his share of the harvest were smaller than they would have been had he given farming his undivided attention. His problems were compounded by the onset of the Depression, when white farmers started defaulting on payment for work they had commissioned. It was later alleged by members of his family that one of Adriaan Nieman's sons, Abraham, had in effect 'robbed' him by failing to pay for the construction of a six-roomed house. The police at Makwassie refused to press charges against the young man, for whatever reasons, which did nothing to ease Mphaka's misgivings about colonial justice.[73]

By the winter of 1928 Mphaka was a beleaguered sharecropper with an ever more tenuous hold on the land, a builder whose skills were under legislative attack from a government committed to improving the lot of 'poor whites' at the expense of black artisans, and a small-scale contractor who was finding it hard to make ends meet in troubled times. This, and a personality markedly less phlegmatic than that of his brothers, meant that he was often more critical and questioning of the social order than they. From that moment in his youth when he had challenged the Elliot boys, Mphaka had been searching for a cause that extended beyond the confines of the African Methodist Episcopal Church. Ironically, when he eventually found it, it was presented to him by a friend of the least overtly political of his brothers. Modise Tsubane, the man who had led Kas on his expedition into the Kalahari, introduced him to the Industrial and Commercial Workers' Union of South Africa (I.C.U.) or, as it was more commonly known amongst SeSotho-speakers, *Keaubona* ('I See You').

*

If at first glance the I.C.U. seemed inappropriate and possibly even misguided in trying to recruit members amongst sharecroppers in the dry southwestern Transvaal, then perhaps it was because its origins were to be found in a more fertile economic environment hundreds of miles farther south. The I.C.U. was, in the first instance, a product of the inflationary climate after World War I that had caused a young and charismatic immigrant from Nyasaland by the name of Clements Kadalie to organise a union for black dockworkers in Cape Town in 1919. Kadalie's initiative was not without parallel: in cities all around South Africa, including those on the industrialised Witwatersrand, the postwar years had seen an upsurge in working-class militancy as the cost of living sprinted away from sluggish wages. A few of the more radical activists in an organisation called the South African Native National Congress, founded in 1913 (the forerunner of the modern African National Congress [A.N.C.]), sought to bypass the slow-moving nationalist movement and incorporate the spirit of industrial resistance into a more broadly based union, which emerged from a specially convened conference held at Bloemfontein in July 1920. At this gathering, where a still inexperienced Clements Kadalie played a minor role, it was decided to extend the new union beyond the obvious constituency of factory workers to previously neglected groups including, among others, women and labourers on white farms.

Having set itself these rather ambitious objectives and having received a fillip during the following twelve months when the dynamic Kadalie assumed personal control over an expanded version of his original organisation, the I.C.U. and its urban working-class leadership nonetheless failed to penetrate rural areas in the next half decade. By 1925, the union was still largely confined to coastal ports and leading cities, where it was involved in disturbances that sometimes ended in violence when the armed forces of an uncompromising white state came into collision with angry, disenfranchised black workers.

Then, suddenly, the I.C.U. assumed renewed significance when the outgoing economic tide left certain black middle-class elements stranded within the ranks of the workers. Unimpressed by its earlier lacklustre performance, and inspired by the newly formed South African Communist Party's avowed aim of organising the black peasantry into supporting a programme for a 'Native Republic,' these highly articulate, better educated, and recently radicalised functionaries forced the I.C.U. to pay more attention to the needs of the countryside.

Starting in 1926, Kadalie and his fiery provincial lieutenants—Thomas Mbeki, Robert Makatini, Doyle Modiakgotla, Keable 'Mote and Jason Jingoes—focused on the Transvaal, the Orange Free State and Natal countryside and systematically set a few tracts alight with a highly inflammable mixture of trade unionism, African nationalism and millennial expectation. Over twenty-four months these fires—fuelled with differing degrees of success by regionally specific economic conditions, traditions of resistance and religious beliefs—spread with great rapidity as the I.C.U. leadership, half-cheered and half-terrified, tried to control them. When the worst of the blaze elsewhere was already dying down in the face of peasant disillusionment, accusations of corruption, and a series of damaging splits within the national leadership, a few sparks of radicalism carried to the tinder box of the western Transvaal.[74]

In 1927, thousands of white diggers and a small army of black workers made their way to the new diggings at Grasfontein near Lichtenburg, where the largely market-determined price of alluvial diamonds at first allowed small-scale producers the luxury of paying wages of between eighteen and twenty-five shillings per week. But soaring production soon exerted downward pressure on the market, and when the state stepped in to restore a measure of economic order with the passage of the Precious Stones (Alluvial) Diamond Act, prices fell by 40 percent between April and June 1928. The diggers tried to salvage their position by reducing wages to twelve shillings per week, but this only succeeded in bringing out thousands of black workers on strike demanding a minimum wage of fifteen shillings a week. At this crucial point, and after the state's attempt at mediation had failed, the I.C.U., which for several months prior to the strike had been holding meetings on the diggings, re-entered the fray. Kadalie and his lieutenants, trapped between the diggers' anger and the workers' militancy, had little room for manoeuvre. When the strike eventually ended and the workers gradually drifted back to wages of around fifteen shillings per week, the I.C.U. had probably played a far less important role in the conflict than that attributed to it by deeply resentful white employers.[75]

But in the minds of most black workers on the farms and diggings of the surrounding districts, the strike at Grasfontein was closely linked with the I.C.U., and news of its success was swiftly transmitted. This was not surprising, given the scale of the exodus from the rural areas to the diggings, the ties of kinship and ethnicity among black workers, and the close link between seasonal agricultural labourers and employment on the diggings. What was noteworthy was that the strike at Grasfontein alerted the I.C.U. leaders to the existence of these conduits into the countryside and helped ease the Union's subsequent entry into the Bloemhof–Schweizer-Reneke–Wolmaransstad Triangle.[76]

Within weeks of the upheaval at Grasfontein, two of the I.C.U.'s most able campaigners—provincial secretary Keable 'Mote and Jason Jingoes, branch secretary at Makwassie—were told to strengthen the Union's shaky presence in the still virgin territory of the southwestern Transvaal. Although still in their early thirties, both 'Mote and Jingoes were experienced enough to cope with the various ethnic groupings found on the diggings, and especially well qualified to deal with the many Sotho-speaking tenants and sharecroppers on the nearby farms. Not only did they both hail from northwestern Basutoland, but they had also gained insight into the problems of wage labourers and labour tenants while organising farm workers in the eastern Orange Free State on earlier assignments.[77]

'Mote and Jingoes arrived in the Triangle with reputations as outspoken radicals in the new black labour movement, but their rhetoric had in truth already lost much of its cutting edge. Yet even if the ideological sparks that flew off their tongues glowed less intensely in the glare of the highveld sun than in the shadow of the Malutis, they had lost none of their zest for organising. During the eighteen months between June 1928 and the end of December 1929, their energy knew no bounds: in addition to setting up regular meetings within the Triangle they also organised other campaigns elsewhere in the southwestern Transvaal. All these gatherings, held against the backdrop of the steadily deepening Depression, were addressed by the most prominent I.C.U. speakers of the day, including the great man himself, Kadalie.[78]

The early I.C.U. Sunday meetings, such as one at Wolmaransstad addressed by district secretary Robert Makatini, attracted hundreds of people who, depending on their station in rural life, had come to town on foot, by bicycle, on horseback, or by horse and cart. Crowds of labourers from the nearby diggings, farm workers and sharecroppers such as Mmereki Molohlanyi and his friends, who made a twenty-mile trip in to Borobalo from Katbosfontein, thronged to hear addresses that had been advertised by circular, poster and word of mouth.[79] As Kas later recalled it: 'All the farm folk went to hear what he had to say. So we also went. There were a lot of people, more than a thousand. Man, the town was full! It was person upon person. And the speaker was above us on a platform so that everybody could hear.'

But not everybody who surged into town on this incoming tide was necessarily black, or buoyed by the same enthusiasms. Like a fleck of foam blown into a remote corner of a beach, a dozen or more white farmers found themselves on the edge of a sea of black faces. At one point Makatini referred openly to their presence, turning briefly toward them to warn them that the I.C.U. was there to scrutinise their treatment of the workers—'Keaubona, I see you when you cheat them.' Redirecting his attention to the black majority, he spoke of the need for higher wages and urged those present to consider strike action if the Boers did not accede to their demands. When this wave crashed, it created yet another spume of white, as policemen, hitherto hidden amongst the ranks of 'very angry farmers,' bent forward to hear and record the speaker's words.

A few weeks later, in October 1928, 'Mote received an equally enthusiastic reception from several hundred excited farm workers and sharecroppers at Bloemhof where he, too, set out to spread the challenge of Keaubona. He succeeded in signing on two hundred men who, for a shilling each, were given the little red card showing membership in the I.C.U. But if the farm workers were hoping for further militant talk along the lines mooted earlier by Makatini, they must have left somewhat disappointed. Although 'Mote had pertinent things to say about the exploitation of farm workers, a relieved white newspaper reporter also noted, ' "the dignity of labour was preached to a very high pitch," ' and ' "the natives were told to obey their masters and work if they were to get an increase in wages." '[80]

This conscious downplaying of the 'wages-and-strike' issue became more noticeable as the Union edged away from the concentrations of militant workers around the diggings, and moved out into the remoter farming districts, where it was conspicuously successful in winning support among sharecroppers. Kas, who attended an earlier meeting at Bloemhof where there was talk about workers demanding a massive increase in farm wages, also went to another meeting at Schweizer-Reneke where Kadalie himself urged farm workers *not* to strike. Although this advice might have alienated the wage labourers, it did nothing to dampen the sharecroppers' enthusiasm for the Union. By April 1929, the Schweizer-Reneke branch of the I.C.U. had more than four hundred members and that at Makwassie, presided over by the wily Jingoes, was at least as successful.[81]

The long-lasting appeal that the I.C.U. had for sharecroppers in the late 1920s was not born solely of conservatism, or because the Union appeared to have backed away from the ugly confrontation that was bound to accompany any strikes in the countryside. On the contrary, in some ways the support of this relatively educated and economically progressive group came because it was capable of articulating other, even more radical African nationalist aspirations with specific appeal to a self-consciously 'respectable' section of rural black society that was in danger of being sucked down into the ranks of wage labour during the recession. Foremost among these aspirations, and of great appeal for the strong-willed Mphaka and his friend Modise Tsubane, was an end to racial oppression. At a meeting in Bloemhof during late 1928, Jason Jingoes spoke out powerfully against the 'South African laws as they affected blacks,' and about 'slavery,' echoing an earlier admonition by Clements Kadalie to an audience in Wolmaransstad—including Mmereki Molohlanyi—that it was 'high time that blacks rid themselves of the chains of slavery.' A member of the Maine family likewise recalled it being said that 'the I.C.U. was going to liberate blacks from slavery.' More than half a century later, this battle-cry still elicited a nod of approval from the radical Motlagomang Maine, who observed, 'Today I can see that those people were the ones who started to liberate us. Before that the Boers treated blacks very badly.'[82]

Mphaka was an extremely enthusiastic convert to the I.C.U. cause. A poorly educated and virtually illiterate man who seldom missed a Union meeting, he developed a close political association with both Keable 'Mote and Jason Jingoes. His sister, Sellwane, remembered that while the family was still based at Vlakfontein, he would often 'come home with I.C.U. documents, although he could not read. After an I.C.U. meeting he would group his people and tell them what had been said at the meeting. He also gave them the Union pamphlets to read.'[83]

Modise Tsubane, whose calling as a herbalist made him well known and trusted in the district, became even more involved in the work of the I.C.U. A former Zionist preacher who had taught himself to read and write and who was something of a rural intellectual, he decided to put his time and literacy at the disposal of the labour movement. He served as a member of the local executive, the Union's eyes and ears on the farms, and was often called on to address meetings in smaller towns. Nor was he the only traditional practitioner in the Triangle to find a leading role in the formal structure of the Union. Across the way, at Makwassie, yet another well-known ngaka, Seabata Koaho, likewise served on Jason Jingoes's local executive committee.[84]

Mphaka, Tsubane and their younger friends responded enthusiastically to all this talk of 'freedom and slavery,' and for many members of the African Methodist Episcopal Church, this message had an old and familiar ring. This familiarity did much to smooth the way for some of the wealthier sharecroppers, such as Mdeboniso Tabu and Nini Tjalempe, to accept the I.C.U. Indeed, on more than one occasion senior A.M.E.C. members left the Sunday morning service in Bloemhof and walked directly across the way to a Union meeting being addressed by Jason Jingoes.[85]

For most of the Maines, especially the women who had long been A.M.E.C. stalwarts, the very familiarity of the I.C.U.'s message seemed to obviate the need for them to join the Union. Sellwane Maine, for example, thought that the two organisations were heading in the same direction albeit on separate religious and secular tracks. She noted of the local ministers that they too, 'talked about freedom. They said that if we prayed wholeheartedly, then God would free us. During A.M.E.C. concerts we sang *Nkosi Sikelela*. That is why I say that those ministers were working with the I.C.U.'[86]

The long-standing presence of the A.M.E.C. in the Triangle and its known link with the United States may well have helped to modulate the reception of another, sometimes highly problematic element in the I.C.U.'s ideology. Whereas elsewhere in the South African countryside it was widely believed amongst I.C.U. supporters that black American liberators would arrive to help usher in a new and more just social order, such open-ended millennial beliefs were harder to find in the southwestern Transvaal. Of course there had been quasi-millenarian flutters around black Americans and the A.M.E.C. during the 1906–8 depression. But men and women like Kas and Sellwane had heard the A.M.E.C.'s Bishop Vernon at Bloemhof during his visit five years earlier; exposed to a real black American, they could arrive at a more sober assessment of the I.C.U.'s often garbled message of liberation.[87]

If they did not have dreams of freedom being ushered in by outside agents, the down-to-earth sharecroppers of the Triangle certainly enjoyed visions of the I.C.U.'s owning property, or of its gaining access to additional land for its hard-pressed members. After half a decade during which white farmers had greatly expanded the grazing available to their own livestock at the expense of their tenants' sheep and cattle, such promises fell on the ears of a land-hungry peasantry like so much gentle rain on parched earth. The prospect of acquiring land, more than anything else, made Kas sit up and take notice of the Union. Always eager to preserve scarce material or political resources, he filed away the Union's offer in his prodigious memory only to dredge it up under more threatening circumstances nearly two decades later.

Sharecroppers joined the I.C.U., he suggested, because they thought the I.C.U. would help them cultivate the land independently. 'The Boers would not allow us to purchase land . . . but the land which they refused to sell us did not belong to them.' Mmereki Molohlanyi, too, recalled a Union booklet (in all probability the I.C.U. constitution) in which it was stated that 'we shall have a right to farm as we want,' while Modise Tsubane was remembered for suggesting that the farms 'be divided in such a way that both the whites and the blacks have their own fields to plough, but that their livestock be allowed to graze on fields held in common.'[88]

These ideological arrows found their way right to the very heart of sharecropping interests. So, too, did the I.C.U.'s effort to convince its members that blacks needed to improve their education if they were ever to 'win their freedom.' As a group, sharecroppers were better educated than most farm labourers and, as we have seen, several of the better-off tenants had already made generous provision for the education of their children. It was therefore from a fairly privileged perch, and with a great deal of sympathy, that they listened to the I.C.U.'s pleas for education, a vibrant message delivered in characteristically part-hectoring, part-pleading tones.

Motlagomang Maine recalled an address that Jason Jingoes delivered to I.C.U. supporters in Bloemhof. The meeting, composed of the usual crowd of blacks and a few white farmers (who themselves were unlikely to have been very well educated), was told that:

> Blacks should blame themselves because, having failed to educate themselves, they succeeded in handing themselves over to the whites. Jingoes said to them, "Look at me, I am educated, I know everything about the world and I am free. I am told, however, that you are called baboons." It was said that when he spoke these words, he turned his back on the whites present and, lifting the vent on his jacket, said, "Look, they say we have tails but do you see one on me?" The whites in the audience bowed their heads in shame when they heard that, and he went on to say that blacks should become educated, free and move out of the darkness.[89]

'Jingoes,' she noted with approval, 'was not scared of whites.'

The I.C.U. spoke the language of progress to sharecroppers, even though some of them, like Kas, did not always appreciate the need for the formal education of children who worked the land. But on other occasions it put out a more ambiguous message, appealing strongly to beleaguered black patriarchs who were in danger of losing their control over family labour. Mmereki Molohlanyi, for example, recalled Kadalie turning to a group of white townswomen at a meeting in Wolmaransstad and saying, ' "We have wives who are working for you instead of working for us, their husbands. How will you feel when they will no longer work for you, and you are forced to go and fetch a bucketful of water for yourself?" '[90]

Like most moderately successful social movements, the I.C.U. could not live by talk alone, and some of its more visible actions won widespread approval in the black community. In many respects it rendered workers on the diggings and farms an invaluable service: for nearly a decade, between 1928 and 1937, Jason Jingoes helped scores of farm labourers to challenge oppressive landlords either through his personal intervention or, even more frequently, by arranging for their legal representation in the small magistrate's courts at Makwassie and Wolmaransstad.[91]

At least as impressive was that the Union intervened with the local police and got them to modify the way they escorted black prisoners into town from the farms and diggings. This was a long-standing grievance in the Triangle, dating back to the period of economic instability after World War I. In 1920, after a serious riot at the Rietkuil diggings during which a worker was killed and two others seriously wounded, the Transvaal Native Congress (T.N.C.), an affiliate of the South African Native National Congress, had complained bitterly to the state's director of native labour about police treatment of African prisoners. Tax defaulters and black men accused of other minor misdemeanours were handcuffed, roped to a white constable's horse and made to scramble and stumble over the twenty-mile haul to the Bloemhof police station.[92] A decade later this practice was again widespread. The I.C.U., like the T.N.C., successfully challenged these practices and in doing so won the lasting admiration of its rank-and-file members. As far as Mmereki Molohlanyi was concerned, it was Clements Kadalie who won the battle: 'He was successful in remedying some of the bad things that took place during those times. The police, after arresting a person, would let him walk all the way to prison while they proceeded on horseback.' For Sellwane Maine the credit was due to Jason Jingoes. 'The I.C.U. helped us because, at the time, we were under the Boers' laws. People were handcuffed and made to walk on foot to Bloemhof while the policeman went on horseback. Jingoes asked them why they should handcuff and treat a poll-tax offender like a common thief.'[93]

These modest victories gave black men and women a new sense of dignity and purpose. Small legal battles won at the magistrate's court linked up with a wider set of ideas and together they gave sharecroppers and farm labourers—the 'monkeys' and 'baboons' of the dominant discourse—enhanced self-esteem and racial pride. As Kas recalled it many years later, 'The white farmers said that we thought ourselves superior ever since we had started following Kadalie.' For the first but not the last time, a form of modern black nationalism presented a meaningful challenge to those who owned and occupied the land. White landlords, accustomed to blacks' bodies bent into submission by the weight of unwritten racial codes, saw little to commend the new posture.[94]

Whenever Jason Jingoes had a meeting at Bloemhof, Mphaka would simply *tell* his landlord on Friday afternoon that he would be away from the farm on Sunday morning. On one occasion, Reyneke dared to probe a little more deeply; he was quickly told that since whites could hold meetings without interference from outsiders, there was no reason why they should concern themselves about black meetings. An essentially amiable soul, Reyneke accepted this rebuff without fuss, but many other Triangle landlords, especially those less well capitalised than he, were not willing to tolerate Union activism.[95]

Precisely because there were differences in the economic standing of various white farmers, the Boers were sometimes unable to agree on how to deal with the threat of the I.C.U. Thus Motlagomang Maine heard of a meeting at Bloemhof where Afrikaner farmers gathered to establish a united front to meet the Union challenge, only to have the assembly end in complete disarray when they started 'fighting amongst themselves' about what strategy to adopt.[96]

The shouting matches between white landlords and black tenants in the summer of 1928 soon gave way to more direct and serious confrontation. By harvest time 1929, scores of discontented black tenants were facing prosecution under the decades-old Master and Servants Act for refusing to see out the period 'stipulated' in unwritten contracts, and many farmers were short of labour at the crucial juncture in the production cycle. At Zorgvliet, young Koos Reyneke was distressed to find that the harvesting teams from Taung failed to appear in their usual numbers, and he and his wife, Anna, were forced to pick mealies 'until our fingers were bleeding.' At Schoonsig, the shortage of labour was so pronounced that, at the start of the following season, J. P. Pienaar went to Johannesburg and, despite the worsening economic climate, bought a steam-driven Ruston & Hornsby thresher for seven hundred pounds—the first such machine in the district.[97]

At this point the state marshalled its counter-attack. In March 1929, the I.C.U. head office in Johannesburg received a spate of complaints from its Bloemhof and Schweizer-Reneke branches, and Doyle Modiakgotla was sent to investigate. As he later recorded, 'On my arrival at Schweizer-Reneke I found that "WAR" had been declared against the workers by both the Local Urban authorities and the Police.' Acting in terms of a contested government proclamation that required all meetings to be authorised by a permit issued by the local magistrate, the police had put a stop to the Sunday morning gatherings. In addition, the town councils, the organs of local government responsible for managing 'native affairs' in urban areas, instructed their Superintendents of Locations, 'to refuse the officials of the I.C.U. permission to reside in the location.'

Over the next two weeks Modiakgotla and Makatini deliberately defied these regulations, and then contested their validity in the Bloemhof magistrate's court, where they found the local police sergeant doubling as public prosecutor. But a Union victory in the courtroom was shortlived out in the field when the police shifted their attention to the I.C.U. rank-and-file. For the next four months the Sunday morning gatherings were disrupted by white policemen demanding that farm workers and sharecroppers produce 'special passes' authorising their presence in what the authorities perceived as 'white' towns.[98]

The 'harvest-season war' between March and June 1929 was backed up by a less organised, but equally effective campaign of terror and harassment deep in the heart of the Triangle. According to Motlagomang Maine,

> Whites became jealous and started looking for people who were attending I.C.U. meetings and killed them. [A lawyer] in Bloemhof warned those attending meetings that they should be careful, and arm themselves with spears in case of any Boer attack on their homes. Mphaka had such a spear.[99]

Considering his friend Modise Tsubane's experience, this was perhaps a necessary precaution. Having addressed an I.C.U. meeting at Witpoort, Tsubane and a party of Union members were on their way home when they ran into a police patrol on the outskirts of Wolmaransstad. The terrified Union supporters 'threw away their membership cards and ran away. When he arrived home Tsubane told us that he would no longer be involved in the I.C.U. because the Boers would kill him.'[100] At Makwassie, Jason Jingoes had an equally unpleasant time. One afternoon he was lured onto a cart and abducted by an angry white farmer and an accomplice. The farmer 'dragged me to the ground and took me to the kitchen of the farmhouse, shouting to Mrs. Fourie to bring him a gun, that he was thinking of shooting a man, a man who had been causing trouble in the district.' Fortunately, Mrs. Fourie was reluctant to help, and when the farmer 'left the kitchen to look for a gun inside the house, she whispered "Go on! Get away!" I got out and ran. My horse was tied behind the cart, and I was on it and off.'[101]

Harassed by the police in town and terrorised by militant whites in the countryside, the I.C.U. never regained its strength after the 'harvest war,' though the branch at Makwassie remained intact for a further eight years under the guidance of the crafty Jason Jingoes. The success of the counter-offensive mobilised by the state and its white populist allies left most of the farm workers and sharecroppers sullen and resentful. But, as was to happen half a century later, the tide of nationalism retreated only until it could flood back into the Triangle. And, somewhere out at Vlakfontein, the Great Laodicean watched the ebb and flow of the political tide with the peasant's customary canniness.

Like most of the Maines, Kas never joined the Union. Sensing the direction the storms would eventually come from, he chose not to make a public issue about joining the I.C.U. and moved instead into the protection of the sheltered area around his older brothers. But when the I.C.U. called its first meeting for farming folk in the Triangle after what was perceived as a 'successful' strike at Grasfontein, he went with those 'riding on horses' to hear Robert Makatini speak while many 'others travelled by cart' to Wolmaransstad.[102]

Kas thought that young Makatini, although not much of a SeSotho-speaker, was undoubtedly a 'clever man' and a 'very educated one,' too. Nor was there much doubt that the I.C.U.'s policy as expounded on that occasion 'addressed itself to the major grievances of farm workers and sharecroppers.' But clever men often preached conflict and did not necessarily 'teach people to understand one another,' a brief which Kas, as a ngaka, took more seriously than other traditional practitioners in the district. In his experience, smart-alecs had a habit of 'starting a war and then disappearing.'[103]

Coming as it did when Kas was enjoying increasing success on the land and getting on well with the landlord and his two older sons at Zorgvliet, Makatini's speech lacked appeal for him. 'I was more interested in my farming. Why did I have to associate myself with the cause they were fighting for? You see, I had never fought anybody, not even the whites.' Even then, he sensed trouble brewing. All that wild talk about 'whites' and 'strikes' was bound to end in tears. 'The Boers would skin you.' Moreover, Kas was strongly of the opinion that strategies of industrial resistance were unsuited to farms, where a paternalistic ethos governed day-to-day relationships. 'How could they call for a strike in a place where they had no social standing? How can you have a strike in another man's home? You can't do a thing like that!'[104]

Kas kept these reservations to himself, but they were shared by a fair number at the meeting. When Makatini started selling the Union's red membership card, 'some took, others did not. You know how it is, people never move as one. Some would strike, others would not.' And Kas noted that once Makatini had left town and the first flush of enthusiasm for a strike had receded, some of the new members became disillusioned and threw away their cards. This was followed by a phase during which Kadalie himself counselled the people to take care and not to embark on any ill-considered strike action. Still later, when the farmers started prosecuting their recalcitrant labourers under the Master and Servants Act, it dawned on some of the remaining loyalists just how powerful was the system that they were opposing. For Kas, pragmatism shaded into defeatism. 'Look, if the law is a law, then obey the law. Don't bother to argue. Can one man deflect the course of a river in spate?'[105]

Though he never joined the Union and his assessment of how the struggle would unfold was remarkably accurate, Kas continued to attend I.C.U.

meetings, in part because of the same deep-seated curiosity that always made him eager to be well-informed. But, given how strongly Mphaka and the other leading sharecroppers felt about the Union and its cause, it also had something to do with his desire for social survival in a potentially authoritarian subculture.

> If I simply sat there I would not have known what was going on. You had to attend in order to see what was happening. If you had the right to vote that was all right, if not then it was also all right. If you had something to say it was best that you said it there, but if you did not attend the meeting you would be branded a rebel. [You would be asked] "Why did you not attend?"

Even then, some in the family could see where his heart lay. His sister, Sellwane, noticed that, while Kas had a residual interest in the I.C.U., 'he was more involved in farming and his medicine.'[106]

It was precisely this involvement in farming that drew Kas back to Vlakfontein in the spring of 1928. The rains arrived at the appointed time and, as usual, he planted his customary crops of maize and sorghum. But elsewhere in the district better-off white farmers were cutting back on land they devoted to grain, not so much because they feared a challenge from the I.C.U. or foresaw difficulties in bringing in the harvest, but because the local market had shrunk so dramatically. By 1929, Bloemhof had only two thousand white inhabitants and there were barely three hundred diggers active in the district. Lichtenburg had drained off much of the Triangle's economic life-blood.[107]

The wisdom of getting the crops in early and concentrating on the types of maize and sorghum that ripened quickly was again evident when the rains over most of the highveld petered out in February. Fortunately, there was just enough late summer precipitation for Reyneke and his partners to watch the crops reach maturity. With the onset of winter, and while several other farms struggled to cope with I.C.U.–induced labour problems (further aggravated by the Taung harvest teams not arriving in their expected numbers), the Maine family helped bring in the harvest at Vlakfontein. At Zorgvliet, at the very far end of the district, Reyneke's sons struggled on.

Normally, black families marked the end of the harvest season by gathering to drink the cattle's legs. But in the winter of 1929 something was amiss, and the beer seemed to slip down less sweetly than it had in the past. As the struggle between the state, the white farmers and the I.C.U. drew to a climax, many farm workers and sharecroppers signalled their dissatisfaction by asking their landlords for a *trekpas* and indicating their intention to move to greener pastures. Even at Vlakfontein, tension and resentment could be sensed everywhere.

The harvest was barely in when Mphaka and Phitise told Reyneke that, for reasons they did not make entirely clear, they intended moving on. Earlier in the season, while working on the diggings at Kareepoort, Sekwala had lost his second wife, Maleshwane, but he had subsequently moved back to Houtvolop. In the interim, the two brothers had visited the diggings often enough to believe it might offer them a viable base for their building, sheep-shearing and leather-working skills. At about the same time, Sebubudi found himself a place on the farm Klippan, a few miles east of Vlakfontein. With Mphaka and Phitise becoming ever closer and starting to look to their children rather than their two younger brothers for practical assistance, the extended family was experiencing minor structural adjustments. The landlord, apparently unconcerned about the exodus of skilled labour from his property, issued Mphaka, Phitise and Sebubudi with the passes that would allow them to relocate their businesses.[108]

Kas was left to oversee the threshing and apply kraal manure to the now deserted fields. But even as the soil was tucked in for its seasonal sleep a strange menacing quality could be detected in the dry winter air. He saw to it that the maize and sorghum were loaded into bags, split into the predetermined portions, and stacked to await transport into Bloemhof where the grain merchants were reported to be under greater pressure than ever from the agricultural co-operative. One morning, while he and Reyneke were supervising the loading of the last of the grain, he ran short of help and had to call in Mmusetsi; he told the six-year-old to open the tap on the overhead tank and fill the dam so the animals could drink before they set off for town; in the meantime, he and Reyneke rounded up the cattle. But, when the oxen caught sight of the wagon, some of them broke rank and ran off into a nearby field. Mmusetsi, forgetting the tap, ran off after the strays, rounded them up and drove them back toward the wagon.[109]

By then the dam was overflowing and the child, sensing trouble, ran to Reyneke to tell him what had happened. The landlord, angered by the loss of water in what had been another dry and demanding year, shouted, 'You bloody big kaffir, you stood by and watched the water overflowing!' and struck the boy. Kas, hearing the commotion, came around the wagon, and remonstrated with Reyneke, pointing out that if the child had closed the tap first, the strays 'would not have had any water to drink.' But the normally placid landlord was in no mood for excuses and merely grunted, 'Agh, he should have first closed the tap.'

Exasperated, Kas hit back in the racist terms that had dominated political discourse in the Triangle all that summer. If Mmusetsi was a 'big kaffir,' then little Fransie Reyneke was 'big' too since the boys were of similar age, and it was high time that the landlord's son was given a set of matching duties and responsibilities. Reyneke retorted, 'For some time now I've realised that you think that this farm belongs to you.' Kas, not to be outdone, replied, 'Oh, now that I have made you rich, you come up with that story.'

Trapped between these rhetorical flourishes, Kas 'off-loaded the wagon' in silent anger and then 'demanded a trekpas so that I could leave and go and search for another place.' When Willem Reyneke put in an unexpected appearance later that morning, he was astounded to learn of his father's behaviour and unwilling to countenance the idea that 'a minor quarrel' would make Kas leave Vlakfontein.

> He said, "Father, you have been cruel. What do you think the little kaffir could have done, the cattle were running away and the water was overflowing? Look, you have expelled Kas. Where will you get another boy like him? We grew up under him and now look at what you have done. It's nonsense!"

Piet did not respond to this but walked away in silence. Later that day, Willem '. . . came to me and suggested that I move across to his place,' Kas recalled. Their friendship, at least, was secure.[110]

But it was too late. The distant rumble of discontent had come to Vlakfontein in a single flash. Perhaps it was because the bolt came from such an unexpected angle that it managed to claim two essentially reasonable men as victims. Or perhaps it was not unexpected at all, and both landlord and tenant had, each in his own distinctive way, been more deeply affected by the I.C.U.'s challenge than they cared to admit. Whatever the underlying cause, an uncharacteristically bitter exchange produced a dramatic ending to an amicable relationship.

With winter drawing to a close and the landlords in the district for once almost as unsettled as the tenants, Kas once again had to find a new home for his wives, three small children and baby Bodule. It is easy to understand why he was tempted to look in the same direction as his two older brothers had and think of Kareepoort. When the family had loaded their possessions onto the wagon and trundled out of the gate at Vlakfontein for the last time, he looked back on seven full seasons during which he had done much to consolidate his career as a farmer, but the road he went down did not lead to a farm. All he could see ahead were the diamond diggings, a deepening economic depression, and certain struggle.

CHAPTER SIX

# Struggle

## 1929–37

Kareepoort, little more than twenty miles due east of Vlakfontein as the crow flies, drew the Maines into a new and unfamiliar world. Although well within the arid Bloemhof–Schweizer-Reneke–Wolmaransstad Triangle, the diggings hugged the scalloped edges of the muddied pools that made up the Bamboesspruit which, in turn, skirted the slightly better-watered areas round Makwassie. With Bloemhof at their backs, the former inhabitants of Vlakfontein looked to slightly more progressive commerce and agriculture around Wolmaransstad to sustain their economic pulse as the Great Depression heaved itself into position over the southwestern Transvaal.[1]

But for all its promise Kareepoort too had probably seen its best days. The original farm of 1870, just over 5,000 morgen in extent, passed through the hands of three Afrikaner families and then, like so many others in the district, to an invading property speculator during the 1890s. In the speculative squall after the South African War, the property was acquired by one of the many land and mining companies that exerted such influence in the western Transvaal. It took the directors of the New Transvaal Gold Farms Company two decades to realise that the future of the property lay not below its surface, but in the gravel so liberally bestowed on the Vaal River flood plain.[2]

After the First World War returning soldiers started taking a more active interest in the Bamboesspruit, and by 1922, Kareepoort, along with a clutch of other farms in the district, was proclaimed as a public digging. Hard-pressed white diggers and desperate black workers invaded the site; hundreds of 'poor whites' were joined by thousands of even poorer blacks drawn from native reserves nearby like Taung, but also from as far afield as Basutoland and the Transkei.[3]

By day, when work on the diggings was presided over by the myopic gods of money and production, those amongst the poor whose skin colour and political power ensured that they had a licence from the state to 'dig' for diamonds mixed with the lesser mortals who did the actual digging;

the shriller sounds of English and Afrikaans climbed above the noise of the mining machinery and triumphed over the resonant tones of Se-Tswana, SeSotho and IsiXhosa. By night, when the sharp-eyed witches of fear and racism ruled supreme, the ranks of the poor were even more segregated, with lighter-skinned men being allowed to stay on in makeshift shanties or portable cabins, while darker ones were made to troop off to peripheral 'locations' whose very names conveyed transience, poverty and squalor—Fly Camp, Velskoen (hide shoe), Vuilkantien (filthy canteen) and Rooistad (red village).

Throughout the 1920s these ramshackle villages were packed tight with thousands of black workers, hundreds of African women and scores of dusky children—living testimony to the few regular and many more irregular unions on the diamond fields. At Kareepoort, where the seasonal splutterings of the Bamboesspruit, or Magoditsane, divided the mainly Xhosa-speaking inhabitants of Rooistad and Vuilkantien on the east bank from the largely Sotho-speaking denizens of Fly Camp and Velskoen on the west, there were well over five thousand men at work by 1923. Similar social arrangements and numbers existed on other farms that boasted diggings—Boskuil, Oersonskraal and Kareepan.[4]

But by its very nature, alluvial diamond mining is only a robber-economy: once the fields of Kareepoort were stripped of their sparkle, the assembled ranks of the poor turned their backs on the pock-marked veld, its mounds of sifted gravel and pits of stagnant water, to march off toward the New Jerusalem. As the battalions of the hopeful and the dispossessed retreated to Lichtenburg in 1927 and 1928, they left behind them the all too familiar casualties of an industrial army of occupation—the old, the sick, the wounded, the idle, the criminal, the abandoned wives and the unwanted children.

Stranded as an urban-like residue by the outgoing economic tide, these industrial outcasts would have experienced even greater hardship had it not been for an influx of marginally better-off black folk from the countryside. Economic refugees from the overextended native reserves, retrenched farm workers, labour tenants expelled from adjacent properties, and retreating sharecroppers all made their way to the public diggings, one of the few places where they could settle legally in what was decreed a 'white man's country.' For the most part younger and more enterprising than the social discards of the former digging regime, the incoming black farmers unleashed a new productive energy in the African villages and, by hiring grazing land for their livestock from nearby white farmers, injected economic vitality into otherwise moribund agriculture.

By 1929, when Kas's family had once again linked up with those of his brothers Mphaka and Phitise, this very modest contracyclical process of social and economic renewal was well under way, and the village of Rooistad was, for most of its black inhabitants, more a farming hamlet than a diamond-digging centre. But this perception was not shared by agents of the state. In a nation where inter-racial sharecropping was illegal, partners across the colour line maintained a discreet silence about their farming practices: to the state, the clutter of inhabitants on the diggings was little more than an ugly proletarian relic of an older industrial dispensation. Outwardly garbed in shabby working-class clothing, the tenants' neater agricultural underclothing remained hidden from official view.

As far as members of the South African Police at Makwassie were concerned, Rooistad and its neighbouring villages were slums populated by beer brewers, prostitutes and men with no visible means of support—people who merely added to the problems of a district already troubled by sporadic outbreaks of illicit diamond buying and stock theft. Families like the Maines, who arrived on the diggings with the dust of the farms still clinging to their rustic ankles, had to adjust to a quasi-urban environment in which the police were constantly demanding passes, searching for small-time confidence tricksters (such as the 'Bull Nines') or raiding the beer parties organised by 'shebeen queens.'

Even here, on the very doorstep of the unknown, the portals of the urban world were guarded by a white man—in this case, a diminutive Briton by the name of Cecil McDonald. McDonald, known to most of the locals as 'Baas Cecil' or, more graphically, as Piccanin, had earlier married into a well-established Jewish family in the district and earned a reputation amongst blacks as a 'good' landlord. When the diggings went into sharp decline during 1928, New Transvaal Gold Farms sold off more than 1,000 morgen of Kareepoort to a landlord from whom McDonald, in turn, hired the land and did some mixed farming.[5]

Unlike the uniformed men at Makwassie, Cecil McDonald was quick to appreciate the scale and potential of the rural influx into the diggings, and when Kas approached him in the spring of 1929, he agreed to provide Kas's animals with grazing at the going price of a shilling a month per head for the larger beasts (cattle, horses and donkeys), and thruppence a head for the smaller animals (sheep and goats). In addition, as a rent-paying tenant, Kas would be free to put as much land as he wished under the plough at no extra cost. Although perhaps inwardly daunted by the prospect of having to find four pounds ten shillings per month in cash just to ensure the well-being of his stock, Kas accepted these terms.[6]

Within days he was at work, ploughing with customary skill and planting a field of maize for his family. The Maine women, in turn, embarked on a programme of domestic reorganisation for what had become a rather lopsided family profile by 1929. Leetwane had to look after three children under the age of six—the boys Mmusetsi and Bodule, and the sister who separated them, Morwesi, while Lebitsa's four-year-old, Thakane, had long since been dispatched to live with her maternal grandparents at Kommissierust.

The imbalance, and the consequent practical problems, were exacerbated at Kareepoort by the fact that Motheba chose to attach herself to Mphaka's household and therefore could not always help with child minding. But the sisters, who had grown much closer in the past five years, resolved the difficulty with elegance and economy, transferring two-year-old Bodule to Lebitsa's care, a common-sense arrangement that not only restored some symmetry to the extended family, but worked out extremely well in practice; the toddler thrived under the new regime. This handing over of a child to stepparents or other kin for nurturing was not uncommon in large farming families on the highveld, nor confined to black families as the Reyneke family history showed. The bonds between the sisters strengthened, at a time when the Maine family was immersed in a new social environment.[7]

The reshuffling of domestic portfolios did not greatly concern Kas, for he believed that such matters were the concern of the women. But cattle raising *did* interest him, and for this reason he paid more attention to six-year-old Mmusetsi who, like other children of his age, was already herding a small and compliant part of the family's livestock. Yet, for all the loyalty he had shown the boy at the time of the clash with Reyneke, the relationship between father and son was not ideal. Mmusetsi was nowhere near as quick-witted as his father would have him, and on more than one occasion the little fellow found himself on the receiving end of a blow or a cuff. This conflict only worsened as time went on.[8]

The likelihood of conflict over herding was greater at Kareepoort, where unlike Vlakfontein, the hired fields were at some distance from the house and the larger animals sometimes went unattended for long periods in remote areas that were traversed by outsiders. Kas asked Phitise to design a portable structure to use as overnight accommodation at a cattle post, and using the corrugated-iron sheets that Hendrik Swanepoel had bought a few seasons earlier, they built a shack that could be readily dismantled, fitted onto the back of a wagon and taken wherever the stock were grazing. Known within the family as the *hloma o hlomolle*, the Put-up and Pull-down, was quickly pressed into service and housed first Kas and, once he was a little older, Mmusetsi, as father and son took turns keeping watch over the family's cattle.[9]

The climatic conditions at Kareepoort forced Kas to pay maximum attention to his crops, a task that was monitored with great interest by his new landlord. McDonald, impressed by his tenant's ability with the plough, was even more admiring of Kas's skill when he noted how the quick-growing Botman and Platpit mealies responded to spring rains and managed to survive the almost inevitable setback of a late summer drought.[10]

In late autumn, when pockets of cold air started gliding down into the few lower-lying depressions that the Magoditsane had scooped out of the surface of the flood plain, members of the Maine household were joined by grandmothers, aunts, nephews and nieces to help bring in the crops. Their collective efforts brought in a sizeable harvest which, once threshed by the horses and donkeys, yielded nearly eighty bags of maize. This noteworthy achievement in a drab season did not escape McDonald's attention. The harvest was barely in when he approached Kas about the possibility of entering into a form of disguised sharecropping for the next year.[11]

McDonald had more reason than most for caution in approaching black tenants, since it was said that shortly before the Maines' arrival he had been prosecuted for inter-racial sharecropping—the only such case Kas ever heard of in his career in the southwestern Transvaal.[12] Perhaps it was because the Kareepoort diggings were so visible from the main Bloemhof-Wolmaransstad road, or perhaps it was simply because bywoners and 'poor whites' resented such inter-racial ventures more and more as the Depression deepened; in any event McDonald took pains to mask his agreement with his new tenant.

McDonald proposed that he would have rights to half of Kas's labour time and equipment, each partner would have access to as much land as he could put under the plough, and each would be entitled to only that portion of the harvest that accrued to him from fields that had been worked 'independently.' According to Kas, this meant that:

> The landlord would not be prosecuted because each partner was tilling his "own" land. Those who wished to inform on the landlord would have difficulty in providing proof because the tenant worked his land separately from the landlord. The landlord, if confronted, could simply argue that the tenant tilled his own land in return for the labour he provided. Even those who wished to investigate the arrangement more surreptitiously would have little chance of proving that the landlord was engaged in "sharecropping." In fact, the very fields allocated to me stretched towards my homestead and not towards his house.[13]

By nature every bit as cautious as the cagey McDonald, Kas drew strength from knowing that Mphaka and Sebubudi both had personal experience of this type of accord and that they approved of the arrangement.

But perhaps the most attractive feature of the new contract was that it allowed Kas to graze his livestock without charge. No longer requiring a monthly cash income, he could contemplate the coming spring more calmly. But any complacency was soon destroyed when Leetwane told Kas that she was once again pregnant. The prospect of having to feed two wives and five children during a depression that each month was becoming more marked meant that there could be no let-up in the struggle to find additional sources of off-season cash income.[14]

The women of the house, accustomed to collecting and drying cow dung for domestic fuel, redoubled their efforts. By covering mile upon mile of brown stubbled veld each day Leetwane and Lebitsa somehow managed not only to keep the Maine home well supplied with fuel, but to stack, dry and sell extra dung to the locals for between one and two shillings per sack. Unlike most of the women in Rooistad, who turned to the supposedly unsavoury practice of commercial beer brewing in order to make a living, the Lepholletse sisters remained morally upright, continued to meet their 'traditional' obligations to an agricultural household in exemplary fashion and, at the same time, acquired small sums of cash that made them less dependent on the patriarch for money. This was, of course, not without appeal of its own, and Leetwane eventually used some of her hard-earned 'extra' income to buy a sewing machine.[15]

Kas was equally quick to appreciate that Kareepoort was a bigger market than that to which they were accustomed. He devoted more of his 'free' time to cobbling, and started repairing and making saddles for some of the better-off whites on and around the diggings. Unfortunately, others, including Phitise, his half-brother, Balloon, and a man called Jacob Lebone had the same idea: before long, at least four cobblers were at work in and around the villages. This competition, as well as his natural curiosity about all things mechanical, led Kas to experiment with Leetwane's sewing machine: he had just learned how to operate it when he heard of an opportunity to reach an untapped segment of the market.[16]

One of the Englishmen on the diggings had married an Afrikaner woman but as far as Kas was concerned this was a risky business that was likely to end in failure. Indeed, the marriage was not a happy one, and after a few months the wife abandoned her husband leaving him with several items for which he had no obvious use and which he resolved to sell. Amongst the bits and pieces which Kas inspected, but in which he showed little interest in order to throw the seller off guard, was a sewing machine. Fearing that he might lose a sale, the Englishman pointed to the machine in desperation, and said 'I am not a woman, don't you want to buy this?'[17] Kas got the machine for three pounds and took it home.

Free of the cultural blinkers that seemed to restrict the vision of many of the white men around him, Kas set about teaching himself how to use the sewing machine and was soon producing shirts and leather trousers cut from hides he had prepared, as well as dresses cut from the cloth that village women bought from Indian hawkers or the local trading store. It was as if the gloom of the Depression served only to sharpen his eyesight and vigilance. 'The people of Kareepoort used to say that I was always on the lookout for new opportunities, a person who lived by his wits.'[18] Even Leetwane was reduced to watching.

For all this diversified economic activity, farming remained uppermost in Kas's mind, and most of the cash that the sewing machine managed to pump out of the diggings' hidden recesses went to acquiring livestock. With money in his pocket and time on his side, Kas found the lure of the stock sales at Makwassie irresistible, and he slipped easily into a seasonal rhythm that came to characterise most of the decade which the Maines spent on the diggings. In midwinter, when man's spirit was at its lowest ebb and the livestock in poor condition, he would set off for the auctions to buy cattle at modest prices. He would fatten the animals on a special diet he devised and then, when the spirit started to rise along with the sap in the small green acacias, resold them at a profit. To this strategy, above all others, he attributed his seeing out the Depression when many of those around him failed.[19]

Working their way along the strands of various economic webs, the three adults in the Maine household extended their social linkages. Leetwane and Lebitsa, drawn together by their shared maternal responsibilities and the long hours spent out in the veld collecting dung, were shielded from the more robust social interactions that surrounded village women who brewed and sold sorghum beer. Indeed, if anything, the commercial brewers were an occupational hazard for them since, to avoid police detection, they sometimes buried four-gallon tins of beer in the veld, returning to collect the supplies later, under the pretext of being dutiful wives who were merely out collecting fuel.[20] Leetwane, who occasionally made beer for her husband's enjoyment, disapproved of such irresponsible ploys, which cast doubt on the reputations of otherwise irreproachable women. But out in the playground of poverty there was no telling who would be found wearing the mask of morality when the bell rang, and she was therefore doubly grateful for the personal support of the small band of committed Christians who shared her and Motheba's affinity for the African Methodist Episcopal Church.

Kas, whose cobbling, tailoring and herbal practice brought him into contact with more people, spent his spare time taking his dog Bles across to Boskuil, where he and Mmereki Molohlanyi would sit and drink beer, talk about farming problems, and discuss the fate of the Industrial and Commercial Workers' Union which, through the efforts of the indefatigable Jason Jingoes, continued to enjoy support around the diggings.[21] With the advent of longer days in spring and the need to prepare the yokes, oxen and ploughs, these social outings tailed off. Yet for all the difference it made, Kas might as well have stayed on at Molohlanyi's drinking beer. The drought of 1930 reached out across the midyear divide and dangled a withered claw in the face of the new season.[22]

As the temperatures climbed under cloudless skies, Kas reverted to ploughing by night, this time working on 'his' land from Monday to Wednesday and then the landlord's fields from Thursday to Saturday. Every morning for more than two months he scanned the east in the hope that a few fat cumulus strays might trundle in over the horizon and stay long enough to slake his thirst. But only in the third week of November, when the first gentle rains came, was his vigilance rewarded. Slowly, stubbornly, and with great deliberation, the drops of moisture worked their way into the dried grass and parched sand and then, in a single magical moment, catalysed it into that joyous, musty, fermenting aroma of life that only a true denizen of the highveld can appreciate. It was—and is—a smell to intoxicate, and for a few days, Kas was rejuvenated.[23]

The quicker-ripening varieties of maize and sorghum that Kas relied on sprinted their way through the sultry weeks of Christmastide and by the beginning of the new year, looked decidedly promising. But Leetwane, a few months in advance of nature, was already slowing down, as she too moved toward completing nature's oldest circuit. Careful to avoid excessive physical exertion, she took steps to make her peace with those around her, especially to ensure that none of her close kin bore her ill-will, an unpropitious circumstance known as *dikgaba*, which, if not set right, might delay the arrival of a child. But this time there was no need to smear the pregnant woman's feet with animal fat to help dispel the dikgaba, or to find a ngaka with herbs to induce labour. All went according to plan and, on 16 February 1931, Motheba simply used a razor blade to cut the umbilical cord when Leetwane was delivered of her fifth child, a girl named Nthakwana. Once the umbilical cord was secured (with a thread drawn from a woollen blanket), and the blade taken out and buried, the baby was lashed with a length of string until it produced the lusty yell that proved that all was well. Hours later, and well before it was put to the breast for the first time, the newborn infant was fed a meal of *momela*, a soft porridge of sorghum sprouts. This too was not always unproblematic: 'We fed the baby by putting food into its mouth with our hands. If the baby did not want to eat we would close its nose, and force it to eat.'[24]

Because the baby arrived on one of the days he was supposed to be at work in McDonald's fields, Kas had to make his way to the house, tell the landlord of her birth, and ask for time off. McDonald quickly acceded to his request, but, because the diggings lay more in the realm of industry than agriculture, he did not offer the customary gift of a sheep. The next morning Kas slaughtered a ewe. Something died, something lived and, before long, the shack was filled with the sounds of relatives and visitors bearing the simple tributes of hard times—a few sixpenny pieces, small bags of mealie meal, a packet of sorghum or a few maize cobs.[25]

Nthakwana thrived, and by the time summer drew to a close, the baby was firmly rooted in her mother's section of the household. Leetwane, however, was stretched by her domestic responsibilities. When the family work party assembled to bring in the harvest that autumn she was less active than usual. And the harvest itself was disappointing. Twenty bags was but a fraction of the eighty the family had reaped in its first year on the diggings.[26]

Despite this, the Maines felt duty bound to host the usual end-of-season beer drink. While Kas and Leetwane entertained a few of their closest relatives and friends in their shack, others, including Jason Jingoes, gathered at Mphaka's home, where the talk again centred on the Union and the way the Depression was savaging the economic lives of Triangle blacks. Not all these midyear celebrations arose from the customary obligations of the sharecroppers and labour tenants who formed the backbone of Kareepoort's 'hidden economy.' Elsewhere on the diggings, larger, noisier and more open-ended gatherings drew together non-farming elements, the less respectable social residue of the former industrial dispensation, and these attracted the attention of the police. Indeed, it was usually just after the harvest that Sergeants Abraham Jonker and Charl Marais called on the white farmers in the district to help them step up the raids on the 'underworld.' By focusing on the commercial brewers, the police, intentionally or unintentionally, widened the gap between the older, poorer 'industrial' inhabitants and the younger, better-off recent immigrants who made their living from livestock or arable farming. Thus Sergeant Jonker was quick to acknowledge that Kas and his wives formed part of the respectable section of Rooistad society, and he was soon on fairly good terms with the family. In the long run this was a mixed blessing.[27]

By winter 1931 even the great entrepreneur himself was finding it difficult to bring forth the smallest trickle of coins. True, his herbal practice and leather work sometimes coaxed a noisy clank out of an otherwise empty tin, but many of the locals could no longer afford to consult Kas about minor ailments, while the number of people traversing thorny farm tracks without shoes seemed to be increasing by the day.

A poor harvest meant that there was less feed to fatten up the cattle Kas had bought from the Kemp Brothers at the Makwassie livestock auctions. And the arrival of the baby meant that Leetwane and Lebitsa had less time to collect cattle dung. As if their plight was not already serious enough, the police chose that moment to launch a new round of raids in which they made the vitally important discovery that Kas's mongrel, Bles, was unlicensed, and so, on 8 September 1931, he was summoned to appear in the Periodic Criminal Court at Makwassie and fined five shillings for having failed to pay 'dog tax.'[28]

It was a miserable winter and Kas looked forward to spring with more enthusiasm than usual. When it came, he experimented with *Rooi Boesman* and *Natal Agtrye*, two hybrids that he had overheard progressive maize farmers discussing in the depot of the South-Western Transvaal Agricultural Co-operative. Fortune favoured the brave, even in a district it seldom chose to visit. The early rains not only arrived but came at such sensibly spaced intervals that even the most hardened cynics believed that the events had been orchestrated by a higher hand. For once it seemed that nature's scale was getting into balance: just as cash in the household was draining away, the moisture level in the fields was rising.[29]

With the assistance of eight-year-old Mmusetsi, Kas managed to get even more land under the plough than he had in 1930, a considerable achievement given the limited time he had. By the end of the year it was clear he would need more hands if the fields were to be kept clear of upstart, moisture-sucking weeds flaunting their preference for the cultivated company of maize and sorghum. Hoeing fields was women's work and, even though Leetwane and Lebitsa were already reeling beneath enormous domestic burdens, Kas put them under renewed pressure either to go out and do the weeding or to come up with an alternative solution. In the end, they decided to call on the services of the same young sister who had helped them at Vlakfontein when illness had reduced Leetwane to crawling about the house on all fours. But six years had passed since then: Tseleng had been through an unsuccessful traditional marriage and was a mature woman in her late twenties with a life of her own, it would be hard to persuade her to move to Rooistad.

How exactly they managed to convince her is difficult to tell, but, early in 1932, Tseleng abandoned the van Deventers' kitchen at Kommissierust and came to Kareepoort, to help in the house and fields. This makeshift arrangement seemed to work reasonably well enough on the domestic terrain, where she had to fit in with the close bonds between her two older sisters, and even better out in the fields where she developed a close relationship with her brother-in-law. Indeed, by the time the summer sun started to lose its sting, it was clear to everyone that Tseleng was carrying Kas's child.

The news of Tseleng's pregnancy was met with stony silence by the Maine wives. Leetwane, with four children of her own to cope with, thought there were already enough mouths to feed, while Lebitsa, sensitive because her marriage had produced but one daughter, saw little reason to welcome yet another child fathered by her husband. These and other more complex arguments about seniority and domestic etiquette were relayed to the three sisters' parents at Kommissierust, and when Kas approached the Lepholletses with the request to marry their pregnant daughter, the usually phlegmatic and accommodating Tshounyane pronounced himself against the idea in unequivocal terms. As Tseleng herself recalled many years later, 'My father told him that he would allow him to marry his second and even the first of his daughters, but not the third. He refused point blank.'[30]

Throughout the summer, in and around the house, there was no evidence of overt conflict amongst the women. The only sharp disagreement arose in a different quarter when Leetwane, sensing that her husband was expecting too much from the slow-witted Mmusetsi, asked that the boy be given time off from his herding duties to attend the informal school in the village and Kas would have nothing of it. What was the point of sending the lad for book learning at a time when domestic labour was already at a premium? Leetwane, who understood but resented this reasoning, blamed herself for not generating enough income of her own to ensure the boy's education, but there were limits to how much dung a breast-feeding mother could collect in a day and to how much work a seamstress could take in from an impoverished community.[31]

Mmusetsi formed part of the work party mobilised to bring in the harvest at the end of the season. Fifty bags, although short of Kas's target, was still enough to allow him to sell some of the new higher-yielding maize. For the sake of convenience he got McDonald to sell a few bags through the co-operative, using the landlord's name. As usual, the seasonal beer drinking in Kareepoort attracted attention from the Makwassie police but Kas, preoccupied with unresolved family problems, took even less notice than usual of Sergeant Jonker and his men.[32]

It was not the first time that Kas's focus on farming and domestic matters left him badly out of step with the thousands of other blacks on the diggings or elsewhere in the Transvaal. Indeed, the endless beer and pass raids of the Depression years twice forced the then inept leadership of the African National Congress to call special congresses—once in June 1931 and then again in July 1932. Even at Kareepoort Jason Jingoes drew on the protest tradition of the old Transvaal Native Congress to continue his struggle against police harassment, mobilising what was left of the once thriving I.C.U. branch at Makwassie.[33]

This new round of rural unrest unsettled the Triangle police, and after a strike by black miners at Leeufontein, Jingoes was summoned to Wolmaransstad on 9 March 1932, where he was handed a letter signed by the Minister of Justice, Oswald Pirow, that forbade him from entering any area in or around the site of active diamond diggings or from addressing gatherings of more than five people at a time for the next twelve months. But state repression held little terror for an organiser who, despite numerous court appearances, had survived the rugged I.C.U. campaigns of 1927–29 without once being imprisoned. Nor did it do much to dampen his enthusiasm for the highly politicised company he found in and around Mphaka and Motlagomang Maine's modest home at Rooistad.[34]

Within weeks Jingoes defied the Minister's orders, boarded a train at Ottosdal and set out for the diggings at Boskuil to address a meeting on a farm adjacent to Kareepoort. But when the train stopped at Makwassie, he noticed that it was joined by two plainclothes detectives whom he recognised as working for Sergeant Charl Marais, better known to the locals as Serasaka Moipeng. By the time the train pulled into the siding at Boskuil, he thought it was dark enough for him to elude his official tail. Uncertain where the farm was, he decided instead to make his way to Mphaka's home, where he was given a place to sleep. As Baefesi, the oldest son, recalled it, 'At that time the whites were against the I.C.U. and Jingoes did not have the right to move around freely. He arrived on Friday in the afternoon and slept. The following day, on Saturday morning—say at about 4:00 a.m.—he was arrested. He was arrested by a policeman from Makwassie named Marais.'[35]

In his memoirs, Jason Jingoes himself tells what happened next:

> The police took me to Wolmaransstad. Before Magistrate A. S. Dunlop again, I suffered this time. I based my defence on the fact that I was not actually at the diggings when I was arrested, and accordingly the court travelled to inspect exactly where I had been arrested. It was found that the farm was a little more than a mile from the diggings. None the less, the banning order stipulated that I should not go on or near any diggings, and I was found guilty. I was sentenced to six weeks' hard labour with no option of a fine.[36]

Serasaka always got his man. What chance did a fugitive have against a man who was said to speak SeTswana so fluently that not even the natives could detect the slightest trace of an accent? How could anyone protect himself against a policeman who had mastered the art of disguise to the point where his blackened face was often mistaken for that of a MoTswana? What sort of a man was it who took the trouble to train his horse to detect the smell of sorghum beer hidden in the veld? Even the streetwise Jason Jingoes believed he had been outwitted by a strange other-worldly man who practised *lithare*, witchcraft.

Jingoes served his prison sentence and a few months later, still undaunted, started a campaign through the columns of A.N.C. President Seme's newly launched official mouthpiece, the *African Leader*, for more broadly based black unity. But it was midwinter in the Triangle, and the signs of ever-deepening economic misery were everywhere to be seen. One day, on his way back from Bloemhof, Kas came across Khayi, an elderly MoTswana, and his sons driving a span of oxen and several dozen sheep to Wolmaransstad. He got talking to the men and discovered they were ex-sharecroppers from Thaba Nchu who had fallen foul of their landlord, had their goods summarily dumped in the road to the farm, and had spent weeks roaming the countryside looking for a place to settle. Taking pity on them, Kas directed them to Kareepoort, where they linked up with dozens of other refugees.[37]

Throughout the Depression, the diggings gave sanctuary to many honest farming folk willing to eke out a living by the sweat of their brows. But the Crocodile lives best where Eland drinks most, and other less respectable elements also went to Kareepoort hoping to survive in less taxing ways. The contest between herbivores and carnivores became ever more localised. Most of the illicit diamond buying deals (or I.D.B.'s as they were called) took place in drinking dens catering exclusively for the black workers working at the few active operations. Hard-drinking labourers who ran up bills that outstripped their meagre cash wages sometimes chose to settle their debts with the 'shebeen queens' by passing on tiny unflawed diamonds—*mulletjies*—they had managed to pick up unobserved.[38]

More sinister, and far more dangerous, were the Bull Nines, small-time gangsters from different ethnic backgrounds who specialised in confidence trickery—passing off bits of polished glass as uncut diamonds to gullible white farmers, 'foreign' storekeepers or anybody else foolish enough to show an unhealthy interest in making a quick profit. Kas, who had no deep-seated moral convictions about a trade that was cut off from legitimate black participation by racist laws, had at least three well-informed sources who kept him abreast of these doings.[39]

First, his nephew, Molefi Maine, who had left school almost as soon as he had entered it, was closely involved with the Bull Nines. When he was only fifteen, two white truck drivers from Kimberley who stopped at the Boskuil siding each Sunday morning to drop off newspapers asked him about the chances of obtaining a few uncut diamonds from the Kareepoort diggings, and a few weeks later he palmed off a 'false diamond' on them for the princely sum of thirty pounds and never again seriously contemplated a life of manual labour. In a semi-successful criminal career that spanned decades, he branched out into Illicit Diamond Buying proper, the selling of 'false gold,' and house-breaking—at which he proved to be less than talented. He eventually married Mmereki Molohlanyi's daughter and drifted around the margins of the diggings.[40]

Second, there was an impoverished MoThlaping friend from Taung who was a frequent visitor at the Maine home. This man took a liking to one of Kas's bulls, which he hoped to buy. Indeed, the man eventually asked Kas to supply him with the 'magical herbs' that would lead him to a diamond large enough to pay for the animal and the 'medicine.' Kas obliged: and, a few days later, the MoThlaping and a friend at work found a mulletjie that the MoThlaping wanted to take to Kas. The friend, fearing he would lose his share of the find if he lost sight of the diamond, instead persuaded the MoThlaping to sell the diamond to Amin Patel, a Wolmaransstad storekeeper. No sooner was this plan in place than the friend lost his nerve and told their employer about what, by then, had become the MoThlaping's find. The digger told the Makwassie police, who raided the shop and arrested Patel. Despite a successful appeal to members of the Triangle's Asian trading community to come up with bail money, the magistrate denied a request for bail and Patel was sent to prison.[41]

Third, since the Bull Nines formed part of black folklore of the diggings, during the slow-paced winter months villagers took delight in recounting tales about them, stories edged with all the ethnically coloured ambiguities that come with social banditry in a racially ordered society. An authentic Bull Nine, so it was said, nearly always disguised himself by posing as a shabbily dressed worker from diggings nearby, never descended to defrauding a black man and, once he had settled on a victim, would move to another district to cover his tracks.

True to spirit, Kas in later life told the cautionary tale of Selatile, a short, fat and exceedingly cunning Xhosa who was based in one of the many villages around Kareepoort. Selatile once went to a remote farm in the district of Delareyville, north of Schweizer-Reneke. There, having identified the true object of his attentions, he persuaded a Xhosa-speaking labour tenant named Mohlejwa that he had a genuine diamond worth eight pounds to sell, demonstrating that the gemstone could not even be crushed by pliers, which he conveniently produced.

But Mohlejwa, who had no money of his own, as Selatile knew, said that he would have to get a loan from his master which he would secure by pledging two much-loved horses as collateral. When the landlord asked the tenant why he was suddenly willing to dispose of horses he had long resisted selling, and got wind of what was afoot, he was overcome with greed and insisted that he, rather than Mohlejwa, would become the principal beneficiary of the sale. The hook was baited. Mohlejwa gave Selatile the money, and Selatile swopped the mulletjie for an identically shaped piece of glass. Mohlejwa took it back to the farmhouse, and by the time the Boer discovered what had happened, Selatile had made good his escape on a bicycle. But this tale had an uncharacteristically unhappy ending, for the irate farmer forced Mohlejwa to forfeit his horses as a penalty for his part in the fiasco.

To Kas, it was the sequel that had a moral for the Triangle's farming folk. In a hostile world, where the shadow of the vulture flickered across the surface of one's consciousness hourly, sharecroppers, labour tenants and wage labourers needed to be doubly vigilant. It was hard-baked peasant wisdom rather than an upwelling of the protestant ethic that drove Kas to work harder than ever before. The signs were everywhere to be seen: it was a struggle for survival. Leetwane, although still a committed Christian, had even stopped paying her modest quarterly contributions to the A.M.E.C.[42]

Winter brought with it a desperate problem for all three Lephollletse sisters as they tried to collect cow dung to prepare their hut floors with, use as fuel, or dry and sell. Resentful landlords were driven to distraction by the lower-than-average rainfall and lack of cash—the Englishman, known to them only as Kgwahla, was no longer willing to tolerate their presence in his fields and ordered them off the property, warning them that not a single cow pat should be touched as the dung was necessary to replenish the soil.

With men and women reduced to contesting the ownership of cow pats, the Maine women took to rising well before sunrise and using the cover of darkness to make two or three forays into the adjoining fields to collect dung which they stacked 'into dumps bigger than houses.' Poverty marked morality, just as surely as tea stained freshly laundered linen, and not surprisingly, Leetwane was later heard to say, 'Even if I found you stealing something from a white man's farm I would not bother you. That was simply another way of making ends meet.'[43]

With three rather than two women 'stealing' from Kgwahla's fields, the sisters did marginally better than making ends meet. By the end of the 1932 winter, which must have been almost literally back breaking, Leetwane and Lebitsa had saved just enough cash to buy two black piglets from a local farmer. Undiscriminating scavengers, the pigs snuffled and snorted their way through the rubbish tips of Rooistad in a depression year, and thrived. For the first time, the Maine wives had overcome the economic isolation of the farms and exploited the limited market opportunities to become livestock owners in their own right. This was an extraordinary achievement, and it was etched on their consciousness. But in a milieu where men rather than women were owners of livestock, the triumph was not unproblematic.[44]

The patriarch saw the approach of the new season as having more to do with planting than with pigs, and he was more interested in making sure that his agricultural equipment was ready when the rains came. But spring did not start off well. Before Kas even thought of getting out into the fields, McDonald took him aside to tell him that he was abandoning Kareepoort and going to hire a farm elsewhere in the Makwassie district. He gave no explicit reason for this unexpected move, but Kas saw it as a by-product of growing white hostility to any form of inter-racial sharecropping; the state was intervening to bolster the position of thousands of 'poor whites,' some of the worst hit being in the Triangle.[45]

But if McDonald had cold feet about inter-racial sharecropping at Kareepoort, he considered the proposed move to Makwassie as a retreat rather than a surrender because he made repeated efforts to persuade Kas to join him on the new property. Clearly, it was the political visibility of the diggings, their proximity to so many 'poor whites,' and his previous conviction rather than any misgivings about the institution itself that motivated McDonald's move. For Kas, on the other hand, the offer came at the wrong time. Already in perilous circumstances, he was unwilling to move to an unknown property just when he was also having personal difficulties that required delicate handling. The whole business left him feeling vulnerable.

As a precautionary measure, Kas asked McDonald to give him a letter of reference that he could pass on to whomever the New Transvaal Gold Farms Company inflicted on him as a landlord. 'He wrote a most helpful testimonial letter to the one who was to take over the farm. He said that the new landlord should keep me on as a tenant, that I was trustworthy, not a thief, that I did nobody any harm, that I was just a good kaffir.' Without the slightest trace of anger or irony, he concluded: 'He wrote a fine letter.' But this could hardly dispel his fears. After all, no landlord ever appeared bearing letters of recommendation from former tenants.[46]

It so happened that when the landlord arrived he brought with him a testimonial that the tenants found far more reassuring than any written document, an amiable MoTshweneng voorman named Johnson Kgosiemang. Kgosiemang acted as intermediary between the landlord—whom they called Sekhofa, or The Owl, because of his bushy eyebrows—and the tenants. Despite his unflattering nickname, Charles Gibson Smith soon won the trust of most of his tenants, and Kas found him 'rich in heart although poor in possessions,' for Sekhofa let him continue with much the same sort of contractual arrangement he had enjoyed under McDonald.[47]

The contract settled to his satisfaction, Kas waited for the rains. But, as seemed to happen so often during the early 1930s, the few clouds gave the plains a wide berth. Motheba had seen the same thing during her time at Kommissierust. Instead of Mother Africa sending soothing bands of tropical moisture slipping down the subcontinent's parched throat, Old Man Kalahari suddenly sat up and spat out a blob of coarse red spittle that came spinning in from the west. By the end of September the women moved inside, along with the dust that crept in under the door, to spend more time with Tseleng. On 5 October 1932, she gave birth to a daughter. The Maines left it to the child's maternal grandmother to come up with a name. Perhaps it was the struggle of the broom against the sand that spring, or perhaps it was the Lepholletses' way of showing their disapproval, but Tshaletseng chose to call the baby Matlakala—'sweepings'—or 'refuse.'[48]

With the basic tension among the families still unresolved, it was decided to hold a *kgotla*, or domestic court. The families agreed that Kas should be made to hand over several cattle to the Lepholletses by way of a fine for his sexual indiscretions. Furthermore, it was decided that the baby would become the responsibility of the Maines, and that she would be reared by Lebitsa along with her own daughter, Thakane, and Bodule. This went some way to restoring the honour of the Lepholletses, and strengthened the symmetry in the household with Leetwane and Lebitsa each assuming responsibility for three of the patriarch's children.[49]

For Kas the fine, along with the failure of the spring rains, meant that the new season started out on an unpromising note. But this time nature was intent on going beyond her usual petulant game of delaying the arrival of the rains and stubbornly refused to allow them to put in an appearance at all. On the few occasions when a passing cloud did release a few drops over the desiccated Triangle, it was not enough even to settle the dust which continued to pile up at an alarming rate. Sand sprawled all over the house, and it seemed that every room, every nook and cranny, was polluted by its presence. And no sooner did you set foot outside the front door than there it was, keeping pace with you, step by step, always with the same disgusting habits—slipping into your shoes, creeping up your sleeves, clogging your nostrils, powdering your hair until you were repulsed by the sight and touch of your own body.

Totally undiscriminating, not even the livestock escaped. As 'spring' crumpled into summer, sand, sometimes six and ten inches thick, slowly rose to cover first the roasted veld and then even the small bushes and trees. Pigs, goats, sheep, horses and cattle, were forced to stretch their necks in strangely contorted postures and to stand on their hind legs as they vied to find wind-dried greenery. It was the Red Dust, *Lerole le Lefubedu*, all over again. The combination of drought and an increasingly maize-mad assault on the vulnerable local ecology ensured that the Kalahari would burst out of its desert confines and push back its eastern perimeter.

The Maine women could not find, let alone collect, cow dung. Leetwane's daughter Morwesi, then all of eight years old and already doing a full day's work, watched as her mother and aunt took instead to walking for miles searching for tumbleweed ensnared in the barbed-wire fences that lined the Bloemhof-Wolmaransstad road. Plaited together, the stems of the one plant that nature had equipped to defeat both wind and sand yielded a last remaining source of natural fuel for cooking.[50]

For members of an extended family clustered on a single property, the drought gave rise to other difficulties and grazing was the most pressing problem of all. Early in the summer of 1933 economically 'strong men' from the northern Orange Free State started moving their stock across the Vaal and into the southwestern Transvaal to what grazing land they could find. The Triangle's black sharecroppers, already hard pushed, suddenly found themselves having to compete for the limited 'extra' grazing resources with white farmers drawn from the more affluent maize belt of the neighbouring province.[51]

The Maines were not without a strategy. From experience gained during the droughts of 1913 and 1926, they knew that part of the solution lay in the temporary dispersal of the extended family. Some would go with a few of the larger livestock and take up work or residence on other farms, spreading the density of the herd. In 'dry cycles' both the family and its cattle tended to move outward, reconvening once the rainfall again approached normalcy and the carrying capacity of the land was restored. This time, though, there was less room for manoeuvre than usual.[52]

Mphaka, for example, had problems with his debtors. These included a local branch of the A.M.E.C. that failed to honour its commitment when he erected a new building for its members. To add to his woes, such construction work as there was kept him away from his livestock, and his oldest son, Baefesi, was away at work on the Boskuil diggings, so there was nobody to take care of his cattle, many of which died.[53]

As part of the thinning-out process that took place during such periods of economic hardship, Mphaka's second wife, Motlagomang, went to stay with her parents at Maokanashomo, the farm Schoonsig, where Kas had bought his two-share Canadian Wonder several seasons earlier. By midsummer the situation was so desperate that it was decided that Motheba would take one of Mphaka's younger sons and four cattle, and go to Prairieflower; at Prairieflower (which the locals rendered tongue-in-cheek as Perdevlei), she and the boy would link up with Sebubudi, who being a younger man had only one wife and two children to see through the drought.

The journey to Prairieflower—the sardonic nomenclature of a Canadian surveyor in a bygone age—like so much of that summer, was awful. Gigantic red sandstorms bowled in across the plain, first enveloping everything in darkness and then lashing, scouring and blasting every inch of exposed flesh or hide. Even the first sturdy little Fords, with their artificial eyes and big windscreen wipers, were forced to stop scuttling around at midday and peer shortsightedly into the gloom, as sand drifts threatened to eradicate every visible trace of a road or roadside fence.[54]

As in 1925, not all of the Maines' livestock escaped the wrath of the storms. Two of Sellwane's bohadi cattle, already badly weakened by the lack of grazing, were suffocated by the thick mixture of sand and mucus that oozed down their nasal passages and clogged their lungs. But it was the suffering of people at Perdevlei rather than animals that most distressed Motheba. A distant relative recounted how up north, at Madibogo, two BaRolong herdboys had sought refuge in a donga only to be suffocated beneath a blanket of fine red sand.[55]

Although there were no deaths at Rooistad, the picture was scarcely more cheering there. Earlier in the season Smith had asked his tenants to reduce the size of their herds—well-meaning advice that flew directly in the face of peasant wisdom. As Kas knew, the rule was *never* reduce or slaughter livestock other than on your own terms and in your own time. Who in his right mind disposed of his banknotes because his purse had shrunk? Culling was the last, not the first, resort of a livestock farmer.

Instead, taking his cue from the invading Free Staters, Kas went to a neighbouring property, Blaauboschkuil, to talk to Mmereki Molohlanyi's landlord. Boskuil, almost wholly owned by the African & European Investment Company, had for several years been let to Piet Ferreira, who by all accounts was a good man. Unlike Kareepoort, the farm had escaped the attention of diamond diggers and that, together with other quirks of nature, had left it with more grazing than most nearby properties.[56]

It seemed like a most unpromising meeting. Kas, without a penny to his name, needed grazing for fourteen oxen, six cows and twelve donkeys—and that at a time when Ferreira had already hired out most of his 'extra' land to the 'big men' from across the Vaal, who could more easily afford his monthly charge of a shilling per beast. But the Boer took a liking to one of Kas's heifers, and he agreed to take in the rest of the animals for the season in exchange for that one beast. This had all the makings of a very good deal. If Mmusetsi were moved across and housed in the 'Put-up and Pull-down,' Kas could stay on at Kareepoort without endangering his agreement with Sekhofa.

When he returned the following morning, however, Kas found that the Boer had had second thoughts, and was coming up with a new proposition: if Kas would supervise the use of his own donkeys to drive the mechanical pump at Boskuil—assuring all the animals on the property drinking water—then Ferreira would exempt Maine cattle from all grazing levies for the rest of the season. While this was less satisfactory than the earlier agreement, Kas could see no other way out of his difficulties. Somehow, he would have to serve two masters. Over the next few months he moved constantly between Kareepoort and Boskuil in one of the worst summers he had ever experienced.[57]

Splitting what was left of the family herd between the properties was not enough to keep the cattle in condition. Well before midsummer most of them were in appalling shape, and Mphaka, in particular, was losing oxen at an alarming rate though both he and Phitise had hired additional grazing for their animals. Kas lost a few horses, but with a heart and mind closer to livestock farming than those of his older brothers, he succeeded in devising a special diet that helped his animals to survive.[58]

> If your animals are starving you can grind maize, mix it with sorghum and sunflower seed and feed them. Even if you have many cattle they will live and you can use a small amount of this mixture to ensure the survival of a large number of animals. For instance, a mere fifty cups of the mixture is enough to maintain a similar number of cattle. Feed it to them, then give them water and they will not die of hunger. But you must remember to add salt to the mixture. After eating it they will drink water and survive until the following day when you can once again feed it to them.[59]

As in the past, a combination of skill and determination won Kas the respect of all who knew him.

Other tenants, some at least as skilled at warding off the challenges of the drought, discovered that the Depression was flushing older, more familiar predators into the open. The unfortunate Khayi family, for example, lost no less than eight draught animals on a single day because, unlike the Maines, they had no sons to man a cattle post. Convinced of foul play, the older brother consulted a ngaka, who informed him that the animals had indeed been stolen by a Kareepoort *maqaqa*, or 'poor white' by the name of Geldenhuys, who in turn had sold the oxen to a Boer who was keeping them hidden on a koppie near Makwassie. The brother went straight to Makwassie, where he accused the Boer of stealing his family's livestock. The outraged farmer sued him for defamation of character and was awarded the sum of fifty pounds. This blow crippled the sharecropping family, the complainant left the diggings, went back across the Vaal and took up work on the Free State farms. The younger brother, who had been at least as convinced by the ngaka's advice as his older brother, lingered and made a point of attending all the stock sales, thinking that sooner or later the farmer would try to sell the Khayi cattle. One day a few seasons later, he recognised a black ox as belonging to his family and reported it to the police; interrogating the Boer they found that the animal had indeed been bought from Geldenhuys.[60]

In the course of his herbal practice, Kas saw and heard of other incidents that revealed the widespread hunger, hardship and humiliation of those clustered on the diggings. Some, like the son of RaMothlabi, came to him with unusual requests, like wanting a potion to prevent hungry people from stealing his food. Kas, ever cautious, told him that although no such herb existed he should get a dog which, with *setlhare* administered to it, would make certain that no strangers entered his home.[61]

White farmers on the properties closest to Kareepoort, men such as Gert Verster and Frans Joubert, would have understood the Mothlabis' problem, even though they had to resort to a different, altogether more dramatic solution. At night dozens of hungry blacks would slip out of the diggings and invade their already bare mealie fields to steal maize cobs—usually to eat themselves but sometimes, more daringly, to sell to others in the village the next morning. The hard-pressed Boers took to getting up at night to discharge their shotguns over their fields, giving intruders the impression that the maize was always under guard.[62]

As Kas well knew, not all the victims of the bitter economic hardship were black, and it was not always the bite of a dog or the sting of birdshot that caused the greatest pain. At Koppie Alleen poverty drove several of Hendrik Swanepoel's younger brothers off the land and into one of the state's many dam-construction programmes for 'poor whites.' Even more awkward to manage was a chance encounter Kas had on the Kareepoort diggings with a badly sunburnt and deeply impoverished failed trader named Norman Anderson who, along with William Hambly, had once owned the general dealer's store at Hessie. 'He did not even want to talk to us. He was embarrassed to talk to people who had known him.' Poverty cut deep but Kas thought that part of Anderson's downfall was due to his having married a Boer woman: it was well known that no Afrikaner, let alone the daughter of a Boer, had the skills necessary to run a store.[63]

A disastrous harvest of eight bags meant that, by mid-1933, the position of the Maine household was at least as precarious as that in most of the other shacks huddled along the scribbled course of a bone-dry Magoditsane. Not even cobbling, saddle-repair work, or the sale of cow dung could help see them through the winter, since nobody had money. Sharecroppers and labour tenants, usually amongst the more privileged elements in the villages, suffered along with the few remaining workers and the ever-growing unemployed. 'There was simply no food. Imagine a three-year-old heifer selling for three buckets-full of mealies!'[64]

There were new, tougher choices to be made. 'I was starving, my cattle were starving and my children were starving.' Kas picked up the few coins he had left, mounted Verloreskaap, and went out to 'buy half a bag of grain for the children.' Even then, at five minutes to midnight, the patriarch and the livestock farmer in him vied for supremacy. 'The horse and the children ate from the same half-bag. I fed it some grain and tethered it at night. Early in the morning Mmusetsi and I awoke to ride the horse to [Ferreira's at Boskuil] to tend the cattle.' There, using his special drought mixture, he ensured that he would not lose a single animal. (Some idea of the magnitude of the drought and the effect it had on even the best capitalised farmers in the district can be gauged from the returns kept by the Bloemhof branch of the Union Creamery. The end of the 1931–32 season had seen an output of nearly 340,000 pounds of butter, by 1932–33 this figure had fallen to 290,000, and 1933–34 saw it dwindle to a meagre 183,000.)[65] Kas's skills were beyond question, but it was also evident that the family paid a heavy price for being presided over by a fiercely committed livestock farmer.[66]

Incredibly, the advent of spring somehow still injected optimism into the mainstream of demoralised village life. Newcomers just off the land, including labour tenants and aspirant sharecroppers, staggered onto the diggings where Charles Smith pushed them into ever more demanding time-sharing arrangements. As Kas later saw it, 'Sekhofa became shrewd.'[67] Late arrivals were contracted to spend as much as four days a week working Smith's fields. Yet, despite this, 'The Owl's' property retained its attraction for blacks, not least because Sekhofa taught his likeable foreman, Johnson Kgosiemang, how to issue his tenants with the highly prized 'special passes' that let them move with relative ease through a district plagued by policemen. Like most tenants, Kas put this illegal facility to good use, although, when time was pressing, he forged passes of his own. That the police accepted his and Johnson Kgosiemang's handiwork as readily as those of 'The Owl's' showed that in the Triangle not much in spelling or calligraphy distinguished the work of a white landlord from that of a black tenant, and that in most cases it was left to a semi-literate black policeman rather than a white constable to enforce the hated pass laws.[68]

Further evidence of 'The Owl's' flexibility was forthcoming when he approached Kas with the offer of a special deal that reduced the number of days per week that Kas would have to spend in the landlord's fields. Smith placed him in sole charge of the best field on the property—a tribute to his ability as a ploughman but one which also signalled Smith's disillusionment with collective work parties that had failed to do good work. By gearing contracts to the abilities of labour tenants and sharecroppers, Smith showed that he was owl-like not only in looks.[69]

Kas, who was growing accustomed to the leeway he was given, accepted 'The Owl's' offer with alacrity, and reverted to ploughing his fields by night. Mercifully, the spring rains materialised and, in a conspicuous show of compassion, the southwestern Transvaal enjoyed better-than-average rainfall through most of the summer.

With a break in the most devastating drought of the twentieth century, 1934 was a year of economic renewal. Although a few swarms of locusts put in an appearance after the drought, they did not wreak much damage, and relieved grain farmers saw the demand for maize push up the price of mealies to levels unknown since 1928. Grazing, too, recovered so well and so rapidly that the Union Creamery anticipated a record output of butter production of close on 500,000 pounds. Livestock producers too had their moment of glory as the price of trek oxen rose to £6 per animal and good slaughter stock fetched as much as £12-10s per head. There was even something for the sorely tested digger: the volume and value of alluvial diamonds recovered from the low point in the industry in 1931.[70]

With richer red blood once again coursing freely through the district's economic arteries, the villages and households along the Magoditsane perked up. While the older Maine children rediscovered the joy of catching catfish and yellowfish in the Bamboesspruit during the rare moments when they were not working, Kas and the women feasted their eyes on the slowly ripening crops of maize and sorghum.[71] The landlord's heavy investment in grain farming meant, however, that grazing was still at a premium around Rooistad, and Kas continued to hire most of his animals' needs from Piet Ferreira at the going rate. The need for a steady flow of cash to pay for all this spurred a new round of tailoring, saddle repairing and cobbling, activities which also benefited from the upswing in the economy.

But it was the revitalised livestock market that attracted most of Kas's attention, and he seldom missed the sales at Makwassie, where he was on good terms with the auctioneers. Across the way at Boskuil, he saw that both Piet Ferreira and his friend Molohlanyi were investing in sheep, for wool prices too were staging a vigorous and sustained recovery. Livestock, always so important in his life, enjoyed a new priority. It was as if, having successfully nursed his animals through the severest test of all, he was determined to reward them by improving the quantity and quality of his herd. The self-confessed 'man of the plough' had eyes only for cattle.[72]

Yet as summer grew to a close Kas, along with others on the farm, was forced to divert his gaze to the fields. The 1934 harvest was greeted with great enthusiasm all over the diggings and celebrated with appropriate gusto not only in the labour tenants' shacks but also on the premises of the Rooistad commercial beer brewers. For the police at Makwassie, this resurgence of drinking heralded a welcome return to the familiar routine of beer raids, and the 'shebeen queens' responded to the challenge by buying 'medicine' from the local ngakas, including Kas, so as to ward off the undesired attention of the much-feared Serasaka and his colleagues.[73]

Eager to repay friends and neighbours who had helped to bring in the harvest, Kas called on Leetwane and Lebitsa to prepare a larger quantity of beer than was usual and store it in the nearest and most conveniently situated of the Maine huts. Preparations resumed the following morning but, round about midday, Kas heard a neighbour raise the alarm which the commercial brewers and their lookouts used to signal the arrival of the police. Normally this would have given him little cause for concern, since most of the members of the Makwassie police knew of his standing as a farmer and as one of 'The Owl's' most respected labour tenants. But the extra beer would arouse unfounded suspicions, and so he hastily prepared a magical potion, applied it to a short stick and stuck it into the thatch roof, pointing toward the advancing men.

The police, who had left Makwassie several hours earlier, were hot and tired when they reached Rooistad and showed little desire to embark on the customary door-to-door search. Instead, they rode up to the Maine homestead, where Kas stepped out to meet two whites and two blacks who were under the command of Sergeant Abraham Jonker rather than Serasaka—'My Enemy.' He immediately told them that he had a lot of beer on the premises, destined for the family's harvest feast. This news was of no interest to Jonker, who merely asked him to get someone to look after their horses and show them where they could enjoy a few hours of sleep.

Later, with the sting out of the sun, Kas woke the men and saw to it that they were given a meal while Mmusetsi went off to collect and saddle their horses. Refreshed, and in full view of the astounded neighbours, Sergeant Jonker handed Kas ten shillings for his trouble before his men remounted and set off. As usual, dozens of 'shebeen queens' were arrested during the raid, handcuffed together, and made to walk ahead of the horses for the ten miles back to Makwassie. The rigours of that march to the police station, rather than the subsequent court appearance and fine, terrified the women. During precisely such ordeals, reminiscent of those the Transvaal Native Congress had complained of so bitterly in the 1920s, Serasaka Moipeng's horse had earned the awful name *'Trap die Kaffer'*—'Trample the Kaffir.'[74]

The sequel to this particular raid was as bitter for the Maines as it was unexpected. Friends who had not been invited to the harvest feast, neighbours who had *seen* money exchange hands during the police operations, and commercial brewers who had used Kas's protective herbs in vain, all had good reason to ask why it was that, in a house rumoured to have been filled with beer, not a single Maine woman had been arrested. The answer, as Kas saw only too clearly in retrospect, was blindingly obvious: 'People said that I was an *impimpi*—an informer.'[75]

Pragmatism and conflict avoidance—social strategies that worked well when dealing with paternalistic landlords on white farms—were obviously differently charged when it came to dealing with the uniformed agents of a repressive state in a turbulent, urbanised community. Perhaps other, less noble feelings also fed the villagers' resentment; the point was soon reached where, briefly, Kas feared for his life. A man who had brought his livestock through the terrible drought without losing a single beast when everyone around him was losing animals left and right must have enjoyed some unfair advantage! What sort of a household was it where the women somehow avoided the hazards of commercial beer brewing merely by collecting and selling *disu*? How had the family managed to survive, when all around them were reduced to unemployment, hunger and destitution? The Maines' conspicuous success seemed almost other-worldly, wizard-like.

Kas perceived this hostility to be most marked amongst the Xhosa- and Zulu-speakers or, as he preferred to categorise them, 'Ndebele' inhabitants of the village. Still, no harm came to him or his family. His strongest critics, like Bles Mpofu, while deeply resentful of his actions were also wary of so powerful a ngaka and did not act rashly. Kas weathered the storm, keeping a low profile and working even harder than usual to raise off-season cash. But here, too, he seemed to be dogged by misfortune. This was especially difficult since some of the troubles could be traced back to the cattle-fever that had beset him ever since the onset of the great drought.

Throughout that winter Kas set aside small sums of cash so that he could make one large-scale purchase of new heifers at the end of the season. The ideal opportunity arose when his favoured auctioneers, Kemp Brothers, announced a Thursday sale at Boskuil. On the appointed day he set off on foot, intent on buying at least a dozen animals, which he hoped to drive back across the track to Ferreira's farm.

At the rail siding his trained eye quickly picked out the animals he thought had the most potential; using an old trick of the trade he tracked down the owner well before the beasts were driven into the ring. Somewhere among the crudely constructed cattle pens, and without the intercession of the auctioneer, the two men agreed on a price for the beasts. The owner then went back to the unsuspecting agent and placed such a high 'reserved price' on his heifers, that nobody else showed much interest in the animals. When the auction was over, he let Kas have the animals at the agreed-on price, thereby cheating the 'middleman' of a commission.[76]

All worked to plan until the moment arrived to pay, when to his dismay, Kas discovered that his purse had come adrift from the belt to which it had been attached by two leather loops. A smaller supplementary fund kept in the safety of a trouser pocket was fortunately still in place, and from this source he managed to conjure up the money for three animals and save a little face. Embarrassment aside, the fact remained that the value of six or more months of hard work now lay in a leather purse at an unknown point on the dusty path between Rooistad and Boskuil. He drove his calves back in fading light at a pace that any responsible herder would have objected to.

Back home he mobilised Morwesi, Thakane, and four nieces from Mphaka and Phitise's households. Under what was left of the sun's obliquely angled rays, he made the platoon trace and retrace his steps between the village and the cattle pens at the rail siding. When poor light forced a halt to their efforts, the girls scurried off into the gathering gloom, and he was left on his own. Then, as the cold silently scaled the banks of the Magoditsane for its nightly rendezvous with darkness, one last glimmer of light illuminated what they had somehow all missed. Suddenly, it was clear to him. 'The ancestral spirits had turned against me.' Amidst the hardship of the drought, the joy of the harvest and his pursuit of cattle, he had neglected the shades. At dawn he made his peace with the Maines who had gone before, by slaughtering a beast in the manner prescribed by tradition.[77]

The spirits, long neglected, seemed propitiated. Shortly after the ceremony, and with Tseleng having left the diggings, Leetwane let it be known that she was once again pregnant—her sixth confinement in fifteen years of marriage. Kas welcomed the prospect of a son who could supplement the increasingly valuable efforts of Mmusetsi and Bodule in the fields, but he was more ambivalent about another development around his first wife.

The little 'disu' piglet that Leetwane had acquired at the height of the Depression had developed into a fine hog, and in yet another flash of entrepreneurial acumen, she had traded it for a promising heifer, named 'Witties.'[78] Now a wife who owned a pig was commonplace, but a woman who laid claim to a cow and its offspring was unusual. But, because the Kareepoort air was still thick with social tension in the aftermath of the beer raid, Kas chose not to make a public issue about ownership of the animal and instead simply integrated it into the family herd.

The wretched heifer was slow to melt into the anonymity of the herd, however, and he soon noticed that Witties had been successfully covered and was already carrying its first calf.

Throughout the blisteringly hot, dry months that followed, the new cow continued to thrive, although several of the older cattle were affected by some of the more common illnesses such as *stywesiekte* or *blaaspens*, wonderfully evocative Afrikaans names meaning stiff-sickness and bloated belly.[79] The animal seemed intent on distinction and because the gestation period for a cow is remarkably similar to that of a woman, the unsettled father watched in silence as Witties and Leetwane set out on the same course, trundled their respective loads through autumn and turned the corner into the safety of winter. Fate seemed as keen to underscore the link between the heifer and its owner as Kas was to have the bond dissolved and the animal absorbed into the family herd.

On 16 May 1935, Leetwane gave the house of Maine its third son, Mosala, and a few hours later Witties, too, delivered her offspring. Told of this happy coincidence, Leetwane seized the moment and regained the initiative in the undeclared war over the ownership of the animal. On an occasion when her husband could scarcely afford to be seen to be acting ungenerously, she made a public display of making a 'gift' of the calf to the baby. For the time being, the cow, no less than the baby, was indisputably hers. But, as Kas knew only too well, a patriarch could afford to be patient, such matters were seldom settled in a day. Time would deliver the boy and the calf into his hands just as surely as it caused a ripe plum to fall to the ground. For the moment, his concerns lay elsewhere.[80]

Since a reasonable spring had been followed by poor summer rains, most of the grazing still had to be found across the way at Boskuil. The monthly fees to Ferreira were a constant drain, and pushed him into more and more tailoring, saddle making and shoe repairing; in the aftermath of the beer raid the herbal practice also slumped. In addition, the midsummer heat burned most of the grain, and the maize and sorghum yields were likely to be down. Everything pointed to yet another hard winter along the banks of the shrivelled Magoditsane. Seasonal beer drinking declined among labour tenants and sharecroppers, while the stronger rhythm of commercial beer brewing reasserted itself. Even there, though, the market was increasingly competitive. Self-consciously 'respectable' women were forced into the few ways left to make a living in a backwater that defied the categories of 'town' and 'countryside' alike. As one of the female residents later put it, 'we grew up under township conditions even though we were on a farm.'[81]

Social distinctions predicated on notions of respectability wore paper thin and generated ironies of their own. Leetwane and Lebitsa managed to hold themselves above the trade in beer by stooping to collect cow dung, while across the way, the younger of Mphaka's two wives joined a circle of brewers that included several women who, like Motlagomang herself, were regular church-goers. Tapping into the uninterrupted stream of malt that flowed throughout the 1930s from two factories in Bloemhof to the stores of proprietors like M. W. Smith at Rooistad, or Solly Meyer in Boskuil, these new entrants to the trade at first brewed their beer alongside that of the established commercial brewers and then hid it in the veld. Later, sorely pressed by the actions of Serasaka Moipeng and his men, they organised a series of 'rotating parties' at which 'stew, cakes and beer' were sold, and gave the proceeds to their menfolk who, in turn, persuaded Smith to intervene with the Wolmaransstad magistrate to get permits that would allow them to brew beer for 'domestic consumption' over weekends.[82]

Out of convenience as much as conviction, members of the Maine household had good reason to steer clear of this trade in beer. After the unfortunate experience the year before and the hostility the raid generated, it would have been adding insult to injury if either Leetwane or Lebitsa suddenly took to commercial brewing. Kas resorted to safer ways of raising winter cash, and as had happened on at least one other occasion, he chose to appropriate an economic niche prised open by his wives. Leaving the boys to look after the livestock, he and Thakane, who had recently rejoined the family, crossed the Vaal via the newly opened bridge and went to the northern Free State, where good rains had ensured an above average harvest. There they purchased a wagon-load of dried mealie cobs from a white farmer for eight shillings. Back on the cold and fuel-starved diggings, the cobs fetched two shillings a bag, to yield a truly handsome profit.[83]

As usual, most of Kas's cash earnings derived from shoe repairing, a trade that showed improvement after the Depression but also signs of serious overtrading. Sensitivity to the increasing competition, and his desire to avoid conflict with the established leather workers in and around the villages, prompted Kas to have some exploratory discussions with the senior cobbler on the diggings, 'Jakob die Skoenmaker.'[84]

Jacob Lebone ('lamp,' 'light,' or 'candle' in SeSotho), a man who in his dealings with whites referred to himself either as 'Jacob Candle' or 'Kers' Lebone (Afrikaans for 'candle'), came from a BaSotho family that had a good deal in common with the Maines. In the mid-1880s they had been active farmers in the Masianokeng district of Basutoland and leading lights in the Paris Evangelical Mission before moving to the Thaba Nchu district of the Orange Free State at the turn of the century. From there they had gone to white farms in the Excelsior district of the eastern Orange Free State and thence to the far northwestern corner of the province, where they became prominent sharecroppers under the Greylings of Kommandodrif. At some point in the early 1920s, a few descendants made the short journey across the river into the southwestern Transvaal where, after a further spell of farming, Jacob Lebone had gone into full-time shoe repairing on Solly Meyer's portion of Boskuil.[85]

During their half-century-long descent from the Malutis across the plains of the African subcontinent, these BaSotho emigrants paused long enough for young Jacob to acquire a few basic skills in reading and writing. Anything he lacked by way of formal schooling was later supplemented by a spell as a migrant labourer in the more challenging environment of Johannesburg after World War I. Employed in a bakery by day, Jacob learned how to repair shoes by night and took a lively interest in the politically turbulent world around him. Back in the Triangle as a sharecropper, Lebone became a staunch supporter of the I.C.U., and long after the departure of Jason Jingoes for Basutoland in the mid-1930s, continued to cultivate the political interests of black and 'coloured' hawkers, cobblers, priests and teachers on the diggings in and around Kareepoort.[86]

In keeping with the tradition of his profession, Jacob the Shoemaker was something of a village intellectual and community leader, and although Kas did not share his appetite for direct political engagement, the two soon established a sound working relationship. By avoiding ideological prescription and concentrating on the practical advantages of co-operating in the trade, Lebone convinced Kas of the folly of unfair price-cutting by local shoemakers such as Kas's brother Phitise and his half-brother Balloon. The two of them agreed on a price determined largely by the cost of the tacks, glue and leather soles they bought from the local stores. In return for getting the Maine brothers to adhere to this new tariff, Lebone agreed to teach Kas some of the finer arts of shoe repairing.[87]

Working from this new economic base line, the winter of 1935 yielded an improved cash return, which was soon taken up on some domestic projects. The arrival of baby Mosala had been followed by the arrival at Rooistad of Leetwane's parents and Thakane. Although a family regrouping was to be expected after the worst of the drought, Kas was not easily reconciled to the presence of the Lepholletses. It was not just that his parents-in-law had refused him permission to marry Tseleng, but that Leetwane's father managed to sit through an entire day without once offering to help out. Since these three extra mouths had to be housed as well as fed, Kas had little choice but to start looking around for more building materials.[88] In mid-June he bought some wood and a few sheets of corrugated iron from a hard-up digger, the only thing making the business palatable being the two scrawny cattle he got thrown into a deal that cost him six pounds. Phitise helped him build a new single-roomed 'house,' and eight weeks later, he drove to Patel's store at Makwassie and spent four more pounds on lime-wash so that he could complete the job to his satisfaction.[89]

The popular General Merchants & Produce Dealers Store situated opposite Neser & Campbell's auction pens was as well known to Kas as it was to most black farmers in the district. Indeed, like many others living off the land, he had to make more than one off-season call there. The proprietor, G. M. H. Patel, like his erstwhile partner, E. S. Seedat, and several other Asian traders in the Triangle, had come to the Transvaal diamond fields as an independent 'Passenger Indian' from Kohlvad, in Gujarat, at some point between 1910 and 1920. After a short spell in postwar Bloemhof, he had shifted his business to the more vibrant farming centre of Makwassie where, with his extended family, he had built up an interest in several smaller shops in the surrounding districts. By the mid-1920s Patel & Co. boasted the largest and most successful store in Makwassie and at one stage employed no less than nine full-time assistants to help customers buying anything from pilchards to ploughs.[90]

The Depression took a toll on his business, but by the mid-1930s Goolam Patel had staged a significant recovery, and was a well-known and much-respected figure in the small Asian trading enclaves to be found along the crescent that linked the farming centres of Potchefstroom in the east to Christiana in the west. Part of his social success was due to the major role he had played in establishing a league for cricket-loving exiles from the British Raj. Yet despite his spats, horse-drawn carriage and new Dodge, Patel, like Daniel Gabbe, his anglicised predecessor in Bloemhof, was part of a generation of emigrants from societies dominated by peasant economies. Whether from Gujarat or Gdansk, trader or *smous*, Muslim, Hindu or Jew, these darker-skinned middlemen had sympathy for Africa's black farm workers, labour tenants and sharecroppers—people clinging to the land that belonged to the political aristocracy of the *platteland* or plains, the lighter-skinned, Christian, Afrikaner landlords.

Kas had witnessed some of this sensitivity when he saw how Daniel Gabbe had made sure that all of his clients' horses, regardless of the owners' colour, were given tethering facilities, water and fodder when they visited his shop. The tall bespectacled Jew also showed a real understanding of the needs of sharecroppers when he extended them seasonal credit, something which the Land Bank did for white farmers but which no bank would do for a black farmer. At Makwassie, many of these same empathetic qualities were to be found in the corpulent, fez-crowned figure of Patel, who had the added advantage of being willing to barter hides, skins, wool, chickens or grain for the small luxuries of life such as sugar, coffee, soap or candles.[91]

The more delicate shades of economic life in the South African countryside often escaped the notice of Afrikaner landlords, who had difficulty in understanding the underlying affinity the *Koelie* or *Joodse winkel* ('Coolie' or 'Jew' stores) had with *plaas kaffers* (farm kaffirs). But those who saw only in black and white were fated to seeing only half the picture. Blessed with a finer vision Kas, like most other sharecroppers, had no hesitation in turning to Goolam Patel when his household suddenly had three more mouths to fill that winter. In the last week of a bitterly cold July he emptied his cobbling tin of coins, took five pounds and drove to Makwassie to buy thirteen bags of sorghum. A week later, having already spent another four pounds and ten shillings on a heifer, he was back at Patel's store to buy a much-needed overcoat on credit.[92]

Buying grain from a produce merchant was nevertheless disappointing, since it meant that most of the heavy labour of ploughing the year before had once again gone to waste, and hard-earned winter cash that should have been used for buying livestock—a renewable productive resource—was being used to pay for grazing or, worse still, food. The sequence of poor agricultural seasons at Rooistad was causing Kas to question the wisdom of arable farming and only further fuelled his desire for livestock. Because drought had impaired the quality of the grazing at a time when wool prices were buoyant, it was probably prudent to start buying a few hardy sheep.[93]

Despite his misgivings about crop farming, Kas found himself once again mesmerised by the onset of the season of renewal. Spring seemed intent on redrawing the Triangle's boundaries in colours so vivid that even Old Man Kalahari could not fail to see them. Like a snake wriggling out of an outgrown skin, the palest hint of hope emerged from the brown veld and then slithered forth in shoots of brilliant green. Dry stubble was mysteriously tinted with silken green, the Old Man checked his movement, and, by September, Kas was working hard at night-time ploughing. The new season had to be better than the one past. It was the iron law of farming; without it, life on the margin made little sense.

Nature, a canny poker player, opened the season's bidding with enough spring rain to keep a glint in the eyes of Kas and 'The Owl' for some weeks. Then, her reputation for inscrutability intact, she used the next few months to provide most of the southwestern Transvaal with above average rainfall. There was no way of winning against so cunning an opponent. All that a man could do was to reshuffle the cards, follow his luck until it had run its course, and hope that he ended up on top.

A normal summer restored predictability to life at Kareepoort and helped to dissipate some of the social static that had built up in the villages. The price of alluvial gemstones crept up, and the industry absorbed some of the unskilled labour that swilled around the margins of the Rooistad community. Labour tenants and sharecroppers, the real economic yeast in the local economy, became more active in the 1935–36 season when beef, maize and wool all moved off the price plateau that had become a worrisome feature in recent years.

Kas went into the off-season with more cash and confidence. For the first time in years he undertook his winter cobbling, saddle repairing and tailoring out of strength rather than weakness. This success, underwritten by the 'Lebone Agreement,' meant that when the agricultural year drew to a close his funds for a new round of livestock raiding stood at twenty pounds. Spring found him with coins in his pocket and notes in a purse securely strapped to his belt. His war chest was exhausted in a short campaign conducted with military-like precision over three days in early November 1936. First, he took over a fine heifer and a good *tollie*—a young ox or steer—from Johnson Kgosiemang which, at three pounds, he considered to be an especially sound investment. Then, having slept on the matter, he marched across to M. W. Smith's Kareepoort Store, where an outlay of £5-15s secured him a 'Hercules Assagai Chief Bike with Balloon Tyres,' an acquisition that would facilitate his trips to the Makwassie livestock auctions. On the third day he caught sight of his real quarry. At Boskuil, Solly Meyer let him have three sturdy young bulls and a very promising heifer for only eight pounds. Each of these transactions was recorded on a receipt and then carefully stored in the large old grain bag that he kept in his room.[94]

The addition of six cattle to the herd in less than a week was a welcome sign of material progress. Kas was especially pleased when not long after, Ntate Masihu, a member of a prominent local farming family, gave him a cow for having rid his grandchild of a 'moving object in the bowel' which caused the infant's rectum to protrude whenever he defecated.[95] Old man Masihu's gift confirmed a trend: during the worst of the Depression Kas's herbal practice had fallen off, and then there had been the unfortunate events surrounding Jonker's beer raid. But now labour tenants and sharecroppers were once again turning to the ngakas for help.

The sudden expansion in the size of the Maine herd was not without its costs, however, since all cattle, regardless of where they came from needed access to grazing. In early spring there was rain enough, but as the pages of Kas's tatty calendar indicated the arrival of 1937, the Triangle assumed its familiar drought-creased grimace. By late summer he and other black farmers on the diggings were once again faced with the sickening prospect of a miserable grain harvest and of having to pay for even more grazing on Piet Ferreira's portion of Boskuil. Kas knew from bitter experience this raised the spectre of famine and a winter in which shoe repairing was a penance rather than a pleasure.

Whatever unease could be detected in the ranks of Kareepoort's labour tenants and sharecroppers was not shared by its clergymen, shoemakers and teachers. Being either self-employed or less directly dependent on the vagaries of nature, they could afford to take a slightly more relaxed view. The early months of 1937 found Jacob the Shoemaker and his lieutenants more interested in politics than in plants—the first time since the heyday of the I.C.U. a decade earlier.

When General J. B. M. Hertzog and his predominantly rural Afrikaans-speaking supporters had been returned to power in 1929, white South African politics had at first seemed only slightly deflected from the course it had followed through the decade. But Hertzog's National Party found that the electoral gains of 1929 meant that it was less reliant on the support of its urban English-speaking Labour Party allies than it had previously been and, by 1931, the 'Pact government' had collapsed. This unexpected development heralded two dramatic years dominated not so much by orthodox party politics as by the stranglehold that the Great Depression exercised over South Africa's fragile economy.

In December 1932, under pressure from Transvaal populists like Judge Tielman Roos and others who favoured a devaluation of the currency (like J. G. P. van der Horst the former M.P. from Wolmaransstad), the Hertzog government was forced to do an economic about-turn and take South Africa off the gold standard. With the economy still spluttering, an election pending and the uncertain political climate, the prime minister and his arch rival, General J. C. Smuts, moved closer and in 1933 formed a coalition ministry in the interests of 'national unity.' A year later, in 1934, Hertzog's National and Smuts's South African Parties joined forces in a United Party whose Fusion government effectively dominated white politics.

Not all of Hertzog's Afrikaner Nationalist friends found the new alignment to their liking, and within months of the formation of the United Party, Dr. D. F. Malan and his supporters broke away to form the Gesuiwerde, or 'Purified' National Party. Hertzog thus became one of the first premiers to occupy what proved to be some of the least defensible terrain in white South African politics—a compromised 'centrist' position exposed to attack from the right by orthodox Afrikaner Nationalists with a strong base in the countryside.

For this and other more deep-seated historical reasons, Hertzog found himself under great pressure to help platteland farmers solve their black labour problems and make some broad ideological advance on what the white electorate saw as the vexed 'native question.' At various strategic moments during the 1930s, he reintroduced a series of controversial segregationist 'Native Bills' whose origins could be traced back to the Pact government of the 1920s.

Of these, three were of particular importance. First, the rather cynically named Representation of Natives Act of 1936 aroused widespread resentment and opposition amongst both Africans and their white sympathisers, for it effectively disenfranchised black voters in the former Cape Colony who, in theory, could still decisively influence the outcome of white elections for the House of Assembly. As part of an exceedingly shoddy 'compromise,' the Act made provision for blacks to elect four white representatives to the Senate. Second, the Native Trust and Land Act of 1936, while marginally extending the amount of land set aside for the exclusive use of Africans in the reserves, helped shape administrative boundaries for yet more territorial segregation. Third, the Native Laws

Amendment Act of 1937 gave the government a whole new arsenal of weapons with which to assist highveld landlords to stem the flow of black labour to the cities of the Witwatersrand.[96]

In practice, the last of these Acts held the most immediate threat for blacks in the Triangle's 'urbanised' communities. But in March 1937, well before the Native Laws Amendment Bill became law, it was the new measures for parliamentary representation of blacks in the Senate that captured the imagination of Jacob Lebone and his politically-minded friends. The election for a senator to represent the 'natives' of the Orange Free State and Transvaal could hardly have offered two more contrasting candidates.

J. D. Rheinallt Jones was the son of a Welsh Methodist minister who had left his native Bangor in 1905 to come to the Cape. In 1910 he had settled in the Transvaal, where, over the next two decades, he established himself as a leading authority on 'native affairs.' Grafting a 'scientific' interest in black society onto a liberal-Christian philosophy he became an influential activist and intellectual and he viewed the signs of growing black radicalism in the cities with some alarm. His interest and expertise in farming matters were less developed.

His opponent, twenty years younger, came from a different world. H. M. Basner, the son of an adventurous non-conforming father who 'ran away to South Africa' in 1895, had been born in the Jewish quarter of Daugavpils, a railway town on the river Dvina, which linked Latvia's heartland with the port city of Riga on the Baltic coast. His earliest years were spent with members of an extended family that included a fisherman, a cabinet maker, a petty state official and a fair number of Chassidic mystics. These were amongst the political and social remnants of various Russian invasions and pogroms that somehow always left just enough 'Jews to carry on the timber trade for the Baltic barons, establish markets for the peasants, keep brothels and shebeens for the railway men and soldiers, and provision the rafts that made the long journeys from the Volga provinces to the Gulf of Riga.'[97]

In 1912, at the age of seven, the boy went with his mother to South Africa, where the rump of the family was reunited with the father, in Johannesburg. The lad disappointed his mother by refusing to become a rabbi and instead went to his sister's home in Los Angeles, California, where he studied law. There, he met a wide range of conservative and radical Russian emigres who had for one or other reason fled their motherland after the Bolshevik Revolution. But it was his chance involvement in a strike of Mexican dock workers that did most to radicalise the young man and determine his future. Shortly after completing his legal studies in 1926, 'Hymie' Basner returned to the country where his heart lay, South Africa.

In Johannesburg he was taken on by a firm of attorneys and, through his legal work on behalf of black workers caught up in the segregationist maelstrom of the Natives (Urban Areas) Act of 1923, was put in touch with most of the prominent political activists of the day. Amongst these was Charlotte Maxeke, a woman who had pioneered several of the early campaigns of the African National Congress but who was ever more deeply involved in her long-standing commitment to the A.M.E.C. Other notables included the lawyer Sidney Bunting and the intellectual Eddie Roux, both members of the small South African Communist Party. Basner, unlike Rheinallt Jones, had some insight into the plight of farm workers, and the penetration of the I.C.U. into the country districts gave him more. Despite his exposure to these more radical political currents and his friendship with committed activists, it was not until 1933 that Basner decided to join the Communist Party.[98]

Given their radically differing backgrounds, it was not surprising that Rheinallt Jones and Basner not only took disparate routes into the southwestern Transvaal countryside during the 1937 election, but also drafted manifestos that differed in important respects. Rheinallt Jones, whose name was familiar to informed blacks on the diggings because he had helped to establish the inter-racial Joint Councils in towns like Klerksdorp and Kroonstad, got access to his black constituency largely through English-speaking United Party supporters. Amongst the most important of these in the Triangle was a local farmer-cum-digger named Moult, for whom Jacob Lebone had once worked, and whose son, J. G. F. Moult, was a candidate for the Provincial Council in the same election. Even more important was Mrs. J. C. Kuhn of Boskuil, an energetic woman with a reputation for always being ready to help blacks at Kareepoort in any community projects relating to church or school.[99]

Basner, although enthusiastically endorsed by the largely urban African political organisations of the day, was comparatively unknown in the southwestern Transvaal in 1937. His allegiance to the Communist Party was hardly likely to open access to the networks that Rheinallt Jones used, nor was his apparent Jewishness an asset in an area notoriously suspicious of foreigners. As much out of necessity as out of convenience, he used potentially sympathetic Jewish traders, shopkeepers and lawyers as his entrance to the Transvaal countryside.[100]

In Wolmaransstad, the key to that particular door lay in the local chemist shop. Leonard and Sonia Wolpe, then in their late forties, hailed from the small village of Shavel, near Vilna, in Lithuania, and had both qualified as apothecaries under Sonia's father in Moscow before emigrating to the United States in 1911. From Philadelphia, the couple had gone to Johannesburg where Leonard had formed a strong personal association with the Labour Party activist Morris Kentridge while practising as a pharmacist in the white working-class stronghold of Fordsburg until the family shop was destroyed in the bombardment of the suburb by Prime Minister Smuts's troops during the 1922 mineworkers' strike.

But factors other than east European origins, American experiences, and a sympathy for the cause of organised labour drew Basner and the Russian apothecaries closer together. Wolpe, like Basner, had cast aside the more obvious elements of his Jewishness and become a remarkably free-thinking man, known to family and friends as an 'atheist,' 'an evolutionist,' a 'Bundist' and one who took an exceptional interest in events around him. Most important of all, it was through the Wolpes that Basner was introduced to E. A. G. Behrman, who became the driving force behind his campaign in the Triangle.[101]

Errol Behrman, five years younger than Hymie Basner, did not readily fit the first-generation Latvian or Lithuanian immigrant mould. Indeed, he did not readily fit into any mould, which no doubt appealed to the iconoclastic Basner. His mother, who like the famous Jewish entrepreneur Sammy Marks could trace her origins back to Neustadt via England, came from a well-educated family of Cape feather merchants; while his father, a trader, had died when the boy was only ten. Behrman was a lively youngster with a quick temper and sharp mind, and when moved to the Witwatersrand to complete his schooling, became a fine athlete and successful boxer. When he matriculated, his mother, hoping to take the sharper edge off a personality that could often be as aggressive and abrasive as it was capable of being charming, arranged for her brother to take him on as an articled clerk in his modest legal practice in Wolmaransstad.[102]

Behrman's partnership with the older, tolerant and more understanding Israel Gordon worked well, although Behrman kept himself aloof from mainstream social activities in the Triangle, preferring instead to go out hunting with a male friend who had the sense to have a beautiful sister. This reclusive manliness masked a potentially explosive temper, however, and on more than one occasion when back in town, Behrman offered to settle a dispute by removing his jacket. But, if the more staid members of the small Wolmaransstad community frowned on such belligerent behaviour, it earned him the undisguised admiration of those beyond the fringes of genteel society. Out on the diggings, among men of all colours hewn from rougher material, it won him the unabashed support of tough elements. Kas later recalled how, in the early 1930s, Errol Behrman was always the Bull Nines' first choice as a defence lawyer in any I.D.B. case.[103]

Despite having the more imaginative electoral team, Basner found that his manifesto failed to convince politically influential members of the black community around Boskuil. Whereas he had to rely on people like Errol Behrman and local shopkeepers like Solly Meyer to get across his message, which was of greater appeal to urban proletarians than cautious peasants, Rheinallt Jones would often arrive on the scene with one or other minor state official and unfold a large map portraying which 'new areas' had been set aside for exclusive African purchase in the 1936 legislation. This differed markedly from the old millennial campaigns conducted for the I.C.U. by black radicals like Jason Jingoes. Sharecroppers and shoemakers needed little persuading about the virtues of acquiring land, and Jacob Lebone soon became a staunch supporter of Rheinallt Jones's. So, too, did the local A.M.E.C. women, reassured by the son of a Welsh Methodist minister at a time when local farmers were muttering darkly about plots by 'Jews and Communists.' And, if there were any lingering doubts in the minds of these privileged elements of black society, they would have been swept away by the knowledge that the 'Bull Nines,' the beer brewers, and the underclasses in general favoured the Behrman-Basner axis. In Rooistad, respectable folk backed the moderate candidate who had the advantage of being endorsed by the local state officials who exercised such a powerful sway over their day-to-day existence.[104]

Predictably, the election elicited very different responses in the extended Maine family. Mphaka and Motlagomang, I.C.U. veteran and beer brewer respectively, were enthused by the political contest and for them, Basner and the Communist Party came closest to meeting the real needs and aspirations of most South Africans. But for Kas, Leetwane and Lebitsa, the 1937 election was of little significance. Smart city politicians who sold stories about black folk in the country getting access to land in return for membership fees or votes had an all too familiar ring. Kas had heard it all before. In any case, his mind was elsewhere. When Rheinallt Jones won by a large majority it had less impact on his memory than a good cattle sale at Makwassie.[105]

There were other, more pressing reasons for such indifference. For the fifth time in eight seasons, drought had destroyed his harvest, and for the second time in as many years, his family faced a winter without food. The odds were becoming too unattractive to keep him in the game. The time had come to abandon the diggings and return to the white farms where, at the very least, he wouldn't have to pay Piet Ferreira grazing fees.[106]

The remains of the grain drained from the bag in Kas's shed as smoothly as sand in an hourglass. Kas looked for a way out. Every day there was one more fold in the bag, one less day in which to find a solution. And Leetwane's changing profile reminded him that soon there would be yet another mouth to feed. Why was it that at Kareepoort hunger always played the role of midwife? Somehow the Boers' farms seemed to have posed different, perhaps even easier questions.

He heard of a white man on the diggings who wanted to dispose of a two-year-old ox at midwinter prices, and notwithstanding the mounting household crisis, he bought the beast for three pounds. Patriarchy entailed making difficult decisions as well as reaping certain privileges, and a sharecropper needed a full span of oxen to convince a Boer to take him on as a partner. Presumably Kas made the correct choice for in a matter of days he heard of an opening with a certain Koos Klopper on the farm Klippan—a property that bounded his old and very successful stamping ground at Kommissierust.[107]

A few weeks later 'The Owl,' unwilling to lose his most talented ploughman, reluctantly issued Kas with a trekpas; it recorded that the Maines were leaving Kareepoort owning three horses, sixteen donkeys and nineteen head of cattle. The 'Put-up and Pull-down' was retrieved from the outpost at Boskuil, brought home and disassembled. On a chilly winter morning in the first week of July, it, along with the planter, a few ploughs, several sheets of corrugated iron, assorted chains, and a few bits and pieces of wire, was stacked high on the wagon. From his huddled perch at the rear, Kas watched with a sense of satisfaction his oldest son coax the herd into making the slow turn westward to Bloemhof. With access to land and a little rain, there was no reason why a man with two wives, six children and a full span of oxen should not make a success of sharecropping. The Triangle was full of such men. He would do them proud.[108]

At Kingswood, Mmusetsi pulled the lead oxen off the main road onto a rutted track that led away from the rail siding north to Schweizer-Reneke. As they trundled through the gate at Klippan, Leetwane whispered a hurried reminder that their supply of grain was at an end. 'I arrived at Klopper's in the year 1937 without any food.'[109]

CHAPTER SEVEN

# Excursion

## 1937–41

Klippan was not a pretty place. A flat stone-strewn depression in the northeastern corner of the property dominated an undistinguished tract of brown and barren veld. Flakes of curled mud lifted off the cracked surface of a desiccated pan which, like some large and demented eye, stared up into a white-hot sun. To the north lay a scattering of *vlei*-like depressions which, it was said, became mildly soggy during good summers, while to the south, the larger salt-encrusted saucer of Sewefontein inclined toward a few gentle folds of slightly darker soil. These disconnected depressions might perhaps have aspired to the collective status of a *spruit* in a very wet season, but like the Magoditsane to the east, it would have been unlikely to reach the Vaal, which mumbled away twenty miles farther south.

As with most properties in the district, Klippan had first attracted the interest of whites in the early 1870s, when burghers proclaiming their unswerving loyalty to the South African Republic had rushed into the Triangle to secure tracts of government-granted land. As in other frontier societies, however, oaths of fidelity were often made with the citizenry's eyes fixed more on mineral rights and property speculation than on agricultural enterprise. When it became apparent that there was no lasting sparkle to Klippan's gravel, the last vestiges of burgher loyalty evaporated beneath the midday sun, and by 1894 the title deeds were in the hands of a firm of Kimberley speculators. The property's interminable economic silences and the unblinking eye of the pan exhausted the patience of even Messrs. Hill & Paddon, and, in 1937, the 3,300 morgen of the original farm were subdivided into smaller portions and sold off to members of prominent local families, which explained how a fair section of Klippan came to be acquired by Mrs. Henry Stern, a widow, mother of the artist Irma Stern and member of a wealthy and influential family based at Schweizer-Reneke.[1]

Like the merchants who had preceded her, Mrs. Stern had no intention of farming the property. Her agents advised her to allow them to find reliable tenants who could work and perhaps even improve the quality of its land. Thus, it was soon made known that several portions of the farm were for hire and could be purchased outright. For various reasons this offer appealed to three local men.

The brothers Hendrik and Piet Goosen, then both still in their mid-twenties, came from a well-established family in the Triangle. In the 1860s, their paternal grandfather, D. P. J. Goosen, had abandoned the Wellington district in the Cape Colony for the excitement of the Griqualand West diamond fields. Later, after a successful spell as a sheep farmer in the Brandfort district of the Orange Free State, he had crossed the Vaal and become a neighbour of the very same Koos Reyneke Sekwala had once worked for.[2]

But Goosen's portion of Rietput did not have enough water for his family's expanding livestock holdings, and in 1924, one of his older sons, P. A. Goosen, sold his share of the original Schweizer-Reneke property to the Sterns and bought a section of the farm Mooiplaas, near London. There he and his young wife, whom he had married as a sixteen-year-old, raised and educated a large family. Their two oldest boys were from a very early age far more interested in farming than in book learning, and therefore chose to serve an apprenticeship under their father rather than devote too much time to formal schooling.[3]

The younger of the two brothers, Piet Goosen, was the first to marry and it was only when Hendrik, too, showed signs of settling down that the collective demands of the rising generation outstripped the capacity of the family farm at Mooiplaas. In 1937, the brothers began looking in earnest for more space. As fortune would have it, J. M. Klopper, one of Dirk Bloem's bywoners on the farm Schoonsig, who was always on the lookout for new ventures, told them about the possibility of renting or purchasing property at Klippan.

At the age of forty-seven, Koos Klopper himself was still waiting for the opening that would enable him to rise into the ranks of those who enjoyed the comparative safety of land ownership. Lengthy and often painful spells as a bywoner, diamond digger and transport rider had all failed to generate enough capital, and had it not been for the modest salary earned by his schoolmistress wife, he and his family would long since have slipped back into the small but stubborn fringe of 'poor whites' who clung to the margins of the diamond diggings. Driven by want as much as by wisdom, his small bespectacled eyes detected in the widow Stern's property perhaps his last chance of extracting a loan from the state's Land Bank.[4]

Kas Maine guarded fragments of information about the Triangle's white inhabitants with the same care and attention that he devoted to storing seed. Knowledge, no less than the possession of an ox and a plough, was an important resource, since who could be certain that yesterday's 'Boer' was not tomorrow's 'Baas'? He had been following Klopper's fortunes on the sharecropping grapevine for nearly a decade, and now he thought that the Boer was likely to be a reasonably 'good' landlord, that a contract with a man who was really still only a bywoner could give him all the psychological space he needed.[5]

Koos Klopper was a man of gentle disposition, which appealed greatly to Kas. Moreover, he had indicated that he intended staying on at Schoonsig so that his wife would not lose her position at the local farm school. With a distance of nearly ten miles separating him from the Kloppers, Kas felt that he'd be protected against most of the irritating day-to-day demands that landlords and their wives could make. The only misgiving he had about their agreement stemmed from the couple's insistence that he provide them with ample domestic labour. He would have to get his two twelve-year-old daughters to spend most of their day working in Elizabeth Klopper's distant farmhouse kitchen.[6] But even this was not without compensation, since the ten shillings that Morwesi and Thakane would earn each month would help the family budget at a time when Kas could no longer profit from ready access to the sizeable market for shoe repairs at Kareepoort.

But, as his brother Sebubudi who worked elsewhere on Klippan had warned him when he had first contemplated moving back to the farms, Klopper's two neighbours were a rather different proposition. Aggressive young men who were self-confident enough to be seen shaking hands with blacks in public, Hendrik and Piet Goosen were eager to cast off the malaise that seemed to afflict many of the local whites and, like J. J. ('Koos') Meyer, whom they took to be a model in such matters, were keen to make the transition to fully-fledged capitalist farming. Like the Kaffir Corn King, Hendrik Goosen had already taken in a white bywoner by the name of Cornelius Fick, as well as several black sharecroppers, whilst Piet, inspired by the same example, had taken to patrolling his fields at night to ensure that no tenant stole maize before the cobs were ready for harvesting. From his own bitter experience at the hands of Koos Meyer, Kas knew what to expect from such ambitious men and resolved to give the two brothers a wide berth.[7]

Only when his wagon finally drew to a halt on Klopper's undeveloped lot did he realise how difficult it would be to effect this resolution, since, as sometimes happened with newly divided properties, the various sections of the farm appeared to be unfenced. Good fences make good neighbours, and the absence of clear-cut or effective boundaries often spelt trouble. But, having committed himself to the move, there was no turning back. Everything suddenly fell silent as the Maines struggled to take in the expanse of their new home.

Kas climbed down from his perch on the wagon and, without saying anything, stared out into the middle distance to take in what looked like some promising tracts of virgin soil. Out of the corner of his eye he also picked out some flourishing patches of *perdestert*, horse-tail grass. That, too, was good; perdestert was useful in bringing cattle to peak condition at short notice.[8] But Leetwane, Lebitsa and the children saw much less and said nothing. Farms always meant hard work, and away from the bustle of the diggings and the social protection of the village at Rooistad, they would be all the more exposed to the sweep of the patriarchal eye. Kas looked out: they looked in.

While the family were contemplating the range of his authority in the new setting, Kas was wrestling with the problem of how to find food for nine people until sufficient maize and sorghum could be harvested in the fields. When Klopper appeared early the next morning he put the question to the landlord; he, running his mind along the familiar conduits of rural credit, could only suggest that Kas try to get mealie meal from the nearest branch of the South-Western Transvaal Agricultural Co-operative. Because the co-ops were usually beyond the reach of black sharecroppers when it came to any advances in cash or kind, Klopper gave Kas a note which, he hoped, would attest to the creditworthiness of his partner. Kas was firmly convinced that Klopper knew all about such things, so he took out the horses, bid the family farewell and set off confidently for the depot at Bloemhof. He was gone for most of the day. For the children, who were famished, it seemed that he would never return. But at dusk they caught sight of his cart coming through the gate, creaking beneath the weight of twelve bags of coarsely ground maize that he would have to pay for by the end of the season.[9]

With the immediate crisis resolved, Kas turned to some long-term planning. For some time it had been clear that wool prices were enjoying a more sustained upturn than those of other agricultural commodities. While he was still at Kareepoort and having to pay Ferreira for his grazing needs, it hadn't been worth his while to have lots of sheep, but now the moment had probably arrived to develop a more substantial flock. He reasoned that if he sold his team of oxen and reverted to ploughing with donkeys—September was upon them, and many grain farmers would already be on the lookout to supplement their draught power for the new season—he could probably afford to buy a few score good quality crossbreeds.

A few days later, he came across Hendrik Goosen's bywoner in the fields. Cornelius Fick was obviously one of the *amaqaqa*, or 'poor whites,' but he had the virtue of being a fluent Sotho-speaker who, like the Maines, hailed from the eastern Free State. Kas took an immediate liking to him.[10] When they next met Kas mentioned that he was eager to acquire a few sheep, and Fick offered to act as an intermediary with the landlord who, he said, needed a span of oxen. Through a process of verbal triangulation it was eventually agreed that Kas would get a hundred of the landlord's sheep for his ten oxen. But on the day of the exchange Kas noticed that the older Goosen had included among the flock a lean and scrawny ewe that had belonged to his brother Piet. When Kas objected to the inclusion of this animal, Fick whispered to him in the vernacular that he should accept the ewe, fatten it up and then sell it to him, Fick.[11]

The sheep, carefully tended by seven-year-old Nthakwana, were an immediate success. Within weeks the animals had their early summer shearing and the wool, sold at one of the Makwassie stores, fetched a good price. Most of the animals improved so rapidly that twice within four months Kas could send sheep to Slabbert & Campbell's stock sales. Young Bodule would leave home well before sunrise to shepherd the animals to distant Bloemhof; later in the morning Kas would take the horses and join the nine-year-old at the auction, where the boy watched while his father disposed of thirty sheep which, between them, brought in nearly eighteen pounds. Klippan was clearly suited to sheep farming. The scrawny ewe earmarked for Cornelius Fick at the original exchange never quite found its way back to the bywoner as planned. Instead, it and several other animals were fattened up, and included in a batch of ten sheep that Kas exchanged for a horse owned by Johnson Xaba, one of several sharecroppers on Piet Goosen's property. The ewe thus eventually, albeit via a rather roundabout route, found its way back to the Goosen brothers' section of the farm.[12]

Livestock farming had its own joys, but it took only the slightest shift in the colour of the grass, the first hint of moisture in the air, for Kas to be drawn back to the main enterprise. Using the donkeys he, Mmusetsi and Bodule opened up a sizeable tract of undeveloped land that Kas opted to devote exclusively to maize. The spring operations went smoothly, and Kas was impressed with the progress made by young Bodule. Under the tutelage of his older but slower adolescent brother, he had shown his mastery of the art of ploughing straight and deep. To those watching around the fire at night, it was evident that the little chap's precocious skill and willingness to work long hours earned him a special place in his father's affections.[13]

The first summer rains and Leetwane's seventh child, their third daughter, arrived at almost the same time. Kas saw an auspicious sign in this happy coincidence, and the child, like Sekwala's first wife who lay buried at the foot of the distant Malutis, was given the evocative name Pulane, 'rain.' When the rains eased Kas slaughtered a ewe and his wives prepared sorghum beer when family members arrived from Kareepoort and surrounding farms to celebrate the birth.[14]

The beer that day was especially good and Phitise, sensing that the *bojwala ba dipitsa* might help ease his way, used the occasion to raise some important family business of his own. For all that, he posed his question diplomatically, even tentatively: years of bitter experience had taught him that his younger brother was almost always reluctant to take time off from farming. But Malefane, his oldest son, was about to choose a wife and BaSotho tradition would have it that the young man's uncle accompany him on a journey to Potchefstroom to meet the prospective parents-in-law and help negotiate the bridewealth.

The caution was unnecessary. True, the beer helped, but other, more deep-seated reasons caused Kas to accept the invitation with alacrity. The move to Klippan and the switch in emphasis to farming small livestock gave him more time and social space than he had at Rooistad, and with the crops already in the ground, the first rains behind them, and Bodule and Nthakwana there to take care of the sheep, he could easily risk an absence of a few days. Besides, the novelty of a train ride and Malefane's paying for the costs of the journey added appeal to what would otherwise be just another family duty.

About ten days later Kas saddled up his horse, took leave of the family and set out for Kareepoort. It was expected that he would spend the night at Phitise's but, as he crossed the Magoditsane spruit and rode to Rooistad, he was struck anew by the noise and shouting at the surrounding diggings, the way the houses clustered together without consideration for decency and privacy, the dangers and delights that were posed by the strange inhabitants. He knew one woman there, Nomahuka, a Xhosa-speaker whom he had seen several times during the hardest days of the Depression and more frequently before going to Klippan. The saddle chafed at his groin, and when he reached the turnoff to Phitise's home, he suddenly decided to ride straight on. He knew where she would be.

Early the next morning he made his way to his brother's house. Malefane rigged up the cart and drove them to the siding at Boskuil where they had arranged to meet Phitise's brother-in-law, the third member of the Maines' negotiating team. The three boarded the train, Kas in an unusually relaxed frame of mind. Even an old dog benefited from the occasional slipping of the leash.

The clan's business—at the place known to Tswana-speakers by its more familiar name of Tlokwe—went smoothly and, as a way of saying thanks to the other two men on the journey home, Phitise slipped out into the back streets of the township and bought a bottle of brandy at a nearby shebeen. The three greybeards then took leave of the kin-to-be, got into a taxi and sped off toward the railway station. But in Potchefstroom, as anywhere else in South Africa, disaster lurked in unexpected places to prey on black men, and when the taxi drew to a halt, they were accosted by a burly policeman who demanded to see their passes.

The passes were in order, but no sooner had they boarded than they were intercepted by a second official, who demanded to see their tickets before allowing them into the compartment. Kas could not find his. The European ticket examiner, deeply unsympathetic, bundled him off the train as it prepared to leave the platform. Kas rushed back to find the taxi, but the ticket wasn't there. He then dashed across to the point where the policeman had forced them to fossick through their pockets for their passes, and to his great relief he chanced upon the missing ticket lying in a pile of old papers. By then the train was moving out of the station, and it was only through the efforts of a second, friendlier examiner that Kas was hauled aboard.

Farming folk, unaccustomed to such traumatic departures, found these events unsettling enough to need something to steady their nerves. Phitise rifled through his possessions and triumphantly produced the bottle that had eluded the policeman's eyes. No one could have asked for a better travelling companion, the brandy slipped into sharecropping company with all the grace and ease of an old friend. By the time the train reached Wolmaransstad, an exercise that had started out merely to calm the rustic mind had developed into a raucous, uninhibited peasant celebration of Malefane's good fortune.

From there on the train seemed to chug along ever more slowly, and when the heavy-breathing steam-engine eventually dragged itself into Makwassie, Kas gradually realised that the figure sprawled across him—still clutching the bottle of brandy—was hopelessly drunk. He persuaded Phitise that it would be irresponsible of them to leave his brother-in-law with the bottle. The man would be going on down the line as far as Christiana, and there was no telling whom he might meet on the train. Seldom had common sense and self-interest blended together so sweetly into a single stream of golden liquid. Before disembarking at Boskuil, the brothers did their duty and relieved their sleeping kinsman of the last bit of liquor. At Boskuil they were met by Malefane and went on to Rooistad for yet another round of celebrations. All of these actions were vindicated when they learned that the police *had* boarded the train soon afterwards but could find no sign of alcohol near the slumped figure. The only minor complication had been when the brother-in-law, who should have got off at Christiana, came to his senses only around Warrenton, by which time the train was already well on its way to Kimberley![15]

*

Back on the leash at Klippan, all thoughts of the kinsman or of Nomahuka were soon buried beneath a welter of demands from the fields, sheep, cattle, donkeys and horses. One morning, not much later, the routine was disturbed by the arrival of a messenger from Hendrik Goosen who asked Kas whether he would let him use his cart for a trip into town. But the vehicle was damaged and he was in the midst of repairing it, so he sent back a message explaining why he could not accede to his neighbour's request.

Out in the field later in the day, Kas happened to come across Goosen who, irritated at not getting access to the cart, made it plain that he thought it was pride, not a logistical consideration, that had motivated Kas's response. Knowing how sensitive many Triangle Boers were to what they considered to be the airs and graces of black sharecroppers, and eager to avoid a confrontation with a new neighbour, Kas backed off and slipped into the appropriately deferential posture prescribed by the platteland's unwritten code of racial etiquette. This ploy, which merely ritualised the form that the disagreement took, defused the tension, but from Goosen's general demeanour Kas suspected the matter would continue to rankle.[16] He found it slightly bizarre that his cart could be the subject of a dispute with a man whom he hardly knew. The cart was nearly twenty years old and in bad shape. Indeed, when he surveyed it along with the rest of his farming accoutrements later that afternoon he realised that much of the equipment he had taken with him from the farms to the diggings nearly ten years earlier needed now to be replaced. With good summer rains behind them and a promising crop in the offing, he needed to repair the cart and buy a new wagon and a double-share plough to replace the old Canadian Chief.

Reinvestment on so substantial a scale required cash, and Kas was already in debt at the co-operative. On the other hand, given that South Africa's spluttering agriculture was at last recovering momentum and that land values in the district were slowly improving, it was probably as good a time as any to take out a loan. The real problem was to whom Kas could turn. An approach to either Goosen or his brother was clearly out of the question. Klopper and Fick were both bywoners, and probably a good deal worse off than most of the black sharecroppers he knew. No, the answer, like the outdated equipment itself, would have to be drawn from the more prosperous pre-Depression era.[17]

The short walk across the road to Vlakfontein was the least difficult and certainly the most pleasant that Kas could have chosen to make. Young Koos Reyneke was, by 1938, a mature farmer of independent means, and he readily agreed to lend Kas fifty pounds until after the harvest—a generous act that Kas greatly appreciated and that further enhanced the standing of the Reyneke family in his eyes. On his way out he found his feet shuffling toward Klippan, but in his mind he was already inspecting agricultural implements in the Bloemhof stores.[18]

But no black man dared venture into town without a pass, and marooned out on the farm, Kas no longer had the services of an understanding scribe like Johnson Kgosiemang. To make matters worse, Klopper failed to put in an appearance for several days. It seemed that landlords, no less than tenants, were never around when one needed them most. Klopper eventually arrived on 18 February and issued him with the written permission that allowed a sharecropper to go about his business. In Bloemhof Kas bought a new cart from Snow's Garage, later making it much sturdier by fitting it with metal wheels he bought at an auction.[19]

The real business of the day was transacted in Wentzel & van der Westhuizen's store. With part of the crop already mortgaged, a wagon was something of a priority, since Kas needed it to market the rest of his grain, but the one he had wanted cost forty pounds, and if he bought it, he wouldn't have enough for the double-share Massey-Harris plough, which cost eighteen pounds. He was on the horns of a dilemma and could not decide; in the end, the genial proprietor came to his rescue. Bill van der Westhuizen, whose livelihood partly depended on his ability to assess the honesty and financial muscle of the Triangle's white landlords and black sharecroppers, suggested that he do an even bolder thing: buy the wagon outright, pay for half the plough immediately and pay for the balance from the proceeds of the harvest.[20]

This agreed to, Kas went home with the same mixed emotions he had experienced on the occasion of his visit to Hambly's store at Hessie nearly twenty years earlier: excitement at the prospects opened up by the new equipment, balanced by an understanding of the awful responsibilities entailed by debt. It had never entered his mind to betray William Hambly back in 1921, and he was equally determined not to disappoint Bill van der Westhuizen in 1939. On balance the move from the diggings seemed to have been justified, and his effort to reinsert himself into the mainstream of economic life on the land had been successful so far.

For several days Kas floated around wrapped in a sheen of optimism until, one morning, Mmusetsi ran up to tell him that their cattle had been confiscated by Hendrik Goosen. The animals, left unattended for a brief moment, had moved into an unfenced area and made merry in a field of ripening maize. Goosen, angered, had instructed his tenants to round up the herd and drive them into his kraal, where they were impounded. Kas knew from experience that there was no easy way round such an awkward and embarrassing problem, but he went directly across the veld to the farmhouse to negotiate the release of his animals. When he got there Sebubudi and the others told him that Goosen had meanwhile left for town and given them no instructions about what to do if anyone arrived to claim the cattle. Under the circumstances there was nothing that Kas could do, so he reluctantly departed for home.

Later that afternoon he made the same long walk to Goosen's house again, but there was still no sign of the landlord. He decided to wait, took up a position near the kraal, and contemplated anew that look of total stupidity which God in His wisdom bestowed on cattle. Viewed from where he sat, his cattle suddenly looked even stupider than most. When Goosen eventually got back he told Kas that his animals would not be released until he had paid a cash fine of three shillings each for thirteen beasts. Kas proposed instead that Goosen consider releasing the animals immediately against a promise of payment at a later date. Goosen wasn't interested, so Kas then suggested that he accept three bags of sorghum in lieu of the two pounds. To his great relief, the landlord agreed.

Kas fetched the grain and went back on his third trip of the day across the veld. Goosen released the cattle and then, interestingly, asked Kas whether he felt the fine had been warranted. Kas acknowledged that it had, but wary of Goosen's intentions and knowing that in Bloemhof law and literacy were blood brothers, he took the precaution of asking the Boer for a receipt. Goosen disappeared into the house, where Kas heard him banging away at a typewriter. When he reappeared he handed him a slip of paper on which was written: 'I fined Kas three bags of sorghum for his thirteen cattle which went into my fields.' Kas, the note in his pocket quicker than a meercat down a hole, took his leave in a way that could give no cause for offence. It had been a long uncomfortable afternoon.[21]

Fortunately the boys kept a close watch on the cattle and the rest of the season passed without incident. By the middle of 1939 Kas was once again in a buoyant mood, the maize and sorghum were living up to their earlier promise. The women and children helped bring in scores of bags of *mabele* (sorghum), and the haul of mealies was even larger. Even after Klopper had walked off with half the harvest, Kas had almost one hundred and fifty bags of maize. Most of this was loaded onto the new wagon and trundled off to the South-Western Transvaal Agricultural Co-operative, where he was quick to settle his debt of twelve bags and sell the remainder.[22]

Kas had reason to feel proud. At a time when most white farmers—who enjoyed privileged access to commercial banks and the state's services to organised agriculture—were producing maize harvests of around three hundred bags, his family, with far more limited financial resources at their disposal, had matched their efforts. Nevertheless, the price of maize was disappointing—something that the Triangle's deeply suspicious populists ascribed to the functioning of the state's newly introduced Maize Control Board. At eight shillings and sixpence per bag, nobody was going to get rich, but Kas was grateful to have exceeded his target.[23]

Kas was a man of honour, and he repaid Koos Reyneke as soon as he possibly could. Good sharecroppers had their own lines of credit to protect. Settling the debt at Vlakfontein left him with a good feeling and, while he still had some cash in his pocket, he took advantage of the seasonal slack to slip across to Kareepoort, where he gave Nomahuka some money for Maqeqeza, their illegitimate daughter. While still at the diggings he heard of a young farmer selling off an ox, and he went home to Klippan driving his alibi before him. He spent the last of his cash on a few items from a newly opened store at Rietpan run by E. S. Seedat, a hawker who had previously operated out of Kingswood.[24]

Not even the lack of money could dampen the prolonged round of postharvest beer drinks that followed. These gatherings were more and more important to the Maines' life on the farms. With grain to spare and beyond the easy reach of the Makwassie police, Kas and other tenants on the nearby farms welded into a tight social grouping dominated by sharecroppers. In the words of a Sotho proverb, they, like members of a growing family, were drawn together by the sharing of sorghum. '*Mabele ke ngwetsi ya malapa ohle*'—sharing sorghum and acquiring a daughter-in-law in like manner engender solidarity.[25]

Before the outbreak of World War II this circle, with a core of ten beer drinkers, usually met at the home of Thloriso ('RaKapari') Kadi, who lived on a section of Kareepan that was both centrally located and had an absentee landlord. Besides his brother Sebubudi, Kas knew RaKapari's other guests well. Among them were the two Moshodi brothers—'Mr. Kwape' and Katse, who were still based at Hartsfontein as they had been when Kas first met them and had now been joined by an uncle named Lerema. Another regular whose friendship Kas could trace back at least a decade was Lerata Masihu, the one-time scribe who had been sent on to the A.M.E.C.'s Wilberforce Institute at Evaton. In ethnic terms overwhelmingly BaSotho, the group included at least two so-called Ndebeles, Hans Mokgosi and Johnson Xaba, who, like so many of the southwestern Transvaal's black immigrants, had come from the Orange Free State. (The two were not really Ndebele but this was the post-*difaqane* BaSotho way of referring to anyone of Nguni origin.)[26]

The underlying social cohesion in the circle came from bonds of Sotho ethnicity, but other factors were at work to keep the two 'Ndebeles' from feeling marginal. All the men were in their late thirties or early forties and most of them presided over large families with one or two sons old enough to be left in charge of farming operations while their fathers were at the 'club.' RaKapari could mobilise no less than five boys to assist him, Kas and Johnson Xaba could each call on the services of two, and Hans Mokgosi could look to the help of an adult son. This reservoir of domestic labour during the slack postharvest period allowed the men the occasional luxury of meeting in midweek in addition to their weekend gatherings. In later years, when most of the boys had become men, they sometimes even met during the ploughing season.[27]

Naturally, much of the talk revolved around agriculture, but they had ample opportunity to chart real and imagined kinship linkages and, where necessary, to contemplate and forge new ones. Thus Kas established that in their long journey from the mountain kingdom of Basutoland into the Triangle, the Kadis, like the Maines, had come through Lehurutshe and the Setlagole Reserve: it was possible that they were distantly related to Mothcba. In the 1940s, the Kadi and Moshodi families did become joined through marriage, and then the core club members were linked not only by class, ethnicity and occupation, but by authentic bonds of kinship.[28]

Afrikaner farmers, whose rudimentary education was powerfully influenced by the Calvinist tradition espoused in the Triangle's churches, saw few of these subtleties, nor could they think of secular arguments to commend these gatherings. They felt hard-pushed to find cheap black labour to help them expand production, and their position in the labour market was eroded by the diggings, which were showing unexpected signs of resilience.[29] So it was not surprising that they directed much of their frustration at the sharecroppers' social gatherings. Four decades later, Kas Maine could recall:

> The Boers used to call us layabouts because, while they were busy cultivating, we used to go and drink beer at RaKapari's. However, our spans of oxen were working. The Boers called us idlers, people who just spent their time relaxing without working. But we told them that we were working, because our children and oxen were gainfully employed.[30]

These ideological contests in the heart of politically volatile farming constituencies did not go unnoticed in official circles, especially in an election year like 1938. In the Bloemhof district, where the local police were under the command of Sergeant 'Ross' Potgieter, Piet Goosen's brother-in-law, the state intervened on the side of the landlords by authorising occasional 'beer raids' on some of the more isolated properties occupied by black sharecropping families such as the Kadis.[31]

But these sporadic forays in farming country could not match the fury which the state unleashed on the inhabitants of Kareepoort and the neighbouring diggings. The Makwassie police, using powers newly acquired under the Native Laws Amendment Act of 1937 and Government Notice No. 1632 of 1938, invaded these 'urban areas' and tried to restrict commercial beer brewing by imposing the provisions of the pass laws on deeply resentful black men and women. So insistent was this onslaught by Moipeng and his men that people in Rooistad and other villages sometimes fled the diggings and sought refuge with friends or kin on surrounding farms. It was partly as a result of these police drives that Kas first met, and later drew into closer association with Jacob Nkosi Dladla.[32]

Hlahla—to give him the name by which he was better known amongst his many Sotho and Tswana-speaking friends—was a Zulu from the eastern Transvaal. Born near the town of Piet Retief in 1904, he came to the Triangle as a young man and, during the 1920s, successfully combined a prestigious traditional medicine practice, as an *inyanga*, with the mundane business of buying up livestock and selling meat on the diggings. A complex man who seemed to have trouble sustaining any relationship with women, Hlahla had just married the third of his eight wives when Sebubudi introduced him to Kas in 1937.[33]

At first Hlahla visited Klippan only occasionally, in order to buy a few sheep that he would take back to the diggings for slaughtering. But then he and Kas discovered their mutual interest in medicine, and soon inyanga and ngaka were having long professional discussions about herbalism; after all, they were meeting on a property where there was no resident landlord to challenge the presence of a 'stranger.' These visits became more and more protracted, and when the police offensive at Leeubos became intolerable, Hlahla and his young wife fled the diggings and moved into the Maine homestead more or less permanently. Although still devoted to his medicine and willing to make trips to Natal to procure rare herbs for himself and Kas, Hlahla merged his livestock with those of his host and quickly became a member of the 'Ndebele' contingent in the local beer-drinking circle.[34]

At about this time, in spring 1938, members of RaKapari's club started noticing that the district's more capital-intensive sector of agricultural production was once again experiencing one of its periodic convulsions—the first since the Great Depression. The price of land had firmed significantly, and several farms had changed hands and been acquired by what an inspector visiting a local bank had been pleased to characterise as 'a progressive and decent type of settler.' But even amongst the longer-established and supposedly deeply conservative Boer farmers, important new developments were taking place.[35]

J. P. Pienaar, the man who had introduced the steam-driven Ruston & Hornsby thresher into the Triangle, had gone on to acquire a Minneapolis Moline tractor that caught the attention of several farmers around Bloemhof. Across the way, at Pienaarsfontein, Koos Meyer lost no time in responding to what he saw as the emerging trend. He bought a dozen tractors and then, early the following season, invested in a 2.5 ton Chevrolet truck that could move no less than fifty two-hundred-pound bags of maize at a time. The Kaffir Corn King had no intention of being left stranded in the wake of any new mechanisation. But these developments had a social cost: as a direct result of these and similar innovations introduced elsewhere in the Triangle, several vulnerable labour tenants were ordered to reduce their livestock holdings and told they would have to work as wage labourers if they wanted to stay on white properties.[36]

Back at the club the spirited advance made by the tractor along this still narrow sector of Triangle agriculture was met with a rather ambiguous response by RaKapari and his friends. As Kas again recalled it, 'Some just laughed and said that it was a load of shit. Others saw it as a good thing.' How on earth were peasant farmers to foretell all the consequences of new technology? In Britain it had been noted on the eve of the Industrial Revolution that 'sheep ate men,' but only much later did the full implication of the enclosure laws become apparent to the citizenry at large. Likewise, how were black tenants to know that the whites' tractors would first develop an appetite for 'surplus' grazing, then move on to consume the oxen that walked the fields and then, in a final gargantuan orgy, go on to devour the very sharecropping families who tilled what little arable soil was left on farms which, up to that point, had always had too much land and too little labour?[37]

No, the tractor did not immediately threaten a good sharecropper. It was known that most of the landlords around Bloemhof were under- rather than overcapitalised, and a hard-working black man with a good span of oxen was more than a match for a white man with a tractor. Anyhow, who would drive the tractors? When all was said and done, it was access to land and labour rather than machines that limited a man's capacity to increase his harvest. Kas and his boys were confident that there was much to commend old skills and familiar equipment.[38]

Their labours did not go unrewarded. The rains again met their commitments, building on a three-year trend, and by October the little maize plants were well established.[39] Then one morning Mmusetsi again burst in on Kas to tell him that their fields had been invaded by sixteen cattle and fourteen donkeys belonging to Hendrik Goosen. Kas ordered the boys to round up the animals, confined them to his kraal, and then sat down to give the matter serious thought. The code of the countryside demanded that he be compensated for the damage to his crop, but, given the history of his relationship with Goosen, this was not an easy message to convey to the farmer.[40]

As Kas saw it, the important thing was to ensure that the landlord was not tempted to kill the messenger who brought the bad news, and the best way of ensuring that was for the communication to go via somebody well known to Goosen, someone who worked for him, someone like Kas's younger brother. So, after much talking, and with great reluctance, Sebubudi and Mmusetsi finally made their way to the Goosen homestead to inform the Boer about the whereabouts of his missing cattle. After what seemed like an eternity Mmusetsi returned with a note that had been carefully composed on Goosen's typewriter. By then it was late afternoon, and Kas had difficulty in coping with the printed Afrikaans. He folded the paper in half and put away the note, hoping that he could persuade Koos Klopper to apprise him of its contents when he came by in the morning.

Keeping a landlord's cattle overnight was risky, and when he told Klopper what he had done, his partner, showing the deference that one would expect of a bywoner, became deeply apprehensive. But once the contents of the note had been read to him, Kas felt reassured and remained convinced that he had done the right thing. Goosen, it seemed, was being very reasonable, for the note merely stated: 'Keep back one of the oxen and I will come across to inspect the damage. Release the remaining animals so that I can use them for ploughing.' He stuffed the note into a bag filled with receipts and other scraps of paper, picked out a large bluish hind-ox by the name of Bloubank and then told Mmusetsi to drive off the remainder of the animals in the direction of the Boer's section of the farm.

He waited for Goosen to assess the damage and reclaim his ox. Days turned into weeks, weeks into months, but there was no sign of the landlord. Was Goosen staying away simply because he had been called on to help his father with the ploughing at Mooiplaas, or because he resented having to pay a sharecropper compensation? Then, on Christmas morning, Kas had just slaughtered a sheep for the feast when one of his guests said that he had noticed on his way to the house that Goosen's cattle had invaded the Maine fields. Kas told one of the boys to take a horse, round up the cattle and bring them back to his kraal.

By the time Mmusetsi appeared with the fourteen offenders, the sky had become unusually grey, as if a spell of uncharacteristically wet weather was about to settle over the sand flats. The cattle were herded directly into the kraal and everyone was grateful to return to the festivities.

It rained that night. Early the next morning Mmusetsi was sent off, this time without the reassuring presence of Uncle Sebubudi at his side, to invite Goosen to come around and inspect the latest damage that his cattle had caused and to recover his livestock. The boy was gone for so long that it seemed as though he must have gone missing; when he got back he was clutching one of Goosen's typewritten notes. This time even Kas could get the drift of a message spelt out in the most colloquial Afrikaans. It was as direct as it was dismissive: 'I will not come to assess the damage, and do with the cattle as you please.' This challenge, so clearly in breach of all conventions, pushed the matter beyond the reach of black diplomacy and called for the intervention of a white mediator.

Early morning of Boxing Day found Kas at Schoonsig. He roused Klopper, explained the position to him and showed him Goosen's latest missive. But his bespectacled partner did not want to interrupt the seasonal tranquillity with unpleasant business. Unsuited to conflict by temperament as well as by status, he was at pains to warn Kas that Goosen was of the 'fighting type,' and his only suggestion was for Kas to take the cattle in to the Bloemhof pound and allow the law to take its course. Kas retrieved his note. It had been a disappointing response and, on his way back to Klippan, he again worked through his options—this time remembering that Sergeant 'Ross' Potgieter was Piet Goosen's brother-in-law. Justice supposedly had no kin, but out in the country it apparently had no shortage of cousins.

Back home, Kas dug out all the notes he had received from Goosen and stuffed them into a pocket. Then, with the help of a friend, Hendrik Lefifi, and Mmusetsi, he rounded up the cattle and set out on the long march to the pound. The roads were deserted, which made herding easier, but at the salt pan near Sewefontein, they were overtaken by a fast-moving cloud of billowing dust. The haze drew to a halt a few hundred yards down the track and lifted slowly to reveal the outline of a truck, stretched by three burly occupants—old man Goosen all the way from Mooiplaas, and his sons, Hendrik and Piet.

Hendrik Goosen, the crown of his brown felt hat fighting the roof of the cab for space in which to express itself, bent his head out the window and called Kas across to the cab:

> I went over to them, they were at the roadside. He said, "What's going on here?" He asked me where I was driving the cattle to and I told him I was taking them to the pound. When he asked me whose cattle I thought they were I told him that I had not the slightest idea to whom they belonged. He said: "You will find out soon enough because by nightfall the cattle will be in the pound and you will be in prison."

With that the truck started up and roared off towards Bloemhof.

But Ross Potgieter was not on duty, and so Hendrik Goosen had to tell his story to another sergeant: about how Kas had instructed Mmusetsi to drive his cattle into the Maines' maize fields and then use the 'invasion' as a pretext for impounding the Goosen animals. When the sergeant asked where the offending parties were, Goosen told him they were on their way to the police station, and since they were in possession of livestock without the consent of the owner, he wanted them charged with theft. The three Goosens then marched out of the police station, rearranged themselves in the cab, and drove off in the direction of Mooiplaas, leaving the sergeant to figure out what had really happened. The farmers' story sounded rather implausible, and so, when Kas and his party appeared at the gate, he asked them to take up a position outside the magistrate's office rather than escorting them into the station.

Kas waited. In the Triangle, as in the rest of the world, peasants, sharecroppers and tenants spent a good deal of their time waiting for officials. When the magistrate—he was R. M. W. Hawes, new to the job—appeared from nowhere some time later and saw the cattle in the yard, he leaned out of the office window and asked what the problem was. Kas, anxious to avoid saying the wrong thing at the wrong time, said only that he had brought in some animals for impounding and then quickly produced the receipt Goosen had given him on the occasion when he had been fined three bags of sorghum for allowing his cattle to stray into the Boer's field. Hawes glanced briefly at the paper and was perplexed by what seemed to be something of a *non sequitur*. Why was it that Kas was turning in Goosen's cattle when it was *his* cattle that strayed into the farmer's fields? But Kas was ready for him and, without saying another word

> I took out the second note—the one that stated that I should keep back an ox until such time as he had had a chance to inspect the damage. He [Hawes] read the document and again expressed some puzzlement, saying "What gives with you?" While he was still confused I handed him the third note, in which it was stated that Goosen would not assess the damage and that I was free to do with his cattle as I saw fit.

The penny dropped, and the magistrate sensed the origin and trajectory of the conflict so carefully chronicled in Goosen's signed and typewritten notes. Hawes summoned a policeman and instructed him to count the cattle, then arranged for the animals to be impounded and issued a receipt. Literacy, or so it seemed, was not always enough to ensure the triumph of the literate over the illiterate, and it paid a man to keep all the documents that came his way. This was a lesson Kas never forgot.[41]

He, Hendrik Lefifi and Mmusetsi rounded up the cattle and nudged them around to the back of the building where a surly-looking policeman guarded the gate. The cattle slowly eased their way forward. Before Kas could say anything, the policeman barked out: 'Speak up, what's your problem?'

> I told him that there was no problem and that we were simply bringing in some stray cattle. He then said: "Hey! Aren't you the fellow who arrived here with Hendrik Goosen's cattle?" I told him to wait, removed the first document from my pocket, and handed it to him.

By the time he had been through the typewritten notes, the policeman had been joined by a few of his colleagues, all of whom seemed to take a great interest in the matter.

> He showed them the documents and they discussed the matter amongst themselves, asking what else I could possibly have done under the circumstances, since the third note made it clear that I had been told to do with the animals as I pleased.

The surly one, seeing no way round the problem, opened the gate and let the animals into the pen. But his colleagues, whose loyalties lay with Ross Potgieter rather than the new magistrate, were unwilling to concede defeat. They crowded in on Kas and fired questions at him from all angles. What was the exact location of his field at Klippan? From what direction had Goosen's cattle come and what was the exact extent of the damage? He answered as best he could, and yet they were not satisfied. There was more discussion between them, and once again Kas and his party were made to wait.

A while later two uniformed men on motor cycles were seen leaving the police station. The 'steambikes' turned toward Klippan. The policemen looked over the fields in question, and then, in search of an 'independent' account of what exactly had happened on Christmas morning, went to interview one of Hendrik Goosen's tenants. But in the countryside justice's cousins are not all equally poorly disposed to the weak, and Ross Potgieter was not the only man with relatives at Klippan. The man they talked to was Serotele, Sebubudi's brother-in-law, who corroborated what Kas had told them. The two policemen went back and told Kas that he and his party were free to leave, provided they returned to the police station the following morning.

The proceedings on the following day were fairly painless. The police acknowledged that damage had been done to Kas's fields and asked how many people had helped to bring the animals to the pound. They calculated the distance between Klippan and Bloemhof as twenty-five miles and paid Kas, Lefifi and Mmusetsi three pounds and fifteen shillings, based on the standard rate of a shilling a mile for each animal impounded. It had taken the police twenty-four hours to do what Hawes had asked them to do.

But Ross Potgieter's men were no nearer to acknowledging defeat than they had been the day before. The magistrate's office and the charge office were separated by a sea almost as large as that which divides law from justice; the police told Kas that because he and his party were still facing a charge of stock theft, within three days they must report again at the Bloemhof police station.

The footprints that can tell us what exactly happened in those three days lie buried beneath the dust that still swirls across the Klippan road on a windy day. Presumably Potgieter paid his brother-in-law a visit and persuaded him to withdraw the charge because, when the Maines next appeared at the charge office, the police simply handed Kas a further ten pounds which, they said, was compensation for the damage done to his fields. Kas was told to take care to pay those who had helped him bring the animals to the pound.

> I took the ten pounds and paid Hendrik Lefifi two pounds, which added to the previous sum I had given to him, amounted to three pounds five shillings. I took the rest of the money and put it in my pocket, and I then asked the police whether it would still be necessary to appear in court. They informed me that the matter had been settled.

It had taken five days and seventy-five miles of travel, a magistrate and several encounters with the police to get a few cattle impounded and to avoid getting prosecuted for theft. The matter had been settled; but settled to whose satisfaction?

A few days later Hendrik Goosen went to Bloemhof where, as a result of the delays brought about by the investigation of the charge of stock theft, it cost him fifty pounds to have his animals released from the pound. This did not please him, and he was still grinding his teeth when, driving back through the gate at Klippan, he saw the man whom he considered to be most responsible for his misfortune. He stopped the truck, got out and walked toward Kas. Kas did not move. Goosen came closer and spat out a declaration that he intended to have revenge for his defeat. Kas said nothing. The Boer suddenly took another step forward and accused Kas of having stolen the one scrawny ewe that had been among the animals exchanged when the Maines had first moved onto the property. Still Kas said nothing. Then, just as unexpectedly, Goosen spun around and walked back to his truck saying that he would be going in to Bloemhof to lay a further charge of stock theft. 'I told him that he should do so.'

But this time Kas, not Goosen, went on the offensive. Long before Goosen could get around to mischief-making, Kas went to the police and warned them that the farmer intended laying yet another charge of stock theft against him. Growing in self-confidence, he then asked to see the magistrate and told him what was happening. Hawes, who appeared most sympathetic, advised him to strengthen his case further by making sure that he quickly returned the hind-ox to its owner. Kas did as he suggested, and left the ox in front of Goosen's house. Goosen, misinterpreting this as a sign of weakness if not of collapse, rushed out and repeated his threat. This time the Boer was true to his word.

Kas went through his papers and found yet another typewritten note recording the exchange of draught oxen for sheep at the end of 1937. The document made no mention of the ewe, but Kas remembered there had been a third party present at the transaction and, since then, Fick too had fallen foul of Goosen, who, he believed, had 'robbed him' of his rightful share of the harvest. He went across to the bywoner's cottage, and by the time he left believed he was ready to make a statement. The endless trips into town were getting him down.

> Now, when I arrived at the police station I informed them of how the exchange had come about and I produced the letter which recorded the original transaction. I also told them that I had a white witness who had seen the exchange take place, and who was willing to testify to my initial reluctance to accept the animal. In fact, it was Cornelius Fick who had persuaded me to accept the sheep on the understanding that I would rear the animal and that I would sell it to him at a later date.
>
> But that sheep had gained in condition and eventually produced a lamb with a reddish head. I included both the ewe and its lamb amongst the animals which I offered in exchange for Johnson Xaba's horse. The sheep which Hendrik Goosen claims to have belonged to him had thus

> found its way back to his younger brother's section of the farm, where it stayed for more than a year. How could he have failed to recognise or retrieve an animal that was supposedly stolen during so long an interval?

The sergeant recorded his words and, when he had finished, Kas picked up the pen and, with that careful and deliberate hand that marked all his writing, signed his 'official' name—Kas Teeu.

The policeman did not seem unfriendly, and Kas got the impression that perhaps even Ross Potgieter's friends were beginning to tire of Goosen's shenanigans. He asked the sergeant what he thought about the whole thing and was relieved to hear the officer say it probably wasn't worth his while to hire a lawyer. The existence of yet another typewritten note would probably settle the matter if it should ever reach the courts. But the sergeant, who by then was talking quite freely, did have one other piece of advice which Kas found less reassuring. Given the conflict with Goosen and the absence of fences it would probably be in his interests to leave the property as soon as it was feasible.

When he got home Kas took a walk through the fields and looked wistfully at his maize. Crops farmed on so large a scale were only really possible on a white man's land, but while his heart anchored him to the fields at Klippan, his head told him the policeman was probably right and the time had come for him to move on. This silent but bitterly contested tug-of-war within him dragged on for weeks; moving now one way and then another. There was no word from Bloemhof, and as far as he knew the police were still investigating the charge of stock theft. Eventually, when he could stand the uncertainty no longer, he made a brave mental move and decided that they should wait for the harvest and then stage a retreat to the diggings.

Life does not always meet you face to face going down the road. Sometimes it sneaks up from behind and tackles you from an unexpected angle. Just as he was beginning to lay plans for a future centred on Kareepoort, someone from Rooistad arrived to tap Kas on the shoulder with a message from his past. Nomahuka and Maqeqeza arrived at Klippan and, from her condition, it was clear that she was about to give birth to a second child. It seemed as if the train from Potchefstroom, much delayed, was about to arrive at a station that had not been scheduled for a stop.

Neither Leetwane nor Lebitsa was enamoured of this development, though it perhaps came as less of a surprise to them than their husband imagined. Through their own mysterious female channels they had known for some time about Kas's involvement with a strange IsiXhosa-speaking woman. But knowledge did not necessarily make for forgiveness or approval, and although wary of incurring the patriarch's wrath about so delicate a matter, they gave Nomahuka a chilly reception. Nor were they more forthcoming when, only a week or two later, the unfortunate newcomer gave birth to a baby boy who was given the name of Nkompi.[42]

In Leetwane's case this diplomatic silence was underscored by other considerations for she, too, was pregnant and, in order to avoid the problems associated with the ill fortune that came with *dikgaba*, it was necessary to avoid all conflict. Kas, misreading this silence as peace, was grateful for what he saw as unexpected domestic tranquillity. The only real pressure would come with the need to feed two more mouths. But then it had always been so. Leetwane had been carrying Pulane when they had arrived at the farm, and she would be pregnant when they left.

Fortunately, it looked as if food would not be a problem. The trend in good summer rains persisted, the crops looked promising and, unlike a field that needs weeding, promising crops had the knack of getting the Boers out of their houses. Klopper's appearances seemed to be more frequent and urgent, and Kas detected in his bywoner partner an almost unseemly haste to get on with the harvesting. But nature's calendar is turned by the hand of God and not man, and Klopper, along with all the other landlords in the Triangle, would have to wait his turn.

Autumn set in very slowly. The leaves gradually tensed in anticipation of worse to come. Then stiffened leaves were suddenly wracked into shapes of tortured brown that groaned in the breeze as they waited for the release that would come with the first icy blast of winter from the Malutis; they would freeze them into patterns of creased whiteness. In the end only stalks in man-made rows, like crosses in a military cemetery, marked the sites of devastation. Amidst so much death only those who knew the land could pick out the promise of life. As he walked through the deserted fields, Kas found his hands darting into the folds of deadened brown so that his fingers could sense the reassuring firmness of the seed that lay hidden in the papered cobs.

In June, realising that the harvest would exceed his domestic labour resources, he spoke to Klopper about the need to hire outside workers to help bring it in. They hired a team from Taung, paying them the usual six bags for every hundred bags harvested as well as feeding them. The Maines helped with the reaping while Klopper hurried back to Schoonsig to negotiate the services of J. P. Pienaar and his son, who between them did most of the local threshing for thruppence a bag. With the harvest coming in so quickly, it was of the utmost importance that the maize be kept dry and threshed quickly so as to keep the seed from sprouting or rotting.[43]

Once Hans Pienaar had installed the Ruston & Hornsby at the central 'feeding point' as agreed upon by the Triangle notables, Kas and his team went across to join the queue of landlords, sharecroppers, bywoners and labourers waiting to have their crops processed. Old Hans and his machine worked on a strictly first-come-first-served basis, and there was no racial ordering in the ranks of the maize producers themselves. Perhaps this seasonal blindness in a society otherwise obsessed with racial divisions caused the sharecroppers to indulge themselves in a moment of class pride, because, as the younger Pienaar noted with irritation, at such moments they would 'play the gentleman,' refuse to assist with any manual labour, and constantly haul out their notebooks to keep a check on the number of bags that they were entitled to.[44] Lower down the scale, however, considerations of class and colour merged to ensure that most of the menial work was left to hired labourers. Thus it was up to the seasonal workers to keep up the pressure in the boiler attached to the thresher's steam-driven engine by supplying it with a steady stream of dried maize cobs. Likewise, it was the hired black workers who had to sew up the heavy jute bags which each contained close to two hundred pounds of maize.[45]

The Maine-Klopper venture yielded nearly five hundred bags, which meant that, even after meeting his share of the cost of the harvesting team, Kas had two hundred bags of maize. Despite the drop in maize prices since their post-Depression high point in 1936, this big harvest meant that he could easily meet his outstanding debts and still have a healthy surplus. Normally the harvest would be moved back to the farm, but Klopper, still showing signs of unseemly haste, 'took his share right away from the thresher,' had it loaded onto a truck and ferried it into Bloemhof, where it was promptly sold. The season's business behind him, the landlord reverted to his reclusive habits and was not seen again for several weeks.[46]

But once the harvest was safely in there was no real need for such urgency. Kas ferried his share by wagon back to Klippan, and the boys helped him stack the maize with the care it deserved. Just contemplating the size of the growing mound of maize was a source of satisfaction even though Kas knew that others in his circle had done even better than he. Around the corner on Piet Goosen's section, Johnson Xaba was basking in the glory of a huge harvest. Still, his own efforts were a source of personal pride and he did not begrudge Xaba his triumph. There was always next season.[47]

A day or two later, shortly after the boys had eased the last bag of maize into its place in the stack, Kas was looking out across Hendrik Goosen's section when he detected a distant eddy of motorised dirt moving slowly toward them. Who could it be? Goosen seldom bothered to drive around the property, and Kas had not seen the man for days. Nor was he expecting anyone from Bloemhof. Unless . . . No, it could not be them, he would have heard something. But, as he watched the dust move closer, it seemed to know exactly where it was going, picking out the way along the indistinct track leading to the Maine house with great precision. It stopped at the gate at the lower end of the property and, when the tell-tale gap opened up between the noisy source of the agitation and its wake, Kas could see for the first time that it was a large truck.

The truck spluttered to an untidy stop, fell silent and then, after a moment's delay, a door swung open to reveal a profile that Kas had seen on the property once or twice before—an agent by the name of Belcher who, it was said, collected rent on behalf of the Stern family. The man got out, said nothing, but took a long look at the mealies stacked ten bags high and then turned to Kas and calmly announced that

> I should not sell any of my bags of grain since Klopper had fallen in arrears with his rent. I said: "*No!* Klopper took his share of the harvest into town, these bags are mine. This is *my* portion, after ploughing on the halves with him." He said: "*No, load the bags onto the truck and come with me.*"

The words came toppling down on Kas, threatening to crush his spirit just as surely as if the grain stack had suddenly given way and snapped his spine. What was he to do? With the theft charge still pending, the last thing he needed was yet another conflict with a white man culminating in a visit to the police station. He called the boys across and together they loaded the grain, every last bag of it, onto Belcher's truck. Bags which once had seemed to weigh almost nothing as he shunted them around the stack so easily seemed awesomely heavy.[48]

The bumpy ride into Bloemhof gave Kas time to reflect on the many ways in which he had known a farmer to lose his crops. Droughts dragged out the agony over an entire season, torturing man and plant alike as relentless exposure to a pitiless sun slowly blistered the life out of the grain. Floods, mocking those who dared farm in an area where water was at a premium, could achieve the same amount of destruction in a week. Locusts could deprive a man of a year's labour in a matter of hours. All these onslaughts took place out in the open, out on nature's battlefield, where one was always aware of if not prepared for unexpected danger. Now, it seemed, one had to add a new item to this list of terrors. An 'agent' could relieve a sharecropper of his grain in minutes—and at precisely the moment he had always considered to be safest, when the crop had been harvested, threshed and bagged!

The more Kas thought about it, the more this seemed to fly in the face of justice and, as they approached the outskirts of town, he persuaded Belcher to give him a little time to speak to Marais before they unloaded the grain. A. J. Marais, a Cambridge-trained lawyer who displayed a set of autographed oars on his office wall to remind him of the rowing Blue that he had earned while studying in England before World War I, was well known for the fact that most of his clientele were drawn from farmers and because he defended both whites and blacks. He was one of the few people who could extricate Kas from this predicament.[49]

They found Marais in his office and Kas outlined his problem, emphasising that the only debt he had personally incurred was for a double-share plough he had bought the previous season. This information seemed to interest Marais, who invited them to come with him to the dealer's store, where his son-in-law, Len Marks, quickly produced documentation that showed that Kas did indeed still owe the firm £9-1s-10d for a Massey-Harris plough purchased in 1938. On the basis of this, Marais persuaded Belcher that Wentzel & van der Westhuizen had a prior claim on Kas's harvest, and that if he wished to protect his principal's interests, he would have to mount an action against Klopper. Since the harvest was already in town, he suggested it would probably be in Kas's interests if he were to leave the grain in the company's warehouse until he wished to sell it.[50]

Kas made his way home, painfully aware of just how close a call it had been. Within only a few hours he had lost and won an entire harvest. He should have been pleased to get back to the safety of the shack, but Klippan had lost what little allure it had. Caught between Hendrik Goosen's wrath on the one hand and the ineffectiveness of Klopper's farming operations on the other, he felt that there was no room left to manoeuvre. He was being tortured by fire and water alternately, a man could only absorb so much punishment.

A week later Kas sold most of his harvest through the co-operative in Bloemhof and took about two score bags back to Klippan to tide the family over the winter. He discharged his debt at Wentzel & van der Westhuizen as soon as he could, and then, at the small post office at Sewefontein, he collected the firm's standard letter of acknowledgment addressed to 'Mr. Kas Teeu':

> Dear Sir,
> We include herewith Promissory Note No: 26902 which has now been fully met and wish to ensure you of our best service at all times.
>
> Thanking you
> Yours sincerely,
> Wentzel & van der Westhuizen

Not surprisingly, in a season in which the written word had figured so prominently in his life, this letter too found its way into his hoard of documents. It was reassuring to know that the end of season would see him free of any outstanding debts. There were others, including Cornelius Fick, who were less fortunate.[51]

Unlike hundreds of other poor whites in the countryside who were starting to haul themselves clear of destitution via the South African government's post-Depression public works programme, Hendrik Goosen's bywoner seemed to be losing the battle for survival. He had approached Kas several times, always with the same request, put to him in fluent

SeSotho: 'My children are starving, could you let me have a pound or two as I need to buy some food? I will let you have the money back as soon as I have been paid at the end of the month.' Kas, who recognised the spectre of starvation from his earlier stay at Kareepoort, had taken pity on him and often helped him out with small loans, which Fick, to his credit, had always repaid. But this time the bywoner's problems were beyond even Kas's reach.

Fick's oldest son had been involved in a shoot-out during an attempted burglary of a store in Johannesburg, and the family needed to raise twenty pounds in bail money. The bywoner was willing to put up two particularly handsome beasts as collateral, but Kas could not help him: 'Unfortunately I did not have the money.' Fick, who owned less livestock than he did, was deeply disappointed and merely observed, 'Man, if you don't help me there is no-one who will.' Nor was Fick the only white man in need of a loan at the end of that winter. Only a few days later, Kas let Dirk Bloem of Schoonsig have six bags of sorghum.[52]

That Kas was in a position to play banker—no matter how modestly—showed that the Maines were at last moving out of the shadow of the Depression. As they left Klippan for the last time, he was moved to take stock of how his family's position had changed over the two years he had spent there. In 1937 he had to borrow twelve bags of grain to ensure his family's survival and, as they left in 1939, he was owed six bags of sorghum. In those two years the boys had matured and become assets in his farming operations, he had reaped two substantial harvests, and he had upgraded his agricultural equipment.

True, the whites were now talking about a new and distant conflict of great proportions, but for the moment the only war he was interested in was the one with Hendrik Goosen which had succeeded in driving him off the property. He would retreat only as far as Rooistad, but felt confident that when he next returned to the farms he would well and truly conquer the soil. Most Boers still needed a good sharecropper.

*

Even though Kas knew what to expect, Kareepoort still came as something of a shock. The hidden economy that centred on the 'shebeen queens' in the villages, the small-time gangsters who hovered around the margins of the diggings, and the sharecroppers who worked the land on the farms were now all of far greater interest to the state and its agents than had been the case two years earlier. The tensions accompanying this growing contest for access to cheap black labour in a reviving economy were evident even in the ranks of the landlords like 'The Owl.'

Charles Gibson Smith was clearly pleased to see Kas, and more than willing to renew his partnership with so capable a crop-farmer, but he was at pains to disguise their old time-sharing agreement and even gave him a document that stated that Kas was a labour tenant and not a sharecropper. This was necessary, he said, because the police had recently attempted to prosecute him for sharecropping and warned him that they had a list of the names of all the sharecroppers on the property.

That list included the names of Mphaka and Phitise Maine, both of whom had been forced to provide some rather imaginative evidence during the course of an unpleasant court appearance.[53] With the threat of a possible court appearance in Bloemhof still hanging over him Kas was relieved to have avoided this conflict, but a discussion with his brothers left him under no illusions. The police were clearly intent on rooting out any resident males who were not *bona fide* wage labourers on the diggings, and it would only be a matter of time before he encountered Serasaka Moipeng and his men. In Kas's words, the police 'held that it was a crime to be idle' in what they had deemed an urban area, and in their zeal to enforce the provisions of the Native Laws Amendment Act of 1937, they 'searched the villages for loafers every Friday and Saturday.'[54]

The expected meeting with the police did not take long to materialise. One afternoon while bringing in the cattle from the fields, Kas was intercepted by a small contingent of mounted men under the command of Charl Marais. Recognising Kas as a former inhabitant of Rooistad, Moipeng asked him to produce the monthly 'tickey' pass that would show he was gainfully employed. Unable to produce the document, he could only counter that he was a self-employed farmer.

> Sergeant Marais questioned the suggestion that I was at Kareepoort solely because I had an interest in livestock farming. "But you do not have a pass, the people who live here work on the diggings."
>
> "Look, *Baas* Marais, I came here to look after my cattle. I left the farms because there was insufficient space for me to graze my cattle on. Here I pay cash in order to obtain the grazing necessary for my animals."
>
> Moipeng then asked me where I obtained the money from, and I told him that I earned cash by repairing shoes and saddles. He then advised me to get my affairs in order or else he would have to arrest me for not possessing a pass. People were not allowed to settle on the diggings simply in order to enable them to herd cattle. Moipeng told me to obtain the documentation reflecting my position as a self-employed person as soon as possible—"Or else I will arrest you. I am Marais. You know me."[55]

Disturbed by this menacing encounter, Kas went back and sought the advice of a few locals, who suggested that he speak to the other Smith, M. W. Smith, proprietor of the Kareepoort Store, who had often sympathised with the villagers in their struggle.

But Kas did not know the other Smith, and so he got his brother-in-law, Sellwane's husband, Legobathe, to come with him to the shop, where he was introduced to the proprietor. He told Smith about his problem and his need for a pass that would reflect his status as a self-employed man who could do cobbling, leather work, and saddle making. But the Englishman, too, was wary of Moipeng and his traps, and so he merely fished around amongst his things, hauled out an old shoe, and gave it to Kas with an instruction that he should take it away, repair it as best he could and then return it to him.

A few days later Kas took the repaired shoe back to the store, fully expecting that Smith would now agree to help him. But much to his surprise, the proprietor merely glanced at the repair and then took out an even older pair of shoes and repeated the instruction. This time around Smith took more interest in Kas's work, which, he said, was at least as good as that of Jacob Lebone, who easily obtained a monthly pass as a self-employed shoemaker. Smith took up his pen and wrote Kas a letter of introduction to Israel Gordon, handed the shoes back to him, and told him to take along all the items to the attorney's office in Wolmaransstad.[56]

Gordon, who along with his nephew Errol Behrman had been on the fringes of the unsuccessful campaign to get Hymie Basner elected to the Senate as a Native Representative in 1937, was becoming more and more familiar with the problems of black folk on the receiving end of the police campaigns. Indeed, so common had the pass raids become that even men of standing like Ishmael Moeng, the 'coloured' hawker attached to Amin Patel's store, had taken to paying the firm a retainer of twenty pounds a year to defend his elderly mother and his wife from what could only be termed police harassment.[57]

Gordon studied the letter Kas gave him, got up and walked out of the office, then reappeared a few minutes later with a policeman. They put a few more questions to Kas, disappeared to exchange a few words in private, and when Gordon next appeared he announced that the police had approved Smith's recommendation and Kas had been granted a pass showing that he was a self-employed artisan. Gordon handed him the document that would put him beyond the reach of Moipeng.

As always, Kas's response to this minor triumph over government bureaucracy was low-key, pragmatic and highly personal. There was no doubting that it was an awful system that required a black man to get police approval in order to follow his calling, but he believed that the law itself was not intentionally wicked, and that in the end an honest man had little to fear. And so he '. . . took the letter and started repairing extensively. I repaired saddles, reins, everything. I repaired reins for those Boers who worked on the diggings. I made good money. Grazing for cattle had to be paid for at the rate of a shilling per beast per month, and I had fifty cattle.'[58]

Between the cobbling and livestock farming, Kas still found time to open up a small field in which he planted maize and sorghum. The spring rains were most encouraging. In fact the closing months of 1939 were almost as auspicious as the previous summer, with its record rainfalls and local flooding within the Triangle. But even with above-average moisture to draw on, the coarse gravel of the diggings was no real match for the richer soils on the white farms, and so Kas monitored the crop's progress with some care. By late November all the seedlings were doing well, the grazing was unusually good, and the shoe-repairing business was still improving when he came home one afternoon to find Sergeant Ross Potgieter and Hendrik Goosen on his doorstep. Had the man from Klippan returned to haunt him? Goosen lost no time in getting to the point:

> He said: "I think that it would be best if this matter between us were dropped." I told him I had not laid a charge, and reminded him that it was *he* who had initiated proceedings by suggesting that I had stolen his brother's sheep. He said that we should forget about the past, and agree to co-operate in future.[59]

It was an offer that Kas could hardly refuse. 'My brother was living on his farm,' and who knew which gates would need opening when he next returned to the farms? He agreed to make peace, and the sergeant quickly drew up a document stating that the matter had been resolved to the satisfaction of both parties. Kas signed it, they shook hands, and the two men left.

This accord, together with his newly acquired pass and his status as a self-employed man, helped to restore a measure of tranquillity to Kas's life. He had been lucky. Others, more especially the women on the diggings, found themselves in far less enviable positions. Already relegated by the police raids to a marginal role in Kareepoort's hidden economy, they now found their position undercut even further by a renewed influx of BaSotho and ochre-smeared Xhosa-speaking women hailing from the most distant corners of southern Africa. This influx succeeded only in attracting more attention from the men at Makwassie.[60]

Moipeng, who had in the past directed most of his energy against the hard core of professional beer brewers on the diggings, now found that 'Almost everyone in the whole village sold beer, there was no woman who lived there without brewing beer.' The result was not only a dramatic increase in the scope and frequency of liquor raids, but the almost daily harassment of unescorted African women, who were expected to produce either 'certificates of good character' signed by a local magistrate in terms of the Precious Stones Act, No. 44 of 1927, or permits signed by Native Commissioners in their home districts authorising them to enter and reside in what had by then been officially proclaimed an 'urban area.' Even Kas, whose two wives and mistress enjoyed a measure of enforced protection as a result of his own privileged position on the diggings, was forced to concede their plight. As Kas put it, 'The women were arrested for either liquor or passes. They formed long queues. *Hey!*'[61]

*

This head-on assault on the rights of wives, widows and daughters, and the sealing off of one of their few independent means of getting cash was deeply resented by women throughout the countryside. In Kareepoort this indignation could be detected in the village churches. Even at village level, however, the church mirrored not only social ties within the community, but some of its class and ethnic divisions. Thus, while the ochre-tinted Nguni traditionalists tended to remain aloof from these more acculturated forms of protest, the Xhosa-speaking Christians from the eastern Cape had no hesitation in turning to the 'Israelites' for assistance—members of the same messianic sect whom the police had mown down during the infamous Bulhoek Massacre at Queenstown in May 1921.[62]

Likewise, while recent BaSotho immigrants may have had difficulty in finding an institutional shield against aggressive police action, the more securely established women—amongst them Mphaka's second wife, Motlagomang, and several other stalwarts—sought refuge under the banner of the African Methodist Episcopal Church. On more than one occasion in the early 1940s the Reverend Kemeng went to the Makwassie police station to secure the release of women arrested for pass offences, alleged to have been committed as they were on their way to one of his Sunday services! On balance, these interventions did little more than ensure a marked rise in the amount of police hostility directed at A.M.E.C. members at the time.[63]

Priests and churches that were of only marginal assistance to the women in their struggle against pass-law enforcement were even less willing to pick up the cudgels, let alone wield them, in any conflict that revolved around alcohol. Church-going women who entered the beer trade in the hope of easing their economic dependence on the men in their lives were therefore in the uncomfortable position of having to go back to their men to mobilise them for a more overtly political struggle. This had a double-edged effect. On the one hand at least the men began to meet after work on Saturdays to discuss the practically inseparable issues of beer brewing and pass raids, but, on the other, it tended to weaken the women's grip on the issue by allowing too much power to devolve into patriarchal hands. As Motlagomang Maine later put it, 'They talked amongst themselves at their meetings and we did not know what they were talking about.'[64]

Who exactly were 'they'? The half-dozen leaders in this group, although from differing ethnic backgrounds, were part of the small aspirant black middle class that somehow managed to maintain a foothold in the shifting gravel of the diggings. Although a few of them, such as Jacob Lebone and Mphaka Maine, had been staunch supporters of the I.C.U. under Jason

Jingoes during its heyday, most of them displayed little evidence of ideological fervour. Underlying social and economic affinities rather than any overt political characteristic gave the group its coherence.

Like 'Jacob the Shoemaker,' most of the men at the informal meetings held behind the shop at Boskuil were either the beneficiaries of a rudimentary education and were self-employed or had jobs as skilled or semiskilled workers. Ishmael Moeng, who had been a sharecropper and a motor mechanic long before he ever became a 'hawker,' and Morris Sello, the local pump operator, were two such men.[65] Mphaka Maine, who had managed to get the magical monthly pass as a self-employed 'builder' after the Wolmaransstad magistrate saw a house he had constructed for a white man at Boskuil, fell into the former category. But, if Mphaka was attracted to the group by the economic status of its members, he was less at home with their religious persuasions because most of them were committed Christians. Lebone, Moeng and George Motjale were all active lay preachers in either the Wesleyan Church or the A.M.E.C., while Morris Sello, like many others on the diggings, was a Lutheran.[66]

Amongst the initiatives taken by this group was a request they directed to the Wolmaransstad municipality for the authorities to use the provisions of Government Notice No. 1632 of 1938 to allow 'respectable women' the right to brew beer for domestic consumption over weekends. But as Motlagomang Maine recalled it many years later, the municipality was deeply suspicious of 'people who were working for themselves and not for the whites,' and the application was unsuccessful. Undaunted, about twenty women 'organised themselves' and launched their own initiative via the man who sold most of the locally made malt, M. W. Smith, of the Kareepoort Store.[67]

Smith's first approach to the municipality was rejected almost as quickly as Jacob Lebone's, but the second time he succeeded in obtaining permits for all the women on whose behalf he applied. In theory, respectable women were restricted to four gallons of sorghum beer, but in practice, the permits opened the door to professional operators. Motlagomang Maine was among those who noted that, although all the brew was supposed to have been consumed by Sunday night, this seldom happened, and if only black policemen were sent to conduct a raid on Monday mornings, 'We gave them some to drink and they left us.' The 'shebeen queens' went even further, adding sugar to the brew to shorten its fermentation period and increase its alcohol content, with the result that 'Those who did not work [for white employers] used to sell beer from Monday to Monday. There was no special day on which beer was sold.'[68]

These developments complicated the task of the Makwassie police. Moipeng and his men, not renowned for their patience or tolerance, now had not only to distinguish between traditional sorghum and 'sugar beer' but to judge whether they were dealing simply with an errant wife, an immigrant MoSotho or a Xhosa woman trying to make money for the first time, or a hardened 'shebeen queen' exploiting the weaknesses of the system on a grand scale. As Ishmael Moeng shrewdly observed many years later, the introduction of the 'permit system' thus saw an increase rather than a decrease in the amount of police harassment of black women on the diggings.[69]

By spring 1939 the demand for 'Certificates of Good Character' and police raids in and around Kareepoort had reached fever pitch, and the women, with no one else to turn to, put renewed pressure on the men who met behind Solly Meyer's shop. Jacob Lebone racked his brains for the name of a broker who could intervene on the community's behalf and then, on 4 August, he wrote to the son of his former employer—J. G. F. Moult, the Cape Member of the Provincial Council, who had stood for office at the same time that J. D. Rheinallt Jones had been returned to Senate as a 'Native Representative' two years earlier:

Boskuil,
Western Transvaal,
4th. 8th. 39

Mr. George Findlay Moult.

*Our Lawship*

We are here by asking an advice of the place we are thereof. Here at Boskuil, though the law of Maquassi is very heavy on us through our wives they are arrested day after day they have got to carry passes. We do wonder much about this arresting this native women therefor we do offer our lawship to give us the light if you know wheather a woman is lawful to carry a pass. If she hav'nt a pass they say she's got to be arrested althrough she say her husband went to work at the diggings. If the Policeman line a native woman in her house, he ask her where is your Husband. If you say my husband went to work they arrest her. The word is you have got to be where your Husband is. Well then our lawship this treatment makes us upsad. It is havy on us we wo'nt hear it to carry the women on our Backs to go to work with them. When ever the Native woman is arrested she has to walk to Maquassi by fect from Boskuil. Will this poor native women bear it to be driven to Maquassi by horses from Boskuil. If you have lost your father and your mother is left with you she has to carry a pass in case she is a widow. If not she got to be arrested. If your daughter is fourteen years old she's got to be arrested she is not married. Well our lawship we do offer an advice from you our Lawship.

We Remain yours
Jacob Candle
Morris Sello
Gabriel Taunyana
Peter Sellodad
George Motjale.[70]

The younger Moult checked the accuracy of these allegations with his father, who was still a resident of the Triangle, and then passed the letter on to Rheinallt Jones.

A week later, on 11 August, Rheinallt Jones wrote to Lebone pointing out that 'the new law in the towns is very strong against women being in towns unless they have homes there,' and asking whether 'it would help if I wrote to the Native Commissioner at Maquassi and asked him to look into your trouble?' When the shoemaker replied to the Senator ten days later he took up an invitation to respond in SeSotho and took the added precaution of getting a scribe to record his thoughts. Rheinallt Jones got the letter translated.

P.O. Boskuil.
21-8-1939

Senator Rheinallt Jones M.A.

Dear Sir,
In replying to your letter, dated the 11 August 1939, which asked about the distress of the women pass laws. 1. This is a very serious matter indeed. It has severely hurt our hearts, through this illtreatment of our wives. When a man is out at duty, on his return, he founds his wife arrested, together with her eldest children, when she has no pass. Farther more they want to know why the girl of 13 years is unmarried. 2. They chase the woman with a horse from Boskuil to Maquassi about 14 miles in distance. We are in great sorrows with this illtreatment of women. This place on which we dwell, we have hired it. We pay for our cattle and sheep. Hon. Jones, here in Maquassi, when the policeman comes to your house at night, he knocks on the door like a savage man. When you ask, who are you? he comes in rushing, and persecute the lady when she has no pass, and she has to leave the family. As you ask, if this will be good if you write to the Commissioner of Maquassi, yes! it will be very good. That's all I have to say.

Remains,
Yours,
Jacob Candle.

Rheinallt Jones did not answer this letter directly, but decided instead to take up the issue with the Native Commissioner when he went to the southwestern Transvaal a few months later.

Long before the Senator had a chance to see his constituents, events on the diggings took a more dramatic turn. In spring Lebone again took up his pen to direct his final appeal to Rheinallt Jones:

P.O. Boskuil
Transvaal.
20-11-1939

*Complaints*

1. We are in great troubles here in Maquassi, in farms and villages. On the 31st. Oct. 1939 there was a marriage in Hartebeespan at Mr. Herman Otto's farm. He had permitted his servant to celebrate the wedding of his daughter. 2. France Marumo's daughter married by Joseph Modisalife's son. Habitually and customarily the lady's mates and friends were all cordially invited to attend the wedding. The father therefore killed an ox or so, and his invited friends helped in cooking, so do the father-in-law's relatives in all details. A tragedy! We were all arrested through the permission of the magistrate of Maquassi. 3. On Tuesday when the wedding commence, all busy cooking. And on the next morning, 1st Nov. 1939, about twilight, where were feasting, there came two police-men and two Police-boys on a pick-up. They arrested France and punished him severely and many others, as well as the ladies, for passes and taxes. They kicked and whipped us I tell you. Oh! It was a very sad sight. The people ran away, and ran over the children's heads through the doings of the police-men. 4. We were taken in front of the magistrate, and no question was asked to anyone of us about her or his guilty. 7 days, 10 shillings was the fine. Through the solicits of the lawyer the fine was then changed to 7 days, H.L. hard-labour. 5. Oh dear sir! Where are we going to get living, we Bantu people as we are much troubled in this country? Troubled by our European friends. Where will our children hold weddings because our commissioner is heartily against these purposes? Aren't the European ladies also get married and doesn't the fathers invite all their friends? Why not possible to ours? Quantity of foods were spoiled that day and our mirth was marred. For an example we were happy for the 2 obedient pairs who respectfully satisfied our will, pity, the police-men & the magistrate marred our joy.

Jacob Candle.

This time Rheinallt Jones acted almost immediately. The Senator got an unknown official to make some informal enquiries about what seemed like a social massacre, and to report back to him.

Notes drafted during the discussions with the local authorities and with the members of Jacob Lebone's group were forwarded to Rheinallt Jones in Cape Town. From them one could see that the story was much as Lebone had told it. Though the host had received permission from the landlord and the manager of the nearby diggings to hold a 'wedding feast,' and though he had the necessary permit to brew beer, Marais and his men had raided because the document had not been 'ratified' by the police and some of those at the gathering had been guilty of defaulting on tax payments, or of being without the appropriate passes. Those arrested had appeared before the Native Commissioner on a range of charges that could no longer be determined with any precision, and Errol Behrman, who had appeared for the defence, had advised his clients to appeal. An unknown Johannesburg-based lawyer had, however, advised against such a course of action and, in the end, several 'young' men and women had been forced to spend a week in the Makwassie prison.

But in the course of making these notes, Rheinallt Jones's unnamed informant also offered a few uncomplimentary asides which showed that, unlike many of the other local observers, he was fully aware of the nature and extent of Kareepoort's hidden economy. The general impression was that 'the community' seemed to be composed of a 'motley crowd' who had settled on the surrounding 'diggings and farms,' and that 'illegal squatting' probably accounted for the women's strong objection to the 'certificate of character.' Most of the wives he spoke to claimed that their men had 'gone away to work or to plough.' 'I did not find out where the men went to plough,' he wrote, but 'possible explanations' were: '1). on W. Tvl. or OFS farms but they don't want their families there—if so are they legally away from these farms? 2). In Reserves (very unlikely) or Basutoland (more likely) and they don't want their women to have to go back or are they not their regular wives? 3). Share-farming somewhere.'

Despite these revealing sideways glances, the official eye remained blinkered by legal considerations. Hastily scribbled 'recommendations' appended to the note only suggested that the local inhabitants be informed of the exact provisions of the Act and that the government be asked 'to make the law better known.' Even Rheinallt Jones could see that this would do little to relieve his beleaguered constituents, but being based in the Cape, he was not in a position to offer them practical assistance.

When eventually he got to visit the Triangle early in 1940, his first stop was at Wolmaransstad, where he apparently met with the local police in an unsuccessful attempt to have them adopt a more pragmatic approach. From there he went to Solly Meyer's shop at Boskuil to meet Lebone's group; he did little more, so Ishmael Moeng was informed, than dismiss the suggestion that the consumption of 'sugar beer' was inherently dangerous and condemn the fact that 'the police used to assault people—women and children—indiscriminately.'[71]

The outcome of this meeting was disappointing for the shoemaker and his supporters, and even more so for the local women. Liquor raids on the diggings and the adjacent farms continued unabated until the police eventually managed to alienate even the local farmers. All of this culminated in a case in which the local magistrate ruled that there was no need for African women in villages adjacent to the diggings to produce 'certificates of character' on demand, but that if the farmers wished to remain within the bounds of the law and have black families living on their properties, they would have to apply for 'permits' under the 1937 law.

Semi-literate farmers found the task of completing the forms which the magistrate handed them so perplexing and time-consuming, that they soon became despondent. 'The Boers were frustrated,' recalls Ishmael Moeng, and desperate to find a solution, they set aside their prejudices and retained the services of an Indian lawyer from Johannesburg, who came to Kareepoort and instructed M. W. Smith in the art of satisfying the state's requirements. Then, and only then, did the farmers succeed in getting the authorisation from Pretoria.[72]

None of this helped the men and women on the diggings, who were ever more disillusioned with what they saw as the ineffectualness of Rheinallt Jones's policies. At this juncture they rediscovered the more radical ideas that had been propagated by the man who had opposed him in the 1937 election, Hymie Basner. How could they fail to be moved when, as Basner reminded them in his 1942 election manifesto, 'the pass laws have not been abolished, the beer raids have not ceased' and 'the conditions of Africans on the farms has not improved'?[73]

This unlikely shift in sympathies—from voting for a 'Christian liberal' to endorsing a 'Jewish communist'—was not easily achieved by the essentially cautious middle-class elements that led black working-class thinking and voting behaviour on the diggings. It took a racist state, intent on forcing a brutal legal code on a small and vulnerable community, to bring about this success. How else could a handful of semi-educated artisans and a half-dozen lay preachers guide sharecroppers, labour tenants or wage labourers—people who were directly drawn off the land and often even more conservatively inclined than they? But if it was the South African state that managed to spin new political militancy out of old middle-class male straw, then it was Moipeng and his men who were responsible for transforming the women's consciousness. Here too there was a set of developments which, at first glance, seemed rather unlikely. The ladies of the African Methodist Episcopal Church—among them Ishmael Moeng's mother, the senior Mrs. Masihu and Mrs. Jacob Lebone—rather than the professional brewers were the most outspoken in protesting the hated 'certificates of character.' These women, whose virtue was beyond question, played a leading part in the women's guild which met at Makwassie each week under the Reverend Tladi, orchestrated the opposition to the pass laws, and helped to forge a new political alliance that was crucial in the 1942 election.[74]

The landlords were predictably disturbed when several blacks started referring openly to themselves as 'communists' (though Hymie Basner himself left the party in protest over the Soviet invasion of Finland). Unable or unwilling to tackle the real issues head-on, the farmers directed most of their anger against the unfortunate Reverend Tladi, making it impossible for him to host public meetings of an overtly political nature. But neither this nor other acts of intimidation could deter the core of the A.M.E.C. women and their colourful supporters, such as Motlagomang

Maine. Even the Israelites, who given their turbulent history had more reason than most to be circumspect, were drawn into supporting the alliance developing around Basner. The Xhosa-speakers made only one concession to the state: they said that they 'did not want to be part of any committee, they only volunteered to help.'[75]

This spirit of resistance, first evident after the disappointing visit to the Triangle by Rheinallt Jones in early 1940, gained momentum in the months that followed and revitalised the meetings behind Solly Meyer's shop. Although dominated by the Christians-cum-communists, the meetings were always open and democratic enough to attract interested onlookers who—like Kas and Mphaka Maine—were usually drawn from the diggings or the surrounding farms.[76]

But, as had happened at the time of the I.C.U., Kas listened and watched a great deal more than he ever participated, leaving most of the serious talking to his brother. It was noted that on the rare occasions that Kas did bring himself to speak, 'he spoke powerfully' about farming matters but could never quite bring himself to join the Basner supporters. In truth, Mphaka's success as a builder meant that he, too, was no longer as politically motivated as he had once been, while Phitise confined his enthusiasm to the A.M.E.C. This lukewarm support for the progressive cause amongst the Maine men did not go unnoticed amongst the 'crossover' communists from the A.M.E.C. Ishmael Moeng, the lay preacher who, by the mid-1940s was sufficiently committed to the 'Communist Party' to take on the task of organising black farm workers on a block of nine properties around Boskuil, later noted of the three or four Maine brothers that they '. . . were not dedicated. Even at church they lacked dedication. Their main concern was business—anything other than business did not hold their interest.'[77]

There was a good deal of truth to this. Moeng saw Mphaka, Phitise and Kas, as well as their half-brother Balloon, in midcareer, at a time when, cut off from the broader expanse of the surrounding properties, they were making extensive use of their formidable skills as artisans to pay for the grazing of their animals and to build up their herds in the hope of eventually returning to the farms. In addition, each of them had other pressing commitments. Kas, for example, was responsible for the well-being of two wives, Nomahuka, and ten children ranging in age from the ponderous seventeen-year-old Mmusetsi right down to the colic-plagued baby Nkompi. Under the circumstances it was understandable, if not always forgivable, that the Maines tended to pay so much attention to their business interests. In the case of Kas, his domestic responsibilities had also undergone a recent extension. In January 1940, Leetwane's mother arrived at the shack to play midwife when her daughter gave birth to Nkamang—the couple's fourth daughter and eighth child. Without a landlord to turn to, Kas gave a ewe of his own to meet the requirements of *leseko*, and the baby was welcomed into the world with a modest family celebration.[78]

*

Shortly after the birth of Nkamang, Kas and the two older boys gave the sheep their midsummer shearing and, for what was probably the last time, sold the wool to 'Ou Vel' Gabbe, who was at the end of his career as a Jewish hawker or *smous* in the Triangle. Much to Kas's delight the wool fetched a much higher price than had been realised the year before and, with the proceeds tucked away in his purse, he set off to honour his only outstanding debt in Bloemhof—the sum of nineteen shillings and six-pence owed to Wentzel & van der Westhuizen for a pair of shears bought before leaving Klippan.[79]

There were other reasons why farming rather than politics preoccupied the Maine brothers in January. Only days after Nkamang's birth, Kas noted heavy clouds moving in from the northeast and, sensing trouble, moved his animals from the low-lying areas around the Bamboesspruit to the higher-lying ground leased from Piet Ferreira at Boskuil. Within hours there was a drop in temperature and, over the next three days, a downpour of three and a half inches of rain transformed the normally sluggish Magoditsane into an angry torrent storming its way to the Vaal. 'It really rained. The fields were waterlogged, houses and bridges collapsed and the valley was flooded. People could not cross the spruit for two days.'[80]

The old corrugated-iron sheets that Hendrik Swanepoel had bought Kas warded off the worst of the water, which meant that the family was spared the danger of collapsing mud walls. Several of the neighbours were less fortunate. Such was the intensity of the flooding that rail traffic on the elevated line linking Kimberley to Johannesburg was disrupted on three consecutive days. Below the rail embankment there was a real disaster when Phitise lost fourteen oxen to the cold and rain. In retrospect Kas could see, 'That was when Phitise became poor.'[81]

Despite his precautions, Kas too lost one of his animals, but this was not a major setback. As the wartime economy gained momentum he found himself with more clients wanting shoes or saddles repaired, and this gave him more cash than he had enjoyed for several years. At about this time he taught himself how to wash and spin wool, which he knitted into the small fluffy caps that found such a ready market in the villages. Once again his access to machinery or material—in this case the wool from the sheep—ensured that a skill which under other circumstances might be defined as 'female' fell into the patriarchal orbit and helped secure the continued male dominance of the domestic economy.[82]

Knitting, and a new round of tailoring for some of the more prosperous blacks in Kareepoort, enabled Kas to visit the stock fairs earlier than usual and, in mid-April, he bought two oxen for cash from Slabbert & Campbell.[83] Yet more money came his way via some patients referred to him by Sekwala, and even more from some 'coloured' shopkeepers who came from as far afield as Kimberley to obtain the 'medicine' necessary to protect their businesses from the ill-will of jealous rivals.

A modest crop meant no need for outside assistance, and it was brought in by the family. This too helped to grease the wheels of finance and underwrite Kas's undiminished passion for livestock speculation. And still the chase for money continued, until one might have wondered whether some undisclosed project was driving such frenetic activity. A few sheets of corrugated iron inserted into the sides and back of the wagon increased its capacity and, with ten-year-old Nthakwana in tow, Kas set off to visit the richer grain-producing areas across the Vaal, where he bought several loads of dried maize cobs from Free State farmers. Back on the diggings, they trundled the cobs from door to door and sold them for eight shillings a bag to beer brewers who had neither the time nor the inclination to spend cold winter days hunting for fuel on a treeless flood plain. Yet again, Kas had set himself up in competition with Leetwane and Lebitsa. How could dung collectors compete with a man who owned a wagon?[84]

Any resentment that this competition might have caused soon paled into insignificance when Kas's wives discovered what his new project was. He was intent on legitimising his relationship with Nomahuka, and no sooner had he paid the bridewealth to her father, Mbewane, than the two Lepholletse sisters turned on his unhappy Nguni-speaking 'spouse.' Leetwane and Lebitsa made a point of casting Nomahuka in the role of the outsider, referring to her only as 'Kas's Xhosa woman,' and contrived generally to have her excluded from the inner circles of the Maine household.[85]

To add to her woes, Nomahuka found that her youngest, Nkompi, was troublesome at night, often crying for hours at a stretch. Trying to understand such fractious behaviour she consulted a traditional practitioner who, sensing the conflicts in the household, suggested that the child was miserable because the Maines had withheld the traditional family welcome, the *leseko*, at his birth. This diagnosis, hard on the heels of the celebration that had followed Nkamang's arrival, made Nomahuka question Kas's commitment to their relationship, and the problem worsened.

Nomahuka was being marginalised by two determined sisters who had earlier seen off the challenge of their younger sibling, Tseleng, and Kas, too, came to see her as the single greatest cause of tension within the family. Before long Leetwane and Lebitsa were accusing her of 'witchcraft.' This led to her summary expulsion from the household, and when she sought refuge with Phitise and his family, Kas found her out and made sure that the 'witch' was also cast from his brother's home. From there she fled to Mphaka's where the oldest of the three brothers, blessed with a well-developed sense of social justice as well as a quick temper, gave her a sympathetic hearing and agreed to take the organisation of a leseko ceremony upon himself.

This time Kas was uncertain about what to do. Mphaka (who had only recently appeared in court for whipping a prostitute who had dared to threaten his son's marriage), presented a sterner challenge than the placid Phitise. His nerve failed him, but he was partly saved by Mphaka's senior wife, who, presumably out of loyalty to her sisters-in-law, refused to have anything to do with the proceedings. In the end, the arrangements for the leseko were made by Mphaka's junior wife, Motlagomang, but by then Nomahuka had clearly had enough of the Maines. In the winter of 1940, she, Maqeqeza and Nkompi were seen for the last time, slowly making their way from Kareepoort towards Wolmaransstad. In later life none of the Maines was ever eager to talk about this shameful business.[86] Indeed, even at the time Kas spoke less than usual and chose instead to concentrate on preparations for the coming spring.

Toward the end of September he took over the debt of one 'Joubert,' who owed E. G. D. Elliot more than twenty pounds for a fine wagon purchased from Isaacs & Gabbe in Makwassie. This helped replace the rather wheel-weary wagon that he and Nthakwana had used earlier on their Free State expedition.[87] He looked over the rest of his equipment and decided that the planter, too, had seen its best days and sold it to a bywoner named Willem Swanepoel for eight pounds. Swanepoel, like most tenant farmers, was hard-pressed for cash, and so Kas allowed him to take the planter for five pounds on condition that he pay the rest after the harvest. Although it might not have appeared so at first glance, this was in fact quite a good deal, given that Kas had only paid five pounds for the planter ten years earlier.[88]

When the first good rains arrived, Kas had second thoughts about the wisdom of having sold it. That magical musty odour of moist spring earth on the highveld produced a yearning for the open expanse of the farms, and he felt constrained by the way that the diggings and their fences hemmed him in. Livestock speculation was all very well, but few feelings rivalled the satisfaction that came with transforming a well-ploughed field into a harvest. No, his pass might give his profession as 'shoemaker,' his wealth might lie in cattle, but his heart belonged to the grain fields.

From the moment the rains fell that spring the desire to get back to the farms never left him. Kareepoort grew smaller with each passing week, Rooistad and its people crowded in on him day by day, and hour by hour one or other official would arrive with yet one more request to produce one more document. When they were not after beer they were after money, and always passes, passes and yet more passes. In mid-February someone from Inland Revenue arrived to demand 'Wheel Tax' and left only after he had been paid twelve shillings to cover four wagons and two bicycle wheels. On the farms one was protected from too many uniforms, from too many white men armed with notebooks.[89]

Soon, with the first of the sorghum harvest coming in, the horses from Makwassie reappeared all along the length of Magoditsane, bringing with them a new and even more vicious round of beer and pass raids. Kas relived the stress of his earlier encounter with Abraham Jonker and, forty years later, he would still recall with some bitterness 'the difficult conditions under which we lived.'[90] Irked by the need to produce his 'tickey-pass' for policemen who now seemed to come in numbers last seen during the locust plague, Kas went off to the Pass Office at Wolmaransstad and made a formal application to be exempted from the pass-law provisions. Mphaka's status as a self-employed builder had earned him an exemption and presumably there was not that much to choose between a builder and shoemaker? All that this achieved was a letter stating that 'the matter [was] still under correspondence.' This spared him a monthly outlay of thruppence, but now he had to carry around the letter if he needed to negotiate his way past the police. In effect, he had been issued with a pass by another name. The joke was on him.[91]

But there was nothing funny about the pass laws. Unbeknown to him, the villagers were about to get some much-needed relief from their enforcement from an unexpected quarter. The community's salvation, it seemed, lay not so much in the political actions mounted by 'modern' men like Jacob Lebone and his friends, but in the mysterious powers of the occult commanded by the peasant underdogs who had suffered most at the hands of Sergeant Charl Marais. Serasaka Moipeng—a policeman 'whose complexion was as dark as if he were a slave,' as one acquaintance put it, and indistinguishable from an ordinary MoTswana when he disguised himself and spoke the vernacular—might have been classified 'white' in a race-obsessed society, but he was also a man who had drunk so deeply from the springs of the other that much of his inner being had become 'black.' Paradoxically, Moipeng's very success at mastering the elements of Tswana cultures was his undoing.[92]

As Kas remembered it, during the course of a raid at Leeubos one Friday afternoon in 1939 Moipeng arrested two BaNgwaketse migrants for 'trespassing' on the diggings. In keeping with the custom he roped them to his horse, *'Trap die Kaffer,'* and assaulted them on the way to Makwassie. At the police station the prisoners were subjected to more gratuitous kicks and thrown into the cells. Before the gates could swing closed on them, they fixed their eyes on Moipeng and told him, 'Marais, today you were a policeman for the last time.'

The following morning Moipeng was taken ill and rushed to the hospital at the place which the BaTswana knew as Matlosana, known to most whites as Klerksdorp. When Moipeng left the hospital two weeks later, he was 'shaking' and over the next twenty-four months his condition steadily deteriorated. In the winter of 1941, weeks before he left Rooistad, Kas happened to bump into him at one of the monthly stock fairs in Makwassie. 'I saw him at a cattle auction, greeted him, and asked him how he was. He told me that ever since he had kicked those BaNgwaketse people he had not been able to stop shaking.' As another long-standing inhabitant of Boskuil, Piet Baraganyi, later recalled it:

> Charl Marais was ill for two years, that is 1939 and 1940. He died in 1941. It was rumoured that he had been bewitched by two Kgalagadi men. They swore that they would get rid of him, and they did.[93]

By the time they buried Moipeng, Kas was already deeply involved in his winter programme. Seldom, if ever, had his cobbling, leather-working, knitting, and tailoring skills been in greater demand. His clientele stretched well beyond the diggings to many whites on the farms, for he enjoyed a reputation as a saddle repairer of exceptional ability. With money regularly flowing into his coffers, he went out and bought himself a copy of *The Rapid Ready Reckoner*, carefully inscribed his name on the inside page, and then protected the slim volume of tables with a stiff brown-paper covering, improvised from a sugar packet bought at Smith's store. A craftsman could use all the arithmetical help he could get, but a sharecropper needed such a book even more.[94]

Ironically, just as he was preparing to leave Kareepoort Kas felt that he was making most progress in mastering some of the arts of urban living. When 'Vrystaat MoTshweneng' failed to pay him the thirty shillings for a saddle he had bought, Kas had no hesitation in going to see Israel Gordon who gave him a letter of demand to drop off at the construction site in Kingswood where the man was working. This strategy worked perfectly, and within days the outstanding sum was handed to the attorney, who deducted ten shillings for representing Kas in the matter of the pass exemption and the action against the MoTshweneng, and forwarded him a cheque for a pound, which he collected from Meyer's store in the last week of May. Out on the farms a sharecropper was largely on his own, but in the towns, as Gordon and A. J. Marais had proved, it was always a good idea for a black man to have a lawyer at his side.[95]

Toward winter's end Kas paid Sebubudi a visit and found his eyes feasting yet again on Klippan's tracts of unworked land. When Hendrik Goosen casually suggested that they consolidate their friendship by entering into an agreement for the new season, he jumped at the chance. By September 1941, the family had once again been assembled for an all too familiar move. As Nthakwana, then still a slip of a girl, patiently explained to one of her bemused interrogators many years later: 'I can say that we grew up in a wagon, trekking from one farm to another.'[96]

All family members were resigned to the inevitability of the move, but some were apprehensive. From past experience the older children knew that the move from Rooistad to the farm usually meant a shift in the demand for labour from father to sons and daughters. The prospect of having to work for Kas and Goosen did not fill them with joy. Life at Klippan could be very demanding for young adults testing the limits of control in a sharecropping family for the first time. But for Kas, nothing detracted from the cheery prospect of re-entering the world of agriculture.

CHAPTER EIGHT

# Re-entry

## 1941–43

After World War I, by its own often rather peculiar standards South African parliamentary politics had remained fairly stable for a number of years. The Afrikaner-dominated National Party, which represented a broad spread of class interests among whites, especially in the countryside, had combined surprisingly well with the more English urban-based Labour Party in the coalition 'Pact government.' Benefiting from the sustained economic growth of the 1920s, it drew whites together and helped to fashion a marginally less exclusive South African identity. But though it tended to take the edge off some of the issues of ethnic identity and class conflict within white politics, the Pact government fell apart during the Depression and the gold standard crisis of 1932–33; by 1934, formal Afrikaner politics was in disarray. Hard-line Republicans under the leadership of Dr. D. F. Malan, broke away from the Pact Nationalists to establish a 'purified' or Gesuiwerde Nasionale Party, while the majority followed General J. B. M. Hertzog into partnership with yet another general, J. C. Smuts, into the new government of National Unity of the United South African National Party.

This 'fusion' government, like its Pact predecessor, got off to a sound start. South Africa began a slow but sustained recovery during the mid-1930s, which helped to consolidate support in both town and countryside: the mining industry prospered, the gross value of agricultural and livestock production almost doubled in the seven full seasons between 1932 and 1939; and wool farmers especially enjoyed a steady expansion in exports to Germany, which helped offset some of the disabilities that Britain's imperial preference system imposed on South Africa. But the outbreak of war in 1939 and the consequent disruption of trade, as well as a resurgence of old and deep-seated Afrikaner Republican ideals, made for real political uncertainty.[1] Anxious to retain the loyalty of their most militant supporters, Hertzog and several parliamentary colleagues abandoned the United Party, and Smuts was left in charge of a wartime coalition ministry. With Hertzog out in the political cold, Afrikaner politics showed new and dangerously fissiparous tendencies. Besides choosing among various party political groupings, red-blooded Republicans could also choose to join any one of a number of extra-parliamentary movements, including several with overtly fascist sympathies such as the Greyshirts, the New Order, or the Ossewabrandwag.

In 1940, the Afrikaner Broederbond, sensing the danger in the situation, succeeded in cobbling together a shaky alliance between Dr. Malan's Gesuiwerde Nasionale Party and General Hertzog's newly formed Volksparty to form a Herenigde Nasionale Party. Though its founding was turbulent, this new party grew in strength and gradually managed to restore a measure of coherence to Afrikaner politics. In the general election of 1943, the Herenigde Nasionale Party under the leadership of Malan showed its power to return Afrikaner farmers on the highveld back to the fold of orthodox party politics.[2]

This political regroupment in the Transvaal and Orange Free State countryside was helped by the surge in the South African economy in general, and the agricultural sector in particular, which grew dramatically between 1939 and 1945, despite farmers' fears at the outbreak of war. Expanded industrial production drew thousands of new workers into South Africa's burgeoning cities, where the demand for food and clothing often outstripped the purchasing power of the workforce; the resulting hike in prices soon had an effect on several important commodities in the booming agricultural sector.[3] Farmers in the southwestern Transvaal, especially those in the marginalised areas of the Triangle, welcomed this surge in demand and increased their production of beef, maize and wool. But wartime opportunities were not confined to these old and familiar stalwarts. With meat and dairy products often in very short supply, manufacturers looked around for alternative sources of vegetable protein such as beans, while production of sunflower seed oil was encouraged in the hope that margarine would make good what seemed like a chronic shortfall in butter.[4]

Many highveld landlords expanded the acreage under cultivation and renewed their efforts to mechanise their farming operations. The number of tractors in the Transvaal, which had increased from a paltry 838 in 1930 to 1,181 in 1937, leapt fourfold to 5,702 by 1946.[5] Whether these were developments from strength, as a positive response to the new opportunities offered by urban markets, or from weakness, having to cope with the exodus of black labour to the cities, they did not bode well for black sharecroppers, whose prosperity depended on their having grazing for their stock and persuading white farmers to use their draught oxen.

Toward the end of 1942, the onslaught of Nazi U-boats on Allied shipping in the Atlantic reduced the supplies of agricultural machinery, equipment, spares and fuel reaching South Africa to a trickle, and the pressure on the land occasioned by the expansion in mechanised production started to ease. By 1943 it was official government policy to encourage white farmers 'to make greater use of animal power,' and thus the great global conflict, which at first had restricted hope for the Triangle's increasingly vulnerable black tenants, came to offer a respite.[6]

Kas could hardly have chosen a more propitious moment to re-enter the world of 'white' agriculture. Not only had he sat out the early years of the war as an artisan-cum-livestock speculator at Kareepoort but, at the age of forty-seven, he could call on the services of two wives and eight children at a time when agricultural producers were once again strongly reliant on manual labour. As physically strong as he was tall and lean, he was making good headway in his career as a farmer and believed he had the necessary experience, equipment and livestock to make a success of his new ventures. He knew that Goosen could be an awkward and demanding landlord, but he hoped their shared desire for material progress would overcome the difficulties in their relationship. If there was enough rain and the markets remained buoyant, he could compete with the best in the district: his arch-rival Johnson Xaba and the others in RaKapari's drinking circle would have to look to their laurels.[7]

But mobilising labour at a time when so many black youngsters were leaving for the cities was not easy. Although Kas could count on more physically mature labourers than ever before, many of his children were keen to use their skills and time in ways that were more compatible with their own dreams and aspirations. Work, up to then mediated by time and season alone, was beginning to be influenced by the chemistry of personality. Mmusetsi, the oldest son at the age of eighteen, should by custom have been both his father's closest companion and greatest help, but although the boy made every effort to please, the family generally agreed that he was abnormally slow-witted, slightly surly, and in need of close supervision.[8] The two oldest daughters, both sixteen and undoubtedly more capable than Mmusetsi, had difficult temperaments. Thakane, Lebitsa's only child, was a quiet, delicate and slightly withdrawn young woman liable to overreact to criticism. She would make a good wife, Kas felt sure, for his favourite nephew, Kgofu, who was turning out to be a good farmer and, like his uncle, had taken up the calling of a herbalist. Morwesi, by contrast, was as tough and independent as she was headstrong and opinionated. Her parents knew that she loved social life and dancing, but what her father did not know was that she had taken to slipping out at night to visit male friends or distant neighbours.[9]

Perhaps because these three were already on the threshold of independence the patriarch sometimes found himself looking down on the younger children with greater affection. Bodule, aged thirteen when the family moved to Klippan, was undoubtedly the star that shone brightest in his father's firmament. The lad was all that the unfortunate Mmusetsi could not be—willing, able and reliable—and, with a head pleasantly free of bookish notions, he was already showing signs of wanting to follow in Kas's footsteps as a farmer.[10] Bodule's efforts were balanced, though never quite matched, by the remarkable talents of Nthakwana. Although only eleven years old, she was already more independent-minded than her brothers and had the knack of mastering any skill, be it basket-weaving or smithying, that her parents exposed her to. In practice she became Leetwane's 'first' daughter, Kas's 'second' son and, not surprisingly, the other children saw her as occupying a privileged position.[11]

Matlakala, Kas's daughter by Tseleng, was nine years old; being an 'illegitimate' child and having always lived with her maternal grandparents at Khunwana, she was not part of the household proper. In some ways her absence put Mosala, then only seven, under marginally more pressure, because his father expected him to take care of most of the smaller livestock alone. This he did well, but the boy was a product of the diggings and already showing worrying signs of 'location' behaviour which, if left unchecked, could become anti-social. Pulane, at five, and Nkamang, at two, the babies of the family, were small enough to occupy the time rather than help to liberate the labour of either Leetwane or Lebitsa.[12]

In October 1941, once the clouds had put on a bit of weight and cast shadows of substance over the plains, Kas and Mmusetsi rounded up the oxen, roused the Massey–Harris from its winter sleep and set off to do the season's battle. Goosen had allocated them several very promising fields, and Kas and his oldest son would leave the shack at four each morning in order to plough as much as possible before the sun was too high. The middle passage of these long and sweaty journeys was sometimes observed by Goosen, who left his parental base at Mooiplaas once or twice a week to put in an appearance on the property he had hired.[13] Kas, who prided himself on being able to make his team of sixteen oxen cut a furrow as deep as it was straight, had little to fear from such casual inspections. Once the fields had been turned, he and Mmusetsi put them under the harrow and waited for the heavier follow-up rains that were bound to come. But even before then the soil was moist enough for the fields to put forth a tell-tale spring fluff of grass and weeds. Kas took this as a signal to proceed immediately with planting the quick-maturing maize and sorghum varieties he was accustomed to cultivating. However, Goosen ordered him not to plant and instead to plough the spring stubble back into the earth. So, shortly after Christmas 1941, the yokes were put back on the oxen and the span sent back into the fields. By the first week of January 1942, Kas had not yet planted any grain. The intervening weeks had not been entirely wasted, however, and he had watched with great interest as Hendrik and Pieter Goosen devoted some of Klippan's other fields to a rather unusual crop.

Union Congo, a Witwatersrand-based firm supplying margarine producers with vegetable oils, had given permission to an agent by the name of Kahn to contract with local farmers for sunflower seed. The Goosen brothers, sensing the chance for earning windfall profits from some of Klippan's hitherto underutilised tracts, had quickly planted several fields of sunflower seed. Kas eventually matched this move and devoted one of his own fields to the sea of bright yellow flowers.[14] Nor was this his only innovation. Like many others he had heard that the price of dried beans had increased by more than 70 percent in twelve months. He decided that this, too, demanded a deviation from the norm and he devoted a second field to the exclusive cultivation of 'sugar' and 'kaffir' beans.

Neither of these departures jeopardised Kas's grain-sharing agreement with his landlord. Indeed, he discussed both developments with Goosen who, although a militant Nationalist as opposed to 'Britain's war' as any of his Boer neighbours, seemed to approve. The sunflower seeds and sugar beans, Hendrik Goosen suggested, was their personal contribution to the nation's war effort.[15]

But the absence of maize and sorghum seedlings continued to worry Kas, even more so when Goosen told him to plough and harrow the field again. This time he felt compelled to question the order, but Goosen gave him a very plausible explanation: given the shortage of commercial fertiliser, it was imperative to plough the grass and weeds back in to increase the soil's nitrogen content and to enhance the moisture-retaining properties of the Triangle's lean Kalahari sands.

Kas found this persuasive. But rationality was a thing of the mind, and neither his son nor his oxen worked by logic alone: only with the greatest effort could he coax them back into the fields. With Mmusetsi under the tongue and the oxen under the whip, he drove them—and himself—to the point where the mixture of dust and perspiration caused their eyes to glow like a desert sunset. The oxen, lacking the love of a son for a father, collapsed first. 'They were muscular and strong but exhausted. To show that they were tired the oxen would simply lie down as they pulled the plough. That was not because they were lean or weak. No! They were tired, man!'[16]

Kas told Goosen about this intractable problem. But the landlord was determined to have his fields ploughed for a third time and refused to take no for an answer; he suggested instead that Kas buy a span of donkeys to finish the job. Kas tried to explain that he had just made a twenty-pound repayment on the wagon as well as a poll-tax payment to the Bloemhof authorities, but Goosen would not hear of it; instead he offered to lend Kas twenty-five pounds. Never before had Kas encountered such resistance: Goosen was at least as stubborn as he. In the end Kas took the loan, bought some donkeys, and then went out to persuade Lerema Moshodi to let him have two oxen in exchange for a plough. With this added draught-power, he put Mmusetsi and the animals back into the fields, which he ploughed and then planted.[17]

Within days the seedlings appeared, but the maize and sorghum fields did not give Mmusetsi the same joy they gave his father. The entire exercise had left him angry and irritable. Nor was he the only one to chafe beneath an abnormally heavy load: the two new crops that Kas had introduced with no discussion in the family were a nightmare for young and old alike. The beans cried out for attention in a way that could only be rivalled by a colicky new-born baby: they had to be harvested frequently and regularly if the pods were not to crack and spoil; no sooner had the women and children worked their way up to the end of the field—genuinely back-breaking business—than they had to turn round and work their way back. Their backs had hardly recovered from this when they were sent out to pick sunflowers, whose deceptively elegant necks were protected by ugly little thorns that punctured one's fingers so that the yellow petals were often stained crimson. From there it was back to the legumes, to pod the beans—yet another enormous task, which in the end yielded only seventeen bags. It was agonising work.[18]

Kas had to pay a high psychological price to keep the women and children constantly mobilised for this intensive labour on unfamiliar crops which, unlike maize and sorghum, lacked any immediate cultural resonance or obvious economic logic within a Sotho household. Sorghum and beer were used in any number of domestic rituals, and everybody knew that a family needed maize to survive the winter, but beans and sunflower seed could only contribute to cash in the patriarch's pocket. Although they did not immediately question his authority, the women, especially, sensed that the game was not worth the candle. So it was all the more ironic that the first clash came among the men, and it had nothing to do with beans or sunflowers but everything to do with a father's ruthless pursuit of success.

From the moment he had walked through the gates at Klippan, it had been apparent to everyone that Kas was going to throw himself into commercial farming. Whether his passion was simply an outpouring of energy accumulated on the restricted diggings, or consequence of his exposure to the competitive male ethos of RaKapari's drinking circle at a vulnerable moment in mid-career, it was clear that he was being driven—perhaps for the first time in his life—by volatile and potentially destructive emotions. His older brothers, whose age and distance allowed them the luxury of irreverence, took to calling him *Ramollwana*, 'box of matches.' Those closer to him and in greater danger of being burned, like Mmusetsi, had to be more circumspect.[19]

Cattle speculation, always close to Kas's heart, became even more important during the war. The price of beef rose by nearly 20 percent in 1941–42 and then surged by a further 14 percent over the next year, and this did not escape the attention of either of the hawk-eyed partners at Klippan. Within weeks of Kas's arrival on the farm Goosen proposed a joint venture in which he would put up the money for cattle and guarantee enough grazing, while Kas would fatten up the cattle on the farm's 'horsetail' grass before selling them at Slabbert & Campbell's monthly stock sales.[20]

The idea appealed to Kas and he visited outlying farms to persuade the Boer landlords to sell their calves at reasonable prices. This was a job he was born to do: he had the trained eye to protect his own interests as a speculator and all the necessary language and social skills to make the Boers feel comfortable during negotiations. He soon discovered, however, that his chance was much higher if he denied any speculative intent and instead plugged directly into the paternalistic ethos that dominated the Triangle's code of race relations, telling the white farmers that he needed cattle as part of an effort to raise bridewealth.

This 'cheating the Boers,' as he later put it, gave him no more cause for concern than did the fate of Nomahuka and her children, and his conduct was the subject of amused comment between himself and Goosen. By early summer several such 'bohadi' cattle at Klippan were being fattened up for the market. Goosen, delighted by this progress, pressured him to ferret out yet more such bargains, but Kas was unwilling to leave the farm when the sorghum crops were clamouring for attention. In May, when grazing fell off, they took the animals into Bloemhof, sold them at a good price and shared the profits. It had been a most satisfactory business. But Goosen, walking away with his wad of banknotes, had no idea of the price that Mmusetsi had been made to pay for this windfall.

Kas had driven Mmusetsi remorselessly all summer. When he was not out ploughing with his father, the lad was busy harrowing, weeding, harvesting, milking or herding the fifty or more cattle on a property that was mostly unfenced. Given the overlapping demands on his time and labour it was hardly surprising that he often got home after dusk to report that one or other of the cattle had gone missing. Kas, who attributed anything that departed from his own exacting standards to mere carelessness, showed little sympathy. The first time it happened he sent Mmusetsi out into a storm and told him not to come home until he had found the missing beast. And when the 'offence' was repeated he took a length of leather to the lad. Mmusetsi resented these assaults, and when they became more frequent he offered increasingly spirited resistance, which only exacerbated his father's anger.[21]

Events reached a climax one evening when Mmusetsi arrived to report that two of the donkeys (purchased out of the twenty-five pounds advanced by Goosen) were missing. He was sent out to search for them, but when by daybreak he had still not recovered them, he came home to milk the cattle. The donkeys did not reappear all day, so again he was sent out on a night search and again he returned alone. Kas lost all vestige of self-control. He grabbed Mmusetsi, tied him to a tree, and used a *sjambok* to thrash the young man's head, neck, back, buttocks and legs. Mmusetsi's screams attracted Thakane and the others: only when they seized his flailing arms did Kas drop the whip and pause for breath.

The sisters untied Mmusetsi, and, still bleeding and shouting, he ran off in the direction of Sebubudi's home, crashing through the acacias. Blinded with pain, he knew exactly where he was going. He found Motheba, as he had expected, in the little mud house that stood beside Sebubudi's. Tall and strong, she was less easily intimidated by her son's ravings than Kas's wives. She called for water, and bathed and cleaned his cuts and bruises, and pacified the most troubled of her grandchildren. Motheba was careful to ask him not to tell his father who had taken him in and cared for him. Ramollwana had gone mad.

Mmusetsi spent several days under the watchful eye of his grandmother, and, when he had recovered, he slowly, almost surreptitiously, eased himself back into the work routine around his father's house. Kas said nothing. If he felt remorse it lay hidden beneath the silence of manly pride, the most worthless of all apologies. An uneasy truce settled over the household. By the time the maize plants put out their first feathery displays, the tension had eased. Even Kas's relationships with his wives, which had not been very sound since the expulsion of Nomahuka, seemed to improve, and early in 1942 Leetwane announced that she was once again pregnant.

Before the maize and sorghum were fully set, Kas sent the women and children back into the fields to help clear the land of *kankeroos*—a weed with a burr-like fruit and sharp hooked spines that not only robbed the swelling grain of moisture, but lodged itself so deeply in the sheep's wool that it became almost impossible to remove. It was a major task to rid the land of this weed but, once it was finished, the family could raid the fields for the first green mealies of the season.[22] 'Stealing' mealies was a risky business, though. Hendrik Goosen started to appear on the property at all manner of strange hours, including the middle of the night, and, taking his cue from the likes of Koos Meyer and other ambitious farmers, he made it known that he considered the removal of any maize prior to the harvest as theft. Neither Kas nor Goosen's other tenants nor anyone else in RaKapari's drinking circle accepted the logic of this argument. The ownership of green mealies or ripening sorghum was the subject of constant disputes between Boer landlords and black sharecroppers, and neither party gave way readily. The idea of sharing crops and yet maintaining separate grain fields always made for uneasy compromises.

At Klippan, the outcome of the battle was effectively decided by a single incident, which was closely observed by all the tenants. When Goosen came across Motsumi making off with a bag of sorghum and challenged him, he was met with rather spirited resistance. Motsumi simply told him, 'Baas, I have not stolen anything from your field. I have removed grain from my own field. Had I wanted to steal sorghum I would have stolen it from your field.' Goosen suggested that he would deduct a bag from the tenant's share of the harvest settlement, but Motsumi simply called his bluff: in that case, he said, the entire matter would have to be resolved in the courts. There were not very many landlords in the district who wanted to risk that, and so, when no postharvest deduction materialised, even the most cautious tenant knew that it would always be easier to take green mealies or sorghum at Klippan than it was at Koos Meyer's or some other farms.

But Kas, who in mid-career showed less patience when entertaining such questions than he did in later life, did not even wait for this implicit answer. Long before Motsumi found out whether his bag of sorghum was safe or not, Kas had taken the struggle a good bit further. In his eyes the exhausting ploughing technique on which Goosen had insisted entitled them to more than a mere half share of the crop. So, well before any outside harvesting teams could be recruited, he waited for a moonless night and then he, Leetwane and Lebitsa slipped into the fields to pick quantities of maize destined never to see the threshing floor.[23]

Two weeks later the family were back in the fields, this time by day and as part of the team a dozen strong that Goosen had asked Kas to organise. The widows, wives and children had not been hard to recruit: Abner and Amelia Teeu—of BaFokeng origin like the Maines but a different clan—lived on a distant part of Klippan (before that they had spent several years working on properties controlled by Koos Meyer). The efforts of Amelia Teeu and her daughters, including a spirited ten-year-old by the name of Dikeledi, were supplemented by those of three outsiders—the widow Mokobo and her daughters, who came to the farm from the location at Schweizer-Reneke.[24]

It took them all nearly ten days to bring in the substantial harvest, and during that time Kas twice slaughtered sheep to supplement their basic diet of mealie meal. The Pienaars did the threshing, and once they had left, Goosen paid each of the team leaders six bags for every hundred bags they had helped bring in. But even after that expense, too, had been met, the Maines still had more than a hundred and fifty bags of grain. Along with the sunflower seeds and the seventeen bags of beans, this amounted to a truly excellent harvest in a year when much of South Africa had been ravaged by drought. Although once again out-farmed by the hard-working Johnson Xaba on another section of Klippan, Kas was nevertheless satisfied with the sense of a job well done.[25]

*

With the most pressing work behind them he and a half-dozen other BaSotho went to RaKapari's new redoubt at Vaalrand, where they were joined by the few 'Ndebele' members of the circle. Undeterred by the hostile comments of impatient white farmers who needed an economically less privileged and socially more compliant labour force, they met throughout the off-season to discuss the problems of the past year, lay plans for the new one, and simply gossip. Hearing talk of the forthcoming marriage of RaKapari's son, Khunou, made Kas think that perhaps the time had come for him to move more boldly in the matter of Thakane and his nephew. He did not want to keep Kgofu waiting too long, for even amongst the Maine clan, a young man had plenty of cousins among whom to choose a bride.

Leetwane and Lebitsa, who had done so much to help shape the profile of the Maine family by refusing to permit Tseleng or Nomahuka to assume wifely status, now took the view that women had no real say in such matters and that, because such marriages were sanctioned by tradition, Kas should proceed as he saw fit. Clearly, 'tradition' was determined almost as much by female resistance or compliance as it was by patriarchal power. So Kas sent a message to the village of Mokgareng near Taung, where Kgofu had been waiting patiently for an answer from his uncle for some time.

Such matters were not always easily settled, and during a visit to Kareepoort to see his father, Kas ran into an unexpected problem. Sekwala, who had effectively lost access to Motheba as a spouse when he had taken a second wife, was more sensitive to the opinions of women in such matters than his son was. Somehow or other he had heard that Thakane was unhappy about the prospect of having Kgofu forced on her as a husband, and so, when Kas asked him for his opinion, he told him it was always unwise for only children—such as Kgofu and Thakane—to wed, since it invariably made for a troublesome union.[26]

Sekwala's views, which soon became known within the family, placed Kas in an acute predicament. Still sensitive about the repercussions of his thrashing Mmusetsi and anxious that Thakane conform to his paternal authority and BaSotho custom, he could hardly be seen to fly in the face of his own father's advice. Not only women, but also children allying themselves with grandparents thus braked the exercise of patriarchal power. But Kas was a wily campaigner on this terrain, and he flanked Sekwala's argument by suggesting instead that Morwesi, who was exactly the same age as Thakane and who had many siblings, should marry his nephew.

But Morwesi was even less enamoured of the idea than her half-sister. On the day the negotiators arrived to determine her bohadi, she fled to Kareepoort, thereby denying them so much as a glimpse of the bride-to-be. Discussions were held in her absence, and as soon as the intermediaries had left to go back to Taung, Kas went to ferret her out of Sekwala's lair and march her back to Klippan. It did no good. On the morning that Kgofu's people were due to reappear with the livestock, Morwesi took

Thakane into her confidence and again ran away, this time to Motheba.

Motheba's strongly held Christian convictions made no allowance for a marriage without the consent of both partners, and when Kas appeared to commandeer his daughter she told him he had no right to force the girl to marry against her wishes. Thwarted first by the traditionalist objections raised by his father and then by the modern A.M.E.C. arguments advanced by his mother, Kas had no room for manoeuvre. He retreated, humiliated, and had to tell his kinsmen that they would have to return to Mokgareng without the wife they had come for.[27]

This was an undeclared war fought between a father and his daughters, between the old and the young, between custom and caprice—a struggle as old as time itself. Kas thought that he could not afford to lose it. At stake was the continued exercise of his patriarchal power and, with it, the very future of the Maine family as a social unit and a functioning economic entity. Precisely because the stakes were so high, Leetwane and Lebitsa found themselves lining up beneath the banners of age and tradition rather than gender and youth when the roll was called for the most decisive engagement.

If Kgofu had not been to Thakane's liking, then there were several other cousins who would willingly pay bohadi for her—amongst them Thobedi Maine, of the Moroke section of the family, who farmed near Schweizer-Reneke. This time Leetwane and Lebitsa were drawn more firmly into the negotiations, and it was eventually agreed that the transfer of sixteen cattle and ten sheep would help bind the families together. But the prospect of a life with Thobedi Maine held no more appeal for Thakane than did a marriage to Kgofu, and she again made her objections known. There was yet more violence, which horrified the younger daughters especially, but, faced by the united opposition of the parents, the enemy lines held and the longer-term preparations for a wedding went ahead in yet another protracted silence.[28]

Sensing that the war within the household was slowly turning in his favour, Kas staged a strategic retreat. He busied himself with off-season activities which had the advantage of keeping him at a distance from his older daughters, who in any case were being called upon to help Leetwane and Lebitsa collect and stack the dung destined for the diggings. Mmereki Molohlanyi and a few other sharecroppers brought in some ploughs for repairing—a pleasant enough winter task that was made all the more enjoyable because Kas could always rely on the help of Nthakwana, who, unlike her older sisters, was seldom the cause of any trouble. Once the warmth of the smithying fire had died away he went out and bought a few pigs. Hogs were always a good investment in cold weather, when the Boers were on the lookout for pork to add to the mixture of beef and spices that helped to make up the coarse farm sausages known as *boerewors*. (A few years later the seasonal edge was taken off this line of business when the first paraffin-fed refrigerators made their appearance.) The bitterness of the domestic struggle was fading from Kas's consciousness when the family went to Vaalrand one weekend to celebrate Khunou Kadi's marriage.[29]

The wedding, unlike the one attended by the unfortunate Jacob Candle and his friends at Hartebeespan a few years earlier, was a great success. The police, always more wary of invading the farms than the diggings, did not show up; the landlord, Piet Labuschagne, was supportive, and all in all everyone—including members of some of the most prominent sharecropping families in the Triangle like the Maines—had a thoroughly good time. When the time came to leave, Morwesi and Thakane asked to stay on, for the party seemed to be taking on new life. Kas agreed, provided they returned to Klippan before nightfall. Morwesi was especially keen on the company of the Kadis' oldest son, Padimole, but the winter sun was as tired as the exhausted dancers, and before they knew it, the last rays of light had gone from behind the nondescript grey ridge that gave the farm its name. When they got home, the blanket of darkness was firmly tucked in around the huddle of shacks at Klippan, only the spluttering of a small dung fire betraying the presence of life.

Their eyes were still adjusting to the light when they heard the sound of a horse coming toward them, and then Kas was suddenly upon them, sjambok in hand. Thakane made the mistake of turning away from the shack doorway and running off toward the pan, screaming. Horse and rider set off after her, and then, from the depth of the darkness, came shrill sounds of pain. 'Oh, you could hear Thakane crying bitterly, feeling the terrible pain penetrating her body, her heart. She sobbed and sobbed,' remembered her aunt Sellwane. The horse wheeled, leaving Thakane to stumble blindly across the stone-strewn pan, and thundered back toward the shacks and Morwesi.[30]

Morwesi met the horse and rider head-on. 'We are tired of running away every day, we are tired of being afraid of you. Do it! Do as you please!' And Kas did. He flogged her until he could no more; but she did not move an inch or utter a sound. As Sellwane recalled, 'Morwesi was very very defiant. My God! Not so much as a tear ran down her cheek!' 'Box of Matches,' too, remained dry. However, when Kas woke the following morning to find that Thakane had still not returned home, he smouldered again. He took out the horse and set out to find her. A teenage daughter who, only days before her marriage, was willing to sleep away from home![31]

Thakane had had little or no sleep. Fearful of the dark and shattered by the assault, she had fled even deeper into the bush and eventually climbed a tree in the hope of avoiding any earthly dangers only to find herself overtaken by a heavenly one. Perched on a branch, she remembered the story of the Boer who had died on the property some years earlier when his horses had taken fright and caused his cart to career into a pole; contemplating his awful death, she suddenly saw and heard the same wild horses coming towards her, driven by the same ghostly Boer smoking a pipe emitting flames. This vision was so vivid and so terrifying that she leapt down from the tree and ran all the way to Sebubudi's homestead, where she was taken in by her grandfather.[32]

Late the next afternoon, still ranting, Kas appeared outside Sekwala's shack shouting, 'Where is she? Where is she? I want to kill her.' Sekwala appeared in the doorway followed by Thakane who, on seeing flames bursting forth from her father's lips, sought refuge in the cavernous folds of Sekwala's overcoat. Sekwala warned Kas not to touch the child, and then in slow and deliberate tones gently eased him into a more reasonable frame of mind, promising that she would be home by nightfall. Two hours later Sebubudi escorted Thakane home but even at that late hour he had to intervene—this time to prevent Lebitsa from assaulting her.

The return home signalled the end of Thakane's resistance. A few days later she—an only child who had been given to the Lepholletses as a baby and returned to the Maines as an adolescent girl—was delivered into the hands of a husband not of her choosing. She left Klippan to start her new life on Delport's farm, near Migdol, to the sound of her sisters' singing. On leaving she caught sight of sixteen cattle and ten sheep staring at her in characteristically stupid fashion. But these were no ordinary animals; they were bohadi, the livestock that would weld two families into one and, so it was said, safeguard her well-being.

*

With Thakane's departure an uneasy calm descended over the Maine homestead. It had been an exceptionally trying time. The usual torments of winter had combined with pervasive social tension to leave everyone exhausted and anxious for the arrival of the season of renewal. It was left to Leetwane to usher in the first signs of spring warmth. Twenty-one years after the birth of her first child at Hartsfontein, she was delivered of a baby named Nneang. Within days of the first of the family's daughters leaving home, a ewe was slaughtered to celebrate the arrival of the last.[33]

The women busied themselves around the house with the new arrival while Kas went about his preseason chores. He got out the cart and drove to Rietpan, where just before the outbreak of war one of the Indian traders from Kingswood, E. S. Seedat, had opened a general dealer's store in an old farmhouse. But when he went into the shop he found that the older Seedat was not there. Instead, he was greeted by the proprietor's son, Solly Seedat, who recognised him from the days he had spent as a hawker around Kareepoort. Kas bought a harness for his donkeys and then decided that, since he was in the vicinity of Vaalboschfontein, he would call in on his old friend 'Maarman' Mokomela, who had once helped him to take care of A. V. Lindbergh's race horses.[34]

On his way home Kas discovered that the landlords were out and about their business in unusual numbers. With the price of maize having risen by nearly 20 percent in a year, many Boers were looking to hire additional land, but with tractors and agricultural equipment in such short supply they were also desperate for draught oxen and labour. As a result, good sharecroppers experienced an unusual demand for their services. Piet Labuschagne not only hired additional land which he put at RaKapari's disposal, but also invited Kas to move across and join them at Vaalrand.[35]

Kas turned down the offer but soon had reason to worry that he had made the wrong decision. Goosen was at least as eager as his neighbours to take advantage of the still-rising prices of maize, sunflowers and sorghum, so more of Klippan was put under the plough in the spring of 1942 than ever before, thereby reducing the amount of grazing. This came at a bad moment for Kas, whose own herd had just expanded with bohadi cattle. Forced into making painful choices, he decided to keep his cattle herd intact but sell all but a dozen or so sheep, which he held back for domestic slaughtering purposes.[36]

Selling the sheep did not please him any more than the increased pressure on grazing for the remaining livestock. His dissatisfaction was further aggravated by Goosen's insistence that again he plough the fields more than once. More work with less food did little for the condition of the oxen, and by the time the fields were ready for planting, Kas had reluctantly concluded that he had perhaps erred in not having moved to Vaalrand. Still, he was sure the family was better off on the farm than they would have been had they stayed on at the Kareepoort diggings.

He got the same impression from listening to the prattle of his sister-in-law, Motlagomang, during one of her visits to Klippan that spring. She reminded the family about the hardships of life on the diggings and the nearby farms and told them how the actions of the police had finally persuaded folk around Boskuil to abandon the liberal Rheinallt Jones and elect the 'communist' Hymie Basner to the Senate in November. But generally speaking, Kas had little interest in Motlagomang's talk of the new and more sympathetic magistrate at Wolmaransstad, or of how Jacob Lebone and the local 'communists' had, with the help of the militant A.M.E.C. women, worked to ensure Basner's victory.[37]

Politicians, even those who talked of making land available to black farmers as the I.C.U. had done, could come and go as they pleased; spring, on the other hand, came but once a year. All that mattered was getting the seed into the earth at the optimum moment, and so Kas planted as much maize and sorghum as he could in the extended fields. The women and children noted that, although the price of legumes had continued to rise, Kas did not make them plant beans. The sunflowers reappeared, however, and, with them, the cuts on the hands as well as painful allergic reactions that affected the lower forearm.[38]

The rains, which had been below average over large parts of South Africa during the previous year, made scheduled appearances in the Triangle with a regularity that tended to mislead those with short memories. By Christmas the countryside sported a thick mane of grass that folded away rhythmically for almost as far as the eye could see beneath a gentle summer breeze. In between, at ankle height, the delicate pastel yellows, pinks and blues of the *Geelblommetjies, Ruikbossies, Hotnotskooigoed* and other veld flowers caught the eye of even the most hardened campaigner. Sometimes even the old Kalahari liked to dress up.[39]

With an abundance of 'blue grass' around the pan, the carrying capacity of the veld held up well, and once the grain was established, Kas could pay attention to the livestock.[40] His first venture for the season, although not involving any tall stories about the need to acquire cattle for bohadi, left no room for sentiment. Without so much as a word of consultation Kas swopped 'Witties'—the cow his wife had acquired via the sale of dung and the purchase of a pig during the family's hardest years on the diggings—for a promising animal belonging to a farmer named Matthews. He called the new acquisition 'Witties,' too, although it was black rather than white like its predecessor. But whatever name Kas chose to give it, Leetwane considered 'Disu' to have been her property and she was bitterly resentful that he had appropriated the animal; she could not forget that when 'Disu' calved, on the day that Mosala was born, she had made a gift of the calf to her newborn son and that Kas had ignored this gift and sold the calf to a Boer.[41]

By January the condition of the herd was deteriorating rapidly. Stranded in mid-season without the resources to ensure the well-being of his animals, Kas reluctantly exchanged a heifer for a cart owned by a Boer who farmed on one of the more distant sections of the property.[42] But this desultory gesture only postponed the need for more serious measures, and Kas's dissatisfaction with Goosen's arrangements mounted as his animals wasted away. Only a month or so later he was forced in the most humiliating way to admit his mistake in not moving to Vaalrand: he had to sell his cattle for cash at a time when there was no immediate prospect of replacing them.

On 16 March 1943, he and Mmusetsi took three oxen, four 'tollies' and a dozen sheep to Slabbert & Campbell's monthly stock sale at Bloemhof. The Boers, many of whom had taken to calling him '*Ou Koop en Verkoop*' ever since he had first succumbed to speculative fever back on the diggings, greeted him with outstretched hands. Kas was as polite as ever, but

'Old Buy and Sell' had no taste for business that day, and the forty-six pounds, fifteen shillings and ninepence in cash that he took home that night were no cause for celebration.[43]

Two days later he was back in Bloemhof at the magistrate's office, this time to meet a regular obligation as well as to lay the foundations for a more ambitious project. The normal rules governing any interaction with a state bureaucracy applied: paying in cash was easy, getting an answer to a simple question almost impossible. Kas handed a sullen white clerk a pound and in return was given the poll-tax receipt, which, in the countryside, also doubled as an identity document. But when he asked whether he might conduct a traditional BaSotho initiation school at Klippan later that year, he was told to wait. It took several hours to get to see a Native Commissioner, who had no clearer answer to the question than the clerk did. Kas left the office without a ruling and was warned not to proceed until he had received written authorisation from the state.[44]

It was irritating. It was a perfectly good way of raising cash, and without it the calves would slip away from him. Luckily the point had not yet been reached where a black man needed a permit to reap his crops. That was one of the features that appealed to Kas about living on the land: the relative freedom from outside interference. Yet, for all that, he would once again be called upon to go beyond the family circle to bring in the harvest. Late in May he undertook the long haul to Taung, where one 'Chief Makgeta' helped him recruit a team of harvesters whom he took back to the Triangle by ox-wagon.[45]

At Klippan they joined forces with the Maine women and children and a few members of Amelia Teeu's family. Yet despite this formidable workforce, it still took several weeks of hard work punctuated only by breaks for meals and sleep, to bring in the harvest. Getting a share of so large a haul of grain appealed to the excited Taung workers, who relished the prospect of going home to their impoverished reserve with the better part of a wagon-load of maize and sorghum. Their enthusiasm was infectious, and a few of the older hands did not fail to notice that fourteen-year-old Bodule became intent on helping Dikeledi Teeu well beyond the usual bounds of collective labour.

They lit fires, roasted meat and drank sorghum beer together while they waited for Jan Groenewald and his threshing machine. The machine easily processed the maize, although many of the Boers complained about the high cost and scarcity of jute bags, imported from India: even in the Triangle there were moments when one felt the impact of the 'Englishman's war.' But Groenewald's equipment fared less well with the sorghum. The grain was unusually large, which meant that, much to the disgust of the onlookers, the thresher tended to spit out much of the kernel along with the chaff. Somebody had the bright idea of hiring Matthews's machine, and in the end the sorghum had to be threshed a second time.[46]

The delay was met with good humour, and when the landlord, his sharecropper and their harvesting team did the final count, they agreed that they were facing a stack of four hundred and fifty bags of grain. Goosen and Kas took about two hundred bags each, the harvesters twenty-four, and the rest was shared by Amelia Teeu and her family.[47]

Goosen was delighted, for his insistence on getting the tenants to double-plough was vindicated. He and his brother, Pieter, sold most of their sorghum to the local malting agents—Eides, Kahn & Kahn, or the Levitas Brothers—who, in turn, passed it on to the municipal beer halls spreading along the length of the Witwatersrand—a route that bypassed and therefore irritated the management of the local agricultural co-operative. Triangle sorghum, which in theory contributed to the national effort at wartime food production, in practice often ended up as so much 'kaffir beer,' by which means the South African state extended the social control it exercised over its burgeoning urban black proletariat.[48]

The bulk of the maize crop, however, was sold through the co-operative at an equally handsome profit. With the national harvest substantially down from 1941 and demand undiminished, the price of a 200-pound bag of mealies had increased by a staggering 26 percent. Mollified by this inflow of winter cash, Kas forgot briefly about the grazing problems at Klippan and eased himself into the off-season with more confidence than he had been able to muster for two decades.[49]

The cash came in handy. In July, Kas got a letter from Sarah Altshuler, executrix of her late husband's estate, requesting that he meet an outstanding debt of £20 which he had incurred earlier in the year. Mendel Altshuler, one of scores of Russian Jewish immigrants who had made their way to the Triangle after the discovery of diamonds, had been the proprietor of Altshuler's Modern Store, which somehow managed to combine the unlikely function of 'Ladies and Gents Outfitter' with that of 'Cartage Contractor' and 'Coal and Produce Merchant.' Sarah Altshuler was one of the few remaining members of Bloemhof's once conspicuous Jewish trading community, the younger generation having abandoned the *platteland* for professional careers in the bigger cities. Kas went into town, settled his debt, and returned home to face the winter's work.[50]

With spares for agricultural equipment and tractors virtually unobtainable during the war, he was in demand as a metal worker. A pair of homemade bellows carefully fashioned from goat skin got the charcoal fire to the point where he could either repair or refit the ploughs, planters and wagon wheels that bigger firms like Wentzel & van der Westhuizen had difficulty in servicing.

Kas's reputation as a blacksmith spread to several of the Boer landlords who already knew him as an able grain farmer.[51] One of his most persistent admirers was Piet Labuschagne. From the very moment that the Maines had arrived at Klippan, Labuschagne had taken a close interest in whatever

Kas did. With Goosen frequently away at his parents' farm, Labuschagne often found the time to lean across the fence and engage his black neighbour in long conversations about farming matters. When Kas happened one morning to complain about the lack of grazing on Klippan, Labuschagne saw his chance. He not only offered to take in all of Kas's livestock, but told him that if he were willing to 'jump the fence,' he could have as many fields as he wanted. 'Labuschagne knew that I was good at farming, so he recruited me to work on his farm.' In August 1943, the Maines moved to Vaalrand, where they linked up with the Kadis and several other prominent sharecropping families. In many ways this was the turning point in Kas's career as a farmer in the Triangle.[52]

CHAPTER NINE

# Advance

## 1943–46

The geographical roots of the Labuschagne family, like those of the Meyers, the Niemans, the Reynekes and many others in the Triangle, lay in the Cape Colony. In the late nineteenth century rural poverty had driven these families north into the Orange Free State, where they turned to wool production, only to move again when the discovery of alluvial diamonds in the southwestern Transvaal had lured them farther west and across the Vaal in the 1890s. Piet Labuschagne's father's history differed from this pattern only because he had been a late starter. His republican sympathies had always been in inverse proportion to his assets, and at the outbreak of the South African War in 1899, he had not hesitated to join the Cape rebels, earning himself a spell in prison without the benefit of a trial. This experience increased his antagonism to British rule, and after the war he moved into the southern part of the Free State, where he took up sheep farming, found himself a wife and raised several sons—including young Petrus Gerhardus, or 'Piet,' as he was almost always known—in the narrow political traditions that lay so close to his heart.

In 1922, at the age of sixteen, and with only a few sheep to his name, Piet Labuschagne left his parents' home and went to the diamond diggings at Grootdorings, where an older brother had established a butcher shop catering to the needs of black workers. Piet worked in the Grootdorings shop for two years, earning three pounds a month, then went to the Free State to raise more capital by selling what little livestock he had left on his father's farm. On his return, he was met at the Boskuil siding by a second brother who told him about exciting new opportunities opening up all along the course of the Bamboesspruit, where 'a Jew' had a concession that gave him exclusive rights to sell meat in the black villages—Fly Camp, Rooistad, Velskoen and Vuilkantien. The Labuschagne brothers were not inclined to respect the 'rights' of 'foreign' traders, so they bought a hawker's licence and a wagon, and then persuaded the owner of Uitkyk, a farm west of Kareepoort, to let them sell meat from his property in return for a monthly rental of five pounds. Each afternoon as soon as the temperature had dropped so the stench was not unbearable, they would slaughter the livestock bought during the day, and then work well into the night cutting the meat into small parcels which they knew would appeal to black customers. They commenced trading at first light, and by adopting a policy of always undercutting 'the Jew' by a penny a pound, succeeded in attracting many of his customers.

But 'the Jew' objected to the siting of a business that infringed his concessionary rights, and he reported the matter to the Makwassie police, who investigated his complaint. The Labuschagne brothers claimed they were 'hawkers' rather than 'butchers' and did not work from a fixed site; thereafter they took the precaution of moving their wagon a few yards every fifteen minutes so as not to break the letter of the law. This ploy left the policemen floundering, and eventually they succeeded in putting 'the Jew' out of business.[1]

Experiences of this sort contributed to the social and political ambiguities that characterised Piet Labuschagne. On the one hand he wanted to succeed, had the capacity to work hard, and could relate easily to folk drawn from most walks of rural life. On the other, he disliked 'alien' traders, developed a special contempt for 'Jews' and 'Coolies,' and generally identified with the Nationalist and populist sentiments that had scarred Triangle politics ever since the Jameson Raid of 1895. By the time he married in the mid-1920s and moved to Muiskraal to try his hand at diamond digging in the Potchefstroom district, he was an ardent Afrikaner Nationalist.

By 1927 Labuschagne had enough money to return to the Triangle, where he and his brother, along with several other farmers, hired portions of the farm Blouboskuil from the famous 'Jewish' firm of Lewis & Marks. With hard work, perseverance and good fortune the brothers survived the Great Depression: by the mid-1930s the older brother was concentrating on diamond digging and Piet was increasingly devoted to sheep farming. When the older brother discovered a diamond worth more than 700 pounds they both abandoned further interest in digging and set their hearts on the farm Klippan, belonging to Messrs. Hill & Paddon of Kimberley. But they did not have enough money to buy the whole property and were outbid by the Stern family of Schweizer-Reneke: as Piet Labuschagne recalled it, 'The Jews bought the property, the whole farm.' This meant the brothers could only hire a small section of it. Piet's need for grazing for his rapidly expanding flock of sheep was eased in 1938, however, when his brother died unexpectedly, leaving him in full control of the hired property.

The timing could not have been more opportune. The price of wool, which had held steady for years, accelerated rapidly during the war, Piet Labuschagne's sheep farming suddenly prospered, and he had even greater access to cash and credit. In 1940, he at last bought a portion of Klippan from the Sterns and then leased two additional parcels of land on the adjoining farm, Kareepan.[2] Two years later he hired most of the 3,000 morgen available at Vaalrand with a view to buying it outright. In 1943, with his eyes firmly fixed on the booming grain prices, he hired a large portion of the farm Prairieflower and got involved in sharecropping ventures.

When Kas Maine and Piet Labuschagne joined forces at Vaalrand in late 1943, South Africa's normally sluggish agricultural economy was expanding at a rate without historical precedent, and both men were approaching the prime of their working lives. Kas, now fifty, owned an impressive array of agricultural equipment, could draw on more than two decades of experience of arable farming, and, if called upon, could put five adults and three children into the field. At the age of thirty-seven Piet Labuschagne was a knowledgeable sheep farmer who owned his own farm and could hire any additional property he wanted. Theirs was as promising a partnership as there was likely to be in a racially obsessed country where the law itself mocked the prospect of genuine equality between a white and a black man.

Their contract contained several familiar clauses as well as a few new ones that cast light on Piet Labuschagne's privileged circumstances. No restriction was to be placed on the amount of ploughing that could be undertaken either at Vaalrand or on the property at Prairieflower. Any crops the Maines harvested at Vaalrand with their own resources would be divided in the customary way, but, in return for the use of two additional teams of oxen and land that Piet Labuschagne would put at Kas's disposal at Prairieflower, any 'surplus' would be divided in the ratio of two to one in favour of the landlord. Kas had perhaps even more than the normal incentives to devote himself to the basic enterprise, since it was further agreed that there would be no individual allocation of acreage—a practice that sometimes left tenant farmers with inferior soils, undeveloped fields or stony lands. In addition, Labuschagne agreed to pay twelve-year-old Nthakwana and eight-year-old Mosala small cash wages for any work they did at Vaalrand—the former in the kitchen as an assistant to Piet Labuschagne's wife, Johanna, and the latter as a herder of his flock of several hundred Merino sheep.[3]

In September well before the first rains, Kas introduced the family to the new routine, which was even more demanding than what they had grown accustomed to at Klippan.[4] The household would be roused long before first light. 'When he called out "time up," ' Mosala recalled years later, 'nobody would be allowed to say another word. The only sound that was heard was that of footsteps running over an earthen floor—patter, patter, patter. His tea had to have been served by 4.00 a.m.' Breakfast was eaten in the dark. At first light the boys had to commence milking and the girls their domestic chores.

With the sky on the horizon only just suggesting a gentle glow, Kas supervised the harnessing of sixteen donkeys and then watched as Mmusetsi and Bodule set out for the fields spread along the grey ridge, which was all that distinguished the property from the plain surrounding it. He stood in the doorway of the shack, eyes narrowed, to watch the silhouettes against the pale pink rays of the sun, as they hitched the donkeys to the plough and then turned to pick up the line of the furrow. When the golden ball of the rising sun was bisected by the line of the ridge, Nthakwana could be seen trotting off toward the farmhouse, while Kas could hear Lerata Masihu's cattle bellowing when Mosala opened the gate and they joined Labuschagne's herd for the day's grazing. Once the children were out of the house and at their work stations, Kas saddled up 'Kolbooi' and moved out into the fields to do his inspection.

Mmusetsi and Bodule had been set a mammoth task. Kas wanted the fields to be as big as possible, but also double-ploughed, after the fashion he had learned from Hendrik Goosen. He would get off Kolbooi, pluck a dried twig from a nearby thorn bush, push it into the ground, use the tip of his boot to mark out a line in the sand, and then tell them that they were not to stop working or outspan the donkeys until the shadow cast by the stick had intersected the line demarcated by his boot. Any unauthorised departure from this routine, would be met with a thrashing. The younger children, out herding, were spared the stick-and-shadow routine, but they too had no room for slacking and on more than one occasion were borne down upon from an unexpected angle by a mounted figure brandishing a sjambok.

Kas's supervisory role, which became harsher over the seasons that followed, was met with both scorn and terror by the younger Maines. Thakane, who had perhaps endured more than most in the months leading up to her marriage, was categorical in her judgment: 'My father was cruel. To him any infringement warranted punishment. He did not work on the farm—he used the labour of his wives and children.' Even Nthakwana, who was sympathetically disposed to the patriarch, felt that after her father had taught her older brothers how to plough, '. . . he became very lazy. Not just a little but very lazy.' Mosala seemed to sense that he would never meet the asking price and reluctantly concluded that his father had '. . . no time for people who were not like him.'

But the young and tearful eye was not always the one that saw most clearly: it is also true that no middle-aged man ever employs his strength and skills in quite the same way he used them as a younger man. Mmusetsi and Bodule, for example, knew that their father was still capable of great physical feats: no sooner was the first seed in the ground at Vaalrand than he joined them to help load three ploughs onto the wagon; the three of them then repeated the entire spring ritual in the soils of Prairieflower.

At Prairieflower, they would erect *hloma o hlomolle,* which would serve as their home for the three weeks they spent on the hired property. Kas harnessed one of Labuschagne's spans to the landlord's heavy three-share plough and then, out of sight of the sceptical youngsters back home, put in a full day's work behind the oxen. He became an instant admirer of this heavy-duty three-share plough. Mmusetsi had Labuschagne's second span to draw the landlord's two-share plough, which left Bodule to deal with the Maine donkeys, which hauled the familiar two-share Massey-Harris.[5]

With his two oldest sons working at his side, each of the men ploughing a field whose size was limited only by physical fatigue and the time available to them, and with not so much as a Boer in sight, Kas was as happy as he was ever likely to be. Part of him positively revelled in the space that had opened up: '[Labuschagne's] oxen were under my control. All his equipment was under my control. I would tell him what type of seed I required and he would oblige. He brought me the seed and then left me to decide how to go about the ploughing.'[6]

But another part of him, the part which would have to give away two-thirds of everything grown at Prairieflower, resented that the landlord only occasionally appeared. There was something inherently unfair in the allocation of capital and labour. 'We ploughed day and night, and he sat and did nothing.' It was a perception not unlike the one that Kas's younger children had of him. In the countryside, anything less than full-blooded physical involvement in the task of the day was likely to be construed as idleness.[7]

More often than not, however, any flash of resentment directed at Labuschagne tended to be lost, given the broader background of Labuschagne's paternalistic practices. As in any good Transvaal thunderstorm, the anger of the lightning was soon forgotten amidst the joy of the rain that followed. In early December the Maine men withdrew to Vaalrand, where for the first time since he and his brothers had done the rounds during the wool boom of the early 1920s, Kas got involved in the summer shearing. Although never able to match Phitise's sixty sheep a day, he found most of his old skills intact and managed to work his way through twenty fleeces a day, which, at a shilling a sheep, brought in about a pound a day. For two weeks this hard work was made more convivial by the liberal lashings of grape and peach *mampoer* which the landlord made available to anyone who seemed to flag—a benign highveld variation on the infamous Cape *dop*, or tot, system which, in turn, cast a fading light on the distant roots of the Labuschagne family. Nor was this the only social relic of the old 'slave south' at Vaalrand. At Christmas time all the sharecroppers and farm workers were invited to a *braaivleis*, or barbecue, on the grass under the huge umbrella-like trees surrounding the big house, where Labuschagne, never afraid of keeping black company, circulated amongst the Maines, the Masihus, the Kadis and others.[8]

On many properties these practices might have constituted little more than seasonal concessions to maintaining good social relations with folk on the family farm: with Piet Labuschagne it was different. At Vaalrand social practices had analogues in productive activities, and these suggested that this paternalistic regime was in fact drawing landlord and tenant into an unusually close relationship. Labuschagne incorporated the few Maine sheep into his own flock, which meant that not only were his tenant's ewes covered by his own quality Merino rams, but they were also routinely protected against diseases such as 'blue-tongue' and 'wireworm.' Kas, in turn, took Labuschagne's cattle into his own herd, which meant that the landlord's animals had expert attention from an experienced livestock farmer and the animals had equal access to whatever quality grazing was available.[9]

These arrangements had obvious advantages for Kas, but they also placed an inordinate strain on the younger Maines, who had to shoulder the responsibility for the well-being of even more stock. Moreover, because the sheep were run at Vaalrand and the cattle kept across the way at Prairieflower, there was a great deal of movement between the properties. The problem worsened as the season progressed, the crops matured, and the fields needed weeding. By midsummer 1944, Kas was away from home for long periods and his visits to Prairieflower became rather predictable.

At dawn the small piebald pony (a direct descendant of the mountain pony that Sekwala had brought back with him from his journey to Basutoland in 1909) was led out of the kraal. He and Kolbooi would then be joined by Baby, an ugly little mongrel which had been given to Kas by Hendrik Goosen's father and which had earned its name through the close attachment it had formed with the youngest of the Maine children, Nkamang. Kas would finish his tea and then man, horse, and dog would set out on the eight- or nine-mile trot to Prairieflower, where they would link up with Bodule and Mmusetsi to inspect the fields and cattle.[10]

These visits, often spread out over the better part of a week, gave Kas a much-needed opportunity to spend the evenings with the older boys, deepening his ties with Bodule and helping to ease the tension between him and the less predictable Mmusetsi. His relationship with his sons, based partially on the exclusive male work patterns, improved at precisely the same time as his relationship with his adolescent daughter, Morwesi, and the younger children was deteriorating. There were limits to just how much physical and psychological terrain a working patriarch could cover in twenty-four hours. Even out at Prairieflower father-son talk sometimes had to make way for the demands of friends, neighbours or passing travellers in search of a bed or a place to keep their animals on an overnight journey. Like most country folk, Kas took seriously his obligation to offer newcomers or visitors some protection from the vagaries of the highveld night. That was how he and his sons met Mokorwane who told them about BaSotho landlords named Motsuenyane who, he claimed, owned a farm near Klerksdorp and were said to be always on the lookout for good sharecroppers. A black landlord on a farm in a Boer district was unusual so Kas made a mental note of the family's name, but having been born and raised in the Triangle, he doubted there would be an occasion to use it.[11]

On the way back to Vaalrand one day he paused to look over a neighbouring property. It served to strengthen his conviction that if the good rains persisted, grain farmers were probably looking at a record harvest. Indeed, in some parts of the Triangle there was already a danger that further precipitation might damage the crops, and somewhat unusually, some of the Boer landlords had been heard to express a hope that the rains would hold off. Kas decided that if it did rain any more his own fortunes were likely to be mixed. The maize and sorghum planted along the ridges of Vaalrand were no doubt safe enough, but the crops on the lower-lying ground at Prairieflower might be flooded. Perhaps the extra effort they had put into the hired property would be in vain. Even so, he was reluctant to complain, since it was obvious that other compensations were going to flow from his increasingly close co-operation with Piet Labuschagne.[12]

One of these was the income he would earn from his sheep. In March, the last month in which sheep could be safely shorn before the Malutis expelled the first icy blast of wintry weather, he saw his decision to allow his sheep to be incorporated into the landlord's flock vindicated. Not only had the number of sheep he owned increased but the two bales of wool they yielded realised more than ten pounds when he took them in to the local merchants. A few other tenants did even better, and RaKapari made enough from his wool sales in Bloemhof that year for his peers to describe him as 'rich.' But Labuschagne, with his intense dislike of any Jewish middleman, sent his bales directly by rail to distant branches of the Farmers' Co-operative Union at the coastal cities of East London and Port Elizabeth.[13]

*

This dislike of 'Jews' and 'coolies'—at a time when Afrikaner-controlled co-operatives were making rapid strides and the prices of agricultural commodities were rising—was not, of course, confined to the ranks of capitalising Nationalist farmers like Piet Labuschagne. At the other end of the class spectrum, where black farm labourers had to pay inflated wartime prices for food and clothing, resentment was no less pronounced. Black prejudice, roundly endorsed by explicit statements of racial hatred directed at Asian storekeepers by Boer landlords, sometimes had horrific consequences. Kas knew of at least one such incident.[14]

At Vaalboschfontein, where he had once worked as a 'stable boy' for

A. V. Lindbergh, Kas's friend Maarman Mokomela and his son Piet had never quite managed to make the transition from being labour tenants to becoming sharecroppers. On more than one occasion while at Klippan, Kas had hired the Mokomelas as part of a hastily assembled harvesting team. Given their availability for this kind of work, and the fact that the younger Mokomela was still working as an ordinary labourer on the Lindbergh farm at the age of twenty, it was clear that the family was struggling to make ends meet. In Piet Mokomela's case these problems came to a head in March 1944, when he was called upon to settle his debts at E. S. Seedat's store, situated in a small building partly hidden from the road by a few bluegums on a track leading to Rietpan.[15]

After work one Friday afternoon Piet Mokomela approached a friend, Johannes Moilwa, and suggested they walk to Seedat's store. Still swinging the leather thongs they used for tethering Lindbergh's horses, the two stable hands set out for the store, which lay about two miles south of the entrance to Vaalboschfontein. It was twilight when they reached it and the proprietor was preparing to close up for the night.

Recognising them, Esack Seedat paid little attention as the two young men stood about waiting for him. When he finished his evening ritual Mokomela took a few paces toward the gap in the wooden counter, carefully removed a banknote from his pocket, handed it to Seedat, and told him to offset some of it against his debt. Seedat took the note and was turning to get change from the cash box directly behind him, when Mokomela suddenly swung the leather *riem* over his neck and proceeded slowly to tighten it. The trader's lifeless body eventually clattered to the floor. The two men then removed the shoes from the body and, trying to leave the impression of a struggle, placed a large wooden spoon in Seedat's bloodless hand. They took the shoes, removed the cash box containing about thirty pounds, pulled the door closed behind them and under cover of darkness made their way back to the farm. At Vaalboschfontein they dug a small hole, eased the cash box into it, and covered their tracks. By then, however, Moilwa's nerve was failing him, and he decided to flee.

In the morning, Seedat's body was discovered by the fourteen-year-old youngster who worked for him, and it was this 'piccanin' who ran to a neighbouring farmer, who informed the Wolmaransstad police about the trader's murder. A detective from Klerksdorp was sent out to investigate the crime. In the meantime, Mokomela, too, had decided that it would probably be best if he left Vaalboschfontein and had unsuccessfully tried to persuade the foreman, Bezant, to lend him a horse. This request, hard on the heels of Moilwa's disappearance, aroused the foreman's suspicions, and when Bezant quizzed Mokomela about Seedat's death, the stable hand invented a story about having seen 'a stranger dressed in green' loitering about the shop the previous afternoon.

Bezant passed this information to the detective, who questioned Mokomela more closely and then left him in the custody of a constable while he went off to make further enquiries about 'the man in green.' But, while the white detective was gone, the black policeman noticed a small blood stain on Mokomela's shirt, and when the detective returned, the groom confessed to the crime and was placed under arrest. Shortly thereafter Moilwa was apprehended and persuaded to provide the police with his account of events at the trading store. At the Circuit Court at Wolmaransstad, in November 1944, Johannes Moilwa was acquitted for his part in the affair and Piet Mokomela was sentenced to death for having murdered Seedat.[16]

Although not unheard-of among the proletariat on the diamond diggings, murders were not an everyday occurrence on the farms. Not surprisingly, the murder at Rietpan left a lasting impression on at least two sections of Triangle society. Amongst Asian traders who were reluctant to admit that blacks often resented them, it was rumoured that Maarman Mokomela was so upset by his son's crime that he refused to visit him in prison prior to his execution. Most black farm workers, including Kas, in sharp contrast, believed that the Indian traders had been so angered by Mokomela's crime that they had raised money to ensure that Moilwa, too, would go to the gallows. In the Triangle, indeed throughout the southwestern Transvaal, the static of class conflict and the crackle of racial antagonism often distorted the messages of community politics. In this sense, the murder of Esack Seedat was a portent of things to come.[17]

The murder was still being actively discussed when Kas had to turn his attention to mobilising labour for the forthcoming harvest. With so many families struggling, as prices outstripped wages, there was not much point in going as far afield as Taung to recruit workers. In fact, there was not much need to look beyond the Mokomelas and yet another family from Vaalboschfontein, the Mokgethis.

As Kas had half expected, the low-lying fields at Prairieflower yielded rather miserly returns, and after he and Labuschagne had shared the harvest in the agreed proportions, the Maines were left with only fifty bags to show for a lot of work. The Mokgethis went home to Kareepoort, only to reappear at Vaalrand a week or two later, where they were joined by Maarman Mokomela and his wife. This time Kas, Leetwane and Mmusetsi helped the harvesters and were more generously rewarded: after Labuschagne and the harvesters had been paid the Maine family had about a hundred bags of grain.[18] But a total of a hundred and fifty bags, although hardly a disaster, was disappointing given the effort and resources they had put into the fields. At the threshing point, Boer landlords and black sharecroppers alike muttered about lower yields, and this simmering discontent was exacerbated by yet another sharp rise in the price of jute grain bags, imported from India and sold through the co-operative or by Asian traders, which went from thruppence a bag before the war, to sixpence a bag by 1942 and then to a whopping ninepence a bag by the winter of 1944.[19]

These escalating overhead costs threatened the profit margins to which the long-suffering grain farmers were only just becoming accustomed, and the continued increase in the price of jute stoked the anger of Afrikaner Nationalists, whose dislike of 'coolie traders' was growing by the year. At the local co-operative, long since a Nationalist redoubt, there were also complaints about some farmers circumventing the institutional marketing arrangements by selling sorghum directly to outside agents—agents who, it was deemed unnecessary to point out, were more often than not Jews.[20]

Piet Labuschagne, a founding member of the local Eendrag, or Unity Branch of the Herenigde Nasionale Party, as well as a loyal supporter of the South-Western Transvaal Agricultural Co-operative, saw to it that *all* the maize produced at Vaalrand and Prairieflower, whether his or the sharecroppers', was sold through the co-operative. The crop was marketed in his name, and on receiving the *voorskot*—the preliminary payment made to farmers which was offset against the end price the grain eventually fetched—he handed Kas his share of the proceeds.[21]

With this tidy sum to see him through the winter and with the agricultural year safely behind him, Kas joined the others in RaKapari's circle for end-of-season celebrations. These social diversions came as a godsend to members of the patriarch's family, who badly needed a break from the grinding routine he had imposed on the homestead. Leetwane and Lebitsa looked around for ways to raise winter cash that would lessen their dependence on their husband, the older children sought out the company of their peers at one or other of the farm 'concerts,' and the youngest Maines whiled away their time in the company of the neighbours' children and animals.

Unlike the bawdy and raucous gatherings that sometimes took place in the Rooistad shebeens and elsewhere on the diggings, most social occasions on the farms were rather modest and took place under carefully monitored circumstances that seldom posed any long-term threat to parental authority. The hidden menace that was integral to the Boer landlords' paternalistic regimes had the unintended effect of tightening the grip that black parents, especially tenant fathers, had on their children. Like many of their contemporaries, most of the young Maine men and women could make no secret of either their destination or the company they intended keeping. It was only Morwesi who would sometimes drift home late or, unknown to her parents, slip out to meet Padimole Kadi, an almost nightly occurrence during the 1944 harvest celebrations.

Once Vaalrand's grassy beard had been lathered by the first foam of winter frost, Kas withdrew from the drinking circle and looked around for new economic opportunities. He was fascinated by the way in which Labuschagne administered large quantities of salt, bought from the Sewefontein works, to his pigs in order to prevent measles and became convinced that this was the key to successful hog raising. 'Stealing the skill with his eyes' and perhaps remembering how Leetwane had used pigs to raise the money for what she persisted in seeing as 'her' cow, he went out and bought a large white boar and a small black sow which, to his great delight, almost instantly produced a handsome litter of ten. By the end of that year he had a thriving colony of pigs.[22]

But he saw the hogs as an investment in the future, and in the interim he had to rely on more traditional pursuits to help raise cash. As luck would have it, Labuschagne arrived at the shack one afternoon with some buckles and hides and asked Kas to make him a set of harnesses for a new team of mules. Kas cautioned him that, in addition to costing five shillings each, the twelve harnesses would take some time to make, and warned him that he would need to have more of his oldest son's time if they were to get the task done in the allotted time. Labuschagne agreed.

Kas and a rather morose Mmusetsi spent several days first beating and then rubbing the salt-encrusted hides, which softened and became supple. It took Kas a day to lay out, measure and cut the treated leather into strips, but it was enjoyable work that inadvertently took him back to his childhood at Mahemspanne, when he had spent many happy hours helping Sekwala cut *karosses* from springbok skins. But the days of hunting springbok and sewing kaross blankets had long since passed, and by the time the awkward young Mmusetsi had children of his own, people would doubtless buy harnesses from the stores ready-made. Until then, Kas would continue to sew as Sekwala had taught him, wait for Mmusetsi to stretch the lengths, and use his awl to pierce the leather and attach the brass buckles.[23]

A fortnight later and three pounds richer, Kas was about to put away the leather-working tools for the season when RaKapari appeared with the remnants of a saddle and the reddish skin of a newly slaughtered pig. Kas put the skin into a solution of saltpetre and alum for several days to dissolve the excess fat, then transferred the hide to a second container for about a week while it darkened. Waiting for this transformation, he stripped down the saddle, set aside its metal parts, and then calculated how much timber would be needed to fashion a replacement. What was needed was some thick pliable wood, not the sort that one could cut from the few acacias still clinging to the ridge at Vaalrand. Fortunately, there was not a spruit on the highveld worthy of the name that did not have at least one or two willows paying obeisance to what passed for a water hole. He found one of the faithful bowing deeply before a nearby *vlei*, slipped in behind it and, while it was still preoccupied, removed what he needed.

He cut, chipped and moulded the willow, then took out the treated pigskin from its place of safekeeping; it had cured to perfection. Using the frame as a template, he cut and stretched the hide and then folded and stitched the leather leaving room for the stuffing. He did the final stitching, and then cleaned and polished the saddle. It was, he thought, the finest saddle he had ever made and, if touted amongst the local landlords, would fetch at least ten pounds. But RaKapari had given him the frame to work on and he was, after all, a neighbour, so Kas charged him only five pounds.[24]

Thloriso Kadi was not the only neighbour to benefit from Kas's skills that season. In 1944, the prices of agricultural implements had risen 60 percent since the eve of the war, and there were few parts, so a steady stream of landlords and sharecroppers beat their way to the Maines' door asking Kas to repair ageing harrows, planters and ploughs. Not surprisingly, the image of Kas as blacksmith, hard at work with his hammer, anvil and forge, left a lasting impression. Forty years after the war Piet Labuschagne could still recall his effecting an innovative repair to a crippled Maxim Planter '. . . as no other craftsman in the world could do it.' Lerata Masihu, schoolmaster-cum-sharecropper, said simply, 'When Kas was not ploughing he was fixing ploughs.' But it was his assistant, thirteen-year-old Nthakwana, who had the most vivid memories of all. 'To tell the truth, I grew up at the bellows—stirring and blowing the fire. On the days that I was not out in the fields minding cattle, I was at the bellows helping the old man.' Most of these repairs were done in exchange for modest cash payments, but as was to be expected in a rural community where many black tenant farmers were financially liquid only once a year, some were paid for in kind.[25]

Amidst the heat, noise and sparks generated by all this 'outside' work, Kas did not forget a project that lay much closer to his heart. The joy of using Labuschagne's heavy three-share plough at Prairieflower was a vivid memory, and he was determined to convert the two-share Massey-Harris into a three-share plough for the new season. When he heard that the first spares manufactured in South Africa were starting to trickle onto the local market, he hurried into Bloemhof to consult Bill van der Westhuizen. Van der Westhuizen was more than willing to sell him parts, but was sceptical that the Massey-Harris could be adapted to hold a third share. Back at Vaalrand, Kas surprised everybody by designing and fitting an ingenious extension that enabled the Massey-Harris to go into the field carrying an extra share.[26]

As usual little if any of this winter cash of Kas's made its way into the family coffers—which rankled with the Maine women, who complained about needing money to clothe as well as pay for the tuition of the children who were attending Lerata Masihu's weekly literacy classes. Leetwane and Lebitsa were left with no alternative but to collect dung, raise chickens and do the laundry of neighbouring farmers in order to generate the cash for items that Kas dismissed as 'luxuries.' Indeed, everyone understood that *all* the patriarch's income would be channelled into farming; even Bodule, who had a privileged position in his father's affections, took to keeping pigeons in the hope that he would one day realise his dream of owning a small Kodak camera.[27]

*

As the days in the Triangle started to lengthen, the outpouring of energy over a warm winter fire died down and eventually stopped. With most of the harvest already disposed of, the merchants were starting to offer better than average prices, trying to coax the last of the black sharecroppers' winter hoards onto the market. Kas and Maarman Mokomela, who had stored most of his grain at Vaalrand, decided to take advantage of this upturn, and one morning in early August, Bodule helped them load maize and sorghum into two wagons destined for Bloemhof. The small convoy left the farm and turned left into the main road, but before it had got very far, the rear axle on the Maine wagon started to screech and groan alarmingly. Kas told Bodule to take Kolbooi, go to the house for a pot of axle-grease, and rejoin them quickly. When the young man eventually caught up with the wagons, however, he was carrying not only axle-grease but disturbing news: Leetwane wanted Kas to know that Nneang had a fever.

Kas was concerned about the toddler, but by then they were so close to Bloemhof that there was no point in turning back. The thought of the sick child continued to niggle away at the back of his mind, however, and he was therefore all the more concerned when negotiations at the grain merchants were interrupted by the unexpected arrival of one of the Kadi youngsters, who had been instructed to tell him that a second toddler, Sebubudi's daughter, had also suddenly taken ill and died at Klippan. He concluded his business in great haste and rushed home; relieved, he found that Nneang's condition had apparently improved.

Kas was reluctant to leave Leetwane and the child, but when there was no evidence that Nneang was worse in the morning, he thought he should go to Klippan and link up with Lebitsa and other family members who were there for the burial of Sebubudi's youngster. But it was not easy: the father in him wanted to stay at Vaalrand, and the brother in him needed to be at Klippan. He picked up his hat and old greatcoat and, as he walked out through the door, he sensed a sudden drop in temperature. He glanced back at the shack only once on his way out. Kolbooi knew the road to Klippan.

Even before the funeral was through he could see sheets of low-lying clouds scudding in from across the eastern horizon, eager, it seemed, to move into the seemingly endless space above the Vaal flood plain. It would be a close-run thing to reach home before they did. Kolbooi was at his best, and they arrived home still dry. But it mattered little, for they lost another important race: death had lured him to Klippan, bolted, and then run on ahead to claim his own child. Nneang was dead.

Kas felt mute. He sent someone across to the farmhouse to tell the landlord what had happened, picked up the bicycle and pedalled in silence all the way to Kingswood, to the Seedats' store. The Seedats knew all about death. He bought a few lengths of plain pine, strapped the timber to the bicycle and rode home, all the way pushing hard against himself. He was dog-tired but used what was left of the light to make a small coffin, something he had done before on the farms, but never for a member of his own family.

The depth of the night, as if sensing the grief that lay huddled within the confines of the shack, refused to yield to the approaching light of a highveld dawn. Kas stood in the doorway and looked out as a dark steel-grey illumination slowly picked out the outline of the ridge. Lingering clouds darkened, showing their disapproval of so feeble an attempt to bring hope. Within the shack not even Bodule stirred, and Kas did not have the heart to set the daily routine in motion. He turned and rolled a soft-spoken sentence into the back of the room before nudging aside a chicken that aspired to a roosting place on his bicycle. He got on the cycle and pedalled away again, hoping he could leave himself behind, but he could not. In Bloemhof he had to wait in turn outside the magistrate's office until he was given the chance to tell an utterly indifferent clerk about his daughter's death. On the way home it drizzled intermittently and by the time he reached the farm gates at Vaalrand, larger drops were beating out a more regular tattoo on his face.

As he pushed the cycle up the path leading to his homestead, Kas caught a glimpse of the Kadi men using shovels to open up the mouth of the grave. The tiny pine coffin was lowered into the earth, and rain, as if waiting for a cue, suddenly pounded down on their heads and ran down their cheeks until there could be no distinguishing tears from water. He and Leetwane stood side by side, contemplating the symmetry of the family they had sought to raise. In 1921, death had taken their first-born, a son at Hartsfontein, and twenty-three years later, in 1944, it had returned to demand their last-born, a daughter.[28]

As spring set in, no one talked of the toddler's death but, as if by common consent, the parents were drawn towards a broadly similar conclusion—that, in their pursuit of material things, the Maines had drifted from their spiritual moorings. Leetwane felt that a distance had been allowed to open up between herself and the church. Kas, whose commitment to Christianity had always been more qualified than his wife's, was slower to acknowledge the existence of an underlying pattern to life such as the Church taught. But he could recall that in the helter-skelter of acquiring livestock he had neglected to honour the shades. This time he had lost more than a leather purse stuffed with banknotes; he had lost a child. Leetwane discussed with the minister the logistical problems of their worshipping in the African Methodist Episcopal Church, and then persuaded Kas that they could set matters right by linking up with a nearby Ethiopian congregation that enjoyed the advantage of close historical ties with the A.M.E.C. Within four weeks of Nneang's death they paid membership fees to the new church and started attending Ethiopian meetings, where Kas, as if trying to put the emotional excesses of Klippan behind him and atone for the immediate past, was baptised under the name of 'Phillip'—a 'lover of horses.'[29]

Any tranquillity transfused into the Maine home via this roundabout route was soon dispelled by a development which not only tested both parents' resolve to the utmost but tended to push them back into practices that drew more on BaSotho traditionalism and Boer paternalism than on Christianity. It had been evident for some weeks that Morwesi had been putting on weight in a way that was most unusual for any nineteen-year-old, and when her mother summoned up the courage to question her daughter more closely, her suspicions were confirmed. The harvest festivities had had an unfortunate sequel. Morwesi was pregnant.

Leetwane was angry, Kas furious. Children born out of wedlock posed important questions of principle, questions which, in this case, were complicated by a set of very messy pragmatic considerations. It was not just that Kas and RaKapari were close friends but that, since the harvest, young Padimole Kadi had been married in a very public and ostentatious white wedding, attended by most of the sharecropping notables in the district as well as their landlord. This, and Kas's prominence on that occasion, heightened everyone's embarrassment. Issues that involved the payment of compensation were always likely to be closely contested, those involving the loss of face perhaps even more so. And so it proved to be.[30]

When Kas told Padimole that Morwesi was pregnant and raised the customary demand for livestock, Padimole denied responsibility for the young woman's condition and took the added precaution of rushing across to tell his father the same story before any of the Maines spoke to him independently. After that, RaKapari steadfastly refused to entertain even the possibility that his son could be the father of Morwesi's child. This intransigence gave rise to great tension between the two patriarchs, and in an effort to defuse the situation as well as to get its basic parameters recorded for any subsequent action in the magistrate's court, Kas got the schoolmaster, Lerata Masihu, to write RaKapari a formal letter in which he spelt out the Maine case and requested an urgent meeting of the families.[31]

The Kadis, hoping perhaps to underscore their social standing in local sharecropping circles, declined to walk across to the Maines' dwelling and chose instead to arrive in a horse-drawn cart. The Maines, not to be outdone, invited them into what passed for a living room. But the meeting was not a success. Arranged at opposite ends of the little room, the parents chose to endorse only those versions presented by their own offspring, eliminating any middle ground and exacerbating the tension. The patriarchs, or so it was later claimed, started out by shouting at one another and ended up by pushing and shoving one another around the shack. Kas then stumbled into an adjoining room only to reappear armed with an *assegai*, whereupon 'Mr. and Mrs. Kadi ran out to their cart and off they went,' as nine-year-old Mosala remembered it later.[32]

Labuschagne heard about the altercation and, anxious to avoid a full-scale explosion along the lower reaches of the estate, summoned representatives of both families to the farmhouse for a hearing. As so often happened on platteland farms, the search for the truth borrowed at least as much from BaSotho judicial procedures as it did from Boer paternalism and the Old Testament. It was one of those many strange and wonderful moments, born of crisis, when the physical isolation of the highveld forced much unacknowledged cultural borrowing to take place; individual identities which could otherwise be fairly readily pigeon-holed as 'black' or 'white,' were subsumed in a new, distinctively South African synthesis. On the appointed day, although not at the nominated hour, Labuschagne took up centre stage beneath a large camelthorn tree where he was flanked by elders drawn of the Maine and Kadi families. There, in the presence of their landlord-cum-chief, Sekwala and the other elders attending the *kgotla*, the charges were once again put to Padimole who, despite being confronted directly by his accuser, again denied that he had had any part to play in her disgrace.

With little prospect of reaching a settlement in the face of such contradictory evidence, Labuschagne, like many a Boer chieftain before him, improvised and temporised. The plaintiffs, he suggested, would have to wait until the birth of the baby, at which point its physical features would doubtless point to the father. With a decision postponed until after the birth of the baby, the judge hoped that passions would subside and the two parties reach an accommodation. He knew Kas and RaKapari to be reasonable men, and there were few problems that reasoning men, as opposed to emotional women, could not solve over a calabash of beer. The problem might have been quintessentially a female one, concerning a woman's unwanted pregnancy, but in a male-dominated universe, its solution was seen to lie in the domain of patriarchal power. A wily old campaigner, Labuschagne knew that, left unobserved, principle and pragmatism could well sidle up to one another. On more than one occasion he had solved problems of his own by calling on his black tenants and sitting down to share in a pot of freshly brewed *bojwala*.[33]

This ruling, which was not so much a judgment as the enforcement of a compulsory 'cooling-off' prior to a more definitive outcome, seemed as if it might work. Labuschagne received no further reports of nasty confrontations behind the ridge. But silence is ambiguous and it was not clear whether the lull in hostilities was due to the wisdom of his ruling or whether the Kadi and Maine households were simply preoccupied with the demands of the new season. In Kas's home, beyond doubt, the family members were being mobilised for another punishing summer routine. All crop production in the 1944–45 season was to be restricted to Vaalrand. But even there, in the heart of the 'family farm,' ploughing was less extensive than the summer before, for the landlord's Merinos made heavy demands on the grazing. Kas, whose hunger for arable land was directly proportionate to the amount of domestic male labour he could put in the fields, found the priority that sheep took over ploughs highly frustrating.

Under these circumstances quantity would make way for quality. He harnessed sixteen donkeys to the newly modified three-share plough and reverted to his old practice of working by night whenever the moon would co-operate. The fields were ploughed and re-ploughed until the furrows ran deep and straight enough to meet his exacting standards. Day shifts were equally uncompromising. By four in the morning the entire Maine household was at work. Kas himself put in four or five hours behind the plough before returning to the shack for a brief rest, then going back to the fields an hour or two later. Only in late morning, around eleven, when Leetwane arrived to feed him and his sons the thick fermented maize porridge known as *polokoe*, could Bodule and Mmusetsi even think of taking a break.[34]

Kas also took a managerial interest in developments across the way at Klippan, where Sebubudi seemed to need help. From what Kas could make out, his younger brother had first got himself into a financial mess by selling off to a distant cousin a team of donkeys that belonged to Sekwala. Sebubudi was effectively bankrupt, and so Kas took him to the livestock sales at Bloemhof, helped him pick out a new span of donkeys, and then advanced him twenty-five pounds to buy the animals. Sebubudi then got involved in a partnership with Johnson Xaba, but no sooner had he got that right than he allowed Hendrik Goosen to talk him into swopping his newly acquired donkeys for an obviously inferior team of draught animals. It was the sort of ineptness that caused Kas to despair.[35]

As a successful farmer, Kas found he also had responsibilities that went beyond the narrow confines of kinship. In early December 1944, he suddenly got a letter from Modise Tsubane, the man who had guided him into the Kalahari shortly after his marriage to Leetwane; Tsubane was stranded at Khunwana, and asked that Kas either forward him twenty bags of Sewefontein salt for trading at the Barberspan station, or lend him cash.[36] How could he refuse a man who was a friend as well as a fellow herbalist? But loans to kin and colleague alike only underscored the need for more hard work, and back at Vaalrand, the family groaned beneath the loads. It was almost as if Kas had decreed that work should become some form of collective family therapy. He knew how much the loss of Nneang had pained them, and he sensed that at such moments Leetwane was especially vulnerable. But was physical exhaustion enough to bury grief? As at the time of Mosebi's death twenty years earlier, he moved closer to Leetwane and thereby to the church. On 4 February 1945, 'Phillip Teeu' and his family were visited by the minister from the Ethiopian Church, R. P. Sidlayi, who baptised Leetwane and then, at her request and with Kas's explicit agreement, also admitted Mmusetsi and four-year-old Nkamang into the ranks of the church.[37] God, along with the three-share plough, had sneaked into the healing process, and the family seemed to be closing ranks.

Five days later Morwesi gave birth to a baby girl who was given the Sotho name Pakiso. Nobody thought of involving Reverend Sidlayi. Death might be coupled to the church and modernity, but life somehow retained its links to paganism and tradition. In another gesture, this time toward the paternalistic ethos that clung to the Labuschagne farm like the mist on a platteland morning, the child was given the additional name of 'Johanna' as part of a tribute to the landlord's wife. Although still convinced about the child's paternity, Kas allowed his demands for compensation to fade, and relations between the Maines and the Kadis slowly improved.[38]

These developments, along with light but well-spaced rains in the Triangle that year, further relaxed tensions along the lower reaches of the estate. The maize and sorghum fields took on an appealing aspect, and everyone around Kas breathed easier. Lazy summer afternoons stroked seasonal work rhythms into gentle wave-like patterns, and for a brief moment, there was an unusual harmony between man and nature, man and man, man and woman.[39]

Kas was out walking through the maize late one afternoon when he heard the landlord's truck purring behind him. It stopped, Labuschagne leant across, wound down the window opposite him, and invited him to join him on a jaunt into Bloemhof, where Arrie, the son of a local white railway worker, was due to take on a young man named van Rensburg, who hailed from across the river at Hoopstad, in a boxing tournament at Robertson's Hall. The invitation was couched in such warm and familiar tones that the thought of refusing never even crossed Kas's mind. He got in, the cab door looped through the thick summer air and thudded closed behind him.

It was twilight by the time the truck had bumped and bounced its way to Bloemhof. They were getting hungry, and Labuschagne drove to a conveniently situated store where he bought some food and then, without announcing his intentions, drove on to the off-sales division of the nearby Commercial Hotel. An ancient relic of the Griqua people wearing a felt hat, a so-called 'Bastard,' sitting on the pavement opposite the shop, watched as the white man disappeared into the doorway and emerged clutching a bottle wrapped in the familiar stiff brown-paper wrapping demanded by the law. Labuschagne coaxed the engine back to life, swung the truck around, and headed back to the outskirts of town. Near to the old town diggings he found a tree hugging the roadside and coasted to a stop. There, half-hidden behind a kindly bluegum—which understood more about the camaraderie of farming life than it would ever be willing to reveal in public—an Executive Member of the Eendrag Branch of the National Party and a black sharecropper partook of a meal and sipped brandy, enough brandy for Kas to start telling the white man about Vlakfontein and young Willem Reyneke. Then they drove back into town and joined the unashamedly partisan Transvaal crowd milling around the hall. Unlike the bluegum, Robertson's Hall spoke only to the public, and so, South Africa's code of racial segregation reasserted its supremacy: friends still glowing from their social intimacy under the bluegum were forced to part company—Labuschagne joining the whites on one side of the ring, Kas the blacks on the other.

Arrie, swarthy darling of the Triangle crowd, entered to wild cheering from both ends of the hall. Van Rensburg, the Free Stater in the other corner, was subjected to a barrage of jeering and catcalls interspersed with bitterly ironic comments from some of the blacks, who reminded him that the local Boers were renowned for their prowess with their fists! And so it proved to be. Van Rensburg took a lot of punishment from the son of the railwayman before Arrie knocked him out. As Kas recalled:

> He stood up, the Bloemhof boy threw a second punch which again sent the visitor crashing into the canvas, and then everybody stood up and shouted: "Hooray! All the way from Hoopstad to be defeated here!"[40]

Afterward, Kas waited for Labuschagne, and together they walked to the truck, chatting about the fight. But the spell of the bluegum had been broken, and their conversation flowed less easily. The truck jolted off into the night, Robertson's Hall in the background scowling its disapproval. Back at Vaalrand landlord and tenant took their leave of one another.

At Vaalrand it was customary to drive all the sheep on the property to the farmhouse on Sunday afternoon for a weekly audit of the herdboys' livestock totals. Occasionally Kas or Labuschagne would note that one of the Merinos or cross-breeds had gone missing—losses which they quickly attributed to theft by the few 'Basters,' 'Bushmen' or 'Coloured' elements who were still to be found on outlying properties. It was as if there was some covert ideological pact that no 'racially pure' Boer or black would stoop to sheep stealing. These minor losses were hardly ever reported to the police. Kas's own flock, along with his knowledge of wool farming in general, grew apace—for which he was more than willing to give Labuschagne credit. The material benefits of this were confirmed in March 1945 when the sheep were shorn for the last time before winter and the price of wool held its wartime value.[41]

The prospect of yet more good returns from the wool clip left Labuschagne especially relaxed. The last days of summer were rolling on into autumn, and the landlord made certain that all his shearers were given generous quantities of peach mampoer. The haze that had hung over Vaalrand for months showed no sign of lifting, but behind the ridge Kas was becoming restless. For those with hearts closer to maize than Merinos, autumn was a moment for mobilisation, not relaxation, and so he was sharp with Labuschagne when the landlord had the temerity to suggest that they pay a return visit to Bloemhof and take in one or other newfangled form of mass entertainment. 'I refused and told him that I could not take that sort of shit—sitting in a tent all day watching others doing their work while we left our own unattended.' Instead of going into Bloemhof, Kas went across to Kingswood, where the late E. S. Seedat's son, Solly, sold him a new wagon for twenty-five pounds in cash. For sharecroppers, autumn was about harvesting, not boxing.[42]

The harvesting team took on what was to become a familiar profile at Vaalrand: three married couples directly under Kas's supervision assisted by Mmusetsi and whichever of the younger Maines could safely be withdrawn from herding duties. Kas and Leetwane themselves worked without direct payment, while the cost of employing the Mokomelas and Masihus devolved on the landlord at the customary rate of six bags for every hundred harvested. This arrangement worked well, but as the pace of the harvesting accelerated through late May and into the early weeks of June, Kas became irritated by Lerata Masihu's laziness, which he attributed to a combination of cunning and book learning. Masihu, in turn, thought Kas's demands were unreasonable and once told him quite bluntly, 'If you were a Boer, I would not work for you.'[43]

The landlord owned his own thresher, so they had no need to wait for either Jan Groenewald or Wikus Pienaar or to pay any other middleman for his services. That was how Labuschagne wanted it. Like most Triangle landlords he had nothing but contempt for 'parasites' who lived off the efforts of Afrikaner farmers while never being directly exposed to the hazards of agriculture. But he was especially annoyed that the price of jute had increased for the seventh successive season (a 200-pound bag now cost more than three times its prewar price). He, along with many others, was of the distinct opinion that this was one more example of how the neglectful United Party government under General Smuts was allowing 'coolies' and other 'foreigners' to hold South Africa to ransom.[44]

But grain needed bags, and so they paid the price. At Vaalrand the harvest was just over two hundred bags, and in order to avoid 'the Jews' Labuschagne arranged for both his and Kas's share to be marketed through the South-Western Transvaal Agricultural Co-operative. This proved to be a good move: in a season of drought when the war had helped to keep demand high, Triangle farmers got prices that had last been reached during the early 1920s. For the first time in its history, the local agricultural co-operative handled more than a million bags of grain in a season.[45]

With more money flowing back into the countryside than at any other time since the Depression, the Smuts government might have been forgiven for thinking that the passions of platteland Nationalists would be tempered by economic success. But in the Triangle it had precisely the opposite effect. Inspired by the formation of the Afrikaanse Handelsinstituut, prominent Afrikaner farmers and political activists such as Commandant Abraham Pretorius, 'Jannic' Wentzel and 'Ampie' Roos approached the chairman of their co-operative, Sarel Haasbroek, with the suggestion that the S.W.T.L. establish a chain of retail stores at Bloemhof, Schweizer-Reneke and Leeudoringstad. All the hallmarks of the Boer Republican tradition would be on display. *Die Handelshuis* would be run by Afrikaners for Afrikaners.[46]

For Piet Labuschagne and Hannes du Plooy, a neighbour from Leeubos and fellow executive member of the Eendrag Branch of the National Party, this commercial initiative formed a part of the Nationalist vision. They and like-minded Afrikaners dreamed of a social order more congruent with an economy belonging to indigenous producers who, although separated by colour, shared the supreme virtue of being authentic sons of the soil who were either aspirant or confirmed Christians. 'Jews and coolies' not only lived off the people by buying cheaply and selling dearly, but were aliens by virtue of their outlandish religions, their intercalary class positions and their indeterminate skin colour. Foreigners by definition had no natural place in South Africa's social, political and economic order. If those at the heart of the nation, such as the farmers, *did* need traders, then they should come from their own ranks—from people who by culture, kinship and work patterns had shown themselves to be part of the social fabric. Until such time, the 'Jews and coolies' needed to be isolated and ostracised as a prelude to their expulsion. There was, they felt, much to commend Die Handelshuis.[47]

In practice, these initiatives often achieved less than their small-town originators hoped for. Xenophobic flares designed to light up the night sky above ethnic minorities did not always shed maximum light on platteland farms, where the realities of everyday life inhibited attempts to mobilise the Afrikaners along purely ethnic lines. Their need for easy credit and competitive prices made many farmers reluctant to abandon the closest 'Jew' or 'coolie' shop, and Nationalists like Labuschagne and du Plooy found it difficult to persuade their Triangle neighbours to fall into line. Only as a result of constant ideological labour by them and other rural activists did this form of ethnic thinking make any headway.[48]

But even as Afrikaner farmers were becoming more set in their views about Jewish and Asian traders their social interactions with black sharecroppers continued to show a measure of openness. Piet Labuschagne, who would no more think of sharing a meal with a Jew or a 'coolie' than of setting fire to his maize crop, saw no contradiction in calling in at the homes of his black tenants to share in freshly brewed beer, saying: 'Come on fellows, what's the problem? Pass me the calabash so I can knock some back.' On secluded highveld farms, race relations continued to be dominated by a paternalistic ethos. The question was how long such quasi-feudal social practices would survive the Boer landlords' accelerated accumulation of capital and the conscious efforts by Afrikaner Nationalists to fashion themselves a more leading role in the modern economy.[49]

Nobody in RaKapari's drinking circle found that socialising with the landlord was strained, unnatural or even uncomfortable. Indeed, on more than one occasion Piet Labuschagne—like Willem Reyneke and Hendrik Goosen before him—lingered on at the Maine homestead long enough for the bojwala to take its full effect and for him to be seen leaving with his head as light as his feet were heavy. Forty years later, when the wickedness of apartheid had worked its way through the system, Kas looked back on his days in the Triangle and observed, 'Those were good Boers who were not unduly concerned [about appropriate codes of social conduct], unlike the modern ones, who are full of nonsense.'[50]

With the season's drinking behind him and little prospect of earning winter cash from repairing shoes or from selling fuel to the few black families on nearby properties, Kas crossed the road to the salt works at Sewefontein, where he earned fifteen shillings a week helping to direct brine into the evaporating pans. He knew that a man had to work, and he did not want to risk becoming like Lerata Masihu; still, he remained convinced that the surest path to financial success lay via livestock farming. When Labuschagne told him about the price that pork was fetching on the Johannesburg market, Kas added a white boar and a fat piglet of his own to the consignment when the landlord next took hogs to the siding at Kingswood. Any reservations about working through unknown agents in distant markets were dispelled when, a week or so later, Labuschagne handed him twenty-five pounds in cash.[51]

Success with the pigs, which owed much to the example set by his wives, was followed by a second triumph, which, if truth be told, was also inspired by initiatives that were traditionally considered 'female.' Shortly after their arrival at Vaalrand, Kas had bought some chickens which, having been allowed to wander freely about the farm, were soon laying as many as three dozen eggs a day. In the past he had sold most of the eggs to storekeepers in Bloemhof, but given the demand for meat in Johannesburg, he decided that the time had come to sell the chickens via the same agents who had disposed of the pigs. To his great delight this too produced a splendid return, and when Sebubudi repaid him the twenty-five pounds advanced for the purchase of the donkeys, Kas found himself flush with cash.[52]

This little bonanza, along with almost all other cash that entered the Maine household, was immediately earmarked for investment in livestock. In the upper-middle reaches of the family, the hunger for cash for domestic consumption or for small luxuries remained as pronounced as ever. Only Bodule, whose primary responsibility was to manage the herd rather than to do domestic labour or grain farming, somehow found the 'extra time' for a hobby, and when Kas started selling chickens he too saw an opportunity. He went with his father to Bloemhof one morning and successfully sold a good number of pigeons.

Kas approved of his entrepreneurial spirit but was much less enthusiastic when Bodule invested the proceeds in a 'Baby Brownie' camera rather than livestock. Bodule surprised him by demonstrating that livestock was not the only investment capable of yielding a cash return: the snapshots he took of members of the younger set on the farm's Sunday afternoon social circuit soon made his new hobby of photography at least as profitable as the pigeons.[53]

His choice of subject matter re-emphasised the fact that, at the age of fifteen, Bodule stood on the threshold of manhood. When Kas asked to see a few of his better pictures, he saw that the boy was much taken with the daughter of 'Ou Kiewiet,' the Griqua. This discovery, coming hard on the heels of Morwesi's rather public disgrace, triggered a response from the patriarch that was almost as enthusiastic as it was premature. Here at last was a relationship that everyone would approve of and that, if it did lead to marriage, would offer a marvellous social analogue to the Maine family's rising economic fortunes. Not only was the young woman very attractive but, as might be expected from someone of mixed-race descent, she had a wonderfully 'light complexion.' Moreover, she was from a 'good' family who had sent her to Bloemfontein for an education, and was known to be very well mannered. In short, here was exactly the sort of daughter-in-law that a respectable family could be proud of.

With all the customary arguments about the need to preserve BaSotho culture, respect tradition and avoid the evils of book learning conveniently shelved, Kas slipped over to the old Griqua's place to sound him out about Bodule's prospects. Ou Kiewiet seemed to be perfectly comfortable with the idea of a marriage, and once this was established Kas thought it was also within the bounds of propriety for him to discuss the subject in RaKapari's drinking circle. But his fellow sharecroppers, sensing that the Maines were in danger of over-reaching themselves in their quest for social respectability, were less than enthusiastic. Kas recalled: 'Man, people were angry when I said that I wanted Bodule to take that woman for a wife! They wanted him to marry one of *their* daughters but I would not hear of it. None of them had decent manners.'[54] At the time, he was undeterred by their opposition. As far as he was concerned, the Maines were only a harvest or two away from acquiring themselves a really fine daughter-in-law.

The prospects for the season's crop farming were enhanced when, shortly thereafter, Labuschagne told Kas that he had decided to hire additional land on Barend Meyer's section of Gansvlei, about twelve miles west of Vaalrand. As before, no restrictions were to be placed on the amount of land ploughed, no individual allocation of fields would be made, and the harvest would be shared on the basis of the number of ploughing teams that each partner brought to the venture. In effect, Kas would manage all the grain farming for a third share while Labuschagne would take responsibility for all the livestock at Vaalrand.[55]

Although working for less than a half share in the grain harvest, Kas was enthusiastic about his contract. Labuschagne would have to buy the seed for the operation on Barend Meyer's property, and, in addition, Kas's sheep would continue to be expertly managed and his cattle enjoy access to the best grazing at Vaalrand. He arranged for Mmusetsi, Bodule and ten-year-old Mosala to load the ploughs onto the wagon and harness the donkeys: together they went to Gansvlei, where they immediately erected the hloma o hlomolle that would protect them against the elements.

Kas had chosen a historically important moment to decamp. The rains, although below-average, the presence of some moisture, the steadying influence of the Marketing Act of 1937, and grain prices that had been climbing steeply for more than half a decade all encouraged Triangle farmers to devote more and more of their land to maize, though the south-western Transvaal had previously been dedicated to more cautious 'mixed farming.' Increased profit margins enabled white farmers to pay back their debts to the Land Bank more quickly and buy the tractors they needed for arable farming. But these mechanical monsters too had to earn their keep, and this in turn necessitated a cut-back in the amount of land available for grazing. Everywhere, the roar of the tractor could be heard triumphing above the bellow of the ox. With imported machinery once again flowing freely into the country after the war, the number of tractors at work in the Transvaal countryside doubled from three thousand to six thousand between 1937 and 1946.[56]

The logic of this structural change and the awful long-term implications that it held for sharecroppers were not always fully understood by black tenants. But in RaKapari's circle they sensed that something important was afoot. 'We also wanted to buy tractors,' Kas remembered. 'We saw that the whites were cultivating more land and that there was less grazing for cattle, so we also wanted tractors.' But they had no capital and, until they were entitled to own property and have access to the banks, they would have to rely on tried and trusted methods. The best way to protect his livestock, Kas decided, was to 'remain on good terms with the landlord.'[57] Fortunately, his relationship with Piet Labuschagne was on a very sound footing.

The work at Gansvlei went well. The three-share ploughs cut long, deep, straight furrows that channelled and conserved moisture, and after a few weeks—the boys deeply grateful that the hardest part of the job was behind them—the Maine men turned to planting. This went smoothly for a while, but several parts of the planter that Kas had bought from old Piet Reyneke all those years before needed replacing. This broke the work rhythm, necessitating a trip into town at a time when Kas was beginning to tire of the bachelor-like existence on Barend Meyer's property. It irritated him that Labuschagne's work allowed him to stay at Vaalrand, and he thought, Why should I spend most of my time away while he stays at home?[58]

As feared, the trip to Bloemhof was a waste of time. Wentzel & van der Westhuizen no longer kept parts for 'The Planter's Friend,' and could only suggest that Kas write to the suppliers in Johannesburg and see if they still had spares. To make matters worse, Bill van der Westhuizen took up a good deal of Kas's time by discussing a new venture he was planning in the well-watered valley of the Mooi River and then inviting him to join him in Natal. Kas heard him out and then firmly but politely declined the offer. It was flattering that a man like Van der Westhuizen should think so well of his abilities as an artisan and as a farmer, but he had more pressing things to contend with. On the way home he gave thought to the problem of how to get a letter written.[59]

A complex communication about spare parts was beyond the reach of Masihu, and Labuschagne, who knew little of things mechanical, was unlikely to be much better. Kas then remembered young Heine Edwards, whom he had met while working at the salt works during the off-season. Edwards, who had already picked up on Kas's reputation as a farmer, was more than willing to help him write to Malcomess Ltd. in Johannesburg. A week later Kas collected the reply at the little post office at Lorena, and took it across to Edwards. Edwards read out a quotation that Kas considered to be unsatisfactory, and then, just as Kas was about to leave, started asking him about a possible sharecropping venture in the coming season. But the irritations of life at Gansvlei paled beside the dangers posed by an unknown landlord, and so he declined this offer, too. Ten days later, he received what at first seemed like a vindication of his decision to remain loyal to Piet Labuschagne.[60]

Around Christmas, when the daily routine eased slightly to accommodate the seasonal movement of family and friends, he was away at Xaba's drinking beer when Mmusetsi came to tell him that their cattle had pushed down a fence, invaded Hans Coetzee's fields and done great damage to his maize.[61] One grey cow, unaccustomed to so rich a diet, had already died of indigestion, but the Boer was so angry that he was refusing to release the remaining animals until he had spoken to Labuschagne. Kas hurried back to Vaalrand, but by the time that he got there it was already dark and Labuschagne told him that they would have to wait until the next day to resolve the crisis.

Labuschagne and his brother-in-law appeared at the Maines' shack at first light. Kas clambered onto the back of the cart, and together they set off for Kareepan. Kas said little, choosing to leave any talking to the two white men. The road was badly rutted, and by the time they reached the Coetzee farmhouse, shards of sunlight were prising open his narrowed eyelids. Kas turned his head from the source of the glare and out of the corner of his eye caught sight of Mr. and Mrs. Coetzee lumbering their way through the kitchen doorway. For a moment or two he did not move. Once Labuschagne and his brother-in-law had eased themselves off the buckboard and onto the ground, he slid around the side of the cart and caught hold of the dangling reins. The parties moved closer to one another and then, like short-sighted rhinos meeting unexpectedly on a narrow path, suddenly stopped. Kas moved in behind the horses and monitored every word, every phrase.

> "More Hans."
> "More Piet."
> "Man," said Labuschagne, "what is this that I hear about you?"
> "About what, Piet?" said Hans.

Labuschagne observed that he had heard that Coetzee had impounded the cattle, and wanted to speak to him. Why, he asked, had Coetzee not at least released the cows with calves?

Coetzee replied that he should not adopt such an aggressive tone, because the cattle belonged not to him but to a kaffir.

Labuschagne lost his temper, told him that the *kaffir* belonged to him! He threatened to punch him so hard that he would end up back in the farmhouse. *Everything* on Vaalrand belonged to him, he told Coetzee. The kaffir owned nothing.

Coetzee pleaded with his neighbour to calm down and talk politely, but that only made Labuschagne angrier.

'Man!' he yelled. 'Am I not talking nicely?' He turned to Kas and said, 'Kas, leave this damned fool and let's go to town, and we will make sure that he also pays for the cow that died in the fields.' Still angry, he repeated that Coetzee should at least have had the sense to keep only one cow and let the remainder return to Vaalrand.

> Mrs. Coetzee told Labuschagne not to talk like that. But Labuschagne was really furious, and asked her whether she had not heard what her husband had been telling him. Mmusetsi had come to claim the cattle

> the afternoon before and her husband should have allowed him to drive the cattle home. "Now, when I get here, your husband tells me that they are the kaffir's cattle." He repeated, *Everything* on the farm belonged to him, even the kaffir.
>
> Mrs. Coetzee then said, "Piet, come on in, let's first have a cup of coffee," and they then all went inside.[62]

The Boers emerged from the farmhouse about half an hour later. Labuschagne told Kas to collect the animals, drive them back to the farm, milk the cows and make certain that all the milk was fed to the pigs. Then Kas should take the wagon and return to Kareepan to collect the carcass of the dead animal. Kas let the reins drop, slid out from behind the horses and moved off.

What was the significance of the exchange that he had witnessed? It left him with decidedly mixed feelings. Kas was grateful for what Labuschagne had done, but there was something deeply disturbing about the landlord's exchange with Coetzee. It was not so much the racist terminology that had shocked Kas—on the highveld such language was freely used by everyone from 'Boers' to 'Bushmen'—but the logic of the landlord's dictatorial attitude. Kas had always seen his and Labuschagne's relationship as a partnership as much as a paternalistic arrangement. But what he had seen at Kareepan that morning left no room for partnership, and paternalism came off no better. Shorn of its Sunday finery, paternalism wore the undergarments of autocracy.

> Yes, 1946–47. Boers! It was then that I realised that Afrikaans was a difficult language. Coetzee had said, "Man, you can't talk like that, they are not your cattle, they belong to the *volk* [people] who work for you." Labuschagne had replied, *"Man, there is no such thing as a kaffir's cattle."*

Kas was gradually coming round to the view that, for all its many virtues in a harsh environment, paternalism was predicated on a structured inequality that required the perpetual adolescence of the junior partner. It left a black farmer with no room for either psychological or economic growth.

The limits of Boer paternalism were again apparent a few days later when the accident-prone Mmusetsi fell from his horse. Kas, examining his son's shoulder, conceded that the young man should consult a general practitioner. He asked the landlord to take the boy into town by truck, but Labuschagne refused, saying, 'Oh, it's just a minor thing that will heal by itself.' Unconvinced, Kas got out the horse and cart and twice over the next ten days escorted his son to Bloemhof, where Henry Hutchinson treated the lad for his badly dislocated shoulder.[63]

Kas found Labuschagne's unwillingness to help out mildly irritating but was not unduly puzzled or perturbed by it. He recalled that, even under Piet Reyneke, there had been moments when he thought he understood the landlord well only to discover that the 'Boer' suddenly did something totally unpredictable. In any case, Mmusetsi's shoulder healed well and soon there was something more serious to worry about. Kas called for Bodule and told him to take the horse and cart and drive across to Soutpan, where the boy's grandfather was living with Phitise and his family.[64]

Sekwala, who was long past his eighty-fifth year, had been unwell most of that summer. Kas had visited him several times, usually on Sunday mornings, and, when that had not been possible, had taken to sending Bodule across to get news of the old man's health. But each report was more alarming than the last, and Kas was deeply concerned. His father's condition deteriorated steadily and Sekwala eventually passed away in the last week of January 1946. He was buried by the Reverend Simon Mokhatsi of the A.M.E.C., which would not necessarily have met with his approval. Despite having been surrounded by committed Christians, Sekwala had persisted in seeing himself as a MoFokeng of old. But he was not around to defend himself when it mattered. Death was indeed about modernity, life about tradition.[65]

Kas realised that with the passing of Sekwala the family had lost its last direct link with Sekameng, the Malutis and Lesotho. The Maines, led out of the mountains on ponies by grandfather Hwai in the 1890s, had over a half century become plainsmen. Suddenly aware that there was no turning back to the mountains if all else failed, Kas re-oriented himself and those around him via his great-grandfather, Seonya, the man who had established himself amongst the BaKubung-BaFokeng of Molote, on the highveld. In times of uncertainty, as when a family lost its patriarch, it was important to know by which historical stars one would steer yet another generation without land of its own.

Sekwala's eclipse had not been entirely unexpected. Even in a family noted for its longevity he had lived to a fair age. Perhaps it was the unarticulated expectation that Maine men simply did not die until they were ninety that made Mphaka's death all the more unexpected. He had been ill for some days after his father's funeral, but all agreed that it was a minor illness. Yet barely four weeks later Mphaka too was dead at the age of only fifty-five. He was laid to rest in the small cemetery at Soutpan where his father and grandfather were buried. Reverend Mokhatsi, displaying the same tolerance that he had extended to the heathens who attended Sekwala's funeral, was on this occasion equally accommodating of the A.M.E.C. cum 'Communist Party' women who came led by Mphaka's widow, Motlagomang.[66]

It was a sad truth that death often brought an ageing extended family closer together than life, and, back at Vaalrand, Kas no longer saw as much of Motheba and his brothers and sisters as he once had. Partly this was a function of distance. Motheba, for example, had been living with Mphaka and his family and after Sekwala's death decided to move to the hamlet of Harrisburg, a railway siding halfway up the track to Klerksdorp, where Sellwane was working as a domestic servant. But the most important reason was the hectic life that Kas led. His own family and work demanded so much attention that he did not have time to devote to what now seemed like more distant kin. The old Maine family—the scion of Sekwala—was disintegrating; the new one coming of age. There was, Kas feared, no other way. When he was not out ploughing or harvesting, he was looking after the needs of his cattle, donkeys, horses and poultry.

*

In March 1946, when his flock was brought in for shearing, it was evident that Kas owned a formidable array of livestock. The immense amount of family hard work, and a postwar boom in agriculture, were bringing him the success denied him in the 1930s. A half-dozen horses met his family's transport requirements and he was more than satisfied with the fourteen donkeys that did service behind his three-share plough. But the really spectacular success was in his flock of cross-bred and Merino sheep. Benefiting from a third season of Labuschagne's professional and personalised attention, Kas's flock had grown to number one hundred and fifty animals. He was quick to give the landlord credit. 'By taking good care of my sheep, that white man [Labuschagne] made me rich.' Kas's cattle, including a span of healthy oxen that grazed on Vaalrand's finest grass, numbered close on fifty. Here again he was unstinting in his praise of Labuschagne: 'He was the man who made me rich.'[67]

Kas had indeed been fortunate with his landlord, but the position of black men in the Triangle who tied up their family's labour in domestic enterprises, farmed grain successfully and accumulated livestock was increasingly anomalous and precarious. White farmers were looking to sharecroppers' sons rather than to entire peasant families for cheaper adult labour, while those who owned tractors frowned on oxen monopolising potentially arable land. Blacks with labour and livestock, and whites with capital and land—men whom a shortfall in resources had pressured into economic partnership and social proximity for half a century—were being pulled apart by forces that neither fully comprehended. And as the moment of their bitter-sweet separation drew closer, it was privileged access to the state rather than mere farming know-how that did most to determine the victor. Afrikaner Nationalists, having spent much of their ideological venom on 'Jews' during the war, were regrouping openly for an ideological assault on the hated 'Coolie Traders' and surreptitiously sharpening their prejudices for the coming encounter with 'rich kaffirs.' The unlikely venue for the first skirmish in this latter war was the monthly cattle sale at Bloemhof. Decades later, Kas recalled:

> That's when they started, in 1946. My father died in 1946. I remember . . . they started those things [efforts to impose stricter racial segregation] on the day I sold an ox named Kolobere. I took it in to Bloemhof, along with all Piet Labuschagne's other cattle. It was really fat! When we got to the auction I asked Labuschagne not to separate out the ox from the lot as a whole, but to sell it along with the rest of his animals. I said, "Do not sell it separately as belonging to me." But Labuschagne insisted that the animal was "light" and that it would only serve to reduce the price his cattle fetched. I then told him to set my ox aside and allow it to be sold separately. Labuschagne sold all his animals and only Kolobere was left. The ox went into the ring on its own. Kemp, the auctioneer, raised, and raised, and raised the price *very much.* It reached the price that Labuschagne's cattle had been sold for, and Koos Meyer eventually bought it at that high price. Labuschagne later came up to me and admitted that, had he left Kolobere in with his lot, it would not have fetched such a good price. "Your ox has obtained a better price than mine," he said. Things like that caused [racial] conflict.[68]

If experiences such as this were difficult for the affable Piet Labuschagne to swallow, then hotheads like Willem Griesel and Willem Nieman choked on them. On the eve of their most substantial political and economic advance ever, a combination of greed, resentment, and racial pride pushed local Nationalists into proposing that livestock auctioneers bar all blacks from attending the cattle sales. When A. J. Marais and his fellow directors at Slabbert & Campbell as well as Kemp refused, the most militant withdrew their patronage and started frequenting racially segregated sales held elsewhere. 'Yes, they sold only to whites and refused to admit black people. But neither Kemp nor Marais refused blacks admission: they held that black folk were free to attend their cattle auctions.'[69]

When a predatory evil stalks the collective good, like a lioness shadowing a herd of impala, a person takes solace in the belief that his own attributes—bravery, cunning, or mere fleetness of foot—will be enough to exempt him from the inevitable strike. And so it was with Kas. Perhaps because they still greeted him as 'Ou Koop en Verkoop,' when they openly referred to all his friends as 'kaffirs,' he failed to appreciate fully the menace. Perhaps because he got on so well with the Kemp brothers, the advent of segregated cattle sales elsewhere failed to arouse greater indignation. Perhaps because his inner being was more attuned to nature's rhythm than history's clock, he noticed only the need to bring in the harvest.[70]

It was easy to assemble a harvest team: for weeks members of the Maine, Mokomela and Masihu families scurried back and forth between

Vaalrand and the hired property at Gansvlei. It was a rewarding experience. Though the Triangle was going back to one of its periodic dry cycles, the rain was still regular enough to ensure a harvest of more than four hundred bags. He and Labuschagne met to do their sums, and at the end of the day, the Maine family had more than a hundred bags to dispose of in a season that saw the sixth consecutive rise in the price of maize. The big white producers, who already had much to smile about, were further cheered by news that the price of Indian jute bags had dropped slightly. Kas however, hardly noticed it.[71]

June drew to a close and, with the night temperatures so low that strangers could be forgiven for mistaking the salt at Sewefontein for snow, he and Labuschagne went across to Gansvlei for the last time to supervise the movement of the grain to the co-operative. Rubbing hands to get blood into chilled knuckles, they stood and watched as the youngsters loaded the two-hundred-pound bags onto the back of the truck and then, eager to get back to the comfort of Vaalrand, decided to lend them a hand. Before long Kas was engaged in some good-natured rivalry with Labuschagne, as he had with Hendrik Swanepoel in a bygone era, seeing who could shift the most bags. But, as almost always happened in the Triangle, any contest that started out as a challenge to masculinity soon darkened; with racial pride at stake, this happened very rapidly. Kas suddenly stopped, for he felt a searing pain ripping through his side, and sank slowly to his knees. It was as if an assegai had been plunged into the lower reaches of his belly. He knew at once that this was a pain which, like that in Mmusetsi's shoulder, lay beyond the reach of herbal medicine. He made his way home and spent a troubled night. The morning found him in Bloemhof.

Kas had known, liked and trusted Henry Hutchinson ever since they had first met during the influenza epidemic nearly three decades earlier. As young men they had spent hours together on horseback, riding to outlying farms, and he found something firm and reassuring in the deft movement of the district surgeon's hands, though Hutchinson was now almost seventy. Hutchinson told him he had torn a muscle and he would have to undergo surgery to set it right: for that he needed to obtain clearance from an adjacent office. Hutchinson scribbled him a note and sent him next door, where someone asked him some questions about how much livestock he owned and then sent him back. Hutchinson told him to go home and get ready to go to the Klerksdorp hospital, but not to distress his wife unnecessarily by telling her that he was to have surgery.

The next morning Bodule drove them back into Bloemhof, this time with Leetwane. Hutchinson took Leetwane aside, told her that her husband would be away from home for about a week and reassured her that all would be well. Leetwane came out of the district surgeon's office, said goodbye to Kas and then, assisted by Bodule, stepped up onto the cart for the ride back to Vaalrand. Kas watched the cart clatter down the road to the Sewefontein turnoff and, when he could no longer hear the sound of the horses' hooves on the hard-baked surface of the dirt road, turned and walked slowly to the railway station. He was suddenly very aware of being alone. Klerksdorp was a sharp angular name for a place in nowhere land, and even its melodious Tswana name, Matlosana, did not sound comforting. He did not know the place.

The train, like all second-class trains in South Africa, was dirty and noisy. The locomotive blew its nose into the cold winter air, showing as little respect for the passing passengers as did a spitting peasant for the niceties of a High Street sidewalk. Men and women shouted to make themselves heard above the din, clutched awkwardly shaped parcels, and scrambled aboard the coaches, which were stained a nondescript brown. The engine cleared its nasal passages and puffed out of the station. Kas, peering through a grime-encrusted window marked with the scoured imprimatur of a crescented springbok, watched as familiar landmarks slipped slowly by—the route followed the course of the Vaal—Kingswood, Boskuil, Swartgrond and Makwassie. Outside Makwassie the train, which up to that point had behaved most responsibly, suddenly became alarmingly independent and swung away from the known world around Wolmaransstad and veered east. Kas felt unusually vulnerable, and when it chugged through Harrisburg, he was reminded of Motheba, their recent meeting at Soutpan, and death.

He decided there was little to be gained by fretting about such matters, and as the train turned once again toward the looping Vaal, he noticed that the grazing around the outskirts of Klerksdorp compared favourably with what his cattle knew at Vaalrand. The engine, spitting a mixture of coal dust and soot, which swirled into the compartment via a broken window, signalled its intention of stopping by whistling, then crawled to a stop. As he disembarked, the sharp pain in his side reminded him of the purpose of his visit: eventually Kas found a stranger who directed him to the hospital.

> Two days after I was admitted to the hospital at Matlosana, Hutchinson telephoned. He was phoning on behalf of Leetwane who wanted to know how I was getting on. They told him that I would be operated on the following day. That was when my wife first learned that I was due to undergo an operation. They told her to remain calm.[72]

At dawn the next morning they wheeled him off to the operating theatre and placed what he could only describe as a 'certain paper' on his face. 'I fell asleep for a long time. When I woke up I was already back in the bed where I had been before the operation.'

A day later Hutchinson arranged for Leetwane to visit Kas at Matlosana. Four days later she found him sitting up and enjoying the company of a 'Ndebele' sangoma who was at least as pragmatic about herbal medicine as he was. Leetwane did not stay for long, and a day or so later, Kas followed her back to Vaalrand, where he showed the children and any curious family members the scar from his operation.[73]

Taking stock of the situation when he returned, Kas was pleasantly surprised to see how well Bodule had done in his absence. There was no doubt that, of the three Maine boys, Bodule was the one who had the makings of a farmer. It was unfortunate that Ou Kiewiet's daughter had suddenly taken ill and passed away. Someone else amongst the cousins would have to be found who was worthy of being a wife to so promising a young farmer. But Bodule himself seemed less than concerned. He was already taking an interest in the well-being of one of the Masihu girls. Now, *that* was a prospect Kas did not relish. Everybody knew that teaching and farming families did not mix![74]

It took a while for Kas to get back into the routine of the farm. He undertook a leisurely inspection of the cattle, carefully examined his heavy agricultural implements, and spent several days making minor repairs to the ploughs. He took to walking the fields, and while he was on one of these little outings he came across Heine Edwards again. Edwards, like most white farmers, wanted to expand the area he devoted to maize. But, like Labuschagne, he still lacked a tractor, and so when he spotted Kas, he once again asked him to 'jump the fence' and join him in a joint venture at Sewefontein. But nothing had changed since they had last spoken, and Kas again declined the invitation.[75]

Fully recovered, Kas set about the last of his winter chores with a will, making certain that when the time came to go back into the fields every animal, harness and plough would be ready. He sensed that, given fields large enough and rains good enough, he now had the resources to produce a really spectacular crop, a harvest that would be counted not in hundreds but in thousands of bags. All he needed was a season when everything went right, and he would enjoy as much security as any black man could without actually owning property.

The optimists in the Triangle were already scanning the horizon for stray clouds heralding the arrival of spring when six-year-old Nkamang took ill. By late afternoon she was bleeding so profusely from the nose that Leetwane became alarmed, and Kas went across to ask Labuschagne to take the child to town. But Labuschagne was unwilling, and so Kas asked him to use his trap—'I used to borrow his cart and drive it.' The reply struck him as a veritable thunderbolt:

> Do you know what he said to me? He said that his cart could not be used by "kaffirs"! I said nothing because my child was ill.[76]

Angry, frustrated and humiliated in equal measure, Kas turned his back on Labuschagne and strode home. He took out the pony, swung into the saddle and then—in full view of the watching landlord—leaped the fence and galloped off in the direction of Klippan. Hendrik Goosen would, for a fee of twelve shillings, drive anyone, especially a sick child, into town to see a doctor. Fortunately he found the Boer at home, and Goosen said that:

> . . . I should go home and prepare Nkamang for the journey. I left immediately, once again jumping the fence at Vaalrand on my way home. Hendrik Goosen arrived shortly thereafter, and together we took the child in to the doctor.[77]

Hutchinson staunched the flow of blood but warned that the child's life would have been in jeopardy had the visit been delayed much longer. They left his rooms and got into the waiting car, as the surgeon reminded them to bring Nkamang back within a week. The car groaned and rattled its way past the familiar names on the farm gates and thirty minutes later, drew up at a point conveniently close to Labuschagne's property. The trio walked together up the path to the shack and, on the way, passed the toddler's grave. They had given one child to Vaalrand, and Kas was determined that they would not sacrifice a second.

Everybody needed time to recover, but Hendrik Goosen's car was barely out of earshot when an irate Labuschagne was on Kas, demanding to know why he had jumped the fence.

> I told him that I had not jumped the fence at night. "I jumped that fence in broad daylight, you saw me! Should I have waited for my child to die? I was forced to jump the fence in order to go and find help for my daughter." I said, "No, man, I was left with no choice. You told me that kaffirs could not use your cart. I only went to Hendrik Goosen because he knew me." Labuschagne said, "Yes, but you shouldn't have jumped the fence." I then said, "No, man, I didn't have time! I haven't stolen anything and I haven't damaged anything. I was just in a hurry to get to Goosen. You wanted me to go via the gate, but it would have taken far too long."

Labuschagne paused, said no more, and then made as diplomatic an exit as the strained circumstances allowed. But despite the silence, all the more marked for the sound of the departing footsteps, the matter was far from resolved. As Kas returned to the sanctuary of the shack to recover his composure, he found himself awash with resentment.

But there was no time to act on this or any other sentiment because Nkamang's nose once again started to bleed. Kas didn't even bother with

Labuschagne this time but went directly to Goosen for help; a second rushed visit to the district surgeon ensued.

On the party's return to Vaalrand, Goosen fell into casual conversation with Labuschagne, exchanging a few pleasantries. He then asked Labuschagne why he had had a problem about lending Kas the trap. Labuschagne claimed that, although he had always had the highest possible regard for Kas as a farmer and while he would gladly allow him the use of the cart for business purposes, he could not sanction its use for social ends. This observation added fuel to the fire raging within Kas. So! Work and life on the farms, hitherto always inextricably intertwined, were to be separated! Yet again, he resisted the luxury of an immediate response. Time would deliver Labuschagne into his hands. Nkamang continued to improve, and then,

> One day [Labuschagne] told me to inspan the cart and fetch some salt for the sheep. I told him that he would have to drive the cart himself because it couldn't be used by a kaffir. He kept quiet and shook his head. Then he asked me whether I was still on about the same old thing. I refused to do that job and told him to give me my trek pass; I told him to sign me off because I couldn't be treated in that way.[78]

Within a week, Kas and the Maine family had moved across the road to Sewefontein.

What exactly had gone wrong? As in the late 1920s, when a new and confident generation of radical black Nationalists in the Industrial and Commercial Workers' Union had challenged the old order from below with the demand that Africans on the land be accorded the dignity and rights that befitted all workers, so, in the mid-1940s, a rising cohort of Afrikaner Nationalists was challenging the old order from above with their demand to be recognised in a society where, for many decades, painfully little had distinguished them from their black sharecroppers.

In both cases it was the notion of paternalism—part chain, part umbilical cord—that was unconsciously questioned and in both cases the resulting wrench was agonising. The betrayal was most brutal where the trust had been most freely given, the pain worst where the pleasure had been greatest, the racism most intolerable where the friendship had been most intense. For Kas the conflicts had always revolved around the well-being of one of his children. At Kommissierust the bonds had snapped when the normally genial Reyneke had slapped Mmusetsi and called him a 'kaffir'; at Vaalrand, when an often generous Piet Labuschagne—struggling to align the emerging philosophy of apartheid with the paternalistic obligations of an older order—had refused to help him with an ailing daughter.

Whether Kas and Labuschagne were fully aware of it or not, both had been drawn into the social backwash occasioned by a shift in the currents within the Triangle economy. No longer either able or willing to rescue one another by using the codes of the older, paternalistic order, and as yet uncertain about the precise rules governing the incoming capitalist dispensation, they continued to pull apart. Even as they drifted apart, their voices could be heard calling out, each acknowledging the prowess of the other. More than forty years later, Kas still spoke with genuine enthusiasm about the progress he made at Vaalrand: 'Piet picked me up.' Labuschagne would with equal relish recall, 'There was *nothing* that Ou Kas could not repair.'

In the spring of 1946, Kas Maine was ready to put any skill he had at the disposal of the Edwards family. But Sewefontein was a troubled farm in turbulent times.[79]

CHAPTER TEN

# Turnaround

## 1946–49

The tedium of the extended plains of the southwestern Transvaal can lull the unwary into complacency. With only an occasional chalk scar slithering across the road to break the monotony, the traveller might be forgiven for slipping into the belief that the whole world is dominated by a succession of dry and dusty depressions. But even here, where nature has allowed the horizontal to triumph absolutely over the vertical, the visitor can be startled by the sudden appearance of the extended pan at Sewefontein.

The largest of ten such shallow evaporating dishes left lying about in the Triangle sun, the three square miles of this heart-shaped depression are gently recessed into the marginally darker soils just north of Vaalrand. Blessed with the 'seven fountains' of its original Dutch name, this unevenly filled saucer of part-brack, part-fresh water is fringed by clumps of bamboo and a cluster of mimosa trees. Situated in what by commonsense criteria is semi-desert, the pan has always been something of an oasis for men, animals and birds who wander from the sweeter waters of the Vaal, which gurgles away some fifteen miles farther south.

When Grandfather Hwai first made his way into this district, back in the late nineteenth century, Sewefontein, when flooded, regularly attracted enough flamingos to paint the entire southern rim the gentlest shade of pink whilst, out in the deeper water, scores of puffy white pelicans marked the centre of the pan. Long before the surrounding acacias had been ripped out to feed the fuel-hungry machinery of the diamond fields, and for some decades thereafter, enormous herds of migrating springbok made their way to the water's edge on their journeys to and from the Kalahari Desert to the west.[1]

The water, along with soil that had more clay in it than that of neighbouring properties, drew white adventurers to Sewefontein even before the speculators started circling the area above the Vaal and Harts rivers in the 1870s and 1880s looking for diamonds. Sewefontein itself, however, had no diamantiferous gravel of consequence, and this, together with its potentially valuable salt deposits, meant that unlike most other properties in the district, the farm seldom changed hands. Indeed, once its first legal owner, the widow Koetze, had sold it to an Englishman named Coleman in 1862, the title remained in the hands of the same family for seventy years.[2]

The first of these Colemans, William Henry, hailed from Colchester, in Essex, the restless scion of a wealthy middle-class family. He had left England in 1837 and, at the age of twenty-one, established himself in South Africa as a sheep farmer in the Witteberg mountains of the Graaff-Reinet district. He was successful enough to be able to spend a good deal of his time in the Eastern Cape, where he mixed freely with the Curries, Ogilvies, Southeys and other leading British settler families in and around Grahamstown. After his marriage to Rebecca Murray in 1844, he abandoned the interior and established himself as a miller at Port Elizabeth.

But by the early 1860s milling had lost its allure for Coleman, and during a visit inland to seek out potential trading sites he met the widow Koetze and bought the farm Sewefontein. Unable to persuade his wife or his three daughters of the virtues of life in the far interior, he moved instead to the Victoria district, where he once again farmed most successfully until Rebecca died unexpectedly in 1870. Without Rebecca to check his restless spirit, he was now able to follow the first exciting diamond discoveries south of the Vaal, and shortly after his wife's death, he and his daughters moved to Kimberley. There he renewed his acquaintance with Sir Richard Southey and turned his hand first to digging and, when he tired of that, to brewing. When the brewery, too, failed to bring him the happiness or success he desired, he called up the last of his ancient East Anglian skills and became a salt manufacturer.

When the Colemans moved to Sewefontein in 1874, they discovered that the game, soil and water around the pan were already supporting more than a dozen other refugees from all over the subcontinent, which was still experiencing the first shock waves of its incipient industrial revolution. Amongst the more prominent families 'squatting' on the property were the van Schalkwyks and the Camilles. 'Ou' Dirk van Schalkwyk was a manumitted slave who, at the age of thirteen, had been given his freedom by the same Cape Dutch farmer who had endowed him with his light complexion. He and his wife, Griet, who took care of the Colemans' laundry, were mostly supported by their son who, armed with an antiquated 'flint-lock fowling piece,' made a living as a professional hunter and supplied the district with venison. In contrast the Camilles (or 'Kameels' as they were known) were crop farmers from the Basutoland plains; they helped to irrigate the maize, sorghum and wheat that was raised in the gardens surrounding the stone-walled house built on the property.

Crystals, crops and carcasses were a formidable combination. In 1878, Coleman invited his son-in-law, Jack Edwards, to join him in the business at Sewefontein. Between them—at a time when 50,000 to 80,000 black workers were making their way to the diamond diggings each year—they helped to supply the Kimberley market with *biltong* (dried meat), fresh vegetables, salt and venison. Many of the migrants, some of whom came from as far afield as the remote northern Transvaal, made a point of breaking the journey south at the Sewefontein watering hole. Their presence, in turn, attracted predators of another sort. Thus, in addition to accommodating several hunters, including the famous F. C. Selous, the Colemans also hosted various labour recruiters hoping to intercept parties of workers on their way to the mines.[3]

Perhaps it was the labour recruiters who first told Jack Edwards about opportunities opening up on the new Transvaal goldfields; in any event he left Sewefontein in 1887 to become secretary to a Witwatersrand mining company. Just two years later William Coleman died, leaving the property to his oldest daughter, Mina, and her two sisters. Lacking the expertise to run so diversified a business, the sisters leased the salt works to the Oceana Company, while the farming side of the venture fell increasingly to Mina's oldest son, Jeffrey John Edwards, whom the local BaTswana aptly referred to as Tetangwane—'the tall one.'

A young man of only eighteen or nineteen when he effectively took control of the farm, J. J. Edwards was at least as capable a manager as William Coleman had been before him. But, unlike his East Anglian grandfather, who had prided himself on his British connections, he was a native of the Transvaal and readily identified with the republican cause espoused by his Boer neighbours, a sympathy that deepened after the Jameson Raid. At about the same time, in 1897, he met and fell in love with Anna Moormeister, the pretty but somewhat vacuous daughter of a German widower. Their courtship was interrupted by the outbreak of the South African War in late 1899; Edwards soon found himself in a prisoner-of-war camp on the island of St. Helena. There, where Napoleon had rotted while sipping Cape wines, this English-Afrikaner spent most of his day carving old animal bones into elegant napkin rings which he dedicated to his beloved Anna and her nine-year-old brother, Walter.[4]

On his return to South Africa in 1904, J. J. Edwards married Anna Moormeister; they moved into the farmhouse at Sewefontein and raised two sons and two daughters. Edwards hoped that his eldest son would eventually take over the farm, but when that young man qualified as a teacher in the late 1920s the task fell to his second-born, the boy whom his wife referred to affectionately as 'Heine.' Heine, who like his great-grandfather Coleman before him bore the names William Henry, took charge of most of the farming operations during the mid-1930s, and when Edwards died in 1937, it was widely assumed that this rather quiet-spoken and thoughtful young man would inherit the property. But he did not. Edwards had made the property over to Anna in the hope that *she* would one day make provision for young Heine. But this move left Heine doubly vulnerable, for it failed to take cognizance of another factor: the reappearance, after many years' absence, of Anna's younger brother, Walter.

Walter Moormeister had eased his way back into his sister's life toward the end of 1933. Anna had always been almost as much a mother as a sister to him, and after her marriage he had left South Africa and returned to Germany. But neither his work nor his marriage had gone well and he decided to turn to his doting sister for assistance and protection. It was a wise move; she never once disappointed him.

Outwardly charming, forty-year-old Walter Moormeister soon reclaimed the privileged position he had once held in the affections of his sister and brother-in-law. However, he soon showed an interest in one of the Edwards daughters that transcended a normal avuncular relationship. Indeed, when his niece left the farm to be a secretary in Johannesburg he left too; along with hundreds of other European immigrants, he found himself a job on one of the Witwatersrand gold mines. Well beyond the purview of the farm and hidden behind the curtains of anonymity in the city, this latter-day Rasputin eventually persuaded his niece to become his 'housekeeper' at an address in the suburb of Bezuidenhout Valley. When J. J. Edwards died, in 1937, Moormeister's hold over the young woman and her myopic mother became ever more pronounced.

All this marginalised Heine Edwards, and the years leading up to World War II were among his most unhappy at Sewefontein. Then, just as he was contemplating leaving the farm to enlist, Walter Moormeister was removed from civilised society for reasons that had nothing whatsoever to do with the outbreak of hostilities. On 22 December 1939, after a bout of heavy drinking, he and an Italian workmate became embroiled in a messy argument about money and sexual favours that culminated in the stabbing of a third miner. Charged with murder, Moormeister was lucky enough to be found guilty only of culpable homicide and was sentenced to four years' imprisonment with hard labour.[5]

On his release, in 1944, Walter Moormeister once again found himself penniless, but this time prevailing anti-German sentiment was so pronounced that he could not get a job even in the mines. Under the circumstances there was little he could do but flee to the farm, where, much to the consternation of Heine and the other siblings, including by now the deeply disillusioned 'housekeeper,' he once again exercised his sinister powers over his sister. Thus, when the Maines moved onto the property in late 1946, they were confronted by a house divided, and entered a situation where the nominal landlord was being pushed aside by a usurper as cunning as he was determined.

Kas, who knew nothing of Walter Moormeister's troubled past, could not have cared less. His mind was focused on a more practical problem. There was no shortage of land at Sewefontein and, but for a shortfall in draught power, there was nothing to keep him from putting three ploughs into the field. He considered buying a second team of donkeys, but good animals were at a premium in springtime. The way round his problem came late one afternoon when, on his return from a remote corner of the farm, he happened to catch sight of Piet Bester making his way through the gates leading to Kareepan.

Kas had known Piet Bester ever since the twenties when the Maines had been based at Kommissierust. A neighbour of Labuschagne's, Bester was better known among the locals as 'MoTswana' Bester because of his exceptional command of the vernacular. The two of them went through the ritual of exchanging a few noncommittal pleasantries and then, as befitted busy farming folk, indicated their willingness to linger and engage in deeper conversation by swopping a few observations about the weather. When Kas happened to mention his difficulty in finding animals to plough with, Bester offered to sell him six oxen for eighteen pounds—half the price to be paid in cash on delivery and half right after the harvest.[6]

With two spans of oxen and a team of donkeys Kas, Mmusetsi and Bodule responded to the seasonal challenge with customary vigour. But it was soon evident that, although most of the Triangle was to be granted a reasonable proportion of its annual rainfall quota, Sewefontein had been singled out for an abnormally dry year. Heine Edwards took pity on them, and in January, when all hope of a reasonable maize or sorghum harvest had been abandoned, he gave Kas three bags of sunflower seeds, believing that he could coax an emergency cash crop out of the unyielding soil. In Johannesburg, where anything was possible, they were still buying seed-oil to manufacture the margarine that was supposed to overcome the continuing postwar butter shortage.[7]

The sunflowers got off to a good start, but Kas was still pondering the consequences of a winter without grain when six of his goats got into Bester's fields and did an extraordinary amount of damage. Not only did it cost him a tidy sum to recover the animals from the Bloemhof pound but, Bester, sensing a crop failure and a possible foreclosure by unknown creditors, demanded an immediate settlement of the debt for the oxen. Kas pleaded for time, but Bester, smarting from the loss of his maize, was in no mood to compromise and threatened legal action. Once again it was left to the gentle Heine to come to Kas's assistance, this time with a cash loan, but by then it was evident that Heine Edwards had serious financial problems of his own. Drought was no respecter of either class or caste.[8]

Walter Moormeister understood the consequences of crop failure as well as anybody. By March 1947 he had inveigled his sister into selling three separate parcels of land of about four hundred morgen each to some maize farmers who, benefiting from the postwar surge in grain prices, were on the lookout for land that could be fed to their greedy little Ferguson and Fordson tractors. This lopping off of eleven hundred morgen from the family farm brought home to Heine Edwards the seriousness of his position; he announced he would soon seek work on the newly opened Orange Free State goldfields that lay over the Vaal River and beyond Wesselsbron.

Kas found the news of Heine Edwards's departure unsettling, but he was partially reassured when Walter Moormeister made a point of asking him to stay on after the harvest. Still, he had reservations, for the land had been bought by the two younger Niemans of Hartsfontein and Willem Griesel of Leeubos: all three, especially Griesel, aggressive men who were poorly disposed toward blacks in general and sharecroppers in particular. They were not the sort whom Kas wanted as neighbours at a time when he was already having difficulties with Piet Bester. Growing self-confidence, based on economic progress, seemed to bring out the worst in some Afrikaner farmers, he believed.[9]

But as we have seen, in these years immediately after the war it was the Asian traders—long resented for their role as creditors and their successes as merchants—who were most directly exposed to this developing Afrikaner arrogance and racism. In the closing years of the war the Smuts government had passed a series of Acts aimed at restricting Asian property rights. When India responded to discriminatory legislation with trade sanctions in late 1946, militant Afrikaner Nationalists hit back with a boycott of 'coolie' storekeepers. In the southwestern Transvaal, where the grain farmers were contemplating yet another huge crop, this took on added meaning when the price of imported jute bags once again rose just as the harvest approached.[10]

Within the Triangle itself the National Party's proposal to boycott and then repatriate all Indian traders (a move co-ordinated by Commandant 'Ampie' Roos) was enthusiastically endorsed by all the members of the Eendrag Branch, which centred on the farm Leeubos. Commandant A. J. Pretorius and an increasingly militant Piet Labuschagne were assisted by Piet Bezuidenhout, the brothers Jan and Hans Coetzee, Ben de Wet, Hannes du Plooy and several others. By early March 1947 the rank and file were out in the field persuading neighbours that it was their patriotic duty to boycott 'coolie stores' at Bloemhof and Kingswood and support the Handelshuis initiative. Anyone who didn't was dismissed as a 'Sap' (a South African Party supporter) and warned that he would get his comeuppance after the next general election. About the same time, pernicious rumours started doing the rounds—such as that a white child had been crushed to death when a rotten jute bag sold to the child's father by an Indian trader had caused a maize stack to give way and fall on the victim.[11]

Fifteen miles north of Leeubos, out at Hessie, where William Hambly once traded his single-share ploughs and leather harnesses for grain produced by bywoners and sharecroppers alike, the proprietor, M. M. Kathrada, was suddenly confronting an increasingly assertive generation of Afrikaner farmers.[12] C. J. Badenhorst, executive member of the local 'Klein Doorns' branch of the National Party, along with C. Meyer, A. J. Liebenberg, S. H. Groenewald and other militants sought to ensure that no whites entered the store premises. 'Sampie' Groenewald, owner of one of the first mechanical threshers to operate in the district, armed himself with a camera, took up a position outside the store and threatened to send a photograph of any Afrikaner patronising a '*koeliewinkel*' to the Potchefstroom newspapers.

In Schweizer-Reneke, where the boycott was enforced with equal enthusiasm, elderly Asian traders became so concerned about the well-being of their customers and their own safety that they hired a former professional boxer from Johannesburg to protect them. In a society where the realities of class often mocked the consciousness of colour, it was an added irony that the boxer was an impoverished Afrikaner. Sadly, not even old Sampie Groenewald could find time to get into Schweizer-Reneke to take the boxer's picture.

Although filled with Nationalist zeal and a desire to help mobilise the *volk* for the coming general election, some of the boycotters had other, less 'public-spirited' reasons for the anti-Asian drive. A. J. Pretorius and C. J. Badenhorst both owned stores at Kingswood where they had to compete with the hard-working Seedat family if they wished to remain in business. Once the campaign was under way, C. Meyer, too, opened a store at Mahemspanne, hoping to undercut the Kathradas of Hessie, while out at Vaalrand, Piet Labuschagne thought about starting up a general dealer's business on a site facing the salt works at Sewefontein. At least a dozen such small-scale Afrikaner enterprises sprang up on farms within the Triangle in the months after the boycott began.

Few if any of these ventures prospered. By the time the maize was in the granaries, the boycott was collapsing. In Bloemhof, Sam Patel organised various long-running sales that counteracted the effects of the campaign, by attracting scores of white customers from the richer farming districts on the Free State side of the river, much to the distress of the Triangle hotheads. In isolated communities where even a boxing match gave rise to passionate rivalry, lines of loyalty could be broken by unexpected variables.[13]

The Triangle firebrands had other problems to contend with. Not everybody was equally enthusiastic about, or willing to abide by, the boycott. By day a few tough-minded English-speaking United Party supporters pointedly ignored it, while at night poorer Afrikaners were let into Indian stores via back entrances. Moreover, the boycott organisers failed to take into account, let alone concern themselves with, the vast majority of the population—the black farm workers. This was not because they thought

Africans would be inherently unsympathetic to the cause. As members of the Seedat family knew, some blacks had strong anti-Asian views of their own. It was because the white Nationalists believed they could do the job on their own. Thus, the Afrikaner campaign against the Indian traders caused not so much as a ripple of interest amongst the sharecroppers, labour tenants and wage labourers on the white farms. Black farming folk, who under normal circumstances were vulnerable to a certain amount of pressure from their landlords, busied themselves with their work and continued to seek out the best available deal at their nearest store. Few remembered the 1947 boycott.[14]

At Sewefontein there was not enough money to buy anything but essentials. As winter approached and the Maines contemplated their first really serious grain failure in more than a decade, Kas decided to do as they had done at Klippan and send the family into the fields at night to collect maize, which he neither declared nor shared with the landlord.[15] A week later Lebitsa, Morwesi and Nthakwana were back in the fields to pick the sunflowers, which had done far better with the lack of rain than had the grains. But harvesting sunflower seed is an unpleasant business, and out in the fields and well away from the patriarch, the women and the younger girls started muttering about having to bring in a crop that was both inedible and unlikely to yield them any cash benefit. Morwesi, aged twenty-two, and Nthakwana, aged sixteen, had reached the point in life where they, like the long-suffering Leetwane and Lebitsa before them, resented that virtually all the income generated from family labour was diverted into farming and almost nothing left for domestic consumption. There were other reasons, too, for disliking harvesting sunflowers. The spiked stems pierced the skin, leaving their fingers and palms aching and covered in blood. They hated the job.

Kas, needless to say, saw things differently. It was regrettable that women working in the fields had minor injuries, but blood nourished the harvest just as surely as rain watered the seedlings. It was part of life. In any case, the head of a household had responsibilities that transcended an irritation that could easily be set right by the judicious application of a little axle-grease. As for setting aside money for domestic luxuries, it never even crossed his mind to do so. He was delighted that Heine Edwards's last-gasp inspiration had helped them salvage the unpromising summer. He sold the sunflower seed and used the proceeds to buy fourteen bags of sorghum from his old friend 'Mr. Kwape' Moshodi. With some maize and a good deal more sorghum to see them safely into the next spring planting, they could all contemplate the off-season with greater equanimity.[16]

By the 1940s the Maines had little reason to fear for their immediate well-being. The herd of twenty cattle ensured at least four gallons of fresh milk each evening. This protein was more than enough for four adults, three teenagers and three smaller children—who lived on a steady diet of sour milk and coarsely ground porridge; excess milk was churned into butter. Throughout the stay at Sewefontein, the Maine family provided all those around them with fresh dairy products. This, together with the chickens and turkeys which the Maine women had taken to keeping, made the family the object of some envy and jealousy amongst their friends, neighbours and even relatives.[17]

Leetwane, always early to sense menacing social undercurrents, understood the dangers. A proper level of subsistence was permissible; it was, after all, what most decent folk strove for, but conspicuous success was always difficult to explain. It invited other, more sinister or highly personalised interpretations that attracted hostility. She sometimes feared her husband's success. 'Most people did not like it. They thought of him as being proud, haughty, even arrogant. The man was rich, he owned a number of cattle . . .'[18]

But Kas had no patience with people who spent more time talking than working. He was as busy as ever, and it was only on 14 May that he found time to take the autumn wool clip into town.[19] Even there, in Bloemhof, the Handelshuis showed how things were changing. There were fewer Jewish traders around, the younger generation having abandoned the bleak and hostile platteland for professional life in the city. In Daniel Gabbe's trading store, where as a young man Kas had been mesmerised by the latest heavy agricultural equipment, old man Gabbe's son, Arthur, now presided over rows of neat shelving filled with cheap commodities. True, the Gabbe family still bought wool, and to that extent they honoured the name of 'Ou Vel' who had more than once spent the night on the floor of the Maines' home. But now if one wanted to buy a plough, one would be referred to the co-operative store. The modern brick-lined premises of the South-Western Transvaal Agricultural Cooperative now held sway over the ageing corrugated-iron stores of 'foreign' traders.

Back at Sewefontein, Kas launched into a round of winter tailoring, leather work and smithying that displayed anew the extraordinary range of his craft skills. Seeing that Nthakwana was becoming a young woman, he took her aside and taught her how to cut and sew dresses. Cloth she would have to find for herself. Workers at a passing roadside camp taught him how to shoe a horse, an invaluable skill he was as quick to exploit commercially as he was to pass it on to twenty-four-year-old Mmusetsi and his younger brothers. Horseshoes along with the hand-hoes that Kas was fashioning from bits of scrap iron not only yielded off-season cash but protected him against the postwar rise in the prices of agricultural implements.[20]

But this money, no less than that raised from the sale of the sunflower seed, was the subject of silent resentment in another quarter. For some years Leetwane had been of the opinion that Kas was not giving her enough money to clothe the older children or to ensure that the younger ones received at least some formal education. It was all very well for him to teach Nthakwana how to sew and cut dresses, but this did nothing for the youngsters, and while the acquisition of craft skills was no doubt laudable, there was no substitute for literacy. But the patriarch's disdain for book learning was undiminished, and only with the greatest difficulty did she persuade him to pay for the instruction that twelve-year-old Mosala received from Lerata Masihu in the small night school held on the farm. The only redeeming feature of Kas's behaviour over domestic expenditure concerned church subscriptions. Throughout the stay at Sewefontein, he saw to it that their dues to the Ethiopian Church were paid regularly.[21]

As always, the advent of the longer spring days marked a change in the family's routine. Time-consuming craft skills, best pursued while the sun was away on its winter vacation, were pushed aside as the Maines abandoned what had once been Heine Edwards's section of the farm and moved into the heart of the property. A new contract, nominally entered into with Tetangwane's widow but in practice negotiated through Walter Moormeister, gave Kas the right to cultivate as many fields as he wished and allowed his livestock to have access to an unlimited amount of grazing. In return, all crops raised on the property were to be shared equally and Kas and Bodule were to assume sole responsibility for the well-being of the Edwards' flock of Merino sheep. Leetwane could, if she so wished, do all the whites' laundry, for which she would receive a little cash, but it was expected that the older daughters, Morwesi and Nthakwana, would take turns as domestic servants in the farmhouse for which they would each receive ten shillings a month. In addition, Mmusetsi had to work on the salt pans during the summer months when his father had less need of his labour; he, too, would earn the normal cash wage paid to any seasonal labourer.[22]

This accord, entered into at a time when black sharecroppers were finding it harder and harder to find land, met Kas's needs admirably. It ensured the well-being of his livestock and the opportunity to reap a substantial harvest, and it helped keep the family together by giving its senior members independent cash-earning opportunities without threatening his overall control of household labour. No patriarch could have asked for more. If Kas had doubts, they were whittled away by the smooth-talking Moormeister.

Though usually to be found with a revolver strapped to his side, Walter Moormeister held no terrors for Kas. Kas saw him only as RaDiphonkgo, an innocuous Tswana nickname whose origin and meaning were lost in time. Invariably courteous and well disposed to those on whose favours he depended for making a living out of nothing, RaDiphonkgo used his broken Afrikaans to persuade Kas that his privileged position as a share-

The attorney A. J. Marais, at his graduation from Cambridge University in 1907
*Mrs. N. Marks*

J. M. Klopper, the landlord at Klippan in 1937-39
*Mrs. J. A. Klopper*

Hendrik Goosen, the landlord at Klippan in 1941-43
*P. A. Goosen*

G. (Piet) Labuschagne, ıe landlord at Vaalrand in )43-46
*G. Labuschagne, Jr.*

Mrs. Anna Edwards, Walter Moormeister's sister, and her husband, J. J. Edwards. The Moormeister-Edwards family were landlords at Sewefontein in 1946-49
*Mrs. E. Lecornu*

Dr. Henry "Hickson" Hutchinson, the district surgeon in Bloemhof in the 1940s
*Dr. H. Hutchinson, Jr.*

Salt pans at Sewefontein, 1989
*S. Mofokeng*

Carts at Vaalrand, 1990
*S. Mofokeng*

Cattle at Molote
*S. Mofokeng*

Kas Maine (right) with his daughter Nthakwana (left) and Mathe (center), in 1949—the earliest known photograph of Kas
*Maine family*

Kas Maine at Dominion Reef in 1951
*Maine family*

Motheba Maine, Kas's mother, with her grandchildren, Sebubudi's twins, Klippan, 1942
*Maine family*

Bodule Maine: after his circumcision; and with his bicycle; both in the 1940s
*Maine family*

(left to right) Mmusetsi, Matlakala, Pulane, Nthakwana, and Nkamang Maine, 1951
*Maine family*

Mmusetsi Maine
*Maine family*

Nthakwana Maine (right) with a friend, Sewefontein, 1951; and with her daughter Ntholeng and nephew Sekgaloane, Bodule's son, 1952
*Maine family*

Mphaka Maine, Kas's brother
*Maine family*

Leetwane, Mokone, Kas, and Lebitsa Maine at Rhenosterhoek, 1951
*Maine family*

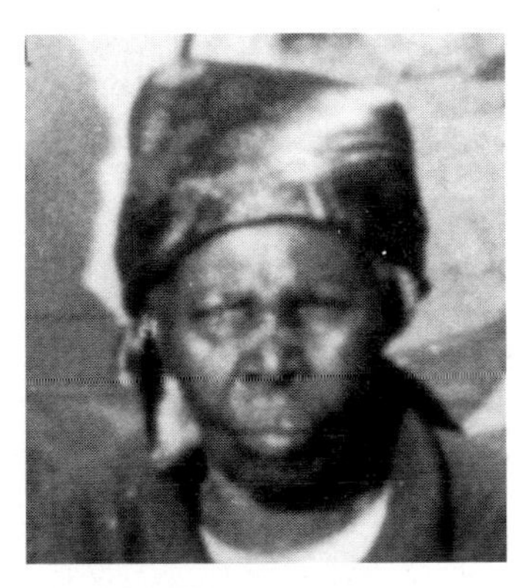

Leetwane Maine
*M. Nkadimeng*

Lebitsa Maine
*Maine family*

Moipone Maine, Kas's half-sister, with her family
*Maine family*

Sabuti Maine with his wife
*Maine family*

Bodule Maine and his wife, Dikeledi
*Maine family*

Two photographs taken by Bodule: his grandmother Motheba Maine; and (foreground) Mosala, Nthakwana and their mother, Leetwane, with (background) children and Kas Maine (seated)
*Maine family*

Johannes RaTshefola, a trader at Molote, 1949
*J. RaTshefola*

L. J. Klue, the landlord at Jakkalsfontein in 1949-50, and his wife
*Mrs. I. M. Pieterse*

J. C. (Cas) Greyling
*Nasionale Pers*

Mrs. Paula Hersch, the landlady at Rietvlei in 1952-57

Nicodemus and (in a photograph of 1989) Moses Ramagaga, the landlords at Molote in 1958–60
*N. Ramagaga, M. Nkadimeng*

cropper owed much to his own intervention with his sister and to his liberal managerial style which, he said, was quintessentially German. 'Look, now, you stay on my farm, you own cattle and sheep, and I do not interfere in your affairs. That is how we live in Germany.' These fragments filtered into Kas's consciousness where they were sorted and filed along with other half-baked notions that dated back to the 1914 rebellion about Germans coming to the rescue of black South Africans.[23] Sensing that these sentiments were well received, Moormeister tried to drive home his advantage by questioning the sort of wartime propaganda Kas had been exposed to at Klippan under Hendrik Goosen. What benefits, Moormeister asked, had blacks derived from the Second World War? None! He suggested that he and Kas, as outsiders, had far more to fear from the Boers than from any supposed 'German' enemy.

But these conversations, which took place over a year, were not in themselves enough to win over a hard-headed realist. As organisers from both the I.C.U. and Communist Party could testify, Kas Maine was not to be persuaded by ideas alone. For him and many others like him, demonstrable material progress outweighed talk, and, set this more demanding test, Kas found the 'German way' at Sewefontein highly persuasive. In the circumscribed world of the farm, where the landlord's power was perhaps the single most important determinant of a sharecropper's economic success, there was always a danger that distinctive individual factors which could be made to account for unusual accomplishments—access to good tracts of land or the optimum configuration of family labour—were compressed and reduced to highly personalised dimensions. And so it was with Kas. Instead of seeing his improved fortunes as linked to the growth of the South African economy as a whole and the spirited response of its agricultural sector during the war, he framed his universe with the farm gates: Walter Moormeister dominated the foreground, personal effort occupied the middle distance, and everything else spilled away toward an ill-defined horizon. Looking through the telescope backward he saw that 'I became a better person when I lived with Germans. I accumulated most of my possessions during my stay with the Germans. It was a German, Moormeister, who told me that we were foolish for having supported the government.' The ethnic preference highlighted in this startling picture offered its own, sad comment on landlord-tenant relationships in the Triangle. It was ironic that a sharecropper as well-disposed to Afrikaners as Kas was should be drawn to this off-beat conclusion. A shrewd observer who had worked with Boer landlords for decades—all the way from the hard-driving Koos Meyer and opportunistic Hendrik Goosen on the one hand, to the religious Piet Reyneke and the hesitantly paternalistic Piet Labuschagne on the other—Kas looked back with greatest satisfaction on his relationship with a gun-toting, Nazi-sympathising killer. Walter Moormeister, it was said, was 'like a brother' to him.[24]

Before the first rains fell, Kas was approached by a diamond digger who had hired a portion of the farm Bosmansfontein, hoping to make a profit from state-guaranteed maize prices. Mindful that the rains had bypassed Sewefontein the previous summer, Kas was eager to take advantage of the offer, and Moormeister raised no objection. A few days later Kas, Bodule and Mmusetsi started work; they ploughed for close on two weeks without once taking a day off. Then, just when they were within sight of planting, the digger told them the farm had been sold to Koos Meyer and they would have to leave without being paid for the work they had done.[25]

With RaDiphonkgo's warnings about Boer treachery and the drone of the approaching Fordsons ringing in his ears, Kas led the boys back to the quiet of Sewefontein, determined to find a way of making good this unexpected pre-season loss. He put twelve-year-old Mosala and ten-year-old Pulane in charge of all herding duties and then got the two oldest sons to help mark out fields on a scale that the family had never before attempted. Bodule, like Kas willing to work the oxen until they simply lay down from sheer exhaustion, took up the challenge with all his customary enthusiasm while Mmusetsi, wandering through the inner mazes of his troubled mind, plugged away at the seemingly endless task.

Slipping into the increasingly familiar role of supervisor, Kas viewed his two sons' efforts with mounting satisfaction. Then, one afternoon, in mid-September, one of the youngsters ran up to him to tell him that Mmusetsi had had an accident and been whisked off to hospital in a car by Moormeister. Apparently the plough had 'jumped' while being manoeuvred out of one heavy clay-lined furrow and into another, and had all but severed his toe. Kas got out the horse and cart and hurried across to Soutpan, where he collected his brother, Phitise, and the two of them went to Wolmaransstad where they tracked down Mmusetsi. The lad was of the firm opinion that the hospital staff had done very little to make him comfortable. Confident about finding his way around a hospital bureaucracy after his recent Klerksdorp experience, Kas managed to get his son discharged, and they all went to Bloemhof, where Henry Hutchinson, then on the very eve of his retirement, treated Mmusetsi at a cost of two pounds and ten shillings.[26]

This accident did nothing to lessen Kas's determination to continue ploughing until the rains made it impossible to do so. So, while Mmusetsi hobbled around the house recovering under the watchful eye of his mother, Bodule was made to shoulder his brother's load as well as his own: his efforts won approval even from the patriarch. With the arrival of the sort of gentle rain that always made for a good start to the season, he and Kas set about planting maize, sorghum and sunflowers. This time, however, they remembered to take the precaution of protecting the bigger fields with the traditional potion that warded off red-billed quelea—a custom they had badly neglected.[27]

Once the seedlings were about three inches high, Kas and Bodule turned away from the fields and back to the sheep. Moormeister, still moving happily down the Prussian road, had neither objected to their absence nor complained about their stand-in replacements, Mosala and Pulane. Bodule was showing real progress in animal husbandry: Kas thought he had all the makings of a fine sheep farmer; and when Kas's old friend Hlahla entrusted him with six sheep for safekeeping, Kas did not hesitate to add them to the already large flock for which Bodule was responsible. This demonstration of fatherly confidence was followed by another, equally important gesture when, on 14 October, he allowed Bodule to negotiate independently the sale of half a dozen sheep at Slabbert & Campbell's monthly stock auction in Bloemhof.[28]

The closing months of 1947 were amongst the most fulfilling that Kas ever spent in the Triangle. His cattle were thriving, the flock of sheep was doing well, and his fields—which stretched from the front of his small shack almost as far as the eye could see—were brimming with three different crops responding to generous amounts of rain and sun. Living on a property owned by a widow always gave an enterprising sharecropper unusual latitude. The absence of a Boer landlord reduced the authority of the farmhouse and enhanced the status of the senior black tenant, whose farming expertise became almost as indispensable to the enterprise as his ability to exercise control over the labour force. At Sewefontein it was only the strange position occupied by the maverick Walter Moormeister that kept Kas from being accorded the unofficial title of 'farm manager.' In practice this deficiency mattered little, because Kas knew that all would be well as long as he and RaDiphonkgo remained close. But by early 1948 worrying signs were emerging that Walter Moormeister had reservations about the amount of physical and psychological space that Kas was taking up.

The first indication that all was not well came when RaDiphonkgo announced he was looking around for a 'farm foreman' and then went on to appoint a rank outsider, Petrus RaMotswahole ('one who comes from afar') to the office. This move was obviously designed to undercut Kas's position and Kas was sure it was done on racial grounds. As a so-called 'coloured,' RaMotswahole had a superior status in South Africa's complex racial pecking order, especially when the Afrikaner Nationalists' demands for the introduction of apartheid were making the country more rather than less obsessed with racial distinctions. Soon RaMotswahole was seen driving a new car that had suddenly appeared on the property; then, adding insult to injury, Moormeister announced that Kas and Bodule were to be relieved of their shepherding duties and that RaMotswahole would assume responsibility for the well-being of the Edwards sheep.[29]

These were painful blows to absorb. Kas was at the very height of his powers and Moormeister's actions not only reduced his role in the day-to-day workings of the farm but tended to undermine his authority within the Maine household. How were his adult children to retain respect for a patriarch whose position had been ceded to an outsider? It was all the more irritating because Kas was still expected to exercise control over his family members though RaDiphonkgo's behaviour was increasingly erratic and unpredictable.

Up at the farmhouse, it was often the Maine daughters who were most exposed to Moormeister's excesses. One morning, when it became clear that RaDiphonkgo and his sister were not going to rise at their usual hour and Nthakwana therefore delayed taking in tea to Walter Moormeister's bedroom, he unleashed such a stream of abuse that she lost her temper and tipped the tray over him. Before he could disentangle himself from his soggy bedclothes, she dashed to the bedroom door and locked him in. From there she rushed outside and from a position next to his bedroom window taunted him for not being able to get out. The noise of this wrangle attracted attention in another quarter of the farmhouse and when Moormeister was eventually freed, he stormed out to his new car, saying he was going to the police station to lay a charge of assault against the girl. But seventeen-year-old Nthakwana, showing the spirit that her sister Morwesi was even more famous for, was not to be outdone. She marched to the other side of the car, got into the seat next to the driver, and told him that she, too, would go to the police to make sure they heard both sides of the story! At this point Kas appeared, and succeeded in calming down RaDiphonkgo by telling him that he would take it upon himself to discipline Nthakwana. Moormeister, who had no real desire for interaction with the police, retreated indoors.

Soon there was more trouble. One afternoon Moormeister got into an argument with Mmusetsi about how the cows were being milked; when the German drew his revolver from its holster, Nthakwana rushed in to shield her brother. More screaming ensued, more than at the time of the tea-tray incident, and once again Kas appeared in the nick of time. But on this occasion RaDiphonkgo was more intractable, and Kas eventually had to warn him that if he did not leave the disciplining of the younger Maines to their father, the family would leave the farm. Neither Moormeister nor his sister, who had put in an unexpected appearance, was ready for that. Anna Edwards reminded Moormeister that it was *she*, not he, who had a contract with the Maines, and in the end this device allowed the warring parties to retreat with dignity intact.[30]

By then it was clear to everyone at Sewefontein and the surrounding farms that Kas would reap a harvest that would be the envy of even a Koos Meyer. But, as always, this prospect did not meet with universal approval; nobody resented Kas's success more than did his old adversary from Kommissierust days, Willem Griesel. Griesel had installed a bywoner by the name of van Rooyen on his section of Sewefontein; but van Rooyen was young and inexperienced, and Griesel had to pay frequent visits to the farm, trips that took him past Maine fields filled with maize, sorghum and sunflowers. Griesel, who had a special dislike of 'rich kaffirs,' made it his business to undermine Kas's relationship with Moormeister. 'This fellow is producing too much,' Griesel told the German, pointing at Kas, 'and it will eventually lead to your arrest and you will end up in prison . . . it will end in a quarrel and, when that happens, this chap will get you into serious trouble. It won't be long before the lawyers appear.' Nothing could have been better calculated to unhinge Moormeister than talk of police and prisons. From that moment on the German started to deal with Kas like an old married man having an affair with a young mistress: one moment overly anxious to please and the next bitterly resentful of his dependence, revelling in secret pleasures but living always with the fear of discovery.[31]

It was not just that his own unfortunate history had primed Moormeister for the favourable reception of Griesel's dark thoughts; the whole political climate was conducive to the transmission of messages of hate and aggression. Griesel merely personalised the general ethnic antagonisms that a generation of Afrikaner Nationalist politicians had been systematically cultivating. Moormeister might have heard Griesel's voice in his ear, but in his head the same small-minded sentiments had a thousand echoes. In May 1948, the very month that Kas Maine contemplated his largest harvest ever, the National Party swept aside Smuts's tired old United Party with a shameful slogan, '*die kaffer op sy plek en die koelie uit die land*,' and formed a new government. Within the Triangle this political victory symbolised the economic triumph of grain farmers who, having benefited from guaranteed maize prices, the postwar agricultural boom, and the large-scale introduction of the petrol-driven tractor, were now indeed ready to put 'the kaffir in his place and the coolie out of the country.'[32]

The simultaneity of these changes and the way they reinforced each other were lost on the Triangle's beleaguered sharecroppers, labour tenants and wage labourers. Like Kas, they lived in a world where the connection between their day-to-day struggles on the farms and those taking place at the national level was not self-evident. And so, while members of the new parliament gathered in Cape Town to start legislating for their apartheid programme, Kas and RaDiphonkgo got into a truck and drove to the 'reserve' at Taung to recruit twenty-five men, women and children to help bring in the harvest.

They had no difficulty in recruiting a team which, so Kas later recalled, came from a dry and overcrowded area where food was chronically short. Hungry parents and ravenous offspring were more than willing to work for six bags of grain in every hundred, let alone the sheep which he and Moormeister offered to slaughter during the harvesting. Enthusiastic BaTswana hastily collected blankets and a few personal items and clambered aboard the open truck for the icy winter ride to Sewefontein, where RaDiphonkgo had erected so many lean-to shelters of loose corrugated-iron sheets that the place 'looked like a location.'[33]

At dawn each morning the tin-town slowly came to life, the shacks extruding a ribbon of men and women wiping sleep from their eyes and strapping babies to their backs. With the teeth of a Maluti wind gnawing away at exposed limbs and joints creaking to the sound of rusted leaves, they spread out into the fields where they bent, stripped, cut, bagged and emptied grain for what seemed like aeons. Relief came in midmorning, when the Maine women, grateful for having been spared the ordeal by sunflower, arrived to serve them with dollops of stiff porridge, sour milk, or chunks of roughly hewn mutton. About this time Kas or RaDiphonkgo would also appear, take in proceedings with a casual eye and then focus on the mounds of maize, sorghum and sunflower that continued to grow, until the weak winter sun was seen to collapse into the field of a distant farm.

After more than three weeks of unbroken effort Walter Moormeister got out the truck and sneaked into Bloemhof to buy the additional grain bags needed for so large a harvest. Wikus Pienaar, too, appeared at last; threshers, it seemed, were always late. The BaTswana workers, rejuvenated by the mere sound of a threshing machine, now fetched, carried and stacked bags with an enthusiasm that at one stage threatened to outpace the mechanical monster which stood about coughing smoke into the limitless expanse of the highveld sky. When the noise of the machine abated, Kas and RaDiphonkgo counted the number of bags and settled their accounts, first with Pienaar and then with the harvesting team. What remained was shared equally between the two men. The huge mound that was Kas's had all the appearances of a presettlement stack: after all his commitments had been met, the Maine family still had a thousand bags of sorghum, five hundred and seventy bags of maize, and more than two hundred bags of sunflower seed. It was a breathtaking achievement.[34]

As Kas savoured his moment of triumph and silently ascribed his success to the efforts of his beloved Bodule, he remained largely oblivious to the historical ironies surrounding him. After a lifetime of sharecropping in the Triangle, his greatest feat had occurred during one of the last seasons in which it was still possible for a black man with a plough and oxen and sufficient family labour, to find the space on white-owned properties. But, as he did not yet realise, the harvest that was his making was to be his undoing.[35]

Kas's most pressing problem was the need to organise the marketing of the harvest. But what he thought would be a minor logistical problem seemed to unnerve RaDiphonkgo, whose advice and behaviour now became increasingly equivocal. On the one hand, unlike most of the white farmers Kas had known, Moormeister refused to market the entire crop in his name, saying that he feared drawing attention to their sharecropping enterprise and the prospect of having to pay tax on such large receipts. On the other hand, he was more than willing to see the crop transported into Bloemhof in his truck and pointed out that, in addition to receiving a large *voorskot* from the agricultural co-operative, the preliminary payment due to all farmers on delivery of the harvest, Kas would also be eligible for an *agterskot* payment, the end-of-season settlement that bridged the gap between the predicted and actual price fetched by the maize. The news of the possibility of a 'bonus' payment on top of the three hundred pounds he got from the co-operative came as a bitter-sweet surprise to Kas, who only then realised how Labuschagne and other landlords had regularly denied him a share of the agterskot that they received in the post each spring.[36]

But, with three hundred pounds in his pocket and the prospect of more to come, who would want to dwell on the past? He drove home with the taste of the beer already in his mouth, determined to enjoy his success. The only difficulty was that Sewefontein was poorly situated for a party. The property lay close to the intersection of at least four minor farm roads and was readily accessible to the police, with whom he had already had two skirmishes. To avoid a third, possibly less pleasant encounter, he asked Walter Moormeister to give him a note allowing him to brew '1 gallon of Kaffir Beer.' The German scribbled something on a piece of paper. It was only when Kas was back in the safety of the shack that he realised the value of a signed but badly written and undated note. He got out the pencil and, in the excessively studied hand that characterises the semi-literate, slowly and very deliberately altered the figure '1' to '4.'[37]

For once all the members of the drinking circle were in full agreement: never had the beer tasted better. While Kas, Thloriso Kadi, the Moshodi brothers, Johnson Xaba and one or two others explored the outer limits of the four gallons so generously approved by RaDiphonkgo, their sons slipped away to take advantage of the shadow cast by the seasonal eclipse. Kas was especially pleased with Bodule's performance and therefore unusually tolerant when the twenty-year-old went out with 'Dira' and the other dogs and seemed to spend hours hunting rabbits, looking after his pigeons or doing the social rounds with his camera. Kas was reconciled to the idea that, although five years younger than Mmusetsi, Bodule would probably get married before his saturnine sibling; he would have been less sanguine had he realised that Bodule was already spending a good deal of his time with a young woman of whom he thoroughly disapproved—Dikeledi Teeu, daughter of the Teeu family, whom the Maines had first encountered during their stay at Klippan but who had since moved to Vaalboschfontein.[38]

*

Kas was still contemplating the coming changes to the profile of his own family when Griesel's bywoner, van Rooyen, asked him whether he might get a ride into town when next the Maine men visited Bloemhof. Kas was pleased to be able to help. Outside of circles dominated by the likes of Piet Labuschagne and his hotheaded Nationalist friends, there were, mercifully, still no hard and fast rules preventing a black man from helping a poor white going into town to collect a wedding suit. A day or two later he and Bodule went with van Rooyen to Goldstein's shop on the outskirts of the 'location' where several women, black and white, laboured side by side to create the raiments that saw respectable Triangle folk from all walks of life through their rites of passage. As he let the young man off at the tailor's, Kas was suddenly overcome by the feeling that he, as father, wished to do the right thing by his own son while there was time and, almost without thinking, he turned the horses toward the bazaar and Ebrahim's Cash Store.[39]

Bodule helped Kas tether the horses, just as Kas had once helped Sekwala tether the horses outside Gabbe's store at the far end of town. The two of them walked in together, bound by unseen ties of love and respect, the one anxious to give, the other ready to share in the proceeds of what had been a truly remarkable year on the land. Bodule expressed a preference for a new bicycle which, after careful inspection, was agreed upon and, when his eye strayed momentarily toward a fashionable three-piece suit, that too was purchased by a patriarch ready to indulge a son cast wholly in his own image. Kas gave Ebrahim fifteen pounds, they loaded the goods into the cart, and then, each content in his own distinctive way, father and son set out on the sandy track that led to Sewefontein.

But love and justice—horses that find it difficult to keep in tandem under the best of circumstances—were often contrary when it came to serving peasant families. The heart and the head bent in opposite directions, causing confusion and dissension. When the bicycle and suit appeared at Sewefontein, they outwardly confirmed long-held inner suspicions that Kas favoured the ploughman Bodule above all others. Leetwane, whose requests for even the most basic domestic commodities had been denied, made her anger known; but it was the one whose blood had been spilt in the field, the one who laboured on the salt pans for derisory wages who was most wounded. Unable to express his anger and disappointment openly to his father, Mmusetsi turned instead on his brother.[40]

The very instant that Kas detected Mmusetsi's anguished reaction, he knew he had erred. Yet he, too, felt constrained by unseen chains. An apology would not only raise the matter of his preference in public but further undermine his authority, and so he chose to ignore his wife's protest and his son's pain. He pretended that all was as it was, hoping that others would help him to bring back order to his moral universe by going along with the illusion. The ploy was as ancient as it was bankrupt. He went about his business as best he could, lending fifteen pounds to Hlahla, who had got himself stranded without money at Zebediela in the northern Transvaal, and attending Slabbert & Campbell's monthly stock auctions. The first outlay netted him two oxen by way of a return, while the second yielded what he considered to be some fine investments, but neither Leetwane nor Mmusetsi would have been cheered by these entrepreneurial adventures.[41]

Undercurrents of discontent were still discernible when Moormeister called up Kas and Bodule to help with the spring shearing. It was tempting to tell RaDiphonkgo to get his coloured 'foreman' to round up the sheep and deal with the Merinos himself, but in the end Kas and Bodule buckled down to the task although not with much grace. RaMotswahole's increasingly privileged access to Moormeister irritated Kas, and once the wool had been baled, he enjoyed considerable *schadenfreude* in telling the German that there were at least thirty sheep missing from the Edwards flock.

Moormeister apparently did nothing about the missing sheep, but then, one morning, Kas awoke to find his shack surrounded by a contingent of Bloemhof police, ostensibly interested in illegal beer brewing. Eager to avoid having to produce the crudely altered note from RaDiphonkgo, Kas readily admitted to having a supply of bojwala brewed for domestic consumption. Much to his relief, Moormeister arrived to confirm his story. But it then transpired that what the police were really interested in were the missing sheep, and when they asked him what he thought had happened to the livestock, he found the old bitterness about ethnic betrayals swilling round his mouth. He told them to direct their enquiries to Moormeister.

> RaDiphonkgo had removed the flock from my supervision and entrusted it to the care of Petrus RaMotswahole. RaMotswahole was a brother of his, whereas I was not. RaMotswahole had long hair, whereas mine was short. RaDiphonkgo had removed the sheep from my supervision and placed them in the care of somebody who looked like him. I told them to ask him. Why had he done that?
>
> The policemen laughed and asked RaDiphonkgo whether it was true, and why he had done so? He admitted that he had handed over responsibility for the sheep to RaMotswahole but could give them no particular reason for having done so. I pointed out that he had also given Petrus the use of his car but that I was not even allowed to ride in it. I was black whereas Petrus was not. RaMotswahole was his—RaDiphonkgo's—brother and I had nothing to do with him.[42]

The policemen turned and left. They were later seen leaving with RaMotswahole, his brother George, and a MoSotho named Ngakana, each of whom was subsequently sentenced to a term of nine months' imprisonment with hard labour for stock theft.[43]

Moormeister's refusal to let Kas use the car had unleashed rancorous memories of Labuschagne's refusal to let him use the cart the time when Nkamang was in danger. Something was happening, something Kas could not put his finger on, something which he persisted in attributing to individual attitudes but which was, in fact, manifesting itself throughout white society. And more evidence of this developing interest in social segregation showed when Willem Griesel's wife, the same fat Jacomina Griesel whom Kas had once been called upon to escort into town from Vlakfontein, died in October 1948.

As befitted someone who had known the deceased, Kas joined the mourners at Leeubos, but it turned out to be a funeral with a difference. Blacks were still allowed to dig a grave, but for the first time in his experience, they were given explicit instructions not to mingle with whites in the funeral procession or to expect to be allowed to handle the coffin. Race relations in the Triangle had never been unproblematic, but in his memory things seemed to have been more relaxed when he was younger. Such incidents soured the atmosphere. Anyway, the funeral had come at an awkward moment in the season, and so when it was over he did not linger over the meat and porridge served to black mourners but hurried back home to the ploughing.[44]

There was good reason for Kas to take special interest in the farming operations. As in the late 1920s, when he had responded to the increasing mechanisation of white agriculture by getting equipment like the planter, so, in the mid-1940s, he had invested heavily in new machinery—including Canadian Tramp and Massey-Harris ploughs ordered directly from the agents in Port Elizabeth, and a second-hand harrow bought from a neighbouring Swazi farmer for five pounds. The early rains had been promising, the ploughing was well under way, and he was looking forward to another good season on the land.[45]

Kas was still busy formulating a planting strategy for the season when RaDiphonkgo, rather unusually, started offering advice about what sorts of crops they should be concentrating on. He thought they should not only cut back severely on unprofitable sorghum production, which took up disproportionate space, but abandon the white 'bread mealies' to which Kas had become accustomed and replace them with the heavier yellow hybrid varieties that the co-operatives were advocating. Kas did not respond to the advice about the sorghum, since there was no shortage of arable land at Sewefontein; knowing white maize to be more drought-resistant than yellow, he questioned the second suggestion. He was irritated by this unexpected intervention and in the end decided to ignore most of the advice of someone whom he classed as a mere novice. But, seen in retrospect, perhaps this episode contained a warning. Having spent four years tucked away in the farmhouse like a lounge lizard, Moormeister was venturing out to offer sharecroppers advice. In his own way, he was giving notice of the imminent change.[46]

On 29 October 1948, Kas, increasingly confident of his ability to deal with the agricultural co-operative on his own, rode into Bloemhof to order fifty pounds' worth of seed. Three days later he used the wagon to go to the rail siding at Kingswood, where he collected thirty bags of white and twelve bags of yellow maize. But a season's work on the land seldom works out as a man plans it: when he started planting early the next morning, the planter broke down, which necessitated yet another time-consuming journey into town. Bill van der Westhuizen arranged for the planter to be repaired but, by the time the machine had been fixed and fetched, it had cost Kas forty-two pounds and the Maines had missed the optimum moment for getting the seed into the ground. In December the seedlings did not look promising.[47]

Even then, the mere size of Kas's fields continued to fuel the enmity of white neighbours, with some of whom he had had unfortunate personal encounters in the past. He and the abrasive Willem Griesel never saw eye to eye, and at one point or another his livestock had invaded fields belonging to Boers like Piet Bester, Hans Coetzee and Hendrik Goosen. He had also clashed with Piet Labuschagne over the increasingly vexed issue of racial segregation. Kas got the distinct impression his universe was shrinking for reasons that he did not fully understand and that his fate now lay largely in the hands of the treacherous Moormeister. Looking back many years later, he recalled: 'The very youngsters who had grown up under my care said that the German had turned me into a white man,' that 'I was getting too rich,' and that another large harvest would see Sewefontein's landlord 'in prison.' But the world was also closing in on the hapless Walter Moormeister. The Afrikaner Nationalists' very public pursuit of apartheid, of racially sanctioned avarice and greed, was dovetailing with his private insanities.[48]

Unfortunately, there was not very much that Kas or anybody else in black farming circles could do to counter a whispering campaign characterised by racial bigotry and double standards. A black man was damned by his success if he worked hard and damned for his laziness if he failed. In any event, Kas found such thinking especially inappropriate during the Advent season and, as Christmas approached and Willem Griesel and his friends busied themselves with making political mischief, Kas directed some of his own energies into the church.

The Maines' switch of allegiance to the Ethiopian Church after the demise of Nneang had marked a small upturn in the family's religious life and had come at a time when the vitality of many local branches of the

African Methodist Episcopal Church was in decline. In Kas's case this upsurge in interest was partly predicated on his relationship with the Reverend Mkwane of Makwassie, who often called on him to organise prayer meetings on some of the outlying farms. This, along with Kas's taking responsibility for paying the church membership fees, left Leetwane rather marginalised—a reversal of the historical trend where she and Motheba had been more dominant in the A.M.E.C.[49]

This meant that at the very height of his powers Kas was not only a match for most white landlords within the district, but also politically, socially and economically unassailable within the family. Omnipresent, he dominated the inner spaces of his home with the same quasi-militaristic vigilance he displayed in the fields beyond it. Those, like Bodule and Nthakwana, who fell in without question and chose to march to the patriarch's beat, found it reassuring, but others who wished to walk more freely, like the teenage daughters Matlakala, Morwesi and Thakane, felt marginalised or, like Mmusetsi, totally exposed. Nobody liked this close-formation marching less, or understood better the cost of abandoning the ranks, than Leetwane. Under constant pressure on the home front and no longer able to derive the same degree of satisfaction from her involvement in the church, Leetwane's health deteriorated and she began to suffer from severe headaches.[50]

Kas, largely oblivious to the long shadows he was casting and the resentment he was causing among the women in the family, busied himself with shearing his sheep, which, despite an unusually hot and dry summer, had grown to number close on two hundred. The expertise which he and Bodule had acquired working under Piet Labuschagne could not have been put into practice at a better time; when they went into Bloemhof to sell the clip, they found that the price of wool had once again increased dramatically and they had yet another handsome cash return. With this money placed back-to-back with the income derived from the spectacular grain harvest of the previous season, it was not difficult to understand why, in later years, sheep farming at Sewefontein was recalled with such fondness.[51]

But an unwritten rule has it that the efforts of the crop and the livestock farmer should never be rewarded equally in the same year. By the time Kas and Bodule returned to the farm, the effects on the crops of the summer-long drought were all too evident, and by autumn the message was unambiguous: as in the 1930s, the maize leaves were prematurely wizened and the cobs showed rows of missing seeds, sporting inane grins, like toothless elders at a village beer drink. Even the proud sunflowers bent their heads.

The Boers responded to nature's dumb insolence by exercising their political muscle, and by March 1949, both Bloemhof and Schweizer-

Reneke had been added to the list of drought-affected districts eligible for state assistance, while black sharecroppers were left to cope as best they could with their poor harvests. White landlords made sure, too, that their own livestock had privileged access to such grazing as there was. Kas found looking at the fields painful. Under the circumstances it made little sense for him and Moormeister to undertake the long journey to Taung to recruit a harvesting team: Lebitsa, the teenage daughters and the younger children were put into the fields to bring in the mealies.[52]

If the harvest at Sewefontein was slightly less disappointing than otherwise, it was because Kas had disregarded RaDiphonkgo's suggestion as to what maize seed he should plant. This truth had been lost on the German, however, and only when it came to threshing the crop did he suddenly notice the marked discrepancy between the amounts of yellow and white mealies. Angered by the shortfall of marketable yellow maize, he was placated only when Kas agreed to forego any claim to a share in it. Kas, aware that the local BaTswana's preference for white maize would always leave him with bartering opportunities, agreed to a share of sixty-five bags of 'bread' mealies. (Later, he exchanged ten bags of these white mealies for yellow maize from a black tenant on a neighbouring property.) The reluctant Maine women were sent back into the fields for the sunflowers.[53]

Kas's disregard of Moormeister's opinion about crop choice may well have produced an echo of the warnings that neighbours had been whispering into the German's ear for several months. Worse, Kas's insensitivity to the preferences of women and children had provoked, unbeknown to him, their first act of collective rebellion. Lebitsa, Morwesi, Mosala and Pulane saw little point in harvesting poor-quality sunflowers when the Maines' farming enterprise was doing so well on other fronts. Even Nthakwana, normally at least as loyal as her brother Bodule, joined the plotters. Many years later, she recalled the revolt against the patriarch:

> I remember the 1948–49 season. There was little rain and my father had ploughed large fields of sunflowers but they did not grow well because of the drought. The plants were thin and the heads so small that you became discouraged merely looking at them. Now, during that harvest season we became so fed up with having to reap such small heads that we devised a way of avoiding having to harvest them at all. We got sticks and beat the sunflower heads until the seeds fell to the ground.
>
> That year he reaped very few bags of sunflower. He could not work out whether the seeds on the ground had been spoilt by birds or by what; only we knew what had been done. He obtained very few bags and he complained, saying the birds had destroyed his sunflowers, but at the same time he turned on us, shouting that we had failed. We told him

that the plants we had left out had had no seeds and if he did not believe us, he could go out and check for himself. *That* is how we got him.[54]

The cost of this hidden challenge from within was soon lost amidst a far more explicit and devastating attack from beyond, an attack that was to change the lives of the Maine family forever, and one in which Moormeister played an important, equivocal role. With most of the harvest already bagged but not yet disposed of, he made it known that a meeting was to be held on the following Saturday to which all the 'rich kaffirs who owned spans of oxen' were invited to hear an official from Bloemhof talk about a new 'government ruling.'[55]

On the day of the meeting Thloriso Kadi, Johnson Xaba, and several others in Kas's drinking circle drifted into Sewefontein early, the sound of the horses and carts barely betraying their presence. As the morning progressed the volume of noise increased, the friends were joined by acquaintances from outlying properties like 'Mokebisa of Mothlodi, Piet Mokgosi and many others,' as Kas remembered it, until 'more than a hundred' black sharecroppers and tenants funnelled their way into the dusty yard behind RaDiphonkgo's house.

Peasants spend much of their lives waiting, waiting for something, waiting for somebody. It seemed that little if anything of significance would happen that day, and they were all reconciled to it. But then they found that they were surrounded at a distance by a number of landlords from neighbouring farms. The Boers got out of their trucks, slamming the doors behind them and, like bees at the edge of a saucer of sugar water, formed and reformed in countless combinations, pointing fingers, waving arms and shuffling feet, mildly menacing in a ritual dance conveying secret information that only they understood. But surely there was nothing to fear. These were the familiar faces of men whom they had grown up with, men whose fathers and grandfathers their fathers and grandfathers had known and worked with for the better part of a century. And yet all was not well.

The Boers moved closer, tightened their circle, and then one of their number stepped forward. The official paused, and the silence of the fields rolled in. For a moment all that could be discerned was the distant sound of a winter breeze stroking away at the greying hair of the surrounding veld. Then the wind rose until the sound of its rustlings and the official's words, man and nature, seemed almost inseparable.

During an unexpected lull could be heard a voice that, despite its soft-toned SeSotho inflexions, contained a needle that slowly, deliberately, released its venom into the very essence of their being. 'They announced that farming on-the-halves was no longer to be practised. Agricultural methods had changed, and tractors had been introduced. Those who owned spans of oxen would have to sell them.'[56] 'Rich kaffirs' who wanted to keep their livestock and ploughs were advised to leave the districts of Bloemhof, Makwassie and Schweizer-Reneke and go to Dithakwaneng, Khunwana, Mogopa or Molote, to 'native reserves' farther north or west, where they were free to buy or hire land near Vryburg, Lichtenburg, Ventersdorp and Rustenburg. All those black tenant farmers wishing to remain in the Triangle would have to dispose of their cattle and sell to the Boers either their own labour or that of their sons.[57]

The landlords turned back to the waiting trucks and most of the crowd dispersed. Kas, a few members of his drinking circle and a small cluster of hangers-on stayed behind, too shocked to move and anxious for more information. Moormeister explained it again:

> He stood up and told us that it was understood that those who owned spans of oxen would not want to sell their livestock. He mentioned places in the districts of Kuruman and Rustenburg where those who owned cattle could go to, but in the Triangle itself, there was no longer sufficient space for them to cultivate. Tenants who wished to stay on with the whites would have to sell their cattle and buy tractors since there was insufficient space for their cattle to graze.[58]

There was dead silence. The black farmers had been made victims of a law which worked to benefit and was propounded by 'those with tractors.' Moormeister had harboured the insect meant to sting them, and there was not much he could do or say to help them. It was an economic death sentence.[59]

Kas responded with customary caution and kept his movements to a minimum. A willingness to be seen to be doing the decent thing often bought a man time with a landlord. And so he drove half a span of oxen to the month-end auction, hoping that if this ploy succeeded, he and his friends would not have to proceed with the more radical plan they had been formulating. He, Kadi, Xaba and several others had 'decided to elect people who would represent us and find farming land for us.' Kas had told them that as a young man working on the roads he had once camped on a large sandy property known as Daggabult, near Kommandodrif, a farm to which the Boers had always given a wide berth. Recent visitors to the river-front confirmed that the land was still unoccupied and everyone agreed that if they could find a skilful intermediary to put their case, the magistrate might well let them settle there. And who better to represent them in dealings revolving around land acquisition than that stalwart from I.C.U. days, Jason Jingoes?[60]

Thus it was that, fully two decades after the Industrial and Commercial Workers' Union had first set the Triangle ablaze with its demand that blacks be allowed to own property, Kas finally caught up with the movement. Knowing how sceptical he had been about the Union, the sharecroppers decided that his brother, Phitise, rather than Kas, should approach Jingoes. But the moment for such an initiative had long since passed. Jingoes, the 'Lion of Makwassie,' had retreated into the Malutis, and Phitise was reluctant to set out on a journey to Basutoland.

When Moormeister noticed that Kas had sold some of his livestock, he cautioned against further sales, suggesting that it might be better to take the remaining cattle and trek north rather than risk losing more; Kas, even at that late stage, chose to interpret this noncommittal advice as a manifestation of genuine concern.[61]

Phitise did eventually manage to establish indirect contact with Jingoes and a group of sharecroppers on the Free State side of the river who had also fallen foul of the 'new law,' but when their request was laid before the Bloemhof magistrate, the inflexible logic that formed the backbone of territorial segregation was again spelled out to them: the farm they had their eye on was designated for whites only. In their pursuit of segregation and ideological purity the Afrikaner Nationalists would rather see marginal agricultural land lie fallow than allow it to be cultivated by a few 'rich kaffirs.'

*

Once Daggabult was ruled out, all Kas could do was go back to Moormeister and try to renegotiate his contract so as to make it marginally more attractive for the white man without necessarily imperilling his interests. But the much-vaunted 'German way' had come to a miserable end: even as Kas was busy formulating his proposal, Moormeister was away at van Vuuren's farm buying himself a tractor that came with an enormous disk-plough attachment. The old order was dead, the tractors had eaten the oxen, and the terms of any new deals would be set by white landlords rather than by black sharecroppers. When the machinery was delivered to Sewefontein a few days later, Moormeister told Kas he would be extending the amount of land under cultivation and reducing the areas devoted to grazing, that 'he would be ploughing up to the very front of my house,' and suggested that 'I should sell my livestock'; he added that 'Bodule could then drive his tractor.'[62]

RaDiphonkgo's 'suggestions' penetrated Kas's heart just as surely as if a dagger had been thrust into his chest. Like the God of the Old Testament, the landlord was not content with the sacrifice of animals, he wanted a son, too! But, desperate to keep Bodule and eager to retain access to the land on which he had been born, Kas continued twisting and turning, trying to find a way of plucking out the knife. He groped about, looking for a counter-proposal to give Moormeister: 'I suggested to him that I would sell my livestock and buy a tractor with which to plough on-the-halves. He said: "No, if there are to be two tractors here where would they find land for both of them to plough?" '[63]

RaDiphonkgo twisted the knife again, suggesting yet again that Kas move north: 'There are farms up there, you had better go. Don't stay here, because the Boers will kill you.' At that moment something in Kas died, just as surely as the miner van Eeden had died by Moormeister's hand one night in Johannesburg ten years earlier. And only then, with the life blood of his existence in the Triangle draining away before him, did Kas manage to set Moormeister's profile against the wider backdrop of events.[64] 'RaDiphonkgo shielded himself behind other farmers. He, too, was jealous of my large harvest. He wished to plough even where my livestock grazed. He also wanted me to give him my children, Bodule and Nthakwana, to work for him. He told me that he would plough for me and that I would have sufficient food. Should I have stayed for food alone? *Aikhona!* I decided to leave.'[65]

The fight to remain at Sewefontein, which occurred over four weeks in May–June 1949 was, in effect, also the climax of a wider, and more drawn-out struggle for mastery of the Triangle. The battle was ostensibly fought between armies of matching strength, with whites who owned land pitted against blacks who had draught oxen and family labour. But the odds were stacked against the sharecroppers. Landlords, by virtue of their franchise, enjoyed privileged access to parliament, the provisions of the Marketing Act and the resources of the Land Bank; they could accumulate capital, consolidate their ownership of the land and mechanise their production techniques in ways that the black tenants could not even contemplate.

Once the balance had tipped in favour of the landlords, as it had at some point between 1943 and 1948, it was simply a matter of time before the tractor—with its exponential demand for land, cereals, and profit—came to displace the ox and the plough. With the productive heart torn out of black sharecropping techniques, all that was left was for white farmers to fossick around amongst the remaining bits and pieces and pick up what interested them. Kas, the very epitome of an independent farmer, would be given 'sufficient' food and Bodule would drive RaDiphonkgo's tractor!

In the end, Kas told Moormeister he would be leaving the Triangle because 'when you hear dogs growling make sure that you get away from them.' Moormeister, eager to avoid legal wrangles, assured Kas that he would try to make the Maines' departure as easy and trouble-free as possible. Kas went across to Thloriso Kadi's to borrow a large wagon; Kadi himself had no need of it because he had decided to take a chance, sell his livestock, invest in a tractor, and stay on in the district.[66]

On the way back Kas gave more thought to the complex problem of where he could find refuge for his family and livestock. In troubled times, one needed the protection of kin; in the end, like Phitise, he set his sights on Byl and Wildebeespan, two 'black spot' farms east of Klerksdorp that were owned by the Motsuenyanes, wealthy BaSotho landowners who, it was said, were always on the lookout for good sharecroppers. If he and some of the family went ahead with a few animals and established themselves during the off-season, the rest of the Maines and their livestock could follow in the spring, when the demand for labour was greatest.

Kas got out the hloma o hlomolle and had it eased into position on the back of the Kadis' wagon where it could serve as a mobile home. The boys saw to it that most of the harvest and some of the smaller possessions were safely secured in the remaining spaces. On a cold morning in June 1949, he told Bodule (twenty-one), Mosala (fourteen) and Nkamang (nine) to round up the Maine livestock: eight horses, a dozen donkeys and sixty cattle. This party took their leave of Leetwane and the remaining siblings: with Mosala and Nkamang perched high up on the wagon and Kas and Bodule astride their horses, the convoy edged its way down the track leading out of Sewefontein, funnelled briefly through the farm gates and then flowed out to take up most of the road leading to Kingswood.[67]

By midmorning they had got as far as Leeubos. But there was no sign of its landlord, and so Willem Griesel was denied the pleasure of witnessing the Maines' retreat. A half hour later they negotiated the turn left into the Bloemhof-Wolmaransstad road, leaving behind them the terrain where four generations of the Maines had worked the land for more than half a century. The family had entered the Triangle in the 1890s as refugees fleeing from the consequences of an industrial revolution sparked by diamond mining. Seventy years later they were still refugees in the land of their birth and still running, this time from a belated agricultural revolution which left whites owning far too much land and a skilled black peasantry hopelessly little.

As the caravan trundled on down the road to Wolmaransstad, Kas's thoughts harked back to those whom they had left behind. Leetwane's headaches had given way to pains in the legs and she was walking with difficulty, an ominous development that was reminiscent of the time at Vlakfontein when he had had to bring in Nini Tjalempe to cure her of *umshosha phantsi.* Bodule, too, was engrossed in a problem; it was one of which only he, his mother and his sisters had knowledge. Before he left Sewefontein he had taken Dikeledi Teeu as his wife, and he still had no clear idea how, let alone when, he would break the news to his father. That night the four Maines bedded down around a fire at Kareepoort, the last port of call before they entered the realm of the unknown.[68]

The first streaks of a cold dawn were still to be seen across the horizon when Kas and Bodule completed their preparations for what they predicted would be a long haul. Once the animals had been watered and the wagon packed, the horses and cattle were eased back onto the generously proportioned grass verges of the road to Wolmaransstad. Moving a herd of sixty animals is an intrinsically slow business, and the pace fell off that

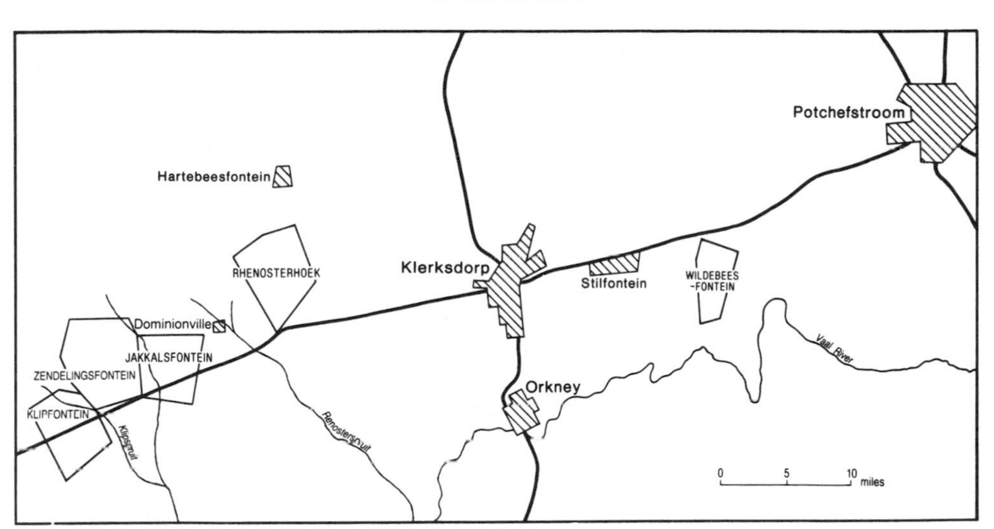


6 Klerksdorp district farms on which the Maine family were economically active, 1949–51

afternoon when they had to negotiate their way through Wolmaransstad itself. Freed from the constraints imposed by the miles of parallel barbed wires lining country roads, the cattle were keen to explore each and every side street and by the time the animals had been escorted out of this macadamised nightmare the winter sun was diving for cover. The men found a corner site into which they could herd the cattle, and then barricaded the animals into this temporary kraal. Mosala and Nkamang clambered into the 'Put-up and Pull-down,' so as not to have their feet bitten by the vicious highveld frost, while Kas and Bodule stretched out beneath the wagon.

The tarred road continued to pull them north and east; having left the flood plain of the Vaal behind it, it wound through a few mild folds that were distantly linked to an elongated ridge flanked by a few stray koppies. By midafternoon they were on the crest of this ridge, and then, leaving a few lazy coils behind it, the road straightened out and slithered down an extended decline with a spruit at its end. Sensing that the animals were tired and thirsty, Kas told Bodule they should risk racing the sun to the spruit and hope that darkness would not find them without a campsite. The Motsuenyane farms were still two days away, well beyond Klerksdorp.

But the spruit was more distant than it had seemed from the height of the ridge. By the time they were three-quarters of the way down the slope it was hard to keep all the animals in sight, and Kas was about to call a halt for the evening when they drew abreast of a gate leading to a farmhouse very close to the road. In the gathering gloom, and out of the corner of his eye, Kas caught sight of a young white man watching them. At moments such as this, a black man, his family and his animals were most vulnerable to the vagaries of a night on the road, and the Boers were most capable of showing that they, too, were the sons of Africa. As Kas knew only too well, the success of any request for help depended on the ethnic and class codings embedded in the highveld's etiquette of race relations.

> So I approached him and said: "*Baas*, can you help me with the cattle? I need a place where they can be kept for the night and it is already very late." He said: "Just open the gate and let them in" and told us where we could find water. We outspanned, made a fire and sat around it—myself, Bodule, Mosala, Nkamang and that "house"—hloma o hlomolle.[69]

A little later the Boer reappeared out of the dark and introduced himself as Pieterse, saying that he was the foreman for one Hotz, the landlord who owned the property.[70]

Pieterse sat down and interrogated Kas at some length—who was he exactly, where had he come from, why had he left, where was he going, how many animals did he own, and how big was the Maine family? Kas responded with customary ease and fluency, sensing that the enquiries might lead to the offer of a contract. And they did. Pieterse offered to find Kas a place on a nearby property, which, he said, his wife would point out to him in the morning, and then disappeared into the night.

At first light the cattle grazed without hindrance, but of Mrs. Pieterse there was no sign. Eventually a car driven by a pleasant-looking Afrikaner woman arrived: after a brief discussion, Kas climbed in and, with her in the front and him in the back, they set off toward the town of Klerksdorp. About five miles along, she pulled off the road onto a dusty track that petered out at a gate where a sign painted in crisp lettering announced, 'L. J. Klue, Jakkalsfontein.' Kas opened the gate.

The farmhouse, which seemed to be in reasonable repair, was only three hundred yards off the main road, but there was not a black worker in sight, and with a furtive sideways glance Kas detected the ruins of what must once have been labourers' mud huts. It was not a promising scene, but the woman's cheerfulness was reassuring, and when they found the Boer she greeted him warmly, saying, 'Dad, here is someone who is willing to abandon a trek if you can offer him a place on the farm.' Louis Klue, a short man in his late fifties, turned from his work to engage Kas. He took over the terrain that had been covered with Pieterse the evening before, and then, without further ado, put to Kas what he considered a reasonable proposition. Kas went over a few details that needed amplifying or clarifying and then said he was satisfied. 'We reached an agreement. I went back to collect my things, and we settled on his farm.'[71]

The next three days were spent building a hut that would fend off the worst of the winter weather. Although keen to ensure the well-being of his family, Kas wanted to get back to Sewefontein to collect the ewes before the arrival of the first spring lambs slowed down the long-distance transfer of livestock. On the afternoon of their fourth day at Jakkalsfontein he told Mosala and Nkamang to collect their belongings, inspanned the donkeys and set off for the southwest. By nightfall they had got only as far as Klipfontein, and although the children were obviously cold and tired, he decided to push on through the icy night. In the early hours of the morning he reached the outskirts of Wolmaransstad where, totally exhausted, he could not even summon the energy needed to outspan the donkeys and fell instead into a light sleep.

But the wooden seat was too uncomfortable to allow for meaningful rest. At first light he got down from the wagon but found it almost impossible to move. He tried walking around the wagon, stopping every few paces to creak and stretch a limb, like a crake patrolling a mudpatch, but eventually gave up. He propped himself up against the side of the wagon, peered into the back at the outlines of two small figures huddled beneath blankets, and decided he did not have the heart to wake them. He hauled himself back on board and belted the reluctant donkeys back into action. By the time the children woke they were well on their way to Schweizer-Reneke. He called a halt at Boskop, assuming the friendly Boer whom they had encountered on the outward leg would not object to their outspanning the animals on his property. Nor did he; Daniel Uys was only interested in finding out how their quest had ended. After they had refreshed themselves, they set off for Soutpan to tell Phitise where exactly in the Klerksdorp district they had settled.[72]

At Soutpan in midmorning they were surprised to be met not only by Phitise and his family, but by Motheba who, having left Sellwane behind at Harrisburg, was on a rare visit to the Triangle. After a miserable night on the wagon the beer, food and warmth of Phitise's home were especially welcome and, while the donkeys had a few hours to graze on 'Ou Willem Coetzee's son's farm,' the brothers swopped notes on the area around Machavie where the Motsuenyanes' farm was said to be located. Before leaving Kas agreed to collect Motheba on his return; Jakkalsfontein was close to Harrisburg, and by then Phitise would want to proceed directly to Byl or Wildebeespan. It was dark by the time they got to Sewefontein.

Up early the following morning, Kas noted that one of the larger fields of sorghum which had been left for Mmusetsi and the girls to harvest was unattended, something he soon set right. Leetwane, unfortunately, was still not well and took no part in either the harvesting or the threshing. There was little point in her staying on at Sewefontein and Kas decided that she should accompany them on the long trek back to Jakkalsfontein with the sheep. Lebitsa and Mosala would stay behind with the remaining livestock and heavy equipment, an arrangement to which Walter Moormeister—his old charming self—raised no objection.[73]

Before that could happen, however, other matters had to be attended to. On 13 July, Kas and the boys loaded twenty bags of maize onto the wagon and undertook what for once seemed like a quick trip to the Co-operative depot at Kingswood. As usual, Kas had no difficulty selling his grain which, despite RaDiphonkgo's earlier misgivings, included several bags of white mealies. The clerk handed him a cheque for the voorskot and Kas made certain that the co-operative was properly informed about his new address at Jakkalsfontein so that he would not be made to forfeit any agterskot. Back at Sewefontein the Maines were met by a suddenly re-agitated Moormeister who, uncertain as to whether he was truly free of the family of sharecroppers, warned Kas not to linger in the district or to wait for any settlement due to him from the co-operative.[74]

The frightened German need not have bothered. Only two days later, and for the third time in a fortnight, Thloriso Kadi's wagon was stacked to the point where the wheels groaned. While the men took care of the heavy manual labour involved in organising the move, Nthakwana (eighteen), Pulane (twelve), and Nkamang (nine) rounded up the eighty-eight sheep—mainly ewes—that had been selected for early relocation at Jakkalsfontein. Kas got Walter Moormeister to issue him with the 'trekpas' and then assembled a column comprising the Kadi wagon, a horse and cart and the flock of sheep.[75] The caravan moved off to collect grandmother Motheba.

Trekking with heavily laden ewes was a slow business. By the end of the first day they had made only about twelve miles' progress when dusk rolled in over the plains. But, still almost within sight of Sewefontein, this did not pose a problem. Kas knew most of the landlords around and easily found a place to camp for the night. He was grateful for the hospitality the Boers showed on such occasions, and that night, seated around the fire, he again reflected on what exactly it was that they were leaving behind them. In abandoning the southwestern Transvaal, he was not only forfeiting his insight into the vagaries of Triangle farming conditions, but also jettisoning an intimate knowledge of local Afrikaner genealogy, a mental directory of Boers based on the experience of four generations of Maine interaction with white farmers. It was all very well for officials brandishing maps of territorial segregation to talk of 'rich kaffirs' buying or hiring land, but even if that were possible, how was a sharecropper supposed to relocate his productive activities without being able to use the insights, contacts and networks that had been developed over a lifetime?

When Kas tried to get moving the following morning, one of the wagon wheels was stuck, so while he and Mmusetsi remained behind to repair it, the girls were sent ahead with the sheep. But as Nthakwana's party approached Seedat's store near Rietpan, the first ewes lay down to lamb, forcing them to call a halt. By the time Kas and Mmusetsi caught up with them, several more animals were giving birth, and to make matters worse, the weather was menacing. The donkeys were outspanned and grazed, ever grateful for a short working day, while the Maines set up a camp. By early evening it was raining steadily.

A steady downpour through the night turned the road's compacted dirt surface into a quagmire that punished the sheep and tortured the donkeys. With the wagon often up to its axles in mud, it took them ten hours to cover the eight miles to Phitise's shack. At Soutpan the animals were again set loose to graze and drink, none happier than the dozen lambs that had been separated from their mothers to benefit from the security of riding in the wagon. The Maines, too, were exhausted, and grateful for the warmth of their makeshift beds. It was one of those magical nights when a sharecropper's shack became a castle.

Kas, first up, checked on the livestock, found the donkeys missing, and woke Mmusetsi. The animals had got out of the camp and, remembering the good times to be had around Seedat's store, had set off back down the road toward Rietpan. Mmusetsi was given the trap and told to go out and fetch the donkeys. It started raining again, and they settled down to wait.

By the time Mmusetsi got back with the donkeys it was past midday. Kas helped get the strays back into harness while Pulane and Nkamang, deliberately violating the boundaries between work and play as only youngsters can when time is pressing, took leave of their cousins. Kas's tongue got the girls moving and the whip got the wheels rolling, but the wagon only sighed and sank slowly into the mud. For more than an hour Kas and Mmusetsi battled to free the wagon from the ooze, but in the end even Kas conceded defeat and agreed to spend a second night at Soutpan. It was Motheba who suggested that, in the morning, he should supplement his span of fourteen donkeys with four of her animals which could revert to her when he chose eventually to return the Kadis' wagon to them.[76]

Even a donkey has some self-respect. With their pride at stake the augmented span shrugged off the challenge of the mud and the convoy swung east toward Makwassie. That afternoon, having made certain that they did not have to take the sheep through the streets of Wolmaransstad, they outspanned within sight of the S.W.T.L.'s main grain elevators at Leeudoringstad. Turning north, the column gained some momentum and they spent another reasonably comfortable night near the farm Borman before cold and cloudy weather again set in around Palmietfontein. The following afternoon the Maines left Motheba at Harrisburg, and then trudged on for several more hours before a steady drizzle forced them to turn off the road and set up camp at Klipspruit. It looked like another wretched night on the highveld.

But the rain suddenly stopped, the girls collected some additional kindling and everyone was soon warming themselves around a roaring fire—a rich reward for bodies that found even sitting difficult after such an extended trip. Before long a young defector smuggled a blanket out of the wagon; within half an hour, most of the family was stretched out on the sand. Kas sat around for a while watching the dying embers, trying to anticipate what problems they might encounter on the final day of their march to a new home. But he, too, was unequal to the struggle, slid beneath his blanket, and turned over. His mind was already doing that slow scan of poorly connected thoughts which characterises the brain's night shift when he thought he heard the distant howl of a jackal.

A minute or two later he was woken by a second and louder howl. This time there was no mistaking it: the sheep had become the focus of a lone marauder. He sat up, tossed a few sticks onto the fire to generate a little light, and then got up to check on the integrity of the thorn barricade they had erected to protect the animals. But the light and movement did not deter the prowler and, throughout the night, Kas had regularly to remind the jackal he was there to protect the flock. By daybreak he was numb with exhaustion.[77]

Late the following afternoon, after more than a week on the road, the column wound its way through the gates at Jakkalsfontein. They were met by Bodule, who told Kas that, with the exception of a visit from an Inspector of Native Labour in search of poll-tax defaulters, nothing untoward had occurred in his absence. Although still exhausted and without sleep, Kas was keen to finish moving the animals, and within forty-eight hours had persuaded Louis Klue to issue him with a pass so that he could return Motheba's donkeys. Only on his return from Harrisburg two days later did he allow himself to relax and start the slow process of gearing up for the challenge of farming in a new environment.[78]

In mid-August Kas received a letter from the South-Western Transvaal Agricultural Co-operative informing him that he was entitled to an agterskot of twenty pounds and thirteen shillings. In an off-season where neither cobbling nor smithying had brought in much cash this was welcome news indeed. He used Bodule's bicycle to get to Klerksdorp, and boarded a train for Leeudoringstad and the Head Office of the S.W.T.L. A clerk was quick to cash his cheque; with the notes stuffed into the pouch on his belt, he made the journey home, where Louis Klue, equally obliging, gave him open-ended written authorisation to brew four gallons of beer for domestic use each week for as long as he lived on the property.[79]

But in life, as in nature, the cat stirs when the mouse is calmest. Bodule, seeing that there was money in the pocket and beer in the belly, seized the chance to tell his father of his 'marriage' to Dikeledi. But the mouse, too, is not without guile: Kas chose not to be provoked and, outwardly calm, simply put a probing question to his son. Yet, within him, his heart pounded and his chest felt about to explode. The name Dikeledi means 'tears,' and Kas had no intention of allowing her to appropriate Bodule. As far as he was concerned they were having an affair: marriage was something different. For the moment, Dikeledi was simply a worry.

Leetwane's condition was another source of concern. Her health had worsened rapidly at Jakkalsfontein, and it was obvious that she was going to need professional assistance. Interestingly enough she showed no desire to consult a Christian faith healer or anyone from the Ethiopian church, in sharp contrast with that painful moment at Vlakfontein years earlier when Kas had taken Lebitsa as a second wife and she had chosen to place all her trust in Nini Tjalempe from the A.M.E.C. Since then, the family had drawn more of its men into its religious practices, but it was almost as if Leetwane's interest in the Ethiopians was in inverse proportion to Kas's involvement with the church. She now expressed a preference for consulting a traditional 'seer,' or *senohe*, her choice settling on Phitise's son-in-law, Mokone Mabe, who had a practice at Boskop, near Potchefstroom.[80]

Mabe, who knew the personalities in the Maine family almost as well as he knew the structure of their household, listened to Leetwane's complaints with rapt attention. After careful deliberation he concluded that she had been 'turning a deaf ear to the spirit of her grandfather, Tsieng,' and that there would be no relief from illness until she responded to the ancestors and took up her calling as a ngaka. This injunction, which emanated from an authority beyond the questioning of mere priests or patriarchs, Kas found impossible to resist. He told Mabe that 'rather than see Leetwane suffer, he should train her as a herbalist.' Leetwane was moving into psychological space which, in the Maine family, had hitherto always been occupied by men. Kas left the novice with Mabe.[81]

On the way back to Jakkalsfontein, Kas found the trees reminding him that spring was approaching, and then realised that most of his heavy agricultural equipment was still at Sewefontein. Moreover, it was becoming obvious that whether he liked it or not, he would have to talk to the Teeus and face the fact that he and Bodule were unlikely ever to see eye to eye about Dikeledi's suitability as a wife. A trip to the Triangle was becoming unavoidable. On 2 October, he got out the horse and cart, helped Bodule harness the donkeys to the Kadis' wagon and together they set off for a farewell visit to Bloemhof.[82]

The journey to what in many ways still seemed like home passed without incident. South of Wolmaransstad the sun was already frying the plain at a temperature that did not augur well. But if Kas had lingering doubts about the wisdom of having left the Triangle, they were dispelled by his discussion with Thloriso Kadi. Kadi's landlord had apparently insisted on provisions in a contract that allowed him to benefit disproportionately from the tractor he had forced Kadi to acquire. It was a sad tale, and Kas (conveniently forgetting that he, too, had once toyed with the idea of acquiring a tractor and working with RaDiphonkgo) thought this development would cause the ruin of his friend.[83]

At Sewefontein, but this time without Kadi's heavy-duty wagon, Kas was appalled to discover that, besides his planter, ploughs and harrow, forty bags of sorghum and an assortment of corrugated-iron sheets also had to be taken to Jakkalsfontein. He decided that it would probably be easiest if he were to have all his farming paraphernalia railed to Harrisburg, and then transport it from there to Jakkalsfontein by road. He hired a wagon from a MoThlaping friend, but even then it required two trips to Kingswood before everything was safely secured beneath a heavy tarpaulin in an open railway truck. Freight charges came to fourteen pounds, which, when added to the sum he had paid the MoThlaping, brought home to him the real cost of relocating his family. If the exercise were ever to be repeated, it would pay him to invest in a truck.[84]

Ten days later Leetwane returned to Sewefontein in the company of her mentor; she had completed three weeks of intensive instruction, her health had undoubtedly improved, and she no longer complained of the blinding headaches that had plagued her. Kas gave Mabe an ox for his services and then, in a further and more public affirmation of his wife's new status as a qualified ngaka, slaughtered another ox for a celebration to which all the Maines, along with Abner and Amelia Teeu, were invited.[85]

The gathering reminded Kas how much he disliked the Teeus' pushy seventeen-year-old daughter. He found it almost impossible to forgive her for appropriating the affections of a son who, if exposed to more elevating influences, was destined for a brilliantly successful career in farming. As the oldest child in the Teeu family, Dikeledi had been sent to school for the first three years of schooling and passed standard one. This was probably unfortunate, and no doubt helped explain why it was that she—unlike the Maine women, who had all been given herbs to circumvent the traditional taboos against mature females working with cattle—had never done any herding. Worse still, ever since the age of fifteen she had had an income of her own from working in a white woman's kitchen. This was the sort of pattern that was likely to make a woman unnecessarily wilful.[86]

Kas balked at the prospect of having to pay bohadi for Dikeledi, and years later, he remained equally forthright about his attitude to the proposed marriage: 'At the time they "eloped" I had no intention whatsoever of arranging a marriage for Bodule.' He therefore deliberately introduced a complication by arguing that 'Bodule's older brother was still around: how could I arrange for him to get married while Mmusetsi remained single?' Was it not more appropriate for Mmusetsi to marry Dikeledi or her younger sister, Disebo, before Bodule's fate was decided upon?[87]

This proposition, which chimed so sweetly with common sense and BaSotho tradition, caused consternation amongst the Teeus, where the prospect of having Mmusetsi as a son-in-law was greeted about as enthusiastically as Kas accepted Dikeledi as a daughter-in-law! But, once the grit had been fed into the works the machinery ground to a halt. With no end in sight, the matter was still unresolved when the Maines finally left the Triangle in mid-October. In private, the lovers took leave of each other, knowing they would meet again at Jakkalsfontein.

PART FOUR

# MATURITY

*

*'Ill fares the land, to hastening ills a prey, where wealth accumulates, and men decay; princes and lords may flourish, or may fade—a breath can make them, as a breath has made; but a bold peasantry, their country's pride, when once destroy'd, can never be supplied.'*

OLIVER GOLDSMITH,
*The Deserted Village*, 1770

CHAPTER ELEVEN

# Retreat

## 1949–52

On the highveld three inches more or less of rainfall a year can make a substantial difference. When the first white settlers crossed the Vaal in the late 1830s they had ignored the Reverend Broadbent's mission station at Makwassie, turned east and pushed on to where the flood plain folded into mildly undulating terrain, the dry Kalahari sands made way for red soils, and the acacia bushes aspired to the status of small trees. Moving on, into an area that had been cleared of most of its indigenous inhabitants by the upheavals that had accompanied the rise of the distant Zulu state, the trekkers established themselves in the districts of present-day Klerksdorp and Potchefstroom.

But the differences between the southwestern and western Transvaal are of degree, not kind. For fifty years these settlements were dominated by the practice of a quasi-feudal mode of agriculture until, in the mid-1880s, they were stirred from their enforced torpor by the discovery of gold on the farm Rhenosterspruit, west of Klerksdorp. For a moment it appeared as if the emergence of an industry and a local market might transform the region but it was not to be. Barney Barnato and Woolf Joel's small Dominion Reefs workings could not rival the emerging Witwatersrand giant, and when mining was eventually abandoned during the 1906–8 Depression, the Klerksdorp district reverted to an agricultural economy that depended heavily on black tenants for labour.

For more than a quarter of a century the Rhenoster Mines, adjacent to the farm Jakkalsfontein, remained dormant. But when South Africa abandoned the gold standard in late 1932, the mines were once again worked, this time under the aegis of the Anglo-French Investment Company, which floated a special subsidiary to manage its venture. Dominion Reefs, employing a few score skilled white miners and several hundred unskilled black workers, enjoyed modest success as one of the few large industrial ventures west of Klerksdorp.[1] Having a mining village at Dominionville did not transform the economic prospects of the region as a whole, but it did help to prime the pump for the more sustained agricultural development that followed World War II and, by so doing, contributed to the proletarianisation of black labour tenants on the neighbouring properties. In 1943, a researcher noted, 'The world around here is becoming more restricted for the "rich" native and there is little hope of his making a living in the district in future. There is a growing need for full-time farm labourers without livestock.' The exact nature of the work demanded of this new generation of black labourers varied greatly, however, from farm to farm.[2]

Farms in the western Transvaal, having more water than those in the southwest and being closer to the metropolitan Witwatersrand, tended to be smaller than those of the Triangle and more self-consciously directed to 'mixed farming.' In the 1950s many landlords close to the towns farmed fruit, vegetables and dairy products for the markets of Klerksdorp, Ventersdorp and Potchefstroom. Farther afield, in a second zone, yet other farmers raised beef, maize, sorghum and tobacco. Only on a few outlying properties in a third zone, did notables such as Jan Wilkens (National Party Member of Parliament for Ventersdorp), J. H. de Kock, P. W. de Villiers and G. J. Kuhn use the 1937 Marketing Act to concentrate on extensive maize production and assemble land holdings greater than the local average.[3]

Jakkalsfontein, twenty miles west of Klerksdorp but conveniently close to the Dominion Reefs mine and the compounds that housed its black workers, straddled the first and second of these zones. The original property, granted to one J. G. J. van Vuuren by the Transvaal government in 1859, had comprised more than 4,700 morgen, but after the turn of the century, and more especially so during the troubled 1920s and 1930s, he and his descendants, like L. R. J. van Vuuren, had sold off land often enough to suggest that they were not very successful farmers. In the wake of the Depression several smaller parcels of land, ranging from 110 to 440 morgen, were put on the market, including one purchased by the Abdurahmans and Pietersens, members of an extended 'Cape Malay' trading family who owned a store on the adjoining property of Sendelingsfontein.[4]

In more remote areas of the highveld, where affluent Afrikaner landlords were firmly entrenched, the sale of land to a 'coloured' family might have raised objections, but around Dominion Reefs, where mineral exploitation, trading rights and urbanisation had already driven the first wedges into farming land, the arrival of the Abdurahmans and Pietersens was not strongly contested. Indeed, these processes of 'development' drew a number of adventurers and speculators to the area, and this in turn encouraged the further fragmentation of Jakkalsfontein.

As with most poorly educated country folk in the 1930s, L. R. J. 'Bully' van Vuuren was deeply suspicious of urban outsiders who, like characters in a Bosman novel, were often confidence tricksters well versed in the art of cheating farmers, *boereverneukery*. But from what he later told Kas, he had discovered early on that not all swindlers were necessarily city slickers or foreigners. According to him, he agreed to sell 387 morgen of Jakkalsfontein to Louis Klue in March 1938 only after he had been plied with so much liquor that he could not make a rational decision. Whatever the truth, for the middle-aged Louis Klue acquiring this small pocket of land was the first secure foothold in a life that had known little but persistent poverty and exceptionally hard work.[5]

The origins of the Klue family, like those of so many other Afrikaners on the Transvaal highveld, could be traced back to one of the great pools of rural distress at the turn of the century: the southeastern Cape Colony. In the 1890s Louis's father had eked out a precarious existence in the George district, where he and his wife raised four boys and two girls under difficult circumstances. In 1910, after years of struggle and the collapse of the local market for ostrich feathers, he had undertaken the long trek to Theunissen, in the Orange Free State, where he and his teenage sons laboured as bywoners on the property of distant kinsmen.[6]

Five years later, in the wake of the 1914 rebellion, the family left the Free State, crossed the Vaal and settled in the Klerksdorp district, where Klue and his sons worked for a landlord during the summer and did manual labour on the Wolmaransstad diamond diggings in the off-season. In 1920, at the age of twenty-seven, and having worked under his father on other men's farms as a bywoner for nearly a decade, Louis Klue went back to the George district and married Susannah Francina Meyer, sister of Koos Meyer, the Kaffir Corn King of the Triangle, himself a refugee from the impoverished southeastern Cape. On his return to the Transvaal two weeks later, Louis Klue became a tenant on the Badenhorsts' section of the farm Rhenosterspruit. Labouring there for the next four seasons, he saved enough money to enter into a joint venture on a farm in the Wolmaransstad district, where he spent the three years of 1924–27.

Only on his return to the Klerksdorp district, in 1927, and at the age of thirty-four, was Louis Klue for the first time able to farm on his own account. Determined to succeed after so prolonged an apprenticeship, he survived the worst of the Depression by planting tobacco on a small hired section of Jakkalsfontein and hauling it to the depot of a co-operative one hundred and thirty miles east, at Parys. Tobacco—that most difficult, demanding and labour-intensive of all cash crops—came to Klue's rescue at a crucial point in his career, and he never forgot his debt to or lost his affection for it. When the Dominion Reefs mine was reopened in 1934 and he won the contract to supply its compound with fresh vegetables, he secured the small but regular income that, four years later, enabled him to acquire the 387 morgen that Bully van Vuuren so resented. Ten years later, by dint of hard labour, shrewd management and single-minded devotion to the production of coarse tobacco, he was well entrenched at Jakkalsfontein.[7]

There was much in the life of Louis Klue that resonated with that of Kas Maine. Both men had spent a long time in the shadows of patriarchy before emerging to marry and set up home on their own; both were familiar with the insecurities that came with landlessness and tenant farming; both knew the meaning of hard work. And they had a good deal to offer one another. But psycho-social symmetries in the personal histories of landlord and tenant are not enough in themselves to ensure the success of the relationship, and in terms of their respective material requirements Kas Maine and Louis Klue met at the wrong moment. As a landlord with restricted access to property, Klue needed wage labourers to work on a cash crop rather than a family of sharecroppers interested in maize production. Kas, with his large investment in livestock and machinery, needed grazing for his animals and land to put under the plough. Their economic needs ran in opposite directions, but for various reasons neither party was in a position to hold out for arrangements better suited to his purpose.

Klue (for some unknown reason Nguni-speakers on the surrounding farms called him Majolo) was a notoriously hard taskmaster who had only recently put to flight a party of Xhosa labourers who had challenged his authority.[8] This encounter had left him short-staffed and, as Kas discovered, helped to account for the vacant huts he had noticed on his first visit to the farm. Kas Maine, for his part, was uncertain whether he would find grazing for all his cattle under the Motsuenyanes at Wildebeespan, and he needed time to take stock of this unfamiliar situation halfway between Bloemhof and Klerksdorp. With neither landlord nor tenant convinced that it could ever be anything more than a temporary arrangement, the two patriarchs reached an accord at the expense of a third party who was unrepresented at their negotiations: the sharecropper's children. It was a compromise that sacrificed the independence of three young adults, a cost that the two principals had met on more than one occasion for *their* fathers.

Klue gave Kas unrestricted access to grazing for his cattle, as well as to a three-acre plot that he was free to cultivate as he saw fit. But since this meant that the landlord's own cattle would be under pressure, Kas had to get rid of his span of donkeys. The landlord also wanted the full-time assistance of Bodule, Mmusetsi and Nthakwana; the young men were to take charge of Klue's herd of dairy cattle, undertake all the seasonal labour associated with tobacco production, and assist in the vegetable plots. In return, Bodule and Mmusetsi would each have three-acre fields, a monthly cash wage of twelve shillings, and an adequate supply of fresh vegetables. Nthakwana, also eligible for a small monthly cash payment, would be expected to assist Mrs. Klue in the farmhouse. It was also understood that other members of the Maine family would help out with seasonal tasks, and would be paid in cash or in kind.[9]

Once these arrangements were in place, Bodule, Mmusetsi and Nthakwana took up their new positions under the Klues, while Kas and fourteen-year-old Mosala, free of direct supervision, ploughed and planted the field allocated to the family. But a three-acre plot hardly exercised a span of oxen that was fourteen strong, and even after Kas had nourished the tobacco-impoverished soil with the first commercial fertilisers he had ever used, the spare draught capacity continued to weigh on his mind. He was grateful when a man by the name of Tutu, one of several IsiXhosa-speaking neighbours, helped him set up an outside crop-sharing arrangement with Bully van Vuuren. Yet he was still dissatisfied, because although the additional field was larger than the one Klue had allocated him, it was still well below what he had had in the Triangle. Worse, the early spring rains were disappointing: the country was locked in one of its drought cycles.[10]

It was all rather worrying. As Kas suspected, neither his grazing requirements nor his ploughing needs could be satisfied on Klue's farm. He was still contemplating these problems when he got a letter from Phitise addressing him as a 'Son of the BaFokeng' and inviting him to attend a feast at Wildebeespan on 5 November. This gathering, and several others like it held on the Motsuenyane farms during the following year, gave him the subliminal inspiration to find a way out of his predicament. An old, rather shapeless notion gradually assumed a more distinctive outline in his mind and helped to facilitate a subtle shift in self-perception over the months that followed.[11]

Phitise, who was distantly related to the Motsuenyanes by marriage, introduced Kas to a family that he had first heard talked about around the campfire at Prairieflower in 1943–44. Sons of Moses Motsuenyane, a distinguished nineteenth-century Orange Free State sharecropper, the five Motsuenyane brothers had had both the political foresight and the financial resources to buy two properties east of Klerksdorp a few years before the passage of the Natives Land Act of 1913. With the assistance of kin and black tenants, their sons had become successful landlords and continued to invest their profits in additional property in the few 'black spots' set aside for African occupation in the Lichtenburg, Rustenburg and Ventersdorp districts of the western Transvaal.[12]

The Maine brothers, although lacking resources like those of the Motsuenyane family, were persuaded by their example that the future lay not in sharecropping on Byl or Wildebeespan—farms which had, in any case, become fairly overcrowded—but in acquiring property of their own. This in itself was hardly a revelation, since the Maines had debated the same notion when they realised that all the 'rich kaffirs' would be expelled from the southwestern Transvaal. But the idea of owning their own property took on new meaning and fresh appeal once they were exposed to first-hand accounts of black farming at places such as Ga-Mogopa, Ga-Motlatla, Khunwana, Molote and Tsetse, all of which were within a ten-day trek of Klerksdorp.

Still, they would need money, and they would have to reactivate their almost quiescent sense of ethnic identity. The few areas reserved for black land ownership in the Transvaal were for the most part under the control of 'tribal authorities,' structures which the Afrikaner Nationalists, with their commitment to apartheid and territorial segregation, were hoping to endow with ever greater political significance as they set about transforming the face of South Africa. Kas, who had survived in the Triangle largely by playing at being a chameleon amongst the Boers, would have to assume yet another cultural hue if, like his great-grandfather Seonya before him, he was to insert himself successfully amongst the BaKubung of the Rustenburg district.

It is instructive to contrast two photographs of Kas taken by Bodule—the one in 1948 and the other in 1951. The former, a snapshot taken after the record harvest at Sewefontein and at the very apex of his career amongst whites, shows Kas sporting a full beard, wearing a hat with a jacket and matching trousers—the portrait of a successful Triangle farmer who, but for the colour of his skin, was indistinguishable from most Afrikaner farmers on the highveld. The latter, taken on an occasion at Dominion Reefs at the start of the Maines' quest to gain access to BaFokeng land, shows Kas closely cropped, brandishing a *knobkerrie*, and wrapped in a full-length blanket beside a prized beast, the very picture of a successful MoSotho peasant with roots deeply embedded in the complexities of southern African black society.

On his way back from the Motsuenyanes, this jumble of ideas slowly took shape, and Kas started formulating a strategy. But when he got home, he was overtaken by events. A letter from his seventeen-year-old daughter, Matlakala, written from Mofufutso, in the Khunwana reserve, reported that both the older Lepholletses had passed away within sixteen days of one another in mid-October. This news was bad enough in its own right, but it also meant that the sensitive Matlakala was adrift in the world. Since Tshaletseng and Tshounyane weren't there to look after her, Kas and Leetwane would have to retrieve her and integrate her into the family at Jakkalsfontein. Kas knew it was a bad time to be away from the farm, but he trusted Bodule with the livestock.[13]

The journey to Khunwana, which pulled the couple north and west toward Lichtenburg, was slow enough by horse and cart and was made even more so because Leetwane insisted on their bringing along a beast for slaughtering. She had taken the loss of her parents badly, and seemed most upset that neither she nor her sisters Lebitsa and Tseleng had witnessed their burial. Kas could sense the deterioration in her health, and under the circumstances, it seemed doubly important that the traditionally sanctioned purification ritual take place without delay. So when she insisted on taking an animal along to Mofufutso, he put up only token resistance. The beast would simply have to tag along.[14]

Mofufutso lay in a congested part of the overcrowded Khunwana reserve. A few stone-strewn dongas were all that separated some badly overgrazed scrub vegetation from dozens of unfenced plots of mealies which, even at that early point in the season, looked doomed. Barefooted herdboys, wielding long switches to head off a dozen scrawny beasts from the maize fields, pointed them in the direction of the Lepholletses' house, where they found Matlakala in the care of her older half-sister, Thakane. The next morning Kas went to find a ngaka willing to conduct the cleansing ceremony, agreed on a time for the ritual, and then returned to the house, where a bewildered Matlakala gave them a detailed account of her life, and of the unsettling events that had led to the death of her grandparents.

When Tshounyane and Tshaletseng had refused Kas permission to marry Tseleng after their affair at Kareepoort, she and her baby had left the diggings and come to Mofufutso, where her brother and parents had effectively raised the child. As a very young girl Matlakala had been put in charge of her uncle's cattle, a demanding task amongst the mischievous herdboys in the reserve and not eased by her inherently nervous disposition. When not much more than eight years old she had fallen into such a deep and extended sleep that a traditional doctor had had to be summoned, and it was he who pronounced her to be in possession of the ancestral spirit that had later pointed her in the direction of 'faith healing' through prayer.

Several years later, Tshaletseng, concerned that Matlakala be fully integrated into BaTswana society before her own death, arranged for the girl to be sent to an initiation school at Madibogo. This proved to be a traumatic experience. After four weeks of traditional rites that culminated in a mock genital operation, the impressionable fifteen-year-old left the school feeling even more vulnerable and insecure than when she had entered it. Back at Khunwana, she had become implicated in a sex scandal which resulted in her and several of her cohort being dragged before a tribal court, the *kgotla*, where they were flogged by an elderly man until their 'clothing was covered in blood.' In the wake of this ordeal and the hardship caused by the 1948 drought, she had fled from the village to a nearby farm, from where she had sent home her food rations, since she feared that her family 'might die of hunger.'[15]

When Tshaletseng realised she was dying she sent for Matlakala and belatedly gave her her blessing, saying, 'Now that I have seen you I can die, and you should continue to pray as you have been doing all along.' Shortly thereafter the old woman passed away and two weeks later, after an equally poignant and pointed farewell, Tshounyane, her grandfather, also died. Isolated, and with no one to help her, she had written to the Maines at Jakkalsfontein, but her half-sister, Thakane, who happened to live close by, had taken her in and agreed to look after her until her father and his family appeared to claim her.[16]

Thakane and Matlakala had a great deal in common besides having Kas as a father. Both were only children, both had been relegated at a tender age to the physical, economic and psychological margins of the extended family and, perhaps as a consequence, were frailer and less robust than either Morwesi or Nthakwana, the daughters of the principal wife in the house of Maine. Thakane, like Matlakala, had seen 'visions' while an adolescent, and had left the Maine home after her traditionally arranged marriage at Klippan, going with her husband to the Migdol district northeast of Schweizer-Reneke, where they had settled on an isolated farm. During the 1948 drought she, too, had sensed famine breathing down their necks and had joined a seasonal harvesting team bound for the northern Free State, in the hope of bringing back enough to see them through the winter. And one day, while preparing a meal for those working in the distant fields, she had fallen into a deep trance during which the spirits of Motheba and Tshaletseng had appeared before her:

> I saw a big grey mountain and my ancestors came out of it. They took me and went back into it. Inside the mountain I was shown a long shiny pipe to which three taps were attached. The first provided only drinking water, the second water for washing, and the third contained 'holy water' capable of healing people.

Such a calling could not be ignored. When the workers returned from the fields, Thakane set about praying for those with ailments, sprinkling the cuts on their hands with 'holy water' and treating their tired feet. The near miraculous cures that followed verified her claims to being a 'faith healer,' and when the peasants in the harvesting team went home, they spread her fame and reputation before them.[17]

Kas considered such quasi-Christian callings to be of a rather lower order than those that emanated from the shades, in time-honoured fashion, and he listened to these accounts in silence. He may have been at the very zenith of the patriarchal arc that circumscribed the family, but he was slowly but surely being nudged into the orbit of more and more independent women who, having been possessed by ancestral spirits, were now—like himself—claiming to have healing powers of one sort or another. Thakane, it was true, was but a distant planet. Her gravitational pull was seldom felt closer to home. But what of Matlakala, who would now join the Maines? And, most pertinent, what of Leetwane, whose influence lay close to the centre of his own social universe? How many traditional doctors, prophetesses and faith healers could one have in a single family without serious confusion? How many people could there be in a peasant grouping whose ultimate loyalty lay not with the head of the household, where it rightly belonged, but with some higher spiritual being? 'These "spiritual" people are a strange lot,' he once observed. 'Once they become possessed by a spirit they seem to forget that they still have elders.' No! 'Any kind of work that you wish to undertake should be sanctioned by your parents, be it church, faith healing, herbalism or whatever.'[18]

Back at Jakkalsfontein Kas confronted an unambiguous, terrestrial crisis. The spring rains had failed to maintain their momentum and the farm's limited grazing was so bad that it was affecting the condition of both his and Klue's animals. As in 1913 and 1933, the solution appeared to lie in splitting up the herd and dispersing the cattle on different properties. This had been easily accomplished when three Maine brothers were living on farms amongst Afrikaners whom they had known most of their lives; it was less easily achieved in unfamiliar surroundings. Nevertheless, Kas did have an idea or two, and he duly put them to Klue, who set out to consult the Motsuenyanes.

While Klue was away, Kas tried to progress in the same direction by discussing his problem with Matsolo, one of the wealthier black tenants on a property adjacent to Jakkalsfontein. It proved to be a fruitful exchange, and, by the time Klue returned from Stilfontein, Kas had found a way to split the Maine herd, optimise the use of his family's labour, and avoid any cash outlay for the additional grazing.

It was a rather complex arrangement. Until the grazing improved at Jakkalsfontein, Bodule was to take all of Louis Klue's dairy herd and half of the Maine cattle and establish them around a cattle post on Byl. Klue would pay the Motsuenyanes grazing fees, and Bodule would stay with the herd. Independently, Kas would take the other half of the herd, and Matsolo all of his, and together they would move the animals to the Stilfontein district, on the other side of Klerksdorp, where they would try to procure grazing on a white farm close to Byl; Matsolo would pay the grazing fees, while Kas's share of the rent would be met by his pledging the labour of Matlakala, the recent addition to his family, to the landlord.[19]

After hastily packing a few personal possessions and some pots, pans, and blankets, Kas, Matsolo, Bodule and Matlakala rounded up the consolidated herd—it numbered more than two hundred animals—and set out. Moving a large herd of cattle along a major road with lots of traffic taxed even four adults to the utmost. Matlakala, waving the obligatory red flag demanded by the provincial authorities, moved ahead of the animals to warn oncoming drivers, while Kas, bearing yet another strip of faded cloth tied to a stick, stayed behind to signal motorists approaching from the rear. In the middle reaches of this unpredictable slow-moving hazard, Bodule and Matsolo did their best to confine the cattle to the grass verges of the road.

At Klerksdorp, as previously arranged with Klue, they herded the cattle onto the common, where the animals could graze and gain strength for the second leg of the journey. Under normal circumstances the common was underutilised, but being a drought year, so many white farmers had brought their cattle into town that the facility was hopelessly overgrazed. The municipal officials were therefore looking to expel any cattle whose presence had not been formally authorised, and in the morning, Kas and Matsolo awoke to the sound of trucks stampeding their already exhausted animals off the property: it took an enormous effort to get the animals to regroup, while Kas tried in vain to persuade hardened municipal officials that they were acting under the instructions of a white farmer. Only later in the morning, when Klue arrived on the scene, was headway made. Klue spoke to the officials, disappeared for an hour, and then reappeared clutching a piece of paper that allowed them use of the common for two days.

On the morning of the third day they steered the herd to a low-water bridge over the Schoonspruit, only to be confronted with a second, and more insistent round of municipal harassment. This time they fell foul of traffic officers. The cattle, daunted by the prospect of having to jump the stream at a point where it narrowed, broke ranks and funnelled back onto an adjacent road, where they stood about blocking traffic and staring at the deeply frustrated motorists. Once again there were frantic efforts to round up the animals. The trek through the outskirts of the town and the road leading beyond it was not much easier. Not surprisingly Kas later recalled, 'It was a really demanding exercise. We had to herd the cattle by day and, at night, we slept in the veld.'

After more than a week on the road the herd split in two: Bodule took his cattle and pushed ahead toward Byl and the Motsuenyanes; Kas and Matsolo took the rest and turned right, past the newly opened 'Pioneer Shaft' of the Hartebeesfontein gold mines and then veered south to the Vaal. At a property named Shenfield, Kas negotiated a short-term tenancy agreement which saw Matlakala going off to serve in a Boer home for a small monthly wage while he and Matsolo moved into two broken-down huts.[20]

Six weeks were spent under the blazing western Transvaal sun, relieved by a single visit from Klue, who dumped a bag of 'mealie meal that we ate with milk.' With only the raucous alarms of crowned plovers to interrupt his train of thought, Kas found that the days gave him ample opportunity to contemplate his uncertain future. Each morning his back reminded him that he was on the wrong side of fifty-six and getting much too old for life at a cattle post.[21]

On Saturdays and Sundays life along the river-front took on a different meaning. Kas and Matsolo took turns visiting neighbouring properties, drinking a little beer and talking to the farm workers. On these excursions through thick thorn bush they sometimes heard the barking of a few emaciated 'kaffirdogs.' These were the front runners of migrant workers from nearby Hartebeesfontein, who, having put behind them the tedium of a purely industrial existence in the compound for the weekend, combed the countryside for springbok, hares and any other small game that survived man's invasion of the highveld. Sometimes they came across small parties of burly white miners trundling their paunches down to the river banks for what started out as quiet fishing expeditions but usually ended up as noisy bouts of drinking when the lethargic carp and yellowfish refused to play their part in the day's entertainment. Once a passing young white miner gave Kas half-a-crown and asked him to find him some dagga. Driven by boredom rather than the desire for profit, Kas found an elderly MoSotho woman who let him have a matchboxful of marijuana; the miner rewarded him with a handsome jersey that he wore for several winters. But such diversions were few and far between. On Christmas Day Klue reappeared and told them that the grazing was good enough for them to return to Jakkalsfontein.[22]

New Year's Day 1950 found Kas back amongst his family and with an old but by now deeply disturbing problem. During his absence Leetwane's legs had worsened, and she was once again having to crawl around the house. Kas took her across to Klipspruit to consult Bulara Sello, a highly regarded MoSotho ngaka whom he had first met while negotiating the family's retreat from Sewefontein. Sello thought that Leetwane's training as a herbalist had been deficient and that, if she were to do justice to the ancestral spirit of her grandfather and her new calling, she should undergo further instruction.

Kas left Leetwane at Klipspruit and he did not see her for ten days, during which time Sello instructed her in the arts of diagnosis and the prescription of herbal medicines. When Kas was summoned to fetch her, Sello was willing to accept but a token fee of one pound from someone he saw as a fellow practitioner. Leetwane's treatment, Kas later recalled, had been 'almost for nothing.' But his wife's condition had only slightly improved; it took the better part of a year for her to recover the full use of her legs. Throughout her lengthy recuperation Kas struggled to find it within himself to pay special attention to the needs of a woman who was still coming to terms with the loss of her parents, and fighting to redefine her role and position within the family. Try as he might, Kas found it impossible to accept that his wife, too, was a ngaka.[23]

By the time Leetwane could move about more freely it was summer, and though Kas had no formal obligation to assist either Klue or his own sons, he found it impossible to get away from Jakkalsfontein and investigate the possibility of acquiring property of his own. It did not take him long to work out that he alone stood between a cursing, demanding landlord on the one hand and rebellious children on the other. As de facto foreman, it fell to him to ensure that the family's labour was fully mobilised for an enterprise that was obviously understaffed: in this tension-filled exercise he was vulnerable to attack from both flanks.[24]

Long before dawn each morning, he would rouse Bodule and Mmusetsi to milk Klue's dairy cattle. After they had loaded the truck and the landlord had set off to deliver the milk, they would move into the tobacco fields, where they and any of the other available Maine children laboured for six and sometimes seven days a week. Tobacco—as Bodule, Mmusetsi, thirteen-year-old Pulane and ten-year-old Nkamang could testify for many years thereafter—was as demanding a crop as any black worker ever came across.

Hundreds of seedlings raised in the shadows of the nursery had to be carefully separated, hand transplanted, and watered individually. Once safely established, the young plants had to be protected from aphids and similar pests by being sprayed with a noxious chemical substance which, so the local inhabitants claimed, not only 'made you cough' but helped to account for the chronic shortage of labour on tobacco plantations. When the tobacco reached maturity after several months of mollycoddling, and when the humidity reached its optimum level, virtually all other work on the farm ceased as the leaves were carefully stripped, sorted by size, given a preliminary grading, baled, cured and then 'sweated,' before being wrapped and packed.[25]

In short, tobacco drained energy from black bodies just as quickly as a plughole sucked water from a high altitude bathtub. After a morning in the fields, Bodule and Mmusetsi then had to move across to the vegetable gardens where they and the children dug, planted, transplanted, watered, picked, packed and loaded vegetables, which their father was expected to deliver to the compound at Dominionville each Tuesday morning. At dusk, having worked for ten hours with only a short midmorning break, the boys reported back to the dairy, where they had to do the evening's milking before going home.[26]

This punishing routine, seldom broken by so much as a Sunday's respite, ground on throughout the long, hot summer. As Kas moved about the property, he sensed that family loyalty was being stretched to the limit, and yet there was neither time nor space to make alternative arrangements. There was nothing that he could do but wait, wait for a harvest that looked distinctly unpromising. When Klue gave him several loads of yellow peaches to sell outside the mine compound at the height of summer, he struck back at what he considered gross exploitation by handing each customer a little less fruit than he was entitled to, and disposing of the resulting 'extra lots' against his own account. The additional cash thus generated he used to buy tea, coffee and sugar at Yelman's General Dealer's Store.[27]

The fact that Kas's children were young adults who had minds of their own only exacerbated the difficulties inherent in any labour tenancy agreement, and made new demands on his diplomatic skills. Kas came home one afternoon to find that not only had a distraught Matlakala absconded from her position in Stilfontein, but she had been followed home by a very angry Jannie Barnard.

At one level it was a familiar story, a tale of racism and poor communications culminating in domestic violence. Like many young white mothers, Mrs. Barnard apparently wanted the privilege of domestic help without forfeiting her sacred racial taboos; she had therefore told Matlakala that no '*kaffermeid*' was ever to touch, let alone carry, her baby. Matlakala, offended at being thus characterised, had pointedly ignored the baby when next it cried. This had infuriated the mother, who picked up a broom and threatened to assault Matlakala. But Matlakala, made of sterner stuff, did not see herself as a victim, and before Mrs. Barnard could hit her, she surprised the woman by grabbing hold of her hair. Mrs. Barnard twisted herself free and burst into tears, then ran off to lock herself in the bedroom. When Mr. Barnard returned from work later that day she told him of these happenings; when Matlakala saw him emerging from the bedroom brandishing a *sjambok*, she ran away and he had followed her.

Here was an unbelievable mess! As far as Kas could see, it might all have been avoided had the white woman possessed even a few words of SeTswana or had Matlakala's newly acquired Afrikaans been a little better. Even as Kas was listening to Matlakala's version of the story—frequently interrupted by observations from a still irate Barnard—he was aware of Leetwane and Lebitsa moving toward the centre of the circle forming outside his shack. Matlakala aggravated the situation by taunting young Barnard—beneficiary of a tenancy agreement entered into by his father —by observing that she, Matlakala, was only working as a servant for his wife because her father happened to own more cattle than he did! Another interjection from Barnard suddenly caused Leetwane and Lebitsa to adopt a very threatening attitude. As black women, they knew what it was like to work in a white woman's house. What did men, any men, know of such things?

Kas sensed that he was losing control of the situation. An incident of crass racism in a far-off household was by now causing a cascade of sparks to fly off a dozen other points of conflict about issues of age, class, gender and patriarchy. It was only with the greatest difficulty that he regained control and delivered his judgment. Barnard, he suggested, had been wrong to try to discipline Matlakala without first referring the matter to him, her father, for a decision. She, in turn, had erred by abandoning her employer and should return to her place of work immediately. This shaky compromise—which studiously avoided the very issue of racism, that had caused Kas to leave Labuschagne—held the day. The Maine women were pleased to note that there had been no unqualified condemnation of Matlakala's behaviour and pleased that the Boer had been reprimanded in their presence; Barnard did not leave the scene empty-handed; and the judge was pleased that his domestic authority was still intact.[28]

It had been a close shave. Jakkalsfontein, which had been meant to give Kas the opportunity to catch his breath before the family pushed on toward the Motsuenyanes, was proving to be anything but restful. Never had it taken so long for May to arrive, and when it came the harvest was bitterly disappointing. The three-acre fields yielded derisory amounts of maize and sorghum, which disturbed Kas even more when he thought of how hard Bodule and Mmusetsi had worked throughout the savage summer for wages of only twelve shillings a month. He found it difficult to look his sons in the eye. Bodule seemed preoccupied, Mmusetsi resentful. Yet his own position was not much better, and had it not been for the twenty-six bags of maize that he had managed to salvage from his venture with Bully van Vuuren, the situation would have been even bleaker. The family needed a place of their own or, at very least, a more equitable partnership, and urgently.[29]

Then, just as he was about to act, Bodule disappeared. One Saturday morning he was simply gone. Kas did not give the matter much thought until the afternoon, when they discovered that though Bodule's hunting dog had reappeared, his bicycle was still missing. At daybreak on the Sunday Kas gave Mmusetsi a horse and told him to scour the neighbouring properties, but nobody had seen Bodule at any of the haunts usually frequented by local youngsters or the black miners from Dominion Reefs. By Monday night everybody was deeply concerned (with the possible exception of Matlakala, who had just rejoined the family after completing her spell at Shenfield). The next morning, Mmusetsi was again sent out and once again returned with no clue as to his brother's whereabouts. Not even Motheba, who had come across from Harrisburg to visit the family, could offer an explanation that Kas found reassuring.

Within the family, however, at least one person *did* know what had happened to Bodule. Matlakala's silence was founded on knowledge and sympathy. Before Bodule's departure he had singled her out as someone who would understand his predicament but not divulge his intentions. He asked her not to tell Kas about his plan to go to the Triangle and return with his lover, since he thought his father would say he was 'still too young and not yet fit for marriage.' But, as not even Matlakala knew, the relationship between her half-brother and Dikeledi had long since reached the point of no return. Dikeledi was several months' pregnant.[30]

On the Tuesday night, after an extraordinarily long and uncomfortable bicycle ride from Bloemhof, the exhausted couple inched their way toward the perimeters of the homestead and then stopped just beyond the light coming from Kas's shack. For several minutes Bodule monitored the movements around the yard, watching as people prepared to settle down for the night; then he caught sight of Motheba. A single whispered word drew her into the shadows. He exchanged a few hurried phrases with his grandmother. She was sympathetically disposed to them and well understood the dangers that her son Kas posed.

Motheba disappeared back inside a hut and then, after what seemed like an eternity, Matlakala walked out boldly into the night and called for Dikeledi. She steered her gently toward the hut where Motheba was; the tension and relief of making that short walk across the yard were almost too much, and Dikeledi was weeping freely. Once safely inside she was surrounded and supported by a phalanx of women including Motheba, Leetwane and Lebitsa. Bodule had done the 'manly' thing and retrieved his lover, but it was the courage of the Maine women that transformed a mere bride into a wife and helped give her a position in the family.[31]

Because his misgivings were so well known, Kas was kept away from Dikeledi and given the task of welcoming home his missing son. But even this duty, which he would normally have undertaken enthusiastically, was now the subject of hidden, perhaps only half-conscious, reservations. In some vague and inchoate way Kas sensed that his relationship with his two oldest boys would never be the same again. Bodule, who had hitherto been at pains to accommodate his father's every wish, had taken the first tentative step toward asserting his independence. How could Kas ever find it within himself to forgive this daughter of the Teeus? And what of Mmusetsi? How could a father dispose of bridewealth for a younger son when he had not yet secured a wife for his oldest? These unexpected, unwelcome developments lacked social symmetry.

Kas said nothing, rummaged through his anxieties yet again, and then decided to call a halt to any further self-indulgence. Personal qualms had to be offset against public duties. Others were involved, including another set of parents, and he had to think beyond his own predicament. The next morning he sent Mmusetsi off to the Teeu homestead on horseback to tell them that their daughter was safe with the Maines, and that he would be visiting them soon to determine how many cattle should be driven to Vlakfontein as bohadi. He went to the kraal and picked out a sheep for slaughtering while some of the women rushed off to Yelman's Store at Dominionville to buy Dikeledi clothes for her wedding. By nightfall she was Bodule's wife not only in his son's eyes, but in the eyes of virtually all the witnesses, and she went to bed determined 'to wake up early in the morning and start working to show her qualities as a daughter-in-law.' But she could work as much or as little as she pleased; there was one person she would never convince.[32]

A few days later Kas set out with Lebitsa and Dikeledi for Vlakfontein. The road back to the southwestern Transvaal was by now very familiar to him, but Kas did little talking, choosing instead to concentrate on a proposition taking shape in his mind. If Abner Teeu would accept eight cattle as half the amount due to him as bridewealth for Dikeledi, then Kas would offer him an additional sum of fifty pounds in cash as bohadi for the younger sister, Disebo, who could then marry Mmusetsi. Dikeledi might then return home for the birth of her child and be taken care of by her family until the Maines had the eight outstanding cattle. When Dikeledi was ready to rejoin her husband, Kas would make the final payment on the bohadi and both sisters could join the Maine homestead.[33]

Such a linked arrangement, if the Teeus agreed, had much to commend it. Not only would it be seen to accord more closely with traditional practices but it would help to restore his troubled relationship with his sons. It could also help to prolong his influence over the boys, and it might even help keep the peace between them. The Teeus, in turn, would be coupled to an economically progressive family and their daughters could keep an eye on each other's well-being. There was something for both families and, more especially, for both patriarchs.

With the assistance of an avuncular intermediary on the bride's side, the menfolk from both families soon reached agreement on the broad outlines of linked marriage contracts for the two Teeu daughters. As Kas had suspected, Abner Teeu could readily see the sense of it, though his wife, Amelia, perhaps knowing that Disebo's heart lay elsewhere, was at first less pleased. This dragged on for several days. Nevertheless, in the end consensus was reached and the Maine delegation returned to Klerksdorp. This time, however, it fell to Dikeledi to remain silent on the trip home. Nobody had bothered to ask her what *she* thought about the prospect of having her younger sister in the Maine household, but, as it happened, she had a very strong opinion on the matter.[34]

At Jakkalsfontein, they were met with the news that Klue had been asking for Kas. The landlord, too, had been giving serious thought to the contract and for rather different reasons, had arrived at a similar conclusion: it was unlikely that they could work together for a second season. While Klue was happy to use the Maines' labour, he did not think a small property dedicated to tobacco cultivation could sustain his tenant's cattle as well as his own dairy cows. He offered to help by introducing Kas to a few neighbouring landlords when the time came, but once spring arrived the Maines would have to move on. For the second time in as many seasons Kas found himself responding to events rather than anticipating them. It had never been the pattern of his life, and he thought it should stop.[35]

*

This time Kas did not hesitate. Bodule was reminded of the complex history of his family's engagement with their BaFokeng and BaKubung forebears, given fifty pounds in cash, put on a bicycle and told to investigate the possibilities of buying land in the northern Orange Free State. But that section of the BaKubung which he had been told to find had long since abandoned the terrain beyond the Vaal and gone back to their ancestral home in the western Transvaal. A week later Bodule was back at Jakkalsfontein reporting on his failure and outlining the need to visit the followers of Monnakgotla's BaKubung, who now lived north of Potchefstroom.[36]

Within days Bodule was on the road again, this time cycling north for about fifty miles, toward the heart of BaFokeng country where late one afternoon he was introduced to a man named Malefo who seemed to know a good deal about the affairs of the BaKubung. Malefo told him that the members of the BaKubung to whom the Maines had historically pledged their allegiance were, along with their old factional adversaries, established on the farm Elandsfontein, near Boons, a small rail-siding south of the town of Rustenburg. With Malefo as his guide, Bodule went to the village of Molote. After being grilled by the chief about his family's ties to the BaKubung, he was allowed to make a cash deposit on four residential stands on behalf of his father, Kas Maine.[37]

The cold print on the receipt that recorded this transaction masked the seriousness of what was at stake. In effect, it had taken the BaKubung one hundred and twenty years to forgive Seonya Maine and his descendants for his having deserted them at the time of the *difaqane* and seeking refuge in the far-off Maluti Mountains. After half a century of picking their way across the chequerboard of the subcontinent, stepping only on white squares assigned to Boer farms, the Maines at last had one foot placed securely on a black square belonging to the BaFokeng. But balancing on one foot is a tricky business. Although four residential stands costing ninety pounds would one day give Kas and his three sons housing sites, the question remained whether or not they would also have access to enough communal land and grazing to sustain their farming operations. Kas could understand why it was that the 'Secretary to the Tribe' who had taken the down payment of fifty pounds from Bodule had told him that the Maines should not hurry to reach Molote.

Back at Jakkalsfontein Bodule scarcely had time to recover his breath before he was again out on the road. The harvest was in, and with Dikeledi's pregnancy well advanced, the moment had arrived for him to take his wife home so that her child might be born amongst her own people. As agreed, Bodule drove eight cattle back to the Triangle as bohadi, and he also handed his father-in-law the fifty pounds that would secure Disebo as Mmusetsi's wife. He did not linger but returned to help his two brothers with Klue's vegetables, which, unlike maize, knew no season. A few weeks later he was delighted to get a message that, on 25 July 1950, his wife had given birth to a son whom they named Sekgaloane.[38]

Bodule, delighted to become a father so soon after becoming a husband, used the excuse of his son's birth to engage in some unbridled celebration. But the news of Bodule's changed status was not greeted with equal enthusiasm throughout the homestead. Kas was correct and polite but hardly overjoyed. For him the birth of the baby, which confirmed that Bodule was no longer solely his son, raised doubts as to its paternity, for he thought it had been conceived under mysterious circumstances. From the moment he first saw the baby Kas was convinced that the child bore a striking resemblance to a Triangle sharecropper, Piet Mokgosi. He said nothing; what could he say? Instead, he watched and waited; when the young boy became an adolescent, he was more and more convinced that Sekgaloane had not only the physical attributes of Piet Mokgosi, but also his 'behaviour and deeds.' Bodule, his Bodule, might have accepted responsibility for the Teeu woman's behaviour, but what other reason could Piet Mokgosi's mother have had for taunting Kas with the observation: 'You are looking after my daughter-in-law'?[39]

Still, most of the family were pleased to have an excuse to celebrate and grateful for the chance to grease the slide into the marginally less demanding routines of the off-season. Leetwane, her health much improved, busied herself looking after her chickens which, along with the occasional payments in cash or kind that came from local women whom she assisted with her services as herbalist and midwife, gave her a small independent income. She was also called upon to 'work' on young Pulane who, having menstruated for the first time, needed to be smeared with red clay so that she could continue to undertake her herding duties without flouting the taboo that most of southern Africa's black societies had about women working with cattle. Lebitsa, meanwhile, used the shorter days to comb the countryside for pieces of dried cattle dung; along with a few thick tobacco stalks, dung was the fuel that held off the highveld winter.[40]

For all the security that came with these seasonal routines, there could be no disguising the truth that the Maines were now well away from their familiar haunts, living through uncertain times and vulnerable economically—a theme Thakane voiced in a letter she wrote to Kas from Schweizer-Reneke, enquiring how they were getting on in 'that strange part of the world.' The black workers around Jakkalsfontein had less money than did sharecroppers in the Triangle and this, together with his having to compete with the services offered in a town like Klerksdorp, meant that Kas was no longer besieged with requests to do leatherwork, shoe horses, or repair damaged agricultural equipment. Indeed, were it not for the pesticide that 'made people cough,' his practice as a ngaka might also have collapsed. He treated a few men on tobacco farms, but here the demand for his services seemed to be sharply down. Poverty bred only poverty.[41]

A decline in off-season income in a year which had also seen a harvest failure was doubly harsh. Moreover, an unknown disease had carried off quite a few sheep. The stands at Molote had thus far involved Kas in an outlay of seventy-five pounds; in addition to this, he had had to find fifty pounds to secure Mmusetsi a wife; Klue had insisted that he sell fourteen donkeys, and he had contributed eight cattle toward the bridewealth for Dikeledi. The retreat from the Triangle had cost the Maines dearly, and the family had lost much of the economic momentum that had carried them through the 1940s. It had all seemed so much easier when the boys were younger and the girls were not yet possessed by ancestral spirits. The future now depended largely on Kas having a good season with a white farmer whom Louis Klue had yet to introduce him to. For perhaps the first time in his life Kas understood why it was that those poorer black sharecroppers who moved each season seldom accumulated much: they were constantly adjusting their livestock holdings and draught power to meet the needs of new landlords.

In late August 1950, Klue drove five miles to Rhenosterhoek, yet another small farm lying just off the main road between Wolmaransstad and Klerksdorp, and introduced Kas to Jack Adamson, a short fellow with a ruddy complexion who said he worked as a winding-engine driver at Dominion Reefs. At first glance Kas saw nothing in Adamson to commend him as a landlord, but he knew that the countryside produced hybrids at least as interesting as those of the city, and as soon as Adamson spoke, he recognised him as one of those Afrikanerised Englishmen who occasionally cropped up on the highveld to remind the unwary of the complexities of South African history. More doubts were dispelled when Adamson told him that he had been hiring a 600-morgen section of Rhenosterhoek from the Koen family for several years and was always looking for experienced black sharecroppers. Six hundred morgen was almost twice the size of Klue's property, so Kas quickly expressed interest in becoming part of the enterprise.[42]

Adamson took care to keep any proposition he made well within the bounds of the law. In a peri-urban area, where cheap black labour was needed by any number of undercapitalised white farmers and men like himself who worked in the mines and on the land part-time, sharecropping was disapproved of even more strongly than in remote districts like the Triangle. So he camouflaged the sharecropping component in the

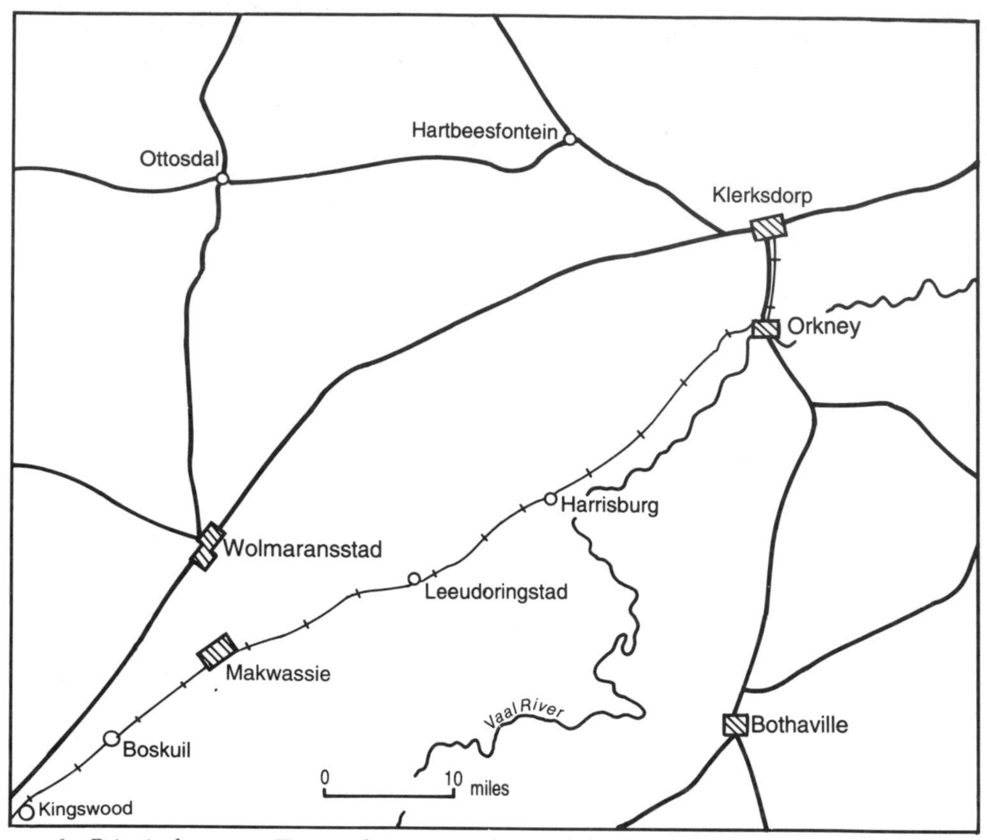


6 Principal western Transvaal towns, northeast of the Triangle

contract in a way that Kas had heard of but never experienced. If Kas allowed him to 'hire' the two spans of oxen that came along with Bodule's and Mmusetsi's labour, and gave him the necessary seed, Adamson would guarantee him two bags of grain for each ox—that is, fifty-six bags of maize. Over and above this, Kas, as owner of the oxen, would be entitled to a half-share of the harvest and allowed to graze his livestock on the property without incurring any further fees. This complex formulation apparently circumvented official charges of sharecropping by meeting the terms of a local court ruling that dated back to several years before World War II.[43]

Kas accepted the terms, and within days the Maines moved to Rhenosterhoek, where they linked up with the three resident sharecroppers named Molapo, Phopho and Nku; who told them that an impecunious white miner with a span of cattle had joined the consortium under virtually the same conditions as the rest of them. The spring rains were promising, and before long, a messenger on a bicycle appeared and started shuttling back and forth between the mine shaft and the farm. This was a truly extraordinary set up: the messenger told the sharecroppers what to do on the land while, five miles away, the landlord earned his wages on a shift in the engine room at Dominion Reefs, moving men and machinery between the 'bank' of the mine and the very bowels of the earth.[44]

No rhinoceros had set foot on Rhenosterhoek for a hundred years; the rhinoceros, like the farm itself, belonged largely to a preindustrial era when the highveld had given man and animal alike space to roam the banks of the Vaal River and live directly off the land. By the twentieth century, the area north of the river had long since been alienated, parcelled up into smaller lots and worked to the point where it struggled to sustain any form of life. The animals, hunted to the point of near-extinction by man, had been the first to go. All that was left now were people struggling to coax a living out of tired soil in the hope that they could stave off the day when they too were driven off the land and forced into the mines and factories of the emerging industrial order.

If the five men who had chosen to hold out at Rhenosterhoek had anything in common, it was an admirable stubbornness, a love of the land, and richly varied cultural backgrounds of the kind found on many western Transvaal farms. Jack Adamson was an 'English' Afrikaner and Kas Maine an Afrikanerised MoFokeng. Paul Molapo, Piet Phopho and Jantjie Nku, respectively a MoTshweneng, a MoTswana and an UmXhosa—as their first names and religious affiliations indicated—were already partly deracinated and Afrikanerised. As usual on highveld farms, time and social isolation undermined notions of cultural purity.

Each of these men longed to own property of his own, but in no heart did the desire to succeed burn more fiercely than in that of Jack Adamson. Like most poorly educated Afrikaners working in the gold mines, his passion for land reached almost obsessional proportions; an endless succession of wage packets was diverted into hired plots, second-hand 'bakkies,' boreholes, pumps, tractors, a few dairy cows, some sheep and the odd goat. The countryside around Klerksdorp was filled with such rootless semi-urbanised Afrikaners, gaunt middle-aged men wearing quaint old-fashioned hats and black leather shoes, who did not have the skills to go forward into an industrial economy or the capital to get back onto the land. This struggle of second- and third-generation Afrikaner males to get back onto the land is one of many still unwritten chapters in South African history. In the 1950s Afrikaners marooned in Klerksdorp's 'Old Town' section were often referred to as *bokslagters*—goat-slaughterers—who, it was believed, lacked the culture and refinement necessary for an English suburban existence. The Adamsons, who ran a small boarding-house in Dominionville, knew many such men, and Jack Adamson understood their dreams and their nightmares.

Jack Adamson's grandfather—Sweinstein Adamson—was an English navvy who had come to South Africa during the late 1880s to work on the railway line from the Cape Colony to the Witwatersrand goldfields;

he had stayed on to fight on the British side during the South African War. After the war his oldest son obtained a position as foreman-cum-bywoner for an attorney who owned a flock of two thousand sheep on a farm near Kroonstad. There he married an Afrikaner woman and raised a family of nine children which included Jack, born in May 1917.

Jack Adamson attended the local farm school until he was twelve and then abandoned the classroom to help his father take care of the attorney's sheep. When his father lost his position as foreman with the onset of the Depression, the family moved to the far northwestern Orange Free State, just south of the Vaal and near a large coalfield. For several months, while two older brothers tried to establish a blacksmith business in Vierfontein, Jack helped his father cut and saw firewood from a nearby bluegum plantation. They sent the wood to the Johannesburg market. But by mid-1930 the bluegums were depleted and father and son were once again without work. At this point the older Adamson brothers decided that the best way to assist the family was to dismiss their 'coloured' assistant in the workshop and give his position to fourteen-year-old Jack.

At the smithy, continuities of class proved to be stronger than brotherly compassion. Jack earned the same thirty shillings a month that his predecessor had. He found this hurtful, but it did not distress him nearly so much as his brothers' unwillingness to pay him overtime. During the rainy season, prosperous white farmers from Viljoenskroon often expected him to work well into the night repairing the implements they needed for their planting, but he got no extra compensation. Disillusioned, he left the smithy and worked in a local trading store owned by a Jewish family. Soon he moved across the river to live with more sympathetically disposed kin in the Klerksdorp district.

In 1933, at the age of sixteen, Jack Adamson, along with two hundred other 'poor whites' from the western Transvaal, enlisted as a navvy earning three shillings a day helping to construct the railway line and cuttings around Boskop, northeast of Potchefstroom. He worked there until the newly opened Dominion Reefs Mine took him on as an apprentice in 1934; he qualified as a hoist-driver in 1937. Throughout the Depression he hankered after a life on the land, but only after the passage of the Marketing Act and the boom in agricultural prices during World War II was he able to hire the land from the widow Koen.[45]

When Kas first met Adamson in late 1950, Adamson was a bespectacled man in his mid-thirties, a 'poor Boer' striving to become a landlord in his own right. Kas, on the other hand, at the age of fifty-six, owned four residential stands and was in possession of far more livestock. The disparities gave him no concern. Had not the Triangle had its fair share of 'rich' black tenants and 'poor' white landlords? Nor was he put out by the fact that while he and Adamson would share the proceeds of the maize crop, Bodule and Mmusetsi would have to make do with a strip of sorghum only three yards wide at the edge of the fields. Good sons worked freely for their fathers. His conscience was clear. Had he not just made provision for the future of all three of his sons at Molote? He was sure that the stay at Rhenosterhoek, which was predicated on sharing grain rather than working in tobacco fields, would be trouble-free.[46]

Throughout the spring of 1950 and the summer of 1951 the Maine men saw little of Adamson. He made brief appearances when he was working night-shifts, or on occasional public holidays, or over weekends. This pattern suited his tenants who, once the crops were established and the visits from the bicycle messenger tailed off, could establish their own routines. The landlord was also most obliging when it came to issuing the passes that allowed them to visit the nearby hamlet of Hartebeesfontein (or, as the Tswana-speakers would have it, Tigane). Nature helped to fashion a pleasant interlude, for the rains fell regularly and the older Maine boys drew satisfaction from their strips of sorghum. Even the youngsters seemed to have new spring in their heels. Freed from having to bend over the stitching on Klue's quilted vegetable patch, they straightened up and seemed to enjoy the few minor chores that came with herding cattle on a small property fenced into camps.[47]

In mid-March, Kas and Bodule paid a visit to Molote, where they were introduced to Samuel Baloe, Secretary for the BaKubung Tribe, taken to see the four stands allocated to the Maine family, and shown the communal grazing available to residents. Baloe reminded Kas that £15-17s-.6d. was still outstanding on the stand earmarked for Mmusetsi, and once again impressed upon him that the Maines should stop working on the white farms only when they were ready to contend with the more congested conditions of what the government and its supporters were pleased to refer to as a 'black spot'—a small black spot on what was otherwise a large white sheet. Kas was more concerned than ever about the pressure on the land that came with communal grazing; still, he raised enough cash for Mmusetsi's stand before sending Bodule back to collect yet another receipt from Baloe.[48]

Shortly after this visit, Adamson appeared with a request that Kas get one of the younger Maine women to help his wife in the boarding-house kitchen. Kas decided, for the second time in as many years, that Matlakala should be the one to help meet the family commitments. Although still smarting from her Stilfontein experiences, Matlakala strode off toward Dominionville to join Mrs. Adamson. Kas and Bodule meanwhile prepared the horses and cart for a long ride back to the Triangle, where they were to collect Dikeledi, her baby, and the woman whom Abner Teeu had pledged as Mmusetsi's bride.

Unbeknownst to them, there was trouble brewing at Bloemhof. In the months following her return from Jakkalsfontein, Dikeledi had thought more about her younger sister marrying Mmusetsi and, the more she thought the less she liked it. Surely it would be inappropriate for her, as the older of the two sisters, to be relegated to junior status within the homestead of her parents-in-law? How could she, a woman with a child of her own, be expected to defer to a younger, newly married sister? Dikeledi went out of her way to paint an unflattering portrait of Mmusetsi and spent a great deal of time and effort convincing her sister of the diplomatic problems that such a marriage would entail. All of this, together with Disebo's own reservations, was bearing its first bitter fruit when Kas and Bodule came to get her. On the morning scheduled for them all to return to Rhenosterhoek, Disebo ran away from home.[49]

Kas, angry and frustrated, demanded the immediate return of the fifty pounds he had paid in bohadi, but Abner Teeu, although highly embarrassed, merely suggested that if the Maine family believed they were owed anything they should recover it from Dikeledi. Kas found this almost incomprehensible: here was a woman who had eloped, her own bridewealth still only half-paid, who had encouraged a sister to abscond, and a father who refused to accept responsibility for the actions of either of his daughters!

On the journey back to Klerksdorp Kas convinced himself that he should refuse to pay the outstanding instalment on Dikeledi's bridewealth until he had recovered the fifty pounds advanced for Disebo. He was not surprised to learn that Disebo had been consorting with 'Zionists'—Christian sectarians whom he associated with licentious behaviour and organised hypocrisy; in his experience they were people whose 'faith healing' made them disdainful of traditional medicine by day but who then approached him by night for herbal remedies. The Teeu girls were obviously a bad bunch. Just looking at Dikeledi's baby was enough to remind him of her dalliance with Piet Mokgosi, and Disebo was probably a sort of 'prostitute.' Mmusetsi was no doubt better off without her.[50]

But the flames of anger were coloured by the fuel of patriarchal guilt. For nearly two decades Mmusetsi had, in his own bungling way, done everything a father could possibly expect of a son—ploughing, planting, harvesting, herding, labouring at his behest at all hours of the day or night, whether on the arid salt pans of Sewefontein in the depths of winter, or in the green vegetable patches of Jakkalsfontein at the height of summer. At the very moment when he should have been rewarded with a bride, his father returned home empty-handed. And all this was made much worse by that ill-considered purchase of the suit and bicycle for Bodule two seasons earlier. Kas had failed, nay, betrayed, his eldest son.

By the time they reached Rhenosterhoek, the fever of anger was making way for the chill of pain, a pain that grabbed at his entrails when he witnessed Mmusetsi's uncomprehending response to the news that Disebo had absconded rather than go through with the marriage: the boy retreated into his innermost thoughts and then, just when it seemed that they would lose contact with him forever, re-emerged from his self-imposed isolation and went about his chores with the same resigned rhythm that had always characterised his work. For Kas it was hard to know which hurt most—Mmusetsi's lingering bewilderment or his willingness to resume work. Then, as if somehow to underscore Mmusetsi's mute loyalty, Kas was plunged into another round of conflict by one of his increasingly articulate, self-confident and rebellious daughters.

Matlakala got embroiled in a kitchen quarrel at the boarding-house and, when Jack Adamson threatened to assault her, packed her bags, made her way back to Rhenosterhoek and refused to return to work. Given the powerful gender solidarity that such disputes aroused in his household, Kas did not pry into the origins of trouble, which he conveniently consigned to the category of 'women's affairs.' But not all men drew the same useful distinction when it came to domestic problems. Adamson unexpectedly materialised, demanding that Kas ensure Matlakala's immediate return to the boarding-house on the grounds that her labour formed part of their original agreement.

Sensing that Adamson was probably just as entangled in domestic complexities as he was, Kas tried to ease the tension in a man-to-man discussion asking why it was that Adamson was so intent on getting involved in 'women's affairs,' but this failed to open up the space he was looking for. So he decided to contest the assertion that Matlakala's labour had in any way formed part of their contract. This vigorous defence raised the temperature of an already heated debate, which ended only when Adamson stormed off to summon the police.

The constable from Hartebeesfontein who came back with Adamson was reluctant to get involved in 'women's affairs.' He was also taken aback when Kas asked him to open a docket so Adamson could confront the Maine lawyer in open court about the precise nature of their contract. This was exactly the sort of situation against which Willem Griesel had once cautioned Walter Moormeister. The very fact that sharecropping was technically illegal made it a double-edged sword. If not handled with discretion, it could damage landlord and tenant alike. Adamson, seeing the arc of the sword and the power of the arm, decided that discretion was the better part of valour. He promptly retreated from his demand that Matlakala return to the boarding-house, and this restored calm. The policeman left, warning Adamson against employing 'rich kaffirs'; it was wealth alone that gave black sharecroppers the confidence to question the authority of white landlords.[51]

Rather surprisingly, this storm did no permanent damage to the communications system linking the winding-engine drivers' compartment at Dominion Reefs to the fields at Rhenosterhoek. Within days the messenger service between landlord and tenant was fully restored and no further mention was made of Matlakala.

With the first reasonable summer since leaving Sewefontein behind them, the cattle had picked up in condition, for which Kas was grateful, given the problems that continued to plague the sheep. Soon after their arrival at Jakkalsfontein the cross-bred Merinos had started dying in large numbers. Kas had never known anything quite like it, but neighbouring white farmers told him it was a common occurrence when an 'imported' flock adjusted to new grasses. Newcomers, they said, were best advised to sell their animals and replace them with sheep reared in the district.[52]

But it was too late. By the time they got round to shearing the sheep for the first time since leaving the Triangle, a flock that had once numbered two hundred and thirty had been reduced to little more than one hundred. And there were other problems with small livestock. It was difficult to keep pigs on a small property fenced off into grazing camps since the animals had no space in which to scavenge; he had to sell the last of the hogs that had been with them since Vaalrand. The destocking that was set in motion by the retreat from the Triangle continued apace throughout the Maines' stay at Jakkalsfontein and Rhenosterhoek.[53]

News from the fields was more encouraging. Once the women and children had brought in the last of the harvest and the landlord had paid for 'the hire of the oxen,' Kas had more than fifty bags of maize and a hundred and thirty-six bags of sorghum, a reassuring profile. But even as his eyes stroked the stack, he was uncomfortably aware that what he was surveying was the outline of his sons' labour. He had met his obligations to Bodule, but Mmusetsi unnerved him; his very presence seemed to remind him of his inadequacies as a father. He resolved to set matters right between them as soon as the harvest had been disposed of.[54]

But before he got to do anything about the harvest, Jack Adamson pitched up out of the blue to tell them that the mine was closing down; the grand old lady had finally run out of gold. In South Africa maize could no more do without gold than a plant could survive without water. A fair number of people were going to be without work, and under the circumstances Adamson could no longer afford to hire the property; the farm would revert to one of the widow Koen's sons. The new landlord had assured him that the tenants were welcome to stay on at Rhenosterhoek, but he had a reputation for cruelty and no sharecropper was willing to work for him; Adamson offered to make an alternative proposal before spring.[55]

For the Maines, already partially mobilised for a possible haul to Molote, the news was less of a shock than it may have been for established residents. Still, even Kas found it gloomy to think that within weeks they would once again be out on the road, on their way to an unknown destination. Pragmatic to the core, he feared not so much the daunting future as the logistical horrors. The disruption and the cost of the move from Bloemhof to Klerksdorp were still recent.

What seemed like a solution to the logistical problems came via an unexpected route. The harvest had been brought in later than usual, and when Adamson had got around to marketing the crop, the neighbouring full-time farmers had filled the limited storage facilities at the Hartebeesfontein depot of the co-operative. The only thing to do was to take his harvest in to the Klerksdorp depot of the Sentraal-Wes Landboukoöperasie (S.W.L.), and Adamson called on the assistance of one of his Afrikaans-speaking friends who owned a truck. It was a rather ramshackle 1937 model Ford; but to Kas it seemed like the answer to all his prayers. He knew nothing about trucks, of course, and so, as with most important matters relating to farming, he first discussed the matter in private with Bodule. Surely at this stage of the family's long trek the purchase of a truck made more sense than of, say, a tractor? Unlike a tractor, the truck could earn its keep throughout the year, and would save them the cost of railing the heavy agricultural equipment to Molote. Kas plucked up the courage to ask the man whether he was willing to sell the truck.[56]

The fellow was more than willing to sell the vehicle, but he warned them that to be relicensed in the name of a new owner, the truck would need a 'certificate of roadworthiness,' in order to qualify for which, it would require professional attention. Kas, more cautious than ever in such uncharted water, made a new certificate a precondition of sale when he submitted a written offer to buy the vehicle for a sum of two hundred and thirty pounds in cash. On 28 August 1951, Kas handed the owner the sum of ten pounds as a deposit and first payment on the truck.[57]

Bodule spent the weeks following these preliminary exchanges in obtaining a learner driver's licence. Once the truck had been repaired, tested, and granted the certificate, he and Kas went into Klerksdorp to pay the balance and take delivery of TZ 225. With Kas beside him, Bodule started the engine, the truck lurched into motion and then clattered off down the road in the direction of Wolmaransstad. They owned a truck! Locked inside the cab for the journey home, the two occupants were drawn even closer together as they shared the delight of being the first in the family to learn what the fourteen-year-old Ford was still capable of doing. At the Rhenosterhoek turn-off they were greeted with squeals of delight as Bodule slowed down to allow Mosala, Pulane and Nkamang to clamber aboard for the short ride back to the homestead. Outside the huts they were met with yet more laughter and noise. But as Kas climbed out and the door of the cab swung closed behind him, he caught sight of half-hidden figures. Mmusetsi cast an anguished glance in the direction of Kas and Bodule, turned, and walked off toward the kraal. Kas remembered that he had promised himself to set things right between them.

Two days later, Kas asked the boys to help him load the sorghum on the truck, and in a gesture calculated to induce a sense of male solidarity, he made a point of asking Mmusetsi to join him and Bodule for the ride into Klerksdorp. His successes at Vaalrand and Sewefontein had done much to dispel any residual fear of distant and impersonal institutions, and he felt sure that he could sell his crops in the best possible market. In town they asked for directions to the premises of a miller whose name had been given to Kas by a farmer near Jakkalsfontein. Bodule drove the Ford onto the weighbridge and, after the usual clerical delay, Kas was given a cheque for more than a hundred pounds, which he cashed right away. On the way back he counted and recounted the notes, mentally setting aside a sum for something he knew had to be done without delay. The spring sun soon faded, and it was dark when they reached Rhenosterhoek.[58]

The following morning, as soon as the family was safely into the daily routine, Kas called for Mmusetsi to inspan the horse and cart, and together they set off for the shop at Rhenosterspruit. Two years had elapsed since that day in Bloemhof when Kas and Bodule had called in at Ebrahim's Cash Store and bought the bicycle; the moment was at hand to make good the damage inflicted on the son whom he and Leetwane had hoped would 'wipe away their tears' after they had buried Mosebi under those bluegums at Hartsfontein nearly thirty years earlier. Inside a small Asian store that catered to local farm labourers and the occasional migrant worker from Dominion Reefs, Kas encouraged Mmusetsi to fossick around in the material treasures so long denied him.

Eventually an impatient Indian youngster hastily compiled a list of the items that had most appealed to the sharecropper's son, did a careless calculation, and handed Kas a scribbled note. Kas bought everything on the list for his oldest son.[59]

| | |
|---|---|
| 6 Yds Material | 1 - 13 - 0 |
| 3 Yds do | 10 - 6 |
| 1 Hat | 17 - 6 |
| 1 Handgloves | 11 - 0 |
| 1 Pair Shoes | 1 - 17 - 0 |
| 1 ½ Yds Nylon | 13 - 6 |
| 1 pair stockings | 9 - 0 |
| Lace | 2 - 10 |
| 1 ring | 3 - 6 |
| Dressmaking | 2 - 0 - 0 |
| 1 pair shoes | 3 - 0 - 0 |
| shirt | 1 - 2 - 0 |
| 1 Tie | 7 - 0 |
| | 12 - 17 - 10 |
| Less | 17 - 10 |
| | £12 - 00 - 00 |

With this act of contrition Kas sought absolution for his failure to provide Mmusetsi with a wife, and forgiveness for the rivalry which, in a thoughtless moment, he had unintentionally precipitated between his sons.

> You see, I was trying to restore peace amongst my sons. When I was still at Sewefontein Bodule had cultivated many fields while Mmusetsi had had to work for the landlord. This had enabled me to produce a thousand bags of grain and I had bought Bodule a bicycle as a present for what he had helped us achieve. Now, by buying these goods, I was hoping to put a stop to the jealousy that existed between my two sons.

But it was too little too late. Somewhere along the long hard road to farming success, Mmusetsi had lost access to his father's heart, lost patience in the struggle to acquire a wife and, most seriously of all, lost hope of ever achieving a place in the family sun. Within days of the visit to the Indian store the new shoes lost their shine, the three-and-sixpenny ring became tarnished, and Mmusetsi '. . . left home and became a wanderer.' Twelve pounds in a trading store could not secure a son's love.[60]

Mmusetsi's departure had been long in the making, but that in itself did nothing to ease his father's guilt and pain. Yet, whatever he felt, Kas said nothing and learned nothing. He looked back but failed to see how distance and discipline had cut whatever ties of warmth and affection he had once enjoyed with his son. Nor did he look ahead and foresee what dangers there might be in the future. Even after Mmusetsi's departure he refused to relax his grip on Bodule and continued to wrestle with his daughter-in-law for the affections of the younger son. Decades later, Dikeledi recalled that during the earliest years of her marriage, 'I had nothing that I could call my own: I depended on what he [Kas] gave me. Even mealie meal, I had to draw from his house.'[61]

One day Jack Adamson appeared and told the tenants he had found a landlord a few miles west of Rhenosterhoek who was willing to take them in as sharecroppers. Most of the families on the property were distinctly interested in this proposition, but Kas, with his eyes still firmly fixed on Molote, was unwilling to trek towards Wolmaransstad and instead expressed a preference to move in the direction of Potchefstroom. To his surprise, Adamson said that there was somebody in the boarding-house who might be able to help him, and promised to arrange a meeting. A few days later he reappeared and introduced Kas to J. J. N. Smit, one of the many white miners who had lost their jobs when Dominion Reefs shut down.[62]

Kas recognised the species if not the variant. Long on initials and short on virtually everything else, Smit reminded him of the hundreds of poor whites who had congregated on the farms and diggings of the Triangle during the late 1920s and early 1930s; rootless men ever searching for that one venture that would give meaning to lives scarred by ignorance and poverty. They seldom found it. Yet, as Kas knew from Hendrik Swanepoel, poverty and honesty were not necessarily antagonistic, and in a society founded unashamedly on racial privilege, Jan Smit had something going for him—a white skin and the legal right to hire property. Amidst the shrinking options of a highveld spring these were important considerations.

Smit's stepson, it was said, knew of a farm section that had been leased to a Jewish trader at Muiskraal, fifteen miles north of Potchefstroom, a small property which, if they came up with fifty pounds in rent between them, could probably be hired from him for a year or more. Although not large enough to plant grain on the scale that Kas had done in the Triangle, he would have enough grazing for his livestock and space for him and Smit each to put a full span of oxen into the field. Any contract the partners entered into with the storekeeper would obviously have to be in Smit's name.[63]

On the face of it, it seemed like a good idea, and Kas was about to agree in principle when he hesitated. For one horrible moment he relived the experience at Klippan a decade earlier, when Koos Klopper's visits to the farm had become ever more furtive as he sank deeper and deeper into debt. There were limits to what a black man should put at risk in interracial partnerships that flouted the law. But the thought of moving to within thirty miles of Molote was appealing, and so he temporised, suggesting that he needed to meet the trader and inspect the property before making a final decision. Smit promised that, in the morning, he would escort him to Muiskraal.

Smit arrived at the farm at dawn in a large heavily chromed Studebaker of the sort that was all but mandatory for white miners in the 1950s, accompanied by a woman who seemed old enough to be his mother. He introduced Kas to the woman, who turned out to be his wife, and then waited while Kas gave Bodule some last-minute instructions about getting a heavy-duty licence so that they could drive the Ford to Molote. Kas eased himself into the back seat of the car and Smit drove off and then swung left, into the sun. The Studebaker, as thirsty as it was powerful, picked up speed and bowled along eastward, roaring through towns which the Smits referred to as Klerksdorp and Potchefstroom but which Kas, working from the ancient cognitive maps of the Tswana, still saw as Matlosana and Tlokwe.

A half-jack of liquor peering out from beneath the driver's seat caught his eye. The bottle and the couple's rather shabby clothing told their own story. Stuck in the back seat Kas got bored and set about prising loose a few more fragments of the couple's personal history. Smit had divorced his first wife and then remarried the older woman, who, being partly crippled, walked with great difficulty. Apart from a sister who lived on the plots at Kafferskraal and a son of hers from a previous marriage, they had no kin to speak of. It made Kas think. Why was it that the farther away you got from the Triangle, and the closer you got to the towns, the poorer the Boers were? Klue, Adamson and Smit were all men who seemed to own almost as little land as they had family.[64]

At Muiskraal they drew up outside a store that could have come straight out of a manual on constructing rural shops between the wars—a small, white, low-slung, mud-splattered rectangular building dominated by an enormous 'Coca-Cola' sign. Beneath it, in smaller lettering, appeared the name A. Cohen. Inside they were met by the proprietor, Kas inferred, who was assisted by a younger black woman who kept out of sight by hovering around the back of the premises. Aaron Cohen took them out through a back entrance and pointed them in the direction of a cluster of twenty-five huts belonging to black tenants who hired the series of small garden lots that lay just beyond. Behind the houses lay a far larger, seemingly undeveloped section of the property.[65]

Kas and Smit walked slowly toward the far corner of the property, checking on the quality of the grass as they went. The veld could certainly provide grazing, but Kas was disappointed to find no sign of successful cultivation. They stopped off at the huts to put a few questions to the tenants, who told Kas bluntly that 'the soil around there was so poor that no arable farming could be undertaken.' Cohen was equally forthright: several white farmers had tried to raise crops on the same section without success, and there was little likelihood of a reasonable maize harvest on such land.[66]

Even as they spoke, however, Kas was busy formulating a strategy. What meaningful challenge could the soils of the Mooi River valley pose to a man who had coaxed a living out of Kalahari sand for three decades? The twenty-five pounds he would have to pay as his share of the rent was undoubtedly worth the risk; in his day he had paid more for a plough. And the property was closer to Molote than they had been before. What did he have to lose? Smit and Cohen discussed a few details relating to the contract and reached an agreement. Kas paid Cohen his share of the rent, and they took their leave of the proprietor for the three-hour journey home. Just beyond Potchefstroom the half-jack again poked its neck out from beneath the front seat, daring Kas to arrange a celebration for later that night. He resisted the temptation. It was dark by the time they got back.

*

The move to Muiskraal, undertaken during the first week of October 1951, was made easier by the truck, even though Bodule had no driver's licence.

Kas hired 'a Boer' to drive them north; everybody at Rhenosterhoek turned out to help them load their belongings, equipment, and protesting ewes, lambs, hens and turkeys onto the back of the Ford; the Maine women and most of the children scrambled aboard the truck and left. Kas and two of the older children would drive the remaining sheep, horses and cattle to Muiskraal by road. Altogether this was an easier and more pleasant experience than the move to Jakkalsfontein had been, and the only difficulty they encountered was on the far side of Potchefstroom, where they had to spend the best part of a day pulling out thorns the cattle had picked up during a night stop on a property belonging to an Afrikaner farmer whom Kas had met along the way.[67]

At Muiskraal Kas threw himself into the spring routine with a passion and commitment last seen when he worked under Piet Labuschagne. At Jakkalsfontein, he later said, he had been 'resting,' at Rhenosterhoek he had left most of the ploughing to the two oldest boys; but at Cohen's the time had arrived to reapply himself to farming because the land demanded special attention. The Maine family once again found itself roused for a working day that commenced long before sunrise and ended only when the last glimmer of twilight had faded in the west. But at the age of fifty-seven, even the patriarch was having to make a few concessions, and Kas later admitted that, for the first time in his life, he had to use an alarm clock.[68]

The women and smaller children were sent out to collect cow dung while sixteen-year-old Mosala took on greater responsibility for the Maines' livestock—an onerous task along the many small tributaries of the Mooi River, where the residents were more heterogeneous and also more densely settled than in the drier Klerksdorp district. Out in the neglected corner of the farm Kas and Bodule cleared, ploughed and then—with the manure that the women had collected—painstakingly fertilised scores of narrow furrows of turned earth, taking care to leave seven-foot wide strips of fallow land between them. By the time the rains came, Kas was sure they would succeed where others had failed.[69]

Smit, who was living in his stepson's home on yet another section of the property, put in only an occasional appearance. But neither this, nor his failure to produce the span of oxen that was supposed to work his section, worried Kas. He took the view that Smit had merely given him cover to hire land he was not legally entitled to rent in his own name. Indeed, as the weeks passed he came to welcome Smit's neglect, and started to give thought to how he could best use the fallow land. He knew that Phitise, for example, would be happy to get away from the Motsuenyane farms and work a patch of his own. He floated the idea past Smit, suggesting that Phitise be asked to plough Smit's section in exchange for grazing rights. Smit, still unfocused, agreed, and Kas set off to speak to his brother.[70]

Phitise jumped at the chance, even though the season was already underway. His agreement with the Motsuenyanes had run its course, and he was eager to move north and find a base from which he could search for land that would secure his children's future. Kas was equally pleased: Phitise was a good farmer, and it would help having him around on a property which, for all its proximity to Molote, was still strange and unfamiliar.

Phitise did Smit's ploughing for him, but the fellow was so disorganised that he failed to come up with the seed in time to benefit from the last of the spring rains. With each passing day it became clearer that Smit was never going to be a farmer; shortly thereafter, he drifted off into another badly paid job, and was only seen at weekends. Kas, with one eye on the livestock and the other on his fields, scarcely heeded the virtual disappearance of his 'partner.' With the seedlings established, he had more pressing problems.

Running the Ford was proving to be a demanding business. Like most old trucks it consumed a fair amount of oil and petrol and, every now and then, required a few minor repairs that had to be done in Potchefstroom and paid for in cash. This took Kas out of the fields and onto the road, where he was reminded of another issue, Bodule's driving, which left much to be desired. It was becoming clear why his son had failed to acquire his licence, yet Kas continued to hope that once he had it the state would give them permission to use the truck to convey farm workers over weekends and public holidays.[71]

Some of these problems, notably those that required a measure of literacy for their solution, he took down to the store, where he often found it useful to talk to old man Cohen. A former farm manager, Aaron Cohen knew how to interact with the various bureaucracies that thrived around city edges, and he understood the practical problems of agriculture. Not that the old man was always easy to get on with; he was a complicated fellow.

Cohen's status as a Jew somewhat marginalised him in a countryside dominated by Christians, and he used the social space that this opened up to move freely between the otherwise discrete worlds of white landlords and black tenants. But this ability to move between different worlds only exacerbated other, deep-seated prejudices and suspicions, thereby further marginalising him. Nguni-speakers disliked him because of his unpleasant habit of rapping the counter when demanding that they 'pay up,' and dubbed him 'Machaya'—'one who strikes.' Others, especially black parents, were frightened by how he had turned several domestic servants into his mistresses. For their part, Afrikaner farmers resented his letting property to 'kaffirs,' whose services, they felt, rightly belonged to them as cheap wage labour.[72]

Aaron Cohen's actions and behaviour were at variance with the official policies of an Afrikaner Nationalist government ever more determined to pursue racial segregation at every level of society, but Kas, who appreciated the art of survival more than ideology, was less censorious. Thus, he allowed Mosala to help out around the Jew's store when he was not needed in the fields, and in return Cohen helped Bodule get a licence for the mongrel he insisted on keeping for rabbit hunting.[73] But above all else Kas admired Cohen for his knowledge of the local property market. Without easy access to their established lines of communication back in the Triangle, the Maine brothers needed to tap into a new flow of information if they were ever to call a successful halt to their costly retreat from the southwestern Transvaal. Cohen's store was an excellent listening post, and it soon yielded them some very promising information.

In November 1951, a MoKgatla by the name of Kwapeng arrived at Muiskraal and told Cohen's tenants of a Pretoria-based agent, Harry Braude, who had subdivided a property at Tsetse into forty-acre lots which he proposed selling to black farmers from the BaKubung and the longer resident BaKwena-ba-Molocoane. These plots, which varied in price from three hundred and fifty to four hundred and fifty pounds each, were located about twenty-five miles away, on the farm Doornkop, just north of the town of Ventersdorp.[74]

The Maine brothers were immediately interested. Forty-acre lots were big enough for arable as well as livestock farming, and these had the added appeal of being located even closer to the four residential stands Kas had already purchased at Molote. Within days the brothers visited Tsetse, where Harry Braude showed them three or four lots which, although not adjacent, seemed admirable for their purposes. After a careful inspection and much discussion, Kas agreed to purchase plots thirty-one and thirty-four for seven hundred pounds. He put down two hundred pounds as a deposit and agreed to pay the balance in smaller instalments over an unspecified period. Phitise, equally impressed, bought himself a plot on basically the same terms.[75]

These two open-ended transactions conducted beneath the summer sun on a stretch of stone-strewn veld outside Ventersdorp were the outcome of thirty years' labour on the land and a lifetime's ambition to own property. On just such a day at Koppie Alleen in 1924, as a young sharecropper dripping with perspiration, Kas had turned on a startled Koos Meyer and warned him that the day would come when he would own a farm of his own, and that he would then hire Meyer and ensure that he, too, was exploited to the point where he dropped. But Kas had long since abandoned thoughts of revenge. Revenge was about the past, about nightmares; the land at Tsetse would be about the future, about dreams. With ownership of the land ensured, Kas could farm to the end of his days and make provision for his three sons.

This vision of retirement bore little relation to reality. The truth of the matter was that the land at Tsetse had come two years too late to form the bridge Kas was trying to build between the past and the future. His agreement with Braude followed on enormous outlays in cash over the preceding twenty-four months—on bohadi for his sons, on the stands at Molote, and on the Ford truck. These three items alone accounted for most of the profit from that one magical year at Sewefontein, and abnormally high expenditure undermined his ability to meet future commitments. Still more distressing, the family's productive power had peaked, and its economic strength was waning.[76]

The patriarch himself had undertaken little of the harder work needed for crop farming while the family was at Jakkalsfontein and Rhenosterhoek, and while he was eternally grateful to Bodule for his efforts, there was no doubt the labour situation had deteriorated once Mmusetsi absconded. Mosala—a wayward youth shaped at an early age by the gambling culture of the diggings—was anything but a farmer. In addition, Kas had to contend with two independent-minded wives and four mature daughters who were less and less inclined to bow before the whip of his authority.

At the time, the thought of surrendering to these shadowy forces didn't even cross Kas's mind. The investment at Doornkop made him more determined than ever to apply himself to the task at hand and to mobilise and direct the family labour to the very best of his ability. As far as he was concerned, two years in the wilderness were drawing to a close, and he could sense the emergence of a new sense of purpose; even the crops in the field showed the Maine family were still a force to reckon with. By Christmas 1951 it was clear that good rains, fallow strips and an abundance of natural fertiliser were yielding fine results. The green patch at Kas's end of Muiskraal became a talking point amongst Machaya's tenants and then, as the cobs started to set, amongst a growing number of white farmers.

Unfortunately much of this talk was not very complimentary, and in versions relayed to him by Cohen and Smit, Kas caught sight of the old green-eyed monster. Looking back, Kas recalled: 'The Boers did not appreciate all the hard work that I had put into that field. They said that they had nourished the soil before my arrival at Muiskraal and claimed, "That kaffir is going to reap a big harvest because of the things that we did." '[77] Amongst the Boers who took a special interest in his field was one P. J. Beyers, an itinerant fertiliser salesman who was also a part-time organiser for the National Party in a neighbouring constituency.[78]

Kas knew that envy posed the greatest possible danger for a black tenant, but, caught in midseason, there was not much that he could do other than to keep a low profile and avoid making enemies. Being a chameleon amongst the Boers could be both costly and insidious. As the crop in his field slowly ripened, Machaya's complaints about Smit's failure to pay his share of the rental of the property became strident. It was all painfully reminiscent of Kas's experience with Koos Klopper. In an attempt to shore up the deteriorating position Kas gave Cohen a further twenty-five pounds in cash, a sum which, given his new commitments, he could ill afford. For the same reason—the need to maintain allies in an increasingly hostile environment—he tried to accommodate his 'partner' when Smit asked that two of Kas's daughters help out as servants in his stepson's home.[79]

This 'request' came at an awkward time. Morwesi, whose daughter by Padimole Kadi was by then nine years old, had just gone out to work—at a liquor store in Potchefstroom, hardly the ideal situation for a single woman, but at least it brought in a little money—so the Maines had already lost the services of one adult daughter. Very reluctantly and without any illusions, Kas called on Nthakwana (twenty) and Pulane (fifteen) to enter the Smit household; they would, at the very least, earn modest wages. As Kas had feared, however, the arrangement soon came to grief. Nthakwana, whose apprenticeship had been served under a gun-toting killer, was unwilling to put up with nonsense from a bunch of ageing poor whites in search of black slaves. And Pulane proved to be a quick learner. When Smit failed to pay them on the last day of their first month they walked out. Nthakwana, along with other SeTswana-speakers saw Smit for what he was—an exploiter—and, from that moment on he was known to them only as 'Dinta,' The Louse.[80]

The problems Dinta caused did not end with his alienating the Maine women. His failure to pay his share of the rent for Muiskraal and to pay Nthakwana and Pulane their wages had other serious, indirect consequences. On 5 January 1952, Kas received a very pointed note from Harry Braude. 'Will you please let us know why we have not heard from you for such a long time? You are now in arrears with your instalments according to your Deed of Sale.' Although Kas could not recall ever having signed, let alone having received, a deed of sale, he did remember agreeing to pay the agent four pounds each month. But he was stranded in midseason with a man who owned less than he did, and he had no customers for his skills as a blacksmith, cobbler or herbalist. What could he do or say? Braude would have to wait either until he had a more regular cash income or until he could pay off everything owed on stand numbers thirty-one and thirty-four. All now depended on Bodule's getting a heavy-duty licence and on their making the Ford pay.[81]

Toward the end of February he and Bodule went into town, where they paid thirty shillings to apply to the Local Road Transportation Board for permission to transport 'Non-European passengers within a radius of one hundred and fifty miles of Muiskraal' over weekends and public holidays. A few days later receipt of this application was acknowledged in writing but, on 11 March the Board informed Kas that his application was being contested by a rival operator, and that the matter would be deliberated at a meeting in early April, which he was free to attend if he wished to make additional representations.[82]

This did not augur well and the truck remained grossly underutilised. But on 7 April, Kas received approval. Inside the large 'Official' envelope he found a 'Motor Carrier Certificate' which, on the payment of a further fee of three pounds, would be valid until June. These certificates were validated for only six months at a time. By the time he got around to paying for the licence it was already 30 April, and in only eight weeks the entire procedure would have to be repeated.[83]

Those eight weeks brought Kas neither joy nor profit. The truck slurped up fuel at an alarming rate, needed yet more minor repairs, and could not be fully utilised while Bodule was still busy preparing to get his licence. Then, just when it seemed that the Ford was at last in reasonable shape, disaster struck: Bodule failed the test yet again. The truck was becoming a major liability and Kas was thinking about getting rid of it when The Louse, in search of the next venture to bring meaning to his life, expressed an interest in it.[84]

Kas was inclined to dismiss this rather silly idea out of hand. How on earth would The Louse pay for the Ford, let alone maintain it? But Dinta had something else in mind. He did not want to *buy* the truck; he wanted to exchange vehicles. Now *that* put a different gloss on it. His Studebaker was smaller and therefore obviously less costly to run. Moreover, the licensing of black truck drivers was a notoriously fickle business and the object of a great deal of bribery; presumably even Bodule could get a licence to drive a car? Even if he didn't, it would always be easier for somebody without a licence to drive a car around country roads than a truck. The more Kas thought about the idea, the more it appealed to him. On 28 June, Dinta handed Kas a document confirming that TX 2408, a re-registered Ford truck, had been legally exchanged for a black Studebaker, registration number TZ 2358.[85] But the car Kas bought was not the car he remembered travelling in on his first visit to Cohen's store. The Studebaker had been badly neglected, and he soon realised that he would have to have the engine overhauled. Bodule drove them to Schoeman's Garage at Potchefstroom, and Kas's heart sank when he heard what it would cost to restore the car to good working order. But there was no turning back; by the time they returned to collect the car, Kas had had to sell four cattle, which irritated him. Disposing of livestock chewed away at his belly, like the cheap brandy that the illicit liquor dealers obtained for their black customers from the bottle store at Rysmierbult. It was a bad sign when cattle, a natural and renewable asset, had to be sold simply in order to maintain a piece of inanimate rubbish like a car. Somewhere along the line Kas had been forced to swop the supposedly primitive draught power of the Triangle for the modern mechanical monsters capable of negotiating the peri-urban fringes of the Witwatersrand. What had it gained him? He had been entrapped in a web of licence fees and a set of costly repairs. Yet what else could he have done? How else did one move heavy agricultural equipment over vast distances?[86]

As the season drew to a close all these pressures started to take their toll on Kas, and on those around him. He thrashed Mosala for not applying himself to his chores. He rebuked Nthakwana and Pulane for not finding jobs. They at least tried to help around the house, which was more than could be said for Matlakala. In her case the spirits of the ancestors supposedly left her unable to do much work, an ailment which, as far as Kas could tell, Lebitsa and the others tolerated to the point of indulgence. To cap it all, Leetwane had become so interested in the little money she made from her chickens and herbal practice that she jeopardised the farming operation as a whole. Nobody seemed to understand that, with Mmusetsi gone, they all needed to work that much harder. Stacked one on another these irritations weighed on Kas's mind. Unable to cope with the burden, he finally snapped and—not for the first time—assaulted Leetwane for one or other minor infringement of the patriarchal code.[87]

But Leetwane had had enough. She left and went to stay with Mphaphatha, a distant kinsman and ngaka who lived on a neighbouring farm. Mosala, Nkamang and Pulane then rebelled, too, and threatened to join their mother, a revolt that was only put down when they were promised a thrashing.

The next counter-attack came from a totally unexpected quarter: Bodule turned on Kas and reproached him for his treatment of Leetwane. This admonishment from his closest and most trusted lieutenant brought Kas to his senses. That night he gave the matter a good deal more thought and in the morning, as Pulane later recalled it, 'he took the cart, followed my mother, and they came home together and sorted out their differences.'[88]

With a measure of calm restored to the troubled household, the Maines mobilised for the harvest, which, given the work that had gone into preparing the field, was a gratifying experience. Dinta allowed them the use of the Ford to take most of their maize into the depot of the Sentraal-Wes Landboukoöperasie in Potchefstroom. Good grain prices, accelerating more rapidly than at any other time in the twentieth century, ensured that Kas received a voorskot of more than thirty shillings a bag for each of the hundred bags.

The irony of receiving a cheque made payable to 'K. Maine' from one of the most powerful institutions underwriting racial domination in the South African countryside was as lost on Kas as it was on the Afrikaans-speaking clerk at the counter. Clearly, the first generation of apartheid bureaucrats could not enforce adherence to strict segregation at every level of the marketplace. Kas took the cheque to a nearby bank where, as luck would have it, he bumped into the same farmer who had helped him find a place to stay the night his cattle had been plagued by thorns. The obliging Boer helped him cash the cheque. Parliament was in the thrall of madmen, but out on the platteland the ambiguities of paternalism often continued to shape inter-personal relations.[89]

The foray into Potchefstroom helped ease Kas's more pressing financial problems, but back at Muiskraal, he went on worrying about the unacceptably high level of sheep losses. The Triangle-bred Merinos were even more susceptible to disease in the valley of the Mooi River than he had bargained for. But Kas hoped to compensate for these losses by doing a little off-season work. Making some money out of cobbling was a distinct possibility, given that for the first time since leaving Bloemhof, he had access to customers—tenants in Cohen's 'location,' labourers on neighbouring farms and workers on the alluvial diamond diggings at Muiskraal (dating back to the 1920s).[90]

As the South African state had long since discovered, policing the market is not easy. Just as the Nationalists found it hard to keep black sharecroppers away from work on white farms, so Kas found it hard to keep his wives and children from peeling off the family enterprise and entering the market on their own terms. Even as he cobbled, the Maines explored how they could squeeze something out of the valley economy for their own account. Lebitsa's collecting, drying and selling of cow dung to the fuel-starved workers on the diggings was hard for Kas to fault, for he had to concede that it was a traditional activity confined largely to the off-season. Leetwane and the older girls' basket making, however, was something new. They disappeared for hours on end, collecting *leodi*, which they dried, wove into baskets and then sold to neighbouring whites. The demand for reeds knew no season, and Kas could see a day coming when some difficult choices would have to be made—between time spent on arable farming and the time spent basket weaving. Patrolling the boundaries of the household economy was almost as fraught an exercise as dealing with the Boers.[91]

Almost, but not quite. Once the last drop of cash earned in transport work had dried up, The Louse crawled out of the seasonal stonework. He had been unable, he said, to 'arrange the necessary papers' to formalise the exchange of vehicles. It would be in everyone's interests, he claimed, if he were to return Kas's truck and reclaim his car. Kas knew there was nothing in this for him. He had already forfeited four cattle in order to repair the Studebaker. Yet Dinta was insistent and Kas was quick to appreciate that, if he could not produce documents proving his ownership of the vehicle, he would never succeed in getting the car relicensed. Who would the magistrate believe if the whole messy business was put before a court, white man or black? No! He had been caught napping. Smit reclaimed the Studebaker keys and drove off in the car.[92]

The Maines' position was now more serious than even Kas anticipated. The Louse had done no maintenance work on the Ford and it required major repairs, which meant yet more money and maybe the sale of yet more livestock. Dinta was sucking the life-blood out of the family enterprise, and growing fatter and more confident. He moved out into the open more often and then, one morning, stood in the doorway of the shack and demanded a half-share of the harvest produced on land hired in his name! That was too much. Kas brushed him aside and strode out of the house, leaped into the cart, and set off at a furious pace for Potchefstroom. As his experiences with Goosen and Klopper had taught him, when all else fails, try the law.[93]

Three days after Dinta laid claim to a half-share of the harvest he was invited to take part in a round-table discussion held in the offices of the law firm of Nel & Klynveld in Potchefstroom. Kas's attorneys argued that not only should Kas be compensated for the amount he had spent on repairing the Studebaker, but that The Louse was responsible for restoring the Ford to a roadworthy condition. Of the disputed harvest no mention was made. As always, inter-racial sharecropping asked too many questions of too many parties to give anybody a clear-cut legal advantage. Dinta got the message. There was no more talk of sharing the harvest.[94]

But the business with the vehicles dragged on, and it had yet to be settled when Kas was reminded that it was only because he was richer than The Louse that the scales of justice had tipped in his favour. Justice might be blind but there was no doubting that, in most states, perhaps especially a racist state, wealth and power—perfumes of privilege—caused her nostrils to twitch. It still came as something of a shock: 'We were sitting at home when we first noticed Boers driving around our field. We were surprised and asked what was happening, only to be told that they were busy inspecting the land because they wanted to buy it.'[95]

Kas had seen this before, including the spring when he had developed new fields at Bosmansfontein only to see Koos Meyer buy the farm before he could plant them. Fields lay barren and neglected; then black tenants cleared, ploughed and cultivated them; then along came a white man, any white man—even someone as 'poor' as Beyers—to buy them. Kas stumbled down to the store to talk it through with Cohen; he was sure he would understand. Hadn't the old man once told Kas that he would give the Maines a five-year lease on the field, provided that no purchaser was found for the whole property in the interim?

How was one to know who was more vulnerable before the arrogance of a triumphalist Christian Nationalism, an ageing Jew with a black mistress or a MoSotho sharecropper with a poor white for a master? The weak wrestling the weak was an unedifying sight. Cohen was very elusive. Unable or unwilling to tell Kas he was going to let Beyers hire the field, he lied and merely said the white man had expressed an interest in buying the property. The feeble had defeated the vulnerable, and the Nationalists could quietly go about their business secure in their knowledge that God had decreed that they should rule the southern tip of Africa. But Cohen took pity on his adversary and promised Kas he would help him find a place on the farm where he had once been a manager.

He told Kas to climb through the fence at the northern edge of the property and keep walking until he came to a cluster of huts where he should ask to speak to a man by the name of Saul Ngakane. Kas did so. Ngakane said that he was the 'foreman,' that he knew all about Cohen, and that the property belonged to a widow Hersch, who lived in Johannesburg. It was obvious that Ngakane was an experienced farmer with oxen and equipment of his own. When Kas asked about the chances of getting involved in sharecropping, Ngakane was careful to check his credentials; eventually he invited Kas to join their small community of tenant farmers. In September 1952, the Maines moved to the farm Rietvlei, their third move in as many seasons.

Kas was determined to put a stop to the costly retreat from Bloemhof.

CHAPTER TWELVE

# Defence

## 1952–56

Kas was exhausted. He needed a fortress, a redoubt from which he could mount a final round of resistance on white-owned land, before retreating, if he had to, beyond the distantly visible folds of the Magaliesberg mountains to the north. Rietvlei, which lay in broken countryside, looked promising. This was a place that was no stranger to hard-fought contests: here, for several millennia, a westward-flung arm of the gold-rich Witwatersrand had sought to prevent the waters of the Mooi River from flowing south to meet the Vaal, but the layered quartzite of its ridges could not overcome the determination of its primal enemy, and the waters of the river had gnawed through the Witwatersrand at the Potchefstroom Gap.

Rietvlei, a roughly rectangular property in rugged terrain, fell into two distinct and unequal parts. A small elevated western section was dominated by the protruding tongue of the Rietvalleirant, or ridge, and rose to its highest point in the flat-topped Tafelkop (Tabletop) mountain; the larger lower-lying eastern section stretched down toward the Mooi River valley. With Tafelkop guarding the approaches to Ventersdorp and the highveld proper forty miles west, and the Mooi River's only real tributary on the northeast forming a conduit that cut its way back into the Witwatersrand for almost eighty miles to reach the outskirts of the mining town of Randfontein, Rietvlei commanded a natural junction of considerable strategic importance, and this was reflected in the complex legal provisions that governed access to the property well into the twentieth century.

But strategic importance is not the same thing as agricultural promise, and for the better part of a century Rietvlei had been something of a Cinderella amongst the local properties. When the first white trekkers abandoned the settlement they had made near Potchefstroom in the mid-nineteenth century, they skirted Rietvlei and showed greater interest in other farms where there was more abundant water and less broken countryside. And in the first three decades of the twentieth century it was the diamond diggings around Muiskraal to the south and Swartplaas to the northwest rather than Rietvlei itself that caught the eyes of undercapitalised white farmers or agents from Witwatersrand mining houses. It is easy then to understand why, when Sender Hersch purchased a 12,400-morgen section of the original farm in 1939, he had done so with the intention of profiting from a speculative investment rather than working the soil productively.

Sender Hersch, like so many South African Jews an emigré from the Baltic littoral, had been born in Zager, Lithuania, in 1888 and emigrated to the Cape Colony as a young man, where he became a successful trader. In 1925 he married Paula Feitelberg, herself a recent immigrant from Pilton in Latvia and the daughter of a rabbi. Shortly after their marriage the young couple settled at Potgietersrus in the far northern Transvaal, where Sender Hersch opened yet another successful venture, the Rand Furniture Store. By the mid-1930s, Hersch had diversified his business interests and invested heavily in property development, a new departure which saw the building of a factory in Johannesburg and the erection of a row of shops at the Venterspost Gold Mines on the West Rand. These endeavours were as successful as his earlier undertakings; by 1939, he was able to retire to Johannesburg and went only occasionally into the countryside to inspect his business interests.[1]

During the course of just such a visit to the western Transvaal in the first year of his retirement Hersch noticed the land at Rietvlei. But he quickly saw that Rietvlei would be a difficult property for an absentee owner to manage. Strangers making their way up and down the valley, legally entrenched rights giving neighbours access to grazing at certain times of the year, and a rich deposit of clear river sand prized by the local building contractors all pointed to the need for constant surveillance if the property was not to be stripped of its natural resources. So he invited the impecunious Aaron Cohen to 'manage' the farm to his own advantage in return for performing certain caretaking duties.

Cohen, at that time still only an aspirant trader with limited capital and very few farming skills, soon found that he would have to call on the assistance of a half-dozen or more tenant farmers, an arrangement to which Hersch raised no objection. Finding black tenants during the war was not difficult: the increase in tractors in the Transvaal countryside and a steady expansion in agricultural output on white farms throughout the war years ensured that there were many black sharecroppers with draught oxen in search of land to put under the plough. During his five seasons on the farm Aaron Cohen assembled a dozen or so exceptionally talented black sharecroppers; amongst the most important of them were Andries Malebo, Aaron Modise, Thomas Mosoeu, David Mogale, Saul Ngakane and Sefudi RaMohano.[2]

This group occupied a relatively privileged position in a world where opportunities for independent black farmers were dwindling with each passing season. For the most part semi-literate, SeTswana-speaking Christians, these sharecroppers also owned some heavy agricultural equipment and plenty of livestock, at a time when destocking had long been the order of the day for most tenants on the highveld.[3]

Aaron Cohen designated forty-year-old Saul Ngakane as farm foreman and, throughout the war, this small and largely self-sufficient community of black farmers at Rietvlei prospered under Ngakane's leadership. A successful sharecropper in the southern Transvaal before the combined impact of the Marketing Act and agricultural mechanisation had driven him into the drier northwest where the Boers were less well capitalised, Ngakane had all the qualifications for success at his job. Besides his obvious proficiency as a farmer, Ngakane was a much-respected Wesleyan who presided over a largish family including five daughters and a son. His technical expertise and his feel for the social problems of farming families earned him the respect of the absentee landlord, 'farm manager' and black sharecroppers alike.[4]

Aaron Cohen, too, profited from this arrangement, and when he left to open the store at Muiskraal toward the end of the war, Sender Hersch saw no reason for making any drastic changes at Rietvlei. Long since convinced of Ngakane's abilities as a manager, Hersch simply did as he had done with virtually all his business interests in the Cape: he left the man in charge with the responsibility for getting on with the job. Saul Ngakane did not disappoint him. He personally supervised and recorded the retrieval and sale of all river sand on the property, and, under his watchful eye, the sharecropping arrangements yielded excellent returns. Indeed, Ngakane did so well that, when Hersch died in 1950, his widow, Paula, decided to allow the estate to continue to operate along exactly the same relaxed lines.[5]

This arrangement between a Johannesburg-based landlady whose family had been forced to flee before the storms of East European anti-Semitism and a dozen BaTswana tenants seeking refuge from the effects of an industrial revolution that hounded out blacks, had something in it for both parties and continued to work well. In a world in which Afrikaner Christian Nationalists proclaimed their privileged and special relationship with the southern African countryside, both parties at Rietvlei were being relegated once again to the status of outsiders—the one on the grounds of religion, the other on the grounds of colour. Having an opponent in common can strengthen underlying affinities, and when Ngakane met Kas in the spring of 1952, there were reasons for close co-operation.

Kas and Ngakane came from almost the same cohort, both being nearly sixty years of age. Their farming experiences had been gained in rather different settings, but both men were battle-hardened veterans who had survived several encounters with teak-tough highveld landlords. Easy rapport sprang up between them, for each had an almost intuitive understanding of the other's position. Ngakane helped ease his new friend into the tightly knit community on the farm, and Kas, for the first time since he had left the Triangle, found himself revelling in the company of progressive and sympathetic blacks. The Rietvlei farmers, like the sharecroppers back in RaKapari's drinking circle, were men of substance with the space to graze their cattle and raise their crops, in marked contrast to the many demoralised labour tenants whom the family had encountered in and around Klerksdorp. Cattle and access to land gave men a dignity and pride in their blackness, just as surely as poverty and landlessness helped to shape many of the more colonised personalities who had to live off the urban margins of industrialising white society.

Yet, for all the continuities in the lives of tenants at Rietvlei during the postwar years, a new menace could be detected in the air. The National Party and its newly appointed Minister of Native Affairs, Hendrik Verwoerd, were beginning to strive more purposively for their goal of total segregation. As early as 1952, Verwoerd, the architect-in-chief of 'grand apartheid,' had given notice to all who would listen of 'how the various Acts, Bills and also public statements which I have made all fit into a pattern, and together form a single constructive plan.' More ominously, as one historian has written, a few months later: 'Verwoerd announced a campaign to eliminate black land ownership in white farming areas, and to get rid of African squatting and labour tenancy on white farms through the conversion of all farm workers to wage labourers, and the revision of the 1936 Land Act to make its anti-squatting provisions enforceable.'[6]

In sprawling rural constituencies with marginal rainfall like Ventersdorp, where there were several areas of 'black land ownership in white farming areas' including Ga-Mogopa and Tsetse, and a considerable amount of 'African squatting and labour tenancy on white farms'—not least of all at Muiskraal and Rietvlei—such policies found instant favour amongst rootless 'poor whites' such as J. J. N. Smit, aspirant landowners like P. J. Beyers, or labour-hungry farmers with political ambitions like J. C. 'Cas' Greyling, who was based on the farm De Beerskraal.

Black tenants viewed these and other even more sinister Nationalist plans with the greatest possible misgivings. Verwoerd's announcement heightened existing tensions between landlords and tenants in the Mooi River valley, and these were further exacerbated when the National Party prepared to defend the gains it had made in the 1948 election. Electoral contests overtly predicated on racist considerations darkened South Africa's troubled political climate; in country districts they polarised fearful poor whites and exposed blacks to the resulting thunderbolts of populist aggression. The Jews, too, were often singled out for special ideological attention, whether they were storekeepers or absentee farm owners.

None of this augured well for the people at Rietvlei. Whether Ngakane had got wind of the Maines' problems with Beyers the previous season, or whether he had simply heard the opening shots in the new electoral campaign more clearly than his colleagues, the precautions he took when allocating Kas his fields at Rietvlei suggest that he was mindful of the larger issues. First, he made sure that Kas, as the newcomer, did not have a really promising tract of land: he marked out fields on the drier section of the property for him to plough. This kept Kas firmly at the bottom of the pecking order and, by hiding him behind the mountain, ensured that his activities were virtually obscured from the attention of nosy outsiders. Secondly, Ngakane and other senior members of the community made a point of warning Kas about the dangers of producing too much maize, which, they said, would jeopardise everyone.[7]

With the speech of the official at Sewefontein still haunting him, Kas needed no reminding. Yet he could not help feeling that the Rietvlei tenants were overstating their case. Perhaps it was they rather than the Boers who feared his competition. In his experience, it was the number of teams of oxen a man put into the fields rather than the size of the harvest that attracted most attention from white landlords. Had it not been the 'rich kaffirs with spans' who had been asked to leave the Triangle? On the Hersch property, there was obviously more than enough space on which to graze cattle and plant crops, and so he tended to disregard what he could only consider as alarmist talk. He was convinced he knew better. As he said years later, 'I always regarded those people as being rather stupid.'[8]

Kas, Bodule, now twenty-four, and Mosala, seventeen, cleared, ploughed and planted the patch allocated to the family, taking care to cultivate the soil as well as possible. As Kas knew, the key to a good harvest lay in the effort one made in preparing and maintaining fields. The other sharecroppers recognised the presence of a real professional in their midst, and with the work progressing well, Kas began to relax and to enjoy farming without the attention of a resident landlord and without having to worry about the uncertainties of dealing with a 'poor white' like Dinta. His confidence, badly shaken by the cost of the retreat from Bloemhof and the unexpected loss of Mmusetsi, returned slowly; as always, he was buoyed up by the moisture in the spring air and the appeal of summer grain farming.

He also expressed this more positive attitude in his dealings with the agent who had sold the Maine brothers the stands at Tsetse. In mid-September 1952, Kas received a circular letter from Harry Braude informing him that the agent whom he and Phitise had met at Ventersdorp had moved office and that henceforth all business would be conducted in the name of 'Harry Braude Limited.' The letter said, 'Anyone who purchased plots at Doornkop [in the district of Tsetse] can now get his Title Deeds provided he has paid for his plot in full.' Kas found this reassuring, but the paragraph that caught his eye read as follows:

> Regarding the farm Boschplaats known as the Thaba ya Batho Agricultural Holdings, we have to advise you that we are now installing an extensive water scheme on this Holding which has been accepted by the Native Affairs Department. As soon as the scheme, which should take about four months to install, is in operation, we shall be in a position to give transfer to all the Purchasers on the Holding who have paid for their plots in full.[9]

Given that most of the land set aside for black occupation on the highveld was in areas with not enough rainfall for commercial grain farming, the prospect of having a plot with a permanent supply of water was alluring. Boschplaats as well as several other possibilities raised in Braude's circular were worth getting excited about.

Kas showed the circular to some tenants at Rietvlei as well as to the Reverend J. P. Jwili of the Ethiopian Church of South Africa. Not surprisingly, Braude's communication was of great interest to them. Jwili, who was based at Ventersdorp but out doing the rounds amongst rural congregants, was especially keen, and largely as a result of his prompting, Kas engaged the services of a local farm school teacher to draft a reply to Harry Braude's circular.[10]

> Rysmierbult
> P.O. Box 22
> Muiskraal
>
> 15–9–1952
> To: Mnr. Harry Braude Ltd
> Dear sir
> I have received your letter which make me so glad to learn and read about so many plots and people here are so many which wants to know of these plots Ruigtesloot which District of which town and Tweefontein which district and give me every plot in which district so that I can tell all that wants to go in for plots and secondly do not send my letters to Potchefstroom as I am now at Muiskraal look on the above address and the one plot which I want to know about it is this one here named Boschplaats known as Thaba ya Batho please write and tell me at which district is this Agricultural Holdings which districts as lot my purchasers here want to know this plots Thaba ya batho I think lot will be sending you their entries fees for the farm Thaba ya batho So far sir I am sending my greetings to you and all at the office
>
> Ta Ta your obedient servant
> Kas Deeu

This plea for help—little more than a message stuffed into a bottle by a handful of peasant survivors hoping to be rescued from the rising tide of capitalist agriculture—floated out toward Braude, who dispatched a pamphlet 'indicating clearly the district of the Holdings and the Purchase price.' But the prices were prohibitive, and with the exception of Reverend Jwili, who somehow managed to scrape together the deposit for a plot in some far-off district, none of the tenants at Rietvlei could follow up the offer. Too 'rich' to be accommodated on white farms and too poor to buy the few smallholdings that did come on the market for purchase by blacks, it seemed that the sharecroppers would remain marooned.[11]

But the situation had not always been so bleak, and Kas was quick to remember how fortunate he and Phitise had been to buy land at Doornkop. Phitise, who had already paid Braude in full, had taken transfer of his property and was busy planning his move to Tsetse. Kas's own situation was far more complicated. The investment in the Ford truck and the repairs on the Studebaker had siphoned off most of his savings, and he was finding it hard to keep up the payments to Braude. Most of these problems could be traced back to his involvement with The Louse, who, despite having entered into an agreement in the lawyer's office, had still not repaid him for his repairs to the Studebaker. Given the demand for freehold property amongst black farmers, Kas knew he risked losing the property at Doornkop, and all of this in turn made the need for The Louse to pay him his compensation just that much more urgent.

Kas undertook a second visit to Nel & Klynveld. The attorneys, however, merely pointed out that the settlement they had facilitated was a voluntary one; if Smit defaulted, the only course of action open to Kas was to bring a civil action against him in the magistrate's court. This ten shillings' worth of advice did nothing to improve the immediate position, and, with the threat of yet more legal fees to come if he were to act against Dinta, Kas was in a quandary. In the end he decided to give Dinta more time.[12]

At Rietvlei, with the first rains well and truly upon them, there was plenty of ploughing to do, and Kas was determined to provide the family with plenty of fresh vegetables. Like most of the tenants he planted potatoes, pumpkins and tomatoes, but then—and much to the dismay of his wives, who remembered their back-breaking experiences at Klippan a decade earlier—he proceeded to open up a patch for beans. But Kas still did not see the problem. As far as he was concerned the beans were just a cash crop, which the older girls could harvest if his wives objected. One thing was beyond dispute: it was far better to grow beans than to raise the dagga plants that he knew Sefudi RaMohano cultivated behind the mountain.[13]

But all crops, even vegetables, had hidden dangers. With the rains falling regularly Kas's demand for labour escalated rapidly, and by midsummer 1952 he was once again pushing the household to the maximum of its ability. On a large property poorly provided with fenced-off camps, where the watering point for the animals was far from the kraal, herding became a full-time occupation that tied up the men in the family for days on end. Unfortunately a wet season also encouraged weeds, of which Rietvlei appeared to have an abundance; the women spent far more time out in the fields than was customary.[14]

In this drive to extract surplus labour power from an ageing family machine, neither the old nor the young were spared. The working week passed without so much as a Sunday break. Kas not only ploughed but took on additional herding duties to supplement the efforts of Bodule, Mosala and twelve-year-old Nkamang, who was often pushed into the role of 'herdboy' despite her age and gender. The two other senior members of the household, Leetwane and Lebitsa, worked tirelessly in the fields, supported by the efforts of five daughters, a daughter-in-law, and a seven-year-old granddaughter. But while everybody was expected to work harder, the women, especially the adult daughters, endured most of the strain. Nthakwana and Matlakala were close to breaking point.

Forced together by circumstances rather than by temperament, the two sisters became firm allies, and each resented the new push for agricultural production for her own reasons. Nthakwana had always been a very capable, hard-working and loyal member of the household team, but at the age of twenty-one, she wanted an independent income and wanted to see something of the world beyond the farm. Matlakala, aged twenty and a recent arrival, was struggling to find her niche, while the call of the ancestors often left her unable or unwilling to take on extra duties. Like Nthakwana, she was eager to establish relationships with men whom her father did not always approve of, and she had embarked on an imperfectly concealed liaison with a fellow named Buti Thale.[15]

The dynamics at work in this rather fraught situation were familiar to both Bodule and Morwesi. But Kas was a prisoner of his personality as well as a system of tenant production that was predicated, in part, on his patriarchal power to keep his family's labour mobilised. And, like a long-term prison inmate, he had so internalised the routines and values of the system that he could no longer see the need for his older children to conduct their first experiments with freedom. In a household where the family was the unit of economic production as well as a social entity, the boundaries became indistinguishable and there was not enough psychological space for boys to become men, or the girls to become women. As Pulane, fifteen years old at the time and already sexually active, recalled three decades later:

> My father would refuse us permission to visit certain places, telling you straight out that you were being denied permission because you were merely seeking a pretext on which to meet boyfriends. He even refused

us permission to attend church. If you told him that you wished to accompany someone to church he would inform you that you should have signalled your intention earlier in the week. You were therefore ordered to stay at home and forget about going to church. When Petrus [her future husband] and I were courting we used to meet at the top of the mountain at Rietvlei. On the way back I would drive the cattle, goats and sheep.[16]

At least one of the other daughters was equally irritated by having to work seven days a week and having no time in which to socialise or attend to personal matters. Many years later Matlakala recalled:

Sunday working did not bother me particularly since I was patient by nature and accustomed to looking after cattle. The person whom it seemed to worry most was Nthakwana. My father would sometimes visit us in the fields just to check whether there were boys with us.[17]

Leetwane and Lebitsa, too, were disgruntled at being denied the chance of generating income of their own through basket making or egg selling, while Dikeledi, always willing to challenge patriarchal authority, never allowed Kas to forget that the issues surrounding the payment of her bohadi had not yet been resolved to her parents' satisfaction. (They, since her marriage, had taken up residence in the Ventersdorp district.)[18]

Everybody was feeling the pressure. Then, with the domestic structures already creaking and groaning, came two letters which made matters even worse. In early January 1953, Braude sent Kas copies of the Deed of Sale for plots 31 and 34 at Doornkop, with a request that he sign them and return them to the office. A mere four days later, Kas got the sort of letter he had been dreading:

Harry Braude Limited,
18 Velra Huis,
2nd Floor.
Bureau Street
PRETORIA.

10th January, 1953.

Greetings,

re: *PLOT 34 DOORNKOP.*

We notice with regret that your last payment on account of your plot was made many months ago. We will appreciate more regular installments. We have very many enquiries about plots at Doornkop and we suggest, in all fairness to buyers, that if you are not anxious to buy anymore or if you find that you cannot keep up with the installments, you must give someone else a chance because plots are very scarce at Doornkop.

Greetings,
H. Braude.
*DIRECTOR.*

Kas found Braude's letters as confusing as they were disturbing. Despite his earlier request, the letters had been sent to two different addresses. While the first letter, with the 'Deed of Sale' conveniently back-dated to 14 February 1952, had mentioned both stands at Doornkop and had been forwarded to him at the Rysmierbult post office, the 10 January letter, sent via Morwesi in Potchefstroom, mentioned only plot number 34.[19] Did this mean that at some point during the intervening period he had forfeited his right to plot 31, or was Braude's administration so disorganised that there had been a mistake?

He knew it would take time and money to sort out the mess and he was running out of both. Where was the justice in this world? Braude was putting the screws on him to pay for land he had spent the better part of a lifetime working to acquire, while The Louse was driving around the district in a car repaired with money raised from the sale of *his* cattle! It was too much to have to put up with. Within a week of getting Braude's second letter he was back at Nel & Klynveld instructing them to institute civil proceedings against Smit. But even here time was not on his side: he was told that it would be weeks, perhaps months, before the matter was raised in court. Neither this nor the ten pounds, twelve shillings and sixpence he had to fork out as deposit pleased him and, under the circumstances, he decided not to consult the same firm about his problems with Braude which were probably best dealt with by attorneys more familiar with land deals in and around Doornkop.[20]

It was hard to get away from the farm at the height of summer and, in the end, four weeks elapsed before he found the time to consult F. B. Gillett, 'formerly member of the Middle Temple, London' at Ventersdorp. Ten shillings gave him some carefully considered advice and a letter to Braude:

24a Carmichael Street,
VENTERSDORP.
TRANSVAAL.
19th February, 1953.

Messrs. Harry Braude Ltd.,
P.O. Box 75,
*PRETORIA.*

Dear Sirs,
*Plots Nos. 31 and 34, Doornkop, Ventersdorp.*

We have been consulted by Mr. Kas Deeu of the farm Muiskraal, P.O. Box 22, Rysmierbult, in reference to your letter of the 6th ulto wherewith you forward for signature by him a Deed of Sale.

He cannot understand why this comes to hand after the lapse of so long a period and he would like a statement showing what the interest on the purchase price of £700 over the period which it will take by way of payments of £4. 0. 0. per mensem to liquidate the capital amount and interest is. Will you kindly let us have this.

Yours faithfully,
F.B. Gillett

The reply from Braude, although perhaps predictable, still came as a shock to him. Even with the best will in the world and another harvest like the one at Sewefontein, it would take a lifetime to redeem the debt of £700 on plots 31 and 34 at 5 percent per annum. The realistic thing to do, Gillett suggested, was to abandon his claim to plot 31 and to concentrate on the marginally larger plot. A second letter from Braude told them that the nineteen morgen of plot 34 would cost £450, and the immediate payment of the sum of thirty-four pounds would secure it. On 30 March, Kas agreed to the new terms of purchase for plot 34 and signed the Deed of Sale which had, in the interim, been amended by Gillett.[21]

But if Kas's unsteady hand had relinquished the right to plot 31 on an autumn day in 1953, his heart had not, and while he could see that raising thirty-four pounds was the immediate objective, he remained inwardly committed to obtaining the money for both plots. He knew it wouldn't be easy. His livestock resources had been depleted by Bodule's bohadi, the debacle with the car and truck, and the ravages of the sheep disease that had dogged them since Bloemhof. His salvation, if it lay anywhere, lay on the land. The family would simply have to redouble its efforts, and Braude, along with the rest of mankind, would have to wait for the harvest. Lower down the rung, he would rely on Nel & Klynveld to squeeze what they could from Smit. If all failed, then his dream of retiring to Tsetse with 'a farm and a large number of livestock' would remain just that, a dream.[22]

The call on a weary household to redouble its efforts came at a very bad moment. The back-breaking chores of bean harvesting and weeding were behind them, the grain was slowly ripening in the fields, so the women were looking forward to a respite when they could pursue a few goals of their own. In fact, Leetwane, Dikeledi, Matlakala, Nthakwana and Pulane had already reorganised their domestic duties so they had time to weave the reed baskets that they sold to whites in Muiskraal.[23]

The older Maine daughters, especially, found basket selling rewarding, not least because their mothers, unlike Kas, laid no parental claim on the cash they earned. For the same reason, the girls were poorly disposed to patriarchal demands for yet more labour in the fields. Kas's view of the primacy of household production on the one hand, and the women's desire for disposable income of their own on the other, set the parties on a collision course and culminated in a spectacular display of psychological pyrotechnics when that 'Box of Matches' of old, Ramollwana, erupted and threatened to burn their baskets and reeds unless they showed more enthusiasm for what he considered the main enterprise. But by their nature such flamboyant 'victories' seldom produce lasting results; the father's intervention did nothing to ease tensions within the household.[24]

Kas found Matlakala especially troublesome. It was not just her sickly disposition and chronic inability to put in a full day's work; her 'visions' were undermining his authority while he was also having to contend with Leetwane practising herbal medicine on her own. Thus, while he was grateful for a Matlakala 'dream' that enabled him to locate a few pigs that had gone missing on Tafelkop, he was unhappy about a 'voice' telling her that 'he would never secure a plot at Doornkop and that he would lose all his money if he persisted in attempting to do so.'[25]

As the season drew to a close the spirits of the ancestors grew ever more lively, and Matlakala's 'visions' and 'voices' seldom reinforced patriarchal authority. From what Kas could determine she lay at the heart of the resistance to agricultural labour and it was she who was leading the hitherto loyal Nthakwana astray. But in fact, it was the other way round. When he reintroduced Sunday shifts of field work to keep away the finches that threatened to devour the family's sorghum, it was Nthakwana who pulled Matlakala aside and told her she was thinking of running away from home.[26]

That very evening Nthakwana and Matlakala sought out Mosala, the only true rebel amongst their brothers, and discussed the outlines of their escape plan. Matlakala wavered, but Mosala persuaded her, pointing out that if she stayed behind she risked being flogged to reveal the whereabouts of her sister. The sisters used the cover of darkness to sneak into Lebitsa's hut and take seven pounds from her hiding place before they retired for the night.

At first light they rose. Questioned by the startled Pulane as to what was happening, they threatened to assault her if she did not keep quiet and go back to sleep. They gathered a few personal items in a bag and then, with the sun still not quite up, set off toward the fields, hoping that if he-who-saw-all caught a glimpse of them, he would be fooled into thinking they had gone 'to chase birds.' It was a calculated risk and potentially the most dangerous moment of the operation. They held their breaths and walked as lightly as they could. One of Bodule's hunting dogs suddenly appeared out of the bushes, wagged its tail and sauntered off.

Beyond the fields, Nthakwana guided them along the path that led to a warren of tumble-down shacks near a diamond digging. They hid there until about 11:00 a.m., then emerged from the tin town and walked to the railway station at Rysmierbult, where they bought two single tickets for the mining metropolis of Randfontein. 'Jobs were not that difficult to find at that time, and there were no "passes" for women.'[27]

By late morning, when Mosala was sent to investigate why the girls had failed to appear for the first of the day's two main meals, the fugitives were already on their way toward new experiences that the extroverted Nthakwana craved and the introverted Matlakala dreaded. At Rietvlei the dominant emotion was anxiety: once Kas, Lebitsa and Leetwane realised that the sisters had absconded, they were genuinely concerned about their well-being; in the case of the patriarch, concern gave way to frequent outbursts of anger. He persisted in believing that this act of rebellion, too, could be traced to Matlakala's malign influence. This time she had risked not only her own reputation and future, but the well-being of the entire family just when, perhaps more than ever, their long-term interests were in jeopardy.

Kas realised that those interests would now be more difficult to defend. A sharecropper's success was in a large measure predicated on the size and quality of his labour force, and, in the last two years he had lost the services of an adult son and two daughters. Despite all his efforts to provide his children with the good life based on the richness of the land, he and others like him were being undercut by a new generation intent on finding and exploiting its own social, economic and political space. Much as he or his unwitting allies in the distant offices of Hendrik Verwoerd's Department of Native Affairs might regret it, each year saw more and more young blacks, men and women, deserting the countryside for the burgeoning cities of the Witwatersrand.[28]

The state's decision not to dam up and then selectively divert this flow of African labour from white farms to factories might have been secretly disappointing to black patriarchs like Kas, but it did not dismay the government, which took the view that its policy of 'influx control' was still in its infancy and would be revised. Within weeks of Matlakala and Nthakwana's unexpected departure for Randfontein, the ruling party was out at the hustings advocating the need to extend and entrench apartheid at every level of South Africa's society and economy. In the National Party's ideal world, access to jobs for all black men would be restricted, and all black women would have to carry passes. In the Ventersdorp constituency, the man who could help turn all the white farmers' dreams into reality in the general election of April 1953 was said to be none other than Johannes Casper Greyling of the farm De Beerskraal, a property five miles north of Rietvlei.

Cas Greyling prided himself on his knowledge of reality. A tough, thickset, middle-aged man with small bright eyes, a weather-beaten face and a mop of grey greasy hair slicked well back, Greyling had a set of pork-sausage fingers that showed that he was no stranger to manual labour. After a childhood on the inhospitable fringes of the Kalahari Desert, he had gone on to the Normal College at Heidelberg, where he had qualified as a teacher and then, through application and sheer hard work, had continued his studies at the universities of Pretoria and Potchefstroom. It was there that he first found the space to express both his physical aggression and his Nationalist fervour. He represented his province at rugby and became actively involved in student politics that consciously revelled in racist authoritarian ideologies, developing the strongest possible antipathy to what he called '. . . sickly, sentimental, post-war arm-chair humanists and philosophers who know nothing whatsoever about the Bantu in practice; those people who idolize the black skin and despise the white skin.'[29]

A self-made man who survived the worst of the Depression by becoming a teacher, Greyling had collected most of the credentials necessary for an aspiring Nationalist politician by the end of World War II. In 1945, he gave up teaching and went to Natal, where he organised a local campaign for Afrikaans Christian National Education. Failing in a bid for a seat in the provincial government, he returned to the Transvaal, and in 1949 took up residence on the small property that he had acquired in the Mooi River valley. Working from his new base as a full-time 'farmer,' Cas Greyling joined various local cultural associations allied to the Nationalist cause and got elected to the Road and School Boards, as well as to the executive of the Transvaal Fresh Milk Association before going on to seek even higher office.[30]

As the newly elected but still relatively impoverished M.P. for Ventersdorp, Greyling showed great interest in local farming conditions. A man with a marked dislike of rural and urban 'black spots,' and with a keen awareness of his white constituents' need for cheap labour, Greyling was especially interested in the practical problems posed by black 'squatters'; Kas had long suspected him of being one of the men behind the Maines' premature departure from Muiskraal. Greyling's election to parliament was a bad omen for black tenants on white properties up and down the valley in general, and for those of Rietvlei in particular.[31]

As night-time temperatures dipped, warning that the approaching chill of winter would soon settle over the valley, there was mercifully no sign of Greyling. The sharecroppers' anxieties slowly subsided and then eventually disappeared as they became preoccupied with preparations for the harvest. It had been a good summer, with above average rainfall, and the fields at Rietvlei were full of promise. Faced with several pressing commitments and still haunted by his daughter's gloomy predictions about his losing his money on Doornkop, Kas was especially grateful: there was more than enough grain to satisfy the landlady and so, well before any public harvesting commenced, he sent his family out on a night harvest that they carefully concealed in his hut.[32]

A fortnight later everybody was back in the fields, this time in full view of the community, bringing in a sizeable harvest. With the grain beyond the clutches of any freak storms, Kas and Saul Ngakane arranged for a local farmer to do the threshing at Rietvlei and paid him in kind; they would be refunded by the landlady. In Rysmierbult Ngakane made a phone call to Johannesburg, and a few days later the corpulent Mrs. Hersch appeared on the property clutching a large black book.[33]

In the three seasons that Kas spent at Rietvlei, Paula Hersch's harvest ritual never varied. She would count the number of bags produced by each sharecropper, assess the size of the tenant's household, and then set aside what she considered a suitable amount of maize for each family to consume until the next harvest. This generous subsistence allowance was deducted from the total number of bags produced, and the remaining harvest divided by two, the results carefully recorded in the black book. The harvest was then loaded into trucks and taken to the co-operative, where it was sold in the widow's name. Once she had her cheque from the co-operative, cash dividends were posted to Ngakane for distribution amongst the sharecroppers.[34]

Kas did well—embarrassingly well if one added in the night harvest tucked away in his hut. Mrs. Hersch had been impressed by what she saw, even without taking into account what he had hidden from her gaze. Thirty years later he could still recall:

> When I arrived at Rietvlei I discovered that no tenant had ever succeeded in producing more than a hundred bags of maize. I ploughed and succeeded in obtaining one hundred and fifty bags. After that, the other tenants started ploughing even more seriously.
>
> MaHersch said to them: "You don't plough properly. Take a look at this man. He has come from a long way off—all the way from Bloemhof—to teach you how to plough. In his very first year here he has brought in one hundred and fifty bags. In all that time you have been with me no one has ever achieved that. In all the years I have known you, the best you could manage was fifty, sixty or seventy bags."[35]

Although this praise singing by Mrs. Hersch was a source of lasting pride amongst the Maines, it was probably less well received by the established sharecroppers at Rietvlei. Why should exactly the same rainfall on similar-sized plots lead to such vastly differing yields? It was known that the newcomer was also a powerful ngaka; what secret powers did he have? And what would Cas Greyling and the neighbouring Boers make of it once they heard the gossip at the co-operative? But if any or all of these

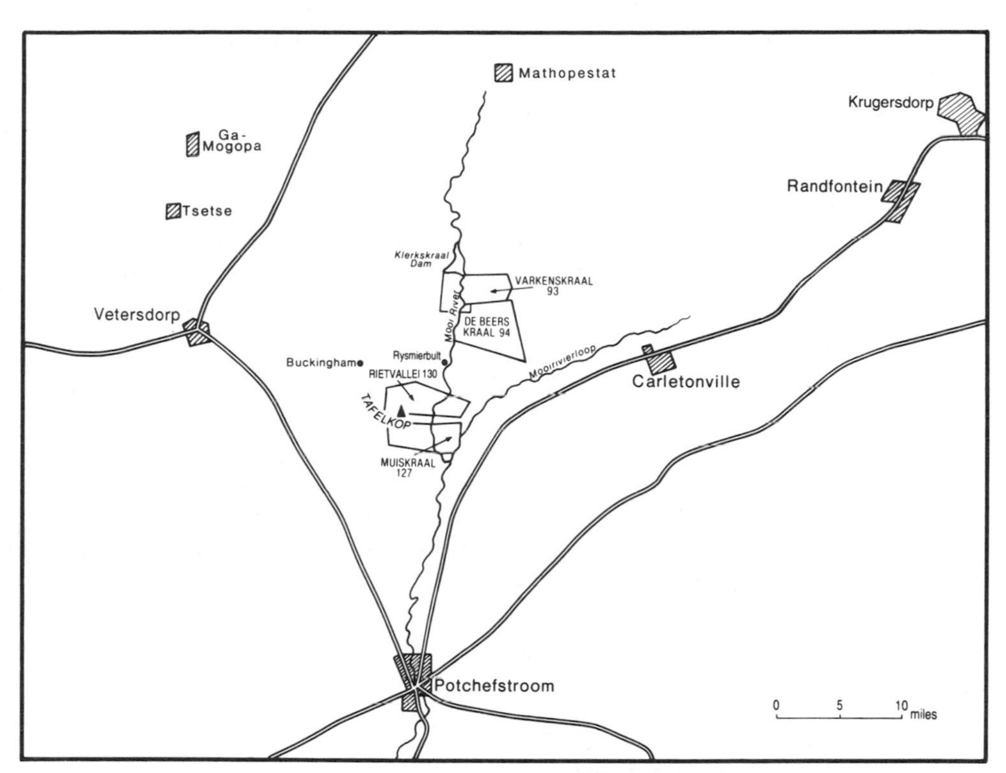


8 Mooi River valley farms in relation to the mining centres of the western Witwatersrand, 1951–56

thoughts did pass briefly through the minds of Ngakane and the others, they were soon forgotten amidst the end-of-season festivities as the men sat down for some serious beer drinking.[36]

The recent loss of Maine labour meant that, sooner or later, Kas would have to make more systematic use of his youngest daughters and perhaps even of his oldest grandchild, eight-year-old Pakiso. But this strategy, replacing the old blood that had been spilt from the top of the family retort with a transfusion of new from below, would have to be pursued with far greater concern for equity if he were to avoid some of the errors of the past. The first thing to do was to acquire a set of donkeys. Children, like old folk, found donkeys easier to plough with than oxen, and it was not long before Kas had acquired a mare and a foal for fifteen shillings.[37]

Unfortunately, not everything in the world was as unproblematic as a pair of donkeys. The ageing Ford, for example, continued to lock up badly needed capital and siphon off a steady trickle of cash as Kas tried to get all the minor repairs done. In theory the truck should have been income generating, but in practice its use was restricted to transporting Bodule and his family to and from his parents-in-law at Ventersdorp. It went against the grain to pour money into unproductive mechanical things, so it was only with the greatest reluctance that Kas decided that he would probably have to sell yet another animal, reducing the herd to fourteen oxen and twenty-six cows—nearly a third down on the number with which he had left the Triangle.[38]

Shedding livestock was worrying, and it was sickening to be reminded of the four animals that he had been forced to sell to pay for the repairs made to Dinta's Studebaker. But the matter was at last coming to a head. Nel & Klynveld had given a date for the hearing. Kas went to the lawyers' offices and then, much to his delight, The Louse suddenly agreed to pay for the oxen before the case went to court. But his joy was short-lived: before so much as a penny of the debt was collected The Louse did as scores of other white miners did up and down the length of the Witwatersrand each year, and slipped out of town to rejoin the circuit of rootless Afrikaner workers further down the track. Cas Greyling knew who such men were, since they formed part of the constituency that had helped to put him and his party in power. They were, he noted during a discussion on the labour vote in the House of Assembly, 'small farmers who have been forced off their farms as a result of climatic conditions and who have sought refuge in the cities.'[39]

Unlike these white men, black sharecroppers were without representation in parliament, and they had to fend for themselves in a system that no longer had much call for their services on the land and, at the same time, denied them the right to seek 'refuge in the cities.' Thus, when MaHersch paid the men of Rietvlei for their share of the harvest in mid-June, Kas hastened in to Ventersdorp to hand Gillett the thirty-four pounds which he believed might still help secure the stands at Doornkop. His attorneys led him to believe that this action would probably satisfy Harry Braude, and Kas also hoped it would help exorcise the ghost of despondency that Matlakala had let out of her bottle of predictions. Plots 31 and 34 would yet be his.[40]

Life seldom yields to counter-attack on all fronts simultaneously. Grown men know that progress is uneven, and so Kas went home ready to face new challenges. The voorskot from the widow Hersch did not go as far as he had hoped, and only two weeks later he had to take an ox to the auction at Muiskraal, where it fetched sixteen pounds. This return to the livestock sales as a seller rather than as a speculator exploiting the vagaries of the market did not go unnoticed.[41]

Dikeledi refused to accept that Kas had the right to sell any animal while her father was still waiting for part of her bridewealth. Kas, in turn, took the view that there could be no further dealings with Abner Teeu until the money that had been advanced to him for Disebo had been refunded. Dikeledi chafed away at the issue of the unpaid portion of her bohadi, never once allowing either Kas or Bodule to forget it. This festering sore cried out for attention, and in order to relieve some pressure on his son, Kas decided to build the couple a small hut, part of a 'traditional duty' activated at a moment when he had good reasons of his own to bind adult labour into the household, which was threatening to disintegrate.[42]

Kas was grateful that Bodule was still around. Most other sons of his age had already gone off to work in the city, but not even Bodule could compensate fully for the loss of the girls. True, Kas had hardly been heartbroken at Matlakala's disappearance, but Nthakwana's departure was different. She had always been a loyal and able companion, helping him with anything from animal husbandry to the spinning of wool, and now he lived in the hope that he might pick up a clue that would lead to her. He went so far as to get a pass from Jacobs, a neighbouring 'poor white,' so he could scour the surrounding villages for news of her whereabouts. But it was all in vain.[43]

Unbeknown to Kas, there was someone whom he both loved and trusted within the family who already knew the whereabouts of the women. But Bodule wasn't talking; there was a new caution, a certain hesitancy creeping into their relationship. He had heard of the sisters' whereabouts in a freakish manner: Matlakala, still devoted to Buti Thale, had sent the young man a letter bearing a Randfontein postmark; this had been spotted by a rather indiscreet third party who had torn the address from the letter and quietly passed it to Bodule. Unwilling to betray his sisters or to expose them to the wrath of their father, Bodule kept the information to himself and then used the cover of the off-season to pay them a clandestine visit. Their story was that of thousands of other black women in the 1950s.[44]

On boarding the grubby third class 'non-European' coach at Rysmierbult, Matlakala and Nthakwana had slipped into conversation with an older woman who was also on her way to the West Rand, where she was employed as a domestic servant. In Randfontein, she took them under her wing and gave them a place to spend the night in the servants' quarters attached to the outbuildings of her employer's home. The next morning she gave them a change of clothing and told them how to set about finding work as 'kitchen girls.' Unencumbered by the '*dompas*' which dominated the lives of black men, and willing to accept the lower rates paid to unskilled African women newly off the land, the sisters soon found employment in a white middle-class suburb. Their new jobs permitted them to visit one another at the end of the working day, which stretched from 6:00 a.m. to 9:00 p.m., or on their Sunday afternoons off.

With only an elementary social network at their disposal Matlakala, now 'Miriam,' and Nthakwana, now 'Rebecca,' at first spent most of their leisure time in each other's company. After a few weeks, Matlakala had a visit from Buti Thale, but anxious to avoid bringing 'disgrace' on her family and still traumatised by her teenage experience at the hands of the kgotla for alleged sexual misdemeanours, she forfeited her bed in the servants' quarters and took to sleeping on the floor in her employer's kitchen for the duration of his visit.[45]

Nthakwana, by contrast, revelled in the space the departure from Rietvlei had opened up. Socially precocious, she had always refused to put up with any nonsense from the men in her life, whether her brothers or her male friends. So brash and self-confident was her behaviour in the new setting that more cautious spirits started to fear for her well-being. Matlakala pulled her aside and offered her some sisterly advice:

> . . . I told her that since we were in town she should avoid that sort of [cheeky] behaviour because she stood the risk of being beaten up. I also warned her against falling pregnant because that would bring further disgrace to our father from whom we had already run away.[46]

After about six months, however, Matlakala noticed that Nthakwana had begun to 'get impatient when I visited her,' and soon she was often to be found in the company of an intelligent MoKgatla policeman by the name of Bernard Msimong, whom they had first encountered on an off-day excursion to the railway station.

When Bodule suddenly appeared in Randfontein one winter's day, the sisters at first feared that he was on a mission from Kas. But once he had reassured them, they were eager to find out all about the family, though they were at pains to say they were not yet ready to consider returning to Rietvlei. Still, the talk of the family softened their hearts, and when the time came for Bodule to go, each of them pressed ten pounds on him for Kas 'so that he might buy a few bags of grain,' but pleaded with him not to tell Kas of their meeting or of their whereabouts.[47]

Bodule was trustworthy, but not everybody else was. Not long after his departure, Nthakwana's life turned a corner. Peering ahead she could glimpse only one possible destination—home! Matlakala was one of the very first to know why:

> One day Nthakwana came to me crying. Well, I had the ability to determine what illnesses people were suffering from, and I used tea leaves to help me in my diagnosis. I examined her and told her that she was in trouble. I asked her whether she was still menstruating and she said that it had been two months since she had last menstruated. I asked her whom she had been sleeping with, and she said that it was the fellow whom we had seen at the station.[48]

After this consultation, visits from the 'fellow at the station' tailed off. Nthakwana tried to find out where exactly the policeman lived, and made the painful discovery that countless women in similar positions have made since the beginning of time: Msimong was a married man.

Alone in a world which only weeks earlier had seemed almost infinite in its space and wondrous opportunities, Nthakwana suddenly sensed that the distance between town and countryside was shrinking fast. Deserted by one man in Randfontein and frightened by another at Rietvlei, she saw no easy way out of her predicament, and in the end it was Matlakala who came up with what seemed like a solution. Matlakala would go home '. . . to see Kas, so that Nthakwana might be welcomed back'—a generous offer which ignored the question of how Matlakala herself was likely to be received.

She waited for a suitable weekend to go to Rysmierbult, creeping into the homestead to tell the women her tale while Kas was out in the fields inspanning the oxen. 'Joo! Joo! When Lebitsa told him that I was there, he called for a knife to kill me with! He accused me of persuading Nthakwana to abscond. I invited him to kill me, but he eventually gave up.' At the end of two tense days Matlakala was grateful to return to the relative tranquillity of suburban Randfontein, where she 'told Nthakwana that while conditions at home were hardly good, it was possible to return.'

Nthakwana needed time to catch her breath. Weeks passed before she was ready to go home: inwardly chastened, she now sought refuge at the very place from which she had once fled. Unlike men, whose urban adventures nature contrived to keep secret, her very profile spoke of her indiscretion. But family demands, which had once constituted the sharp rectangle of prison life at Rietvlei, assumed the rounded contours of home as she entered the last months of her confinement. Lebitsa, Leetwane and the younger sisters were unfailingly understanding and supportive. Kas, always more ready to forgive those able and hard-working siblings who had been 'misled' by outsiders than those who were slow, ponderous or merely 'stubborn,' positively welcomed her return. This willingness to forgive the favoured daughter extended to her offspring: when Nthakwana gave birth to a daughter, Ntholeng, she, too, assumed favoured status in the eyes of her grandfather. Her virtues were compounded, so it was said, by the superior intelligence which she happened to have inherited from her father, Msimong the policeman.[49]

With spring only weeks away, Kas found an upturned tin under a nearby tree and got to work on a pile of shoes belonging to friends and family. On this relatively isolated property, with a clientele of sharecropping families and some workers from the diggings at Muiskraal, cobbling was once again a viable option for winter cash. Closer to the cooking pots and the shacks, several Maine women wove baskets, while in the distant background Leetwane collected eggs which she either sold to neighbours or took across to Cohen's store. As Kas's bone-hard hands forced the awl through the strips of leather and his elongated fingers drew the waxed thread through the holes, his mind ran through the last of the off-season chores. When Bodule took command of the new ploughing team, he would pay his brother a visit. Phitise had left for Doornkop months before, and the time had come for him to assess for himself the conditions at Doornkop.[50]

But first he would have to ask Jacobs for a pass for yet one more trip into Potchefstroom. That great bureaucratic monster, the Local Road Transportation Board, needed to be stroked yet again before it would let them use the Ford to transport a few passengers over weekends and on public holidays. It would be left to the likes of Cas Greyling and his friends to decide whether or not one black man's truck posed a threat to the interests of the South African state or of competing white businessmen. Fortunately, the responsibility for pronouncing on Bodule's driving skills lay elsewhere, with the distant Transvaal provincial traffic authorities.

When it came to helping blacks with passes, Kas thought Jacobs was obliging, like many 'poor whites' he encountered in the Triangle. It was one of those strange things that just happened in a racially obsessed order: every now and then the static of class affinity would jump the gap that was supposed to separate the electrodes of colour. Jacobs gave him a pass scribbled on the corner of a page torn from a school exercise book, and on 1 September, Kas handed a deeply disaffected clerk in Potchefstroom thirty shillings to cover the cost of the application. A few days later he got a printed notice that the matter would be considered only three weeks later; on 25 September, the entire procedure had to be repeated. Jacobs might give him a pass, but there was no telling what those South African Police patrolling the streets would do once you got into the streets of 'white towns.' The whole business was unbelievably tedious.[51]

At Rietvlei, Kas mobilised the new ploughing teams which now included his thirteen-year-old daughter, Nkamang, and Morwesi's eight-year-old girl, Pakiso. In essence, this involved putting into the field two teams of differing age, strength and experience, employing differing combinations of technology. While the boys used the oxen with a heavy plough, opening up the virgin soils Ngakane had allocated them, Kas and the girls used the donkeys to draw the lighter plough. This prompted him to reflect on how, as a newly married young couple at Kommissierust, he and Leetwane had gone out into the fields on their own to do battle with the light wooden-beamed 75 plough while now—at the age of nearly sixty—he was out in the fields with the first of his grandchildren.

Kas made few if any allowances for differences of age or gender, and any slacking or deviance from his instructions met with swift rebuke. And yet, beneath the dry and slowly wrinkling skin of his weather-beaten face, one could sometimes detect the hint of a softer and more thoughtful attitude. Perhaps deep down the departure of three children had slowly evoked in him greater tolerance of human frailty and an understanding of the need to show appreciation for a person's efforts. Or was it simply a little old-fashioned grandfatherly indulgence? In either case, the girls did well, and once the ploughing was completed, he took Nkamang and Pakiso aside and planted them each a patch of their own, from which, he assured them, they alone would benefit.[52]

The early rains were up to the task. Once the first maize and sorghum were safely in the ground, Kas was ready to go and see for himself what the promised land at Doornkop looked like. He got yet another tatty pass from Jacobs and, on 16 October, he and Bodule left, going first in the direction of Ventersdorp. Bodule knew the road well because of frequent visits to his parents-in-law, but as they drove out past the elegant stone kraals that Phitise had constructed, Kas was relieved that it was his brother rather than Dikeledi's parents whom they were going to see. There was not much love lost between the Teeus and the Maines.[53]

They had no real trouble in finding their way to Doornkop or in getting to Phitise's place, even though it was dark by the time they arrived. Not only did Kas remember more or less where his brother's property was located; there were several locals who could point them in the direction of the energetic newcomer who had settled in their midst. Around the fire that night they talked about old times in the Triangle, and Phitise spoke of his plans to sink a borehole and build a shop. There was no doubting that the place had great potential, and Kas was comforted by the thought that he and Phitise could help one another as they grew older. He went to bed thinking about all the foolishness of Matlakala's 'prophecies.'

But the voices of the ancestors are not to be trifled with. In the morning Kas saw his dream of retiring to Doornkop disappearing before his very eyes: he found someone else, a complete stranger, already living on the very land he had set his heart on! *'The plot that I had bought had been sold to me and to the other man.'*[54] Agent, bureaucrat, clerk, swindler—who cared? What did it matter? Like raindrops vanishing in the dust they spoke only of what might have been. The plough and the soil had been betrayed by the quill and the ink; perhaps it had always been so. Braude had contrived to sell one plot to two people.

Kas was filled with remorse. He had driven the family to the point of disintegration with his demands, he had struggled to earn the additional thirty-four pounds that Braude had demanded, and he had spent a great deal of time and money on getting legal advice—only to find it all a mirage. Even if he did establish his ownership, how could he possibly meet Braude's demands for monthly payments? Harvests read seasons, not calendars. He had tried and lost, so he decided 'to drop the whole matter.' Three days later he went back to Gillett and instructed him to recover such money as he could from Braude.

Farming, by its very nature, gives a man very little time for self-indulgence. In any case, it was not in Kas's nature to dwell on what might have been. The loss of the land nevertheless tasted like bitter aloe and it was several days before he was ready to take consolation from the fact that the stands at Molote were paid for. That at least was a secure accomplishment in a world in which the options open to a sharecropper declined with each passing year. It was important to take life season by season, and this year, thanks partly to the efforts of Nkamang and Pakiso, the fields were looking good.

Kas's strategy of using those on the bottom rung of the family ladder to replace the labour of those leaving at the top was running into a familiar problem: Leetwane, who was once again fretting about the girls not being well enough educated, pointing out that the family had access to a nearby farm school. 'Bantu Education' was intended more to provide a white electorate with a supply of cheap black labour than to generate an educated workforce, but Mr. Letebele did at least teach children how to read and write. Their mothers' relentless campaign for Nkamang and Pakiso to attend school was successful: despite Kas's opposition to the idea of educating children who were going to be neither priests nor teachers, he consented to their attending classes and paid their school fees on condition that they continue to do herding jobs during the summer afternoons.[55]

As fourteen-year-old Nkamang soon found out, even this curtailed roster could be very demanding, for in addition to the unknown disease that continued to stalk the sheep in the valley there were many jackals on Tafelkop. One afternoon, while Pakiso was working in the fields, Nkamang was entrusted with the flock which included a ewe that was about to lamb. She drove the sheep up the mountain path to a familiar grazing ground, and when the time came to take them home and the number tallied with the one she had started with, she drove them back down the hillside. But at the kraal gate Kas detected the ewe, no longer pregnant, and asked why it had returned without the lamb. When Nkamang admitted that she hadn't noticed that the ewe had lambed, he took a switch to her and sent her back out into the dark to recover the missing animal.[56]

Frightened at the prospect of walking over such a large expanse on a moonless night when the jackals were afoot, she simply walked a short distance into the bush, waited for a while, and then returned to the shack to tell him that her search had been unsuccessful. Kas exploded, and only the physical intervention of Leetwane and the others saved Nkamang from a full-scale thrashing. To placate him, Mosala and Nthakwana eventually agreed to go with her on yet another foray into the dark. But that, too, was futile. At first light he was still raging on about the lost lamb. Nkamang, in desperation, took the flock back along the path they had followed the day before and the mother somehow found the lamb for itself. Even Kas was impressed by this; by the end of the season, the patriarch would proudly introduce Nkamang and Pakiso to interested outsiders as 'my herders.'[57]

The problem of getting the children to attend to their farming duties was not confined to the junior ranks of the family. Shortly before Christmas Dikeledi had given birth to a second son, Ratshilo, so she was preoccupied and although Bodule was still far and away the best and most able of his helpers, Kas could not help noticing that he too sometimes paid slightly less attention to detail than he had done in the past. Perhaps it was the burden of parenthood or perhaps Dikeledi, who had yet to be pacified on the bohadi issue or to find herself an appropriate niche in the Maine household, was sapping her husband's commitment to collective enterprise. It was difficult to tell.

The second week of January saw yet more paper-chasing as Kas once again had to go to the Local Road Transportation Board and then buy third-party insurance for the truck. All this cost him slightly more than five pounds over four weeks—and he was still responsible for feeding two wives, six adult children, a daughter-in-law, and three grandchildren. To make matters worse, the Ford was once again consuming spares at a frightening rate. Unfortunately, not much of the expense could be offset against the small sums that the occasional hire of the truck brought in, and its driver was not always around to help, either. In mid-February Bodule, who was showing ever growing confidence in his ability to deal with the greater Witwatersrand transport network, took several days off to visit friends in Johannesburg and Vereeniging.[58]

At Rietvlei the fields of ripening grain gave Kas the compass by which he steered his family. The double-ploughing technique learned under Hendrik Goosen, along with careful weeding, ensured that he would once again achieve a higher yield of maize per acre than most of his neighbours, and this meant that once again he could organise a 'night harvest' and once again please the landlady. So despite the setback with Braude, Kas thought he had much to be grateful for. Ever since the family had left Jakkalsfontein and moved farther north and east, the rains had held up well, and at Rietvlei he was blessed with a supportive foreman and no landlord. White widows like Maria Coetzee, Anna Edwards and Paula Hersch were often a black farmer's best friend—at least until an older son or some new man entered their lives to help run the farms. As in most matters, the answer lay in seizing the moment. On a cold night in May several of the Maines slipped out of the shack for some stiff-fingered harvesting and set aside a reserve of twenty bags in the back of Kas's hut. A fortnight later, with the winter sun struggling to bring warmth to shuffling legs, the women brought in the 'official' harvest, which Ngakane once again went out and arranged for Bouwer to thresh.

Saul Ngakane mouthed the magical words into the telephone at the Rysmierbult post office, and, a few days later, MaHersch materialised along with her black book. The Maines, it seemed, had produced an astonishing two hundred bags of maize. This time, not content with making a short speech lauding his virtues as a farmer, Mrs. Hersch in full view of the other sharecroppers gave Kas 'an extra ten bags because you put so much effort into your ploughing.' But exhorting tenants to emulate a man's work with the plough was one thing; paying him a bonus was another. This posed dangers of which the landlady could only have been vaguely aware. Back in Latvia the loathing of Jewish successes had come from outside the community, but in Africa it was the enemy within that people feared most. People went home taking silent exception to an act that singled out an individual at the expense of the community.[59]

In the Mooi River valley, there were others who resented the entire community of black tenants. White farmers, many of whom were benefiting from an extraordinary rise in grain prices that accompanied the outbreak of the Korean War, were searching high and low for yet more ground to put under maize. For the Afrikaner Nationalists amongst them, the very existence of sharecroppers in their midst was not only economically unpalatable but a political affront at variance with Verwoerd's avowed policy to rid white properties of independent black producers. The Hersch property became the subject of increasingly envious or resentful glances. Cas Greyling, newly elected Member of Parliament for Ventersdorp, repeatedly approached Percy Stabin, the lawyer representing the Hersch family, with requests to hire or purchase the farm. But Paula Hersch saw no reason for disposing of the property or for reneging on what was a convenient and profitable agreement with her black tenants.[60]

But in a racially ordered society, one woman's 'tenant' was another man's 'squatter.' The situation was fraught with danger, though the foreman and sharecroppers at Rietvlei were unaware of any behind-the-scenes manoeuvres. Kas, in particular, was interested only in enjoying the spoils of the season: the part of the harvest he had openly shared with the landlady; the illegal 'night harvest'; his bonus; the family's official subsistence allowance for the following year; and, perhaps best of all, a further sixteen bags that came from the two patches prepared for Nkamang and Pakiso. The ancestors had obviously smiled on his efforts to make special provision for the youngsters, which helped to account for the generous rainfalls. All that was now required was to fulfil the promise to give the girls the full benefit of their fields.

Kas got out the horse and cart, proudly escorted the young ladies to the local trading store, and there told them to choose whatever they fancied; a privilege previously granted only to Bodule and Mmusetsi. Who knows what thoughts passed through his mind or what debts he was endeavouring to settle on that blustery July day in 1954? Kas stood in the doorway as the girls scurried around picking out and putting aside so many items that the storekeeper worried whether they could pay and asked loudly whose children they were. Kas stepped out of the shadows and, in ringing tones, accepted responsibility for 'the herders of my sheep and cattle.' The girls had selected several practical items, such as blankets and shoes, but they were most taken with several lengths of prettily printed cloth. They took everything home, where the versatile Nthakwana fashioned the cloth into dresses in the style of the day. The whole exercise was an unqualified success, and gave them all many happy memories.[61]

With the cash proceeds of the harvest still to follow and the property at Doornkop no longer making demands on his purse, Kas entered the off-season in a more relaxed frame of mind than usual, which eased tensions throughout the family; while he engaged in some cobbling, herbalism and smithying, the women spilled out onto the adjacent farms and worked in makeshift harvesting teams. Most of the maize or money earned through these efforts found its way back to Kas, but one or two of the younger daughters, such as Pulane, made a point of giving their earnings to Leetwane, and, in addition, she went on selling eggs and baskets. Nthakwana, accepting responsibility for the baby, Ntholeng, took to knitting sweaters which found a ready market amongst workers around Muiskraal. With the exception of Dikeledi, who never let up on her campaign for the settlement of the bohadi debt, most of the family seemed to breathe more easily.[62] It appeared that everybody was nodding off to the slower beat of winter's metronome—everyone except Dikeledi.

Realising that her arguments about bohadi were making little headway with Kas, she changed tack and suddenly started badgering Bodule to accept moral responsibility for his father's 'debt.' This hazardous ploy, which involved chipping away at her husband's loyalty to his father, produced results more rapidly than even she dared hope for. Before the winter was out, Bodule suggested to Kas that he might leave Rietvlei for a while and work in town.

If this was a declaration of independence, then Bodule was at pains to disguise it. He bent over backward to reassure his father that the search for work would take him no farther than forty miles northeast of the valley, to the nearby mining town of Carletonville; that he would make certain that most of his earnings would come back to Kas and help with the bohadi debt; and that he would return to the farm on weekends whenever possible. How could Kas refuse? His son was merely asking for his freedom as an adult, which some of his daughters had already taken without even asking. Had he not done much the same thing when, as a young man, he left Sekwala to work on the roads? Bodule had worked as hard and as faithfully as any father could wish for, he was the father of two children and, as Kas was the first to acknowledge, had never once placed his personal interests before those of the family as a whole. As spring came to the valley Bodule left Rietvlei for Carletonville, where he became a plumber's assistant.[63]

It was an amicable departure, and yet, as with so much in life, the very banality of an everyday occurrence screened matters of profound material and psychological importance. It's often only the curtain of the commonplace that hides the tears of the victim from the unwanted stares of passers-by. All that was happening was that a farmer's son was going to town to become a plumber's assistant, but, at a deeper level, the Maine family was losing its grip on white-owned farmland. A sharecropping family without the services of an older son was like a wagon without a wheel. With Bodule's departure Kas lost ready access to his most trusted adviser, his closest friend and his warmest confidant. For a laconic man like Kas, whose desire for self-reliance sometimes pushed him toward introversion, life would never be the same. And, as if to confirm the private doubts to which he dared not give public utterance, the new season got off to a disastrous start.

It was not that the remaining family members were unwilling to work at ploughing and planting, but had to do with the whims of nature and an indefinable tension in the dry Rietvlei air. Mosala, Nkamang and Pakiso all helped him, and although the girls especially worked hard, their efforts were of limited value when the rains were so paltry. With too much heat and too little moisture the crops withered prematurely or grew into ugly little plants with leaves twisted into grotesque shapes. Almost as mysteriously, the tenants started quarrelling and looking askance at their neighbour's fields.[64]

This uneasy atmosphere worsened when the Department of Native Affairs suddenly decided to clear the valley of black tenants, as advocated by Verwoerd. As a first step in this coercive process the so-called Native Commissioner at Ventersdorp issued instructions that all black workers in the district report to his office to register and collect new 'reference books.'[65]

At the Native Commissioner's offices, Kas and Mosala joined a long queue of, in the insidious paternalistic idiom of the day, ' "boys" from the farms.' There, the sixty year old and his nineteen-year-old son rubbed shoulders with hundreds of other 'boys,' black men who, regardless of age, dress, income, physique or status, had their thumbprints taken like so many common criminals before being issued with a 'dompas' that, overtly, recorded the date and place of their birth and their ethnic identity and covertly determined where, when and at what rates they could sell their labour in the land of their birth. Under these fraught circumstances it was hardly surprising that the sharecroppers at MaHersch's wondered who was responsible for bringing down this bureaucratic plague upon them, and how long it would be before they would be asked to make way for the white farmers who wanted Rietvlei.

On the way home Kas and Mosala each reflected privately on what the passbook was meant to achieve. For Mosala it was a clerical ball-and-chain, a document designed to keep him in the countryside and deny him access to the lucrative labour markets of the Witwatersrand. Indirectly it made him subservient to patriarchal authority. The cautious and pragmatic Kas saw it differently. For him the book, like virtually all government documentation, had the potential both to imprison and to free. The Native Commissioner, on learning Kas had been born before the rinderpest, had told him he might well be exempted from the poll tax and be eligible for a state pension. For Kas and others like him, the dompas might well be the key to hitherto unknown benefits. Time alone would tell.[66]

Unfortunately, not everything in life could wait for time. At Rietvlei the maize, sorghum and sunflowers continued to suffer from the drought. Kas and Saul Ngakane, who were spending more time together since Bodule's departure, discussed the situation, but there was nothing to be done without rain. Neighbouring farms were no better. Visitors up and down the valley spoke gloomily about the coming winter, pessimistic talk that served only to unsettle people further.

Not all the visitors to the farm were bona fide guests of the inhabitants. Complex legal provisions allowed strangers to funnel through sections of the property at all sorts of odd hours of the day and night. This made the tenants responsible for guarding the livestock and other possessions all the more tense. Yet this unwanted traffic rendered an unexpectedly pleasant surprise.

One afternoon in November 1954, Kas and Mosala were out in the fields with the truck when a white man drove past, stopped and then turned around to take another, more careful look at the Ford. Kas watched, motionless, as the Boer got out of his car and moved very slowly toward them. He edged within speaking distance and then introduced himself as J. C. Marx from Cyferbult, near Buckingham.[67]

The seventeen-year-old Ford was a complete liability. It was a good time to get rid of it. The only problem was that Marx, like so many of Cas Greyling's peri-urban constituents, was yet another economically marginalised Afrikaner with no capital. In the end, after a great deal of preliminary negotiation Kas decided to take a chance: the Ford was exchanged for an ox-wagon and forty-six pounds in cash payable in December. Marx recorded the transaction for them in the unsteady hand characteristic of most rural folk whether black or white. Marx, unlike Dinta, was an honourable man, and he paid his debt. Afrikaners, like everyone, came in all shapes and sizes. Despite his joy at getting rid of the Ford, Kas knew that all was not well. The exchange of a truck acquired in a time of economic strength for an ox-wagon in a season of financial weakness epitomised the fragility of his defences. He was being left behind in the great race for progress, a race in which he had always been amongst the very keenest competitors, and there was very little he could do about it. He felt vulnerable, and over the next weeks he felt worse still.[68]

Early one morning shortly after the advent of the new year, Kas took stock of what was left of the remaining maize, sorghum and sunflowers in the field. He was in a sombre mood. Then it slowly dawned on him that perhaps it was not just lack of rain that was destroying the crops. Had he not lived through droughts before and survived? No, it was unresolved tensions within the community that accounted for his failure! 'There were fights between the blacks at Rietvlei and they made use of potions to impoverish one another. My crops died without any apparent reason. I realised that the crop was bewitched.'[69]

Yet again he was being victimised for his virtues rather than his vices. Back at Sewefontein he had had to endure the envy and resentment of neighbouring whites; at Rietvlei it was the blacks. 'They had once heard Mrs. Hersch remarking, "Good Heavens! You have been outfarmed by Maine. He has only just arrived and yet he has produced so many bags." It seems they became jealous.'[70] Success and failure were, in the end, two sides of the same counterfeit coin.

Yet not even hidden enemies exempted a man from the need to feed his family and, as the battered grain struggled to set, Kas detected a new problem: thousands of small insects, *makgwaba*, were invading the maize and sorghum. He did his best to minimise the damage by snaring a large hadedah and placing the outstretched wings of the bird over the fields. But eventually he conceded that he was using the charm far too late, for most of the damage had already been done by the time he got around to killing the bird. The plague shrank the harvest even further.[71]

Just as the last nibble of autumn gave way to the first bite of winter, the blight that black sharecroppers feared most of all suddenly reappeared. The Member of Parliament for Ventersdorp and a few vociferous supporters started visiting the property at odd moments, complaining about 'Jews being made rich by kaffirs.' Cas Greyling told Maine that the issue of 'squatting' had been discussed in parliament and that '. . . there were many whites who did not have a place' to farm while others said the police were doing nothing to enforce the law.[72]

Without a landlord as a buffer, the tenants at Rietvlei were made uneasy by these pronouncements, but nothing more happened and they regained their composure. The Maines collected a modest 'night harvest,' carefully stacked in the back of the shack. Kas relaxed a little. A week later, on 20 June, he visited Phitise, and while at Doornkop went across to the BaKwena-ba-Mogopa farm at Zwartkop to visit Thakane, whom he had not seen since leaving the Triangle. It was a pleasant reunion but a costly one, because she borrowed thirty-eight pounds from him: Kas took the precaution of getting her to sign an I.O.U. witnessed by Phitise and two of her male friends.[73]

Cas Greyling reappeared. This time, in addition to the usual menacing talk about Jews, squatters and lawyers, he expressed interest in buying two of Kas's oxen. Kas was not averse to the idea—the loan to Thakane and the prospect of a poor harvest meant that cash would be at a premium—but he was mindful of whom he was dealing with. He told Greyling he would need the written permission of the farm owner before he could dispose of livestock, and Greyling agreed to wait. When MaHersch's son, Oscar, chanced to visit the farm on 28 June, Kas got him to draft a note to assure any prospective buyer that the animals belonged to the tenant and were not stolen.[74]

As far as Kas could see, the landlord at De Beerskraal was a 'poor white man.' Cas Greyling could not have found it easy to do business with an independent black farmer; not only was it personally humiliating to have to deal with black sharecroppers whose demeanour distinguished them from the cowed farm labourers over whom he and his constituents presided, but their presence in the valley was an affront to the theory and practice of racial segregation which he championed. When he called to collect the two oxen a few days later he was in the arrogant and belligerent mood that characterised most of his parliamentary performances. The old regime, which allowed people like Mrs. Hersch to turn 'black folk into bosses,' he announced, was on the way out. He, Cas Greyling, was about to hire Rietvlei so that it could be cleared of 'squatters' and farmed by white men.[75]

The prospect of having Greyling as a landlord was unthinkable. Kas rushed to Saul Ngakane to find out what was happening, but the foreman knew of no plan to let the property. The two of them went to the post office at Rysmierbult to phone the landlady, who reassured Ngakane that she had not even been approached about the matter. It seemed like just another false alarm in a bad year. Cas Greyling reappeared a week or two later to tell Kas that the oxen that he had bought had sickened and died, but he made no further mention of hiring the property.

Kas's share of the harvest was so small that he did not even bother committing the sum to memory. Mrs. Hersch came, bustled about with the big black book and then left for Johannesburg. Shortly thereafter Kas went with Mosala to the police station at Rysmierbult, where the Bantu Commissioner, remembering their earlier encounter at Ventersdorp, gave him and several elderly men from other farms certificates exempting them from the poll tax. Kas felt vindicated. Sometimes he found it difficult to understand why it was that the young made such an awful fuss about the reference books.[76]

But the Afrikaner Nationalists knew what they were doing. For the likes of Hendrik Verwoerd, Cas Greyling and a growing army of bureaucrats, *all* Acts, Bills and instruments at their disposal 'fitted into a pattern and together formed a single constructive plan,' apartheid. And Greyling had been far from idle. Once again there was talk of action being taken against 'squatters.'[77]

> Mrs. Hersch then arrived and gave us the same story that I had once heard from Walter Moormeister. "My boys, we have always got on well, but the Boers are now threatening me with committees and lawyers, saying that this farm is occupied by kaffirs while their own children have no land to cultivate. I will find some land for you at Vermeulen's." She chose Ngakane and me because we produced more than the other tenants.[78]

There was not much more to be said. They would have to leave.

In 1955 as in 1949, the Afrikaner Nationalists, after careful and conscious deliberation, were putting the Maines and black families like them under the political lash. Twice within six years whites had used their power and influence to attack and render inoperative such paternalistic structures as remained in the countryside so that their demands for an exploitable pool of cheap black labour could be readily met and so that they might benefit from the apartheid regime's twinned objectives of racial segregation and capital accumulation. Only the cruelest twist of fate dictated that Kas should have chosen to construct his last major defence of peasant enterprise on the very site of an enemy outpost. A black sharecropping family fleeing across the face of inhospitable white farmland and already riven within by generational, gender, and personality conflicts was no match for the organised strike power of the Boers. The Maines prepared for yet another move.

Even at that late stage, an experienced campaigner could still find ways of hiding his loved ones from the pursuing troops, putting his head down and surviving for just one more season. Here and there, if one only knew where to look, a few tumble-down paternalistic structures could be found dotting the increasingly bleak and devastated highveld social landscape. Kas put his faith in MaHersch.

Paula Hersch knew the names of several farming families in and around Klerkskraal, six miles north of Rietvlei. The Herschmanns and Rosenthals, for example, had lived in that district for many years, and not only occasionally sub-let sections of their own property, but also knew the names of other landlords who might be willing to take in a few sharecroppers. It was through this network that she had first heard of the farm Varkenskraal (Tswana-speakers preferred to call it Mamanthana) where a friend of hers had hired land to a certain Vermeulen. She got out the black book, made a few phone calls, and then told Kas and the foreman where they might find Hans Vermeulen, who was waiting for them.[79]

Vermeulen did not actually live on the property, which was always an advantage, and the farm was close enough to the upper reaches of the Mooi River, which flowed to the nearby Klerkskraal Dam, for the livestock to have easy access to water. Grazing, too, seemed adequate. The only problems were the many poisonous snakes and the fact that the land set aside for arable farming had never before been put under the plough. Kas and Ngakane exchanged notes in the vernacular and then told Vermeulen that they would join him at Mamanthana as soon as they had wound up their affairs at Mrs. Hersch's.[80]

At Rietvlei Kas told the family to prepare for a short move north which would leave the Maines only twenty miles short of their ultimate refuge, the stands at Molote. His wives and daughters packed up their modest collection of household goods, while he and Mosala assembled their formidable array of agricultural equipment, leather goods and hand tools. On 30 August, only days before leaving Rietvlei, Kas and Mosala took ten bags of white maize and six bags of sunflower seeds in to the Buckingham branch of the Sentraal Westelike Koöperatiewe Maatskappy Beperk (S.W.K.M.B.), and sold them for twenty-one pounds, ten shillings and sixpence.[81]

Not having to retreat into the distant wildernesses which had once been known as 'native reserves' and which the word-wizards of the regime were now calling Bantu Homelands, the Maines were amongst the last to leave Rietvlei. They left behind them Bodule's first home, the hut Kas had built for his oldest son. Leaving the farm for the last time, the family passed several other mud houses with gaping holes where fleeing tenants had plucked door and window frames as they ran before the advancing nationalist storm. Where sheets of corrugated iron held down by rounded stones or giant pumpkins had once protected peasant living rooms, the summer sun now scorched deserted earthen floors. The unknown white farmers to come, seeing only the mud shells left by squatters, would demolish what had, for a decade and a half, been the homes of a self-sufficient sharecropping community. The Nationalists had already shown they could eat up the ghetto homes of South Africa's urban blacks for purposes of segregation rather than social upliftment, and now they were developing what would become in due course a truly gargantuan appetite for rural homes. As Verwoerd told anybody who would listen, 'Everything fitted into a pattern and together formed a single constructive plan,' apartheid. In another world, in another place, Oliver Goldsmith had once witnessed a not dissimilar phenomenon and written:

> One only master grasps the whole domain,
> And half a tillage stints thy smiling plain . . .
> Far, far away, thy children leave the land . . .
> Where then, ah! where shall poverty reside
> To 'scape the pressures of continuous pride?

At Varkenskraal, his mind still as far away from abstracted generalisations as it had been on the day three decades earlier when he was first approached by the organisers of the Industrial and Commercial Workers'

Union, Kas had urgent immediate tasks: to find a suitable site for the homestead, to build a kraal and some shacks. Then he and Ngakane paced out and marked plots of their own and made a preliminary survey of all the bushes, trees and stones that would have to be removed before they could start ploughing.

Mosala helped Kas clear Vermeulen's field, a task which was less tiring than they had feared, since the soil had been turned so often in the past. As Kas suspected, the real problem was the new fields, which had been allocated so as not to be, among other things, too visible to the likes of Cas Greyling, whose own farm, De Beerskraal, was only a few miles south. Without his two oldest sons to help him, Kas found working the new fields very taxing. Mosala did his best, but even he had to admit that the family was desperately short of muscle power. A little way off Ngakane, too, battled to clear virgin bush by day. By night, around a fire, the men reassured one another that it was all worthwhile, for the soil looked so promising and the rains were on schedule.[82]

This up-beat assessment was vindicated over the next weeks, and by the time the plumber's assistant returned from Blyvooruitsig in early December, the first sowings were showing signs of promise. Bodule came home to enjoy the statutory break that employees in the building industry were allowed, but he decided to stay on for the birth of his third child, who was due later that month. He was disappointed that not more had been accomplished in his absence, and asked Kas to let him open up another patch of land so that he, too, might capitalise on the rains. It seemed a good idea, all the more so because none of the fields was in any way comparable to those at Rietvlei.[83]

Right from the outset, Bodule let it be known that he was now the head of a household in his own right: his relationship with Kas had changed. This was evident not so much in what he said—for he, too, was a quiet man who often preferred the company of his hunting dogs to that of people—as in what he did. He chose a patch far away from that ploughed by Kas and Mosala, cleared it all by himself, and then signalled his long-term intentions by building a separate grain store and a new hut for his wife. After six years of married life she would no longer have to draw mealie meal from her father-in-law's house.[84]

Kas never overtly challenged these outward manifestations of changing inner convictions. As far as he was concerned, they were part of the life cycle and had been mercifully slow in coming. Yet Bodule's declaration of independence gave rise to minor resentments and tension. Dikeledi, always quick to detect a change in Kas's behaviour, had the impression that her father-in-law sometimes felt that the process had not been taken far enough, that maybe Bodule and his entire family should move right out of his orbit. Bodule, however, was more upset by the harsh words exchanged when his father simply slipped back into his old habits and assumed that his son would weed his field after spending hours working on his own.[85]

This argument left a gash in their relationship, but it was largely forgotten when, on 23 December, Dikeledi gave birth to her third son, Mokentsa. The arrival of the baby, who like all the other grandchildren was ushered into the world by Leetwane, was in itself an occasion for celebration, but it became an even bigger one when Morwesi appeared. She had been away working at the bottle store in Potchefstroom for a long time, but she wanted to spend Christmas with her parents and her daughter, Pakiso.

The composition of the Maine family was changing rapidly. Nthakwana had once again left home, this time to be near Matthews Moate, a new man in her life who worked on Faan Naude's farm near Rietvlei. Mosala had slipped away to visit his lover, a woman named Motshidisi Mogwase. As Kas surveyed this gathering he could not but see that his children were moving out into a wider world, and were being replaced by the five grandchildren.

Yet not even that pattern was clear or constant. The children (with the notable exception of Matlakala and Thakane, daughters of the 'second' and 'third' wives) tended to drift in, and out of the household. As with most young adults the world over, the first break from the family was hardly ever complete. The only permanent loss appeared to have been Mmusetsi. Bodule and Nthakwana shifted around, depending on circumstances. Indeed, later that month Morwesi suddenly told them she would not be returning to Potchefstroom. She had been offered work as a domestic servant at Klerkskraal, where her employer was to be none other than Sergeant Jonker, the very policeman whom Kas had got to know so well during the years at Kareepoort.[86]

A job in Klerkskraal meant that Morwesi would be closer to Pakiso who, at the age of ten, was a major help to her grandparents. Like Nkamang before her, however, she was soon subjected to many pressures that came with working for her grandfather. He would send her out late at night to find stray sheep or goats even though she had a great terror of owls; he was impatient when she shied away from dealing with an ox that had pulled the plough clear of the furrow—he shoved and pummelled her towards the unruly beast. Once, after he had taken a leather thong to her for dithering over her herding duties, she had called him Ramollwana as her aunts and uncles had done before her. At moments like that, when the Box of Matches was in full flame, she became strangely quiet, as if she were keeping a private account that she would one day settle with her grandfather.[87]

Yet, despite their being a rather odd couple—an irascible old man and a timid young girl—they got on well enough and spent many happy hours together. Sometimes the work cheated them of the success they deserved.

Despite their best efforts, sheep accustomed to the diet of dry scrub in the Triangle seemed incapable of adjusting to the lusher grasses of the Mooi River valley. Kas discussed this with Bodule and they decided to sell all but ten of the sheep. Luckily the cattle did better: at Rietvlei the herd numbered twenty cows and twenty-eight draught oxen belonging to Kas and Bodule and the landlord was not too concerned about how many animals they kept at Varkenskraal.[88]

Indeed, Vermeulen didn't seem concerned about anything. His visits to the farm were infrequent, and he tended to confine his attentions to his maize field, which, thanks to above average rainfall, did well. The only time he noticed the two new fields opened up by his tenants, he merely expressed surprise at how well the crops were doing on land he had considered unsuited for them. Of other Boers there was very little sign. The only remotely suspicious outsider to visit the property was an official from the Transvaal Provincial Administration who appeared in mid-February to issue dog licences and collect a wheel tax. Having already lost two dogs to snake bites, Kas paid him a pound for the right to keep the two surviving dogs and a further ten shillings for the ox-wagon.[89]

An outlay of thirty shillings was no great burden on a man who owned close on forty cattle, and Kas did not begrudge the state its income. As with many sharecropping patriarchs, however, it was not so much state taxes he feared as taxes from within. A fortnight or so after the visit by the T.P.A. officer, Mosala told him of his wish to get married and asked him to arrange for the customary discussion about bohadi with his prospective bride's family who lived on a farm near Muiskraal. This predictable development was something of a disappointment to Kas. He had always hoped that his sons, like himself, would marry in their mid- rather than early twenties. But there wasn't much he could do about it, and so he arranged for an intermediary drawn, like the prospective parents-in-law, from the Reverend Jwili's congregation. There could be no room for misunderstanding. The on-going struggle with the Teeus was reminder enough that bohadi was always a potentially explosive issue.

By the time the parties had agreed on a price of ten cattle, the worst of the summer's heat had been drawn and Kas's thoughts were on harvesting. Kas watched as Leetwane and the younger girls brought in the crops from what he still considered the 'family' field, while, out on the smaller patch, Bodule worked with Dikeledi and his 'mother,' Lebitsa, to bring in the grain for his new, separate store. Kas was watching the dissolution of the old household: Dikeledi was participating in the birth of the new. It hurt.

Even without Bodule's help, the old firm managed to produce more than two hundred bags of maize. Farther down the track Saul Ngakane, too, did well. When Vermeulen arrived to divide the harvest, he was satisfied with his tenants' performance and again expressed admiration for what had been achieved on the newly developed fields. Bodule's smaller harvest was exempted from sharing. A private arrangement with the property owner allowed him the right to his crop in return for looking after some of the owner's cattle, said to be destined for the Johannesburg meat market. Indeed, he soon drove the cattle to the abattoir, a forbidding prospect. But the privilege of not having to share his harvest only rubbed salt into old wounds and did nothing to ease underlying tensions.[90]

Soon preparations for Mosala's wedding were well advanced. Kas was told his son's marriage would have to be formally registered at the magistrate's offices in Ventersdorp by an appropriately empowered official. The state, it seemed, heard of everything. It could also do as it pleased. On a clear spring day back in 1939, the police had invaded the wedding reception that Joseph Modisalife had organised for his son, turned it into a riotous tax-and-pass raid, overturned the beer and food, and then arrested the guests. In South Africa, not even a well-meaning black patriarch could ensure his son a trouble-free wedding. He tracked down Vermeulen, obtained verbal permission to brew beer for the guests and then got the police sergeant on duty at the police station in Klerkskraal to give him a permit. Even pragmatists could get scars.[91]

But there was no trouble. Mosala's wedding was a perfectly enjoyable occasion, attended by members of both families as well as close friends like the Reverend Jwili and the Ngakanes. Kas soon built the couple a house of their own, a gesture that had been too slow in coming in Bodule's case but in this instance, was especially important since Motshidisi would soon produce her first child. As Kas aged he began to realise that tact and diplomacy which, in his youth, he had tended to reserve largely for dealing with landlords and outsiders, had as much place within the family as beyond it.[92]

With the harvest and wedding behind them, life fell easily and naturally into the rhythm of the off-season. Kas spent time drinking beer with Saul Ngakane; occasionally, he would be interrupted by a former client from Rietvlei in search of herbal remedies and wanting to chat about the fate of MaHersch's tenants. There was cobbling, harness making and leather work to be done, and, as always, winter was the only time to repair planters, ploughs and harrows, which, along with their owner, were beginning to show their age.

And, of course, the women of the household looked to earn a little extra cash making baskets, collecting and drying cattle dung, helping out in the kitchens of local farmers, and, in Leetwane's case, providing sage advice and practical assistance as a midwife. Unlike the winters at Muiskraal and Rietvlei when the spectre of losing the property at Doornkop had cast a pall over their efforts, there was no real conflict about these activities at Varkenskraal, and, until Matlakala reappeared, the family seemed pleasantly relaxed.

Matlakala was not well. She turned up late in the winter and for several weeks spent most of each day in bed. While the women, Nkamang especially, were helpful and sympathetic, Kas was largely indifferent. As far as he could see, the woman had spent most of her life avoiding work, and when one suggested she should be taken to see a doctor, he said the idea was a waste of time. Whatever the truth of this allegation, there could be no doubting that it was *his* approval, above that of all others, that she craved. After being denied the services of a doctor, 'I told my father that the ancestors had said that I should have a sheep slaughtered for me and that I would recover after that.' But Kas was as deaf to voices speaking of sheep slaughtering when he had sold most of his sheep, as he was deaf to wives who chirped on about the need for doctors. 'My father refused to slaughter an animal, saying that he had struggled too hard to raise the sheep he still owned.'

Kas's second rejection cast Matlakala into darkest depression.

> In the morning, after everyone had gone off to work, I forced myself out of bed. I went out to search for a piece of wire that I could use to hang myself. I found it and went back into the house. I then started to pile the chairs on top of one another and fastened the wire to the roof. I knew that nobody would disturb me since they only returned from the fields at about four o'clock. I was going to tie the wire around my neck, kick away the chair that I was standing on, and that would have been that.
>
> Then, just as I was about to do it, I heard the voice of an angel calling me. I really have seen an angel with my very own eyes! The angel said: "Miriam, God needs you since you still have much to do for your people. You will perform miracles and heal people." The angel said that I should leave home rather than kill myself. The angel then took away the wire and I was left standing and looking where it had been. That wire was never found in the house. The angel lifted me and placed me on my bed, put one hand on my forehead and the other on my chest, and prayed. The angel then left but, as it was about to go out of the door, stopped and said, "You will find a present beneath your pillow that will help you leave home."
>
> I left home the very same night, there were strong winds blowing around Rysmierbult. I looked under the pillow and found four pounds. I travelled all night until I reached Grasmere station where I bought another ticket for Park Station at Johannesburg. At Park Station I bought a ticket for Vereeniging.[93]

Knowing nothing of this the Maines simply assumed, when they found no trace of her the following morning, that Matlakala had vanished as mysteriously as she had arrived. But Matlakala's was a deeply troubled spirit. There was always psychic turbulence in her wake, and her unattended demands for sacrifices and her subsequent disappearance were bad omens.

With spring almost in sight Hans Vermeulen suddenly appeared, as if from nowhere, to inform his tenants that unnamed white farmers had complained to the owner about black sharecroppers on the property. The section they were farming, moreover, had been hired out to a neighbouring Boer, C. G. J. Geldenhuys, for the coming season. Vermeulen seemed genuinely distressed to relate this unexpected development and offered to find both the Maines and the Ngakanes another place to farm on. But they had had enough. As Kas recalled with great bitterness many years later:

> I was sick and tired of being allocated a field and then being evicted once it had been cultivated and the soil proved fertile. You were chased away as soon as they discovered that you could produce a good harvest from soil that had previously been considered useless. You tamed the land and they got rid of you.[94]

Muiskraal, Rietvlei, Varkenskraal, what did it matter? In the Mooi River valley every track seemed to lead to a Boer doorstep.

And yet, Kas hesitated. He was a son of the highveld, that dry, open, unlimited expanse of land which challenged human ingenuity to find a productive link between the countryside's wet east and dry west. He, his father and his grandfather before him had all worked on farms that conquest and the law alone rendered 'white.' But for him and others like him, black sharecroppers were as much a part of white farming as grass was of grazing. So, while Ngakane was quick to say he would move north to find a place amongst the BaKwena-ba-ga-Mogopa who lived beyond Doornkop, Kas delayed telling the family what he had in mind. Maybe he did not know. He had no wish to be forced into early retirement at Molote. How could a man with a family and cattle wrest a living from communal grazing and a residential stand? And how could he, a loner, live and farm among the hundreds of other BaKubung at Mathopestat?

He waited. They waited. The grass, tardy as ever, put out a few tentative shoots to test the spring air. And still Kas vacillated, poised uneasily between a known past and an uncertain future. Nobody prompted him, nobody questioned him. The patriarch knew his lines. Or did he? Bodule was the first to break ranks. He separated out his oxen and announced that he was going to take occupation of his stand at Molote. A day or two later he, Dikeledi and the three grandchildren were gone. It was quieter around the homestead, and then, as if to emphasise the silence, an unusually gentle spring drizzle set in over the valley, releasing that familiar musty smell that told a man he should be planting.[95]

The Ngakanes took their leave, and at twilight the first swallows were seen patrolling the upper reaches of the Mooi River. Kas poked about the yard, like a chicken trying to scratch its way back into yesterday's morsels. But there *was* no way out, and, if truth be told, he now needed Bodule more than Bodule needed him. His problems were not those that a man discussed openly, let alone with women and children. On 16 August, he and Mosala put six bags of sunflower seeds into the ox-wagon and took them to the co-operative depot at Buckingham. Slowly, stubbornly and with the greatest possible reluctance, like an ox going into the slaughtering pen, Kas was being pushed to a conclusion. Five days later he made up his mind. On 21 August 1956, he took seven bags of maize to the co-operative and, on his return, announced to the family that they, too, would be moving to Molote.[96]

The Maines went through the routine of preparing for the journey, and once everything was ready, Kas got out the horse and cart and set off to find the landlord. For a black man, even a journey into oblivion had to be authorised by a white master. He came back clutching a trekpas, a simple enough note scribbled on cheap stationery, meant to reassure nervous whites about livestock movements rather than to serve as a record of a sharecropper's possessions:

> 4th Sept. 1956.
> P.O. Rysmierbult.
>
> Pass for native Kas to proceed to Boons with 20 cattle, 15 sheep, 10 goats, 2 horses and 2 donkeys.
>
> J. Vermeulen.[97]

Behind this drab piece of paper lay the richness and complexity of an extraordinary man's life, and the odyssey which had begun in hope at Kommissierust in 1921 and ended in resignation at Varkenskraal in 1956. It took thirty-five years and fifteen farms for the likes of Hendrik Verwoerd and his supporters to get the Maines where the Nationalists wanted them. The trekpas did not show how a man who but six years earlier had possessed 8 horses, 12 donkeys, 60 cattle and 220 sheep had now been reduced to owning less than forty animals. The Maines, who had entered the Mooi River valley on a Ford truck with the chance of acquiring freehold property of their own, were leaving on an ox-wagon for a residential stand on a communal farm in a 'black spot.'

Kas was on his knees, a posture neither comfortable nor natural. At Molote, once safely with the BaKubung-ba-Monnakgotla of his forefathers, he would show the world what he was still capable of.

CHAPTER THIRTEEN

# Defiance

## 1956–59

The heartland of the BaKubung people lay seventeen miles north of Varkenskraal, not a very daunting journey for a family that had been cut off from its Fokeng cultural roots by hundreds of miles and more than a century in time. Yet for all its proximity to Vermeulen's hired property, Monnakgotla's Location was in a different social and political universe. In purely geographical terms, the divide separating the white farms of the Mooi River valley from the communally owned land presided over by Chieftainess Catherine Monnakgotla lay along an east-west arc of elevated ridges: this is the Witwatersrand outcrop, rising above the surrounding countryside by a few hundred feet, part of the watershed between the highveld to the south and the slightly lower regions to the north. Rain on the southern flanks of the ridge drains into the Vaal River, which joins the Orange and flows west toward the Atlantic, while that on the northern slope feeds into the Crocodile River and goes on to join the Limpopo, curving east toward the Indian Ocean. Nature charged this ridge with responsibility for separating the temperate highveld of the south from the warmer 'bushveld' climate to the north.

Despite its pivotal position in South Africa's destiny, the Witwatersrand is reluctant to inflict its importance on the surrounding landscape, deferring instead to the seemingly more impressive Magaliesberg some ten miles farther north, a parallel arc of more formidable proportions. On the gently curving plain between the two, the Maines found the small clusters of agricultural communities that together formed Monnakgotla's Location.

This apparently happy congruence between the work of God and man, which the government's Department of Native Affairs strove to reinforce for reasons of its own, had, however, long since been divided by the fissiparous tendencies of BaTswana politics. An ancient BaKubung site, reconstituted from the farms Elandsfontein and Palmietkuil with the aid of missionary help in the 1880s, Monnakgotla's Location by the mid-1950s was home not only to the many followers of the formidable Chieftainess

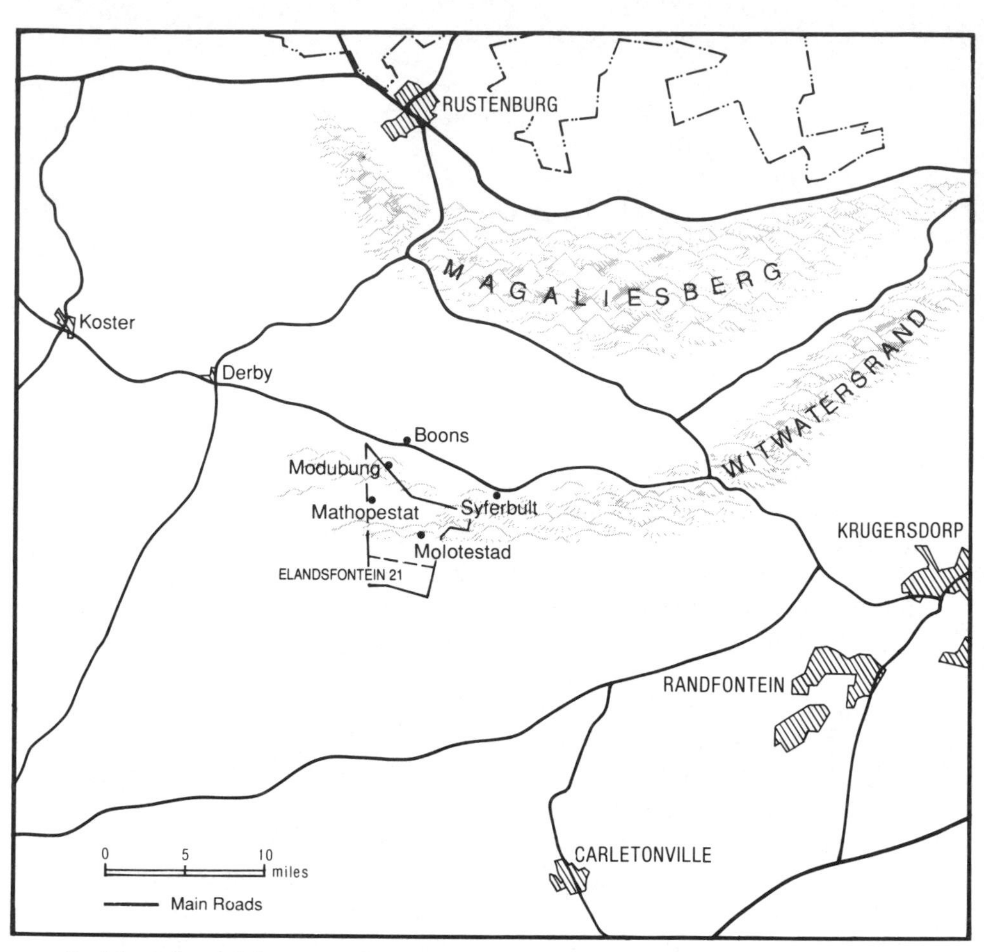


9 Monnakgotla's Location, 1956–67

Catherine Monnakgotla but to the BaKubung-ba-Mathope, the older and smaller scion which disputed her right to the throne of the BaKubung-ba-Monnakgotla.[1] The century-long division between the Monnakgotlas and the Mathopes was partly reflected in the layout of the location and its three principal villages—Modubung, Molote and Mathopestat: the inhabitants of the first two were Lutherans and Monnakgotla stalwarts; the residents of the last, a village on the western perimeter, were supporters of the Mathope family (as the name implied) and affiliated to the Anglican Church. In practice these differences seldom if ever gave rise to open conflict, and to the outside eye the villagers lived in harmony and seemed to co-operate readily enough in farming.[2]

Life in Molote, the largest of the clusters and the one for which the Maines were bound, had much to commend it. Situated on the highest point of a gently sloping spur at the eastern end of the valley, the village boasted a church, a school, two stores, hundreds of neat mud houses and scores of solidly constructed brick and corrugated-iron homes built in the bold unpretentious style of the 1920s and 1930s. These, along with the bluegums lining its dusty streets and the many peach trees in its back yards, were the signs of an exceptionally prosperous peasant community by mid-twentieth century South African standards. Below the village lay the maize fields that formed the backbone of the local economy and, beyond, the dam that provided the community with water.

Molote's apparent self-sufficiency, like that of its neighbouring villages, was circumscribed by the wider world and the structures of the South African state, which rested largely on the interests of a white electorate. A few miles north, on the tarred road connecting the distant platteland town of Derby to the mining centre of Krugersdorp, lay the white hamlet of Boons which, with only a fraction of the population of Monnakgotla's Location, boasted a thicket of provincial road signs, a post office, a telephone exchange, a rail-siding and a police station. And beneath this tough administrative hide the soft organs of white South Africa drew sustenance from the lifeblood of the countryside: warehouses for the regional branch of the agricultural co-operative, a mill run by members of the original Boons family, a filling station that sold more diesel to farmers than petrol to motorists, and two more trading stores.

For Lebitsa and Leetwane, as well as those of the Maine children who had spent most of their lives in the comparative isolation of the south-western Transvaal, Molote suggested the possibility of a rather more exciting life. But for Kas, whose earlier experiences on the diamond diggings had left him with few illusions about the supposed benefits of urban life, the prospect of being confined to a residential stand in a location where the cattle had to survive on communal grazing was not appealing. His last request to Bodule before he left Varkenskraal was to find him a property close to Molote where they could continue to farm in the old manner.

In his first week amongst the BaKubung Bodule came on an isolated house on the very outskirts of Molote occupied by an old man named Nkwane. Nkwane told him that one of the last portions of the farm Elandsfontein remaining in white hands had been recently sold off to a firm named Cyferbult Investments and to the 'South African Native Trust' in the hope that it could one day somehow be incorporated into the adjacent Monnakgotla's Location as part of the state's grander scheme for territorial segregation. But given that neither Chieftainess Catherine nor her many peasant followers had been able to raise the capital to purchase the thousand morgen comprising 'Elandsfontein Number 21,' the property had been subdivided and sold off to various individuals. The southern border of communally owned BaKubung land was therefore flanked by strips of freehold property belonging to relatively wealthy black farmers.[3]

This historical anomaly, which predated the triumph of D. F. Malan and his supporters in the 1948 election, held out the possibility that Kas could enter into a sharecropping arrangement with an independent black landlord in the heart of the Transvaal at a time when the Nationalists under their new leader, J. G. ('Hans') Strydom, the Lion of the North, were busy extending their ideas of territorial segregation within the National Party. As Bodule discovered soon enough, Nkwane himself did not own property. Even the patch his house stood on belonged to a distant kinsman who held title to the 500 morgen of land on the flank of the northern Witwatersrand. Nkwane was merely the farm caretaker. Bodule was interested by this information and he became positively enthusiastic when Nkwane told him that his kinsman, interested more in the long-term speculative potential of the property than in its productive capacity, had given him permission to take in tenant farmers. After several more discussions, Bodule and Nkwane, the one acting on behalf of his father and the other for a distant kinsman, agreed that the Maine family settle on Elandsfontein No. 21 in exchange for the normal set of sharecropping rights and obligations. When most of the Maines arrived at Monnakgotla's Location in the spring of 1956, they did not disappear into the anonymity of residential Molote—as Cas Greyling, Beyers, and their political henchmen might have wished—but instead took up an economically defiant position on a thin strip of farming land belonging to one M. H. Ramagaga.[4]

Born in Thaba Nchu in 1904, Moses Nehemiah Ramagaga was a prominent member of the small but tightly integrated rural aristocracy that dominated independent black farming and landowning on the South African highveld. A grandson of Mongane Ramagaga, who had been granted a farm for his political services to the BaRolong court of Chief Moroka in the late nineteenth century, the young 'Musi' Ramagaga had been educated at St. Paul's in Thaba Nchu, then in Natal at Edendale, and later at the Reverend John Dube's prestigious Ohlange Institute at Inanda. After completing a course at the agricultural college at Tsolo in the Transkei he had taken up a position as a 'demonstrator' in the Department of Agriculture before being seconded to the Nongoma, Witsieshoek and Vryburg districts. In the late 1930s he decided to return to Thaba Nchu to help steer the family farm at Leeupoort, which he did through the agricultural boom precipitated by World War II.[5]

The success at Leeupoort meant that Musi Ramagaga had no wish to expand his farming into the western Transvaal, but early in 1952 he was persuaded by three kinsmen—Stephen Monametsi, Ismail Motsuenyane and Ismail's son, Andreas—who were keen to invest in additional land at Molote. Between them they purchased virtually all the land which Cyferbult Investments put on the market, and for an outlay of £7,000 Ramagaga acquired himself a parcel of 5,000 morgen, which fringed the ridges of the Witwatersrand and ran on into the valley toward the stream which the people in nearby Mathopestat knew as Mokgadi.[6]

Kas was relieved to find that Musi Ramagaga seldom ventured beyond Thaba Nchu and was in no great hurry to dispose of his land at Molote. Ramagaga family members seemed interested more in exchanging small parcels of land amongst themselves than in selling them to outsiders. These transactions were usually designed to rationalise existing holdings, not to alter the character of the property as a whole. To protect his everyday interests, Ramagaga asked a younger brother based in Soweto, Nicodemus, to take charge of his affairs at Molote. Although Orlando East was only three hours away from the Magaliesberg, Nicodemus found it hard to visit the property and persuaded yet another kinsman by marriage, Benet Nkwane, to settle on the farm and act as the caretaker-cum-supervisor.[7]

Undaunted by an arrangement in which he was legally subject to an absentee landowner residing in Thaba Nchu, required to share his crop with a putative landlord living in Soweto, and expected to respect the wishes of someone else living on the premises in Molote itself, Kas set about finding a suitable site for his homestead. The physical configuration of sites and homes that emerged from some hasty deliberations held amongst the Maine menfolk mirrored the power structure of a sharecropping family presided over by an ageing patriarch.

At the heart of the complex, perched on a rocky platform at the base of the ridge, Kas built three corrugated-iron structures designed to accommodate himself, his two wives, and the few remaining members of his inner household—Nkamang, Pakiso and Nthakwana's baby, Ntholeng. A short distance away, at the same elevation and so angled that it, too, managed to capture the first shafts of the morning light, he helped to build a similarly crude lean-to for Mosala and his wife, Motshidisi. Five hundred yards farther down the slope, on a site marked by a lone acacia, Bodule erected a shack that turned its back on the patriarchal complex. Clearly, neither Bodule nor Dikeledi was ready to yield so much as a single inch of the psychological space that had opened up between them and Kas.[8]

In their own peculiar way, these metal shells, clinging to the lower part of the ridge like snails to the foot of a garden wall, gave Kas a sense of achievement and marked a personal triumph: the impermanence of the accommodations signalled his unflagging commitment to sharecropping, a willingness to move on and search out new opportunities if things did not work out, and also a devotion to life on the land. These were deeply held values, which the more sedentary inhabitants of Monnakgotla's Location may not have understood, but the habits acquired during thirty-five years along the highveld road died hard, and, at the age of sixty-two, Kas thought of Molote as but one more place to find grazing for all his animals and to plough.[9]

Kas paced out his fields beyond Bodule's place so as to avoid the scree that had tumbled down from the surrounding koppies over the years. Bodule marked out a small field of his own, which he ploughed in great haste before leaving for yet another round of migrant labour in Carletonville. Kas got Mosala to round up the oxen and then, calling on each of his favourites by name—Basterman, Verneuk, Wildeman, Senkwasie, Fransman, Jammerein, and Donker—drove them into a corner where Nkamang and Pakiso helped to harness the draught animals into their accustomed positions.[10]

The maize and sorghum fields at Elandsfontein, although smaller than those of the surrounding white farms, were nevertheless a formidable challenge for sixteen-year-old Nkamang and her eleven-year-old niece. The ground was fairly free of stones and reasonably level, but loam at the southern end of the farm gave way to clayey soil at the point where Ramagaga's property dipped toward the banks of the Mokgadi. Once they reached this lower end, the oxen tired and the plough sometimes got stuck in the clay. Kas made few allowances for such problems. Like Mmusetsi, Bodule and Mosala before them, the girls were soon reminded, if they needed reminding, that he was still an uncompromising taskmaster and that the supposed privileges of gender counted for little with him.[11] The girls made excellent progress and, before long, the seed was safely in the ground. This early success set Kas thinking about the need to retain a full team of draught oxen. With Bodule away and Mosala, too, having gone off to work in one of the Witwatersrand's satellite towns, the family's need for fourteen highly trained animals was reduced. But any speculation in cattle would require Bodule's consent, because there were certain animals in which they had a shared interest, so Kas decided not to act until the security of the new position had been firmly demonstrated.[12]

Summer unfolded well. The regular rains vindicated Kas's decision to cling to the soil on the side of the koppie rather than descend into the bowels of the location below, and by mid-December, when the two girls were ordered into the fields to do the hoeing, the Maine maize was already being admired by Benet Nkwane, who put in frequent appearances. Another enthusiastic observer was Kenneth Tembani Nkaba, a school principal from Vryburg who was acquiring from Ismail Motsuenyane the narrow strip of 125 morgen just west of Musi Ramagaga's property.[13]

Ever since his experience with Lerata Masihu back at Vaalrand in the early 1940s, Kas had entertained the gravest possible doubts about the practical usefulness of teachers. Like Masihu, most of them seemed only to combine 'cunning and laziness' in equal proportions. But Nkaba was different. Kas took an immediate liking to the man, and while they differed markedly in age, education, ethnic background and professional expertise, the two soon developed a friendship characterised by trust, tolerance and a respect for one another's abilities.[14]

Kenneth Nkaba's father had been a Xhosa-speaking immigrant who had taken up residence amongst the BaTswana and had started out life as a transport rider and worker on the Cape railways before turning to sharecropping in the Schweizer-Reneke district. From there the young Nkaba had been sent to the Eastern Cape for his schooling. On leaving Lovedale College in 1936, he had taught in various towns in the northern Cape before being offered the principalship of the primary school at Vryburg. But, unlike Masihu, Kenneth Nkaba had never quite shaken off the desire to live on the land, and when Kas first met him in late 1956, he was trying to buy the property at Molote, hoping to leave teaching and take up farming full-time. A teacher who had his roots in the Triangle and who could see the advantages of abandoning a mortar board for a maize field was the sort of man whom Kas found easy to respect. What else would you expect from a sharecropper who for three decades had earned much more from farming each year than a Lovedale graduate did from teaching?[15]

But at the age of forty-two, Kenneth Nkaba was still an inexperienced farmer. This, along with the fact that his long-standing base at Vryburg was more than a hundred and fifty miles away, meant that his new enterprise was exposed and vulnerable. So he was delighted when Kas gave him valuable tips about farming and offered to keep an eye on the activities of the sharecroppers he had 'inherited' along with the farm. In the complex web of economic relationships that linked tenants and subtenants in Monnakgotla's Location, Kas was a 'supervisor' as well as being supervised.[16]

In practice, Kas's offer required almost no effort, and once his own crops were established, most of his time was spent making certain that Nkamang and Pakiso were herding properly, which wasn't easy on a property bordered by communally-owned land and short on fencing. This need for vigilance meant more work for the girls, which reactivated a long-standing dispute with some of the older women in the family about the importance of education. Leetwane and Morwesi were both insistent that their daughters spend as much time as possible in the classroom at Mathopestat, while Kas wanted the youngsters to look after the livestock. This tug-of-war was a variation on the struggle that had gone on during the trek across the highveld, but at Molote it took on a new dimension once Morwesi decided to underwrite the cost of Nkamang's and Pakiso's education from her wages. This unexpected intervention in the micro-economics of the household, which came in the form of monthly remittances, tipped the scales against Kas at a moment when his access to family labour was already severely circumscribed, and it significantly heightened the domestic tension. For the first time in his life Kas was unambiguously on the losing side when it came to mobilising and directing female labour. Age siphoned off social strength with the same insidious zeal that it sapped physical prowess, and it left the patriarch bruised and sensitive.[17]

Kas was forced to turn to Mosala for help, not only with the cattle, which needed constant supervision, but with the horses, which were showing the effects of the warmer, less sympathetic bushveld environment. Most men could rely on help from a twenty-one-year-old son, but Mosala was no ordinary son. Right from childhood the boy had not taken kindly either to discipline or hard work, and even as a newly married man, he showed few signs of settling down. In fact, it was hard to know whether it was better to have him hanging around Monnakgotla's Location, where he was always wandering off to seek out beer and undesirable company, or to have him away in the mining towns, where his attempts at casual labour produced equally feeble results. Almost everybody agreed: Mosala was 'a troublesome son.'[18]

To add to Kas's problems, the price of wool, having peaked in 1950, slumped to a postwar low in 1956. Animal husbandry was increasingly onerous and time-consuming, and Kas needed to reorganise his livestock holdings to make them more congruent with the labour at his disposal. But again he hesitated. Draught animals embodied the struggle of a lifetime and could not be disposed of easily. With the chill of autumn breathing death over the last of the maize, he turned his attention to the fields.[19]

Fortunately, it was still possible to organise a team for the harvest without major conflict. The Maine women did not dispute the need to bring in the grain: the logic of securing a supply of food always trumped their misgivings about the cash finding its way into the patriarch's pocket. Leetwane, Lebitsa, Nkamang and Pakiso, the core of the Maine labour detachment, turned out readily enough, and they were assisted by the more disaffected Dikeledi, Mosala and Motshidisi. But even then there were signs that the Maines could not sustain sharecropping indefinitely. Lebitsa, who had been coughing in a most alarming way for months, was now also experiencing difficulty in using her fingers: she was little more than a passenger as the harvesters sailed up and down the neat channels of maize that demarcated the extent of Musi Ramagaga's property.[20] In fields so much smaller than those on the white farms, her halting efforts were, however, more than compensated for by the quick darting movements of Pakiso: the red blood of youth oiled the back, shoulders and hands in a way that the pallid liquid of senescence could never hope to match. With Benet Nkwane already hovering about like a crow at a picnic site, Kas noted how, lower down the valley, where there was more clay, the locals were using sledges to drag their maize across to the threshing points. Higher up, on the flanks of the ridge, where there were more rocks to negotiate, he would have to find a different solution. He needed a truck, and decided to speak to Johannes RaTshefola.[21]

RaTshefola, the owner of a small store midway between Molote and Mathopestat, was an interesting fellow. Like dozens of other storekeepers in rural South Africa, his domestic arrangements harked back to an era before Afrikaner Nationalists had enshrined their racial obsessions in law. RaTshefola was a MoTswana who had married the daughter of the local Indian shopkeeper, David Mohammed, and then gone on to manage a family store on the outskirts of the location. Unfortunately he could not help Kas but suggested instead that he speak to his father-in-law, at the principal store in Molote.[22]

Maine and Mohammed were both drawn from a cohort with an almost instinctive ability to read the contours of honesty that lined most weather-beaten highveld faces, and they took an instant liking to one another. Mohammed showed great interest in Kas being a herbalist as well as a sharecropper. After a lengthy digression exploring this issue, they got around to establishing the terms on which a truck could be hired. After that, Kas found it easy to move the grain to the threshing point, and he and Nkwane, acting on behalf of the Ramagagas, shared about a hundred bags of maize.

Fifty bags was not a bad return in a season which at one stage had threatened to see Kas relegated to a 'residential stand.' Before returning the truck, he arranged to take most of his share a few miles down the road, where thanks to the efforts of the Maize Board, he received a handsome *voorskot* from the Koster Koöperatiewe Landboumaatskappy Beperk. Stable or improving maize prices remained a priority for a government eager to please its rural supporters; ironically, the few remaining independent black grain producers on the highveld also profited.[23]

Welcome as the money was, it was nevertheless much less than what Kas had grown accustomed to in the trek across the Transvaal, and he could not contemplate the winter with equanimity. He scoured his mind for possible alternative sources of income and then remembered how, two years earlier, the Native Commissioner at Ventersdorp had softened the pass issue by suggesting that he was of an age where he qualified for a state pension. However, when he and Leetwane approached 'the tribal authorities at Molote' for assistance they were turned down on the grounds that Kas's ownership of a span of oxen made him ineligible. He was caught in a bureaucratic double-bind: too poor to make a living on his own and too 'rich' to qualify for help.[24]

Leetwane, whose inclination was to keep quiet and avoid conflict, had special reason to question the logic of this means test. As the original owner of Disu-cum-Witties, she believed that most of the Maine cattle were descended from a cow that had once belonged to her and had since been appropriated by her husband. Denied independent access to or rights over the cow's offspring, *she* was now being denied a pension on the grounds that her *husband* owned cattle. It was as if her cattle had been stolen, only to have the theft confirmed and be told that the thief's wealth rendered her ineligible for assistance!

Regardless of the logic, the fact remained that the older Maines faced a winter with not enough cash. Kas did some cobbling and a little herbalism; Leetwane collected, dried and sold dung, which enabled her to pay her dues to the Reverend W. Matabane when he abandoned his Ventersdorp base for a few days in order to visit some of the more far-flung members of the Ethiopian Church.[25]

But whereas in the not too distant past such off-season activity had permitted the Maines to buy 'luxuries,' at Molote the 'marginal' winter monies had to prop up the basics of the household budget. The pace of this decline was at least partly offset, however, by the remittances from adult children working beyond the family enterprise. At various times Bodule, Nthakwana, Pulane and Morwesi all sent small sums in cash to either Kas, Leetwane, or Lebitsa.[26] The social links binding the older children to the family nucleus were still intact. But as Kas knew only too well, the day was not far off when the economic demands of the grandchildren would take priority. The Maine children were trying to make their way in the world under their own steam, but this distancing process was often incomplete and fraught with difficulties, and it sometimes involved Kas in extraordinarily complex and even painful matters. Pulane's rather tangled personal affairs were a case in point.

Shortly after the family's move to Rietvlei, it had become clear that the Maines' adolescent daughter was becoming sexually involved with Dikeledi's younger brother, and not long thereafter, she had run off with Mojalefa and settled in the Ventersdorp district. The wisdom of getting married at sixteen was in itself questionable, but the issue then became embroiled in the problems surrounding Dikeledi's unpaid bohadi which, in turn, harked back to Kas's failure to secure Mmusetsi a wife. To make matters worse, Pulane's marriage was not a happy one, and this further aggravated the long-standing hostility between the two families. The acrimony between Kas and Abner Teeu increased, each of them believing that the other owed him at least half a dozen cattle for outstanding bohadi payments.[27]

The anger over this thicket of issues became more rather than less serious with each passing year, since both patriarchs were becoming older, less productive on the land, and therefore more inclined to mount a challenge for such income as could be derived from beyond the sphere of domestic production. By the winter of 1957 Kas was muttering openly about the need for a legal solution to his difficulties with the Teeus, and he would no doubt have instituted proceedings had he not been preoccupied with the question of negotiating bohadi payments for Nthakwana.

After her adventures in Randfontein, which had culminated in the birth of Ntholeng, Nthakwana had rejoined the family for a time before striking up a promising relationship with a MoTswana son of the soil named Matthews Moate. Matthews Moate was said to come from good farming stock in the Ventersdorp district, and he showed every sign of becoming a responsible son-in-law. He and Nthakwana produced their first child, a boy named Bona, and the Moates, unlike the Teeus, were aware of their obligation to raise bridewealth for Nthakwana. Preliminary discussions with the groom's parents were already underway, and Kas was sure that the whole business could be settled within months rather than years.[28]

But the patriarch was aware that he could not afford to limit his search for bohadi cattle to the most promising instances, especially so when both his cash income and livestock holdings were rapidly declining. With age stalking him and his two wives ever more purposefully, even the most headstrong and troublesome of young women would have to be brought back into the family fold. Thus, unaware of Matlakala's unhappiness and the attempt at suicide which had preceded her mysterious disappearance from Varkenskraal a few months earlier, Kas persuaded Bodule to escort him to Carletonville in the hope of tracking down the missing daughter. But though they diligently explored several quarters in the township of Khutsong, there was no sign of Matlakala. Like Mmusetsi before her, she had disappeared without trace.[29]

The days were lengthening, and at Molote nature's clock stopped Kas's mind from dwelling on Matlakala's 'foolishness' and encouraged him to face the problems of the coming season. The difficulty the cattle had in getting on and off the rocky ledge on which he and Mosala had constructed the kraal had long since convinced him of the need to move their homestead lower down the ridge. So at his insistence they re-erected the cluster of tumble-down shacks at a spot farther down the slope which gave them easier access to the fields and watering points for their livestock.[30]

Kas was out repairing a plough when Nkaba suddenly appeared at his side. His neighbour made a few complimentary remarks about his abilities as a blacksmith and then, rather tentatively, asked the old man whether he would be willing to go with him to the Rustenburg district, where he wanted him to take a look at a property he was thinking of buying. Kas was not very keen, but did not want to let the man down. Rustenburg (or Tlhabane to give it its Tswana name) was far away; but then so too were the rains, and in the end he agreed to accompany Nkaba.

In the animated exchanges that rose above the whine of the car engine, Kas learned that the farm near Rustenburg which Nkaba was interested in belonged to Aaron Motsuenyane, a member of the same formidable family that the Maines had encountered in their sweep through the

Klerksdorp district. But once Kas examined the property, he decided that it had not been one of the Motsuenyanes' better investments. He saw dangers in its heavy clay soils, which, he explained to Nkaba, not only created problems for grain, which needed good drainage, but also tended to stick to the oxen's feet, causing the hooves to rot. Nkaba, pleased to have received these insights before laying out additional capital, let his option lapse.[31]

During the mostly silent journey home, Kas was left to ponder how it was that he had been reduced to giving advice to aspiring property owners when only a few seasons earlier he had been within striking distance of acquiring a farm of his own. For sharecroppers the dividing line between success and failure was always paper thin, and somehow, for reasons he could not quite put his finger on, he seemed to have ended up on the wrong side of the line. In his heart he felt it had something to do with the progressive fragmentation of his family, who were no longer wholly committed to farming. Still, there was no point in dwelling on what might have been; he just had to push on. Back at Molote he told Nkamang and Pakiso to withdraw from the classroom and, much to the dismay of their mothers, once again made them round up cattle and help with the ploughing.

The oxen, accustomed to more formidable challenges than those posed by Musi Ramagaga's modest fields, made light of their task, and Kas again asked himself whether it was wise to keep a full span of draught animals when the price of beef was scaling new heights. When they finished planting he had decided to sell the oxen, cut back on labour-intensive grain farming, and expand his other cattle holdings. But how would Bodule respond? Kas still saw him as the son of a black sharecropper with a future on the land, not as a migrant worker destined to disappear down the gizzard of industry.

But his sensitivity was misplaced. The seductive rhythms of weekly wages and urban living had already done much to undermine Bodule's remaining commitment to the uncertainties of rural life. The potentially disturbing idea of disposing of the oxen and buying younger beasts fell on Bodule's ears almost as gently as did the summer rains on their newly planted fields. Kas put the proposition to Bodule during one of his weekend visits from Carletonville, and by the middle of the following week the local agent, 'Freedman the Jew,' had given Kas twenty-five pounds a head for each of the twelve beasts they had known and worked by name for more than a decade. In the end, dispatching the animals to the Johannesburg abattoir troubled Kas more than it did Bodule.[32]

Notes were beginning to count for more than nostalgia, so Kas did not waste time fretting about the loss of the oxen. The three hundred pounds he made on their sale put him in a strong bargaining position, since the livestock market, like most markets, favoured cash purchasers. Within three weeks he acquired twenty-six calves and set aside some cash savings. But whereas once his skill as a livestock speculator would have been publicly endorsed by an admiring son, on this occasion it was met with no outward sign of enthusiasm. Bodule merely thanked him for his share of the spoils, the fourteen calves, and led his animals away. There was no bitterness or recrimination, only indifference, which Kas was having trouble detecting.[33]

The calves, taken care of by Nkamang and Pakiso during the afternoons, took readily to their new surroundings. Although the summer rainfall was less generous than it had been the year before, there was enough grazing for them to pick up in condition, and even in autumn, the veld was sufficiently vital to allow Kas to acquire yet another cow. The patriarch's new venture was showing signs of promise.[34]

Unfortunately, the same could not be said of the enterprise in the fields. The 1958 maize harvest disappointed landlord, tenant and harvesting team alike. In his postmortem at the end of the season Kas decided that this was not only because the rains had not always come propitiously, but because the soil at Molote was tired. Next spring he would invest in some of those artificial fertilisers he had seen men like Jack Adamson put to such good use on Klerksdorp farms.[35]

*

Winter slid down the hillside and took control of the valley and its inhabitants. The flowing movements of collective labour that characterised the harvest made way for the more stilted manoeuvres of off-season tasks. Lebitsa moved with the greatest of difficulty. Even those who had become accustomed to the rasping sounds that emanated from deep within her, were struck anew by the harshness of her cough. She did not complain and it was difficult to know how much of her infection could be attributed to her age and how much to winter. While she languished within a corrugated-iron structure on an exposed northern ledge of the Witwatersrand, Kas got out a chair and moved out into the watery sunshine to pursue his calling as a herbalist.

His reputation as a *ngaka* had started percolating down into the valley, and he was becoming better known amongst the people of Molote and Mathopestat, who at first had been uncertain as to what to make of a man who had self-consciously relegated himself to the margins of the community. At odd moments one or two more determined patients picked their way up the path to his shack. But then, especially after he had introduced himself to David Mohammed at the trading store, there had been an increase. These visits, which Kas found especially agreeable when there was less to do in the kraal or fields, gave him cash that others, especially his wives, found hard to keep track of.

The trickle of patients continued as Kas became better known and more integrated into the growing community of former agricultural wage labourers, labour tenants and sharecroppers, all shoe-horned into Catherine Monnakgotla's crowded stronghold at Molote. These new economic refugees did not always have real or imagined links of kinship with the older BaKubung residents, and by 1958 the inner dynamic of Monnakgotla's Location was changing. Jonas Maponyane, secretary to the Chieftainess's council, or kgotla, was amongst the first to detect these changes and to point out how the established crop farmers, who enjoyed privileged access to land in the valley, were being drawn into a variety of complex sharecropping agreements with *matsenelwa*, or outsiders, who, newly off the farms, sometimes owned more livestock than did the resident BaKubung.

In essence, there were no formal agreements. It was understood that the outsiders were there only on a temporary basis and would eventually have to move on. However, many of the BaKubung families did not have oxen, yet owned fields. They would then enter into a variety of agreements with outsiders who cultivated crops on BaKubung land. The harvest would be shared in accordance with the specific agreement entered into by the parties. For instance, an outsider who had grown crops on a field owned by a MoKubung, because he owned the oxen and equipment and performed the labour, would get a larger share of the crop. If he harvested twenty bags, he would keep fifteen and give five to the owner of the field.[36]

*

Not all these tenancy agreements worked out amicably, which further complicated the byzantine politics of the location. Catherine Monnakgotla, already embroiled in a dispute surrounding her accession to the throne, gave silent assent to the influx of matsenelwa in the belief that their presence could help bolster her numbers in any conflict with the Mathope faction. But because the newcomers sometimes got into difficulties with her more senior supporters, they occasionally undermined the solidarity of her own faction. In short, the sudden influx of outsiders caused her power-base to start fracturing along economic as well as political lines.[37]

From his position high up on the hillside, Kas first observed and then analysed these developments. Later, when he became more adept at negotiating the stony path that led down into community politics, he occasionally ventured forth and began to align himself openly with Catherine Monnakgotla's cause. From a purely personal perspective, these rather tentative forays into the valley made good sense: they protected his interests as a potential resident in Molote and gave him leeway with the matsenelwa. His instinctive ability to make the adjustments necessary to ensure his family's survival—the distinctive feature of his thirty years on the highveld under white landlords—was not about to desert him now that he found himself amongst black landowners.

Kas's feel for domestic politics, too, remained largely intact, although with so many of the children grown up and his material resources gradually declining, it was harder for him to have the final say when opinions differed. But the backbone was unyielding. When Dikeledi told Bodule that he should recognise the inevitable, sell his livestock, and move permanently into Carletonville, Kas was quick to point out the folly of selling cattle and cutting all ties with the countryside. And, as the winter nights slowly gave way to spring days, he had another minor domestic success when the Moate family delivered five of the dozen cattle the Maines were owed for Nthakwana's marriage.[38]

The new season started badly, however. For the first time in several years there were no early spring rains, and by mid-November, farmers and tenants up and down the valley were so unsettled that Catherine Monnakgotla sanctioned the holding of a traditional rain-making ceremony. But the molutsoane, the first communal hunt Kas had participated in since his days under Reyneke at Vlakfontein, was slow to bring about the desired result; only in December were his fields fully planted and the seedlings fertilised. By then, Bodule had long since gone back to town, and left behind a field even smaller than the one he had cultivated the previous season.[39]

As anticipated, the commercial fertilisers spurred on the growth of the maize, but even so, the absence of regular rain meant that the crops did not make sustained progress. Nevertheless, by early January there was still a chance of reaping two or three hundred bags if only the heat eased off and the cloud cover allowed the soil to retain its moisture. The pressure did not get to Kas. He had always known that arable farming was a gamble, and he never resented nature stacking the odds against the farmer. On the contrary, the challenge brought out the best in him, and in its own mildly addictive way merely served to heighten the expectation and tension in the months before the harvest. All this made a blow from the hand of man, rather than an act of God, totally unexpected.

Late one afternoon Benet Nkwane barged in to inform Kas that Ramagaga had sold the farm to Andreas Motsuenyane without even going to the trouble of telling his brother, Nicodemus, about the transaction! This meant that soon the Nkwanes, the Maines and another sharecropping family farther down the slope, the Mogakabes, would all be asked to leave the property. The new landlord, Nkwane had been told, was an active and wealthy young farmer in Lichtenburg, who was likely to take a far greater interest in the property than Musi Ramagaga had.[40]

Kas was startled by this bad news but refused to panic. It was well known that tenants with standing crops had the right to bring in their harvest, and he thought it was unlikely that the new landlord would ask them to leave before the end of the season. But only a few days later all hopes were dashed when Motsuenyane appeared and told them to vacate the farm at once.[41] Even then, Kas's nerve held. Somehow, something would happen and he would find them another place on which to live and graze their cattle. Had it not always worked out that way in the past? But nothing did happen and, in the end, it was left to Kenneth Nkaba, who thought that he should repay his unofficial 'foreman' for his many kindnesses, to invite the Maines to move across to his land.

In a perverse sort of way, this act vindicated Kas's faith: as in the Triangle, a policy of good neighbourliness carried with it its own dividends. Within a matter of hours, the family's shacks were re-erected on the schoolmaster's property. But then, just as Kas was about to pick up the threads of his farming operations, the unthinkable happened. Three tractors from Lichtenburg suddenly appeared, and before he or anybody else could do anything about it, his and Mogakabe's crops were ploughed into the ground. Motsuenyane had decided to plant a crop of his own on the very fields the tenants had prepared, fertilised and cultivated! In his lifetime Kas had seen a harvest carried away by creditors because of a third party's debts, but he had never, not even in the Triangle, where the Boers were as tough as their *sjamboks*, seen a landlord plough a man's seedlings into the ground before his eyes! 'It showed one something about the rich. Whites were rich, but rich blacks were also shits. Motsuenyane was just a shit. A rich black man will trample a black tenant underfoot just as readily as a rich white man!'[42]

The Maines were living through a nightmare. Kas and the girls scrambled about assembling a makeshift team of donkeys and mules, and tried to salvage something from the season by doing some last-minute ploughing while the gods, smiling on the usurper, allowed Motsuenyane to complete his mechanised operations even before the Maines had finished planting anew. By the time they had finished, they had missed out on the best of the rains, and Kas watched helplessly as Motsuenyane's seedlings flourished and his own struggled. The loss of a potentially good crop was bad enough, but his blood boiled at the thought that it was his fertiliser and fields that were now supporting the efforts of an interloper. Such a monstrous injustice should not, indeed could not, go unchallenged.[43]

At his instigation he, Mogakabe and Nkwane met to discuss the situation, and Kas suggested they consider taking legal action to obtain compensation. Mogakabe was persuaded easily enough, but Nkwane, whose status as supervisor meant that he had been less directly affected by the invasion, was more cautious. Still, Nkwane escorted them to Orlando East, where Ramagaga's younger brother would advise them as to which lawyer to consult.[44]

Nicodemus Ramagaga, irritated by his brother's failure to tell him about the sale of the farm and embarrassed by the manner in which Nkwane had been rendered homeless, was a sympathetic listener. He referred them to a prominent lawyer in Rustenburg who, after making an appropriate preliminary charge, promised to make further enquiries on their behalf before taking the matter to court.[45]

But the lawyer was outgunned by the firepower of the Lichtenburg firm that Motsuenyane retained. On the sharecroppers' second visit the lawyer warned them that he did not believe that they would enjoy any success at law. Musi Ramagaga had apparently neglected to disclose his partnership with them when he sold the farm to Motsuenyane, and the new landlord had therefore been within his rights to plough what he perceived to be the previous owner's crops into the ground. As usual, law and justice were not good bedfellows, and sometimes not even the efforts of experienced pimps could bring such reluctant partners together.[46]

It was a difficult blow to absorb. Anger smouldered within Kas, and sometimes he felt as though his belly would burst into flames. His entire family was now faced with a winter of real hardship, for it was clear that little would come of his late planting at Nkaba's. The first frost of autumn slithered in along the valley floor and then reared up to savage the few unsuspecting ears of corn that lingered in his fields. By May, when a farmer should be thinking about the composition of his harvesting team, Kas was looking at a barren field. As usual man rather than nature was the sharecropper's biggest enemy.

By the same token, it was to man that Kas had to look for help. The news that the Maine family was in trouble filtered into the communications network of the extended family, and just as the local farmers were preparing for the harvest, Moketla, one of the patriarch's nephews, arrived from the Triangle with what looked as if it might be an answer to their problems: Sebubudi had entered into a crop-sharing agreement with a white farmer and needed labour to bring in what looked like a bumper harvest.

Kas suggested that the younger women in the family be the work party and that they be accompanied by Mosala, because harvesting teams, beer, and the presence of strange men all posed dangers for girls in an unfamiliar setting. This, together with the stipulation that the harvesters be paid in cash rather than in kind, was readily agreed to by his nephew. Moketla, Mosala, Dikeledi, Motshidiso and Nkamang set out for Wolmaransstad, each with a different degree of enthusiasm for the task ahead.

Soon their reservations gave way to the less discomforting feeling of nostalgia, as they came to familiar terrain and recognised the larger properties around what the whites called Wolmaransstad, but they always knew as Borabalo. Collective labour was more agreeable than individual work, the harvest *was* surprisingly large, and it was good to renew acquaintance with kin with whom they had lost contact. Moketla allocated more than forty pounds in cash to the Maine work party; when the moment to leave arrived, Mosala, wishing to stay on for a day or two longer, suggested that the women hand him their earnings for safekeeping and go back to Boons by train. This they did without incident, although by the time they reached home, Dikeledi was complaining of not feeling well.

The foray into the Triangle seemed to have been successful; the women went back to their work routines around Molote. But within days relief turned to despondency and then to anger: when Mosala eventually appeared and the women asked him for their earnings he told them that on boarding the train at Makwassie, he had been robbed of his purse by a group of unknown men. With the exception of his wife, Matshidiso, nobody gave this story credence. Dikeledi and Nkamang spoke openly of Mosala having stolen their money; they thought he should be made to go out and borrow the funds necessary to pay them back. But Mosala had long since slipped beyond the reach of patriarchal authority and nothing came of their demand.[47]

The atmosphere around the homestead soured. Matters worsened when Dikeledi became sicker and had to ask her father-in-law to hire a car to take her to the Krugersdorp hospital, thirty miles away. She and Kas had been at loggerheads for almost as long as anybody could remember, and in a season when every penny counted double, nobody was much surprised when he refused to entertain an idea he considered wholly unnecessary. As in the case of Matlakala, he did not seem much concerned about the health of the younger women even when they asked him for help. But Dikeledi's condition suddenly deteriorated, and Kas had a message sent to Bodule, who arrived to find both his wife and his 'mother,' Lebitsa, ill. During the next week Dikeledi staged a slow recovery but Lebitsa's health remained at a low ebb. The winter, her arthritis and, above all else, her cough were killing her.[48]

Bodule went back to work, but things were never the same. To Dikeledi, Kas's refusal to help her in a moment of crisis was the last straw. As soon as she had recovered sufficiently she announced she was going to Carletonville to consult a doctor; when she got there she told Bodule she no longer wished to be part of her father-in-law's homestead. But Bodule, unwilling to cut his last link with Kas, pleaded with her to return to the farm; she made her way home with great reluctance.

Back at Molote and still not well, it took Dikeledi but a day or two to realise that she had made a mistake. Once again she proclaimed the need to consult a doctor, but this time, rather pointedly, she hired a car herself and persuaded Lebitsa to come with her to Carletonville. There she arranged for them both to consult a doctor; although he did not determine the exact cause of Dikeledi's own illness, he said that Lebitsa was suffering from tuberculosis and that it was far advanced. Dikeledi, determined to assure their future in town, went out and found herself a job as a domestic servant in a white suburb. Through a series of boldly executed moves, she had thus out-manoeuvred not one but two Maine men. After a struggle lasting more than a decade she had at last managed to free herself, and her husband, from the clutches of her father-in-law. But her moment of triumph was darkened by the shadow that hung over Lebitsa, who was weaker than ever.

Bodule realised that he had lost the battle to maintain the last link with the disintegrating household at Molote. His wife crafted a plausible story about her supposed need to see a doctor three times a week, and then, on the first weekend they could get off work, they took Lebitsa back to Boons, where Bodule explained to Kas that Dikeledi would be staying on in town until she had regained her health. Kas, cut off from any line of retreat by his earlier behaviour, offered no resistance. Bodule had finally been captured by his wife.[49]

Kas scarcely had time to assimilate the implications of this development when he was laid low by a second, even more vicious blow. On Thursday, 9 July 1959, Lebitsa died at the age of only fifty-three. She left the world much as she had lived through it, quietly and without a fuss. Kas had lost a close companion and a much-loved wife. The body was taken into Boons, and because the death had occurred when Kas was strapped for cash and the family could not afford the prohibitive mortuary fees, the interment had to be organised at short notice and very few family members were at the funeral.

Early that Saturday morning a few neighbours trickled into the homestead and then sat around waiting. Funerals in rural areas were always surrounded by a great deal of uncertainty about working people's commitments and transport arrangements, and it was never clear until the last moment who exactly would be attending. By midmorning they were joined by Phitise and his wife, who came across from Doornkop by horse and cart. Thakane arrived by car from Schweizer-Reneke and, even more impressively, Motlagomang, Mphaka's widow, who had somehow managed to get on a train from Taung at very short notice. There was another long wait and then, in the distance, they heard a motor cycle chugging its way up the path to the house. Members of the Ethiopian Church recognised it as belonging to the minister based at Ventersdorp. He, too, was welcomed, and soon the procession wound its way down a stony path leading to the grave, dug on a hillside site a little above 'Ramagaga's property.'

At Leetwane's insistence Lebitsa was laid to rest after a short service led by the Reverend Ndzeku, even though Lebitsa, unlike most of the other women in the family, had remained a pagan throughout her life. It was left to Kas to guard tradition by making certain that her grave was dug in a manner befitting a MoFokeng, properly aligned with the rising and setting sun. By the time the mourners got home the short winter twilight was spreading gloom up and down the length of the valley. The neighbours peeled off the path and the family was left to contemplate its lot in darkness and silence.[50]

Lebitsa's absence was difficult for Kas to come to terms with. For several days his movements were ponderous. Emotional quicksands seemed to surround the homestead on every side. Death, seeking to pull along the living, grabs not only at the heart but at the hands and feet, numbing everything it touches. After a struggle lasting a week, he wrenched himself free and found enough energy to fossick around amongst his tools and equipment. Mustering all the care and precision he could, he fashioned a curved wooden mould for the cast of a tombstone. He poured concrete into the mould, waited for the mixture to stiffen, and then suddenly remembered there was a point where craft skill had to give way to literacy: he called for Nkamang, who scrabbled around for a stick and, with Kas standing at her shoulder, slowly engraved an inscription dictated to her by Leetwane and the others. The completed monument embodied the skills of an elderly craftsman with those of a literate young woman—a pragmatic mixture of which Lebitsa herself would have approved.[51] Members of the family set aside the tablet to harden the stark and cheerless message of Hymn 111: 'Death and Judgment.'

Kas spent most of his 'free time' attending to the needs of local people who found their way up the hillside to ask for help with minor ailments, not to mention others from farther afield directed to the house by David Mohammed at the trading store. All this brought in cash, and it was supplemented by the proceeds from cobbling undertaken for some neighbours. Meanwhile the 'real' work of the season was done: making the harnesses for the donkeys and mules that Kas would put into the fields in the spring.

After he finished the harnesses, Kas made a few social calls in Molote, which gave him the chance to catch up on some of the village gossip and monitor the bickering between the valley's two dominant political factions. He found that the normally low-keyed debate had taken on a new, sharper edge as a result of an unexpected development: the National Party government, intent on pursuing to its conclusion its insane dream of territorial segregation, had instructed its Department of Bantu Administration and Development to inform Catherine Monnakgotla that, at some point in the not too distant future, all the inhabitants of 'black spots'—including those in her location—would be given notice to abandon their homes and incorporate themselves into the appropriate 'tribal homelands' which the state was in the process of consolidating.[52]

The insecurity which such 'official notices' engendered in South Africa's long-settled and relatively stable black communities was serious enough in itself, but around Boons, it could unleash devastating social, political and economic forces. The threat of removing peasant families from their established farming sites to ethnically defined 'homelands' injected new venom into the succession disputes between the Monnakgotla and Mathope factions. It also exacerbated divisions between BaKubung insiders, who enjoyed access to land, and immigrant matsenelwa from many different cultural backgrounds, who had been pushed into tenancy relationships with their hosts. Removal threatened the very fabric of this society.

Ramagaga's hasty disposal of the property under the ridge took on new meaning in this context and prompted an even more disturbing train of thought in Kas's mind. Had Ramagaga perhaps got wind of something that Motsuenyane did not know about when he sold the farm? How vulnerable was the thin strip of freehold land bordering Monnakgotla's Location? Where did all these developments leave Kenneth Nkaba, and more pertinently still, where exactly did the Maine family's political fortunes lie—with an adventurous independent landlord like Nkaba, or with a more cautious traditional authority like Catherine Monnakgotla? The answers to these questions were, to say the very least, hidden. What was certain was that from now on Kas and the rest of the inhabitants in the valley would be farming under the sword of Damocles.

# PART FIVE

# SENIORITY

*

*'When a chameleon moves from one position to another it changes colour. Likewise, when I moved from one farmer to another I had to adopt a strategy that would meet the new circumstances.'*

KAS MAINE,
1983

CHAPTER FOURTEEN

# Threat

## 1959–67

Each season now seemed to require more effort than the one preceding it. With virtually all the older Maine children now working in the towns and only Nkamang and the granddaughters around to help Kas in the fields, he had to keep adjusting the combinations of draught animals and ploughs to declining labour resources. Youth and old age could be equally harsh for a sharecropper. Kas, like all black farmers coaxing a living from the soil, needed an adult son, but Mosala was not to be relied upon.

Nor could the women be commanded with the old authority. Nkamang, nineteen, and Pakiso, sixteen, were both of an age where they wished to take control of their own destinies, and their plans did not centre on spending lengthy spells in the fields. But somehow or other Kas managed to wheedle yet one more season's labour out of them, while Mosala was sent off to the Boons Station Cash Store to buy half a ton of commercial fertiliser.[1]

Once the seed was in the soil the young women breathed more easily, since the progress of the maize then depended more on God than it did on grandfather. Kas, too, perked up when, shortly before Christmas, the Moates arrived with twelve cattle and a horse to honour their bohadi debt. A few days later Nthakwana married Matthews Moate in a ceremony at the home of the Reverend Bothma of the Nederduitse Gereformeerde Kerk in Ventersdorp. Kas did not attend, which was probably just as well since some aspects of the proceedings would have offended his traditionalist sensitivities. In a radical departure from family custom Nthakwana chose to get married in a long white dress. Nor was Kas much impressed when, some months later, Bodule and Dikeledi went through a similar ceremony in Carletonville in order to become eligible for municipal housing under the government's apartheid policies. The family structure was splintering under the pressures of ageing but also from those of the state: and several of its fragments were becoming embedded in an unfamiliar urban culture with new regulations and laws.[2]

Kas had other reasons for being irritated with Bodule and his wife. Dikeledi had suggested to him that he consider disposing of his cattle and, deeply offended, he had told her that while she was at liberty to sell all of Bodule's animals, he had no intention of selling his own livestock. Bodule then put in an unexpected appearance over a weekend, sold Nkaba two of his better beasts, and went on to ask the schoolmaster to buy him a few heifers when next he was in Derby. Nkaba did so and the heifers were added to the Maine herd.[3] Nkamang and Pakiso, both resentful that neither their education nor career prospects were being enhanced under a regime requiring them to split their working day between school and farm, now had even more livestock to contend with. The animals were neglected, wandered off into other fields and got impounded, with the result that Kas had to spend time and money retrieving them from the pound at Klerkskraal.

When Bodule next appeared, Nkaba rather pointedly drew these problems to his attention and offered to buy all the younger man's cattle for his property at Vryburg.[4] Bodule took exception to the suggestion that he was abusing privileges extended to his father, and despite the lateness of the hour, demanded that Nkaba issue him a permit to move his livestock to Carletonville, where he could sell the animals at his convenience. Angrily, he got out the animals and set off in a southeasterly direction, but he was being matched in pace by the onset of twilight. By nightfall he was back at Molote, put the cattle back in the kraal, and told Kas that as soon as he returned to Carletonville he would arrange for someone to come and collect his cattle. He never did.[5]

In January 1960 the crisis in livestock management deepened when Kas got a letter from Jacob Dladla. The last time he had seen 'Hlahla' had been in 1947, when the herbalist had entrusted six sheep to his care while the Maines were still under Moormeister at Sewefontein in the Triangle. Time, tractors and the Afrikaner Nationalists' drive for territorial segregation had caught up with Dladla; like thousands of others, he had been pushed off the farms and was looking for a place where he could keep the seven cattle and thirty sheep he still owned. Ties of friendship, bonds of professionalism, and the fact that he owed Dladla the equivalent of six sheep and their offspring, meant that Kas was hardly in a position to refuse a request for help. The animals were forwarded to him by rail from Bloemhof.[6]

But loyalty to one friend had to be offset against exploiting the trust of another; Kas knew that by taking in yet more livestock he was stretching his friendship with Kenneth Nkaba to breaking point. The landlord sensed the old man's predicament, however, and, mollified by the prospect of sharing in what looked like a good crop, he maintained a diplomatic silence. The pressures on Kas were eased slightly when Nkamang and Pakiso suddenly announced they would not be returning to school at the start of the 1960 term and would instead go into town to seek work—this was something of a mixed blessing since while it increased his short-term access to labour it only exacerbated the long-term problem. He advised against it. He persuaded Nkamang to stay on for a few more months, but Pakiso soon took off for Carletonville where she intended to use Bodule and Dikeledi's home as her social conduit into the urban labour market.[7]

It took all of Kas's accumulated wisdom and still formidable physical strength to see him through the 1960 growing season. When Jacob Dladla arrived to reclaim the sheep he had left with him thirteen years earlier, Kas persuaded him to settle for a cow, a light-brown ox, and grazing rights for the two animals by way of compensation, which seemed reasonable. Always a man of mysterious movements, Jacob Dladla then disappeared into the alleys of nearby Modubung where, despite the disadvantage of being a mother-tongue Zulu-speaker, he joined other 'newcomers' seeking refuge in Catherine Monnakgotla's location.

Kas, driven by nervous energy and eager to divert his anxiety, planted lots of watermelons, which did very well in the field closest to the shacks looking out over Mathopestat. But the maize was his primary concern, and in late summer his worst fears were realised when the biggest field was suddenly invaded by the same grain-eating insects he had encountered at Rietvlei. He and Mosala tried to ward off the *makgwaba*; but how could four arms hold off four million adversaries? The harvest did not even meet the family's domestic requirements.[8]

For once the magnitude of the crisis seemed to penetrate even Mosala's head; he, with Motshidisi and Nkamang, was sent off to the far western Transvaal, to Madibogo, to search out the kin and work that might enable them to get the Maines through the winter. This harvesting party was markedly more successful than the one that had returned from the Triangle the year before. Nkamang soon left for the mining town of Stilfontein, where she employed the same tactic that Pakiso had used at Carletonville and got Morwesi to get her a position as a domestic servant.[9] Prime Minister Verwoerd's government might have wanted to limit black rural-urban migration, but during the early 1960s its success rate with African women was only marginally better than that which Kas and Mosala enjoyed against the makgwaba.

Nkamang's remittances helped to ease the strain on the domestic budget, but Kas was also interested in a tale that she and Mosala brought back with them from Madibogo, a tale that Bodule, too, told to him with so many significant similarities that he felt he had to give it credence. For the first time in a decade, Kas was getting meaningful clues as to the possible whereabouts of Mmusetsi.

According to his brothers, Mmusetsi had left Rhenosterhoek for Carletonville where he had met up with a sharp-witted white man who, impressed by his butchering abilities, had taken him on as an assistant in his butcher shop. The shop, most of whose supplies came from a farm on the Transvaal's northwestern border with Bechuanaland, offered meat at unusually competitive prices. Mmusetsi and the man got on well, and some months later, the white man decided he should take Mmusetsi with him when next he visited his farm, which was situated in country made famous in the cattle-rustling tales of South Africa's greatest story-teller, Herman Charles Bosman. And indeed, up in the border badlands Mmusetsi soon learned how the butchery maintained its competitive edge.[10]

On Friday evenings the white man would link up with a gang of BaTswana cattle rustlers and agree on a price for livestock stolen from farmers in the neighbouring territory. The next night the butcher and a few trusted assistants would take a truck to a designated point on the border, flatten the barbed-wire fence with a length of canvas, and send Mmusetsi deep into Bechuanaland to link up with gang members. Mmusetsi would take charge of the stolen animals and drive the herd back over the border where they were loaded onto the truck for the journey to Carletonville.

These arrangements worked well for several months, but then, on a trip back to the farm one autumn, the butcher decided he needed more help in bringing in the maize and asked Mmusetsi to stay behind to assist the seasonal labourers. During the harvest Mmusetsi met a woman whom he subsequently married, and as far as his two brothers knew, the couple had settled on the property, where they were still to be found.[11]

Kas pondered this story for several days, but he either could not or would not act on it. He had already conducted one unsuccessful search for a lost child, on the occasion when he and Bodule had failed to establish Matlakala's whereabouts. His relationship with Bodule was now at a low ebb, and he did not want to negotiate the mining towns on his own, while the prospect of conducting a search 'somewhere along the Bechuanaland border' for a butcher whose name no one knew was equally unappealing. The bizarre tale was perhaps best forgotten, he decided. Better that he make use of such off-season energies as he had to raise cash for his next effort on the land.

Looking out over what had once been Nicodemus Ramagaga's fields and down across the valley toward Mathopestat, Kas could see the last of the harvesting teams winding up their seasonal efforts—a cruel reminder of how, between them, man and nature had robbed him of two successive harvests. He was contemplating this scene when a few of Andreas Motsuenyane's workers came to tell him that their landlord's business interests had delayed him in Gaborone, and that they were having great difficulty in getting the harvest moved to the threshing point. They were anxious not to lose their place in the queue for the thresher and thus delay their departure for the reserves, so they asked Kas to help them transport the harvest to the threshing point. He was quick to point out that he was 'not on good terms with Motsuenyane,' but the workers said they were 'the ones who were suffering' and persuaded him to help them. He borrowed Kenneth Nkaba's wagon, got out his horses and then spent a good part of a day moving maize to the threshing point. There was something wonderfully reassuring about being seated high up on a wagon behind two large horses plodding their way towards a distant mountain of maize. The solid rhythms of transport-riding, the unexpected reappearance of Jacob Dladla, and talk of Mmusetsi all took him back to his days in the Triangle. He was lost in reverie; the season seemed filled with echoes. He regained full consciousness when the grateful harvesting team gave him four bags of grain drawn directly from their quota.

When Motsuenyane arrived he was pleased to find that the harvest had been fully processed, and he walked over to thank his neighbours for their assistance. But Kas, smarting beneath the injustices of two successive poor seasons and in no mood to be patronised by the rich and powerful, told him that the excellence of his harvest was directly attributable to the fact that his fields had been fertilised by Maine crops which had been ploughed into the ground by greedy Motsuenyane tractors. Motsuenyane, who had come in peace, was not to be deflected, and he said: 'Maine, let's not talk about that. I will try to make things easier for you. You complain that I once took your fields. Take them back, plough them again, and let us then share the proceeds.' Kas, who could no more avert his eyes from gaining access to land than a sunbird could ignore an aloe in bloom, paused and then relented. 'We resolved our differences.' They parted company, having reached agreement on a broad strategy to pursue when spring came again.[12]

*

Winter had barely set in, but in Kas's mind it was already August. The prospect of being able to cultivate not one but two sets of fields conveniently close to his little cluster of shacks on the hillside revitalised the inner core that drove the outer being. He set about cleaning, modifying and repairing ploughs with a vigour reminiscent of his Vaalrand days.[13] He was becoming more familiar with the vagaries of the local environment, and he believed that, if he were given but one more chance, he could still produce a bumper crop. Just one lucky break, like the one at Sewefontein, would enable him to offset the damage suffered in recent seasons. Surely it was not too much to ask?

Four years in Monnakgotla's Location were starting to yield other, less obvious, benefits. For the first time since leaving the Triangle Kas had been in one place long enough for the locals to get to know him and his reputation as a ngaka was spreading. Even the Boer who owned the café in Boons, a man known to him only as 'Piet,' was referring white workers from the surrounding mining towns. Afrikaner women, themselves only recently off the platteland, often consulted him about the difficulties they experienced during pregnancy and he supplied them with appropriate herbal remedies. As usual in such cases, he accepted payment only once the couple reported on the success of his treatment.[14]

But this system of deferred payment had its drawbacks, especially in a season when cash was at a premium. The financial squeeze, talk of Mmusetsi, and Bodule's threat to sell his cattle forced Kas to trawl through his list of debtors and, almost inevitably, helped to dredge up his long-standing dispute with Abner Teeu. While willing to admit that Dikeledi's bohadi had not yet been fully paid, he was of the opinion that this was only because her father had refused to return the cash that had been advanced to him for the hand of Disebo. The mess had become even more complicated once Pulane had run off with Dikeledi's brother, Mojalefa, but this too was surely a 'marriage,' and one for which the Maines had not even received partial payment! By his accounting, the Teeus owed the Maines far more than the Maines owed the Teeus, and he was angry and frustrated by their debt. Visits from the Carletonville-based Maines became tense. Once, when Bodule put in a weekend appearance, Kas told Dikeledi that he intended laying a charge of theft against her father when the families met later that month at the Bantu Commissioner's office in Ventersdorp to formalise Mojalefa and Pulane's marriage. This news was greeted with total incredulity when Dikeledi told her father. 'No,' said Abner Teeu, 'I don't believe that a brother would have me arrested.' How was it possible for the parents of children who were so intermarried to end up confronting one another in a white man's court?

On the day of the wedding Dikeledi, who had done more than anyone else to ensure that storm clouds would put an end to Mmusetsi's marriage to Disebo, suddenly found herself attracting lightning from both patriarchs. 'The one would turn on me accusingly, saying, "Your father did this and that," and then the other would turn and say: "Your father-in-law said this and that." ' She pleaded with them to settle their differences amicably and shuddered to think what would happen once they entered the Commissioner's office, but much to her relief, 'the signing ceremony went without a hitch.' Mojalefa and Pulane were allowed to leave the building without having to witness a major public confrontation. Perhaps Kas's nerve had failed him after all.[15]

But Dikeledi had underestimated Kas's determination. As the unfortunate business with Andreas Motsuenyane had showed only too clearly, age and faltering agricultural production increased rather than decreased his appetite for litigation. Her father-in-law had already had a word in private with the magistrate. As the families were about to disperse, the Bantu commissioner-cum-magistrate suddenly emerged from his office, called the two patriarchs aside and conducted an informal hearing. He listened to complainant and defendant, expressed his surprise at elderly men who were related by marriage needing to have recourse to the law, and sketched out the terms of what he considered to be a fair out-of-court settlement. He proposed that any outstanding debts arising from bohadi for Dikeledi or Pulane be swiftly settled in the customary way. Then, hinting at what he thought was the origin of the problem, he suggested that Teeu pay Kas ten pounds a month until the amount that had been advanced in expectation of Disebo's marriage to Mmusetsi had been liquidated. These street pronouncements did not carry the full weight of the law, but they were offered with enough conviction to persuade the parties to agree. Both men left Ventersdorp relieved—Abner in the knowledge that he had avoided arrest, and Kas in the belief that he would recover his money.[16]

Back home, with an assured income of ten pounds a month Kas could afford to slacken the pace of his off-season activity, but chose not to do so. Cobbling, his work as a ngaka and smithying continued to take up most of his time, and only now and then did he venture down into the villages. In the valley much of the talk centred on the government's threat to resettle the people of Monnakgotla's Location. Kas should perhaps have been more concerned about this, but a lifetime on the land had taught him that, even with the best eyes in the world, one could seldom see further ahead than a season at a time. His own eyes were no longer as sharp as they had once been, and there was no point in fretting about something which, if the Boers had already decided upon it, was inevitable.

This phlegmatic approach did not commend itself to most folk, and in the valley, the tension and uncertainties surrounding the threat of removal were causing the community to split along old and new lines of stress. Mathopestat residents clung to all their deep-seated reservations about the inhabitants of Modubung and Molote, and the inhabitants of these two villages, in turn, harboured more and more suspicions about 'newcomers'; Catherine Monnakgotla's once united supporters fragmented into smaller and ever more conspiratorial groupings. As long as the social bases and political programmes of these groupings were unclear, however, Kas could see no practical purpose in people withholding their support from the chieftainess. In politics as in life, the two cardinal virtues were patience and pragmatism. Had not time and circumstance delivered first Motsuenyane and then Abner Teeu into his hands?

With spring fast approaching the Maines assembled a span of oxen and a team of horses, the former to work under Kas on the smaller field at Nkaba's, and the latter, under the guidance of Mosala, to plough the larger area that had once been Ramagaga's farm. It took but a single day for Kas to realise that this was a mistake. Without Nkamang or Pakiso to help him he did not have the strength to get up at dawn, collect and inspan the oxen, put in a full day in the field, release the animals for grazing, make certain that they were all watered, and then round them all up before returning them to the safety of the kraal before nightfall. At the age of sixty-six, he reluctantly admitted that he was getting too old to do this on his own, and when Nkaba offered to let him have a team of donkeys and a full set of harnesses in exchange for the draught oxen, he was unusually quick to agree to the swop.[17]

The donkeys, slower and less powerful than the oxen, were easier to handle. Working at a more sedate pace, Kas made steady progress along the upper slopes. Lower down the hillside Mosala, working on the fields that had earlier been exposed to the weight of the tractors, found that the horses made light of the same clay soil that the girls had once struggled to plough. Kas decided he would plant maize on Nkaba's property, where he could personally keep an eye on the crop. Motsuenyane's fields would be planted with sorghum which, being more robust than the maize, could survive Mosala's casual supervision.

The spring rains were not very good, but despite this, the seedlings in both fields were soon bursting with promise. A thorough ploughing and his earlier investment in commercial fertiliser were reaping dividends. By Christmas 1960, Kas knew that if the rains continued into the first months of the new year he would be looking at a fine crop. Yet something within him warned him against optimism. Landlords, like locusts, were unpredictable. He found himself more sympathetically disposed towards valley inhabitants trying to cope with the threat of removal. What if the 'Boer' authorities chose to resettle everybody before the harvest was even in? But his fears proved groundless and the summer months slipped by without incident.

Cooler weather in April attracted its quota of inquisitive landlords who, like the quelea on the telephone wires, suddenly appeared in flocks, taking a seasonal interest in the proceedings. God and the government posed problems of one order, but it galled Kas to think that after successive seasons of hardship landlords were entitled to fully half of all the grain he surveyed. On the upper slopes there were potentially at least a hundred bags of maize, and lower down there was even more sorghum. He bore Nkaba no ill will—indeed, the schoolmaster had come to his rescue at a crucial moment and was always willing to lend the Maines a hand but Andreas Motsuenyane still conjured up less charitable thoughts: 'I decided to punish him because of his wealth. I wished to reduce his strength. He was a man who did not even live on the farm and yet he gave me a lot of shit.'

So, when Motsuenyane was away attending to his primary enterprise in the far western Transvaal, Kas arranged for Leetwane, Mosala and Motshidisi to swoop down on the sorghum at the first moment of ripeness. They commandeered a full trolley-load of grain, diverted the wagon up the hillside and then, in the privacy of the shack, carefully decanted it into seven bags. On reflection, Kas decided that seven bags was insufficient compensation for the anguish and inconvenience he had been made to suffer at Motsuenyane's hands. Mosala was sent out on a second raid and returned with ten more bags of grain. The psychological balance sheet had been adjusted.[18]

The remaining sorghum was stacked on the lower slope to await Motsuenyane's arrival, and the harvesting team moved higher up the slope to tackle the maize. Nkaba, generous as ever, offered to help them, and as expected, once the harvest had been divided the Maines's share amounted to about fifty bags. Kas considered this satisfactory: after all, lower down the hill there was more to come. Or was there? He was mildly apprehensive about Motsuenyane's impending visit. Landlords might be knaves but they weren't fools.[19]

Motsuenyane took in the size of the grainstack that greeted him on the lower slopes of the property with a single glance. Kas scanned his landlord's face for signs of unease, but Motsuenyane did not flinch. On the contrary, after inspecting the stack more closely he expressed his delight and said how pleased he was that Kas was entitled to a hundred and fifty bags of sorghum. Motsuenyane was convinced that he had made his peace with his elderly neighbour, and he left for Lichtenburg happy in the belief that he had righted an historic wrong. But men sometimes celebrate the right thing for the wrong reason. As the Afrikaner Nationalists heralded the advent of the South African Republic under their premier, H. F. Verwoerd, on 31 May 1961, and Andreas Motsuenyane drove back home basking in a feeling of well-being toward his fellow man, Kas sat down to contemplate a harvest of more than two hundred bags of grain.[20]

The bonanza could not have come at a better time. It gave Kas a cushion for the winter, and it helped restore his faith in his own productive powers. For a man in his mid-sixties that was important. Indeed, as he reflected on the season and analysed the reasons for his success, the outlines of a new and radical plan took shape in his mind. But it involved so sharp a break with past practice and entailed so many risks that he needed more time to think it through.

*

With the harvest stored and an adequate food supply secured for the first time in years, the family settled down for a winter that bore some resemblance to those of years past. Instead of sending out a work party into distant kinship networks in a last-minute bid for maize and money, or extruding ever more sons, daughters and grandchildren into the bottomless pits of ugly mining towns, the farm and the family were once again a safe haven for those who were lost, weak, or vulnerable. This new-found strength did much to restore Kas's self-confidence, dignity and pride. Sixteen-year-old Pakiso certainly sensed it, because Molote presented itself as an obvious bolt-hole when she discovered she was pregnant.

As the illegitimate daughter of Morwesi and the granddaughter of a patriarch who had himself fathered three illegitimate children, Pakiso was perhaps fated to produce an illegitimate child. Like her mother and Nthakwana before her, the shells of age, attitude and experience that enclosed her were not hardened to withstand the pressures of the transition from rural to urban life and the onset of early adulthood. She had met Lucas Mogosi while working at a hospital in Carletonville, and the nearby township of Khutsong had given them a base for what appeared to be a satisfactory relationship. But unlike Kas, who had made the honourable offer to marry or take in the two 'outside' women who had entered his life, Lucas Mogosi had simply rejected the young woman once he had learned of her condition.[21]

Maybe it was merely the passage of time or an advance in the status of women, but the family no longer had much room for anger, recrimination or tears in such matters as they had had when Morwesi had been seduced by Padimole Kadi in the 1930s, back at Vaalrand. Morwesi and Nthakwana's earlier experiences had helped clear the way for such 'unforeseen' developments. For members of a self-confident younger generation like Pakiso, unquestioning acceptance of pregnancy outside marriage might have been part of an agreeably 'modern' attitude, but for members of an older generation immersed in the strictures of the A.M.E.C. and notions of respectability, like Motheba, it spoke of social decay.

But great-grandmother Motheba had the good grace to keep her misgivings to herself. That winter, already well into her nineties, she and Kas's younger sister, Sellwane, called in at Molote on a journey designed to re-establish links with sons whom she had last seen when they were still farming in the Klerksdorp and Potchefstroom districts. Leetwane the church-goer had always occupied a special place in the old woman's affections, and Motheba spent several pleasant days in her company before moving on to Doornkop, where she hoped to spend some time with Phitise and family before going back to Harrisburg.[22]

That was the last time Kas saw his mother alive. A few days after her departure for Ventersdorp he received word that the old lady had passed away. He and Leetwane undertook a painful journey to Doornkop where he, Phitise and Sellwane were joined by Sebubudi and, together for the first time since Triangle days, the siblings laid their mother to rest in the only way imaginable—in an uncompromisingly Christian ceremony presided over by a minister from her beloved African Methodist Episcopal Church. In death as in life, she seemed to be marked by an inner strength and magisterial dignity. She was a truly formidable woman.[23]

The children each mourned her departure in their own way. For Kas she epitomised those qualities he associated with the older, most respected farming families in the Triangle, and on the way home he reflected on her compassion, discipline and loyalty to kin. It was she who, by her example as much as her prayers, had done most to convince him of the existence of God, and even during the closing days of her life, she had somehow still found the strength to take leave of her children in a dignified fashion. For women like Motheba, drawn from a generation that had witnessed the first major shock waves accompanying the transition from a respectable peasantry to a vulnerable proletariat, the sanctity of family life was paramount. For younger women like Morwesi, Nthakwana and Pakiso, along with the insect-like men who flitted through their lives, pausing only long enough to implant their seed before flying on to another chance union, it apparently meant much less. An industrial revolution had corroded the values of the old rural order just as surely as rain rusted the shares of an uncovered plough.

Perhaps it was this subliminal glimpse into the process of decay, or guilt at having failed to pass on appropriate values to the next generation, or, more simply still, the stark juxtaposition of Motheba's departure and Pakiso's arrival with what would be his first great-grandchild, but Kas returned to the farm in a distinctly sensitive frame of mind. He was therefore outraged to encounter there a new and even more damaging round of complaints about Mosala's drunkenness, sexual promiscuity, and penchant for wife beating. The spectre of putrefaction materialised before him, and in the horror of the moment, he lost his moral footing. He grabbed a knobkerrie and flailed away at his son's crapulous frame until Mosala lay bleeding from a gash on the head.[24]

Kas realised later that he had gone too far. The lesson, if indeed there was one, was almost totally lost on a drunk. Mosala's head healed with a scar and the incident was largely forgotten as Kas began to focus on the opportunities that came with spring and new production strategies.

❀

The first move was made in early July. Kas asked neighbouring Boers whether they had any good second-hand implements for sale, and then, one day, he came back from the farm Onverwacht with a heavy disk-plough, which he cleaned and modified to meet his new requirements. A month later he and Mosala put seventy-six bags of sorghum into Nkaba's wagon and took them to the Boons depot of the Koster Koöperatiewe Landboumaatskappy Beperk, where they fetched 223 rands and 44 cents of South Africa's new republican money,* or, as he and most of the people

* The issues surrounding central banking, currency and legal tender in the four former British colonies that formed the Union of South Africa in 1910 constitute a maze of historical problems for the specialist. To all intents and purposes, however, the lives of the Maine family between 1900 and 1960 were shaped most fundamentally by the value of the British pound. The establishment of the Republic of South Africa in 1961 saw the abandonment of the sterling block and the introduction of the rand with an initial exchange rate of two rands to the pound. Since 1961 the value of the rand has depreciated steadily and, by mid-1995, it took R5.80 to purchase a British pound.

in the valley still thought of it, about a hundred and ten pounds sterling.[25]

But Kas was still well short of his intended target. Only after a long session with Nkaba during which he managed to persuade his landlord to lend him £150 was he ready to take the next step. He sent a message to Phitise asking for the help of his brother's son, Malefane. When Malefane arrived Kas, Mosala and he wound their way down across the valley to Easton's Garage at Syferbult, where the real object of Kas's attention was housed.[26]

Malefane, who knew about such things, assured him that the tractor was a good investment. The Eastons were a well-established family with a good reputation, but even so Kas found it hard to part with £300 for a mere mechanical thing. In his mind's eye he could still see the outlines of the old Ford truck and The Louse's wretched black Studebaker. But he had worked through these misgivings a thousand times before and always reluctantly concluded that there was no other way round his problem. He had either to mechanise his production techniques or abandon serious grain farming. Twelve years after first contemplating the idea at Sewefontein under very different circumstances, Kas became the owner of a tractor. Malefane drove them home, Mosala, at his cousin's shoulder, anticipating the moment when Kas would allow him out into the fields on his own.[27]

Back at Molote the men met with a strangely muted reception. Not only were there no squeals of delight from excited children, as there had been on the day when Kas and Bodule returned to Jack Adamson's farm with the Ford, but the women in the household seemed positively sullen. Within days Leetwane broke ranks and openly criticised Kas's willingness to spend money on a tractor at a time when she and the rest of the family were living in the most rudimentary domestic circumstances. Equally dark mutterings started to emanate from the rest of the Lepholletse clan. But Kas was not unduly concerned: jealousy was a common enough human failing. What hurt more was his wife's consistent lack of support for what, in his eyes, was a necessary, if risky, venture.[28]

Despite Leetwane's opposition, Kas believed that financial decisions about the family were his personal prerogative and that he would have been failing in his duty to them had he not bought the tractor. Farming and food production were the responsibility of the head of the household, and in the Maine home that made it a man's business. Ignoring the pointed comments within the shack and the whispering campaign beyond it, he sat through Mosala's first awkward attempts at fully mechanised ploughing with no outward signs of fear or uncertainty.

On 8 September 1961, and with the ploughing in full swing, Pakiso gave birth to a daughter named Mpho. Inside the small corrugated-iron structure the baby would wake by night and demand the milk that soothed her until dawn; while out in the fields, the tractor would occasionally fall silent at midday and refuse to move until the patriarch replenished it with the diesel and oil that sustained it until dusk. And, just as surely as Pakiso learned that demands made of the breast could be physically draining to the point of exhaustion, so Kas realised that the leviathan's appetite was unrelenting. Like Pakiso, Kas managed to get through the moments of doubt by focusing on the expectations of a longer-term reward.[29]

On balance Kas was satisfied with the tractor's performance and pleasantly surprised by how quickly they managed to do the planting. While Mosala rounded off operations on the lower slopes he cultivated vegetables and watermelons in the smaller field nearer the shacks. The tractor's time- and labour-saving abilities were not lost on Mosala. No sooner were they through with the planting than he announced his intention to leave for the West Rand, where he would work for a firm of contractors that had been actively recruiting unskilled labour in the valley for some weeks. Kas pleaded with him not to leave him with sole responsibility for the fields, but the last battle for Mosala's loyalty had been fought and lost during their winter conflict. By November Mosala was gone, leaving behind only a vague offer to come home before the harvest.[30]

A short spring gave way to a long hot summer, and by the time 1962 had arrived, there was no longer much reason to lament Mosala's absence. The rains tailed off disappointingly, but that meant fewer weeds, so the demand for family labour eased off. But the drought also meant that the grain was slow to pack and swell. As the season dragged to a close, emaciated maize cobs cracked open to reveal grins that mocked all Kas's efforts. Worse, Leetwane was always on hand to remind him of the folly of the tractor.

With the grain reserves depleted and no sign of a decent harvest, Kas was under renewed pressure from two quarters. First, Kenneth Nkaba, realising that he might lose the gamble they had taken in devoting more land to arable production than to grazing, became touchy about the strains Bodule's cattle put on his resources, and he was also concerned about recovering his share of the money paid for the tractor. Secondly, the women in the household started insisting that the fruit and vegetables from the field near to the shacks be used solely for domestic consumption, thereby putting paid to any idea of Kas using them as cash crops. Kas tried to meet this domestic demand by bringing home small quantities of inferior-quality produce, but the women soon tumbled to this trick, and Pakiso and the others resorted to 'stealing' maize, pumpkins and watermelons when he was not around. His priorities might have included the purchase of a tractor but, for mothers and daughters, the primacy of food production for domestic consumption was not negotiable. They had a point.[31]

When harvest time came around there was no sign of Mosala, though

Leetwane, Motshidisi and Pakiso turned out without complaint. Years later, in philosophical mood and with all the benefits of hindsight, Kas still found it hard to conceal his disappointment at the size of that harvest:

> No, the introduction of the tractor did very little for me. I managed to produce only twenty bags. It would have been better had I ploughed with oxen. Even if I had only four donkeys I would have harvested more than I would have with the tractor.
>
> Look, farming is a funny business. When oxen plough they are working in harmony with God and with nature; a tractor cannot. It restricts you with its unnatural demands. I could produce more with four donkeys than I could with a tractor because, with animals, you do not make a cash outlay for grazing—they don't need diesel and oil. The cost of the donkeys remains hidden.[32]

It had been a deeply frustrating season, to put it mildly.

But there was no thought of turning back. Kas had made the investment, committed himself to mechanised farming, and the only way out was to redouble his efforts. Grain farming, like gambling, was addictive, and somewhere in the future there had to be just one more year like that magical one at Sewefontein. Courage and craziness were separated by a poorly defined frontier, and who was to say where a sixty-eight-year-old man stood without access to a tractor driver? One thing was certain: if there was to be any improvement in yields, then the fallow ground would need turning. But where on earth was Mosala? By the second week of June Kas could wait no longer. He gleaned a clue or two from Motshidisi and then set out for the West Rand in the hope of finding the son who now separated hope from success.[33]

Kas found Mosala in Roodepoort, at an address which the son had sometimes used when sending cash to the post office at Boons. There seemed good reason for his having failed to reappear at Molote in time for the harvest: like thousands of black men seeking work in South Africa's 'white' industrial areas, he had run foul of the government's increasingly repressive use of the pass laws, and he had spent four months in prison. According to Mosala, the police had warned him that '. . . if he were ever again arrested for a pass offence he would have to spend the remainder of his life in prison.'[34]

The truth of the matter, which remained concealed from Kas for at least a decade, was rather different. Even before his departure for the West Rand, Mosala had been spending time in poor company, which abounded in the mining towns. During his recent excursion into Roodepoort he had been befriended by 'Ishmael,' a MoTswana, and by Thabo, a MoSotho who hailed from the Transkei. After an evening's heavy drinking in the local shebeen, the three had spotted a client tendering notes of a large denomination and decided to waylay the man when he left the premises. The victim was ambushed and, when he attempted to resist, Ishmael and Thabo stabbed him to death. Mosala had then helped drag the body to a nearby spruit, where it was tossed into a trickle of urban effluent.[35] The three had then split up, with Mosala taking the added precaution of changing his employer and adopting the new 'Venda' ethnic identity of 'Samuel Mondau.' The two principal offenders, however, when arrested decided to implicate their accessory and led the police to the firm of fencing contractors at Roodepoort where Mosala was working. In the court hearings, Ishmael and Thabo, who appeared without the benefit of aliases, were found to have had several previous convictions and were each sentenced to fifteen years' imprisonment. Young Samuel Mondau, without previous convictions, was found guilty on two charges: one, a pass offence, earned him four months in prison and the other, being an accomplice to a murder, a sentence of fifteen years, conditionally suspended for five years. It had been a narrow escape.[36]

Kas was delighted to find how easy it was to persuade his son to come back to Molote and get involved in a spell of tractor driving. But their return to the maize fields was marked by such bad luck that he was tempted to believe that perhaps it was due to Leetwane's unrelenting hostility to the tractor. Mosala let the tractor engine overheat and catch fire, and this so badly damaged it that Kas could not even think of repairing it himself. His lack of expertise when it came to diesel engines frustrated him, and he knew that the cost of repairs would drag him even more deeply into debt. He was bogged down in the cross-cutting logic that surrounded arcane notions like depreciation and vehicle maintenance; with great reluctance he acknowledged the need for professional advice.

Kas went to Syferbult hoping to encounter the drizzle of reason; instead, the solution that Easton proposed struck him with the force of a highveld thunderstorm. Why spend lots of money repairing an old machine when trading in the damaged tractor for a later model would secure the twin advantages of reliability and better resale value? Easton was willing to let him have a hundred pounds for the old tractor. The idea of acquiring a new model was not without appeal, but £500 was an enormous investment to make at a time when he already owed Nkaba money and the entire farming community was living under the threat of forced removal. Kas needed time to think it all through.[37]

Try as he might, he could see no way of retreating from his initial investment without enormous financial and psychological losses. He would have to either meet the enormous costs of repairing a machine in which he had lost confidence, abandon the idea of farming with a tractor, thereby vindicating Leetwane's scepticism, or go forward with Easton's proposed deal. But how was he supposed to raise £400? Like most black sharecroppers he had only his harvest and his livestock. Suddenly, he saw the glimmerings of a solution.

On 27 June 1962, Kas and the landlord had yet another long meeting at which it was agreed that Kas would sell Nkaba twenty-two oxen for close on £500 payable over nine months. Virtually all these animals would come from the herd Bodule had left in Kas's care when he and Dikeledi had gone to Carletonville. This deal, the largest Kas had ever been involved in, eliminated the landlord's grievance about Bodule's supposedly 'free grazing,' settled Kas's debt for the amount he had advanced for the first tractor, and at the end of the day left Nkaba still owing the Maines three hundred and fifty-five pounds. It was the closest that Kas could get to solving his most urgent problems.[38]

Nkaba had long coveted Bodule's animals, and he gladly took the oxen to his base at Vryburg while Kas went back to Syferbult to discuss terms for a new tractor. Easton agreed to Kas's paying off the balance over several months, and then, with the sun still hovering low in the winter sky, Mosala drove the new tractor back to Molote. Kas knew he had taken a calculated risk and he relaxed only when he got around to telling Bodule what he had done with 'their' stock. This, too, had been a risk worth taking, for Bodule raised no objection.[39]

The security of returning to off-season work patterns was something of a relief after such weighty matters. Between sporadic bouts of cobbling Kas found the time to tackle several more challenging projects with typical vigour and determination. Kenneth Nkaba, no less than Piet Labuschagne before him, was awed by Kas's ingenuity with things mechanical. The old man might still have been learning about tractors, but there was no doubting his versatility with the standard repertoire of farm machinery. When some stray cattle damaged the carburettor of Nkaba's paraffin-driven engine at the borehole, Kas cut and inserted a copper template into the damaged part, a temporary repair that lasted while they waited for a replacement to reach the store at Boons. Equally impressive was the way he commanded the bellows and hammer to modify a seemingly derelict planter that the landlord had picked up during one of his rambles through Monnakgotla's Location.[40]

In the villages, where there seemed to be a lull in talk about resettlement, older men such as William Lephadi and Malebje were more impressed by Kas's skill as ngaka. When Lephadi's son got married the family hired Kas to safeguard the festivities against any manifestations of malevolence or ill-will—a wise precaution since a half-witted woman suddenly appeared midway through the celebrations and handed out chunks of dried dung as if they were so many pieces of bread or cake. Such was the protective strength of Kas's herbs, however, that the unfortunate woman disappeared almost as mysteriously as she had appeared and the guests were 'miraculously' protected from undesirable after-effects.[41]

Income from winter work and occasional cash remittances from children in the mining towns helped to balance the Maines' household budget during a season characterised by a net outflow of labour into the urban labour markets; this later was temporarily halted when the Moates left their employer at the farm Duisterpoort and returned to Boons. Nthakwana, by now expecting the Moates' fourth child, needed her kin to help with her family while her husband, Matthews, wanted a spell of relatively independent labour under his father-in-law rather than at the beck and call of a Ventersdorp farmer.[42]

With Mosala there to drive the tractor and Matthews around to help with the livestock, Kas cheered up. But rural production was cursed: no sooner did a man have one set of variables under control than God decreed that another should slip beyond his grasp. Thus, just as labour ceased to be a major problem, the rains failed. The total precipitation recorded in the district declined for a third successive season.[43]

Not everyone was equally concerned about what was happening out in the fields. Sensing that the time was approaching for the arrival of her baby, Nthakwana left the Moates' house in Mathopestat and moved into the shacks on the hillside so that she, like Pakiso with Mpho, might be near her mother. Leetwane had lost none of her skills as a midwife or child minder. Indeed, it was she, rather than the sixteen-year-old Pakiso, by now reduced to being family laundress, who looked after Mpho. On 11 October 1962, Leetwane helped deliver Nthakwana of a son, to whom the Moates gave the name Johannes.[44]

Leetwane's life had become more rather than less onerous with the move to Molote. In an ageing sharecropping family with static or declining income the patriarch risked losing his sons to urban labour markets, but the matriarch was forever acquiring new charges as her daughters returned home for access to a social support system that nurtured grandchildren. Just when the patriarch was most predisposed to invest in expensive labour-saving machinery in order to help prime the faltering pump of grain production, the matriarch was likely to need money for domestic consumption and comfort. It was probably inevitable that Kas and Leetwane should hold conflicting views about the priorities that new tractors and tin shacks should enjoy in a marriage.[45]

Leetwane strained even more when, not long after the birth of Johannes, Pakiso's fourteen-month-old baby suddenly took ill. At first Leetwane thought Mpho had flu and did her best to nurse her back to health (which unintentionally made the baby's teenage mother feel even more marginal than usual). When the baby failed to respond to traditional remedies, Leetwane and Pakiso took her across to the clinic at Modubung where she was examined by a visiting West Rand doctor. The doctor, whom Pakiso had first encountered during her adventures in Krugersdorp, had little sympathy for adolescent mothers and directed his diagnosis at the child's grandmother. They went back to the shack on the hill with Pakiso still uncertain as to what exactly ailed her child. Over the next few days Pakiso watched helplessly as the baby grew ever weaker and then, on 15 November, died.[46]

A cloud descended over the Maine homestead. For Kas, the events were reminiscent of the episode at Hartsfontein forty years earlier. The following morning he went into Boons, and, as he had done with Nneang, bought a few lengths of pine from which he fashioned a tiny box coffin. With the nearest A.M.E. Church some distance away it was hard to get a minister at short notice, and in the end it was Nteke, a church elder, who led the small procession of close kin down the hillside towards the poorly fenced plot south of what had once been Ramagaga's land. In a simple service, Mpho was laid to rest next to 'grandmother' Lebitsa, and on the way home the family closed ranks to comfort and protect the child who had lost a child.[47]

Mpho's funeral reminded Kas and Leetwane how time and distance had separated them from the African Methodist Episcopal Church. Nthakwana was quick to point out that any elderly couple needed easy access to a church, and she suggested they consider joining Reverend Ndabas's theologically conservative Nederduitse Gereformeerde Bantoekerk at Molote. But it was only after they had received the blessing of the A.M.E.C. minister at Ventersdorp that Kas and Leetwane agreed to join the Moates at services that sometimes also attracted Afrikaner landlords, such as Gert Bosman. Slowly, painfully, members of the family slid back into the accommodating rhythms of the summer routine.[48]

In December, when friends and family used the seasonal lull in activity to move about the countryside more freely, Kas put the tractor and its trailer at Morwesi's disposal to transport to Boons some second-hand furniture she had bought from a white woman in Stilfontein. This was not a trouble-free exercise. White provincial traffic officers whose political sympathies lay with the National Party were zealous when it came to dealing with black road users. Hidden behind the shields of their sunglasses they accused Mosala of overloading a trailer that was not properly licensed. Kas had no difficulty in forgiving his son when it came to the unpleasant experience of transgressing unashamedly racist practices, but he was less understanding when he discovered that Mosala was letting the village boys drive the new tractor while he lazed under a tree drinking beer.[49]

Out in the fields the heat of the bushveld summer seduced Mosala into a bout of profound idleness, but down in the valley it made for a far more distressing problem. One Thursday afternoon in February 1963, three-month-old Johannes suddenly developed severe diarrhoea. Nthakwana took him to see the herbalist in Mathopestat, but when the ngaka seemed unable to do anything, she decided to take the baby up the hill to seek her parents' advice. Kas thought the baby was suffering from *phuana*, an affliction that caused the membranes of the child's fontanelle to pulsate alarmingly; before they could do anything, the baby died.[50]

Nthakwana spent the night in the shack and in the morning Kas went off to Boons to get pinewood from which he fashioned his second coffin in three months. Someone was dispatched to Carletonville to tell Bodule and somebody else to tell the minister of the N.G. Church. On Saturday afternoon—the time most favoured by black migrants for moving across the highveld—a crowd of about a hundred family and friends joined the Moates to witness Johannes being laid to rest beside Lebitsa and Mpho. The Maines, who had wandered through the South African subcontinent for the better part of a century in search of their roots, had come home only to bury their youngest in the very heart of BaFokeng country.[51]

The rest of that summer was lost in a haze that seemed timeless. The maize and sorghum battled to retrieve what moisture they could from soils which, although well ploughed and generously fertilised, cried out for rain that never came. By the time the cooler weather set in in April, it was apparent that, although the harvest could feed the family for another year, there would be no surplus to market at the Koster Co-operative. If the economic momentum of the household was to be maintained, it would have to be a busy off-season.

In autumn Matthews and Mosala went out and earned several bags of grain by attaching themselves to harvesting teams in the valley. Then they turned to the local building contractors for wage labour, and made the painful discovery that opportunity, like Satan, often mocked ordinary folk by donning the most outrageous costumes. The only unskilled work available was in Boons itself, where the state had authorised the construction of a new police station and a modern lock-up facility. Maybe Mosala, himself only recently out of prison, sensed the irony of working to enhance the repressive capacity of a racist government, but it would have passed Matthews by.[52]

Kas earned most of his winter cash from rather more orthodox and predictable pursuits. As usual, some of it came from sitting in the sun and cobbling and, on colder days, from working with the bellows to repair planters and ploughs, or from shoeing the villagers' horses. He was seldom idle and, when not physically active, could be found listening attentively to his clients' problems or sorting through his collection of dried animal parts, fats, and herbs as he prepared medication for his patients. But, as he explained to a nervous schoolmaster who had come to see him, not all evil in the world could be dispelled by herbal remedies.

Mr. Lesika, who had taken up a position in the school at Mathopestat, hailed from the area around Potgietersrus, in the northern Transvaal, the playground for half the nation's witches. A recent marriage to a local woman ensured that he would enjoy access to valley society, yet despite this, he wished to establish a permanent base amongst the Ndebele with whom he had grown up and, to this end, had gone back the summer before and built himself a house near the Zebediela Citrus Estates. No sooner had the structure reached a reasonable height than some unknown force had singled it out for a mysterious lightning strike, razing it to the ground. Lesika wanted the newly rebuilt house protected by a powerful ngaka before he risked sending his bride back to it, and who better than Kas?[53]

Kas explained that he seldom left the farm, even for short spells. But Lesika was insistent, and when he offered to pay their rail fares north as well as a suitable fee, Kas agreed to help him. A few days later they disembarked at Zebediela station and went to Patrick Kekana's home at Moletlane where they were put up for the night. In the morning they worked their way along the banks of the Nkumpi River to the point where it meets the even smaller Mogoto. At the village of Ga-Mogoto, Kas performed the necessary rituals to protect the teacher's home from anti-social elements, and a clearly relieved Lesika slipped him a ten pound note. They spent a second night deep in the heart of Ndebele country and the next day made the dirty, punishing 'second-class' rail journey back to Boons.[54]

This trip, which took Kas farther north than he had ever been before, bore testimony to his growing reputation as a ngaka, and he found his experience as an 'outside' consultant amongst the Ndebele both interesting and rewarding. Lesika's home, suitably protected against witchcraft, suffered no further damage from lightning, and this gave yet more credence to his powers as a herbalist, a deeply satisfying development at a time when his strength behind the plough was on the wane. Perhaps nature sought to compensate man for the ravages of age by allowing him to shift the burden of his labour from the physical to the mental. Back at Molote, he was delighted to find several patients eagerly awaiting his return.

But as Kas knew only too well, life has a habit of jumping the tracks at unexpected moments, and one seldom knows for certain where it is going. In his absence officials from the Department of Bantu Affairs, mandated to extend and entrench the Nationalist government's obsession with territorial segregation, had started negotiating with the independent black landowners on the southern fringe of Monnakgotla's Location; this was the prelude to tougher resettlement initiatives that were to be directed against the valley's more densely populated area. Their discussions with Nkaba had some bearing on the future of the Maine family, and Kas listened with great interest to his landlord.

The state wanted to buy rather than expropriate Nkaba's farm and had, as an opening bid, offered him a respectable price for it. But in a nation where race placed artificial limitations on the amount of land available for purchase by blacks, Nkaba knew that it was better to be compensated in land rather than cash: when he countered with this suggestion, the government offered to let him have 900 morgen of what had formerly been a 'white' farm in the Kuruman district of the northwestern Cape. Nkaba was uncertain about farming successfully in a notoriously dry part of the country, and in another counterproposal had offered to hire the property on a trial basis for two seasons before relinquishing his rights to the land at Molote and making a final decision. This was acceptable to the B.A.D. officials. The landlord now proposed shifting to Kuruman when spring came.[55] He made it clear that he was willing to extend their partnership, which reminded Kas that they were living in a shrinking world where the political will of Afrikaner Nationalists was still encroaching on the black farmers' already limited space. Yet as a seasoned product of Triangle farming, where sharecroppers had never enjoyed much security of tenure under the Boers, this new tremble in the sword of Damocles frightened him less than it did some of the more independent and longstanding valley residents.

In October 1963 Nkaba railed part of his herd to Kuruman, where the hired property, although more extensive than his holdings at Molote, offered less grazing; most of the livestock remained on at Molote under Kas's direct supervision. A new contractual arrangement, which saw the landlord sacrifice his right to a half-share of his tenant's crop in exchange for the full-time management of his herd of cattle, was made possible by certain important changes that had taken place within the Maine family structure. To be precise, Kas now had access to the labour of a new generation of young herders—his grandchildren.

Ten-year-old Ntholeng and her cousin, Nkome, a daughter of Thakane's who had been sent from the western Transvaal to live with her grandparents, were old enough now to be of real help in managing the cattle. They, like Pakiso and Nkamang before them, were therefore entering a troubled period during which they would be dogged by the conflicting demands of work and school, with Kas struggling to keep them on the farm while the older women battled to see that their education was not neglected.[56]

But in the spring of 1963 this battle was only just beginning to take shape. The girls were not yet old enough to give voice to their own preferences and school was not yet making heavy demands on their time. Moate and Mosala used the tractor for fertilising, ploughing and planting in time for the summer rains (which, in the end, proved to be only marginally better than those recorded in the year before) but even with the girls' help, management of Nkaba's animals was a full-time business.

Nkaba owned more than fifty oxen and close on three hundred sheep, which showed the financial muscle of a man who, in addition to the property he hired at Kuruman and the farms he owned at Molote and Vryburg, held the dominant share in three trading stores managed, with indifferent success, by his sons. Sensing that the grazing resources at Molote were overextended, Nkaba went out and hired from the state a portion of the neighbouring farm, Booyskraal, thereby extending still further the area under Kas's supervision.[57]

With the lines of communication between landlord at Kuruman and tenant at Molote stretched almost to breaking point, Nkaba delegated ever more powers to Kas, who now, at the age of seventy, began to function more like a farm foreman than a tenant. The first recorded evidence of such devolution of power came in January 1964, when Nkaba gave Kas written permission to take into custody all stock trespassing on the property and to arrange for it to be impounded at Klerkskraal.[58] In an area poorly served by fences, such a concession was not granted lightly and called for the exercise of real diplomacy and tact. On several occasions Kas did indeed arrange for cattle to be removed from his maize fields, but before sending them to the pound, he would talk to the owner and somehow agree on compensation before anything else needed doing. Only once, during one of Bodule's weekend visits, did he send stray cattle to the pound, and even then an agreement was reached at the eleventh hour and someone was sent out on a bicycle to intercept the herd before it reached Klerkskraal.[59]

Anticipating the day when the state would evict the villagers and the inhabitants of Monnakgotla's Location would be forced to sell their livestock at the depressed prices inevitable with an artificially swollen supply, Nkaba told Kas to begin disposing of his sheep in small lots. Nkaba could not have found a better man for the job had he scoured the Transvaal from end to end. With a lifetime's experience to draw on, 'Ou Koop en Verkoop' was in his element. In the next twelve months Kas seldom missed a livestock auction, and his interventions yielded exceptional results for his principal. Indeed, he was so successful that Nkaba rewarded him several times with payments amounting to between fifteen and twenty pounds sterling.[60]

Kas never once betrayed Nkaba's confidence. Just as he was ready to hit back at those who crossed him, he was reluctant to abuse the trust of those who proved their friendship. By autumn he was actively engaged in selling Nkaba's cattle, and this, too, was of financial advantage to both parties. Indeed, these successes as a livestock agent let Kas contemplate yet another disappointing harvest with some equanimity. The tractor, it seemed, was not earning its keep. The harvest was enough for subsistence purposes, but there was nothing left to market. Kas resolved to think through the problem yet again before embarking on the spring campaign.[61]

But, for the time being, he had to deal with the problems of a white woman from a nearby mining town who had been referred to him by the local Indian storekeeper. The woman and her black servant had apparently parted company on bad terms, and when the maid had walked out, she had left a *tokoloshe* to haunt the premises. The presence of this short, ugly creature with an enormous penis not only caused the dishes in the kitchen to shake and wobble, but, perhaps more importantly in an area where underground mining operations made sharp earth tremors a frequent occurrence, made it impossible for the white family to obtain the services of any other black servant.[62] As in the case of the teacher from Zebediela, Kas could not deal with problems of this nature at a distance. He took up an invitation to visit the woman's home where, after bleeding her and her husband, he used powerful traditional herbs to fumigate the premises. This stopped the crockery from rattling and allowed the couple to engage a new maid. The husband, relieved to have his wife freed from the sexual attentions of the tokoloshe, sent Kas five pounds via the Indian at Boons.[63]

Assignments like this, along with the management of Nkaba's livestock, gave Kas pleasure. The work, inherently satisfying, relied heavily on skills acquired over a lifetime, and that he was being paid for it made it doubly rewarding—especially since his fields were not yielding the hoped-for returns. This, in turn, helped to secure a period of domestic tranquillity, and the family seemed to knit together more tightly than it had for some time. Even Bodule, reconciled to his position as a semi-skilled assistant to various building, plumbing or painting contractors around Carletonville, was becoming more helpful and supportive during his weekend visits to the farm. The presence of Nkome ensured that the Maines were kept informed about Thakane's life as a faith healer in the western Transvaal. With the painful exceptions of Mmusetsi and Matlakala, Kas was in direct or indirect communication with virtually all his children.

Bodule told Kas that he heard that Matlakala was a member of a group of faith healers operating in Sharpeville, a BaSotho enclave in a part of the southern Transvaal which provided much of the manual labour for the heavy industries near Vereeniging. Kas was anxious to re-integrate Matlakala into the family and asked her brothers to go to the township, find her, and persuade her to come home with them. The police at Sharpeville, accustomed to assisting men from distant areas searching for women who had abandoned the farms and reserves, helped track her down. But Matlakala had no intention of returning to Molote to live in her father's shadow. When the police tried to persuade her to go with them to the office of the Bantu Commissioner—it was clear that society was being increasingly shaped by apartheid considerations—she refused and denied even knowing Bodule or Mosala. The two brothers returned home without her.[64]

This rejection so obviously hurt Kas that Bodule took it upon himself to pay Matlakala a second, clandestine visit in order to discuss the situation with her further. Matlakala became more forthcoming, and told Bodule what had happened to her after the aborted suicide attempt at Rietvlei some years earlier.[65]

She had gone by train from Rysmierbult to Park Station, Johannesburg, where she had changed trains and, still somewhat bewildered, eventually found herself in Vereeniging. On alighting from the coach she fell in with two BaSotho women who went with her to a café near the station, where she sat down and bought herself a pint of milk. But, before she could finish the drink, she had fallen unconscious and, so she was told later, the BaSotho women had then ordered a taxi and taken her to their home in Sharpeville, where she had remained unconscious for three days and nights. Then the women had escorted her to a nearby house where they entrusted her to the care of a local 'prophet' and faith healer named David Pule. Pule, having been paid for his services by the BaSotho women, suggested that if Matlakala were willing to undergo treatment and accept her calling as a faith healer, she would eventually recover from her mysterious illness. Anxious to regain her strength, she had stayed on at Pule's and, a few days later, had excreted a small, lizard-like animal, which she referred to as a '*mampharane*.' Once fully recovered she had thanked the women for their assistance and told them she had agreed to become a 'prophetess' in David Pule's faith-healing church.[66]

Matlakala had been working for the 'Prophet' for several months. She was 'making a lot of money' for Pule who, although he did not pay her, gave her food and clothing. But, being something of a free spirit and either unable or unwilling to live in the shadow of male authority, there were signs she was tiring of being an 'assistant' to the 'Prophet.' The ancestors had already suggested it was time for her to move on, to take up an independent calling, to do her real work—which lay with 'her people' at Khunwana. From what Bodule could gather, there was little likelihood of her returning to the family home. The person for whom Matlakala had the greatest affinity was her half-sister, Thakane, with whom she was already in contact. They were Kas's only children by Tseleng and Lebitsa and they were both faith healers. Bodule, it seemed, was fated to leave Sharpeville alone.[67]

There were other reasons, too, why it was probably not the optimum moment to contemplate returning to Molote. The very idea of 'home' itself was under siege. Nkaba heard renewed talk that winter about the possibility of the Monnakgotla community being moved to Saulspoort in the Rustenburg district; mindful of Kas's achievements over the past season, he offered him a place on the hired property at Kuruman, which, he said, showed signs of great promise. It was a tempting thought, and Kas wavered. When Nkaba offered to take some of the Maine livestock back to Kuruman for safekeeping, Kas accepted; he told Mosala to load six cattle and ten goats onto the truck for the journey west. Nkaba was a man he could trust.[68]

But when there was no move on the villages from the 'Bantu Authorities' in the next few weeks, Kas had second thoughts about having banished his animals to the Kalahari littoral. Perhaps the future really did lie with Catherine Monnakgotla and her people. Kas was, after all, the owner of several residential stands in the village and he was bound to receive compensation in cash or in kind should the community ever be resettled. At Kuruman he might be caught flat-footed in unfamiliar surroundings. In any event, he had already committed himself to buying yet another machine, and that alone bound him to Molote for some months.

In May Kas had been back to the garage at Syferbult for a third visit to the house of his dreams. This time it was a second-hand Fordson Diesel costing R450 that had caught his fancy. Easton admitted that the tractor needed a good deal of work before it could be sent out into the fields with any confidence, but he also managed to convince Kas that it was a sound investment. Easton offered him R200 for the old tractor and, with the winter behind him, Kas was ready to go across and settle the small amount still owing against the opening balance of two hundred and fifty rands.

On 17 August 1964, he paid Easton the final instalment of thirty rands, and Mosala drove them home perched high up on the Maines' third tractor. This machine, like its predecessors, was a source of pride. But Kas's investments in tractors, like those in trucks and land before them, were probably too little too late. The gods delighted in confining him to the middle ground—too skilled and too wealthy to be a mere subsistence farmer, and too black and too poor to become a capitalist—no one knew exactly where the boundaries lay. But it was barely spring, and the gods were just emerging for their seasonal play.[69]

The Fordson was put to work almost at once, and throughout September, Mosala and heaven knows who else used the tractor to prepare the fields for the coming rains. But the rains never came. The 1964–65 summer rainfall was the worst recorded in the district for nearly fifteen years, and the countryside slumped to the lowest point of another 'dry cycle.'[70] Fate went on to deal Kas another blow: by the end of that month the engine required extensive repairs, and he had to have the tractor towed to the nearest garage at Boons. Two weeks later, on 14 October, Kas was back at the garage, to be confronted with a bill for R225. He gave the proprietor R80, took back the tractor and a week later reappeared to hand over a further R90. The outstanding balance of R55, and the realisation that his fields were going to yield little grain, sent a shiver of doubt down his spine. Perhaps he had erred in not taking up Nkaba's offer. Ten days later and by then flat broke, Kas railed eight donkeys to Kuruman, leaving the ever-understanding landlord to pick up the bill for transport at the other end.[71]

It was a truly wretched summer. While Ntholeng and Nkome spent their days moving Nkaba's cattle to find grazing, Kas, Moate and Mosala took turns weeding and guarding the family's grain fields. Even more time was spent trying to avoid the merciless highveld sun, which threatened to incinerate man and beast alike.

Kas could pay more attention than usual to his herbal practice that summer. His fame as a 'doctor' continued to spread and, on New Year's Day 1965, a representative of the African Dingaka Association, an umbrella organisation seeking to represent the interests of all highveld herbalists, asked him to consider joining the body. Sceptical about the benefits of collective action, Kas nevertheless saw little harm in an organisation whose aims did not go much beyond obtaining 'recognition of the Association by the Government in the Republic of South Africa.' He became a member and got a certificate that specified that it was 'not issued within the Medical, Dental and Pharmacy Act 13 of 1928.'[72]

Since virtually all Nkaba's sheep had been sold off, there were few sources of cash income around. Whatever Kas earned was usually spent at the local stores for domestic necessities. What of the debt at the Boons garage? In February Kas paid off R10, but it took until May before he could hand over a further R20; on 7 June 1965, he eventually paid the outstanding R15.10; even then the proprietor insisted on his paying an additional cent for the cost of a revenue stamp. Kas stuffed the receipt along with the others into the old maize bag which he kept at the back of the shack.[73]

Hyenas, jackals and vultures, along with nature's other opportunists, thrive on hardship. With many in the valley suffering from a shortage of cash after the drought and an awful harvest, state officials now arrived to issue the inhabitants of Monnakgotla's Location with an ultimatum: they should either move to a new site voluntarily within twelve months, or face expropriation and forced removal. As an indication of its 'goodwill,' the government would give the villagers a choice of three sites in districts it was busy consolidating into a larger 'Tswana homeland,' but if the state and the people failed to reach agreement, this choice would lapse and the government would simply allocate them the site it considered most appropriate. The already tenuous political ties that held the valley's inhabitants together unravelled further, with groups degenerating into 'collaborating' or 'resisting' factions. Kinsmen were beginning to turn against kinsmen.[74]

While most villagers were still actively debating the pros and cons of various tactics and arguing about the likely outcome of the ultimatum, Kas made up his mind. There *could* be only one possible outcome from such an unequal contest. Events at Walter Moormeister's and at Mrs.

Hersch's left him convinced that the Boers would get their way. Conservative by nature he might well have been, but his pragmatism was also deeply rooted in the acid soils of personal experience. The inhabitants of Monnakgotla's Location were in a novel situation, but for Kas there was something sickeningly familiar about the business of being hounded out of a district by Afrikaner Nationalists, whether at the behest of a group of the party faithful, as at Sewefontein in 1949, or at the hands of an individual like Cas Greyling, as at Rietvlei in 1955. No! It was time to prepare for their final season at Molote. The livestock would be sold, since the animals were unlikely to survive the ravages of heartwater (a tick-borne disease of the bushveld) away from the highveld, and he would have to see to it that his assets were as portable as possible.[75]

The approach of spring gave further shape to these priorities. In early September Kas heard of a short course in tractor maintenance being run at a local garage by a leading oil company. This gave him an opportunity he had been longing for ever since purchasing his first tractor from Easton four years before—the chance to develop a knowledge of internal combustion engines so that he would be less reliant on Mosala, Moate or Bodule. A week later he emerged from a dirty old garage in Boons clutching a certificate testifying that he had attended the Mobil Oil course on tractor maintenance.[76]

Using this new lens he went over the old Fordson and found it wanting. It was obvious that the tractor had a limited life and, despite the recent overhaul, would have to be replaced sooner rather than later. If he was going to sell his livestock anyhow, it made sense for Kas to use the proceeds to buy another tractor: not only could he now maintain it, but it could be moved readily and put to profitable use at a new site. He paid a fourth visit to Syferbult and set his heart on an even newer model, costing in excess of a thousand rands.[77]

As usual, Easton made him a generous offer for the older machine. But perhaps that was also part of the problem, which remained hidden from Kas: the way the four trade-ins in five years whittled away at the value of the original sum invested in an inflationary climate. By then, however, Kas was long set on his chosen course of action, and during the first two weeks of October he sold enough livestock to hand Easton at least two hundred and forty rands in cash, and then took delivery of his new tractor.[78]

Mercifully, nature chose not to oppose this development openly. The rainfall in the district during the 1965–66 season was generous by any standards, and for several weeks Kas, Moate and Mosala had their hands full as they prepared and ploughed the fields on Nkaba's property for possibly the last time. Ntholeng and Nkome were equally hard-pressed because, although there was by then a visible shrinkage in the size of their grandfather's livestock holdings, Nkaba still had a sizeable herd. This, and the escalating demands for attendance at school, brought them into renewed conflict with the old man, who remained unconvinced about the virtues of a formal education.

Resentful of having to pay local tax for a school building, Kas was further irritated when, one night around the camp fire, his granddaughters ventured the opinion that the younger generation, which knew about complex things like 'gases and oxygen,' was smarter than older folk who had never had the advantage of a book education. For weeks thereafter he mocked them as the 'children of gases and oxygen,' mere kids who thought themselves clever enough to take him on in an argument.[79]

He did, of course, have formidable skills of his own. As the summer progressed and the livestock picked up in condition, he chose his moment to enter the market and sell the animals, knowing that if the government made a decisive push in its ongoing negotiations with Catherine Monnakgotla, it was likely to do so after the harvest, and when resettlement became a reality, it would precipitate panic selling and cause prices to drop. Early sales brought in a steady income throughout the summer; during February and March 1966, he gave Easton another R100.[80]

The last haze of summer sidled off down the valley and made way for the crisper images that come with cooler autumn air. Organising a harvesting party was harder than it had been in 1965 because the Moates, unsettled by the talk of removal, had moved back to the white farms around Duisterhoek shortly after the ploughing was completed. It was therefore left to Leetwane, Mosala, Motshidiso and the 'children of gases and oxygen' to help bring in the harvest.[81]

It was not only the inadequacy of the harvest that was galling but the realisation that in a matter of weeks the state would put the villagers under intense pressure to 'agree' to the move to Rustenburg. When the assault commenced, the first blows were aimed at the children: the government ordered the summary closure of the Molote school. What better way to get at the adults than to deprive their children of education? The Afrikaner Nationalists were anything but subtle, and since the school at Mathopestat was left alone, it was clear that the initial targets for resettlement were the villages of Modubung and Molote.[82]

This crude message convinced Catherine Monnakgotla and her followers that they would have to accept any reasonable offer if they wanted a prompt and peaceful resettlement. Loyalists, led by one Ramokogo, asked each supporter to contribute R1.50 to a fund to buy additional land to supplement the grant the government was offering at a site thirty miles northwest of Rustenburg. Kas, along with about thirty others, was in the first party sent to inspect the site at Ledig, on the southern slopes of the circular Pilanesberg hills. He and the other heads of households who agreed to being resettled were each allocated residential stands in exchange for the ones at Molote and, along with other contributors to the new fund, promised access to an additional four morgen of fields for either cultivation or grazing.[83]

Within the valley this deal failed to meet with anything like universal approval. In Mathopestat, where the BaKubung-ba-Mathope had always questioned the political legitimacy of BaKubung-ba-Monnakgotla society, the very act of entering into an agreement with the government was perceived as betrayal and strengthened the community's resolve to resist resettlement. In Modubung and Molote, too, elements within BaKubung-ba-Monnakgotla society—mostly wealthy land-owning male notables—questioned not only Catherine Monnakgotla's legitimacy but the wisdom of the deal struck. This group opposed any attempt at resettlement, banded together to form a so-called Sofasonke faction ('We all die together'), and denounced those who had agreed to be removed to Rustenburg as mere 'outsiders,' 'newcomers' or 'tenants.'[84]

Accusations and counter-accusations, based on doses of fact and fiction administered in equal proportions, raised the political temperature in the valley to fever pitch. In more traditional communities, such tensions sometimes precipitated bouts of witch-hunting, but in a totalitarian society like South Africa's, it was the Security Police who were most eager to sniff out 'agitators' and 'enemies of the state.' Leading members of the Sofasonke faction who opposed Catherine Monnakgotla and her followers' decision to move were arrested, charged and subsequently convicted on charges of terrorism.[85]

Such state-induced terror added fuel to the fires of recrimination and persuaded the chieftainess and her admirers to make even greater haste as they scurried to flee a troubled past and seek refuge in a problematic future. Within weeks of the harvest, the 'G.G.s'—'Government Garage' vehicles—rolled into the location and by August Modubung was deserted, with only a few mud shells and budding peach trees left to testify to what had once been a thriving peasant village. In Molote, scores of homes stripped of their corrugated-iron roofing told much the same story: only a few structures still occupied by Sofasonke members were left to remind curious outsiders of a community disembowelled by the insanity of segregation. Sixty miles north, perched on an open hillside in the Pilanesberg, six hundred families battled to erect makeshift accommodation to shelter them from the last gasp of the winter wind and the first storm of spring.[86]

But not everybody suffered these traumas in equal measure. Kas's dual economic citizenship—as residential stand owner in Molote and tenant of Nkaba's on the neighbouring freehold property of Elandsfontein—put his family in the curious position of being both within and beyond the reach of the painful events occurring in the valley. Quick to exploit any ambiguity, Kas used the resulting space to maximise his economic opportunities and minimise the cost, inconvenience and pain of relocation.

The Maines had avoided having their shack doors daubed with crudely painted white numbers when Bantu Affairs officials moved through the villages during late July and early August. The exercise did give them advance warning, however, of the arrival of the 'G.G.s,' and Kas used the next few days to make sure that his interests at Ledig were protected when the government vehicles did move in. He told Mosala and his family to go with the villagers to the new site and to protect the three stands the family had been allocated, along with the shacks and some of the heavier agricultural equipment. Kas, Leetwane and some of the youngsters stayed on to look after the remaining livestock and, more importantly, to extract yet one more harvest from Nkaba's property before it was finally transferred to the government-controlled Bantu Trust.[87]

While Mosala manned the outpost at Ledig, Kas hired a tractor driver to get the ploughing done in a season that again had a better than average rainfall. As the planting drew to a close a white man from the Bantu Trust told him that Nkaba—on learning that the removals were under way and that property belonging to any black resisters was likely to be expropriated by the state—had telephoned him with a message for Kas. He wanted Kas to sell his remaining livestock as quickly as possible at a reserve price of R40 a head.

Kas gave this message considerable thought. It was not so much the authenticity of the communication that caused him to hesitate, but his landlord's sense of timing. The auction pens around Boons were swollen to bursting point with animals belonging to villagers who had agreed to undertake the move to Ledig, and prices were being pushed down. But he felt he owed it to Nkaba to do precisely as the man asked, so he rounded up the cattle. With the better-fed animals from Booyskraal assembled it occurred to him that it would probably pay to secrete ten of his own animals in the herd, for a few animals in poorer condition could be well hidden amongst Nkaba's fatter beasts and benefit from the higher price.

The rumour that cheap pickings were to be had around Boons attracted several Afrikaner farmers from the Rustenburg district to the auction, as well as a fair number of professional speculators from farther afield, including one known to Kas only as 'a certain Jew.' Sensing that the big crowd might drive up the bidding, Kas was mentally readjusting Nkaba's reserve price from R40 to R50 a head when he was approached by the speculator about the possibility of a private sale. But 'the Jew' was an experienced dealer, and when he realised what Kas was trying to do, he put him under great pressure 'to separate the big oxen from the small ones.' Kas, equally good at bluff and counter-bluff, resisted this, and insisted on selling the herd as a single lot at the uniform price of R50 a head. In the end, and to his great delight and relief, 'the Jew' agreed to his terms. By taking the initiative, increasing the asking price by 20 percent and not losing his nerve, he had secured an additional R600 for

Nkaba, a fitting reward for someone who had been a loyal friend and a good landlord. When Nkaba called in toward the end of that month Kas handed him the speculator's cheque for R3,000; Nkaba, delighted at this handsome outcome, showed his gratitude by giving him R30 for his trouble. Kas added this to the R250 he had earned from the sale of his own animals and, for the first time in several months sensed the comfort of a modest financial cushion at a time of great uncertainty. For a moment, the pressure was eased; in a rare gesture, Kas made Leetwane a gift of R15.[88]

Even as Kas counted his cash and lamented the loss of his last oxen, he could sense a finger summoning him to Syferbult. He had last visited the garage just after the harvest, when he had given Easton R40, and the time had arrived for him to make another payment. On 31 August, he went across and gave Easton another R40. Clearly, it was going to take a long time before he could safely say that he owned the diesel-driven machine with which he had chosen to replace his livestock holdings.[89] He had recurring doubts about the wisdom of investing in things like tractors and trucks which unlike animals, could not generate wealth through natural reproduction. Still, he was prompted to think through the virtues of owning a car. The stands at Ledig were at least ten miles away from the hamlet of Boshock which housed the nearest garage and trading stores. The tractor, he knew only too well, was constantly needing fuel, and what would be the point of taking it on a round trip of twenty miles in order to get it filled with diesel? With the immediate prospects of obtaining a crop uncertain, the women would have to use the trading stores at Boshoek so as not to become prisoners of the shops in Ledig with their inflated prices. Acquiring a vehicle for the luxury of private transport made little sense when he still had access to a pair of horses and a cart, but a motor car would make life a whole lot easier.

At first Kas looked around for a vehicle in rather desultory fashion, but once the ploughing was over he decided to speak to Fanuel Kgokong. He had long admired the small black Austin that Kgokong used to chug up and down the valley's dirt roads. Could he persuade Kgokong to part with his car? Buying a second-hand vehicle was a risky business, all the more so for a man who was more familiar with cattle than with cars. Kas moved slowly. Kgokong showed him a list of parts that had only recently been replaced. Kas was reassured and, on 10 September, he acquired the Austin for R100.[90]

For Leetwane, whose reservations about the purchase of vehicles of all sorts were well known, the acquisition of the Austin was a source of additional puzzlement. Kas could not even drive, yet he seemed almost mesmerised by motorised transport. He was constantly shedding oxen, wagons and carts, which one associated with tenants, in his unsuccessful effort to catch up with tractors, trucks and cars, which characterised the world of landlords. And, even when he *did* manage to corner one of the monsters, he had to rely on others to tame it. Mosala had to drive his tractors, and with the Austin he had to turn to Mongoaketse Baloyi for assistance.[91]

Kas neither wanted nor needed a full-time chauffeur. He simply summoned Baloyi, the son of a village friend, whenever he wanted to go into Boons or Koster on business, and would sometimes allow him to use the Austin for personal business without charging him for fuel. This arrangement worked well enough, but it left Kas vulnerable to some of the abuses he had experienced when he had trusted Mosala with the tractor. Around Christmas, he lent the young man the car to visit his family in Ledig, only to be told later that the engine had broken down and that it would have to be repaired before the Austin could return to Molote.[92] Kas sent Baloyi a note authorising a repair, but when Baloyi had not reappeared a week later, he became perturbed. Concern about the car, and also a desire to make certain that the headman at Ledig had allocated Mosala the agreed-on sites persuaded him to go to Ledig. Well before dawn on 30 January 1967, he got out the bicycle for a midsummer's ride into the heart of the bushveld. His eyes were not as good as they had once been, but his strong sinewy frame did not flinch before a fifty-mile challenge. A modest appetite, a diet of sour milk and maize porridge, plenty of fresh air, and a lifetime behind the plough had its compensations.

The first fifteen miles were exceptionally hard. The little dirt track from Boons wound its way north through the foothills of the Magaliesberg toward Steinfurt. By the time he was pedalling up the long incline at Olifantsnek, it had been light for several hours, and the sun was making serious demands of anything left exposed or uncovered. At the top of the pass he paused and then pulled his straw hat down low over his brow before coasting down to the place which most whites knew as Rustenburg but which for any SeTswana-speaker would always be Tlhabane. Klerksdorp, Potchefstroom, Rustenburg and Ventersdorp—amongst the very first places where the Boers settled after crossing the Vaal River in the mid-nineteenth century—might form part of segregated South Africa; but the very names of these places, Matlosana, Tlokwe, Tlhabane and Tsetse, posed a challenge to Afrikaner Nationalists. In the end the word, the mind and a sense of belonging were stronger than the pen, the map and the politician.

Rustenburg was not a hospitable place for black men. Kas stopped, refreshed himself, and then set out on the next stage of his journey, which took him past the sprawling centre of Phokeng and the surrounding chrome mines, which helped to ensure that the local BaFokeng enjoyed a standard of living significantly higher than that of most of their neighbours. Just beyond Phokeng a groggy signpost pointed him in the direction of Paul Kruger's house at Boekenhoutfontein, one of Afrikaner Nationalism's most hallowed political shrines. But like most black folk Kas had little reason to mourn the passing of a man whom the BaFokeng remembered only as an especially uncompromising landlord, not as a president of the Zuid Afrikaansche Republiek. Instead he pushed on to Boshoek. He got there at ten o'clock and, although the Pilanesberg mountains were already clearly visible in the distance, it took him another two hours of steady pedalling in blistering heat to cross the bone-dry plain that separated the small commercial outpost from the hillside settlement at Ledig.

He found Mosala's shack easily enough, and spent the rest of the afternoon sheltering in the mottled shade of a thorn tree. The following morning he checked on the progress that Baloyi was making with the car and asked Mosala to direct him to someone who could write a letter on his behalf. He located the scribe and, using the name that Leetwane preferred when dealing with church matters and the written word, dictated a letter to her:

Bakubung P/bag,
P.O. Box 1100 Ledig
Rustenburg
31st Jan 1967

Marta Teu Maine,
I travelled safely by bicycle and arrived at Ledig at 12 o'clock. I found everybody in good health. I don't know when I will be back as I am still struggling with the car. I am not certain whether it will be fixed or not because the engine is still being repaired and has yet to be returned.

Go and collect some money from Matshitse, the shopkeeper, and pay Ranti. Tell the people at the Garage that I will see them when I get back home.

I do not have any more news, greetings.

I am, Kas Teu Maine.

P.S. Please, give the horses two measures of salt and paraffin on Saturday. Please do as requested.

The scribe got someone to take the letter to Molote, which pleased Kas greatly. It meant that the horses would not be neglected.[93]

Over the next few days Kas assessed the grazing available around Ledig and compiled a mental inventory of the most readily available herbs. He told Mosala to take care of the bicycle, clear the site on stand 282 where he and Leetwane would erect their shack, and prepare for their arrival as soon as he had brought in the harvest at Nkaba's. When Baloyi eventually appeared Kas took leave of Mosala and his family, wound his slightly stooping frame into the front seat of the Austin and told the driver to go to Tsetse. His mealies in the fields back in Molote were marking time until autumn, and Kas wanted to use this break to see his older brother.

In some ways Kas's visit to Doornkop was painful. Phitise—now in his mid-seventies, cautious, talented and increasingly religious—had put the property he had bought from Braude to very good use. He used just two donkeys for ploughing, but the quality of his maize showed that he had lost few of his skills. Most of the capital generated from farming had been put back into the property in the form of improvements such as fencing and a borehole, and the family talked of opening a small shop. Owning land clearly made a difference to a farmer. Kas suddenly became anxious to get back home. He wanted to check on the health of his horses and he needed to pay Easton a visit. Always competitive, the brothers took their leave of one another by boasting about the size of their prospective harvests.[94]

Back in Molote Kas surveyed his fields, did a rough calculation and realised he might not harvest more than his fraternal rival. His tractor had opened up a bigger patch than the donkeys, but that also made him more dependent on labour to loosen the soil and weed the fields. When half his potential labour force was stuck on a hillside in the distant Pilanesberg, this comparative neglect of midsummer maintenance showed up in the yields. His prospects were good but not outstanding.

On 20 February he had Baloyi drive him across to Syferbult. He gave Easton a further instalment of R20 because his tractor would soon need a major service if it were to get them and all their possessions, shacks and heavy agricultural equipment to Ledig. He and Mosala took it to the Boons Garage; when they returned to collect it, Kas was handed a bill for R227.19; together with the existing debt of R53.85, he owed the garage so much money that he could not hope to liquidate the debt at once but gave the proprietor R20 in cash and promised to make a further, bigger payment directly after the harvest.[95]

But even in Boons garages' memories are organised in months rather than seasons and, as the twenty-fifth of each month came round, Kas received a written statement reminding him of his debt. These notices nagged at him, and he worked away at his crafts to raise money. By autumn, and with the harvesting about to commence, he was within sight of the end. As always, his farming had been called upon to pay the biggest price, and Kas found it difficult to conceal his disappointment with the harvest. 'That was the year that Phitise and I were in competition. He used two donkeys to plough with and obtained fifty bags. I used the tractor and got seventy. Now, *you* tell *me* who harvested more!?'[96]

Kas sold most of his grain through the local co-operative, and on 25 May, he paid the man at the Boons Garage a final instalment of R80; when there was no receipt book at hand, he asked the proprietor to make certain to forward a receipt to him at Ledig. Over six years he had spent R2,500 on tractors and tractor maintenance: on the smallholding overlooking Mathopestat, hardly an ox was to be seen.[97]

The next two weeks were especially hectic. Kas, Leetwane and Bodule's oldest son, Sekgaloane, who had been sent from Carletonville to help his grandparents, loaded everything from pots and pans to ploughs onto the trailer. Baloyi went to Ledig to collect the family tractor driver, and a few days later Mosala started up the engine for the long haul to the Pilanesberg. A day later Leetwane and the smaller children were shoe-horned into the back of the Austin, and they, too, left for Ledig. Kas and Sekgaloane, along with the few Maine animals, remained behind for a final meeting with the landlord.

Reliable to the very last of their eight-year friendship, Nkaba arrived at the appointed time. He and Kas walked over the farm one last time before the property accrued to the South African Bantu Trust. That night Nkaba scribbled Kas the one note that could assist an elderly black tenant moving through a countryside haunted by the spectre of stock theft.

> Pass to Kas Teu to drive from Elandsfontein No. 19 to Ledig, Rustenburg: four horses, one mule, two cows and two calves—the property of Kas Teu who is now trekking.
>
> K.T. Nkaba,
> Elandsfontein No. 19,
> Boons
> 15.6.67.[98]

At dawn the next morning Kas and Sekgaloane mounted their horses, rounded up the animals and set off for Steinfurt. Kas's earlier cycle ride had been a great reconnaissance exercise; as the afternoon sun faded, they pitched camp on the property of a friendly white farmer he had met on the roadside near the Olifantsnek dam. Dried acacia twigs were nudged into a smouldering fire that helped to keep their hands warm well into the night, but the following morning the chilly southerly breeze never ceased to practise scales up and down their vertebrae as they moved off to Tlhabane and far-off Phokeng.

The days were short. The cows and calves struggled to keep up with the horses and mules; by nightfall, they had got only as far as Boshoek. They stopped on the roadside and a scramble through a barbed-wire fence produced yet another nest of acacia branches. Streaks of orange flame sliced at the cold night air, and in the distance, the Pilanesberg was outlined against the southern sky. Kas suddenly felt very tired. He had been on the road for more than half a century, and it was time to call a halt. In the morning, they would be at Ledig.

CHAPTER FIFTEEN

# Reflexes

## 1967–78

The Pilanesberg mountain range is something of a geological oddity. Set on an extended plain of clay that separates it from the Magaliesberg mountains to the south and the bushveld proper to the north, the hills are almost perfectly circular and some fifteen miles in diameter. The concentric circles of rock that make up the range are, in fact, the eroded remnants of an extinct volcano. Within the dust-clogged crater, pitted protuberances more than a million years old reluctantly release small quantities of water that trickle down into boulder-strewn saucers where the acacias and grass have to stand on tiptoe to peep into neighbouring depressions. Pressed up against the outer wall of this natural fortress is the place where Catherine Monnakgotla's followers found themselves in the middle of 1967.[1]

Known to members of the Department of Bantu Affairs as Ledig—a word which, depending on one's mood and worldview, can be defined as idle, otiose, sterile, vacant or simply worthless—the site that the BaKubung had selected was on the southern flank of the most conical of the hills. Beneath it lay a large stretch of barren veld dotted with thorn trees gradually falling away to the grandiosely named Elands River, which, in truth was little more than a seasonal watercourse in those years when the rain gods chose to pass that way. As far as any outsider could see Ledig might as well have been situated in Satan's back yard since both its climate and surroundings seemed only to challenge the human spirit to triumph over the odds posed by natural adversity. But, while officialdom looked out on an 'idle' landscape, the erstwhile inhabitants of Modubung and Molote looked in on their uprooted communities and, in the first flush of hope and reconciliation, named their cluster of shelters Kopano, meaning 'fellowship' or 'union.'[2]

With his instinctive longing for the desolate Vaal flood plain, Kas did not feel in need of either fellowship or union. Decades behind the plough and the desire to be left alone to look after his livestock had only deepened his preference for isolated sites, and as at Molote, he chose residential stands on the very perimeter of the settlement. His property, like that which he selected for Bodule and Mosala, was at some distance from his neighbours, and gave access to grazing on the nearby mountainside. (He had owned four stands at Molote, but Catherine Monnakgotla was willing to allocate him only three at Ledig because he was unable to produce evidence of the existence of his third son, Mmusetsi.)

With the help of Mosala and two of his children, eleven-year-old Tsetse and nine-year-old Morapedi, they erected the shack so that it looked out over the surrounding plain and some stray koppies. Kas realised that the place would not qualify as home until they had also put up the hloma o hlomolle, the tiny shelter that had once been used as a cattle post but that, in recent times, had been pressed into service as a tool or grain shed. Structures that had once accommodated the peripatetic lives of a share-cropping family were, with the passage of time, slowly changing function and becoming increasingly emblematic of the world they had lost. True, tin shacks were unsuited to the needs of an elderly couple, but (as Leetwane would never understand) for Kas, the corrugated-iron structures signified not only home but freedom of movement in a shrinking universe. If things went wrong, he wanted to be able to move on to more promising fields.[3]

Twinges of doubt were reinforced by the unfamiliar sights and sounds of the new surroundings. Grey loeries, their raised crests making them look like the advance guard of a war party, made noisy landings in the tree beneath which Kas had erected the enclosure for his animals, and used this lofty perch to shriek out warnings in a strange tongue. Reading these omens, and mindful of advice he'd been given about the susceptibility of 'imported' animals to local diseases, Kas resolved to wait until spring before deciding what direction his livestock farming should take. Planning sensibly for crop farming was equally chancy. Catherine Monnakgotla and her advisers were finding it difficult to allocate the plots of 'additional' land purchased with the fund they had established on leaving Molote.[4] This delay was bad enough during the end of winter, while Kas was still busy around the shacks, but once spring came, the enforced idleness of man and machine became intolerable. The tractor was only used to service the water-run between the 'river' and the house, but the need to buy fuel necessitated frequent trips into Boshoek. Kas eventually established that, like most white farmers in the district, he was eligible to have fuel delivered to his house, but for this to work effectively, he would have had to install a costly storage tank—and this at a time when his farming was bringing in hardly any cash.[5]

During one of these visits to Boshoek Kas heard about a Boer who, exploiting the market that the state had obligingly dumped on his doorstep, was selling goats at what sounded like very reasonable prices. The logic of farming goats was so obvious that it surprised Kas that he had not thought of it before. Goats were not susceptible to the diseases that affected larger livestock and would thrive on the sparse hillside grazing. Moreover, not only could he and Mosala's youngsters do what little herding was required, but the goats would also find a ready market as slaughter-animals amongst the inhabitants of Ledig and the surrounding Tswana-speaking villages at Chaneng, Mahobieskraal and Mogwase.

Kas bought his first two goats from the Boer on 31 July and then like the 'Ou Koop en Verkoop' of old, bowled through the rest of the spring and early summer months buying, swopping and trading goats wherever and whenever he could with the local landlords or their labour tenants, mostly for a little cash. His largest and most enduring source of supply lay within the Pilanesberg crater itself, where, on the remote and somewhat isolated farm of Welgeval, years before the state had settled the followers of Chief Ofentse Pilane's BaKgatla.[6]

The men of Welgeval, some of whose predecessors had benefited from the presence of the Nederduitse Hervormde Gemeente's mission that had once operated from the neighbouring property of Doornhoek, were enterprising livestock farmers with well-established trading links that stretched back all the way to the Bechuanaland border. Like Catherine Monnakgotla's BaKubung, Ofentse Pilane's BaKgatla were not ethnically homogeneous; Kas had no difficulty in identifying with men like J. P. Sefara, Henry Moloto or Johannes Mbatha—they, too, being peasants from various walks of rural South Africa who, in the twilight of their careers, were being confined to tribal enclaves by Afrikaner Nationalist politics and the accelerated development of capitalist agriculture on the highveld. A member of this group, Johannes Molefe, early in November 1967 sold Kas the eight goats and twelve kids that became the core of his rapidly expanding herd.[7]

Kas's decision to concentrate on small rather than large livestock was vindicated as the summer wore on. Although similar to Bloemhof in temperature, the bushveld's greater cloud cover made for increased humidity levels, and Ledig often sweltered for days on end. By March 1968 heat and horsesickness had claimed the mule and the four horses he had brought with him from Molote. His modest cattle holdings were safe only because, within days of his arrival at the Pilanesberg, he had exchanged the two cows and their calves for six so-called 'salted' or immunised animals which he got from Mr. van Vuuren, a Rustenburg-based butcher with a large farm on the southern banks of the Elands River.[8]

These half-dozen animals were sufficient to rank Kas, along with the Maseko and Mngomezulu brothers, among the ten most prominent livestock owners in the village. But the Maine family's underlying economic strength was more evident from the remarkable growth in the size of Kas's goat herd tended by Tsetse and Morapedi, despite occasional losses to jackals that stalked the more rugged sections of the crater. As at Molote, when Pakiso had come to Kas's rescue, there was somehow always a grand-child to turn to just at the moment when Kas seemed to have reached the limit of his labour resources. By late summer 1968 he could boast of a herd of more than one hundred goats.[9]

The success with the goats was reassuring because soon the tractor seized up, like the Austin before it, and Kas had to decide whether to spend money on repairing it or to cut his losses and abandon it entirely. The decision was complicated since it was still unclear to her followers when exactly Catherine Monnakgotla and her advisers would allocate the fields. After careful thought Kas abandoned the idea of having the engine reconditioned, dismantled it, and sold the parts to the garage in Boshoek. This brought in some cash but left him with a new problem: how would he move water from the river to his homestead?

He scrabbled around amongst the scrap iron he had collected during his travels, found a car chassis and then, with customary ingenuity, modified the rear axle so that it formed the framework for a robust little water cart that could be drawn by a pair of donkeys. Before long this service was extended to neighbours and other villagers at a fee of a rand a time; in the course of a full working day, Mosala, or whomever else Kas could find to help him, would make four such round trips. It was hardly transport-riding as in the days of yore with his old friend Hendrik Swanepoel, but it did bring in some cash.[10]

In autumn other skills were dusted down and put to use in the new environment. Kas set up a forge and did some work for those fortunate enough to own their own agricultural equipment. He also made a saddle for Jacob Mokebeswane, which brought in a further R14. This, together with a little cobbling and the fact that a day seldom passed without someone needing his services as a herbalist, gave his economic life familiar and comforting features. Moreover, all around him, as the erection of new shacks showed, the size of the market for his skills was still growing.[11]

That winter dozens more families moved into the Pilanesberg in search of residential stands and patches of land on which they could keep a few animals. Some of these were BaKubung stragglers from Modubung and Molote, but many more were *matsenelwa*. These Nguni-speaking 'outsiders,' having long since been pushed onto the highveld, were now being expelled from white farms there and, cut off from their lusher eastern habitats, were forced to look for refuge in the dry 'Bantu homelands' of the western Transvaal. At Ledig men like the Maseko and Mngomezulu brothers were directed to Catherine Monnakgotla, who was more than willing to sell them residential stands in a bid to bolster her power base.[12] More matsenelwa families established themselves in the 'Ndebele section' of a BaKubung settlement which the Afrikaner Nationalist government, in turn, hoped eventually to incorporate into a greater Tswana-speaking state. In this rural area, as in many others, and as in the vast majority of

South Africa's sprawling cities, the social, political and economic realities of everyday life made a mockery of the supposed ethnic symmetry of apartheid planning.

Although Catherine Monnakgotla's policies in these and other matters did not always meet with the approval of the Bantu Commissioner, who lived in the Pilanesberg, she nevertheless allowed the same pragmatic considerations to shape the composition of her Tribal Council. During the first three years of community life at Kopano anybody, including the matsenelwa, could serve in a kgotla, which at one stage included Abel Maseko and Scotch Mngomezulu. Acting in the spirit of 'fellowship,' the council mobilised the community's limited cash and labour resources to construct the tribal offices, post office and the BaKubung Primary School.[13]

For black families living on white farms and buckling under the combined impact of agricultural mechanisation and a fall in their real income, access to a school, even a Bantu Education school, often tipped the scale in favour of migrating to the nearest rural slum. At Ledig, where the BaKubung Primary was later supplemented by a more substantial Ratheo Middle School, the provision of even this rudimentary form of education increased the inflow of 'outsiders' from the distant platteland, as well as of 'insiders' with direct claim to linkages with BaKubung society. It was the possibility of their children going to school that motivated Matthews and Nthakwana Moate's decision to abandon a relatively secure position on a farm in the Ventersdorp district and rejoin the rest of their family on the slopes of the Pilanesberg on 28 September 1968.[14]

While Kas had taken care to make sure that two of his three sons were given stands at Ledig, he had made no such provision for his daughters. He took the view that married women were the responsibility of their husbands and that unmarried daughters either would or should get married. Mosala and Bodule both had access to residential property, but Nthakwana had to buy a site as though she were a mere 'outsider.' Because the settlement had expanded since the family's arrival fifteen months earlier, the kgotla could only offer her a stand on what was by then the very outer margin of the BaKubung sector. The Moates would have to live much higher up the hill, isolated from most of their family.

If Nthakwana had doubts about the wisdom of erecting a home on this remote site, her father did not share them. Indeed, Kas took such a liking to the setting that he eventually persuaded her to swop stands. This helped him to regain the quiet life he hankered for, ensured the continued access of his goats to hillside grazing, and gave him more space for Kholoro, a wild and undisciplined dog he had got to protect his livestock and property.

In October, with spring well advanced and still no sign of a definitive allocation of fields for arable farming, Kas helped to dismantle the shacks and re-erect them on the site that was to become the closest to a permanent home that he and Leetwane were ever to know. The corrugated-iron hovels served as home to himself, Leetwane and Nkamang as well as to Mposhoza Shongwe, a youngster whom he took in as a herdboy to help Tsetse and Morapedi with twenty sheep that he had bought at Welgeval at the start of that season. Kas developed a genuine liking for the lad, the first of two surrogate sons he adopted during the closing years of his life. But, as the still vacant lot lower down the hill showed, Kas was never fully reconciled to the permanence of Bodule's urban commitments and went on hoping that one day his son would return to assume his rightful place.

Until he got more support from his own sons, however, he, Mposhoza, and his grandsons had to raise the sheep and goats on their own. The new arrangements worked well for the first three seasons at Ledig. At dawn, amidst much barking and snarling from Kholoro, he would let the animals out of their small barbed-wire enclosure and slowly walk them to the top of the mountain for the day's grazing. After school Tsetse and Morapedi would link up with Mposhoza and together the three would bound up the hill to collect the animals. Good summer rains meant that grazing was seldom a problem, and the animals multiplied fast enough for Kas to dispose of the natural increase at good prices without reducing the size of his herds.

Toward the end of the winter of 1969, this arrangement began to fray. Late one afternoon the herdboys reappeared with the goats and then, with the settlement already wrapped in the chill of the evening gloom, told Kas that they had been unable to find any of the sheep. This dereliction of duty infuriated the old man, who ordered them not to return home until they had recovered the flock. But the boys were unwilling to go out into the cold and dark and instead went to Mosala's place, where they spent the night in hiding. An hour later seventeen of the sheep returned to the kraal of their own accord and were followed, at first light, by two more stragglers. Only then did the herdboys set out to find the single remaining sheep. But it was too late: the animal had ventured down into the crater and been taken by jackals; all they could find was a badly torn fleece.

These events left everybody unhappy, but later Kas tried to reflect on them dispassionately. What had happened, he thought, was that the youngsters had lost some of their boyish charm and enthusiasm, stumbled into adolescence, and falling into bouts of more questioning and resentful behaviour had become less reliable and vigilant. These transitions were hard enough to manage in the confines of the home, where the reins of Kas's authority were firmly held; beyond it, with only grandsons and an outsider to work with, it was impossible to enforce discipline. Kas reluctantly conceded that the youngsters were 'no longer interested in herding' and it would probably be wisest to get rid of his small livestock gradually, and instead have a few larger animals which required less attention.[15]

There was no difficulty in finding a market for sheep and goats: Tswana- and Zulu-speaking inhabitants placed great store on meeting their ritual obligations, and there was always a lively demand for slaughter-animals. In a matter of weeks Kas had sold several goats and sheep at good prices. Later that winter demand increased with the arrival of another two hundred families from Molote—people who had been driven from their homes along the Witwatersrand by a concerted campaign of police intimidation and state harassment.[16]

The arrival of these latecomers, composed for the most part of diehards and resisters to relocation, gave rise to new problems, for the group included people who had not only entertained serious doubts about Catherine Monnakgotla's legitimacy as a ruler, but had not contributed to the fund for purchasing land at Ledig. This disqualified them from the full range of resources available at their new home, though they gradually mellowed and rediscovered the kinship ties with the BaKubung who they said had collaborated with the authorities in the move to Ledig. However, this melding of the two Tswana-speaking sections was achieved at the cost of making the Nguni-speakers feel more marginalised.[17]

Besides the economic impoverishment that went along with the implemented policies of territorial apartheid, the very act of resettlement itself reactivated old political agendas and caused new class, community and ethnic tensions to flare up at the points of departure and arrival. At Ledig the re-emergence of previously submerged ethnic tensions jeopardised the 'fellowship' that had characterised the earliest deliberations of the tribal council at Kopano. Conflict within the community manifested itself most dramatically in sabotage of the concrete water tanks erected in the BaKubung section of the Ledig settlement. In Pilanesberg, the Bantu Commissioner started to doubt that Catherine Monnakgotla could act as an agent of local government in an independent homeland.

Amidst this social turbulence, Catherine Monnakgotla and her advisers made their first land allocations. These, although fairly equitable in their own terms, had the unintended effect of keeping ethnic tensions within the community simmering. In part, this was because the 'Tswana' and 'Ndebele' brought different resources to working the land: the BaKubung, although grain farmers by tradition, did not have enough draught animals since most of them had sold their livestock in Molote; the 'Ndebele' or Nguni-speakers, by contrast, were traditional cattle keepers but, having only recently been expelled from white farms, tended to own more oxen and have less knowledge of crop farming.[18] These mismatched resources were partly reconciled through sharecropping contracts made between Tswana and Ndebele patriarchs, but whereas at Molote the incoming matsenelwa had often shared the crop fifty-fifty with the resident land-owning BaKubung, at Ledig the terms of trade swung in favour of the owners of draught animals: the Nguni-speakers came to insist on retaining as much as 75 percent of the crops which they ploughed, planted and harvested in BaKubung fields. These changing economic fortunes on the land, exacerbated by the South African government's programme of forced removals, served to sharpen social divisions. This, in turn, contributed to a gradual decline in the political influence of the 'Ndebele' men who had formerly been important in the chieftainess's kgotla. Men like Scotch Mngomezulu who, although of matsenelwa origins, had once been full members of the tribal council, were marginalised when the Pretoria government encouraged a growing sense of Tswana consciousness.[19]

Meanwhile, the apartheid state's social engineers were at work on another project which, supposedly designed to serve the interests of an emerging Tswana 'homeland,' was destined to have a most ambiguous effect on Ledig. A Potchefstroom-based team of 'development experts,' commissioned to find ways of enhancing the economic viability of an 'independent state' based almost exclusively on political and ethnohistorical considerations, hit upon the idea of transforming the Pilanesberg crater into a game park.[20]

One could hardly conceive of a project less likely to generate economic development for men and women who had just been removed from fertile, private property bordering on the Witwatersrand industrial labour markets and dumped, like so much human rubbish, on the dry slopes of a bushveld hillside. A game reserve, besides being unable to generate many jobs, held obvious danger for Catherine Monnakgotla's patched-together community: any scheme to enclose the crater would curtail access to grazing. Care was taken to ensure that all talk of the project was kept away from the BaKubung notables and strictly confined to official circles.

So while academics trickled their ideas into the slow-moving streams of state bureaucracy, members of Catherine Monnakgotla's community struggled to find an economic foothold on the slopes of the Pilanesberg. For Kas, the historical ironies were doubly unfortunate: just as he was about to sell the last of his sheep and goats, he got a message from Mahlangu's Thanda Bantu Store, the village shop, informing him of the arrival of an item postmarked 'Stilfontein.' He was expecting a letter from Morwesi, who had been largely silent since her move to the western Transvaal. When he got to the store—set like most such shops in a dusty compound surrounded by a six-foot-high security fence—he was surprised to be handed an envelope addressed in Pakiso's hand rather than that of her mother.[21]

Kas had not seen Pakiso since shortly before her marriage to Serame Tantane at Molote; she was experiencing a difficult pregnancy, she wrote, and wanted help with herbal remedies. Although difficult as a child, Pakiso had been of great help to him at Rietvlei, and Kas had always taken pride in this granddaughter who had served an informal apprenticeship under him as blacksmith and mechanic and gone on, rather unusually, to work first in a panelbeater's shop and then later in the spare parts division of several garages on the West Rand. She, in turn, though resentful at what she considered to be exploitative behaviour on his part, was always quick to acknowledge the influence that her grandfather had had on the development of her attitudes and skills. Kas made up his mind to go and see her.[22]

The skeletal grid formed by the rutted streets of Khuma township, just beyond Stilfontein, was new to Kas although he was familiar with the countryside, for the township was located on the very site of what had once been the Motsuenyane farm; two decades earlier, he had had to spend six weeks on the neighbouring property, Shenfield, when the summer rains failed at Jakkalsfontein. But whereas then he had used the Vaal River and a tall bluegum tree located on the rise near Machavie to serve as natural beacons, the street urchins in Khuma used the concrete headgear that marked the surrounding gold mines to direct him to the Tantanes' home. The distinctive mud huts used by Motsuenyane's sharecroppers had long since made way for row upon row of drab working-class housing. If this somehow constituted progress, as 'the children of gases and oxygen' suggested, then why did it come so heavily disguised? A good thing should surely be more obvious, Kas mused.

Despite the uniformity of the matchbox houses Kas easily found Pakiso's home. She and her young husband—who on first meeting seemed to be an agreeable sort—welcomed him fondly; they mentioned that it might be a while before Morwesi, whose movements were unpredictable, put in an appearance. By nightfall there was still no sign of Morwesi, but nobody seemed to mind much. In the morning Kas attended to Pakiso's complaint, which fortunately was not difficult to treat; the real shock came later that day when her mother eventually shuffled through the door of the house clad in a dirt-encrusted maid's uniform and a pair of cheap, rapidly disintegrating slippers of indeterminate hue.

Although only forty-four, Morwesi had aged beyond recognition. Her face, creased like that of a veteran twice her age, was gashed by a disproportionately large mouth hanging open to reveal gaps in her yellowed teeth: her speech was halting and deliberate. Her demeanour was that of a drunk or a tramp, and as she moved forward to embrace him, Kas detected the tell-tale sour smell of sorghum beer. Like thousands of other unskilled and uneducated women pushed off the land of their birth, and unable to secure a niche in the urban economy, she had been pulled into the ghetto underworld. As always, alcohol had started out by posing as a friend and then very quickly become a master.[23]

Father and daughter had little to say to one other. She mentioned some of the difficulties she had experienced and he gave her a brief account of life at Ledig. Together, they managed to crank out a few sentences about Pakiso's pregnancy. Then, suddenly, everything about the place, its people and their problems, the very air they breathed, made Kas feel claustrophobic. The next morning he fled back to the Pilanesberg. Given the choice of living in a rural or an urban slum he would always opt for the countryside.

Back at Ledig he confronted a string of complaints about the bloody dog. But he also had more time to think. He found it easier, although painful, to reflect on how his offspring tended to cluster at the ends of the spectrum of success and failure. Those who as children or young adults had been perhaps most prone to question his authority at crucial moments—Mmusetsi, Morwesi and Mosala—had all faltered when they tried to cut their family ties and chart an independent life course; Mosala had been implicated in robbery and murder, Mmusetsi involved with a gang of cattle rustlers in Bechuanaland, Morwesi was on her way to becoming just another township derelict. By contrast, those like Bodule and Nthakwana, men and women who had internalised the ethic of hard work and delayed their move out of the patriarchal orbit, seemed to be coping better with the stresses that came with moving off the land. The remaining children, too, showed this pattern. Thakane and Matlakala were governed by their spiritual callings, while Pulane and Nkamang seemed uncertain of the centre of social gravity in their lives. Kas had fathered saints and sinners in equal measure. His children seemed either to be shattered by his demanding standards, or to become tough, well-integrated individuals. It was not easy to be a Maine child or grandchild. But it was no easier being the head of a rural household amidst an industrial revolution in a racist society. There was no place left to hide.

The dog Kholoro had become a menace. On several occasions he had savaged other animals in the neighbourhood and, during Kas's absence in Stilfontein, had killed a neighbour's dog. He had all the instincts of a hunting dog and needed the run of a farm, far more space than Kas had. He was like the animals Bodule used to keep for rabbiting back in the western Transvaal, and this gave Kas an idea: he would make a present of the dog to the BaKgatla-ba-Kgafela when next he visited Welgeval.

The BaKgatla, who were having difficulties with the few wild animals still living in the crater, were as pleased to have a hunting dog as they were to do business with Kas. Between August 1969 and January 1970, Kas had visited John Sefara at least five times and, for an outlay of nearly R400, acquired a cow, a donkey, four calves, four oxen and nine sheep. (He sold the sheep at a handsome profit when returning migrants funnelled into Ledig for the festive season.) Even before the last of these transactions had been finalised, a MoKgatla showed Kas the remains of skulls and bits and pieces of three baboons that Kholoro had caught and shredded. It was a truly vicious dog, the man said with a smile, and he knew that a ngaka could put such parts to good use.[24]

No question, the baboon bits would come in handy. Animal parts, especially wild animal parts, had been hard to obtain ever since they had left the farms where most of the Boers had guns. But when spring came around the most important patient Kas had to treat was himself. In a move designed more to emphasise his access to a portion of the land purchased by Catherine Monnakgotla than to engage in serious cultivation, Kas had hired a tractor to plough his plot. One afternoon, on returning from the fields, the driver stopped at a gate, and when Kas got down to open it, the link between the trailer and the tractor snapped, the trailer lurched forward, and Kas's leg was trapped against the rear of the tractor. There was an awful crunching sound, Kas let out a scream of pain, and amidst yet more shouting, he was lifted back onto the tractor and rushed home.[25]

By the time they got to the shack Kas had decided that there was no need to call a doctor. His leg had not been broken but only badly bruised and, not long after applying the appropriate herbs and poultices, he was back at work in the fields. This dramatic recovery by a seventy-six-year-old man added unintended lustre to a reputation that had already received a boost earlier that winter, when his prowess as ngaka had been cited with approval by the professional association of herbalists during a radio programme heard nation-wide.[26]

Indeed, the secretary of the African Dingaka Association of South Africa, by then based at Rustenburg, was at pains to ensure that Kas remained a paid-up member of their association. In March 1970 S. C. Mcgobje called on him and urged him to renew his membership. This he did, since he was comfortable with its 'non-political' nature. At a time when Afrikaner Nationalists were inflicting apartheid on every facet of black life in South Africa, there was a small measure of comfort in possessing a document that called on 'all members of the South African Police and the Authorities concerned' to recognise the right of the member to 'travel to any part of the Republic of South Africa on his or her business'—though in all but name, the membership card was an extension of the hated pass system.[27] It all depended on how one read the document. One man's badge of oppression was another's emblem of craft distinction. Less compliant herbalists, like his old 'Zulu' friend Hlahla, who had abandoned his native Natal for a life amongst the Tswana of the Transvaal, probably objected to belonging to an organisation that appealed to their racist oppressors in such pleading terms, but Kas carried the badge lightly. In any case, Hlahla had died that autumn. The 'Tribal Authority' at Ledig designated a burial place in the 'Ndebele section,' thereby inflicting on Jacob Nkosi Dladla the sort of label in death that his very life had mocked. It was not just Afrikaners who were interested in patrolling ethnic borders.[28]

Kas's harvest that year was hardly worthy of the name. Five miserable bags of maize were almost meaningless and certainly did not warrant having gone to the expense of hiring a tractor. Luckily the drawn-out sales of small livestock produced better results than expected, and once he was down to six sheep, Nthakwana offered to buy the remaining animals, which she resold at a profit. Her skill as an entrepreneur was beginning to distinguish her from other inhabitants of the economic backwater: it was increasingly evident that she was the daughter of 'Ou Koop en Verkoop.' It was just as well, because Matthews had become sickly since leaving the farms and was having trouble getting work around Ledig.[29]

At the other end of the spectrum Leetwane and Nkamang were finding it almost impossible to generate even a trickle of cash. The barren Pilanesberg environment was especially hard on women: without cow dung to collect from within the fenced confines of a Boer farm, or reeds to make baskets with, such as on the well-watered properties around Rietvlei, they had no value-added commodities to sell. Leetwane sometimes took Nkamang with her when she attended pregnant women, but most of the locals were too poor to pay for a midwife, and this was more of a 'community service' than a commercial venture. In any case, even the modest returns that came from this had to be divided three ways once Morwesi, recognising the extent of her drinking problem, had fled Khuma township and returned home. Poverty meant that Leetwane had to look to her granddaughters, especially to Ntholeng, to pay for her most intimate requirements, which was deeply humiliating.[30]

If Leetwane entertained silent reservations about the return of her oldest daughter to the parental fold, her husband had even more explicit cause for complaint. Morwesi helped little around the house, went on drinking and was given to unpredictable and unpleasant bouts of irascible behaviour. Most of the time Kas managed to cope with these irritations, but he found it virtually impossible to forgive her for one thing.

One weekend afternoon, shortly after Morwesi's return, Pakiso and Serame drove up the rocky path leading to the shacks in a brightly coloured second-hand car. Ostensibly there to visit their mother, the real purpose of their visit emerged when Pakiso urged Kas to buy the car from them for R600. The Tantanes had put down a substantial deposit on the vehicle but after one or two monthly payments could no longer afford the instalments and faced the prospect of losing both car and deposit. This was a familiar working-class predicament—having just enough cash to embark on what seems like an essential project and then not having enough to see the business through. Now they tried to palm off the vehicle on Kas.[31]

But Kas was restocking his herd and explained that he had little use for a car, let alone one that cost R600. This cut little ice either with Pakiso or Morwesi who was hovering around in the background. They linked up with Serame and stalked off to Morwesi's room, muttering and shaking their heads. Soon the youngsters re-emerged, and drove off to Potchefstroom. Only months later did Kas learn what must have been discussed in Morwesi's bedroom.

After the visit to Ledig, the Tantanes went to Lichtenburg where, without disclosing their business, they called on Thakane and asked her to direct them to Nkaba's farm. There, Pakiso gave Nkaba a note, allegedly dictated by her grandfather, authorising him to sell five cows and all the goats which Kas had entrusted to him when the two men had parted at Molote. A week or two later the Tantanes collected the cash, and settled their debt with the car dealer at Stilfontein.[32] No longer bound by the stuffy conventions of kin and countryside, Pakiso had swept in to recover what she considered to have been the real cost of her labour at Rietvlei. The 'children of gases and oxygen,' part of a poorly educated first generation of proletarians churned out under the Verwoerd regime, had spotted the vulnerability of a nearly illiterate old man and exploited it. But whether she knew it or not, the fraud to which Pakiso had been party extended well beyond her grandfather to all the Maines, who, like most black families being pushed off the land, had one foot in the countryside and the other in the city. The excursion to Kuruman started a feud that smouldered for a decade and a half.[33]

Kas sensed that open discussion of this anti-social behaviour could be destructive, and he played it down. A patriarch had to see farther than a grandchild, and the way out of the mess was for him to concentrate on restocking the herd so that he could ensure that his own children received what was due them. Thus, even though the Mngomezulu brothers and a few of the better-off cattle owners started buying tractors so as to benefit from the increasingly lucrative sharecropping agreements to be negotiated with the BaKubung, Kas remained committed to his new plan for livestock farming.[34] Because he had no interest in hiring the Mngomezulus' tractor, he allowed others to plough and harvest his fields on the same demanding terms to which so many of the resident families were subjected. The prospect of recovering only a few bags of grain was less important than the need to be seen to be working the fields. In a way this avoidance of crop farming was regrettable, because the first decade that the Maines spent at Ledig they enjoyed good rainfall. Yet instead of helping to farm grain, Morapedi and Tsetse were asked to help their grandfather with the cattle, though from their point of view this was better than having to look after small livestock in the crater.[35]

Throughout 1971 and the summer of 1972 all of Kas's old skills were in evidence as he tracked developments in the livestock market in and around the Pilanesberg. Most of his deals centred on trading cattle and donkeys which he got from farmers such as the BaTlokwa-ba-Kgosi, who were struggling to eke out a living near the crater. Perhaps the most telling moment came when he persuaded W. J. Grobler of Boshoek to let him have twelve cattle in exchange for six young oxen he had bought earlier from his old friends at Welgeval. In five years, Kas had built up the herd from two cows and their calves to well over a dozen animals—a significant achievement for a man in his late seventies—and he was justifiably proud.[36]

Livestock speculation, like herbalism, was dependent more on skill and knowledge than on the treacherous combination of chance and strength that dominated success in the fields. And of course, with the passing of the years Kas was becoming more reliant on his mental than his physical prowess. In his youth he had viewed the off-season largely as an irritating period of enforced idleness during which one challenged the natural order of things by engaging in physically demanding activities like transport-riding, but now he was becoming reconciled to the rhythms of nature. Indeed, the milder climate of the Pilanesberg was a relief after the rigours of the highveld, and he enjoyed the time that he could devote to his herbalism.

When patients from Ledig and the other villages came to his shack, the consultation usually took place beneath the overhanging branch of a favourite tree. Irregular bowel movements, constipation, diarrhoea, headaches and urinary tract infections accounted for most of the complaints of these rural slum dwellers. Seldom, if ever, did Kas send away a patient without giving him or her a handful of herbs for which he made a small charge. There was, however, a second, more profitable side to his business—the demand for the exercise of mystical powers. Many of these requests, too, were rather mundane, but others, such as those which came from black policemen, were often interesting.

Throughout the 1970s South Africa's armed forces were expanded as the Nationalists found apartheid policies questioned vigorously at home and abroad. A racist state, high on police and low on justice, put more and more black men in uniform, but these recruits did not know what to do to gain promotion in a bureaucracy so tightly controlled by their white masters. Kas, who recognised people more readily than policies, helped them in the same pragmatic way that he helped anyone else. The only difference was that policemen paid more.[37]

His services were also in great demand by storekeepers—some from as far afield as Madibogo, Mafikeng, Pietersburg, Potchefstroom, Taung and the Witwatersrand—who wanted potions to protect their businesses from bankruptcy, ill-will, jealousy or just plain misfortune. A 'coloured' client by the name of Msimang drove him to and from Kimberley to protect his premises, and retained his services to help with an application to the notoriously unsympathetic Local Road Transportation Board to operate a bus service. In a country where the obsession with the need for territorial segregation was putting ever greater distances between black workers and their families, transport was an expanding market.[38] True, in some remote and rather obscure way he was servicing what young black radicals would later call 'the system,' but, as with the would-be sergeants, who would deny an old man in a resettlement camp the chance of survival? The climate reeked of moral decay and the stench of opportunism penetrated every level of society. Nobody escaped contamination. If policemen paid more than peasants, well then, storekeepers tended to pay even more than policemen. It was that sort of world.

Spring rains helped to clear the air. The moisture coaxed forth a musty odour from the countryside, and it invigorated men and animals with new hope for the work ahead. The rains put a stop to off-season diversions and, evoking memories of a functioning rural economy, somehow managed to make the villagers more purposeful. The last quarter of 1972 saw the best rains since the BaKubung had been forced out of their ancestral pastures. Overnight the Pilanesberg was decked in bright greens, and when Mosala's boys came to collect the cattle to graze on the common, entrepreneurial matsenelwa were already sending out their tractors to work the fields of the poorer BaKubung. Kas remained curiously unmoved. He had seen it all before: nature was a bitch; what was born of hope often died in drought.[39]

But the rains persisted well into the summer. By March 1973 the veld and fields around the village were full and fecund, and for the first time since his arrival in Ledig, Kas genuinely regretted that he was not doing arable farming. Easy pickings were to be had from sharecropping contracts in the villages around the crater. Whether it was the moisture in the air or the image of the maize stack at Sewefontein that floated into his mind was not clear but, as soon as Jan Mngomezulu told him about a 'coloured' at Boshoek who was selling a good second-hand tractor at a reasonable price, Kas knew he had to see the machine.

There was something familiar about the social profile of the posse that picked its way across the veld towards Shadrack Bessit's garage—two older men with money to buy a tractor, and a younger one to assess its mechanical worth and drive it. Now, though, Kas was relying on the help of friends. While he and the older Mngomezulu debated the merits of the International, the youngster took the tractor out for a test drive. Everyone agreed it would be a fine investment, but Kas needed to raise R800: in ten days he disposed of six cattle at a good price, and by 10 April he was ready. He and the two Mngomezulus went to Boshoek, the purchase was made—Kas and Jan Mngomezulu signed a voucher in halting script—and the younger Mngomezulu drove them home, parking the tractor under a tree in the Mngomezulu yard, well away from Mosala. In the following months the younger Mngomezulu drove the tractor to haul sand to building sites, bring in firewood from the bush, and transport water from the Elands River.[40] He took such good care of the machine that Kas eventually let him use the tractor for 'private' business without charging him. Kas's confidence was not misplaced, and with their shared passion for farming, he and Jan Mngomezulu, the one a matsenelwa and the other a Mo-Fokeng, became close friends. From his earliest days in the Triangle, Kas's most intimate relationships had been based more on shared interest than on common ethnic identity.[41]

In spring Kas had young Mngomezulu plough the Maines a small patch of land. Morapedi and Tsetse got out the oxen to plant maize. With only the boys around to weed the fields, Kas saw no point in extensive crop farming, and it was probably better to hire out the tractor for ploughing at a guaranteed income of R5 per morgen than to sharecrop, where the potential profit was greater but the risk of losing so much higher. The recurring dream of one more big harvest was gone. A man approaching his eightieth year was entitled to be cautious.[42]

The tractor worked well throughout that summer. By the first quarter of the new year, Kas was once again ready to search for livestock. Welgeval remained his first port of call; in late February, he acquired two donkeys, a mare and a bull from Sefara. The donkeys could be used to transport water and a man needed a horse if he wanted to send a son or grandson out into the fields. But, as always, cattle were the most prized possession of all; a month later, Kas was back at Welgeval to collect six heifers bought from Joseph Ramoufi. Later he ventured to Swartruggens to buy donkeys from the BaPhiring. All in all, 1974 saw a significant increase in his live-stock holdings.[43]

But the harvest was bitterly disappointing. In retrospect he could see exactly why. He had hired out the tractor so often at the beginning of the season that by the time it got to his own fields, it was too late to benefit from the rains which, in any case, were not as good as the year before. If truth be told, the hiring out of the machine was always going to be a better idea, for he simply didn't have the labour. A man on the land without sons was nothing, and the urchins around Ledig showed no interest in working at the rates that local black farmers were willing to pay.[44]

In midwinter, Thakane arrived to spend several days of combined business and pleasure with him and Leetwane. Her skills as a healer were in great demand in Mooidorpie, the small black township attached to the white town of Lichtenburg, but she sometimes faced unusual problems or needed rare herbs. Kas was of great help to her, and their confidential professional discussions drew father and daughter together.[45]

At about the same time, Kas's relationship with Matlakala also improved. Operating out of a Zionist Church based at Matlapaneng, near Taung, rather than at Winterveld, Matlakala remained dedicated to her calling as a prophetess, or *senohe*, but she no longer self-consciously shunned the family. She came to several family gatherings, and on one occasion Kas was pleased to assist her with a loan of R700 so that she could buy a panel-van. Relegated to the psychological margins of their father's affections for so much of their lives, Lebitsa's and Tseleng's daughters were, at last, finding a place in the family fold.[46] But ironically, those at the very core of the structure, Kas's children by Leetwane, became increasingly detached. True, Nthakwana was moving in her father's orbit, and Nkamang and Pulane both made the effort to stay in touch with the family between their spells of domestic labour in the cities, but most of the children were distancing themselves. Mosala and Morwesi were submerged in the local *demimonde* in Ledig, and Mmusetsi—the first sibling to show signs of wanting to get some distance between himself and the family—was lost to the farms of the northwestern Transvaal. But, as always, the distance between Kas and his middle son pained the father most: true, Bodule visited them a few times, but it was clear that the boy still marched to the beat of Dikeledi's drum. It was not so much what was said or what was done between them that hurt, but what was left unsaid and undone. Once, when Kas invited everyone in the family to attend the slaughtering of a sheep in order to honour the Maine ancestors, even Mosala noticed Bodule's hurtful absence.[47]

Though there was always someone in need of advice and herbs, a pair of shoes or a harness to be repaired, a piece of metal to be cut or sharpened, or an animal needing attention in the barbed-wire kraal that Kas had erected in the southern corner of his plot, he had more time for reflection. He had never been a great sleeper and the nights now seemed to him like a giant chasm which he tried to bridge by pondering on personal matters, recalling past achievements or, more often than not, simply running through the strategies for the days that lay ahead. He still rose well before anybody else in the house and was virtually the only one in the family to see the first shafts of weak morning sunlight as they slid off the backs of the noisy, glossy starlings that had moved into the tree that towered above his metal toolbox.

Spring came around and Kas's tractor was in even greater demand than before. A steep increase in the price of white maize following on the drought whetted the speculative appetites of some of the matsenelwa notables, while the subsequent rise in the price of maize meal forced even the poorest members of Ledig to plant mealies for cheaper food. As before, it was only much later in the season that the young Mngomezulu found the time to plough the Maines' fields.[48]

Kas, his focus still firmly on his cattle, was not unduly concerned, especially when better rains came and the grazing improved dramatically. There was another reason for his self-confidence. He had secured reliable assistance: Nthakwana's seventeen-year-old son, Bona, along with another youngster in the village, Richard Maseko, would come after school to saddle up the horses and go into the crater to recover any cattle that had strayed during the morning.[49] Yet more help appeared from a most unexpected quarter when Bodule suddenly arrived at the homestead, unaccompanied by his wife or children. He had been laid off work and, wanting to re-establish links with the family as well as avoid the harassment experienced by the unemployed in Khutsong township at the hands of the South African Police, had retreated to the relative safety of Ledig. Verwoerd's dream of 'white' cities that extruded unskilled black labour 'surpluses' into the 'homelands' until they were needed again at the core of South Africa's industrialising economy might not always have come true but, in the case of the Maines' extended family, it shaped their life and struggles for nearly twenty years.[50]

The welcome return of the prodigal did not, however, transform Bodule's troubled relationship with Kas. Two decades of chafing cannot be soothed by a single summer. But the fact that he was at Ledig without Dikeledi helped; it gave father and son several weeks of unmediated interaction. Their discussions, which came at a time when Bodule no longer needed to contest household income, acted as a balm and restored some affection and suppleness to their relationship. When autumn came and Dikeledi put out the call for her husband's return, Kas did not contest his son's departure.[51]

Bona and the other grandchildren brought in the little maize that stood in the field. Kas had never expected much of the harvest, but the fact that it grew in an unfenced area some distance from the village did little to improve the yield, for the number of outsiders expelled from the white farms and forced to seek refuge in Ledig seemed to increase with each passing year, and most of the newcomers were locked into a struggle for survival. How could one blame them for taking a few green mealies when their children were starving? Yet, at the end of the day one couldn't do much with five bags of grain.[52]

It is impossible to know what the BaKubung as a whole felt about their yields or about the drop in the price of maize that marked the 1975 season. The theft of mealies, the cost of tractor hire, or simply having to share the harvest on ever more disadvantageous terms with matsenelwa notables all contributed to dissatisfaction. But it was clear that the BaKubung were becoming disenchanted with the rule of Catherine Monnakgotla and her advisers. The never-ending stream of refugees from white farms, the declining living standards, and the return of a 'Sofasonke' element who hadn't been able to get access to purchased land all contributed to the alienation of the faithful at a time when Pretoria was singing the praises of an 'independent' Tswana state. The shift from the spirit of 'fellowship' that characterised the earliest settlement to the 'sterility' of an emerging slum was palpable. Names have meaning, and the Kopano of eight years ago had long since been forgotten: the people now lived in Ledig.[53]

The first rumblings of discontent had become apparent during 1973, when several influential BaKubung had urged Catherine Monnakgotla to stand down in favour of her son, Solomon Ratheo. She had ignored the request, but when the movement gained popular support during late 1974, she reluctantly informed Bantu Commissioner Steytler that she would hand over the chieftainship of the tribe to her son on 8 January 1975. But Solomon Ratheo was destined never to rule. On New Year's Day, a week before his installation, he went to visit an aunt in Mohlakeng near Randfontein, got involved in a quarrel with a former school friend, and was stabbed to death. A knife in the ghetto had renewed his mother's rather tenuous lease on political life at Ledig.[54]

The BaKubung notables returned to the fray from another, more radical angle of attack. They demanded that Catherine, whom they increasingly portrayed as a stooge of the apartheid administration, make way for the very man whom they had ousted in her favour back at Molote, Lucas Monnakgotla, her brother-in-law and a Sofasonke resister. In mid-1975 she relinquished her authority to this man, who had never courted 'outsider' support and who was therefore beyond the reach of the matsenelwa cattle keepers. The internal dynamic of ethnic politics and the external logic of apartheid, thus combined, allowed the BaKubung to reassert their 'Tswana' identity at precisely the moment that the Afrikaner Nationalist government was unfolding its plans to launch the state of Bophuthatswana.[55]

Men like Kas and Scotch Mngomezulu found it difficult to account for their patron's sudden abdication. Indeed, so reluctant were some of the cattle keepers to believe it that it was said that she had arranged for her own son to be murdered rather than turn her back on her clients! Nor were the matsenelwa reassured when, within months of her abdication, she suddenly took ill and died: as Kas and others saw it, the usurping faction had resorted to witchcraft to eliminate a woman whom they considered a foe.[56]

If Lucas Monnakgotla did not embrace the 'outsider' cause, he was equally careful not to alienate those amongst the matsenelwa who had formed such an important part of Catherine Monnakgotla's constituency. After some initial turbulence the community was restored to a more even keel. By late 1975 the political temperature had dropped significantly, and Kas was once again seeing a stream of patients in need of his help and advice. In between consultations, he busied himself with the usual winter routine working with leather, repairing agricultural equipment and machinery.

The one piece of equipment he was no longer willing to repair was the trailer he had lent the Mngomezulus. In fact, his arrangement with them was beginning to sour. The father and son made constant use of the trailer to ferry fuel and water or bricks and sand to the village for commercial purposes, and yet, whenever the vehicle broke down or a tyre punctured, Kas had to pay for the repairs. When the trailer broke down again that winter, he suggested that the Mngomezulus take it over. But they had no wish to own a vehicle, and so Kas arranged for it to be towed back to his yard, where it was dumped beside an abandoned planter, some broken ploughs and a formidable pile of scrap metal. The cobbler kept his needles and thread in a box, the ngaka stored his herbs in a collection of bags, bottles and pouches, but it was space, above all else, that the blacksmith needed to pursue his profession.[57]

The arrangement for the tractor also fell gradually into disfavour. When the price of maize sank back to more familiar levels, interest in working the land declined amongst both the tractor owners and the BaKubung. So, after the tractor had been hired out for the spring ploughing (at the reasonable fee of R5 per morgen excluding fuel), Kas asked the younger Mngomezulu to take it home and park it in his yard.[58]

Throughout the spring of 1975 and well into the summer of 1976 the rainfall remained above average, but without ready access to secure fields, sufficient adult labour, or consumers with money to spend on fresh fruit and vegetables, few commercial farmers could emerge along the Pilanesberg. BaKubung villagers who had once proudly hawked their produce along the entire length of the Witwatersrand had been reduced to slum dwellers. It was as if the official name, Ledig, 'idle' and 'worthless,' had become a self-fulfilling prophecy. It was, as Kas said with bitter irony, just a '*kaffer plek*,' a mere nigger joint.[59]

Perhaps because the economic landscape was so bleak and cheerless, members of the homeland administration scheduled to take office in the following year were open to the advice of white 'development experts' who, it will be recalled, had had their eyes on the tourist potential of the crater and its surrounds since the late 1960s. In mid-1976, Chief Lucas Mangope—the man destined to become the first state president of a country that, like the original Pakistan, comprised distinct parts separated by hundreds of miles—paid Lucas Monnakgotla a visit. His purpose was to persuade the BaKubung chief to underwrite a scheme even more ambitious than that originally envisaged by the planners from the University of Potchefstroom. The scheme was also already firmly in place in the minds of one of South Africa's most powerful entrepreneurs and the members of what would soon be the government of the 'Republic of Bophuthatswana.' The crater, Mangope argued, was too valuable an asset for the Tswana nation as a whole for the government to allow it to be used exclusively for the needs of stock farmers like those at Welgeval, or the cattle farmers of Ledig. A site in the southeastern corner of the crater should be leased to a South African company that would develop a luxury hotel and gambling complex there capable of attracting thousands of visitors to the area. 'Sun City' itself would be surrounded by the long-mooted Pilanesberg Game Reserve, which would, in turn, feed off tourists attracted to the casino and sporting complex.[60]

Sun City was a scheme conjured up from beyond the realms of the normal imagination. For the venture to succeed, the Bophuthatswana administration would have to clear the crater of the black farmers living there and secure the hills on which the men from Ledig grazed their cattle. But the development generated by the linked projects would, it was suggested, hold obvious benefits for local communities: jobs for youngsters struggling to find employment close to home, and a market for local agricultural produce. In addition, the Mangope government promised to compensate cattle owners with new pastures at nearby Mahobieskraal and Sandrivierspoort.[61]

There was not much for Lucas Monnakgotla to negotiate. Mangope's government was already deeply committed to the scheme, and given that Ledig was to be incorporated into the new state, it made little sense for him to alienate its authorities, all the more so since the BaKubung were bent on rediscovering their wider ethnic identity. The hotel and casino would, of course, give rise to some construction work and, thereafter, jobs for cleaners, gardeners and waiters. The loss of grazing was regrettable, but it would affect the matsenelwa more than the BaKubung, and in any case there was the promise of other land in compensation. Monnakgotla took Mangope's proposals back to his kgotla where they were duly endorsed. Progress, it seemed, was intent on taking a walk through the Pilanesberg even though it had made a rather unusual entrance.[62]

Kas, along with the older men and the matsenelwa, was sceptical but his mind was elsewhere. It was spring. The tractor was hired out, but the early rains were not promising. By early in the new year it was clear that the season would be abnormally dry, and Kas's cash income fell. The harvests were down, too, and the first few months of that winter saw yet another influx of people from the farms, along with widows and children who could not find a social toe-hold in the 'white cities' of an apartheid-mad state.

While younger men in the village went off to the construction site at Sun City to try their luck as unskilled labourers, Kas relied on craft skill and mental agility to pull him through the winter. His ability as a ngaka was undiminished, but he was having difficulty repairing shoes, his eyes watered excessively and felt unusually sensitive. On days when they were painful, he took to wearing a pair of unfashionable sunglasses that he had found in one of his boxes. The white desert of salt at Sewefontein and a lifetime's exposure to the ultraviolet light radiated by the remorseless highveld sun were exacting their toll. His eyes were clouding over with cataracts.

The problem with his eyes exacerbated his difficulty with reading as he scrabbled about amongst his letters and receipts, peered down at personal documents, or browsed amongst discarded magazines in search of information about agricultural equipment or patent medicines.[63] But as had happened in the fields back at Molote, it was the grandchildren who extended his reach at a time when his physical powers faltered. One of the 'children of gases and oxygen,' Ntholeng, came to his rescue—and this, ironically, at a time when black children in Soweto and elsewhere were in open revolt, questioning adult authority, demanding a decent education and an end to the apartheid policies of a cynical and destructive government.

Ntholeng, who worked in Tembisa, on the East Rand and visited her mother over weekends, lent Kas her eyes, put her literacy at his disposal, and won his trust. Given his ingrained caution and his earlier misfortune at the hands of another granddaughter, this spoke well of the young lady who became his banker, clerk, and valued adviser during the closing years of his life. Kas remained dismissive of the supposed advantages of formal education, but the truth was that both he and Leetwane now relied on the sharp eyes of a holder of a 'Standard Six Certificate' to help them penetrate the fog of modernity that had settled over parts of the Pilanesberg.[64]

During Ntholeng's spring visit Kas raised a problem that had been stalking him for years. Ever since the move from Molote, Leetwane had pleaded with him to build a permanent structure the Maines could call home. After a decade at Ledig he was ready to acknowledge that it was likely that they would spend the rest of their lives there, but this admission came at an awkward moment. The tractor was no longer operative and Ntholeng would have to advance him money if they were to abandon the shack and ask Mosala to build them a house from the coarse cement blocks used in the village.

Ntholeng was willing to help, but she was preoccupied with a more serious problem. Matthews Moate was ill and had not worked for months—a persistent cough was sounding more and more like that of Lebitsa's before she had died—and Ntholeng was anxious to see her parents through a period when the family was without a male breadwinner. She could not say when she'd be able to offer financial assistance. For Kas the delay was mildly irritating but he was willing to make do. When Bophuthatswana became independent, on 6 December 1977, and the building of Sun City continued apace, the Maines were living in the same corrugated-iron shack that had served as a home for more than fifty-seven years.

Leetwane was less patient. Her pointed silence signalled her dissatisfaction, and when the cooler weather set in Kas returned to the fray. Ntholeng urged him not to worry, telling him that she was working on a way for him to buy the cement and make bricks of his own. At the end of May she reiterated her desire to help, but there was still no sign of any money.[65]

By June the matter seemed to slip from Kas's consciousness, and by the time the Bophuthatswana government got around to paying him and Leetwane the first instalment of a modest state pension four weeks later, all thought of buying cement had gone from his mind. With the coldest part of the year behind him, he walked across to Mdluli Street, in the 'Zulu' section of Ledig, and bought a horse from 'Christinah's son.' Livestock was always a good investment, and a second horse would allow them to use the cart to haul water to the house. From the doorway of the shack Leetwane watched in resentful silence as her husband guided the brown mare into the yard and tethered it. In his clouded vision tractors and horses had always been more important than houses.[66]

The rainfall for 1977 was the lowest recorded in ten years. The matsenelwa were uncertain as to what to do in yet another dry season, but Kas had no doubts: it was time to think of sunflowers. They were well suited to dry conditions and of no temptation to hungry folk who raided fields once the sun was down. The Maines' new neighbour, Hika, who owned a tractor as well as a taxi, gave him a good rate for doing the ploughing.

Kas donned a tight-fitting straw hat and walked down to the fields, where he directed operations from a seat under a stunted acacia. The heat was stifling; by midmorning only the odd smartly dressed crow, like a lost soul searching for the casino, risked moving through the bush. But Kas soon realised that he had been too pessimistic. As the year ended, the rains picked up and Kas was certain that if they persisted, he'd get at least one and perhaps two crops of sunflowers. He relaxed slightly, looked up a friend, and then settled down to enjoy a little *mokoko* (a supposedly traditional 'Tswana beer' made from a mixture of brown bread, sugar and yeast).[67]

On Christmas Day Kas had a good deal more mokoko. Late that afternoon his nose started to bleed. Then he noticed that, even though there was no cut to be discerned, he also seemed to be bleeding from his tongue. He tried to staunch the flow but couldn't and by the time someone was sent to summon the neighbours, even Kas was concerned.[68]

Hika helped him into the car and they set off for the hospital. There was blood everywhere. The stuff never stopped coming. It streamed into the cloth Leetwane had given him, dripped down his arms and legs, and eventually formed a semi-congealed mess on the car floor. Even at Saulspoort, where the doctor insisted that he be admitted as a patient, it took them time and effort to staunch the flow. Loss of blood and concern about his blood pressure meant that Kas needed to spend time in Moruleng Hospital regaining his strength, but to him this was an idea without appeal. The sunflowers were ready for harvesting. Too weak to offer real resistance, he agreed to stay on only after he had given Leetwane and Nthakwana detailed instructions about what to do with the sunflowers and how to go about organising a second planting.[69]

The instructions were a necessary precaution, for the days at Moruleng turned into weeks. After a month he still felt weak and dizzy. But by then he was alarmed by something even less flexible than a ploughing schedule: if he wasn't present at the 'tribal office' on the last day of the month he would miss the January 1978 pension payout. 'I told them: "I want to go home." They said, "No." I then pleaded with them: "Please discharge me so that I can go home to collect my pension."' In the end, and much

Kas Maine and his sister Sellwane, 1982
*M. Nkadimeng*

Kas Maine and his wife, Leetwane, with a daughter-in-law and grandchildren: (left to right) Motheba; Ratshilo's wife, Masidiye; Bodule's wife, Dikeledi; Lebitsa's daughter Thakane; Leetwane and Kas; Thakane's daughter Dikeledi; Ramosele, and (in front, squatting) Tsetse
*M. Nkadimeng*

M. Nkadimeng

D. Goldblatt

M. Nkadimeng

Kas Maine working at Ledig, 1981-82

M. Nkadimeng

Kas's sons Mosala and Mmusetsi, 1982
*M. Nkadimeng*

Kas and Leetwane's daughters Morwesi (with Mpsana), Pulane, and Nkamang, 1982
*M. Nkadimeng*

Kas and Lebitsa's daughter, Thakane, 1985
*M. Nkadimeng*

Kas's common-law wife, Tseleng Lepholletse, and their daughter, Matlakala, 1985
*M. Nkadimeng*

Kas's grandson Sekgaloane, 1982
*M. Nkadimeng*

Kas's sister-in-law Motlagomang, Mphaka's wife, 1985
*M. Nkadimeng*

Kas's niece Mamfuna, daughter of Motlagomang and Mphaka
*Maine family*

Kas's brother Phitise, 1982
*M. Molepo*

Kas's brother Sebubudi Maine, 1985
*M. Nkadimeng*

Lerata Masihu, 1983
*M. Nkadimeng*

Kas's nephew Baefesi, Mphaka's son, 1983
*M. Nkadimeng*

Members of the Maine family at Ledig, 1985
*M. Nkadimeng*

Kas's daughter Nthakwana, Kas, a child, and his son Mmusetsi
*M. Nkadimeng*

Mequatling
*S. Mofokeng*

Ledig, looking south toward Mahobieskraal
*S. Mofokeng*

against their better judgment, they relented and the family arranged for Kas to be taken home.[70]

On the way back to Ledig Kas's mouth started bleeding again, and Hika had to turn off into Mogwase, where a black medical practitioner ran a surgery from the rooms behind his house. 'Jesus! Trying to stop the blood!' Eventually, in desperation, the doctor got out a pair of forceps and started fiddling about in Kas's mouth. The bleeding appeared to ease and then stopped. The doctor wanted him to stay around for a while, but Kas asked for vinegar which, he explained reassuringly, was guaranteed to prevent a recurrence. Black doctors are as hard to convince as white ones, though, and it was some time before the vinegar was forthcoming. When it arrived Kas thanked the doctor, filled his mouth, and went out to the waiting car.

Back home Kas reflected on his experiences at the hands of the doctors. Sloppy diagnosis and indifferent success at Moruleng and Mogwase had done nothing to enhance their standing in his eyes. Had they not told him quite openly that they were unable to determine exactly what the cause of the bleeding was? Something that the young fellow at Mogwase had mentioned, however, had stuck in his memory, and that was a warning about the need to avoid alcohol. He decided it was probably the yeast in the mokoko that accounted for impurities in his blood. That, along with the social and moral problems evident in his own family and elsewhere in the village, was enough to convince him to give up drinking all alcohol. It was not a great sacrifice. He had never been a big drinker or, for that matter, a great eater.[71]

A lack of reserves in the form of body fat slowed down Kas's recovery. Age, too, meant that he was frailer than he had been, and stiffness seemed to have set into his lower limbs. Still, with the help of herbs, which doctors in white coats would never know anything about, he gradually removed the remaining 'poison' in his blood and by avoiding all alcohol made certain that there was no recurrence of the nose-bleeding. He felt stronger each day, although not strong enough to harvest the second crop of sunflowers when the time came.

The successful switch from grain to oil-seed was certainly an experiment worth repeating. The sunflowers brought in a little cash, which he hoped to supplement with the midyear pension payout, but Kas had yet to learn what black folk elsewhere in South Africa had long since discovered: one could never be sure of the actual amount a pensioner received. When the state officials responsible for the payout did not actually steal all or part of the money, clerks in the tribal office imposed levies on residents for the erection of schools, or fined those who had failed to attend meetings of the kgotla. Apparently a little coercion was sometimes necessary to maintain popular involvement in traditional structures, which, so African Nationalists were otherwise wont to argue, were inherently and spontaneously democratic. Kas found the levy on education especially irritating.

Why should he be made to pay for a service that could be of benefit only to members of the next generation?[72]

All this meant there could be no slacking off in off-season labours, and Kas took in some more cobbling. Misshapen shoes needing new soles were thrown on an untidy pile growing beneath a nearby tree. Dry and cracked, like the faces of owners only recently ejected from the apartheid state's cities and farms, the shoes spoke only of poverty. As with his herbalism, Kas adjusted the price of his cobbling to meet the client's pocket. The only difference was that the rising cost of leather imposed a lower limit that he neglected at his peril. A man had a responsibility not only to his family but to his community.

As the spring approached Kas's children and grandchildren, sensing the return of the warmth, crept out of their urban lairs to enjoy a weekend on the slopes of the Pilanesberg. Matlakala, Thakane and Ntholeng all came to see how the old man was getting on and, in between attending to his planters, ploughs and yokes, he dispensed advice, guidance and support. Ntholeng took the opportunity to press a few notes on him and her grandmother. It was reassuring to have the support of the younger family even though it was noticeable that the daughters rather than the sons were shouldering most of the load. Kas knew that his sons had families of their own to attend to, but he longed to see Bodule.

The Transvaal spring is always short on patience, and it did not have time to indulge the whims of an old man. The early rains were unpromising and Kas again asked Hika to help with the ploughing. The tractor made short work of a modest patch and, with the help of Nthakwana's children, the field was devoted to sunflowers. But the early summer rains were as intermittent as they were light; the crop was slow to show signs of progress.[73] By December Ledig lay sweltering beneath a blanket of red dust. It made breathing unpleasant and people thought twice before venturing about the stone-strewn, rutted streets. The day started with a few modest stirrings and the muffled sounds of early-morning preparations, but then everything fell silent until late afternoon, while man and beast alike sought refuge from a white-hot heat sent from hell. Kas perched on a three-legged stool beneath the acacia in the yard and sipped tepid water from a sand-scoured Coke bottle. Beads of sweat trickled down his brow, skirted his cheeks, congregated around his chin and then fell to his chest where they blended into the damp stain spreading slowly across his heart. Everybody and everything was waiting, waiting for relief from the heat. But nothing ever happened. Every hour or so he would rise, shuffle across to the metal drum that stood behind the shack, fill the bottle and make another attempt to douse the fire that smouldered in his belly and ravaged his throat.

A few days before Christmas cooler air moved in from the east. That evening, as he peered out toward the Magaliesberg, Kas could see the night sky pierced by the spiked aggression of forked lightning over the Witwatersrand. A thunderstorm on the horizon should have made him feel better, but he went to bed filled with foreboding. However, the morning air was pleasantly cool and moist. He got out the stool, eased his tools into position and confronted chores that had been crying out for attention.

Time slipped by. About midday he became aware of the groaning of a car trying to come up the storm-washed channel to the shack. It was hard to see where exactly it was going, but, once it turned in at the gate, he recognised it as belonging to one of Bodule's sons. Ratshilo got out, let the door swing shut, walked slowly towards him, and then said, '*Grandfather, I am here to tell you that Bodule is dead.*'[74]

Kas took a pace or two toward the boy. Half unbelieving, he cried out: 'Your father is dead?' Ratshilo answered, 'Yes.' Still reeling, Kas tried to steady himself by asking how it had happened, but Ratshilo's eyes filled with tears and he could not bring himself to speak. Kas sought to comfort the boy, and eventually learned that Bodule, his Bodule, had been killed at the mines. Head bent forward, tears poured down his cheeks, missed the chin and the stain on the vest, and disappeared into the dust. By the time Leetwane found them, Ratshilo was sufficiently composed to provide them with more details. But what did it matter? Details were like teardrops in the sand; they served only to confirm a man's helplessness.

On his return to Carletonville Bodule had found work with a firm of contractors responsible for the maintenance of headgear and surface installations on the local mines. Although painting was not the best-paid work in a town where most men went underground in a cage rather than up a ladder to earn a living, it at least offered continuous employment. This, along with the chance of occasionally earning overtime rates, allowed him and the family to sustain a dreary proletarian existence in Khutsong township.

The previous Tuesday Bodule had set off for work earlier than usual. It had been a cool, overcast morning. At the mine he let himself into the storeroom well before the official commencement of work and donned his overall before collecting paint and a brush and going back out into the mist. What happened next was a mystery. A minute or two later he was struck by a train of hoppers, *makalanyane*, the heavy metal 'cocopans' that conveyed ore from the mine shaft. A worker found him lying beside the track and raised the alarm, but by the time the first-aid assistants arrived he was dead. His neck had been broken and the right side of his head mangled.

Workers moved his body to a makeshift mortuary and the contractor sent someone to the house at Khutsong, where, shortly after eight, Dikeledi and her six children were told of Bodule's death. She and the two oldest boys, Sekgaloane and Ratshilo, were taken to the mine where she identified the body. The following morning, somebody from the contractors arrived with an envelope containing about R100. The family had lost its major breadwinner and, in return, had been given the equivalent of a week's wages. There was, it was said, no prospect of additional money from the state's Workmen's Compensation Fund because Bodule had been killed before the officially sanctioned starting time. In the swirl of a highveld mist, apparently only the contractor's lawyers could see justice, and who knew what the state of their eyesight was?[75]

Dikeledi had never been one to shy off a challenge, so she sent Ratshilo off to see a lawyer of her own and make certain that the family interest was protected by lodging an appropriate claim. Then she told Ratshilo to go to Ledig, tell her parents-in-law the bad news and bring them back to Khutsong. Kas was taken to the mortuary at Carletonville to view the body, but he insisted on visiting the spot where the accident had taken place; the next morning, Ratshilo drove him to the mine to show him where the hopper had destroyed his greatest love. As they got back into the car Kas made a resolution: he would take his son home, home to the fields and countryside where he belonged.[76]

Dikeledi disagreed. She thought that if the contractors were to be forced to pay for the coffin, the funeral, and compensation, Bodule would have to be buried in Carletonville. Kas pursued her and the two oldest children from every angle he could possibly think of:

> I asked repeatedly about relatives who might not have heard of Bodule's death and who might not be able to get into the township as readily as they could enter a village such as Ledig. I was defeated by the argument that they would not receive compensation from the contractors if Bodule were not buried in Khutsong.
>
> I then gave them an example of somebody who, although he had died in Natal, had been buried at Kuruman. Why then should Bodule, who had died so close to his parents' home, not be buried at Ledig? I did not mean to imply that they would have to meet the cost of transport to Ledig. No. All the expenses would be met by me, if only he were buried at my place.
>
> But Dikeledi would have nothing of it. *She* was the one who objected most strongly. In the end I gave up and I told them to do as they saw fit.[77]

In death as in life, it was Dikeledi who stood between Kas and Bodule. More than ever, he wanted his son by his side, at home, at Ledig. It was time for them to be fully reunited. But two days later they buried him at Carletonville, a place even more meaningless than the bluegums at Hartsfontein where he and Leetwane had laid their first-born to rest half a century earlier. They made their way home, Kas leaving behind him the hope of one day sharing a field with his favourite son. The towns seemed to be claiming everything. Even the corpses.

CHAPTER SIXTEEN

# Survival

## 1978–85

They took the northerly route back from Carletonville, meeting the circular road that brushed the fringe of the crater at Heystekrand before swinging south for the short drive home. At Mogwase, where electricity, well laid-out streets and modern houses marked Bophuthatswana's emerging administrative elite, the road straightened out, and for the next five miles it pointed unerringly toward the sprawling mess that was Ledig. About halfway through this section Kas noticed a team of workmen erecting lengths of game fencing, but before he had the time to assimilate this information, Ratshilo said that they were approaching the turnoff to Sun City.

The casino complex was screened from the eyes of passing BaTswana by the outer walls of the Pilanesberg. In Bophuthatswana, as elsewhere, the poor could see the rich at work but not at play—a glimpse of the former carrying with it a positive and elevating message, the latter provoking only negative and undesirable comparisons. Gambling without guilt was to be part of the Sun City experience. A little farther on, opposite the Thanda Bantu Store, the car turned up a rutted track past the vacant lot meant for Bodule, and then negotiated the entrance to the yard, and drew slowly to a halt outside the shack.

Kas and Leetwane were exhausted by the experiences of the previous week, she perhaps even more than he, and he was concerned about her silences, which seemed to run ever deeper. Kas tried to pick up the thread of farming operations, but while the rains stayed away not much could be done in the fields, and sitting around at home only allowed time for painful memories to surface. Trapped in limbo, he pottered and shuffled about the yard hoping for a sign that would give new purpose to his efforts. But things seldom happen that way, and when a sign did materialise, it called for a practical rather than a metaphysical response.

A few months earlier a line of lime-washed stones had suddenly appeared along the southern lip of the crater, early warning to BaKubung cattle keepers of the boundary of the proposed game reserve and the eventual arrival of a fence. At first such unnatural markings reminded the matsenelwa of their anger at Lucas Monnakgotla's 'betrayal' of their cause to the Mangope government, but when nothing untoward happened, the stones gradually faded in significance and the Maine, Maseko, Mngomezulu and other families continued to send their cattle out into the hills and valleys. Governments, in their experience, were slow-moving things.[1]

But a few weeks after Bodule's death, the herdboys suddenly raised the alarm about a giant game fence snaking across the Pilanesberg, and Kas then recalled seeing the workmen near Sun City. Within days the fence denied BaKubung cattle access to the reserve, and while Bophuthatswana government officials attached to 'Operation Genesis' busied themselves with ferrying in leopards and other wild animals to supplement the few still in the crater, the matsenelwa had to remove their livestock and retreat to the more distant grazing allocated them by the Mangope administration.[2]

Even for those who owned transport and had herdboys, the move to Mahobieskraal was problematic: the quality of the grazing was poorer and the fields, adjacent to human settlement, were prone to veld fires. The cattle, far from any mountain spring, were separated from the nearest watering point by a long haul bisected by a busy road. As if this were not enough, at night the animals had to be left amongst virtual strangers: predictably enough, stock theft soon increased.[3]

For someone like Kas, who was without transport or young sons, a move to Mahobieskraal was potentially disastrous. He was determined not to be stampeded into any ill-considered action. Had he not staged an effective retreat when the Boers at Sewefontein had given the 'rich kaffirs' an ultimatum to get rid of their oxen or leave the district? Through careful planning, he had managed. But now there was a difference: instead of being threatened by advancing tractors and sprawling maize fields, which he could understand even if he did not like them, his oxen were having to make way for baboons, leopards and ostriches! It defied comprehension. He told Bona to pull the livestock back to the hill behind the shack.

Even this rather tentative move exacted a cost. Autumn was a bad time to restrict grazing, and equally inopportune for selling animals that were losing condition. One or two stock holders were panicked into premature action when a leopard started taking animals in the hills behind the township, but Kas's nerve held. He decided that it was best to bring in the sunflower seeds and use the off-season to come up with a longer-term strategy.[4]

The eight bags of seeds the grandchildren harvested supplemented the income from other activities, but even this modest sum exacerbated a problem that was more pronounced with each passing year: the misperception on the part of those around him about the exact extent of Kas's

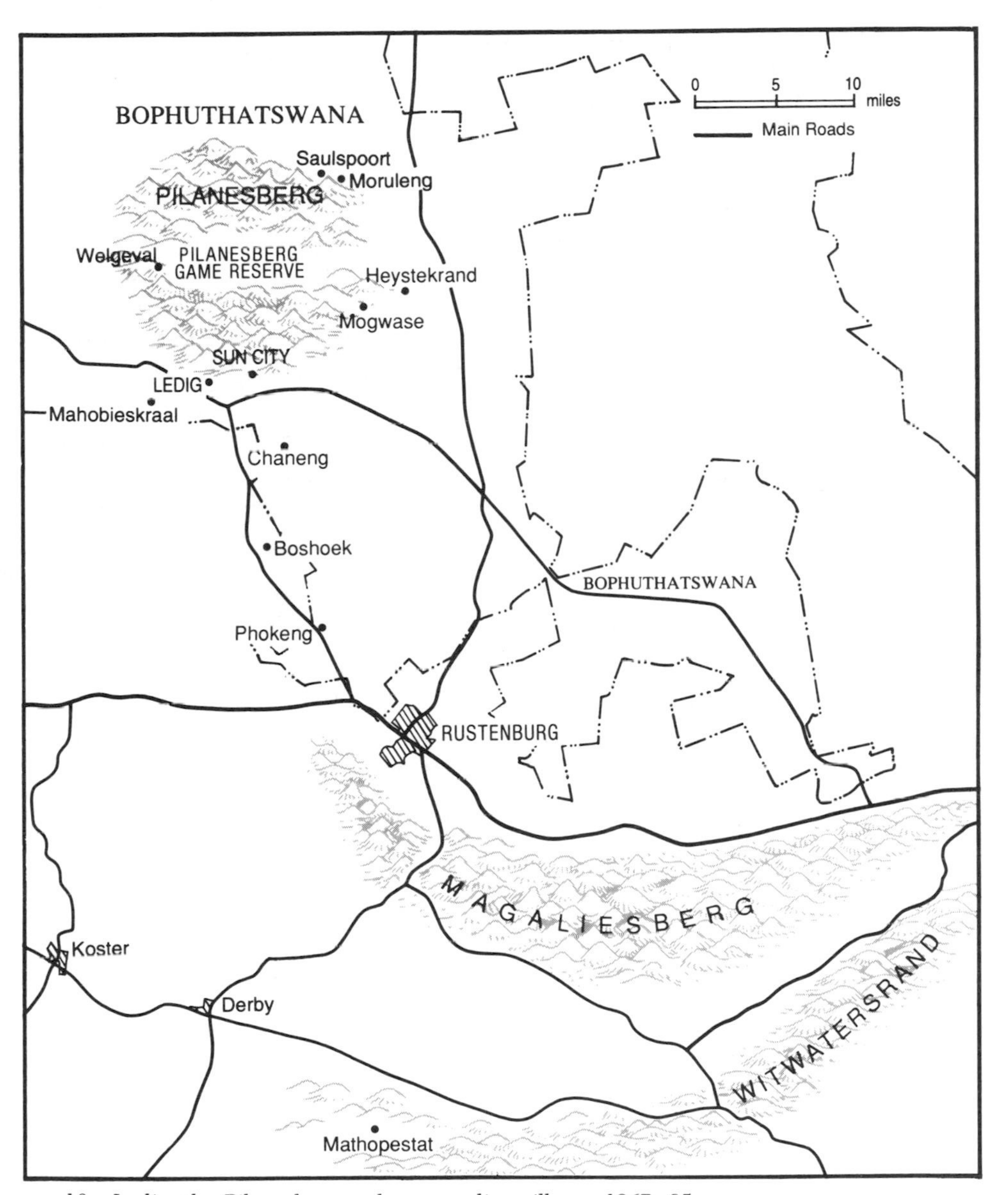


10 Ledig, the Pilanesberg and surrounding villages, 1967–85

financial resources. The family persisted in seeing him as a man with assets, including the broken tractor that stood in his yard, and failed to appreciate just how much his income had declined since the move to Ledig. This misunderstanding was compounded by his frugal lifestyle and by the fact that, even when he had had money in the past, he had seldom been willing to give it to his wives. Past performance and current practice merged to give people the impression that he was a tight-fisted old man, when he simply wanted to make a firm distinction between productive and unproductive investment.

In winter 1979, shortly after the midyear pension payout, Leetwane renewed her plea for a home that could give them comfort and warmth. But Kas was as unwilling as ever to commit resources to a project that seemed so marginal to their overall economic well-being. This time, he directed the problem back at Leetwane, suggesting that if she were willing to hand over her pension, he might reconsider the matter. But this was impossible. Nor was he more forthcoming when she asked him for the rail fare to Taung so that she might see her family for the first time in fifty years. In the end, it was Ntholeng who took pity on her grandmother and gave her the money which allowed her to take leave of her kin.

These were painful blows for Leetwane to absorb, all the more so because she remained convinced that most of Kas's livestock holdings could be traced back to Witties, the cow that she had brought to the family herd with her money from collecting, drying and selling dung during the Depression. Fearing his wrath when these issues were raised openly, she simply kept quiet and only mentioned them to a young friend when she was sure she would receive a sympathetic hearing. 'I am the one responsible for buying all the household goods,' she told him. 'I buy relish, wood, water and many other things; he does not spend a cent on things like that.' Quizzed about her husband's supposed wealth, she was equally critical: 'I just don't know. He acts as if he has no wife, children, or even an older child whom he could trust. In whose name are those savings registered? If he should die who would inherit that money?'[5]

On this issue, so close to the heart of patriarchal power, Kas Maine and his wife were condemned never to find one another. His secretive behaviour and her reluctance to express her resentment underscored a mutual silence that he mistakenly attributed to her grief at the loss of Bodule and she, equally mistaken, took as a sign of his indifference. The awful truth that simultaneously bound and separated them was rather different: except for a few hundred rands which he kept about the house for an emergency, there *were* no cash savings. All Kas had was his skill as a craftsman, the few stands at Ledig, a tractor in need of repair and a few beleaguered animals. It was not that he was unconcerned, only that he saw their collective well-being tied up in things which Leetwane thought less important than he did. In the greater domestic universe, patriarchy and matriarchy moved in distinct orbits, which at various points in the history of the family converged or diverged but were fated never quite to touch.[6]

So, while Leetwane fretted about a house Kas agonised about cattle. By August 1979 it was clear that the livestock could not be confined to hillside grazing and that Mahobieskraal was not a viable alternative. Deep down, Kas reluctantly acknowledged that he would have to sell the cattle.

This chilling thought set him along a different chain of logic: he would have to revert to grain production; though at his age tending maize and sorghum was no more feasible than herding cattle. Try as he might, there was no escaping the difficulty; if they were to survive, Kas had to choose between farming at Ledig and Mahobieskraal.[7]

He made the choice knowing that if he followed it through all its twists and turns, it would culminate in the purchase of yet another tractor and open wounds that might never heal. Perhaps it was this awareness, the consciousness of yet more domestic pain, that accounted for an uncharacteristic slackening in pace. The cattle would have to go, but he would delay selling them until the last possible moment and then take the added precaution of phasing in the new farming operations over two years. During the first season he would make do with draught animals mustered at short notice and then, if necessary, use the second summer to make the risky switch to mechanised power.

On 20 November 1979, he went across to Simon Modisane and bought three sand-coloured donkeys and, the very next morning, set about cleaning and repairing ploughs and planters that had not been pressed into service for years. For all the problems there was something satisfying about returning to his never-ending fray with nature.[8] The very sight of the first cumulus clouds waddling across the southeastern horizon filled him with the same expectant excitement he had felt as a boy when, shortly after the South African War, he and Sekwala would go out and stalk springbok, Koranna-style, at Mahemspanne.

But the rains were late and sparse, and it was hard to get youngsters to help with the ploughing. The older boys in Ledig, if not already at work in the casino complex, hankered after the modest wages to be earned from fringe employment in the Pilanesberg Game Reserve, which Lucas Mangope opened in December. Kas was surrounded by a generation of youngsters who frowned on agricultural labour, and this not only increased his dislike of Sun City and all that it stood for, but rekindled his contempt for the one person whom he felt should have been helping him, Mosala. But Mosala's life was taken up with alternating spells of labour on local building sites and bouts of heavy drinking, hardly the sort of assistance that an eighty-five-year-old farmer could rely on. Kas was reluctantly coming to the conclusion that his youngest son was 'a shit.'[9]

Kas slowly abandoned the idea of a bridging season and turned instead to a quicker mechanical solution. He hired Hika's tractor and, with the help of a man named Motaung, opened up a patch which he hoped would be large enough to be profitable. But given the costs of the tractor and the driver's services, the whole business was delicately poised; this strengthened his resolve to acquire a functioning tractor of his own.[10]

The new year arrived determined to remind everyone of the harsh realities of bushveld farming. The Pilanesberg lay curled beneath a ball of fire that incinerated anyone foolish enough to venture out after ten. Animals stood rooted beneath thorn bushes whose tufts of dust-laden greenery offered the only glint of hope to those still interested in the annual struggle for survival. Out in the fields the sunflowers, poised between life and death, were uncertain whether they could summon up the energy to raise themselves above dwarf-like stature.

For weeks it seemed that fire would triumph over water, but then, at first almost imperceptibly, the heat lost its edge and the acacias started absorbing more moisture from below than they lost through the foliage above. A passing thunderstorm dislodged a torrent of water that careered off mindlessly in the direction of the Elands; for a few hours it gloried in the name 'river.' The veld slowly changed colour; first the animals, and then the people, started to move about more freely. Kas was amongst the last to abandon his refuge beneath the tree in the yard. The old skills were still much in evidence but the champion of all seasons was slowing down. His reflexes were less certain, his walk stilted. Yet the fight was not over; there were too many things to do.

Somebody had told him about a 'Ndebele' living on one of the platinum mines around Rustenburg who was selling a horse and, on 6 January, he sought out Abram Magagula and returned with a brown mare that had cost eighty rands. The mare would, in time, yield a foal which could be readily sold in a place where many folk still relied on horse and cart for transport. The unsympathetic bushveld environment meant that the mare and the other horse which he already owned needed constant attention; when they developed ugly cuts on their shanks, he was quick to apply a mixture of motor oil and paraffin to their wounds.[11]

Breeding was the ostensible reason for acquiring the mare, but there were other, more deep-seated reasons for buying horses. The truth was that Kas had never lost his love of the animals with which he had first worked as a young stable-hand back at Vaalboschfontein. The mare was an investment, but like the cattle, it was a comforting link between the world of the Triangle and the present. Two weeks later he went out and bought a cow from Stephen Sechele. The fight had to go on.[12]

The sunflowers promised to deliver about a dozen bags. This was hardly a great return, but it at least justified his asking the grandchildren to help him bring in the crop, which in the end was only ten bags. The bridging season had not been great but at least he had not lost ground, his medium-term plans were unchanged, and he started making enquiries about tractor prices.[13]

In autumn Kas got a note from Phitise agreeing to meet his travel costs if he would visit him to treat him for a stomach ailment. Kas hesitated, for Phitise was no longer based at Doornkop: along with hundreds of other independent black farmers around Tsetse, he had been driven out of the

Ventersdorp district by the government's segregationist policies and forced to relocate his family at Uitkyk near Groot Marico. His brother's request was, of course, not without irony. Had not Mphaka and Phitise always been ambivalent in their responses to his success as a ngaka? Still, he bore him no ill will and it was fitting to help Phitise and catch up on news of the family.[14]

Phitise, who had once owned a small farm with a borehole, a fine house, and a shop, was now living in a mud hut in a resettlement camp. Like so many others he had received inadequate compensation and his pension from the Bophuthatswana government was much smaller even than the one their younger brother, Sebubudi, received from the South African authorities. Kas found the whole question of pensions most perplexing. How was one to make sense of a world in which governments gave a woman the same pension as her husband, and paid a younger brother more than his older sibling? Such practices mocked the natural order of things.[15]

More disturbing was the decline in his brother's mental health. There were clear signs of senility. Phitise's mind focused with obvious difficulty, and much of his time was spent in apparently meaningless repetitive talk of God. Admittedly he was eighty-eight years old, but to Kas it seemed like only yesterday that they were all engaging in friendly rivalry at Soutpan to see which of the four Maine brothers could produce the largest harvest. But bitter-sweet memories of the Triangle did nothing to solve the problems of the day. He left Phitise with some herbs and returned to the Pilanesberg, uncertain as to when or where he would next see his brother. Life was short.[16]

A month or two later Nkamang, having completed a spell of labour as a domestic in Rustenburg, rejoined the family, moving into the shack at Ledig and sharing a bedroom with Morwesi. This was only a minor inconvenience in a climate where much of life was lived outdoors, and having one more person to feed was more than offset by Nkamang's willingness to help with the domestic chores. It was nice to have somebody reliable around the house, and Leetwane benefited from the presence of a daughter whose quiet-spoken and placid disposition complemented her own.

*

On a Friday in late July 1980, Nkamang, Morwesi, Nthakwana and Mosala joined Kas and Leetwane for a drive to Carletonville where *ho ntsha diakobo tsa mofu*, 'the distribution of the clothes,' was set to take place at Dikeledi's home. If Kas had any reservations about a journey filled with painful memories, or about the suitability of an urban venue for a ceremony set to honour a son of the soil, as he still thought of Bodule, he did not show them. He was past quarrelling with Dikeledi; even if only for the sake of Leetwane, they should all proceed in the manner sanctioned by Sotho custom.

Early on Saturday morning they were joined by Matlakala and Thakane, who, along with other kin who had come from the more remote corners of the Triangle, had travelled through the night to meet their obligation. Kas asked Mokentsa to slip out and secure the sheep they had purchased the day before, and then he announced that the task of purifying the garments and distributing them amongst the family would fall to Leetwane and Matlakala. Dikeledi went into the house, came out with Bodule's clothing and laid it out on a blanket spread on the ground; beneath a tree, the men slit the unhappy sheep's throat. Kas used a large penknife to separate the second stomach from the intestines and then emptied greenish chyme into a small enamel basin, which he handed to his wife. Leetwane and Matlakala sprinkled the *letlale* over the garments, and then handed out the worn and crumpled items to grateful kinsfolk, who were more than ready to press old shirts, shoes and trousers into new use on the distant circuits of western Transvaal farm life.[17]

People slowly drifted off to collect a plate of *pap* and meat and find a seat beneath the tree. The muffled sounds of amiable social chatter rose from several huddles of kin as folk relaxed after the exertions of the week. It was an unusually tranquil moment. The day had passed without so much as a hint of unpleasantness between Kas and Dikeledi. Moreover, the patriarch had made a magnanimous gesture by involving in the ritual not only his wife but also the daughter who, in the past, had so often been relegated to the outer realms of her father's universe. His daughter, always sensitive to the slightest change in the psychological climate, sensed this, sidled up to her father, bent down, and speaking so softly that at first he misunderstood her, said: 'Bodule has just appeared before me in a vision and asked me to tell you that he is pleased with what you have done.'[18] At that moment they were perhaps closer than they had ever been. For the first time in his life Kas left Carletonville feeling at one with himself.

Back at Ledig he eased himself into the slipstream of a winter that seemed to have rushed by with all the disdain of a modern motor car. The approach of spring meant any amount of repairs had to be done to old agricultural implements that would be put to new use with a tractor. The bigger tasks were easy enough to manage, but when it came to focusing on fine work such as shoe repairing, Kas was having difficulty. His eyes, brimming with moisture, struggled to penetrate the film of white tissue spreading slowly across the pupils. He tried to shrug off the problem by making light of it. A man might not be able to thread a needle, but he could see people and objects like tractors as clearly as ever. He went across to Boshoek to see what machines they had on offer.

During late September and early October his eyes were the cause of so much discomfort that he twice consented to having a friend take him to see a doctor in Rustenburg. On each occasion the advice was the same: check into a state hospital at Pretoria or Johannesburg just as soon as possible, and have the cataracts removed by a specialist at a nominal charge. But Kas was unconvinced. Doctors, trapped in their rooms, knew nothing about nature. Spring was already well advanced and the early rains had been encouraging. His eyes were fixed on rural life and processes that not everyone could discern. The doctors would be the last to spot the difference between a man with sight and a man with a vision.[19]

On 25 November, at the age of eighty-six, Kas acquired his fifth tractor. L. C. van der Merwe of Boshoek Garage & Boumateriaal accepted the old International he had bought from Bessit as a trade-in, and together with cash raised from the sale of most of the cattle, it covered R2,300 of the R3,000 needed for a small red Ferguson tractor. He asked an acquaintance, Fly Ndawene, to drive the tractor back to Ledig. This time there would be no lending it to third parties: Kas was already formulating a plan to lessen his dependence on predatory drivers like Mngomezulu or Motaung.[20]

Ndawene parked the tractor beside the shack, where its rounded and coloured contours contrasted sharply with the squat, angular, rusting structure of the Maine home. The pride that Kas derived from owning this machine far transcended its utility as a productive resource. Much as a herd of oxen gathered in a kraal signalled a young man's status in the community, so the tractor indicated the presence of a patriarchal power in a productive household. The problem was, Kas no longer presided over a socially cohesive farming family in a functioning rural economy. He was living in a resettlement slum surrounded by the neglected fields of a dying peasantry whose labouring lifeblood had long since drained away to the cities. Leetwane looked across at the tractor, but her eyes saw only the spectre of a house.

In the morning Kas still rose long before anyone else. He peered at the tractor through the filtered rays of the angled sun and checked on the drum containing the two hundred litres of diesel they had brought back with them from Boshoek. Prices were running away from people faster and faster; he had been shocked to discover that the diesel cost R83.80. The price of fuel, the need to pay a driver, and the rising cost of seed all meant that he would have to drive a hard bargain with anyone willing to let him sharecrop on their fields. But there was no turning back. On 5 December he was back at Boshoek to buy a ploughshare. He then spent several days, using Motaung as his intermediary, to negotiate contracts with families on the plain near the eastern Pilanesberg who, through want of productive resources of their own or a lack of knowledge and skill, had abandoned hope of farming their allocations themselves.[21]

But the countryside was not a factory. Men made contracts: only God could ensure the conditions that allowed them to be met. Nature, which in the past had never hesitated to punish a Maine enterprise if she disapproved, apparently took a benign view of the new venture, in which Kas stood to gain no less than three out of every four bags harvested. By January 1981, he had more than thirty-five morgen under maize, sorghum and sunflower on plots lying in an arc from Moruleng in the north across to Ledig in the south. Even though the proceeds from ten of the morgen would accrue to the tractor driver, and the yields were likely to be low, since the fields were badly neglected and needed fertiliser, the success of the venture seemed assured.[22] Manual labour was undertaken by Motaung and a youngster whom Kas hired (hoping that eventually he would replace the adult driver at a more competitive rate), while Kas accepted responsibility for all the necessary support services. Throughout that summer he monitored contracts, organised fuel purchases, selected seed, repaired implements and travelled up and down the length of the Pilanesberg plains supervising the ploughing, planting and harvesting. He was in his element. Despite scepticism in certain quarters, the tractor was proving to be a good investment, and the whole exercise seemed to rejuvenate him. He looked back on his heyday in the Triangle as a pleasant memory, and, although the prophecy he had made to Koos Meyer about owning a farm had never come to fruition, he could nevertheless survey his life's work with some satisfaction. He *knew* how to farm grain.[23]

Organising the harvest was quite an exercise. The heavy manual labour once again fell to Motaung and his young assistant, but Kas made sure that a representative of the family who 'owned' the land was present to witness the harvesting. Once the crops had been threshed he exercised his right to the maize stalks, which he fed to the last of his cattle on the hillside. The animals were being fattened for selling so that he could pay his tractor debt at Boshoek.[24]

He sold off the cattle—a slow business, because there were too few animals left to warrant his undertaking a long haul to the nearest livestock auctioneers—and by the end of May 1981 he had R1,000 in cash, a sum that reflected both the volume of his livestock sales and the decline in the value of money: the government never hesitated to print money for hundreds of unproductive apartheid projects. On 29 May, Kas went across to the Boshoek Garage and handed van der Merwe the money he still owed him. He folded the receipt into a small green leather wallet that he had taken to carrying around with him and went home.[25]

The following morning Kas attended to the needs of a few patients. At about midday he was interrupted by the arrival of Motaung, who had come to deliver maize. The trailer was unloaded and the maize placed in the hloma o hlomolle, by now serving as makeshift grain store. He and Motaung exchanged notes about what remained to be done to conclude the harvest, the afternoon shade slid down the hill to warn that the day was drawing to a close, and Motaung left.

By the time Leetwane summoned Kas for the evening meal, it was chilly, with the last of the sun's rays only just penetrating the pink haze over the Pilanesberg. He got out a chair, positioned it so that it looked out across the plain, and sat down to eat. The meal finished, he sat on, taking in the sounds of early evening and waiting for Leetwane to reappear to collect the metal dish. But the gap between delivering the food and her reappearance was unusually long; tired of waiting, he turned to the shack and called out her name. There was no reply. He stood up, walked across, and put his head through the doorway and again called out her name, this time more firmly. Still no sound. He went down the passage and found her slumped awkwardly across the bed. He bent down and whispered to her, but she was unable to respond and made only the feeblest attempt at communicating by moving her hands. Alarmed, Kas stumbled back down the passage, staggered across the patch that separated them from Hika's property, and called out to the neighbours.

The Hikas came rushing across the darkened yard, dodging harrows, planters and ploughs. They, too, made repeated attempts to rouse Leetwane but she could only raise her hands in weak signals that no one understood. Kas asked Hika to drive her to the doctor at Boshoek, but by then several other neighbours had appeared and everybody agreed that it would be pointless, since the doctor would already have gone home to Rustenburg. Somebody said that Leetwane should be taken to Moruleng, but Kas balked at the idea; he knew about the hospital at Saulspoort and had little faith in it or its staff. Had he not almost bled to death under their very noses? If Leetwane were taken there she would be made to lie around in cold corridors for hours before a doctor would be found to attend to her. If that was to be her fate then it would be better for her to spend the night in the comfort of her home surrounded by her family. She could go to Moruleng in the morning. And so, stranded between the shoals of patriarchal compassion on the one hand and the rocks of peasant pragmatism on the other, Leetwane spent the night listening to the winter silence.[26]

In the week that followed Kas visited her at Moruleng on several occasions. She had suffered a stroke, but nobody took the time or trouble to explain this to him. In hushed tones, crouched over her prostrate body, he made increasingly desperate attempts to communicate with her, only to be confronted by an occasional incoherent phrase, a puzzling hand signal or, most frustrating of all, one of those lengthy silences that had characterised their relationship ever since Bodule's death. In the end, drawing on an explanation that was closer to his own heart than to hers, he concluded, 'It was the death of her son that weighed so heavily on her.' It was that, he said, which 'rendered her unable to speak.' It was as good an explanation as any: anything else would simply have caused both of them further, probably needless, suffering.[27]

But the comfort of this diagnosis did nothing to dispel his reservations about the Saulspoort hospital. During the first day or two he made repeated attempts to determine what exactly had happened to her and what the family could expect by way of recovery but was fobbed off with replies that were either manifestly incomplete or pitched beyond his comprehension. As the hands on the ward clock ticked slowly by to complete their mindless circuits of time suspended, Kas's frustration mounted about the inadequate attention that his wife was receiving. By midweek it was clear to him that he either had to get Leetwane discharged and examined by a private practitioner of his choice, or get her admitted to hospital in Rustenburg. There was nothing radical in this approach; hadn't he got Mmusetsi out of the hospital at Wolmaransstad when his toe had been almost severed by the plough? And hadn't he seen to it that the boy was treated by Dr. Hutchinson with excellent results? But his inter-personal skills, so successful when it came to negotiating the social intricacies of Triangle life, were almost useless when it came to dealing with the bureaucrats of a modern state. They refused point blank to release his wife, and spoke instead of an 'improvement' in her condition when it was obvious to him that there was none to be detected. Again and again he wrestled with them over this question, and by the Friday, exhausted, he retreated to the shack to reconsider his options.

He was still pottering around the yard late on Sunday morning when Ratshilo arrived to take him to the hospital to visit his wife. He found his hat and clambered into the car and they set off for Saulspoort. Ratshilo parked the car in the grounds and together they walked towards the main entrance.

> As we approached the gate one of the attendants asked whether we had come to visit Leetwane. We said: "Yes." They told us that she had died at midday.
>
> They took us to the mortuary and showed us the corpse. Until the moment that she had died, she had never uttered another word.[28]

Kas was stunned but remained composed throughout the long car journey home. The immediate struggle was over. 'God had called her to Him and there was nothing more that could be done.' It was 7 June 1981.

A week later, a minister from the Nederduitse Gereformeerde Kerk mission at Saulspoort came south to conduct a funeral service attended by dozens of kin from Taung and the Triangle, as well as scores of friends and acquaintances from Phokeng and, of course, Ledig. Leetwane's role as *sage femme*, always played unobtrusively so as not to detract from Kas's calling as ngaka, had gained her a following amongst the disadvantaged of that deprived community. She was laid to rest among the thorns, stones and weeds of a neglected cemetery, the distinctive east-west angle of the grave betraying her BaFokeng origins. (In the sprawling resettlement camp, even burial was a privilege, because those whose financial affairs were not in order at the kgotla were often denied burial in the community's rough barbed-wire enclosure.)[29]

Nkamang and Morwesi picked up the domestic duties that had once been their mother's, assisted by Nthakwana who, despite the deterioration in her husband's health, made a point of walking across to the shack each day to see how her father was getting on. Kas took the network of female support largely for granted; when he complained about lack of assistance from his children, as he sometimes did toward the end of his life, it was the absence of *male* labour he had in mind. It was galling to him that when his new tractor could be opening new economic frontiers, he could not rely on the help of a son. Mosala was drinking more than ever, which meant that Kas was ever more reliant on his hired driver and locally recruited young assistants. His contempt for Mosala's debauched lifestyle grew daily, while Motaung and his boy brought in the maize. Their work in June and July mounted to close on ninety bags of yellow mealies stacked beside the little shack beneath the hill.[30]

*

For anyone to have reaped such a large harvest on the depleted soils of a Pilanesberg labour reserve in the 1980s was an achievement. For an octogenarian with faltering eyesight and declining physical powers to have done so without the assistance of male offspring was extraordinary. But, even though chance, design and talent had elevated him to a relatively privileged position amongst struggling folk, Kas continued, as always, to assess his economic performance against his own exacting standards. Ten rands was not a very good price for a bag of maize, and he remembered that he owed Motaung ten bags. The harvest could have been better, and there was room for improvement. He gave the matter more thought when spring returned to the bushveld. Kas's new plan centred on getting supplementary equipment for the tractor to extend its usefulness beyond the agricultural season, which would improve his return on investment. But before he could act on these ideas he was sucked into the preparations for two domestic gatherings with a logic and urgency of their own. First, he had to find an animal to slaughter for *ho tea lejoe*, as the family gathered on the first weekend in August to honour Leetwane and propitiate the ancestors. Given escalating stock prices and the limited funds at his disposal, this was harder than he had anticipated. In the end, a friend, whom Kas had once helped out of a similar predicament, took pity on him and came to his rescue; even then the animal cost R180. Secondly, in early September, the unfortunate Matthews Moate finally succumbed to tuberculosis. It was the family's second funeral in three months.[31]

In between these family commitments, Kas found time to do a little cobbling and to help his patients. He performed his duties as herbalist as conscientiously as ever, but knew he was doing leather work almost as much by touch as by sight. Through all of this, his commitment to farming never once wavered. He was 'a man of the plough,' as he noted and the faintest trace of moisture in the air brought on the old longing to be out in the fields looking for the big harvest.[32]

Kas persuaded a friend to let him have a panel-van and, in mid-September, they transported thirty bags of yellow maize to the depot of the Magaliesbergse Graankoöperasie at Rustenburg. Although separated by more than half a century in time from his first visit to such a 'modern' institution at Bloemhof, the excursion yielded what Kas calculated as only 'one hundred and fifty pounds,' or R300. He was disappointed, but farming was an unpredictable business, and who could tell when next there would be a year like that magical one at Sewefontein? The anticipation was heightened by the notes and messages he got from prospective sharecropping partners, and he busied himself buying, modifying or repairing agricultural machinery.[33] The early rains boosted his optimism. In the three weeks before Christmas, Fly Ndawene went twice to Boshoek to buy diesel fuel (the suppliers for some reason or other had stopped delivering to outlying areas). Kas found the need for such frequent visits to Boshoek worrying. Besides costing about R50, every expedition took the tractor out of the fields for part of a working day and he was always dependent on intermediaries like Hika and Ndawene. Everything pointed to the need to act on the plan he had formulated earlier.[34]

The first two weeks of 1982 again saw Kas out under the bushveld sun instructing his schoolboy tractor driver about the maize fields he had contracted to plough. But even Kas's hopes faded when, like so many times before in the Transvaal, the crucial midsummer rains failed to materialise. By the end of January the fields around the crater were locked in the grip of numbing heat and unrelenting drought; conditions in the adjoining Pilanesberg Game Reserve were no less appalling. Angry farmers, explicitly forbidden by the Bophuthatswana officials to kill marauding animals, stood by and watched helplessly as baboons scaled the outer game fences, invaded their fields, and raided their maize. By the end of that season the rainfall at the Saulspoort Hospital station was the lowest recorded since the arrival of the BaKubung from Mathopestat.[35]

Kas did his best to salvage what he could from the season. In February, he paid an undisclosed sum to a hawker from Boshoek named Marais for a second-hand panel-van for transporting diesel to the fields. This liberated him from the tyranny of distance and the inefficient use of the tractor, but did nothing to reduce his dependence on 'outside' drivers. He then had the idea of hiring a second youngster not only to do the diesel run into Boshoek, but to help take kraal manure and water to the fields, whose tired soils were impoverished by years of neglect.[36] But the vehicle was not suitable for this: soon he decided it would be better to have a big trailer that could be coupled to the tractor and used to haul the maize harvest as well as manure. He made some tentative enquiries. Before he could buy one, though, he happened one morning to trip over a length of wire in the yard and, in falling, landed on a protruding metal post that dealt him a painful blow to the ribs. The accident occasioned a rare visit to the nearby BaKubung Clinic, but the significance of the incident lay not so much in the horrid bruise, as in the shadows spreading across his eyes. Kas was going blind.[37]

Eventually, Kas bought a trailer for R100 and got the schoolboy to haul water or maize whenever necessary. He told Motaung to take the tractor to the outlying villages where it was used to power threshing machines at a charge of R200 for every hundred bags of grain. In short, Kas dealt with the financial crunch caused by the acquisition of the tractor by increasing his investment, a bold and imaginative move, but in truth, his reliance on hired labour and low agricultural output limited his returns. He was losing.[38]

The distinctive economics of mechanised farming and heightened uncertainties about rainfall in a 'dry cycle' called for some critical thought, which under normal circumstances Kas would have given them during the off-season, but in the winter of 1981 he was drawn into several punishing psychological encounters that drained him of energy, and left him no time to think about technical issues. In the third week of July he got a message from Phitise asking him to come to Marico to help sort out a squabble that was out of hand. Phitise, whose powers were failing, had lost control of a domestic dispute and needed help to keep it out of the Bophuthatswana courts. Kas arranged for a young friend from Johannesburg, Malete Nkadimeng, to drive him to Uitkyk. But by the time Nkadimeng got to Ledig, Kas had decided he would also go farther west, to Slurry, to follow up on a snippet of information Mosala had brought home a few weeks earlier. In the end, Kas, Nthakwana and Mosala all squeezed into the car for a ninety-minute drive to Uitkyk.

The story Kas was told on arrival at Uitkyk captured important elements of the social, economic, and moral decay that had crept into many respectable black sharecropping families over the preceding twenty years. First they had been forced into overcrowded 'black spots' and then they were pushed out into the political oblivion of ethnically defined Bantustans, relegated to the national trash heap. Poverty picked away at the remaining social fabric, and conflicts about money, liquor or labour turned fathers against sons, mothers against daughters, and brothers against sisters.

Phitise's daughter, Matshidiso, had been going around the village telling folk that her brother Malefane had been defrauding their parents of their pension and using the proceeds to educate his children. Malefane angrily confronted his sister, who lost her temper and struck him: he retaliated, and a full-scale fight ensued. The next day she went to the police and laid a charge of assault against her brother.

Kas summoned the parties to hear their versions of the conflict. While he was listening, the police arrived and told Malefane to come to the police station to make a statement. When he returned, the hearing resumed until Kas felt he had all the facts. His 'judgment' ignored the substance of the charges and countercharges and he focused instead on the procedures to follow in order to resolve the issue. He made it clear that he thought that the police had erred in accepting a charge before the family had fully discussed the dispute. The proper thing was now for Matshidiso to withdraw the charge of assault as a prelude to another family hearing. The siblings agreed to act on their uncle's advice.[39]

This compromise, reached shortly after midday, pleased Kas, not least of all because it left his own party with enough time to push on to Slurry, the site of a large cement factory fifteen miles east of Mafikeng. There were good reasons for his wanting to undertake a diversion that would take them virtually to Botswana before looping back to Ledig. About a month earlier Mosala had met a labourer who had told him about a chance encounter with a certain 'Mmusetsi Maine' who lived on a farm in the Slurry district. The geography of this story tallied closely with earlier rumours of cross-border cattle smuggling, and it led Kas to believe that the man must have been the same child who had gone missing at Rhenosterhoek all those years ago. At the age of eighty-eight Kas had no time to lose. He and Matlakala had been reconciled, and now he wished to make his peace with all those children whom had been emotionally marginalised under his old patriarchal regime during the sharecropping years.

Just beyond the cement factory, they turned off the main road and followed a small dirt track leading north to a cluster of traditional mud houses belonging to some labour tenants near Benadesplaas. Mosala and Nkadimeng got out, climbed through the barbed-wire fence and approached a hut where a small group of people were gathered outside the doorway. They enquired about the whereabouts of a Mmusetsi Maine, but nobody seemed to have heard of such a person. Then, suddenly, as they were about to move off, a middle-aged man who until then had appeared lost in thought spoke up. There *had* been somebody by that name who had once worked on that very farm under the landlord RaMae, 'Father of Eggs.' But RaMae had died, leaving the property to a son who had sold the farm to another Boer, who had expelled most of RaMae's former tenants. The Maine person had moved on to Vlakpan,

Gert Pretorius's farm, about ten miles north along the very same road.[40]

The road was poorly signposted, and they overshot the entrance to the Pretorius property by several miles. Nkadimeng turned around and retraced their route. All of this took time, and the winter sun was hanging low in the west when they reached Vlakpan. They turned in at the gate and found the outline of a track across the veld leading to an isolated hut. Kas got out and walked round towards the front of the house, where a man and woman were seated on upturned tins, the pale shafts of fading daylight playing across their faces. Their legs were wrapped in twilight. As he approached, the woman stood up and took a tentative step toward him. The man remained seated.

Kas shuffled forward to the seated figure until, despite his clouded vision, he thought he could detect the profile of his oldest son. He paused, poised awkwardly between joy and fear, and then asked the man if he knew who *he* was. Puzzled, the man glanced up and said, 'No.' But even as that single syllable pierced the evening gloom, Kas knew he was looking at 'the replacement'—the boy who had come 'to wipe away their tears' when Mosebi had died at Hartsfontein fifty years earlier. He leaned forward and asked the man who his father and mother were and whether they were still alive. Turning his head to avoid the last rays of sun, the man repeated his parents' names and gave Kas the name of the farm where he had last seen them. All doubt was banished from his mind. Kas pressed forward and again asked the man whether he did not know him, and again came the answer, 'No.' Desperate, he turned to Mosala, tugged at his sleeve, pulled him forward until he, too, stood opposite the seated figure. Did the man not recognise the new face? Still came the answer, 'No.' The man remained perfectly motionless.[41]

The woman, sensing Kas's frustration, beckoned them to one side. She introduced herself as Ntsholo, wife of Mmusetsi and, coincidentally, a distant kinswoman of a family of the same name that had once farmed at Hartsfontein under the Niemans. Kas remembered the family. She confirmed that the man sitting on the bench was indeed Kas's son, and explained that Mmusetsi had been traumatised by certain events that had left him even more hesitant, confused and insecure than before. If she might explain what had happened over the past five years, and they gave him more time to appreciate their questions, she was sure Mmusetsi would come to realise who they were. Unable to speak directly to a son who had failed to recognise his own father, the old man sat down to listen instead to the daughter-in-law whom he had only just met.[42]

She and Mmusetsi had married on the border farm belonging to the butcher from Carletonville. But Mmusetsi had tired of working for the cattle smuggler and had found more regular employment with RaMae, Father of Eggs. RaMae, who in addition to his poultry business also kept cattle, appreciated Mmusetsi's hard work and loyalty, gave him a horse, and placed him in sole charge of the cattle on a farm where most of the hard manual labour fell to immigrant workers from Botswana.

Mmusetsi's elevation to the rank of foreman and the enthusiasm he showed for his work had not met with universal approval however. In the small world encompassed by farm fences, a man who 'would round up the cattle early in the morning while others were still asleep' undermined the deliberately sluggish and self-protective work rhythms developed by poorly paid manual labourers. This, along with his fierce temper and great physical strength, earned Mmusetsi the enmity of the BaTswana workers, amongst whose ranks there was said to be a powerful 'witchdoctor.' Underlying tensions were aggravated when RaMae's relations with his employees deteriorated: he shouted at the workers and assaulted the herbalist for 'talking too much.'[43]

The ngaka swore revenge and used his herbal potions to bewitch Mmusetsi. One morning, Mmusetsi's horse reared: he was thrown from the saddle and landed heavily on his back. When he tried to get up, his legs failed to respond: from that moment, he was both paralysed and incontinent. The man before Kas was seated not out of disrespect but out of necessity. Mmusetsi spent his waking hours out of doors for reasons that were as obvious as they were humiliating. Poorly endowed in mind but always willing in the flesh, Mmusetsi now found his troubled soul trapped in a sadly defective body. RaMae, alienated both from farming and from his workers, had arranged neither for medical assistance nor for compensation for his employee. Not long thereafter he died from causes the BaTswana attributed to the ngaka. Mmusetsi was left to rot outside the door of his hut; when RaMae's son sold the farm to a third party and the new owner expelled most of the tenants, Gert Pretorius had taken pity on Mmusetsi and his wife and given them a place at Vlakpan. The world was filled with good as well as evil men. Pretorius had also done more than any of the other Boers, Ntshohlo said, by arranging for treatment at the local hospital, where Mmusetsi had been given crutches. Pretorius also saw to it that Mmusetsi got a small state disability allowance.[44]

By the time she finished this account, to which the seated figure hesitantly contributed, Mmusetsi had come to realise he was talking to Kas, Mosala, and Nthakwana. They, in turn, outlined the fate of other family members. But Mmusetsi was too bewildered and upset to take it all in. With darkness already upon them, Kas and the others took leave of the couple, got into the car and set course for Ledig.

Kas returned to his empty shack at about eight o'clock. It had been a painful, long day, and sleep did not come easily. The following morning, his head was still aswirl with ideas and his heart pounding with a bewildering array of emotions ranging from fear and anger to guilt and sorrow. When a friend passed by, he called him in and poured out the tale of his son's plight. On learning of the evil-doings of the ngaka and the Ba-

Tswana, the man comforted him, offering to introduce him to an even more powerful 'witchdoctor' in Phokeng, who could cause 'all those who had bewitched Mmusetsi's legs to die.'[45]

It was a possibility. Ngakas, not least of all those from Botswana, sometimes had access to mystifying powers. Had not Moipeng himself told him all those years ago at Kareepoort how the BaNgwaketse from the far-off Kgalakgadi had bewitched him? Was the village not full of stories about the mysterious disappearance of an eighteen-year-old youth from Ledig? Were shopkeepers not asking ngakas for human skulls which, it was claimed, could protect their businesses from ruin? And had not the black freedom fighters in 'Rhodesia' won the war against the settlers by employing animal fats that caused bullets to turn to water? There was often more to witchdoctors than met the eye.[46]

But Kas's anger slowly subsided, and he started looking around for other explanations as to why Mmusetsi, of all people, should have been the object of such misfortune. Then another development took place which ensured that Mmusetsi's troubled past was never far from his mind. Towards the end of winter Dikeledi left Carletonville and took over the stand Kas had bought for Bodule at Ledig long before, when hope and despair had been balanced more equally in his life. The joy of his oldest son's support had been denied him, and now, ironically, he found himself in the company of the very person whom he considered to have done most to poison his relationship with Bodule and, indirectly, with Mmusetsi.[47]

Dikeledi had reasons of her own for wanting to leave Khutsong of course. Having succeeded in winning financial compensation from the contractors for whom Bodule had been working at the time of his death, she now had a small monthly income. That money would go further in the countryside than in the town, where life was more expensive by the year and where she had to share a tiny township house with her son Ratshilo and her daughter-in-law. Always sensitive to changes in the emotional climate, she noted, 'I felt that for them to have a peaceful and happy relationship, I should leave. If I lived with her I would have seen her mistakes or she would point out mine and this could have been a source of conflict.'[48] Dikeledi's sensitivities were largely lost on Kas, who could never bring himself to forgive her for what he considered highly irresponsible behaviour. How could a wife encourage her husband to sell the last of his oxen as a ploy to get him rooted in urban soil? Yet that was what Dikeledi had done at Molote. And had she not told her sister, Disebo, not to marry Mmusetsi? The tragic consequences of that advice were now plain for all to see.[49] What Kas and the family needed was a period of reconciliation, and one of the best ways of facilitating the healing process was, he thought, for everybody to go about their business as usual. Much of his own day was taken up with helping a youngster from Chaneng complete his training as a ngaka. Here, as in the case of the herdboy whom he had earlier 'adopted,' Kas became something of a father to the lad in the hope that one day he would help him as a driver or in some other practical way.[50]

But, as with Mposhoza Shongwe, this attempt at establishing a quasi-paternal relationship did not work out as planned, and some family members accused the young man of using the panel-van too much. Pakiso went so far as to tell Kas that he was being cheated by both his drivers, and that he should sell both the tractor and the van. Pakiso had little moral authority when it came to the question of deceit, and she soon learned that 'you were not supposed to say anything.' How could a child of the townships possibly understand the needs of a man without sons in the countryside? And, even if Kas's eyes were not as sharp as they had once been, the changing chant of the birds reminded him that spring was fast approaching. What could they teach him about farming?[51]

No sooner had he taken down a harness from a branch on the tree outside his shack to repair it, however, than he was forced to put it back and attend to yet more painful family business. Death had developed a taste for spring rituals: on 15 September, it claimed Phitise, who passed away peacefully in his sleep. A week later Kas journeyed to Groot Marico to help bury the sibling who, in his passion for farming, was perhaps closest to himself in temperament. An active and deeply devoted member of the African Methodist Episcopal Church until the end of his days, Phitise had left a son, six daughters, thirty-seven grandchildren and fifty-seven great-grandchildren to bring honour to the name of the farming Maines of Schweizer-Reneke.[52]

Kas found it hard to summon up the strength for the new season. For as long as he could recall he and Phitise had enjoyed a friendly rivalry to produce the larger harvest. Had it not been for 'the Boers,' they would have continued this competition to their dying days. But Uitkyk, like Ledig, was a place 'where people who had nowhere else to go went to live,' and their offspring would never again farm in the way of their forefathers.[53] The farm had given way to the factory, and values centring on family and production had been replaced by an ethos of self and consumption, an ethos that slithered about the townships and rural wastelands of apartheid's Bantustans like a puffadder in search of prey.

Kas remained in the countryside as the willing prisoner of an order in which ritual and tradition offered a logic and a comfort of their own. The moment for Motheba's children to wash and purify themselves could no longer be delayed if they wished to avoid the polluting effects of death. To accommodate his late mother's kin, who came largely from around Taung, the ceremony would be held at a western Transvaal venue. Kas eventually decided that the family should be summoned for *ho hlatsuwa* at Sellwane's home, near Delareyville, in late October.

The decision to hold the ritual at his sister's home made for complicated logistical arrangements, and Kas, as oldest surviving son, was expected to provide the beast for slaughtering. Not wishing to be captive to distant livestock speculators, he started looking around Ledig for an animal that could be transported to Delareyville by truck. In the end it was only late on the Friday afternoon before the event that he found a heifer; fortunately Nkamang's part-time position in the 'Tribal Office' meant that the permit necessary to move cattle around the countryside could be obtained at the last moment.[54]

This last-minute scramble was worsened by the late arrival of the truck, for they had to wait for a friend who could only use his employer's vehicle at the end of the working week. It was well after eight o'clock that night before they set out on the four-hour drive to Gannalaagte—a dry, barren, treeless resettlement camp to which Sellwane had been removed when expelled from Tsetse. They reached Delareyville, about twenty miles north of the Triangle, shortly after midnight. Kas supervised the tethering of the heifer, staggered into the house, and fell asleep on the floor.

The next morning people arrived in dribs and drabs from the rural west, from farming centres like Mofufutso, Khunwana and Madibogo; after lunch they were joined by folk who had come from the urban east, from the Witwatersrand; by nightfall, more than a hundred men and women from across the face of the Transvaal had been shoe-horned into Sellwane's yard. Later that evening the elders gathered in the larger of the two rooms where they were joined for a prayer and the singing of a hymn before retiring to various nooks and crannies for some broken sleep.[55]

On Sunday Kas was up long before the rest. Shortly after 6 a.m. he supervised the slaughtering of the beast. The heifer was slit open and the partially digested grass removed from the first of its stomachs and blended with the contents of the gall bladder and some herbs and water to produce a thick dark-green liquid, *moshwang*. At Kas's request, the animal's head was put to one side to be taken back to Ledig while the carcass was removed and butchered into more manageable portions. Shortly thereafter, a long wooden pole was produced, a few of the men commenced roasting selected cuts of meat, while others propped up the pole as a makeshift seat for those who were about to be 'washed.'

Kas, Sellwane and her son by a first marriage, Michael Abrams, stripped down to their undergarments, re-entered the yard, and took up positions on the pole. Two nephews, Kgofu and Hwai, both practising ngakas, appeared with bowls of moshwang, which they poured on the heads of the three on the pole and rubbed it into their hair; more moshwang was used to wash their feet. The ngakas then produced a bottle containing a mixture of animal fat and herbs, which they rubbed into their torsos. The three family representatives were then taken aside and offered smoked liver; this completed the ritual and signalled everyone present to partake of the meat, sour porridge, samp, vegetables and sorghum beer that the women had prepared.

On the way home Kas thought approvingly about these rituals, which not only propitiated the ancestors and protected the children, but drew together those who followed, and heightened the family's kinship and solidarity in a world intent on dissolving their bonds. It was disappointing that his younger brother, Sebubudi, had not met his transport at the appointed time and place in Bloemhof and had therefore forfeited the chance to honour their mother. And, it would have been nice if Bodule and Mmusetsi had been there to join the other grandchildren at Gannalaagte, but the one was dead and the other sat paralysed on a distant farm. Such distancing from the ancestors and the family could possibly account for his son's misfortune: perhaps Mmusetsi had failed to honour them at a critical juncture in his life. Kas resolved to find out when next he saw his son and daughter-in-law.[56]

The rains that spring were late and sparse. They did nothing to lift the malaise that seemed to have settled upon him from the Pilanesberg. At times, he would rage inwardly about Mosala's drunkenness and incompetence; at other moments, he doubted his own commitment to grain farming in what was, after all, his eighty-ninth year. He made a few half-hearted attempts: in mid-November he gave Hika R45 to buy some diesel at Boshoek and then sent Motaung out on the tractor to work a few of the arid fields around Ledig. Of his panel-van, and the schemes to fertilise the fields, there was no sign. Maybe Pakiso had been right and the drivers were cheating him, but it was so hot and dry that even if they were defrauding him, they could not have got much for their efforts.[57]

By January 1983, Kas had given up hope of getting any serious return from the few sharecropping contracts he and Motaung had entered into with the alienated plot holders whose hearts were on the farm but whose heads were in the factories. As had happened the year before, the tractor would have to earn its keep later in the season when he could hire it out to drive the threshing machines of those lucky enough to have something to harvest. Instead, he devoted himself to cobbling and herbalism. But, while his fame as ngaka continued to attract many clients, the deterioration in his eyesight made shoe repair difficult. His eyes watered continuously and eventually became so troublesome that he even consented to visit the clinic at Saulspoort.[58]

Predictably, the doctors at the Moruleng hospital gave him no satisfaction, and he went home reconciled to the prospect of being progressively more reliant on hearing and feeling than seeing. He sensed the darkness closing in on him with no self-pity, accepting it as simply one more burden that came with age, like no longer being able to urinate freely or efficiently; he knew that others, like his son, were shouldering greater loads. As if to remind him of this, only a few weeks later he received a message that Mmusetsi wanted to visit his father and see those living at Ledig.

But Slurry was a long way off, and it would take time to find someone patient and understanding enough to transport a feeble-minded, incontinent man over such a distance. In the end the task fell to Mokentsa, by then a genial giant of a man who had recently qualified as a minister in the Dutch Reformed Church and who, besides living in the Groot Marico district, not far from Gert Pretorius's farm at Vlakpan, owned a car.[59]

By the second week in April Mmusetsi had been installed in the shack on the hill, and for the first time in his life was enjoying untrammelled emotional access to his father. Even then he felt vulnerable, and he soon asked Mokentsa to fetch his wife and child from Vlakpan—they duly joined him at Ledig. He and Kas spent their days seated in the winter sunshine outside the shack talking, consciously and unconsciously rediscovering the personal strengths and weaknesses that had kept them apart for the better part of a lifetime. They would never enjoy the complementarity that Kas and Bodule had once had, but they drew comfort from their new closeness.

Throughout these seemingly interminable discussions with his slow-witted son, Kas's razor-keen intellect never once deserted him; he was constantly probing, pushing and searching for that one elusive clue which, he believed, would account for Mmusetsi's defences having been breached by the forces of evil. Eventually he found an explanation that both satisfied him and exonerated his son. During his marriage to Ntshohlo, Mmusetsi had apparently fathered two children who had died in infancy. In neither instance had the mother arranged for the parents to undergo the appropriate purification rituals after the burial. Mmusetsi's vulnerability to misfortune had less to do with malign BaTswana herbalists and wild horses than with his wife's failure on this score.[60]

This belated discovery did not assuage Kas's frustration at being unable to do anything about his child's suffering. As a father he longed to intervene, but as a ngaka he thought the 'illness' had progressed too far for it to be treated successfully. There were also ethical questions about dealing with members of one's own family, which troubled Kas greatly. To whom did one turn? The hospital at Saulspoort had always failed miserably and the one at Rustenburg was not much better. Only when Nkadimeng visited him in April did Kas think of getting Mmusetsi seen to in Johannesburg.[61]

Nkadimeng, who lived in Soweto and knew his way around the local health facilities, believed it was not an insurmountable problem. But Mmusetsi was not certain about leaving his family, communications between Johannesburg and Ledig were difficult, and, as always, it took time to penetrate hospital bureaucracies. Mmusetsi's health was deteriorating rapidly, so Nkadimeng took them to a friend's home in Soweto; there Mmusetsi, too agitated and embarrassed to stay inside the house, spent a midwinter's night seated on a drum in the garage. The following morning he was admitted to the Johannesburg General Hospital.

Kas went back to Ledig and, without Mmusetsi to take care of, tried to attend to farming even though his eyes were troubling him more than usual. As he feared, the arrangement with Motaung brought precisely no grain, so he negotiated a new contract with a man from Moruleng named Anthony Makiti. Makiti was willing to take over the tractor as well as the cost of its maintenance in exchange for an annual cash rental of R80 and a modest percentage of the harvested grain. It was hardly a lucrative deal, but it freed Kas from organising a driver, arranging for fuel supplies, negotiating individual contracts and ensuring that the vehicle was properly maintained. He had contracted out his management problems, always a risky business, but could see no other way out of his predicament.[62]

Kas continued to see patients. Those who came from far afield, such as the car full of middle-aged men and women from Lichtenburg who consulted him during the last week of May 1983, suffered from familiar ailments, and it was easy to conjure up the appropriate herbal remedies. But those who came from the Witwatersrand townships or the sprawling labour reserves in the countryside had distinctive problems that showed the sordidness and violence they were forced to endure and were difficult to treat. The adults were preoccupied with their bowel movements, the condition of their blood, and problems of fertility; the youngsters wanted charms to protect them against 'bullets' or 'stabbing,' or sought strange herbs that would ensure them jobs at Sun City.[63]

When Nkadimeng appeared at the end of May and Kas again complained about problems with his eyes, it was suggested that they combine a visit to Mmusetsi with an appointment at the St. John's Eye Hospital in Soweto. At the clinic Kas was given familiar advice about needing surgery to remove the cataracts, but he could see no point in a man of his age undergoing an operation, and it made him think that perhaps the doctors in Johannesburg were not all that much better than those at Moruleng. His reservations about the efficacy of modern medicine only increased when he was taken to Mmusetsi. He was shocked: his son had lost the will to live.[64]

Mmusetsi lingered for six weeks. In mid-July he was transferred to a second hospital for a further round of tests; on his return to the General Hospital, he contracted pneumonia and died on 3 August. Young Nkadimeng had to tell the family of his death; a few days later Kas arranged for Nthakwana, Dikeledi and a man who owned a truck to go to Johannesburg and collect the corpse. Friends contributed to the cost of the coffin and the transport, while the ministers from the Nederduitse Gereformeerde Kerk in Afrika at Saulspoort were equally obliging. On 13 August Mmusetsi was laid to rest in the cemetery at Ledig where Kas had once wanted Bodule buried. Those who had been the closest companions in life were sometimes most distant in death, while those who had been remote in life were sometimes brought closer together in death. The Lord had a logic of his own.[65]

Kas found it harder than ever to get back to farming. A month after the funeral he purchased a cow for R140 but spring seemed to pass him by. Cobbling and herbalism brought in some cash, but it did not linger in his purse for more than a day or two before finding its way into the open-mouthed till in Mahlangu's Thanda Bantu Store. Tea, coffee, sugar, salt and washing powder—items that Kas had very much taken for granted when Leetwane was alive—were now necessities he had to purchase for himself.[66]

Even small amounts of money were harder and harder to come by. Kas's daughters often complained that all the years they had been made to spend in the fields on white farms had robbed them of the chance of an education and decent employment. They were of some help to him but it was limited. Morwesi still spent on liquor most of the money she earned as an occasional women's help and child minder, but she sometimes managed to bring home a bite or two to eat. Nkamang, who earned a modest but regular wage for part-time work in the 'Tribal Office,' was more of a help, but she was pregnant, and if she approved of the child's father, she might eventually marry and move out of the house. Nthakwana, the most highly motivated and best organised of his children, ran several informal enterprises, including an illegal shebeen, and could always be relied upon for a meal or a little cash.[67]

This increasing dependence on his children at the age of ninety irritated Kas, and, in the late summer of 1984, he started thinking about new enterprises to ease his reliance on cobbling, herbalism or a state pension. He was taken with the idea of hiring unskilled labour to manufacture the coarse cement bricks that found such a ready market amongst the hundreds of homeless refugees who poured into Ledig from town and countryside each year. So he explored ways of raising capital from a friend to purchase a front-end loader that could be used to move sand and cement. Nor were his dreams of a more integrated domestic life at an end: when Tseleng came to visit him shortly after Mmusetsi's funeral, he suggested that they set up home together. Daughters, no matter how well-meaning, were no substitute for a good wife. Leetwane, he now readily acknowledged, had been a very good wife.[68]

As the chill of autumn slid in and shivered its way along the floor of Kas's shack, the new vision taking shape beneath the tree in the front yard blossomed when Makiti arrived to tell Kas that his share of a favourable maize harvest at Moruleng amounted to sixty bags. This was the first good news to come in from the fields in three years, and it lifted the old man's spirits. It vindicated his decision to hand over the tractor to Makiti, and moreover, the harvest would put a deposit for the brick-making scheme within reach. But then, as he stood up and prepared to move on, life tripped him up.[69]

On the Tuesday following Makiti's visit he gave Nkamang, by then five months' pregnant, some money and asked her to go to Mahlangu's store for paraffin. On her way back she became dizzy and was about to put down the bottle of paraffin to recover her breath when she fell forward, striking her head against a rock. Passers-by rushed to her aid and, when she had recovered a little, took her to the nearby BaKubung Clinic. There she was examined by the nurse on duty, who advised her to rest for a day and then go to the Moruleng hospital for a thorough examination. But it was a chance that Nkamang was to be denied. She passed away during the evening of 5 June 1984.[70]

The death of Nkamang, at the age of forty-three, caught Kas by surprise. Sickly as a child, the girl had been shielded from physical labour by Leetwane and—a concession from her father—had been permitted more time for school than any of the other children. Apart from her attachment to the Zion Apostolic Church of South Africa, which Kas held in contempt, she was a likeable young woman whom Kas knew had been very good to her parents; the most domesticated and least troublesome of all their daughters. Her death now robbed him of his most reliable help around the house, and he would be thrown on the tender mercies of Morwesi.[71]

It took the better part of a week for Kas to raise the money for his share of the mortuary fees, the cost of a coffin, transport and the three sheep that the members of the Z.A.C.S.A. felt would meet the earthly needs of those attending the wake. Late on the afternoon of 15 June the family members based on the Rand started arriving at the shack and were joined during the evening by more kin from the Triangle. At about ten o'clock Bishop Matlhatane and the Reverend James Dube from the Z.A.C.S.A. appeared along with a few followers dressed in the distinctive red, white, blue and green uniforms of their church. The arrival of the Zionists was the cue for the Maine women to serve everyone with food and coffee, after which the two clergymen led the congregation in the hymn singing, interrupted occasionally from the floor by short speeches in honour of the deceased woman.[72]

The Zionists sustained this punishing ritual throughout the night, breaking it only once. Shortly after midnight a blast on an ox-horn summoned all able-bodied men attending the gathering to the cemetery for duty as *diphiri* (hyenas), or gravediggers. The kgotla, in a ruling that was designed as much to prevent mischief as it was to uphold 'tradition' in an ever more deracinated and unruly community, had ordered that all graves were to be dug only a few hours before burial. So some of the younger men, including Mosala's son, Morapedi, left the warmth of the wake and made their way to the cemetery, but it was a cold night and the diggers were few. The funeral procession had been scheduled to leave the shack at 7 a.m., but it was delayed when, shortly after dawn, it was announced that the grave was not yet ready.

Almost two hours later the van bearing the coffin eventually inched its way out of the yard, but even then the mourners were greeted by truly extraordinary scenes when they entered the cemetery.[73] The diphiri had long since abandoned their shovels and were using various sticks and whips to thrash Morapedi, who had absconded under the cover of darkness, gone to a neighbouring shebeen and returned drunk. Only when they caught sight of the priests did the diphiri stop, by which time Morapedi was up, hurling abuse at the spectators and threatening to kill his assailants.

This behaviour on the part of a grandson embarrassed Kas, but there was worse to come. Once the diphiri had dug the grave to the satisfaction of the elders, the rites were read and the coffin lowered into the grave. But when it came to the scattering of soil, Reverend Dube asked only his Zionist followers to partake. Kas shuffled forward and intervened, telling the minister that custom dictated that his family, too, participate, but the people in charge ignored his objection and went on with the service. Quiet embarrassment gave way to rancorous silence. Once the coffin was covered the Reverend Dube invited someone from Nkamang's burial society to say a few words in honour of the deceased, but no one stepped forward. Looking around, he asked anyone who had worked with Nkamang in the Tribal Office to say their piece, but this, too, was met by silence. Undeterred, Reverend Dube blundered on, asking a representative of the BaKubung chief to respond. Only at the very end did he turn to the Maines, but the family, sensing Kas's unhappiness and unwilling to betray him, locked themselves into the silence of the grave, which they contemplated with downcast eyes.

The mourners gradually dispersed, moving back up the hill to the house. The funeral had been a shambles. Back at the shack they waited their turn to wash their hands in a small enamel-coated basin before partaking of the main meal of the day. Kas disappeared for a moment, and a few minutes later a whispered message was passed along the family grapevine: after the visitors had left there would be a meeting of the family addressed by Kas's favourite nephew, Kgofu. This development, anticipated by the elders and feared by the juniors, sent a surge of energy through the network and reactivated the tension that had been so evident earlier in the day.

Ngaka Kgofu, a bearded giant, rose to tell them that Kas wanted it known that the rites performed on the grave of one newly deceased, ho tea lejoe, would be performed in a month's time. No one objected, so it was agreed that the family would next meet at the shack on 15 July. Kgofu then slowly worked his way round to telling them, in understated and euphemistic terms, that the patriarch was exceedingly unhappy not only with the morning's events but with the way that the family in general, and the young in particular, were ignoring 'his advice about the customs and traditions of the people.' Kas, irritated by what he perceived as an excess of diplomacy, interrupted. Kgofu should call a spade a spade, he said. Morapedi, recognising his cue amidst all the static, rose and tried inconspicuously to leave. Kgofu grew more explicit, specifying the Reverend Dube and his church's many shortcomings, pointing out the deficiencies in Morapedi's behaviour, and criticising the way in which the younger generation were giving their children names without first referring the matter to the elders for advice and guidance. This last breach of etiquette and custom was especially serious, and Kas's misgivings were warmly endorsed by all the fathers present. Consensus having been reached on this convenient but tangential issue, the meeting slipped slowly into more relaxed chatter; then people who had come from far away started collecting their possessions for the journey home.

The scars left by these events and the loss of Nkamang pained Kas, and there were more upsetting developments to follow. A few weeks after the funeral Morapedi appeared at the shack and, trying to repair the damage to their relationship, entrusted Kas with the safe-keeping of R300. Unfortunately, this transaction was witnessed by the older brother, Tsetse, who later sneaked into the shack and stole the money. By the time Kas discovered the theft, Tsetse had been on a week-long binge and nothing was left of his younger brother's earnings. Such things, unheard-of amongst respectable sharecropping folk in the Triangle, underscored Kas's gloomy observations about the ravages of shebeen liquor and the implosion of morals in the Maine family.[74]

A slightly more hopeful note was struck when his nephew, Hwai, came across from Mothlabe to present his apologies for being unable to attend the ho tea lejoe. It was nice of the young man to have taken the trouble, but, frankly, he need not have bothered. When 15 July came around the local livestock farmers broke their promises and there was not a beast in sight. With no animal to slaughter there could be no proper ritual; instead, the family arrived for yet another Zionist-dominated affair. Chronic disorganisation, fast becoming one of the hallmarks of life in the 'homeland' slums, left Kas deeply disappointed and the anger of disappointment turned into depression.[75]

> What has happened is that I have been overtaken by tragedy. I have been overtaken by a series of tragedies in rapid succession, and they have weakened me. A mother died, a child died and another followed; and all of these deaths have involved a considerable outlay in cash . . . each funeral costing R400, R300 or R500. All of these things break one's spirit

> and life drains away. This is what has depressed me. Even now, as I sit here talking to you, I constantly think through these things, brooding.[76]

His physical vigour, too, was draining away. 'Where will I get the strength from to continue?' But his intellectual powers were undiminished, and an alert mind patrolled its bodily limits like a caged jackal. '*My mind is not old, it's only the body that is old.*' This zestful part of Kas's spirit never stopped probing and questioning, always looking for a way out of his predicament even if it meant thinking the unthinkable and betraying Leetwane. On more than one occasion he contemplated building a house in the hope of attracting a wife to help shore up his collapsing domestic life.[77]

Robbed of Bodule and cheated of Nkamang, Kas directed his anger at his surviving children, a torrential anger that sometimes swept all before it in an ill-directed outpouring of resentment. 'My children are stupid, they are bad, they don't want to help me. They forget that they have left me in the wilderness and they prefer to indulge in alcohol.' By the end of winter 1984 his resentment had grown so acute that it reactivated deep-seated sharecropping reflexes, and Kas gave careful thought to abandoning Ledig and his family, and going farming full-time with Makiti.[78]

> I want to leave this place, that's why I don't build a house. I want to go to the village behind the mountains, it's better there. It has been suggested that I should trek from this place and move to Moruleng where my tractor is. I could live on the farm there and look after everything. I could use the tractor, buy cattle and keep them with me. It means that I would have cattle as well as cultivate crops.

For Kas this was an appealing prospect. Behind the mountain, there at Moruleng, he would reconstruct the universe he had grown up with, a world where agriculture and the social order were reciprocally bound together.

> There are many unemployed people of good character who would willingly work for me, they would labour in my fields. I would feed them, compensate them for working in the fields and even buy clothing for them. They would work for me in good faith because I was kind to them. If they had families I would feed them and pass on gifts to them. I would become a foreman who would give instructions to which they would respond positively and in good spirit. They would be grateful because I would not be harsh with them. They would be unlike those who complain about the fact that they are abused and over-worked, those who have become the victims of excessive drinking.

But only reverie could turn the Pilanesberg into paradise, and as spring battled to make itself known, Kas was rudely awakened to a world that demanded that he come to grips with the realities of Ledig and patriarchal responsibility. He told a visitor that if Makiti was going to go on using the tractor, he would have to assemble a team of donkeys to plough the field on the Elands. 'I wish that I could get out and buy some donkeys, but I am having trouble walking. If only I were capable of walking, I would long since have had the donkeys ploughing.' While the rains stayed away he could live with this frustration, but when a few promising showers fell in the first week of November, the smell of the moisture on the parched soil outside the shack drove him to distraction. When Nkadimeng happened to call on him, Kas tried to persuade him to take him to Hwai's place at Mothlabe, where there were always good donkeys for sale. But Nkadimeng was in a hurry to get back to Johannesburg and suggested instead that they go to Nthakwana's for lunch. During the meal Kas returned to the attack, this time looking to his daughter to help get him to the donkeys at Mothlabe. Nthakwana discouraged him. She told him that she could see no reason for his buying donkeys since all of his children were old enough to support themselves. This suggestion, which made light of his responsibilities as a father, angered Kas. 'He told her that he was going to plough for them.' But Nthakwana, who could be equally forthright, would have none of it. They had no need of his crops, she replied. If they needed food, they could always get it at the shops.

The shops! A son of the plough buying maize from the shops? What men had to put up with from their children! Imagine it, going to a shop to buy maize! Kas went home still muttering, and was not to be deflected.[79] He made enquiries from neighbours who were more sympathetic than daughters in such matters, waited for his pension to be paid, and then, on 30 November, persuaded David Maseabi to sell him three donkeys. But nature, like the donkey, is a cussed thing and no sooner had he acquired the animals than the rains stayed away. As the old year gave way to the new, Kas realised he probably couldn't get down into the fields that summer, and despite his virtual blindness, he reverted to cobbling and the herbal practice. Toward the end of that summer, Anthony Makiti called in to pay him the fee for the hire of his tractor and confessed that he, too, had been unable to extract a harvest from the fabled fields behind the mountain.[80]

The failure at Moruleng did not trouble Kas unduly. He knew, perhaps better than Makiti ever would, that farming was a hard, even treacherous business. He was already looking ahead, concerned that the ceremony to distribute Mmusetsi's clothes should proceed without the problems that had beset recent family occasions. It would be attended, he decided, only by the immediate family and Bodule's children.

On 9 July, he gave Mosala R60 and told him to buy a sheep for slaughtering. Mosala returned later with an animal that had cost only R45; it pleased Kas no end that Mosala retained a good eye when it came to small livestock purchases. Perhaps there was time yet for the boy to come to his senses?

That afternoon, while the family were preparing for the ceremony, there were several interruptions, including one by a woman and her teenage son who had been told to obtain the protection of a ngaka. Kas had become all too familiar with this sort of township problem. The boy had apparently 'lost' no fewer than four pairs of trousers and a jacket to the local dry-cleaning agency over the past few months; ill fortune on this scale pointed to the work of sinister forces, and the mother pleaded for his help. Although not much moved by such a trivial matter, Kas felt unable to turn them away; and with Mosala hovering in the background, he sat them down, took out the bones, cast them into the sand, read the message and then gave them some advice. The two departed, satisfied.[81]

In the morning Kas was first up, long before dawn, supervising the slaughtering of the sheep, thinking his way through the day. As the first spikes of sunlight pierced the heavens high above the Elands River, he sent Morwesi into the shack to retrieve Mmusetsi's clothes, which were laid out on the ground in front of the house. Then, in an intentionally challenging gesture, he called on Mosala to assume responsibility for ridding the clothing of lingering evil spirits. Mosala sprinkled chyme over the ragged collection of shirts, trousers and jackets, while one of Bodule's youngsters, Ramosele, used a stick to beat the apparel, all the while chanting: *'Ha di ye,'* 'Let them go.' The clothing was distributed among the kin, Nthakwana served everyone with liver, while some of the men grilled the rest of the meat for the midday meal.[82]

Kas was relieved when the day drew to a satisfactory conclusion. Mmusetsi had been a troubled soul, and it would be reassuring for those who came after him to know that ho ntsha diakobo tsa mofu had been conducted in accordance with the customs of the people, and with no untoward manifestations or unseemly behaviour. He had done his duty and could now look forward to the new season.

*

In keeping with his recent more thoughtful mood, Kas gave much time to searching for the origins of his current predicament. He acknowledged that the failure to invest in suitable property when he was much younger had been a major mistake.

> As my mind dwells on these things I realise that I am old now, but when I was younger I was in a position to do things for myself. I now regret

> the stupid mistakes that flowed from lost opportunities. I should have bought land while I was still working. Had I done so, I would be secure in my retirement.[83]

The initial error had been compounded by the ill-considered move to Ledig, so unsuited to arable farming and therefore unsuited to a progressive grain farmer like himself.[84] The corollary was obvious: the time had come to explore new horizons.

> As you know, everyone, black or white, has a gift. Some men are gifted livestock farmers. In my case it's tilling the land. My survival depends on that. If I had access to land, as I had on the white farms, I could get a good harvest and could purchase a plot of my own. A plot costs anything between eight hundred and a thousand pounds. This is the reason for my wanting to grow beans. Today, a bag of beans fetches around twenty-five pounds. Fifty bags of beans are worth about a thousand pounds. There is a suitable plot for sale near Kameel for which the owner wants eight hundred and eighty pounds. At the moment I don't have the money but, should I get a good harvest, it could be mine. If you plant beans in October, you could be harvesting them by February.[85]

'If you plant beans in October, you could be harvesting them by February.' Even at ninety-one there was always just one more season, just one more plan, just one more way of cheating adversity. And the way to ensure seed in the spring was to work your way through the winter. Just one more client to help with herbs, just one more pair of shoes to be repaired. Small sums set aside for beans or plots were tucked into the green leather purse, only to be taken out again for a necessary visit to Mahlangu's store or handed over to the Bophuthatswana official who arrived to demand wheel tax for his water cart or grazing fees for the three donkeys tethered in the corner of the yard.[86]

August dragged by as it always did, an inherently indecisive month. The days seemed to hover about pointlessly, like giant cumulus clouds waiting for the wind to stir and give direction to their lives. By early September, the spiked acacia thorns were carefully rewrapped in the bright green of a bushveld spring. Visitors from the east spoke of rain on the highveld, and deep within the recesses of the shack, there were signs of movement—slight and faltering, it's true, but movement nonetheless. If you plant beans in October, you could be harvesting them by February. A day or two into September, and Kas took down the harnesses for repair, but a mild bout of diarrhoea slowed down his progress even more than did his failing eyes. He persevered for a while, but then dug around amongst his herbs for a trusted remedy. The diarrhoea was persistent, and it sapped his strength. Rather uncharacteristically, he felt the need to go to bed.

Morwesi and Nthakwana fussed about him and suggested that he see a doctor, but there seemed little point in going all the way to Boshoek for so minor a complaint. In summer Ledig was filled with small babies and old folk who suffered from diarrhoea: it was part of life in the homelands, one of the gifts of the Elands.

Within days his condition had deteriorated so markedly that it gave cause for genuine alarm, and when Kas asked for Thakane, Nthakwana hastened to get a message to Lichtenburg. Thakane and a nephew, Mokawane, came quickly, and they, too, made unsuccessful attempts to persuade him to see the doctor. By now his mind, like his emaciated body, was uncertain as to whether it should slip back into the past or make one last determined effort to push on into the future.

The toxins of age and illness spread through Kas's body until it felt as if they had oozed into every crevice of his limp frame and sucked up his last drop of energy. At sunset on 13 September, he asked Nthakwana to gather the children round him. He was about to die, he told them, and they should each say a prayer for him in the privacy of their homes. But the following morning, when Nkadimeng arrived for one of his visits, Kas the patriarch was less resigned, and when the young man suggested that they go to a doctor, he at last agreed. Nthakwana and Thakane propped him up and helped to dress him, but by the time they had finished he was too weak for a two-hour car journey; he promised Nkadimeng he would try again in the morning. Nkadimeng, distressed, drove to the nearest chemist, some distance away, and bought patent medicines and fresh fruit juice, hoping they would help see the old man through the night. He was confident they would get him to a doctor. There would be no talk of Moruleng; they would go to Rustenburg.[87]

Kas did not disappoint him; he never had. On Sunday morning he allowed them to carry him to the car and put him in the back seat. It took time and effort to find someone willing to see the old man. The Indian doctor agreed that he was suffering from diarrhoea and dehydration but also drew their attention to Kas's irregular heartbeat. He was given an injection, and they were given a prescription. Nkadimeng drove through the deserted streets of Rustenburg until they found an emergency chemist, bought the medicine and then drove back to Ledig. Back in the shack, Nkadimeng explained to the daughters how to administer the medicine, and left for Johannesburg.[88]

For a while the medicine seemed to work. Two days later, however, Nthakwana again took fright at Kas's condition and summoned a nurse from the BaKubung Clinic, who told them he would have to be taken to Saulspoort in the morning. But the mind was not old, only the body, and, while everyone was clucking about, Kas busied himself putting the final strokes to the great plan. He had been thinking about it for weeks. Anyone who went to Moruleng needed a will and that evening he got Nthakwana to assemble the children for the last patriarchal statement. His work was nearing its end.

Morwesi, Mosala, Nthakwana and Thakane—the offspring of Leetwane and Lebitsa—along with Dikeledi, who had become an integral part of the house of Maine, as well as the kinsmen Hwai and Morosane Mokawane all congregated in Kas's room to hear and bear witness to his wishes. Some came to the shack that night expecting much. The only surviving son of the first house might have the right to his father's machinery, so that he could carry on the work of the family. Others, who thought they were marginal to the patriarch's life, such as the daughter of a second wife, expected little. These issues were never easy to decide on, and a wise man had to find ways to ensure that the weak were not without hope, that the strong were not discouraged, and that the goods acquired during a lifetime of struggle found owners who would keep them productive. It was all a question of balance, a matter of judgment.

Dikeledi and Nthakwana were strong women and relatively well provided for. Dikeledi, who owned a matchbox house in Khutsong, had a small monthly cash income and already occupied the stand in Ledig set aside for her late husband; Kas was confident that she could provide for Bodule's children. Nthakwana and her children had been provided for by the late Matthews Moate; she was exceptionally capable and she owned the stand that had been given to her. Kas was mindful of the unstinting help that she had given to her parents during the closing years of their lives. Somewhere out there in Chaneng was a young man with a panel-van that belonged to the Maines. Maybe she could recover it for herself?

Pulane and Matlakala, the latter a love child by Tseleng, were seemingly well-adjusted young women; they had drifted away from the family since their marriages, and they would have to be taken care of by their husbands.

This left Morwesi, Mosala and Thakane. The first two, children by his first wife, and sources of recent great anguish, physically close but emotionally distant, had given up work and caring for their families and had instead degenerated into drunken idleness. But they were his children, vulnerable souls in a world stalked by evil, and deserving of his protection. Kas had to find a way to give them resources that could not be sold off for alcohol but could give them security while they struggled to find themselves.

Thakane, Lebitsa's only child, was married, like Matlakala, and her husband should therefore provide for her, but it was only right that she, too, be given something substantial. It was important that there be peace between these two branches of the family, and in such matters considerations of gender took second place. So Kas wanted the tractor to go to

Thakane and to the grandson who showed most promise as a family man, the Reverend Mokentsa Teeu, son of Bodule.

Morwesi and Mosala's disappointment was palpable. What was left? Had they been forgotten? But Kas had not finished.

There was the stand on which the shack stood. Morwesi, despite the strength of character she had shown as a child, had become a weak woman without a husband. This, and the fact that she had only one child, Pakiso, meant that special provision would have to be made for her: Kas wanted it known that, if she should so choose, she was entitled to live in the house on the hill for the rest of her life. But Morwesi could not be trusted with property, so he had decided that the stand should be left to one of the granddaughters. Who better than Ntholeng? Not only had she done most to help him and Leetwane during the period of their greatest need, but she was an illegitimate child, the daughter of the policeman who had abandoned Nthakwana.

This left Mosala—the last surviving male of his generation in his branch of a great house—son of Kas, grandson of Sekwala, great-grandson of Hwai, great-great-grandson of Lethebe and great-great-great-grandson of Maine. What was there that a father could do for this wayward son of the soil who followed in the tradition of cobblers, farmers, ngakas, stonemasons and thatchers? In a gesture pregnant with ambiguous meaning, Kas willed his son his divination bones, the *diakola*, his hand tools, and his farming implements. Mosala's past as well as his future lay in the palm of his hands. His father had given him everything and nothing. Had it not always been so?[89]

In the morning they took Kas to Moruleng, where he spent nearly ten days struggling to find a way out of the prison that they said cared for people. As the month inched to a close, he hit upon a ploy that had worked once before. He asked to be released for a day or two in order to go to Ledig to collect his pension. Bureaucrats understood this sort of reasoning only too well, and the children brought him home. In the morning he went to collect the pension; when they wanted to take him back to Moruleng, he complained about the heat and asked to be given just one more day at home before returning to the hospital. He had completed his business. Kas died in his shack that afternoon, 25 September 1985. At twilight Mosala brought the donkeys in.[90]

*

Kas Maine's funeral, ten days later, drew together family, distant kin, friends, neighbours, a sharecropper or two, former clients, and a few curious outsiders. At ten o'clock in the evening, the Reverend Selepe of the Nederduitse Gereformeerde Kerk at Saulspoort presided over a vigil attended by more than four hundred people. In keeping with the spirit of the great pragmatist himself, the proceedings harked back to the past as much as they pointed to the future and were as 'modern' as they were 'traditional.' A man of so many parts, known to so many people drawn from so many different walks of life, needed to take leave of his friends and loved ones in different ways. After all, did not a Christian, a MoFokeng and a ngaka speak to men and women of the world in different tongues?

The diphiri were mustered at midnight and returned from their nocturnal labours about three hours later, led by a prancing young ngaka with a leopard-skin hat who was clutching an ox-horn in one hand and a pouch containing diakola in the other. A blast on the horn summoned Reverend Selepe and the men to take up the coffin, leaving the women in the shack behind them, they joined the ngaka in a noisy procession down the hillside. In the cemetery, where the grave was aligned in the distinctive east-west manner that marks the resting place of a true MoFokeng, two men leapt into the grave to receive the coffin handed to them. An elder from the N.G. Church, Mr. Melk, led the hymn singing and read several short passages from the Bible. After the coffin had been covered with soil, three kinsmen were nominated to stay behind and guard the grave until sunrise, when a second service would take place.[91]

Shortly after 7 a.m. the evangelist Letheo began an orthodox N.G. Church service with a prayer before handing over proceedings to the Reverend Mokentsa Teeu. Mokentsa made several general announcements before inviting members of the family to step forward and speak in honour of his grandfather. A tall white man, one of Kas's friends, slipped into the back of the gathering, and he, too, spoke. Then there were more prayers and hymns, before a second and larger procession wound down the hill to the cemetery, where several wreaths were placed on the grave. The service concluded with yet more hymns and prayers. Most of the crowd then went back up the hill to the shack.

But not everybody left. The tall white man stayed behind and waited until the cemetery emptied. A slight breeze had sprung up; it was an October day. He reached down, picked up a handful of earth and walked across to the grave. With the wind gently spraying the grains of sand that trickled from his hand, he looked down, made a silent promise and then turned to leave.

On the way back to the Witwatersrand he noticed the first heavy clouds of the new season rolling in from the south, and in a small field beside the Tlhabane road, there was a man planting beans.

# Epilogue

In Giuseppe di Lampedusa's masterpiece *The Leopard*, a follower of the Sicilian court, Don Pietro Russo, warns his Prince of imminent revolutionary changes to the old order: 'Your Excellency knows that we can stand no more; searches, questions, nagging about every little thing, a police-spy at every corner of the street; an honest man can't even look after his own affairs. Afterwards, though, we'll have liberty, security, lighter taxes, ease, trade. Everything will be better; the only ones to lose will be the priests.' The Prince and his nobles give the matter some thought and come to the conclusion that 'Unless we ourselves take a hand now, they'll foist a republic on us.' 'If we want things to stay as they are,' they reason, 'things will have to change.'

On 2 February 1990, State President F. W. de Klerk, opening parliament, outlined a radical new political dispensation for South Africa. Nelson Rolihlahla Mandela, leader of the banned African National Congress and a prominent lawyer who had worked in opposition to the regime in the 1950s and 1960s, having served twenty-seven years of a life sentence was released from prison, and long-standing bans on political parties opposed to white minority rule and apartheid policies were lifted. The country was prepared for the emergence of a non-racial democracy.

In the summer of 1992, the BaKubung-ba-Ratheo, under the leadership of township-dwelling Monnakgotla elders, approached three Witwatersrand-based liberal pressure groups—the South African Council of Churches, the Black Sash and the Transvaal Rural Action Committee—to help them get lawyers to submit a claim to the Commission on Land Affairs to regain title to their former property around Boons. A year later, on 1 December 1993, most of the BaKubung still resident at Ledig learned, as a Johannesburg newspaper reported, that 'the government had decided to give them back their Western Transvaal farm, twenty-five years after their forced removal.' By then the BaKubung, manifesting old fissiparous tendencies, had split into two factions—those eager to return to the sites of Mathopestat, Molote and Modubung, and those wanting to stay on in the Pilanesberg. In March 1994, only weeks before South Africa's first non-racial election, the historic vote that led to the installation of a 'Government of National Unity,' a small group of BaKubung were re-established on the farm they had once owned.

In August 1993, the BaKgatla-ba-Kgafela, who thirteen years earlier had been removed from the farm within the crater from which they had sold livestock to the people of Ledig, at last started receiving financial compensation from the Bophuthatswana Parks Board for land incorporated into the Pilanesberg National Park. They then became interested in buying a large farm adjacent to the area from which they had been removed. By July 1994, and with the assistance of the Mazda car company and Bophuthatswana's Agricor division, they had turned the property into a small game reserve which they called Lebatlane. 'The project will generate money for the BaKgatla through hunting, ecotourism, ethnotourism (visitors will be accommodated in traditional villages) and the sale of craftwork,' noted a Johannesburg newspaper. It would also 'be the clan's contribution to the government's Reconstruction and Development Programme.'

Kas Maine's children, their roots prised clear of the soil, chose to stay on in the Pilanesberg rather than return to Boons. Mosala Maine along with his wife and children lives in a tumble-down mud shack in Ledig on the stand his father bought for him, eking out a living from occasional construction work in and around the township. His movements are unpredictable and he is often away from home for days and weeks at a time. Across the way, Nthakwana Maine, now a pensioner and a reasonably successful small-scale entrepreneur, lives in a well-constructed house supplied with electricity and draws her water from a borehole on the property. Her impeccably clean and neat lounge boasts a television set, a solid dining table, several small glass vases, along with a huge colour portrait of Nelson Mandela and a smaller black-and-white photograph of her father. Morwesi, after another bout of self-destructive behaviour during a stay with Pakiso in Stilfontein has, of late, been living with Thakane in the far western Transvaal. Her health is said to be 'much improved.' The Maine siblings at Ledig occasionally see Matlakala, who continues her work as a faith healer.

Of Kas and Leetwane Maine's tin shack on the conical hill at Ledig there is now no outward sign. The yard has long since been cleared of rusting ploughs, scraps of metal, bits of fencing and broken-down carts, and where the shack once stood, there is a modest two-roomed breeze-block cottage which Ntholeng—with Mosala's help—is gradually adding to on her occasional visits to Ledig. It is hard to open and shut the back door, its metal frame having been badly bent during careless building operations. At present the cottage lacks a ceiling, and the roof, fashioned from the very corrugated-iron sheets that once framed Kas and Leetwane's home, has rusted through. As the sun pours in through scores of tiny holes, it leaves an irregular mottled pattern on the grey concrete floor. Shades amongst the shades.

Those who knew Kas well look at the tree under which he sat working and can almost sense his presence—pondering the real and imagined changes around Ledig, and coming up with a balanced formulation weighted to perfection with the wisdom of a peasant who had survived several different political dispensations.

Back in Sicily, the Prince reflected on the outcome of the revolution Don Pietro Russo had warned him of and concluded, 'I belong to an unlucky generation, swung between the old world and the new, and I find myself ill at ease in both.'

# A GUIDE TO PRONUNCIATION OF SOME OF THE NAMES IN THIS BOOK

| | | |
|---|---|---|
| Bae'fesi | bah-eh-feh-see | [bae'fesi] |
| Bafo'ke'ng | bah-foo-keng | [bafʊ'ke'ŋ] |
| Bahu'rutshe | bah-hu-ru-tsee | [bahu'rutshi] |
| Baku'bu'ng | bah-ku-bung | [baku'bu'ŋ] |
| Bathla'pi'ng | bah-tlah-ping | [batłha'pi'ŋ] |
| Bo'dule | boo-du-leh | [bʊ'dulɛ] |
| Bo'dumo | boo-du-moh | [bʊ'dumɔ] |
| Bo'hadi | boo-hah-di | [bʊ'hadi] |
| Bophutha'tswana | boo-pu-tah-tswah-nah | [bʊphutha'tswana] |
| Difa'qane | di-fah-qah-nee | [difa'!anɪ] |
| Dike'ledi | di-keh-leh-dee | [dike'ledi] |
| Di'phiri | dee-pee-ree | [di'phiri] |
| 'Disu | di-soo | ['disu] |
| 'Dladla | dlah-dlah | ['ɮaɮa] |
| Ho tea 'lejoe | hoo-teh-ah lee-jwe | [hʊtɛa 'lɪdʒwɛ] |
| 'Hwai | hwah-ee | ['hwai] |
| 'Kadi | kah-dee | ['kadi] |
| Kasia'nyane | kah-see-ah-nyah-nee | [kasia'ɲanɪ] |
| Keau'bona | kee-ah-oo-boh-nah | [kɪaʊ'bɔna] |
| 'Kgofu | kgoh-foo | ['kxofu] |
| 'Kgotla | kgoo-tlah | ['kxʊtła] |
| Khu'nwana | koo-nwah-nah | ['khunwana] |
| 'Kolo | koh-loh | ['kɔlɔ] |
| Ko'pano | koh-pah-noh | [kɔ'panɔ] |
| Le'bitsa | lee-bee-tsah | [lɪ'bitsa] |
| Le'bone | lee-boh-neh | [lɪ'bɔnɛ] |
| Lee'twane | lee-eh-twah-nee | [lɪɛ'twanɪ] |
| Lephol'letse | lee-phool-leh-tsee | [lɪphʊ'letsɪ] |
| Le'rata | lee-rah-tah | [lɪ'rata] |
| Le'rema | lee-ree-mah | [lɪ'rɪma] |
| Le'thebe | lee-teh-beh | [lɪ'thɛbɛ] |
| Ma'chaya | mah-chah-jah | [ma'tʃhaja] |
| Madi'bogo | mah-dee-boh-goh | [madi'bɔxɔ] |

| | | |
|---|---|---|
| Magodi'tsane | mah-goh-dee-tsah-nee | [m'axɔdi'tsanɪ] |
| Makao'te'ng | mah-kah-oo-teng | [makaʊ'te'ŋ] |
| Male'fane | mah-lee-fah-nee | [malɪ'fanɪ] |
| Male'shwane | mah-lee-shwah-nee | [malɪ'ʃwanɪ] |
| Ma'luti | mah-loo-tee | [ma'lʊti] |
| Maqe'qeza | mah-qeh-qeh-zah | [ma!ɛ'!ɛza] |
| Matla'kala | mah-tlah-kah-lah | [matɬa'kala] |
| Matshi'diso | mah-tsee-dee-soh | [matshi'disɔ] |
| Mequa'tli'ng | mee-qoo-ah-tleeng | [mɪ!ua'tɬi'ŋ] |
| 'Mmer'eki | m-meh-reh-kee | ['mer'eki] |
| 'Mmu'setsi | m-mu-seh-tsee | ['mu'setsi] |
| 'M'ngome'zulu | m-ngoh-meh-zoo-loo | ['m'ŋgɔme'zulu] |
| Moka'wane | moo-kah-wah-nee | [mʊka'wanɪ] |
| Mo'kentsa | moo-kehn-tsah | [mʊ'kentsa] |
| Molo'hlanyi | moo-loo-hlah-nyee | [mʊlʊ'ɬaɲɪ] |
| Mo'lote | moo-loo-tee | [mʊ'lʊtɪ] |
| Mo'ate | moo-ah-tee | [mʊ'atɪ] |
| Modia'kgotla | moo-dee-ah-kgoo-tlah | [mʊdia'kxʊtɬa] |
| Modi'ehi | moo-dee-eh-hee | [mʊdi'ehi] |
| Modu'bu'ng | moo-doo-boong | [mʊdu'bu'ŋ] |
| Mo'gwase | moo-gwah-see | [mʊ'xwasɪ] |
| Moi'pe'ng | moo-ee-peng | [mʊ'ipe'ŋ] |
| Mongwa'ketse | moo-ngwah-keh-tsi | [mʊŋwa'ketsɪ] |
| Mo'nna'kgotla | moon-nah-kgoo-tlah | [mʊ'na'kxʊtɬa] |
| Mora'pedi | moo-rah-peh-dee | [mʊra'pedi] |
| Moru'le'ng | moo-roo-lehng | [mʊru'le'ŋ] |
| Mo'rwesi | moo-rweh-see | [mʊ'rwesi] |
| Mo'sebi | moo-seh-bee | [mʊ'sebi] |
| Mose'lantja | moo-see-lahn-chah | [mʊsɪ'lantʃa] |
| Mo'shodi | moo-shoh-dee | [mʊ'ʃodi] |
| Mo'sotho | moo-soo-too | [mʊ'sʊthʊ] |
| Mo'theba | moo-tee-bah | [mʊ'thɪba] |
| Motlago'ma'ng | moo-tlah-goo-mahng | [mʊtɬaxu'ma'ŋ] |
| Motlaka'ma'ng | moo-tlak-kah-mahng | [mʊtɬaka'ma'ŋ] |
| Motshi'disi | moo-tsee-dee-see | [mʊtshi'disi] |
| Motsue'nyane | moo-tsweh-nyah-nee | [mʊtswɛ'ɲanɪ] |
| Mpo'shoza | m-poh-shoh-sah | [mpɔ'ʃɔza] |
| Nde'bele | n-deh-beh-leh | [ndɛ'bɛlɛ] |
| 'Ngaka | ngah-kah | ['ŋaka] |
| Nga'kane | ngah-kah-nee | [ŋa'kanɪ] |
| Nkadi'me'ng | ngkah-dee-mehng | [ŋkadi'me'ŋ] |
| Ntha'kwana | nthah-kwah-nah | [ntha'kwana] |
| Padi'mole | pah-dee-moo-leh | [padi'mʊlɛ] |
| Pa'kiso | pah-kee-soh | [pa'kisɔ] |
| Rama'gaga | rah-mah-gah-gah | [rama'xaxa] |
| Ramol'lwana | rah-mool-lwah-nah | [ramʊ'lwana] |
| Ratshe'fola | rah-tsee-foo-lah | [ratshɪ'fʊla] |

| | | |
|---|---|---|
| Sebu'budi | see-boo-boo-dee | [sɪbu'budi] |
| Seka'me'ng | see-kah-mehng | [sɪka'me'ŋ] |
| Se'kwala | see-kwah-lah | [sɪ'kwala] |
| Sel'lwane | seel-lwah-nee | [sɪ'lwanɪ] |
| Se'mpane | seem-pah-nee | [sɪ'mpanɪ] |
| Se'onya | see-oh-nya | [sɪ'ɔɲa] |
| Sera'saka | see-rah-sah-kah | [sɪra'saka] |
| Sofa'sonke | soh-fah-sohng-keh | [sɔfa'sɔŋkɛ] |
| Ta'u'ng | tah-oo-ng | [ta'u'ŋ] |
| Tja'lempe | chah-leem-pee | [tʃa'lɪmpɪ |

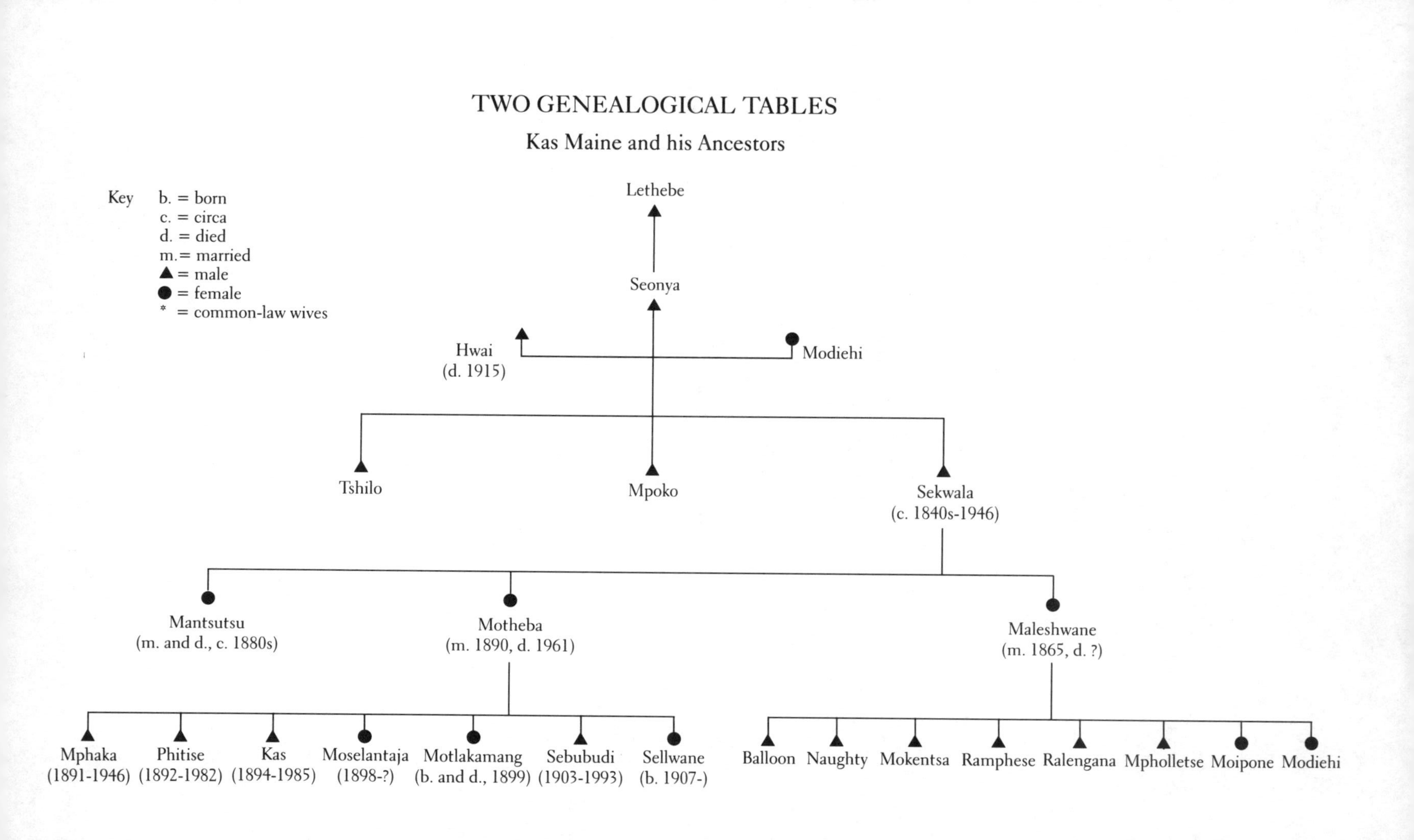
TWO GENEALOGICAL TABLES
Kas Maine and his Ancestors
Key
b. = born
c. = circa
d. = died
m.= married
= male
= female
* = common-law wives
Lethebe
Seonya
Hwai
(d. 1915)
Modiehi
Tshilo
Mpoko
Sekwala
(c. 1840s-1946)
Mantsutsu
(m. and d., c. 1880s)
Motheba
(m. 1890, d. 1961)
Maleshwane
(m. 1865, d. ?)
Mphaka
(1891-1946)
Phitise
(1892-1982)
Kas
(1894-1985)
Moselantaja
(1898-?)
Motlakamang
(b. and d., 1899)
Sebubudi
(1903-1993)
Sellwane
(b. 1907-)
Balloon
Naughty
Mokentsa
Ramphese
Ralengana
Mpholletse
Moipone
Modiehi

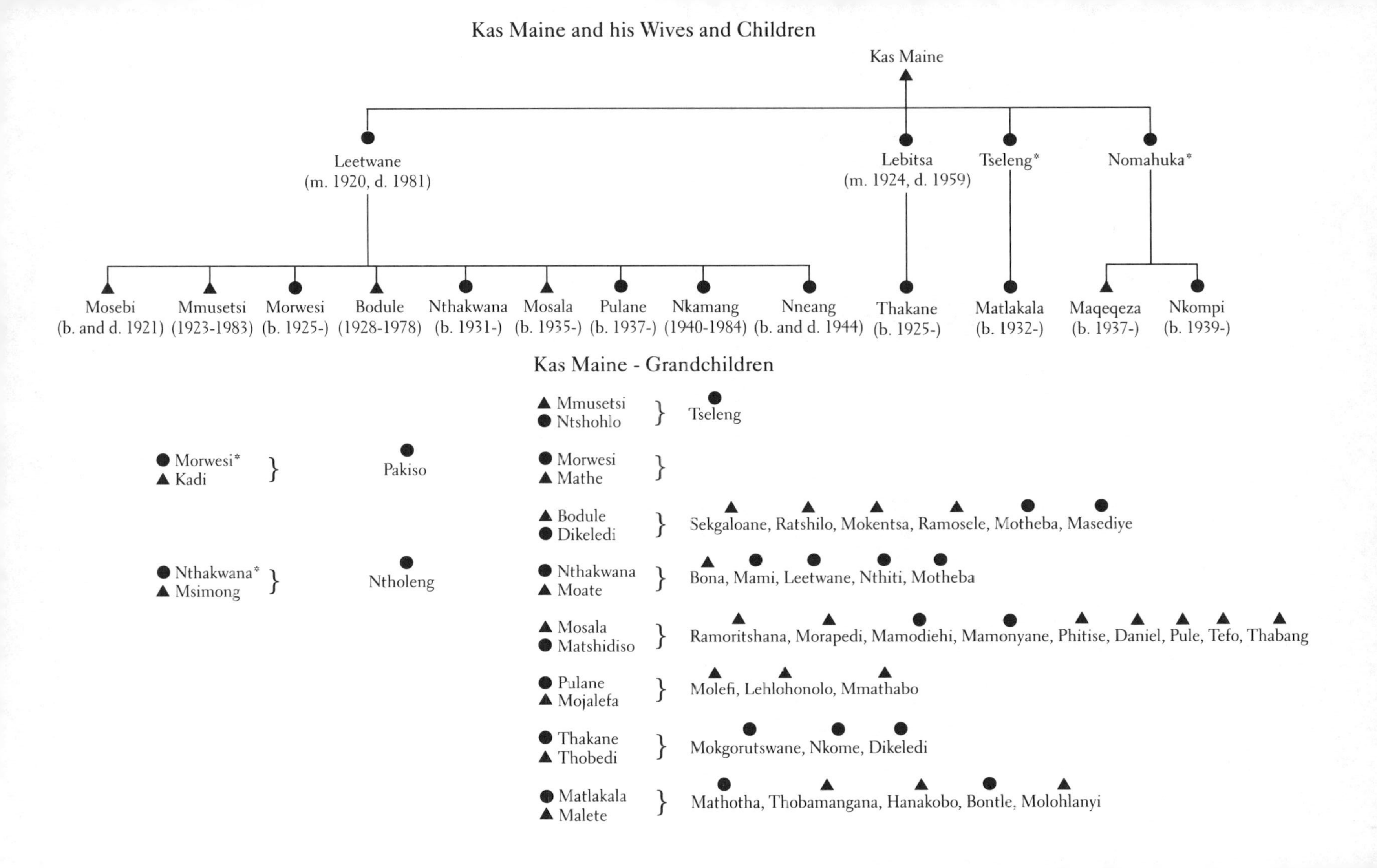

Kas Maine and his Wives and Children
Kas Maine
Leetwane
(m. 1920, d. 1981)
Lebitsa
(m. 1924, d. 1959)
Tseleng*
Nomahuka*
Mosebi
(b. and d. 1921)
Mmusetsi
(1923-1983)
Morwesi
(b. 1925-)
Bodule
(1928-1978)
Nthakwana
(b. 1931-)
Mosala
(b. 1935-)
Pulane
(b. 1937-)
Nkamang
(1940-1984)
Nneang
(b. and d. 1944)
Thakane
(b. 1925-)
Matlakala
(b. 1932-)
Maqeqeza
(b. 1937-)
Nkompi
(b. 1939-)
Kas Maine - Grandchildren
Mmusetsi
Ntshohlo
Tseleng
Morwesi*
Kadi
Pakiso
Morwesi
Mathe
Bodule
Dikeledi
Sekgaloane, Ratshilo, Mokentsa, Ramosele, Motheba, Masediye
Nthakwana*
Msimong
Ntholeng
Nthakwana
Moate
Bona, Mami, Leetwane, Nthiti, Motheba
Mosala
Matshidiso
Ramoritshana, Morapedi, Mamodiehi, Mamonyane, Phitise, Daniel, Pule, Tefo, Thabang
Pulane
Mojalefa
Molefi, Lehlohonolo, Mmathabo
Thakane
Thobedi
Mokgorutswane, Nkome, Dikeledi
Matlakala
Malete
Mathotha, Thobamangana, Hanakobo, Bontle, Molohlanyi

# A CHRONOLOGY OF THE FARMS WHERE KAS MAINE WORKED AND THEIR LANDLORDS

| *Year* | *Landlord* | *Farm* | *District* |
|---|---|---|---|
| 1920–21 | J. van Deventer | Kommissierust | Bloemhof |
| 1921–22 | W. A. Nieman | Hartsfontein | Bloemhof |
| 1922–23 | W. A. Nieman | Kommissierust | Bloemhof |
| 1923–24 | P. A. Reyneke | Koppie Alleen | Schweizer-Reneke |
| 1924 | J. J. Meyer | Koppie Alleen | Schweizer-Reneke |
| 1924–27 | P. A. Reyneke | Vlakfontein | Bloemhof |
| 1927–29 | P. and W. Reyneke | Zorgvliet/ Vlakfontein | Schweizer-Reneke |
| 1929–31 | C. McDonald | Kareepoort | Wolmaransstad |
| 1931–32 | C. G. Smith | Kareepoort | Wolmaransstad |
| 1932–37 | P. Ferreira | Boskuil | Wolmaransstad |
| 1937–39 | J. M. Klopper | Klippan | Bloemhof |
| 1939–41 | C. G. Smith | Kareepoort | Wolmaransstad |
| 1941–43 | H. Goosen | Klippan | Bloemhof |
| 1943–46 | P. G. Labuschagne | Vaalrand | Bloemhof |
| 1946–49 | H. Edwards/ W. Moormeister | Sewefontein | Bloemhof |
| 1949–50 | L. J. Klue | Jakkalsfontein | Klerksdorp |
| 1950–51 | J. Adamson | Rhenosterhoek | Klerksdorp |
| 1951–52 | J. J. N. Smit | Muiskraal | Potchefstroom |
| 1952–55 | Mrs. Paula Hersch | Rietvlei | Ventersdorp |
| 1955–56 | J. Vermeulen | Varkenskraal | Ventersdorp |
| 1956–59 | J. Ramagaga | Molote | Koster |
| 1959–67 | K. T. Nkaba | Molote | Koster |
| 1967–75 | Chieftainess Catherine Monnakgotla | Ledig | Bophuthatswana |
| 1975–85 | Chief L. Monnakgotla | Ledig | Bophuthatswana |

# BIBLIOGRAPHY

## Abbreviations

| | |
|---|---|
| C.A.D. | Central Archives Depot, Pretoria |
| C.P.S.A. | Church of the Province of South Africa, Archival Collection, U.W. |
| FAM. | Maine family interviews, M. M. Molepo Oral History Collection, I.A.S.R., U.W. |
| GES | Department of Health |
| GNLB | Government Native Labour Bureau |
| I.A.S.R. | Institute for Advanced Social Research, U.W. |
| JUS | Department of Justice |
| K. | Interviews with Kas Maine, M. M. Molepo Oral History Collection, I.A.S.R., U.W. |
| K.O.D. | Kas [Maine] Original Document, M. M. Molepo Oral History Collection, I.A.S.R., U.W. |
| LDB | Department of Agriculture |
| MCK | Mining Commissioner Klerksdorp |
| MNW | Department of Mineral Affairs |
| NTS | Department of Native Affairs |
| O.R.D.M. | Office of the Registrar of Deeds, Mafikeng |
| O.R.D.P. | Office of the Registrar of Deeds, Pretoria |
| P. | Interviews with Kas Maine's peers, M. M. Molepo Oral History Collection, I.A.S.R., U.W. |
| S.B.A. | Standard Bank Archives, Johannesburg |
| SNA | Secretary for Native Affairs |
| S.W.T.L. | Suid-Westelike Transvaalse Landboukoöperasie |
| T.A. | Transvaal Archives, Pretoria |
| UG | Union Government |
| U.W. | University of the Witwatersrand |

## I. Manuscript Sources

### A. Official

1. Central Archives Depot, Pretoria
   Estate, no. 2648/1943
   Estate, no. 1676/1944
   GES, Vol. 698, File 328/13
   GNLB, Vol. 87, File 3694/12/27
   JUS, Vol. 148, File 102/14/77
   JUS, Vol. 245, File 3/1007/20
   JUS, Vol. 257, File 1/307/19
   JUS, Vol. 275, File 1/307/19
   JUS, Vol. 275, File 1/307/119, Schweizer-Reneke 1919
   JUS, Vol. 304, File 1/811/21, Bloemhof 1919
   JUS, Vol. 367, File 1/554/23
   JUS, Vol. 429, File 1/667/27
   JUS, Vol. 517, File 6175/29
   LDB, Vol. 1378, File Nos. 44042 & 4308
   Makwassie Periodic Court Criminal Register, 28 October 1930–7 January 1933
   MCK, Vol. 1/53, File 1789/16
   NTS, Vol. 93, File 1/10/3 & File 4/17/2
   NTS, Vol. 1243, File 12/214
   NTS, Vol. 2478, File 13/293
   NTS, Vol. 7645, File 48/331
   WLD, Criminal Register, 1/2/1937–27/10/42, Case No. 95 of 1940
2. Johannesburg City Health Department Archives
   Vol. 159, File No. 11/1
3. Office of the Registrar of Deeds, Mafikeng
   JQ, Folio 91/1, 7 April 1899 & 10 August 1911
   JP, Folio 171/1, 10 July 1933
4. Office of the Registrar of Deeds, Pretoria
   IQ, Folio 94/11/1
   HO, 55–65, Folio 59/1 and 60/4/1
   HO, 80, Folio 80/1
   HO, 184–191, Folio 188/1
   HO, Folio 210/1/1
   HO, Folio 210/2/1
   HO, Folio 210/1
   HO, 197–254, Folio 243/1
   HO, 208–231, Folio 211/1
   HO, 208–231, Folio 212/1
   HO, 208–231, Folio 216/1/1
   HO, 208–231, Folio 216/2/1
   HO, 208–231, Folio 217/1
   HO, 208–231, Folio 221/2/1

HO, Folio 229/2/1
HO, Folio 231/6/1
HO, 232–251, Folio 240/6/1, 240/7/1 & 240/8/1
HO, 233, Folio 233/1
HO, Folio 234/1
HO, Folio 234/1/1
HO, Folio 234/6/1
HO, Folio 340/1
HO, Folio 349/1
IQ, Folio 21/9/12, 21/22/1 & 21/21/1
IQ, Folio 93/4/1
LG, Folios 323/1, 323/1/2, 323/9/1 & 323/12/1
LG, 835/61, N.R.D.A. (2)
5. Transvaal Archives, Pretoria
MCK, File 525/4
MNW, Vol. 257, File WL. 565/14
SNA, Vol. 457, File NA. 410/10
SNA, Vol. 457, File NA. 412/10
TPS, Vol. 45, File 6089

## B. Unofficial

1. Standard Bank Archives, Johannesburg, Inspection Reports of the Bloemhof Branch
INSP. 1/1/193, 13 September 1919
INSP. 1/1/290, 17 September 1929
INSP. 1/1/299, 27 December 1930
INSP. 1/1/318, 30 April 1932
INSP. 1/1/336, 6 June 1933
INSP. 1/1/351, 30 November 1934
INSP. 1/1/361, 24 October 1936
INSP. 1/1/337 & 1/1/377, 21 February 1939
2. University of Cape Town Archives
W. G. Ballinger Papers
3. University of the Orange Free State, Institute for Contemporary History Archives
South African Political History Collection, 'Apartheid and Guardianship: Short Summary of HNP Policy' (n.d.)
Vol. IG 6/5/5
PV2: N.P. Transvaal, File 191, Ventersdorp
4. University of the Witwatersrand, Church of the Province of South Africa Archives
Rheinallt Jones Papers, Senatorial Correspondence, Makwassie, 1939
Saffery Papers, AD 1178, B3, 1929
5. University of the Witwatersrand, Institute for Advanced Social Research Archival Collection
a) Correspondence with author

| | |
|---|---|
| Basner, M. Mrs. | 16 February 1987 |
| Behrman, G. | February 1988 |
| Coetzee, J. S. | September 1985 |
| Greyling, J. C. | 4 July 1985 |
| Hambly, F. M. Mrs. | 27 May and 24 July 1985 |
| Hutchinson, H. | 11 March 1985 |
| Jonker, H. D. Mrs. | 28 January 1992 |
| Klopper, M. J. A. Mrs. | 10 September 1985 |
| Wolfinsohn, W. | 26 July 1985 |

b) Manuscript, 'Grandmother Remembers' by Clarkson, A. B. (n.d.)

c) Original documents, Maine Collection. KOD 1–KOD 524, period covering 18 March 1922–5 October 1985

## II. Printed Primary Sources

### A. Official Records

*Annual Report of the Department of Justice*
1910, UG 51-11
1910, UG 53-11
1911, UG 56-12
1910–1912, UG 44-13
1914, UG 28-15
1916, UG 39-16
1917, UG 36-1918
1918, UG 36-19
1919, UG 35-20

*Annual Report of the Agricultural and Pastoral Production of South Africa*
1925, UG 13-27
1926, UG 24-28
1929, UG 16-30

*Annual Report of the Native Affairs Department*
1909–1910, U.15-1911

Breutz, P. L., *The Tribes of the Marico District* (Pretoria 1953–1954)

———. *The Tribes of the Mafeking District* (Pretoria 1955–1956)

City of Johannesburg, Non-European Affairs Department, *Report on questionnaires relating to the 'Kaffir Beer Profits Enquiry, 1945'*

Department of Agriculture, *Handbook of Agricultural Statistics 1904–1950* (Pretoria 1961)

Department of Native Affairs, *General Circular No. 22 of 1959*

*House of Assembly Debates* (Hansand), 1 June 1954–21 May 1959

Judicial Commission of Enquiry, *Causes and Circumstances Relating to the Recent Rebellion in South Africa, Minutes of Evidence* (UG 42-16 and UG 46-16, Cape Town)

Oliebeheerraad, *Beknopte Historiese Oorsig van die Oliesaadbedryf in Suid-Afrika* (Pretoria 1968)

*Reports of the Central Board of the Land and Agricultural Bank of South Africa*

Year ended 31st December 1915, UG 12-16
Year ended 31st December 1916, UG 17-26
Year ended 31st December 1921, UG 10 and UG 16
Year ended 31st December 1922, UG 10 and UG 16
Year ended 31st December 1923, UG 16 and UG 17
Year ended 31st December 1924, UG 17
Year ended 31st December 1925, UG 24
Year ended 31st December 1926, UG 12
Year ended 31st December 1927, UG 18
Year ended 31st December 1928, UG 18
Year ended 31st December 1950, UG 17
*Report of the Drought Investigation Commission 1922* (UG 20 of 1922)
*Report of the Native Affairs Commission*, G.P.S. 4620–1942/43
*Report of the Native Economic Commission*, 1930–1932 (UG 22 of 1932)
*Report of the Native Farm Labour Committee 1937–1939* (Pretoria 1939)
*Report of the Native Land Committee, Western Transvaal* (UG 23 of 1918)

## B. Newspapers and Periodicals

*African Leader*, 3 September 1932, 15 October and 22 October 1932
*The Diggers' News*, 6 February 1914 and 13 March 1914
*The Klerksdorp Record and Western Transvaal News*, 1 November 1918
*The Star*, 3 December 1935
*Saturday Star*, 16 March 1991
*The Times*, London, 13 May 1977
*Umteteli wa Bantu*, 27 June 1931
*The Workers' Herald Weekly News Bulletin*, 5 May 1929

# III. Secondary Sources

## A. Select Books

Amery, L. S. (ed.), *The Times History of the War in South Africa, 1899–1902*, Vol. IV (London 1906)
Basner, M., *Am I an African? The Political Memoirs of H. M. Basner* (Johannesburg 1993)
Beinart, W. and Bundy, C., *Hidden Struggles in Rural South Africa: Politics and Popular Movements in the Transkei and Eastern Cape 1890–1930* (Johannesburg 1987)
Beinart, W., Delius, P. and Trapido, S. (eds.), *Putting a Plough to the Ground: Accumulation and Dispossession in Rural South Africa, 1850–1930* (Johannesburg 1986)
Boshoff, S. P. E. and Nienaber, G. S., *Afrikaanse Etimologie* (Pretoria 1967)
Bosman, D. B. and Nienaber, G. S., *Tweetalige Woordeboek* (Cape Town 1984)
Bosman, H. C., *Jurie Steyn's Post Office* (Cape Town 1971)
Bozzoli, B., *Women of Phokeng* (Johannesburg 1991)

Bradford, H., *A Taste of Freedom: The I.C.U. in Rural South Africa, 1924–1930* (London 1987)
Bret, M. R., *Pilanesberg–Jewel of Bophuthatswana* (Johannesburg 1989)
Brown, T., *Setswana Dictionary* (Gaborone 1980)
Campbell, J., *Songs of Zion: The African Methodist Episcopal Church in the United States and South Africa* (New York 1995)
Cawthorn, R. G., *The Geology of the Pilanesberg* (National Parks Board of Bophuthatswana, Mafikeng 1988)
Davenport, T. R. H., *South Africa: A Modern History* (Johannesburg 1987)
Desmond, C., *The Discarded People* (Johannesburg 1969)
Department of Agriculture, *Handbook for Farmers in South Africa* (Pretoria 1929)
Donald, K. (ed.), *Who's Who of South Africa* (Johannesburg 1939)
du Boulay, S., *Tutu: Voice of the Voiceless* (London 1989)
Dubow, S., *Racial Segregation and the Origins of Apartheid, 1891–1936* (London 1989)
du Toit, A. and Giliomee, H. (eds.), *Afrikaner Political Thought: Analysis and Documents, Vol. 1 1780–1850* (Cape Town 1983)
Dvorin, E. P., *Racial Separation in South Africa: An Analysis of Apartheid Theory* (Chicago 1952)
Edgar, R., *Because They Chose the Plan of God: The Story of the Bulhoek Massacre* (Johannesburg 1988)
Elphick, R. and Giliomee, H. (eds.), *The Shaping of South African Society, 1652–1840* (Middletown 1989)
Federasie van Afrikaanse Kultuurvereniginge, *Afrikaanse Kultuuralmanak* (Johannesburg 1980)
Flynn, C. L. Jnr., *White Land, Black Labor: Caste and Class in Late Nineteenth Century Georgia* (Baton Rouge 1983)
Fox-Genovese, E., *Within the Plantation Household: Black and White Women of the Old South* (London 1988)
Frederickson, G. M., *The Arrogance of Race: Historical Perspectives on Slavery, Racism and Social Inequality* (Middletown 1988)
Genovese, E., *The Political Economy of Slavery* (New York 1967)
——— *Roll Jordan Roll: The World the Slaves Made* (New York 1976)
——— *The World the Slaveholders Made* (Middletown 1988)
Guest, H., *Klerksdorp's Fifty Years of Mining* (Klerksdorp 1938)
Gutman, H. G., *The Black Family in Slavery and Freedom, 1750–1925* (New York 1977)
Hahn, S., *The Roots of Southern Populism: Yeoman Farmers and the Transformation of the Georgia Upcountry, 1850–1890* (New York 1987)
Hammond-Tooke, W. D. (ed.), *The Bantu-speaking Peoples of Southern Africa* (London 1974)
Hexham, I., *The Irony of Apartheid: The Struggle for National Independence of Afrikaner Calvinism against British Imperialism* (New York 1981)
Hindson, D., *Pass Controls and the Urban African Proletariat* (Johannesburg 1987)

Jones, J., *Labour of Love, Labour of Sorrow: Black Women, Work and the Family, from Slavery to the Present* (New York 1986)
Keegan, T. J., *Rural Transformations in Industrializing South Africa: The Southern Highveld to 1914* (Johannesburg 1986)
Kriel, T. J. (ed.), *Popular Northern Sotho Dictionary* (Pretoria 1966)
Kriel, T. J., *The New SeSotho-English Dictionary* (Johannesburg 1950)
Kruger, D. W. (ed.), *South African Dictionary of Biography*, Vol. III (Cape Town 1977)
Lemann, N., *The Promised Land* (New York 1991)
Lewis, I. M., *Ecstatic Religion: An Anthropological Study of Spirit Possession and Shamanism* (Harmondsworth 1971)
Lowrey, T. K. and Wright, S., *The Flora of the Witwatersrand* (Johannesburg 1987)
Marcus, T., *Modernising Super-Exploitation: Restructuring South African Agriculture* (London 1989)
Maree, H. P., *Die Geskiedenis en Ontwikkeling van die Dorp en Distrik van Schweizer-Reneke, 1870–1952* (Roneo 1952)
Marks, S. and Rathbone, R. (eds.), *Industrialization and Social Change in South Africa: African Class Formation, Culture and Consciousness* (London 1982)
Mauss, M., *The Gift* (London 1954)
McMillen, N. R., *Dark Journey: Black Mississipians in the Age of Jim Crow* (Chicago 1990)
Mohlamme, J. S., *Forced Removals in the People's Memory: The BaKubung of Ledig* (Johannesburg 1989)
Mokgatle, N., *The Autobiography of an Unknown South African* (London 1971)
Morgan, B. H., *Report on the Engineering Trades of South Africa* (London 1902)
Murray, C., *Black Mountain: Land, Class and Power in the Eastern Orange Free State 1880s–1980s* (Johannesburg 1992)
Natrass, J., *The South African Economy: Its Growth and Change* (Cape Town 1981)
Oakes, J., *The Ruling Race: A History of American Slaveholders* (New York 1982)
——— *Freedom and Slavery* (New York 1983)
O'Meara, D., *Volkskapitalisme: Class, Capital and Ideology in the Development of Afrikaner Nationalism 1934–1948* (Johannesburg 1983)
Pakenham, T., *The Boer War* (London 1979)
Paroz, R. A. (ed.), *Southern Sotho-English Dictionary* (Morija 1983)
Perry, J. and C. (eds.), *A Chief is a Chief by the People—The Autobiography of Stimela Jason Jingoes* (Cape Town 1975)
Posel, D., *The Making of Apartheid 1948–1961: Conflict and Compromise* (Oxford 1991)
Robertson, A. F., *The Dynamics of Productive Relationships: African Share Contracts in Comparative Perspective* (Cambridge 1987)

Russell, M. and M., *Afrikaners of the Kalahari: White Minority in a Black State* (Cambridge 1979)
Shillington, K., *The Colonisation of the Southern Tswana, 1870–1900* (Johannesburg 1985)
Simons, H. J. and R. E., *Class and Colour in South Africa, 1850–1950* (Harmondsworth 1969)
Smith, C. A., *Common Names of South African Plants* (Pretoria 1966)
Tinley, J. M., *South African Food and Agriculture in World War II* (Stanford, Ca. 1954)
Tyson, P. D., *Climatic Change and Variability in Southern Africa* (Cape Town 1986)
Warwick, P. (ed.), *The South African War* (London 1980)
Warwick, P., *Black People and the South African War, 1899–1902* (Johannesburg 1983)
Wheatcroft, G., *The Randlords* (London 1985)
Willan, B., *Sol Plaatje: South African Nationalist 1876–1932* (London 1984)
Worster, D., *Dust Bowl: The Southern Plains in the 1930s* (Oxford 1979)

### B. Select Articles and Chapters in Books

Ambrose, D. P., 'Maps of Lesotho and Surrounding Areas, 1779–1828.' *Mohlomi, Journal of Southern African Historical Studies*, Vol. I, 1976
Armstrong, J. C. and Worden, N., 'The Slaves, 1652–1834' in R. Elphick and H. Giliomee (eds.), *The Shaping of Southern African Society, 1652–1840* (Middletown 1989), pp. 109–183
Bozzoli, B., 'Marxism, Feminism and South African Studies.' *Journal of Southern African Studies*, Vol. 9, No. 2, 1983, pp. 139–171
Clynick, T., 'Romance and Reality on the Vaal River Diggings: Race and Class in a South African Rural Community, 1905–1914.' *Canadian Papers in Rural History*, Vol. 9, 1994, pp. 401–418
Cohen, J. S. and Galassi, F. L., 'Feudal Residues in Italian Agriculture.' *Economic History Review*, 43, 4, 1990, pp. 646–656
Davenport, T. R. H., 'Some Reflections on the History of Land Tenure in South Africa, Seen in the Light of Attempts by the State to Impose Political and Economic Control.' *Acta Juridica*, 1985, pp. 53–76
de Klerk, M., 'Seasons that will Never Return: The Impact of Farm Mechanisation on Employment, Incomes and Population Distribution in the Western Transvaal.' *Journal of Southern African Studies*, Vol. 11, No. 1, 1984, pp. 84–105
Delius, P., 'Abel Erasmus: Power and Profit in the Eastern Transvaal' in W. Beinart, P. Delius and S. Trapido (eds.), *Putting a Plough to the Ground* (Johannesburg 1986), pp. 176–217
Dugmore, H., 'The Rise to Power of the Monnakgotla Family of the Bakubung, 1830–1896.' *Africa Perspective*, (Johannesburg) New Series, Vol. 1, Nos. 3 & 4, 1987
Feierman, S., 'Struggles for Control: The Social Roots of Health and Healing

in Modern Africa.' *African Studies Review*, Vol. 28, Nos. 2/3, 1985, pp. 73–127

Feinberg, H. M., 'The 1913 Natives Land Act in South Africa: Politics, Race, and Segregation in the Early 20th Century.' *The International Journal of African Historical Studies*, Vol. 26, No. 1, 1993, pp. 65–109

Genovese, E., ' "Our Family, White and Black": Family and Household in the Southern Slaveholders' World View' in Carol Bleser (ed.), *In Joy and In Sorrow: Women, Family and Marriage in the Victorian South, 1830–1900* (New York 1991), pp. 69–87

Giliomee, H., 'The Eastern Frontier, 1770–1812' in R. Elphick and H. Giliomee (eds.), *The Shaping of South African Society* (Middletown 1989), pp. 421–471

——— 'Constructing Afrikaner Nationalism.' *Journal of Asian and African Studies*, XVIII, 1–2 (1983), pp. 83–98

Hammond-Tooke, W. D., 'Urbanization and the Interpretation of Misfortune: A Quantitative Analysis.' *Africa*, Vol. 40, 1970, pp. 25–38

Isaacman, A., 'Peasants and Rural Social Protest in Africa.' *African Studies Review*, Vol. 33, No. 2, 1990, pp. 1–119

Keegan, T., 'The Sharecropping Economy on the South African Highveld in the Early Twentieth Century.' *Journal of Peasant Studies*, Vol. 10, Nos. 2/3, 1983, pp. 201–226

Kirby, J. T., 'Black and White in the Rural South 1915–1954.' *Agricultural History*, No. 58, July 1984, pp. 411–422

Minnaar, A., 'The South African Maize Industry's Response to the Great Depression and the Beginnings of Large-Scale State Intervention 1929–1934.' *The South African Journal of Economic History*, Vol. 4, No. 1, March 1989, pp. 68–78

Morris, M. L., 'The Development of Capitalism in South African Agriculture: Class Struggle in the Countryside.' *Economy and Society*, Vol. 5, 1976, pp. 292–342

Murray, C., 'Land, Power and Class in the Thaba Nchu District, Orange Free State, 1884–1983.' *Review of African Political Economy*, Vol. 29, 1984, pp. 30–49

Patterson, O., 'Slavery.' *Annual Review of Sociology*, Vol. 3, 1977, pp. 407–449

Rogerson, C., 'A Strange Case of Beer: The State and Sorghum Beer Manufacture in South Africa.' *Area*, Vol. 18, No. 1, 1986, pp. 15–24

Sahlins, M., 'The Sociology of Primitive Exchange' in M. Banton (ed.), *The Relevance of Models for Social Anthropology* (London 1965)

Swartz, L., 'Transcultural Psychiatry in South Africa' Part 1. *Transcultural Psychiatric Research Review*, Vol. 23, 1986, pp. 273–303

Trapido, S., 'Landlord and Tenant in a Colonial Economy: The Transvaal 1880–1910.' *Journal of Southern African Studies*, Vol. 5, No. 1, 1978, pp. 26–58

——— 'From Paternalism to Liberalism: Cape Colony, 1800–1834.' *International History Review*, 12, 1, 1990, pp. 76–104

Tyson, P. D., 'The Great Drought.' *Leadership S.A.*, Vol. 2, No. 3, Spring 1983, pp. 49–57

van Onselen, C., 'Race and Class in the South African Countryside: Cultural Osmosis and Social Relations in the Sharecropping Economy of the South-Western Transvaal, 1900–1950.' *The American Historical Review*, Vol. 95, No. 1, 1990, pp. 99–123

——— 'Reactions to Rinderpest in Southern Africa.' *Journal of African History*, XII, 3, 1973, pp. 473–488

——— 'The Reconstruction of a Rural Life from Oral Testimony: Critical Notes on the Methodology in the Study of a Black South African Sharecropper.' *The Journal of Peasant Studies*, Vol. 20, No. 3, 1993, pp. 494–514

——— 'The Social and Economic Underpinnings of Paternalism and Violence on the Maize Farms of the South-Western Transvaal, 1900–1950.' *Journal of Historical Sociology*, Vol. 5, No. 2, 1992, pp. 127–160

## C. Select Unpublished Articles and Dissertations

Anderson, R. C., 'Keeping the Myth Alive: Justice, Witches and the Law in the 1988 Sekhukune Killings,' B.A. (Hons.) diss., Department of History, U.W., 1990

Bonner, P., ' "Desirable or Undesirable Sotho Women?" Liquor, Prostitution and the Migration of Sotho Women to the Rand, 1920–1945.' Paper presented at the I.A.S.R., U.W., 9 May 1988

Booth, C., 'Liberalism in the Countryside: J. D. Rheinallt Jones, the Joint Councils and the Farm Labour Question in the 1930's and 1940's,' B.A. (Hons.) diss., U.W., 1987

Bottomley, J., 'The South African Rebellion of 1914: The Influence of Industrialisation, Poverty and "Poor Whiteism." ' Paper presented at the I.A.S.R., U.W., 14 June 1982

Bradford, H., 'The Industrial and Commercial Workers' Union in the South African Countryside 1924–1930,' Ph.d. diss., U.W., 1985. [Published as *A Taste of Freedom*.]

Campbell, J., 'The Social Origins of African Methodism in the Orange Free State.' Paper presented at the I.A.S.R., U.W., 22 March 1993

Claasens, A., 'Who Owns South Africa,' Occasional Paper No. 11, Centre for Applied Legal Studies, U.W., February 1991

Coetzee, J. M., 'Lineal Consciousness in the Farm Novels of C. M. van den Heever.' Paper presented to the Department of English, University of Cape Town

Dugmore, H. L., 'Land and the Struggle for Sekama: The Transformation of a Rural Community—the BaKubung, 1884–1937,' B.A. (Hons.) diss., U.W., 1985

Du Toit, A., 'Deel van die Plaas: Notes on the Contradiction of Antagonism on Western Cape Fruit and Wine Farms.' Paper presented to the Conference of the Association of Southern African Sociologists, June 1991

Eales, K. A., 'Gender Politics and the Administration of African Women in Johannesburg 1903–1939,' M.A. diss., U.W., 1991

Faison, M. L., 'Pixley Ka Isaka Seme, President-General of the African National Congress 1930–1937: A Study of the Impact of his Leadership and Ideology on the Congress,' M.A. diss., University of Columbia, U.S.A., June 1983

Hexham, I., 'Modernity or Reaction in South Africa: The Case of Afrikaner Religion.' Seminar, University of British Columbia, Consultation on Modernity and Religion, 15–18 December, 1981

Hughes, H., 'Black Mission in South Africa: Religion and Education in the African Methodist Episcopal Church in South Africa, 1892–1953,' B.A. (Hons.) diss., U.W., 1976

Hyslop, J., 'The Representation of White Working Class Women in the Construction of a Reactionary Populist Movement: "Purified" Afrikaner Nationalist Agitation for Legislation against "Mixed" Marriages 1934–1939.' Paper presented at I.A.S.R., U.W., 24 May 1993

Martin, W. G. and Beittel, M., 'The Hidden Abode of Reproduction: Conceptualizing Households in Southern Africa.' Paper presented at Workshop on Conceptualizing the Household: Issues of Theory, Method, and Application, Harvard Institute for International Development, 2–4 November 1984

Millar, D. J., 'To Save the "Volk"! The 1947 Consumer Boycott of Indian Retail Traders in the Transvaal.' Seminar paper presented to the Geography Department, U.W., 1987

Nasson, W. R., 'Black Society in the Cape Colony and the South African War of 1899–1902: A Social History,' D. Phil. diss., Cambridge University, 1983

Roth, M., 'The Natives' Representatives Council, 1937–1951,' Ph.d. diss., U.W., 1987

Segal, L., 'A Brutal Harvest: The Roots and Legitimation of Violence on Farms in South Africa.' Seminar Paper No. 7, presented in Department of Applied Psychology, U.W., Project for the Study of Violence, 26 September 1990

Trapido, S., 'Poachers, Proletarians and Gentry in the Early Twentieth Century Transvaal.' Paper presented at the I.A.S.R., U.W., 12 March 1984

——— 'The Emergence of Liberalism and the Making of "Hottentot" Nationalism.' Paper presented to the Seminar on Southern African Societies at the Institute of Commonwealth Studies, University of London, 23 February 1990

Venter, P., 'Die Staatsversorging van 'n Tipiese Boeregemeenskap in S.W. Transvaal,' M.A. diss., University of Potchefstroom, 1944

## IV. Interviews

All recordings of interviews, unless otherwise indicated, are housed in the M. M. Molepo Oral History Collection, I.A.S.R., U.W.

### A. with black farm owners

Motsuenyane, A. B. M. Interviewed by Mrs. B. Motau, No. 563 Macheng Street, Mohlakeng, 6 March 1984

### B. with Kas Maine

K. 1 M. T. Nkadimeng, 21 November 1979, Tape Nos. 82 & 83
K. 2 M. T. Nkadimeng, T. Couzens and G. Relly, 2 July 1980, Tape Nos. 204 & 205
K. 3 M. T. Nkadimeng and T. J. Matsetela, 2 December 1980, Tape Nos. 130 & 131
K. 4 M. T. Nkadimeng and T. Couzens, 15 July 1980, Tape No. 224
K. 5 M. T. Nkadimeng and T. Couzens, 15 July 1980, Tape No. 223
K. 6 M. T. Nkadimeng, 24 July 1980, Tape Nos. 229 & 230
K. 7 M. T. Nkadimeng, 27 July 1980, Tape No. 231
K. 8 M. M. Molepo, 17 September 1980, Tape No. 234
K. 9 M. M. Molepo, 20 January 1981, Tape No. 276
K. 10 C. van Onselen, 24 February 1981, Tape No. 264
K. 11 M. M. Molepo, 28 July 1981, Tape No. 300
K. 12 C. van Onselen and T. Couzens, 2 November 1981, Tape No. 298
K. 13 M. M. Molepo, 1 December 1981, Tape No. 295
K. 14 M. M. Molepo, 13 May 1982, Tape No. 325
K. 15 M. T. Nkadimeng, 29 July 1982, Tape No. 338
K. 16 M. T. Nkadimeng, 1 November 1982, Tape No. 402
K. 17 C. van Onselen and T. Couzens, 2 November 1982, Tape No. 301
K. 18 M. T. Nkadimeng, 13 January 1983, Tape No. 355
K. 19 M. T. Nkadimeng, 23 January 1983, Tape No. 357
K. 20 M. T. Nkadimeng, 3 March 1983, Tape Nos. 367 & 371
K. 21 M. T. Nkadimeng, 3 March 1983, Tape No. 381
K. 22 M. T. Nkadimeng, 9 March 1983, Tape No. 370
K. 23 M. T. Nkadimeng, 10 March 1983, Tape No. 368
K. 24 M. T. Nkadimeng, 20 April 1983, Tape No. 386
K. 25 M. M. Molepo, 25 February 1982, Tape No. 314
K. 26 M. T. Nkadimeng, 29 May 1983, Tape No. 387
K. 27 M. T. Nkadimeng, 2 November 1983, Tape No. 403
K. 28 M. T. Nkadimeng, 29 November 1983, Tape No. 400
K. 29 M. T. Nkadimeng, 30 November 1983, Tape No. 399
K. 30 M. T. Nkadimeng, 24 January 1984, Tape No. 406
K. 31 M. T. Nkadimeng, 26 January 1984, Tape No. 407
K. 32 M. T. Nkadimeng, 26 January 1984, Tape No. 408
K. 33 M. T. Nkadimeng, 17 April 1984, Tape No. 462
K. 34 M. T. Nkadimeng, 17 April 1984, Tape No. 463
K. 35 M. T. Nkadimeng, 16 May 1984, Tape No. 464
K. 36 M. T. Nkadimeng and E. Msimango, 29 August 1984, Tape No. 486
K. 37 M. T. Nkadimeng, 9 July 1984, Tape No. 478
K. 38 M. T. Nkadimeng, 10 July 1984, Tape No. 479

K. 39 M. T. Nkadimeng, 1 October 1984, Tape No. 490
K. 40 M. T. Nkadimeng, 2 October 1984, Tape No. 491
K. 41 M. T. Nkadimeng, 2 October 1984, Tape No. 492
K. 42 M. T. Nkadimeng, 30 October 1984, Tape No. 501
K. 43 M. T. Nkadimeng, 6 November 1984, Tape No. 502
K. 44 M. T. Nkadimeng, 6 November 1984, Tape No. 504
K. 45 M. T. Nkadimeng, 7 November 1984, Tape No. 505
K. 46 M. T. Nkadimeng, 7 November 1984, Tape No. 503
K. 47 M. T. Nkadimeng, 15 February 1985, Tape No. 508
K. 48 M. T. Nkadimeng, 16 February 1985, Tape No. 509
K. 49 M. T. Nkadimeng, 1 March 1985, Tape No. 511
K. 50 M. T. Nkadimeng, 2 March 1985, Tape No. 512
K. 51 C. van Onselen, 13 March 1985, Tape No. 513
K. 52 M. T. Nkadimeng, 29 March 1985, Tape No. 516
K. 53 M. T. Nkadimeng, 20 March 1985, Tape No. 517
K. 54 M. T. Nkadimeng, 20 April 1985, Tape No. 518
K. 55 M. T. Nkadimeng, 20 May 1985, Tape No. 519
K. 56 M. T. Nkadimeng, 20 May 1985, Tape No. 520
K. 57 M. T. Nkadimeng, 14 June 1985, Tape No. 521
K. 58 M. T. Nkadimeng, 14 June 1985, Tape No. 522
K. 59 M. T. Nkadimeng, 13 June 1985 (notes)
K. 60 C. van Onselen, 22 July 1985, Tape No. 533
K. 61 M. T. Nkadimeng, 9 July 1985, Tape No. 531
K. 62 M. T. Nkadimeng, 10 July 1985, Tape No. 532
K. 63 M. T. Nkadimeng, 26 June 1984, Tape No. 477
K. 64 H. Mohlabane, 20 March 1984, Tape No. 417
K. 65 M. T. Nkadimeng, 27 August 1982, Tape No. 351
K. 66 M. T. Nkadimeng, 28 November 1983, Tape No. 401

### C. with Kas Maine's peers

P. 1 Masihu, L. M. T. Nkadimeng, Molelema, Taung, 8 September 1983, Tape No. 395
P. 2 Modise, T. H. Mohlabane and B. Motau, Forest Town, Johannesburg, 22 February 1984, Tape No. 410
P. 3 Molohlanyi, M. M. T. Nkadimeng, Wolmaransstad, 19 April 1984, Tape No. 468
P. 4 Nkaba, K. M. M. Molepo, Tlapeng, Vryburg, 23 January 1982, Tape No. 310
P. 5 Padimole, R. M. T. Nkadimeng, Vaalrand, Bloemhof, 13 July 1983, Tape No. 393
P. 6 Kadi, S. M. M. Molepo, Makwassie, 5 August 1986, Tape No. 534
P. 7 Baraganyi, P. M. M. Molepo, Boskuil, 5 August 1985, Tape No. 535
P. 8 Molohlanyi, M. M. T. Nkadimeng, Wolmaransstad, 28 June 1985, Tape No. 530

| | | |
|---|---|---|
| P. 9 | Tabu, I. | M. T. Nkadimeng, Letlhapong, Taung, 24 April 1986, Tape No. 547 |
| P. 10 | Tsubane, P. K. | M. T. Nkadimeng, Barberspan, 1 June 1986, Tape No. 549 |
| P. 11 | Baraganyi, P. | M. T. Nkadimeng, Boskuil, 21 April 1987, Tape No. 551 |
| P. 12 | Moeng, I. | T. Matsetela, Oersonskraal, 12 March 1987, Tape Nos. 561 & 562 |
| P. 13 | Moeng, I. | T. Matsetela, Oersonskraal, 16 January 1987, Tape Nos. 565 & 566 |
| P. 14 | Moeng, I. | T. Matsetela, Oersonskraal, 12 May 1987, Tape Nos. 567, 568 & 569 |
| P. 15 | Moeng, I. | T. Matsetela, Oersonskraal, 19 May 1987, Tape No. 570 |
| P. 16 | Moeng, M. | T. Matsetela, Oersonskraal, 19 May 1987, Tape No. 571 |
| P. 17 | Lephadi, G. | T. Matsetela, Ikageng, Potchefstroom, 27 May 1987, Tape No. 572 |
| P. 18 | Moeng, I. | T. Matsetela, Oersonskraal, 8 July 1987, Tape Nos. 575 & 576 |
| P. 19 | Moeng, I. | T. Matsetela, Oersonskraal, 9 July 1987, Tape No. 577 |
| P. 20 | Lephadi, G. | T. Matsetela, Ikageng, Potchefstroom, 11 August 1987, Tape No. 582 |
| P. 21 | Moeng, I. | C. van Onselen, Oersonskraal, 29 November 1988 |
| P. 22 | Molohlanyi, M. | E. Kgomo and M. T. Nkadimeng, Wolmaransstad, 1 July 1988, Tape No. 634 |
| P. 23 | Ngakane, B. | E. Msimango, Ikageng, Potchefstroom, 2 March 1989, Tape No. 510 |
| P. 24 | Ramagaga, M. | M. T. Nkadimeng, Thaba Nchu, 3 February 1989, Tape No. 679 |
| P. 25 | Patel, S. | M. T. Nkadimeng, Schweizer-Reneke, 4 July 1988, Tape No. 639 |
| P. 26 | Monnakgotla, M. | E. Kgomo and M. T. Nkadimeng, Ledig, 13 July 1983, Tape No. 392 |
| P. 27 | Ramagaga, N. | M. T. Nkadimeng, Orlando East, Soweto, 10 February 1989, Tape No. 681 |
| P. 28 | Molohlanyi, M. | M. T. Nkadimeng, Wolmaransstad, 28 June 1985, Tape No. 530 |
| P. 29 | Maponyane, J. | S. Lebelo, Ledig, 13 November 1990, Tape No. 712 |
| P. 30 | Ratshefola, J. | M. T. Nkadimeng, Wedela, 22 November 1990, Tape No. 714 |
| P. 31 | Malebje, P. | M. T. Nkadimeng, Ledig, 7 March 1991, Tape No. 717 |
| P. 32 | Lepedi, W. | M. T. Nkadimeng, Ledig, 8 March 1991, Tape No. 719 |

| | | |
|---|---|---|
| P. 33 | Monnakgotla, G. | M. T. Nkadimeng, Meadowlands, Soweto, 22 April 1991, Tape No. 720 |
| P. 34 | Maseko, R. | S. Lebelo, Ledig, 9 August 1991, Tape No. 746 |
| P. 35 | Maponyane, J. | S. Lebelo, Ledig, 9 August 1991, Tape No. 749 |
| P. 36 | Mngomezulu, S. | S. Lebelo, Ledig, 8 August 1991, Tape Nos. 750 & 751 |
| P. 37 | Maseko, M. and others | S. Lebelo, Ledig, 18 August 1991, Tape No. 753 |
| P. 38 | Mngomezulu, S. | S. Lcbelo, Ledig, 13 November 1990, Tape No. 713 |
| P. 39 | Mngomezulu, J. | S. Lebelo, Ledig, 8 August 1991, Tape No. 756 |
| P. 40 | Mokgethi, T. & Tlhakanye, H. | M. T. Nkadimeng, Rietvlei, 5 July 1989, Tape No. 699 |
| P. 41 | Baraganyi, P. T. | C. van Onselen, Boskuil, 16 July 1985 |


## D. with doctors, landlords, lawyers, tenants, traders and others by C. van Onselen


| | |
|---|---|
| Adamson, J. | Kafferskraal, Klerksdorp, 15 February 1984 |
| Becker, M. Mrs. (b. Hersch) | Norwood, Johannesburg, 6 July 1985 |
| Du Plooy, J. H. | Leeubos, Bloemhof, 14 March 1985 |
| Du Toit, A. Mrs. (b. Coetzee) | Van Riebeeck Street, Wolmaransstad, 16 July 1985 |
| Goosen, P. A. | Bapsfontein, 29 August 1985 |
| Goosen, P. | Kareepan, Bloemhof, 1 October 1985 |
| Gordon, A. Mrs. (b. Zilibowitz) | Rosebank, Johannesburg, 5 March 1987 |
| Grant, J. | Holfontein, Bloemhof, 13 June 1984 |
| Hafejee, M. Mrs. | 2 Honeysuckle Street, Lenasia, 5 June 1985 |
| Humphreys, G. E. Mrs. | Ficksburg, 12 April 1988 |
| Isane, Z. Mrs. & Nanabhay, A. Mrs. (b. Patel) | 83 Crown Road, Fordsburg, Johannesburg, 11 November 1987 |
| Kanareck, J. | Hillbrow, Johannesburg, 3 September 1985 |
| Kathrada, M. M. | Hessie, Schweizer-Reneke, 15 March 1989 |
| Labuschagne, J. Mrs. | Sewefontein, Bloemhof, 29 November 1988 |
| Labuschagne, P. G. | Vaalrand, Bloemhof, 25 February 1981 |
| Labuschagne, P. | Sewefontein, Bloemhof, 25 February 1981 |
| Lecornu, E. Mrs. (b. Edwards) | Bryanston, Johannesburg, 20 March 1987 |

| | |
|---|---|
| Lindbergh, F. M. | Vaalboschfontein, Wolmaransstad, 12 June 1984 |
| Mandelstam, P. Mrs. (b. Wolpe) | Johannesburg, 24 February 1988 |
| Marks, N. Mrs. (b. Marais) | Rosebank, Johannesburg, 12 September 1988 |
| Meyer, J. J. | London, Schweizer-Reneke, 11 June 1984 |
| Meyer, J.J.P. | Wolmaransstad, 16 July 1985 |
| Meyer, R. S. | Devon, Eastern Transvaal, 15 March 1984 |
| Nieman, A.C.G. | Bloemhof, 16 July 1985 |
| Nieman, R. J. | Brits, 12 June 1985 |
| Nieman, M. Mrs. (b. Griesel) | Bloemhof, 16 July 1985 |
| Pienaar, L. T. | Schoonsig, Bloemhof, 11 June 1984 |
| Pieterse, I. Mrs. (b. Klue) | Klippan, Klerksdorp, 29 March 1984 |
| Reyneke, J. | Olifantshoek, 21 May 1986 |
| Swanepoel C. | Kimberley, August 1985 |
| Swanepoel, H. F. B. | Houtvolop, Schweizer-Reneke, 21 February 1985 |
| Wentzel, J. J. | Bloemhof, 15 March 1989 |

### E. with members of the Maine Family

| | | |
|---|---|---|
| FAM. 1 | Maine, Leetwane | M. T. Nkadimeng, Ledig, 27 July 1980, Tape Nos. 232 & 233 |
| FAM. 2 | Maine, Leetwane | M. T. Nkadimeng, Ledig, 17 September 1980, Tape No. 235 |
| FAM. 3 | Maine, Sebubudi | M. T. Nkadimeng and M. M. Molepo, Klippan, Bloemhof, 18 September 1980, Tape Nos. 236 & 237 |
| FAM. 4 | Moate, Nthakwana | M. T. Nkadimeng, Ledig, 4 November 1980, Tape No. 239 |
| FAM. 5 | Deu, Petrus & Pulane | M. M. Molepo, Khuma, Stilfontein, April 1982, Tape No. 316 |
| FAM. 6 | Tantane, Johannes | M. M. Molepo, Khuma, Stilfontein, 5 April 1982, Tape No. 320 |
| FAM. 7 | Moate, Nthakwana | M. M. Molepo, Ledig, 13 May 1982, Tape No. 326 |
| FAM. 8 | Teu, Mosala | M. T. Nkadimeng, Ledig, 22 July 1982, Tape No. 334 |
| FAM. 9 | Maine, Thakane | M. T. Nkadimeng, Lichtenburg, 28 July 1982, Tape No. 348 |
| FAM. 10 | Deu, Nkamang | M. T. Nkadimeng, Ledig, 24 August 1982, Tape No. 347 |
| FAM. 11 | Legobathe, Sellwane | M. T. Nkadimeng, Gannalaagte, 27 August 1982, Tape Nos. 349 & 350 |

| | | |
|---|---|---|
| FAM. 12 | Maine, Molefi, Maine, Abraham & Maine, Sellwane | M. T. Nkadimeng, Gannalaagte, 24 October 1982, Tape No. 352 |
| FAM. 13 | Maine, Sellwane | M. T. Nkadimeng, Gannalaagte, 24 October 1982, Tape No. 382 |
| FAM. 14 | Maine, Sebubudi | M. M. Molepo, Klippan, Bloemhof, 25 November 1982, Tape No. 323 |
| FAM. 15 | Maine, Baefesi | M. T. Nkadimeng, Mareetsane, Mafikeng, 6 September 1983, Tape No. 398 |
| FAM. 16 | Teu, Mosala | M. T. Nkadimeng, Ledig, 24 January 1984, Tape No. 405 |
| FAM. 17 | Lepholletse, Tseleng | M. T. Nkadimeng, Taung, 19 April 1984, Tape No. 467 |
| FAM. 18 | Maine, Hwai & Maine, Kas | M. T. Nkadimeng, Motlhabe, Rustenburg, 25 July 1984, Tape No. 483 |
| FAM. 19 | Teu, Mosala | M. T. Nkadimeng, Khuma, Stilfontein, 15 November 1984, Tape No. 507 |
| FAM. 20 | Abrahams, Abie | M. T. Nkadimeng, Boitumelong, Bloemhof, 21 April 1984, Tape No. 466 |
| FAM. 21 | Malete, Sellwane | M. T. Nkadimeng, Matlapaneng, Taung, 25 June 1985, Tape Nos. 525 & 526 |
| FAM. 22 | Maine, Kas | M. T. Nkadimeng, Ledig, 23 October 1985, Tape No. 537 |
| FAM. 23 | Maine, Sebubudi | M. T. Nkadimeng, Vaalrand, 20 November 1985, Tape No. 538 |
| FAM. 24 | Maine, Ramphese | M. T. Nkadimeng, Warrenton, 21 November 1985, Tape No. 539 |
| FAM. 15 | Maine, Molefi | M. T. Nkadimeng, Boitumelong, Bloemhof, 27 June 1985, Tape No. 524 |
| FAM. 26 | Maine, Sellwane | M. T. Nkadimeng, Gannalaagte, 27 March 1986, Tape Nos. 543 & 544 |
| FAM. 27 | Maine, Sebubudi | M. T. Nkadimeng, Vaalrand, 26 March 1986, Tape No. 541 |
| FAM. 28 | Maine, Baefesi | M. T. Nkadimeng, Mareetsane, Mafikeng, April 1986, Tape No. 545 |
| FAM. 29 | Deu, Ntholeng | M. T. Nkadimeng, Tembisa, 20 May 1986, Tape No. 548 |
| FAM. 30 | Maine, Sellwane. | M. T. Nkadimeng, Gannalaagte, 2 June 1986, Tape Nos. 550 & 554 |
| FAM. 31 | Maine, Sellwane | M. T. Nkadimeng, Gannalaagte, 18 June 1986, Tape No. 553 |
| FAM. 32 | Maine, Sellwane | M. T. Nkadimeng, Gannalaagte, 18 June 1986, Tape No. 553 |
| FAM. 33 | Maine, Motlagomang | M. T. Nkadimeng, Hertzogville, 23 October 1986, Tape No. 555 |
| FAM. 34 | Maine, Motlagomang | M. T. Nkadimeng, Hertzogville, 25 November 1986, Tape Nos. 556 & 557 |

| | | |
|---|---|---|
| FAM. 35 | Moate, Nthakwana | C. van Onselen, Ledig, 10 March 1987 |
| FAM. 36 | Moate, Nthakwana | C. van Onselen, Ledig, 30 March 1988 |
| FAM. 37 | Moate, Nthakwana | C. van Onselen, Ledig, 26 July 1988 |
| FAM. 38 | Teu, Dikeledi | M. T. Nkadimeng, Ledig, 17 January 1989, Tape No. 676 |
| FAM. 39 | Teu, Mosala & Morwesi | M. T. Nkadimeng, Ledig, 17 January 1989, Tape No. 677 |
| FAM. 40 | Teu, Mosala & Moate, Nthakwana | M. T. Nkadimeng, Ledig, 18 January 1989, Tape No. 678 |
| FAM. 41 | Maine, Leetwane | M. T. Nkadimeng, Ledig, 3 November 1980, Tape No. 238 |
| FAM. 42 | Moate, Nthakwana | C. van Onselen, Ledig, 30 April 1989 |
| FAM. 43 | Deu, Ntholeng | M. T. Nkadimeng, Tembisa, 20 May 1986, Tape No. 548 |
| FAM 44 | Moate, Nthakwana | E. Kgomo and M. T. Nkadimeng, Ledig, 30 March 1989, Tape No. 680 |
| FAM. 45 | Maine, Thakane | M. T. Nkadimeng, Itsoseng, Lichtenburg, 28 June 1985, Tape No. 529 |
| FAM. 46 | Lepholletse, Tseleng | M. T. Nkadimeng, Taung, 26 June 1985, Tape No. 527 |
| FAM. 47 | Moate, Nthakwana | M. T. Nkadimeng, Ledig, 17 July 1990, Tape No. 711 |
| FAM. 48 | Tantane, Pakiso | M. T. Nkadimeng, Stilfontein, 23 November 1990, Tape No. 715 |
| FAM. 49 | Moate, Nthakwana | M. T. Nkadimeng, Ledig, 14 November 1990, Tape No. 716 |
| FAM. 50 | Moate, Nthakwana | M. T. Nkadimeng, Ledig, 8 March 1991, Tape No. 719 |
| FAM. 51 | Teu, Ramoritshana | M. T. Nkadimeng, Ledig, 18 January 1989, Tape No. 618 |
| FAM. 52 | Maine, Mmusetsi | M. T. Nkadimeng, Ledig, 20 April 1983, Tape No. 385 |
| FAM. 53 | Moate, Nthakwana | M. T. Nkadimeng, Ivory Park, Tembisa, 27 August 1991, Tape No. 752 |
| FAM. 54 | Maine, Mosala & Maine, Morapedi | M. T. Nkadimeng, Ledig, 17 September 1991, Tape No. 754 |
| FAM. 55 | Moate, Nthakwana | M. T. Nkadimeng, Ledig, 18 September 1991, Tape No. 755 |

# NOTES

## Introduction

1. *Sunday Star*, 29 May 1993.
2. Quoted in C. Murray, *Families Divided: The Impact of Migrant Labour in Lesotho* (Johannesburg 1986), p. xi.
3. See T. J. Keegan, *Rural Transformations in Industrializing South Africa* (Johannesburg 1986), pp. 15–19.
4. See M. Lacey, *Working for Boroko: The Origins of a Coercive Labour System in South Africa* (Johannesburg 1981).
5. See C. van Onselen, 'The Reconstruction of a Rural Life from Oral Testimony: Critical Notes on the Methodology Employed in the Study of a Black South African Sharecropper,' *The Journal of Peasant Studies*, Vol. 20, No. 3, April 1993, pp. 494–514.

## 1. Origins

1. See FAM. 23, pp. 1–3.
2. I am greatly indebted to David Ambrose for generously providing me with factual information relating to the geography and history of pre-colonial Lesotho.
3. See A. Mabrille and H. Dieterlen (eds.), *Southern Sotho-English Dictionary* (Morija 1950), pp. 110–11. On the difaqane see, for example, W. F. Lye and C. Murray, *Transformations on the Highveld: The Tswana and Southern Sotho* (Cape Town 1980), pp. 28–39.
4. For 'myths and realities' surrounding this early northern offshoot of BaFokeng society, see B. Bozzoli, *The Women of Phokeng* (Johannesburg 1991), pp. 29–34. Seonya's movements between the Malutis and the highveld and the birth of Hwai, are reconstructed from K. 1, p. 6; K. 6, p. 32 and K. 35, pp. 1 and 10–12. For a slightly different version, see FAM. 3, p. 4 and FAM. 11, p. 24. On the complexities of Sotho-Tswana culture see Lye and Murray, *Transformations on the Highveld*, pp. 15–19.
5. It is possible that Seonya—descendent of the BaPhoka who had as their

totem *phoka* (dew)—sided with Madubane, since the BaFokeng (of which the BaPhoka are part) were linked to the BaHurutshe through chiefly marriage, see P. L. Breutz, *The Tribes of the Marico District*, Ethnological Publications No. 30 (Pretoria 1953–54), p. 27, para. 75. The Maine totem is alluded to in FAM. 4, p. 4 and the BaFokeng linkage in K. 46, p. 7. The complex role that Lesele played in BaKubung politics is outlined in H. L. Dugmore, 'Land and the Struggle for Sekama: The Transformation of a Rural Community—The BaKubung, 1884–1937' (B.A. [Hons.] diss., University of the Witwatersrand [U.W.] 1985), especially pp. 11–13.

6. See K. 1, p. 6; K. 6, p. 32 and K. 35, pp. 1 and 10–12.
7. See C. Murray, *Families Divided: The Impact of Migrant Labour in Lesotho* (Johannesburg 1981), pp. 10–11.
8. See K. 1, p. 6; K. 3, pp. 2–3; K. 6, p. 7; K. 35, p. 6; K. 46, p. 3; FAM. 11, pp. 4–5 and FAM. 24, pp. 1–2. See also C. Bundy, *The Rise and Fall of the South African Peasantry* (London 1979).
9. See Murray, *Families Divided*, pp. 11–12.
10. On Mequatling see R. C. Germond, *Chronicles of Basutoland* (Morija 1967) or T. R. H. Davenport, *South Africa: A Modern History* (Johannesburg 1987), p. 148. Kas Maine, who as a young boy first heard this told to him by his grandfather Hwai, later—as a nonagenarian—recalled the landlord's name as 'Vloermann.' See K. 42, p. 11 and K. 46, pp. 1–2. The title deeds show that in the century that followed its closure as a mission station by President Brand in 1870, Mequatling had only three owners—C. G. Radloff (1874–89), J. M. L. Woldmann (1889–1946) and M. A. A. M. Geertshen (1946–67). I am indebted to Karel Schoeman and the late Mr. Woldmann's daughter, Dr. M. A. A. M. Geertshen, for biographical details relating to her father. In particular, I have drawn on her unpublished manuscript, 'Mequatling' (n.d.).
11. The brothers Tshilo and Sekwala, in accordance with Sotho custom, married cousins—in this case they also happened to be sisters, the daughters of Moabi, a kinsman of Hwai. See K. 35, p. 9; FAM. 11, p. 10; FAM. 23, p. 5 and E. Msimango, Field Report, Ledig, 25–26 October 1985. See also K. 35, pp. 1–4 and K. 48, pp. 26–27, 30 and 32.
12. On Winburg see T. J. Keegan, *Rural Transformations in Industrializing South Africa: The Southern Highveld to 1914* (Johannesburg 1986), p. 42. See also Dugmore, 'The BaKubung, 1884–1937,' pp. 34–54.
13. Following the logic of this argument, their most likely hosts would have been the BaHurutshe-ba-ga-Gopane, since it was from amongst the people of Chief Sebogodi that Madubane, Lesele's mother, had originally been drawn. See Dugmore, 'The BaKubung, 1884–1937,' p. 13 and Breutz, *Tribes of the Marico District*, p. 102, para. 276.
14. A considerable number of BaHurutshe, a group that also enjoyed connections with Basutoland, were living there, too. See P. L. Breutz, *The Tribes of the Mafeking District*, Ethnological Publications No. 32 (Pretoria 1955–56), p. 24, para. 57. On the dispossession of the BaRolong, see K. Shillington, *The Colonisation of the Southern Tswana, 1870–1900*

(Johannesburg 1985), p. 103. Kas Maine referred to the family being taken in at 'Verlof' by the 'BaTshweneng'—presumably referring to the period spent in exile at Tswaneng by this section of the BaRolong. See K. 37, p. 1; K. 49, p. 2 and Breutz, *Tribes of the Mafeking District*, pp. 185, para. 534 and 187–88, para. 541.

15. The Mokawanes, living in the village of Bodikela around Madibogo under the headman Lebenya, were members of the Modibedi clan who originally came from the village of Phahameng, in the Morija district of Basutoland. See K. 19, pp. 1–2 and, more especially, K. 35, pp. 13–14. In the mid-1940s a sizeable community of southern Sotho speakers was still resident there. See Breutz, *Tribes of the Mafeking District*, Table No. XII, p. 40. Motheba Mokawane's *kwena* (crocodile) totem is noted in K. 37, p. 9.
16. The period between Sekwala's first and second marriages is noted in K. 48, p. 32.
17. Decades after these events, Kas Maine could still offer a detailed description of the four 'ponies' lost. See K. 35, pp. 15–16 and C. van Onselen, Field Report, Ledig, 7 June 1984. On the historical context of rustling see Shillington, *Colonisation of the Southern Tswana*, pp. 138–39.
18. See Shillington, *Colonisation of the Southern Tswana*, pp. 129–36.
19. See K. 1, pp. 13–15; K. 12, p. 19 and K. 37, p. 4.
20. The title deeds indicate that half of Rietput was made over to J. C. Reyneke and his son, a minor, Petrus Albertus Reyneke, on 4 February 1904. Office of the Registrar of Deeds (O.R.D.P.), Pretoria, Farm Register HO, Folio 60/1. Details relating to Reyneke and his early career are taken from a letter to the author, 28 January 1992, from Reyneke's only surviving child, Mrs. H. D. Jonker, then 85.
21. Letter to the author from Mrs. Jonker. On Scribante see G. Sani, *History of the Italians in South Africa, 1489–1989* (Edenvale n.d.), p. 72.
22. See K. 12, pp. 19–22.
23. See K. 1, pp. 13–15; K. 5, pp. 4–5; K. 12, p. 19; K. 26, pp. 1–2 and K. 37, p. 4. On the 75 plough see B. H. Morgan, *Report on the Engineering Trades of South Africa* (London 1902), pp. 135–36.
24. K. 56, p. 26.
25. FAM. 23, p. 9.
26. See Shillington, *Colonisation of the Southern Tswana*, p. 140. O.R.D.P., Farm Register HO, Folios 79/1 and 80/1 show that Gerrit Jacobus van Niekerk was given title to Niekerksrust and Zorgvliet on 9 and 15 August 1893, respectively. Van Niekerk was also the local Native Commissioner; see H. P. Maree, *Die Geskiedenis en Ontwikkeling van die Dorp en Distrik van Schweizer-Reneke, 1870–1952* (roneo 1952), p. 21. For the coercive role that such officials sometimes played in the rural economy see S. Trapido, 'Landlord and Tenant in a Colonial Economy: The Transvaal 1880–1910,' *Journal of Southern African Studies*, Vol. 5, No. 1, 1978, pp. 26–58.
27. See K. 1, pp. 8–10; K. 5, pp. 8–10; K. 12, p. 24; K. 36, pp. 19–20 and

K. 37, p. 3. See also C. van Onselen, 'Reactions to Rinderpest in Southern Africa,' *Journal of African History*, Vol. XII, No. 3, 1972, pp. 473–88.

28. The return to Reyneke's property is noted in rather desultory fashion in K. 36, p. 20 and K. 37, pp. 4–6. Kas Maine remembered a smattering of Kora words which were almost certainly also shared by local SeTswana speakers. See K. 6, pp. 2–7 and Breutz, *Tribes of the Mafeking District*, p. 77, para. 201.
29. See K. 1, p. 8 and K. 6, p. 25.
30. O.R.D.P., Farm Register HO, Folio 59/1 shows that J. C. A. Henderson acquired the property from J. W. Fletcher on 19 September 1891 and sold it to Henderson Consolidated on 20 August 1896. The farm was apparently something of a centre for 'kaffir-farming' in the 1890s. See, for example, M. T. Nkadimeng, interview with E. M. Maine, Molelema, Taung, 8 September 1983 (Tape No. 394, M. M. Molepo Oral History Collection, Institute for Advanced Social Research [I.A.S.R.], U.W.). This reputation persisted into the twentieth century—see Transvaal Archives (T.A.), Secretary for Native Affairs (S.N.A.), Vol. 457, NA. 410/10, S.N.A. to Native Commissioner, Christiana, 7 February 1910. Niewoudt's presence as landlord is noted in K. 36, p. 20 and K. 37, p. 6.
31. C. van Onselen, interview with H. F. B. Swanepoel, Schweizer-Reneke, 21 February 1985. The formation of this class of white tenants is noted in Trapido, 'Landlord and Tenant in a Colonial Economy.'
32. See K. 20, p. 8 and K. 40, pp. 27–28.
33. See K. 57, p. 17.
34. On Swanepoel see K. 20, p. 17 and K. 57, p. 17.
35. See T. Pakenham, *The Boer War* (London 1979), pp. 9–99.
36. On the role of the 'agterryer' see W. R. Nasson, 'Black Society in the Cape Colony and the South African War of 1899–1902: A Social History' (Ph.d. diss., Cambridge University, 1983), pp. 148–54.
37. See K. 12, p. 25; K. 37, p. 13; K. 42, p. 1 and K. 49, pp. 2–4. The wider context can be traced in Pakenham, *Boer War*, pp. 200, 329–30, and 341–42. The position assumed by the Bloemhof Commando at Magersfontein is illustrated in P. Warwick (ed.), *The South African War* (London 1980), p. 80. Clearly Tollie de Beer was not amongst those who surrendered at Paardeberg since he was still actively involved in the war in September 1900; see L. S. Amery (ed.), *The Times History of the War in South Africa, 1899–1902*, IV (London 1906), p. 56.
38. This strategy of discarding Sotho names was later recalled by one of Balloon's sons, Ramphese; FAM. 24, p. 4. On the use of 'captive balloons' at Magersfontein see Pakenham, *Boer War*, pp. 205–6.
39. See K. 1, pp. 1–9; K. 12, p. 28 and K. 37, p. 6. Also L. S. Amery (ed.), *Times History of the War*, IV, p. 224.
40. See K. 1, p. 9 and K. 8, p. 7.
41. See K. 1, p. 10 and K. 12, pp. 28–29.
42. See K. 1, pp. 9–12; K. 8, p. 7 and K. 37, p. 14. But see also K. 12, p. 38, where the brother is referred to as 'Moshawa.' Also P. Warwick,

*Black People and the South African War, 1899–1902* (Johannesburg 1983), pp. 33–34.

43. See K. 1, p. 10; K. 6, pp. 29–30; K. 8, pp. 6–7; K. 12, pp. 27–29 and K. 37, pp. 14–15.

## 2. Foundations

1. K. 53, p. 26, and more generally, S. Trapido, 'Landlord and Tenant in a Colonial Economy,' *Journal of Southern African Studies*, Vol. 5, No. 1, October 1978, pp. 44–45.
2. See K. 1, p. 12; K. 8, p. 7; K. 12, p. 28 and K. 37, pp. 14–15.
3. Office of the Registrar of Deeds, Pretoria (O.R.D.P.), Farm Register Vol. 12, HO 208–31, Folio 212/1 (Zoutpan), and Folio 217/1 (Mahemspanne).
4. See K. 43, pp. 25–26; K. 27, p. 36; K. 43, p. 26 and K. 61, p. 17.
5. See K. 1, pp. 3 and 11; K. 5, pp. 6–8; K. 6, pp. 2–3 and K. 49, p. 6. It is worth noting that Kas used the Kora terms *diduku* and *ditshepe* to describe pits and springbok rather than the SeSotho nouns *lengope* and *phuthi*. This points to a Koranna influence at the turn of the century which is difficult to detect in subsequent decades. For the numbers of Koranna living in the district at the time see, for example, Transvaal, *Annual Report of the Native Affairs Department for the period 1st July 1909 to 31st May 1910* (Pretoria, U.G. 15-1911), Appendix No. 3, p. 80. For springbok in the district during 1900–1910 see Union of South Africa, *Annual Report of the Department of Justice 1910*, p. 59, or *1912* (U.G. 44-13), p. 148. The Maines also practised BaSotho hunting and rainmaking rituals such as *molutsoane*. See K. 43, pp. 15–17, and J. and C. Perry (eds.), *A Chief is a Chief by the People: The Autobiography of Stimela Jason Jingoes* (Cape Town 1975), pp. 33–34.
6. See K. 36, p. 20 and FAM. 3, p. 2.
7. See K. 1, p. 12; K. 26, p. 3 and K. 29, p. 2. At the time the farm was run by G. H. Meyer on behalf of his widowed mother, who had lost her husband during the South African War. This arrangement lasted until the death of the widow in 1912, when the property was sold to W. A. Nieman. See O.R.D.P., Farm Register Vol. 12, HO 208–231, Folio 216/1/1. See also C. van Onselen, interview with J. J. P. Meyer (son of G. H. Meyer), Wolmaransstad, 16 July 1985.
8. See K. 23, pp. 15–16 and K. 41, p. 8.
9. See K. 55, pp. 18–19 and FAM. 14, p. 2.
10. See K. 1, pp. 11–13; K. 5, pp. 13–14 and K. 26, p. 3. In K. 26, p. 5 mention is made of a cow acquired from Magistrate Kemp at Schweizer-Reneke after the war.
11. See K. 5, pp. 11–12. But see also FAM. 3, p. 8; FAM. 14, p. 2; K. 42, p. 1; K. 43, p. 36; K. 44, p. 11 and K. 50, p. 1.
12. O.R.D.P., Farm Register Vol. 12, HO 208–231, Folio 221/1 and C. van Onselen, interview with H. F. B. Swanepoel, Schweizer-Reneke, 21 February 1985. On Stern, a noted Republican sympathiser during the war,

see the obituary 'Mr. L. Stern. A Personal Friend of President Kruger,' *The Star*, 3 December 1935.

13. See K. 2, pp. 1–2; K. 26, pp. 3–4; K. 28, pp. 7–8; K. 30, pp. 8–11; K. 38, p. 11; K. 40, p. 31; K. 43, p. 36 and K. 50, pp. 1–2.
14. See Transvaal, *Annual Report Native Affairs 1st July 1909 to 31st May 1910*, Appendix No. 11, Stock 1909–1910, p. 107 and M. T. Nkadimeng, Field Report, Ledig, 28 February 1983.
15. See K. 6, pp. 6–20 and 52–55.
16. See K. 6, pp. 30–32 and 36–37; K. 55, p. 24 and FAM. 11, pp. 17–18.
17. Transvaal Archives (T.A.), Mining Commissioner Klerksdorp (M.C.K.), Vol. 1/72, File 525/4, 'Memorandum as to Alluvial Diamond Diggings in the District of Bloemhof, Wolmaransstad and Potchefstroom, Transvaal Province,' 1911.
18. BLOEMHOF DIAMOND FIELDS—OUTPUT, 1909–1918

| Year | Output in carats | Value (£) | Price per carat (£) |
|---|---|---|---|
| 1909 | 783 | 2,886 | - |
| 1910 | 1,542 | 7,412 | - |
| 1911 | 37,382 | 196,777 | - |
| 1912 | 76,661 | 378,114 | 4-18-6 |
| 1913 | 81,780 | 420,015 | 5- 2-6 |
| 1914 | 36,645 | 136,619 | 3-14-6 |
| 1915 | 31,977 | 121,838 | 3-16-3 |
| 1916 | 45,558 | 239,519 | - |
| 1917 | 68,833 | 415,575 | 6-10-0 |
| 1918 | 75,144 | 410,792 | 5-10-0 |

Standard Bank Archives (S.B.A.), Johannesburg, Annual Inspection Reports, Bloemhof Branch, 31 January 1912, 15 February 1913, 26 March 1914, and 13 September 1919.

19. See S.B.A., Johannesburg, Inspection Report of the Bloemhof Branch, 31 January 1912, p. 5. See also T.A., Transvaal Provincial Secretary (TPS), Vol. 45, File 6089, Petition addressed to His Honour, The Administrator, December 1911 and undated departmental memo entitled simply 'Bloemhof.'
20. O.R.D.P., Farm Register Vol. HO, Folio 340/1, shows that Kareefontein was owned by three companies between 1902 and 1922—the B.N.B. Syndicate (1902–1920), The African Farms Ltd. (1920), and South African Townships and Mining Corporation Ltd. (1920–1922). It is possible that Wood's brother facilitated the hiring of the property. See also T.A., TPS, Vol. 45, File 6089, M. G. Wood to Resident Magistrate, Christiana, 20 February 1911 (and accompanying 'Report on Bloemhof and Commonage'); and M. G. Wood to the Administrator, Transvaal Province, 29 March 1911.
21. See Union of South Africa, *Annual Report of the Department of Justice 1911* (Pretoria U.G. 56-12), p. 115 and 1910–12 (Pretoria U.G. 44-13), p. 130, where the complaint is repeated.
22. See K. 37, p. 6 and K. 43, p. 27.

23. See K. 36, p. 15; K. 43, p. 27; K. 46, p. 10; FAM. 11, p. 3 and p. 1.
24. See FAM. 23, p. 4 and K. 49, p. 34.
25. FAM. 3, pp. 34–35.
26. See K. 49, pp. 34 and K. 57, pp. 50–51.
27. See K. 43, pp. 27–28 and FAM. 14, p. 4. This practice reeked of the pre-war coercion sometimes practised by local Boer notables and, in another context, Kas Maine once noted that, 'The British Magistrates were brought after the Anglo-Boer War, but the law remained that of the Boers.' K. 26, p. 5.
28. T.A., S.N.A., Vol. 457, NA 412/10, Sec. for Native Affairs to the Native Commissioner, Christiana, 7 February 1910. This thirst for labour was partly fired by a 25 percent increase in land values in the Bloemhof district between 1908 and 1910. See Union of South Africa, *Annual Report of the Department of Justice 1910* (Pretoria U.G. 53-11), p. 52.
29. T.A.D., S.N.A., Vol. 457, NA 412/10, Res. Magistrate, Christiana to S.N.A., 14 June 1910. See also Trapido, 'Landlord and Tenant in a Colonial Economy,' p. 43.
30. K. 26, pp. 3–5. In 1909 the Bloemhof district received a miserly 11.84 inches of rain over forty-nine days, which helps account for the poor harvest in 1910. See Union of South Africa, *Annual Report of the Department of Justice 1910* (Pretoria U.G. 51-11), p. 53. Kas, however, often incorrectly referred to 1906 as having been the 'Year of the Star.' For an example of this rare failure of his memory, see K. 47, p. 16.
31. See K. 26, pp. 3–4.
32. See D. W. Kruger (ed.), *South African Dictionary of Biography*, Vol. III (Cape Town 1977), p. 525.
33. O.R.D.P., Farm Register No. 10, HO 184-191, Folio 188/1, shows that A. V. Lindbergh first acquired a one-third share in the property (together with A. Mackintosh and E. C. Jamieson) in April 1908, and that Lindbergh subsequently bought out his partners' shares in the property between June and October of the following year. See also C. van Onselen, interview with F. M. Lindbergh, Wolmaransstad, 12 June 1984. For the wider context see S. Trapido, 'Poachers, Proletarians and Gentry in the early Twentieth Century Transvaal' (Institute for Advanced Social Research [I.A.S.R.], University of the Witwatersrand [U.W.], 12 March 1984).
34. See, for example, Union of South Africa, *Annual Report of the Department of Justice 1911*, p. 115 or *1912*, p. 130.
35. See K. 1, p. 36; K. 6, pp. 38–42 and 62; K. 43, p. 29 and FAM. 3, p. 10.
36. On sheep farming in the district at the time see Union of South Africa, *Annual Report of the Department of Justice 1911*, p. 100.
37. See K. 57, pp. 35–38; K. 61, p. 20 and FAM. 11, p. 22. The Maine brothers' reputation as sheep-shearers is noted in C. van Onselen, interview with A. C. G. Nieman, Bloemhof, 16 July 1985. On gal-lamziekte see Ministry of Agriculture, *Handbook for Farmers in South Africa* (Pretoria 1929), pp. 216–17.

38. See K. 1, p. 18; K. 6, p. 50 and K. 50, pp. 1–2.
39. On these and the following rituals, see K. 9, pp. 11–13; K. 22, pp. 19–22; K. 30, p. 21; K. 50, pp. 10–11 and K. 57, p. 34. On the communal hunt see J. and C. Perry (eds.), *A Chief is a Chief by the People*, pp. 33–34.
40. See K. 6, p. 23; K. 55, pp. 6–7 and FAM. 23, pp. 9–10.
41. See K. 46, pp. 4–5. On the expansion of the A.M.E.C. see *Annual Report Native Affairs 1st July 1909 to 31st May 1910*, p. 11. Since Khunwana was home to several of Motheba's kin, it is possible that it formed one source of her enthusiasm for the A.M.E.C. See also FAM. 11, p. 42.
42. See K. 29, pp. 1–2.
43. See K. 38, p. 16 and K. 47, p. 8. Biographical details relating to Anderson and Hambly drawn from letters to the author from Mrs. F. M. Hambly (daughter-in-law of the late W. H. Hambly), 27 May and 24 July 1985.
44. K. 43, p. 28. On the Fencing Act see, for example, I. Hofmeyr, 'The Spoken Word and the Barbed Wire: Oral Chiefdoms versus Literate Bureaucracies,' I.A.S.R., U.W., 2 March 1992.
45. See K. 6, pp. 54–55; K. 26, p. 5; K. 39, p. 9; K. 43, pp. 28–29; K. 49, pp. 34–35, and C. van Onselen, interview with F. M. Lindbergh, Wolmaransstad, 12 June 1984.
46. See K. 6, pp. 43–48; K. 11, p. 19; K. 47, p. 17 and K. 53, p. 36.
47. See K. 6, pp. 49–52; K. 17, p. 10; K. 48, pp. 1–4; K. 52, p. 20 and FAM. 14, pp. 10–11.
48. See K. 6, p. 53; K. 49, pp. 6–7 and K. 55, pp. 13–14.
49. See K. 46, p. 4.
50. The demand to cull the family herd is recalled in K. 61, p. 22. The triple impact of gal-lamziekte, drought, and herds of invading springbok in search of grazing in the Bloemhof and Wolmaransstad districts is noted in Union of South Africa, *Annual Report of the Department of Justice for 1912*, pp. 128 and 148.
51. See K. 6, pp. 39–41.
52. See K. 6, p. 56.
53. See K. 6, pp. 56–58.
54. See K. 47, p. 9; K. 53, pp. 36–38; K. 55, pp. 6–7 and K. 57, p. 11.

## 3. Preparation

1. See K. 52, pp. 17–19 and FAM. 14, p. 12.
2. See K. 8, pp. 2–3.
3. On Plaatje see B. Willan, *Sol Plaatje: South African Nationalist 1876–1932* (London 1984), pp. 161–62. For the broader context of the 1913 Land Act see T. R. H. Davenport, 'Some Reflections on the History of Land Tenure in South Africa, Seen in the Light of Attempts by the State to Impose Political and Economic Control,' *Acta Juridica*, 1985, pp. 53–76.
4. K. 11, p. 19.
5. See K. 61, p. 21. Office of the Registrar of Deeds, Pretoria (O.R.D.P.),

Farm Register Vol. 12, HO 208–231, Folio 212/1, records H. F. E. Pistorius as having acquired title to Soutpan on 11 March 1898. For Kas's references to 'Stories' (i.e. Pistorius) see, for example, K. 26, p. 5 and K. 61, p. 21. Note, too, that it was around Bloemhof that Plaatje '. . . encountered a large number of African families with their stock, who had travelled from the Free State, thinking that the Land Act was only in operation in that province.' Willan, *Sol Plaatje*, p. 162.

6. See K. 52, p. 18.
7. In its narrowest sense, lekgolwa refers to 'a satisfied person.' See T. J. Kriel (ed.), *Popular Northern Sotho Dictionary* (Pretoria 1966), p. 66. But the word is also used to refer to a migrant worker who has left home for so long that he or she loses contact with the family. See, for example, FAM. 14, p. 11.
8. See K. 52, p. 19 and K. 61, pp. 26–27.
9. The collapse of the Mooifontein diggings is noted in Standard Bank Archives (S.B.A.), Johannesburg, Inspection Report of the Bloemhof Branch, 13 February 1913, p. 15. Transvaal Archives (T.A.), MNW Series, Vol. 257, W.L. 565/14, 'Notes of Meeting held at Bloemhof on 17 August 1914.'
10. See K. 6, p. 58.
11. K. 6, p. 55. See also K. 32, pp. 3–5 and C. van Onselen, Field Report, Ledig, 22 July 1985.
12. See J. Bottomley, 'The Orange Free State and the Rebellion of 1914: The Influence of Industrialisation, Poverty and Poor-whiteism' in R. Morrell (ed.), *White but Poor: Essays on the History of Poor Whites in Southern Africa, 1880–1940* (Pretoria 1992), pp. 29–39.
13. See K. 8, p. 11. Although Kas referred to 'Holl*iday*' it was more likely to have been Holl*oway*. Charles Holloway was one of the first owners of Soutpan, having title to the property between 1859 and 1864. See O.R.D.P., Farm Register Vol. 12, HO 208–231, Folio 211/1. Throughout interview K. 8 Kas referred to the activities of Veld-Kornet 'van Rooyen' who, he said, was active in the district. From circumstantial evidence it seems likely to have been Veld-Kornet P. J. K. van Vuuren. Compare this with van Vuuren's account in Union of South Africa, *Judicial Commission of Enquiry into the Causes of and Circumstances Relating to the Recent Rebellion in South Africa, Minutes of Evidence* (Cape Town U.G. 42-'16), pp. 191–95.
14. See K. 8, p. 11.
15. K. 8, pp. 4–5 and pp. 1–12.
16. See K. 8, pp. 10–11.
17. See K. 8, p. 12; FAM. 11, pp. 47–48 and FAM. 14, pp. 5–7.
18. In K. 8, p. 12, Kas Maine suggested that the rebels went to the farm Makouskop; this is possible, although the meeting was undoubtedly held on the adjoining farm, Vleeskraal. See Union of South Africa, *Report of the Judicial Commission of Inquiry into the Rebellion*, p. 42.
19. See Union of South Africa, *Report of the Judicial Commission of Inquiry into the Rebellion, Minutes of Evidence*, p. 194 (evidence of P. J. K. van

Vuuren). Also H. P. Maree, *Die Geskiedenis en Ontwikkeling van die Dorp en Distrik van Schweizer-Reneke 1870–1952* (roneo 1952), p. 22, and Union of South Africa, *Report of the Judicial Commission of Inquiry into the Rebellion*, p. 57.

20. K. 8, pp. 13–15. In the original version Kas incorrectly refers to 'de Wet' rather than to Beyers.
21. This was, of course, also the period during which blacks and whites in the far western Transvaal reported seeing other mysterious illusions such as 'aircraft' flying overhead, notably in the neighbouring Taung and Vryburg districts. See Union of South Africa, *Annual Report of the Department of Justice 1914* (Pretoria U.G. 28-'15), p. 9.
22. See K. 6, p. 26 and K. 57, p. 9. See also Union of South Africa, *Annual Report of the Department of Justice for 1914*, p. 108, and Union of South Africa, *Report of the Land and Agricultural Bank of South Africa for the year ended 31st December 1916* (Cape Town U.G. 26-'17), pp. 3–4. Commercial banks, too, were pessimistic about agricultural prospects in the district; see S.B.A., Johannesburg, Inspection Reports of the Bloemhof Branch, 11 March 1914 and 23 October 1915.
23. See K. 6, pp. 58–59; K. 7, p. 5; K. 29, pp. 7–9; K. 57, p. 17 and K. 61, pp. 27–28.
24. See K. 6, p. 4; K. 26, pp. 4–5 and FAM. 11, p. 23.
25. See S.B.A., Johannesburg, Inspection Report of the Bloemhof Branch, 31 December 1917; Union of South Africa, *Report of the Land and Agricultural Bank of South Africa for the year ended 31st December 1916*, pp. 3–4 and Union of South Africa, *Annual Report of the Department of Justice for 1916* (Pretoria U.G. 39-'16), pp. 81–83.
26. As Kas recalled it in K. 5, p. 2: 'Look, if you did not get on well with him you would leave him, but you would come back to him again and tell him that you had been overwhelmed by anger. He would say that you could come and stay with him again.'
27. See K. 6, pp. 54–55; K. 7, p. 13; K. 30, pp. 32–34; K. 63, pp. 10–11 and FAM. 11, p. 52.
28. See FAM. 11, p. 52 and C. van Onselen, interview with A. C. G. Nieman (b. 1906), Bloemhof, 16 July 1985.
29. See K. 23, pp. 12–13.
30. See K. 10, pp. 10–12.
31. See K. 11, p. 26.
32. See Office of the Registrar of Companies, Pretoria, 'London Diamond & Estate Company Limited'; Central Archives Depot (C.A.D.), Mining Commissioner Klerksdorp (MCK), Vol. 1/53, File 1789/16, W. T. Lloyd and Mrs. Lloyd to MCK, 4 August 1915 and an advertisement for 'Kaffir Eating-House,' London Farm, in *The Diggers' News*, 13 March 1914.
33. See the report on the annual general meeting of the company in *The Diggers' News*, 6 February 1914. The grievances of the London diggers are in C.A.D., MCK, Vol. 1/53, MCC 315/14, W. T. Lloyd to MCK, 9 August 1915, and W. T. Lloyd and others to MCK, 12 April 1916.
34. See K. 11, p. 24. This produced an understandable labour shortage and

the inevitable request for the state to provide cheap contracted labour —on this occasion from the Zoutpansberg, where it was said that the government was feeding 'starving natives.' See C.A.D., MCK, Vol. 1/53, W. T. Lloyd and others to H. Rees, MCK, 12 April 1916.

35. See C.A.D., MCK, Vol. 1/53, Petition signed by S. Davis and several others to G. J. Brink, Claims Licensing Officer, Bloemhof, 1 April 1916; W. T. Lloyd and fifteen others to G. J. Brink, 5 April 1916 and W. T. Lloyd and others to H. Rees, MCK, 12 April 1916.
36. On the mohlathe system, see K. 11, pp. 21–23; K. 27, pp. 20–24 and FAM. 15, pp. 12–15. As late as 1917 the system was still widespread enough in the Bloemhof district for the state to contemplate the passage of new laws to curb the 'problem.' See Union of South Africa, *Report of the Department of Justice for 1917* (Pretoria U.G. 36-1918), p. 53.
37. See K. 27, pp. 20–21 and K. 50, pp. 9–10.
38. This vigilance was, in the first instance, intended to prevent the theft of diamonds. At a time when the mohlathe system was being widely resorted to, however, it also served to reassure other whites that the digger was not working on-the-halves with blacks, the reasoning being that the only digger who could afford to absent himself from a claim for any time was one who was certain that the remaining workers would police each other in order to guarantee their share of any find. See K. 11, p. 23.
39. See K. 31, pp. 1–3.
40. See K. 11, p. 24 and K. 31, pp. 3–4. Also C.A.D., MCK, Vol. 1/53, Petition by W. T. Lloyd and others of London Diggings to G. J. Brink, Beacon Inspector, Bloemhof, 5 April 1916.
41. C. van Onselen, interview with F. M. Lindbergh (son of the late A. V. Lindbergh), Wolmaransstad, 14 June 1985; it was suggested that A. E. Bezant replaced van der Walt in 'about 1919.' According to Kas, however, this change must have taken place at least two years earlier. Bezant spent several years on the estate and was still delivering crop reports to the Department of Agriculture in 1921. See C.A.D., R.1074/5/21, Chief, Division of Economics and Markets to A. E. Bezant, Vaalboschfontein, P.O. Kingswood (1921).
42. K. 6, p. 61.
43. See K. 6, p. 62—contrast this with the practice under van der Walt as outlined in K. 6, pp. 41–42.
44. See K. 29, p. 15 and K. 58, pp. 31–34.
45. See K. 7, p. 14; K. 21, pp. 17–21 and K. 46, p. 17. On complaints by Bloemhof farmers about diggers driving up the price of black labour see, for example, Union of South Africa, *Report of the Department of Justice for 1917*, p. 60.
46. See O.R.D.P., Farm Register, HO 341–379, Book 17, Zoutpan, Folio 349/1, entry No. 1 and HO 197–254, Kareepan, Folio 243/1, Entry No. 1. Also, C. van Onselen, interview with Mrs. A. du Toit (b. Coetzee), Wolmaransstad, 16 July 1985 and letter to the author from J. S. Coetzee ('John,' son of the late Johannes Stephanus Coetzee), September 1985.

47. It is not clear which of Sekwala's siblings followed him to Kareepan nor is it known exactly where they were located. Kas implied they were on Kareepan proper under Hans Coetzee, while his younger brother, Sebubudi, thought that most of them were on a neighbouring property belonging to Hendrik Coetzee, the oldest Coetzee brother. See K. 6, p. 61 and FAM. 3, pp. 15–16.
48. See K. 5, p. 3; K. 6, p. 62; K. 7, p. 5; K. 26, p. 6 and K. 47, p. 22.
49. See K. 26, pp. 5–6 and K. 42, p. 12. Peter Tyson has suggested that the South African climate is characterised by alternating wet and dry cycles each of approximately nine years' duration. His data suggests that a 'dry cycle' during 1906–15 was followed by a 'wet cycle' in 1916–25. See P. D. Tyson, 'The Great Drought,' *Leadership* S.A., Vol. 2, No. 3, 1983, pp. 49–57. On the breaking of the 1906–15 'dry cycle' in the southwestern Transvaal, see Union of South Africa, *Annual Report of the Department of Justice 1918* (Pretoria U.G. 36-'19), p. 102.
50. See K. 43, pp. 31–32; K. 57, p. 49; FAM. 3, p. 15 and letter to the author from J. S. Coetzee. See also Union of South Africa, *Report of the Land and Agricultural Bank of South Africa for the year ended 31st December 1915* (Cape Town U.G. 12-'16), p. 7, for the trend toward substituting donkeys for cattle as draught animals as a result of gal-lamziekte.
51. See K. 35, p. 19 and K. 63, pp. 15–16.
52. See K. 6, p. 62 and K. 61, p. 27.
53. See K. 27, pp. 27–28 and K. 61, p. 27. For documentary evidence of the points raised see C.A.D., MNW, M.M. 2333/18, Report on the Alluvial Diamond Diggings of the Western Transvaal by the Inspector of White Labour (D. Steyn), presented to the Superintendent and Chief Inspector of White Labour, 2 July 1918, and JUS, Vol. No. 357, File 3/16701/2, Police Enquiry, Statement by Major A. E. Court, Bloemhof, 18 January 1923.
54. See K. 7, pp. 9–13. The line from the song is taken from a fuller version recalled by Leetwane Maine—see FAM. 1, p. 20. A brief account of the flu epidemic at London and Kameelkuil is in the *Klerksdorp Record and Western Transvaal News*, 1 November 1918 under the headline 'The Flu Epidemic Bloemhof.'
55. Kas, who knew that the influenza virus was brought to South Africa by troops 'who went to Europe in the Kaiser's War,' thought his bout of flu had been contracted from a horse which, he suggested, was imported from Europe. See K. 7, pp. 11–13 and K. 63, p. 11.
56. See FAM. 11, pp. 48–49.
57. See K. 34, p. 4; K. 41, p. 19; K. 58, pp. 20–21 and FAM. 11 pp. 48–50. Also letter to the author from J. S. Coetzee. Further details about both Abraham Pretorius and his sister, Maria Magdalena Pretorius (Mrs. Hans Coetzee), are in O.R.D.P., Farm Register HO, Folio 231/6/1, entries 3 and 4.
58. See FAM. 11, pp. 48–50 and K. 41, p. 19.
59. Details of Hutchinson's career are in the records of the Registrar of the South African Medical and Dental Council in Johannesburg. Other bi-

ographical information supplied by Henry Hutchinson's son, H. Hutchinson Jnr. of East London, in a telephone conversation with the author, 6 August 1984.

60. For Kas's views of Hutchinson (referred to as 'Hickson') see K. 15, p. 48; K. 34, p. 4; K. 55, p. 1 and K. 61, p. 3.
61. See K. 15, pp. 14–15; K. 49, p. 19 and FAM. 11, p. 51.
62. See K. 15, p. 15; K. 21, pp. 2–3 and K. 26, p. 6.
63. See C.A.D., JUS, Vol. 275, 1/307/19, Annual Report of the Resident Magistrate, Bloemhof, for the year 1919; JUS, Vol. 304, 1/811/21, Annual Report of the Resident Magistrate, Sub-District of Schweizer-Reneke, for the year 1919, p. 3 and JUS, Vol. 275, 1/307/119, Annual Report of the Resident Magistrate, Wolmaransstad, for the year 1919, p. 2. For confirmation of these trends see also Union of South Africa, *Annual Report of the Department of Justice for the year 1919* (Pretoria U.G. 35-'20), pp. 94–95, and Suid-Westelike Transvaalse Landboukoöperasie Beperk (S.W.T.L.), *'n Halfeeu van Koöperasie, 1909–1959* (Leeudoringstad 1959), pp. 51 and 64.
64. See K. 29, p. 17; K. 34, p. 4; K. 41, p. 19 and K. 49, p. 18.
65. C. van Onselen, interview with J. J. P. Meyer and Mrs. H. H. de Waal (son and daughter of the late G. H. Meyer of Rietfontein), Wolmaransstad, 16 July 1985.
66. O.R.D.P., Farm Register HO, Bloemhof, Folio 234/1, entries 1–7. On Nellmapius, see H. Kaye, *The Tycoon and The President: The Life and Times of Alois Hugo Nellmapius, 1847–1893* (Johannesburg 1978), or G. Wheatcroft, *The Randlords* (London 1985), pp. 75 and 124–25.
67. See K. 18, p. 3; K. 34, pp. 5–6; K. 37, p. 10; K. 53, p. 38 and FAM. 3, p. 16. It should be noted, however, that in K. 1, pp. 16–18, Kas implies that he had a sharecropping arrangement with Meyer. For several reasons—including the payment of rent, the absence of draught animals or a plough of his own, and his lack of access to family labour—this seems unlikely, and I have therefore chosen the dominant interpretation emerging from the documentation. But even that contains certain minor contradictions. For example, there is some uncertainty as to whether the rent at Kommissierust amounted to four pounds or five shillings per annum—see K. 18, p. 3 and K. 34, pp. 5–6. Given the economic context I have opted for the higher figures.
68. See K. 46, p. 7.
69. See K. 35, p. 21; K. 46, p. 7 and FAM. 12, p. 10. The issue of names and naming under racially oppressive regimes is a well-developed theme in the literature. Under slavery see, for example, E. Genovese, *Roll Jordan Roll* (New York 1976), pp. 443–50.
70. See K. 20, p. 5. On the differences between a herbalist and a diviner see, for example, W. D. Hammond-Tooke (ed.), *The Bantu-speaking Peoples of Southern Africa* (London 1974), pp. 342 and 349.
71. See K. 6, pp. 32–33 and K. 20, pp. 3–4.
72. See K. 2, p. 26; K. 6, pp. 33–34; K. 15, p. 22 and K. 20, pp. 2–5. Although Kas always referred to himself as a ngaka, or herbalist, it is clear that his

skills went beyond this since he made use of divining bones; in traditional terms, that was probably closer to being a *selaoli*. As Hammond-Tooke notes, in *The Bantu-speaking Peoples of Southern Africa*, p. 349, 'The traditional South Sotho diviner is called selaoli and uses bones in his diagnosis and treatment. The skill is acquired from another selaoli who is paid for his instruction.'

73. See K. 20, pp. 6–8.
74. See K. 12, p. 2.
75. See K. 12, p. 3 and K. 38, pp. 13–14.
76. See FAM. 1, pp. 1–8 and K. 43, p. 23.
77. See K. 15, pp. 6–7.
78. See FAM. 1, p. 23 and K. 15, pp. 9–11.
79. There is some ambiguity about the extent of bridewealth. In K. 7, pp. 14–15, Kas suggested that the bohadi was twenty cattle. Leetwane, however, remembered that ten cattle, a horse and bodumo were involved—see FAM. 1, p. 25. In later interviews with Kas he made no further mention of the additional ten cattle. Bodumo is said to be a customary BaSotho payment of ten sheep to the bride's family over and above the bohadi; see K. 15, pp. 13–14.
80. In K. 15, p. 12 and K. 18, p. 4, Kas mentions a farmer whom he refers to as 'Jan Griff' or 'Griffith.' Circumstantial evidence indicates that it might have been Johannes Erasmus Greef, who farmed an adjacent portion of Kommissierust during the early 1920s. See O.R.D.P., Farm Register HO, Bloemhof, Folio 234/6/1, Entry No. 1.
81. For an account of the Bloemhof stock fairs see C.A.D., JUS, Vol. 275, File 1/307/19, *Annual Report of the Special Justice of the Peace, Bloemhof, for the year ended 31 December 1919*, p. 4. The same report notes the increase in livestock prices between 1918 and 1919.
82. See K. 15, p. 13 and K. 18, pp. 3–4.
83. See K. 15, pp. 12–13 and K. 27, pp. 16–18. Also, C. van Onselen, interviews with R. J. Nieman, Brits, 12 June 1985, and Mrs. A. C. G. Nieman (daughter of the late W. Griesel Jnr.), Bloemhof, 16 July 1985.
84. See K. 15, pp. 15–16. In SeSotho, a married woman is sometimes said to be somebody who has 'taken a lere.'
85. FAM. 1, p. 21.
86. K. 15, pp. 17–18.
87. See K. 35, p. 19. A similar observation is offered in K. 15, p. 5: 'All of my age group were already married by that time. I was already old by then, I was already thirty years of age.' In fact, Kas was probably closer to twenty-five when he married Leetwane.
88. See K. 5, pp. 16–17. More serious locust invasions occurred later, but the first appearances of the swarms were in 1920. See, for example, C.A.D., JUS, Vol. 367, File 1/554/23, Annual Report of the Magistrate, Sub-District of Schweizer-Reneke, 1923, where it is noted: 'The last three years have been a continuous battle with drought and locusts.'
89. See K. 20, pp. 23–24; K. 36, p. 18 and K. 38, p. 14.

90. Kas recounted these events, with minor discrepancies, twice—K. 5, pp. 16–17 and K. 38, pp. 14–16.
91. K. 15, p. 18.
92. See K. 57, pp. 40–41. It should be noted that in K. 57, p. 41, Kas suggests that it took three years for Leetwane to become pregnant. This is almost certainly incorrect: K. 15, p. 18 offers a more convincing account, which dovetails with the circumstantial evidence, that she failed to become pregnant in 1920.
93. One suggestion as to the origin of these exotic farm names is to be found in L. van der Post, *The Voice of the Thunder* (London 1993), p. 92.
94. See K. 15, p. 18; K. 36, pp. 6–9; K. 41, p. 15 and K. 57, pp. 40–41.
95. See K. 1, pp. 17–19.
96. See K. 11, p. 18.
97. See K. 42, p. 5 and FAM. 1, p. 27.
98. O.R.D.P., Farm Register HO, Bloemhof, Folio 234/1 notes that Jacob Jacobus van Deventer purchased a 400-morgen section of the farm from Northern Minerals on 15 February 1922. In all probability this was the person to whom Kas referred as 'Jaap van Deventer' since elsewhere he suggests, 'We once worked with him. The place where we lived was taken over when Kommissierust was bought. I worked with him a short while and decided no . . .' K. 43, pp. 45–46.
99. In K. 1, pp. 16–17 and K. 42, p. 5 Kas suggested that he was involved in sharecropping *prior* to his marriage. This is most unlikely, however, since other evidence shows that the property at Kommissierust was hired from Gert Meyer, that the rent was paid in cash, and that other whites complained about the farm having been 'hired for kaffirs.' None of this squares with sharecropping. It would seem that Kas in his later years spoke of sharecropping as being of a piece with his first attempts at independent cultivation.
100. See K. 43, p. 47.
101. On the acquisition of the wagon see K. 5, pp. 18 and 20.
102. See K. 46, p. 8.

## 4. INDEPENDENCE

1. C. van Onselen, interview with A. C. G. Nieman (oldest surviving son of the late W. A. Nieman), Bloemhof, 16 July 1985.
2. C. van Onselen, interview with R. J. Nieman (son of W. A. Nieman), Brits, 12 June 1985.
3. Office of the Registrar of Deeds, Pretoria (O.R.D.P.), Farm Register No. 12, HO 208–231, Folio 216/2/1, entries 1–3. See C. van Onselen, interview with R. J. Nieman.
4. On van der Horst see Federasie van Afrikaanse Kultuurvereniginge, *Afrikaanse Kultuuralmanak* (Johannesburg 1980), p. 73. See also C. van Onselen, interview with R. J. Nieman.
5. On Hersch Gabbe and wool sales, see C. van Onselen, interview with

J. Kanareck (b. 1899), Johannesburg, 3 September 1985. Jack Kanareck emigrated to South Africa from Russia in 1929, was at first employed by the firm of Gabbe & Isaacs, and later married Sarah Gabbe, daughter of Wolf Gabbe.

6. See Standard Bank Archives (S.B.A.), Johannesburg, Inspection Report on the Bloemhof Branch, 13 September 1919, p. 4; Central Archives Depot (C.A.D.), Pretoria, JUS, Vol. 367, File No. 1/554/23, Magistrate's Annual Report on the Wolmaransstad District, 1923, p. 4; and K. 58, pp. 3–4. Kas remembered fifty horses at a time being driven to the dealer Smith; see K. 58, p. 4. On Nieman's distaste for Jews like Gabbe see C. van Onselen, interview with R. J. Nieman. Nieman was not alone in his dislike of 'foreign' middlemen—see D. O' Meara, *Volkskapitalisme: Class, Capital and Ideology in the Development of Afrikaner Nationalism 1934–1948* (Johannesburg 1983), p. 163.
7. See C. van Onselen, interviews with R. J. Nieman and A. C. G. Nieman. M. T. Nkadimeng, Field Report, Ledig, 28 February 1983; K. 57, pp. 43–45 and K. 58, p. 8.
8. See K. 42, p. 4 and K. 49, pp. 19–20.
9. For the view of a black sharecropper who knew Tjalempe during the mid-1920s, and a white farmer whose father worked with him in the 1930s, see Mmereki Molohlanyi's comments in M. T. Nkadimeng, Field Report, Western Transvaal, 24–29 June 1985, p. 3 and C. van Onselen, interview with R. S. Meyer (son of J. J. Meyer), Devon, 15 March 1984.
10. See K. 2, p. 9; K. 7, p. 35; K. 38, pp. 6–10; K. 48, pp. 46–48 and K. 57, pp. 43–45.
11. See K. 47, pp. 18–19.
12. FAM. 1, pp. 28–31; see also, K. 15, pp. 2 and 19.
13. K. 15, pp. 20 and 29.
14. For a general statement on the agricultural economy at the time see, for example, Union of South Africa, *Report of the Land and Agricultural Bank of South Africa for the year ended 31st December 1921* (Pretoria U.G. 16-22), pp. 7–8.
15. See K. 29, p. 10; K. 40, p. 28 and K. 49, p. 24.
16. See K. 15, pp. 20–24 and FAM. 1, pp. 28–31.
17. On *lekweba* (SeSotho) or *motlhaba* (SeTswana) see K. 15, p. 20.
18. O.R.D.P., Farm Register Vol. 3, HO 55-65, Folio 59/1, Entry No. 7, August 1921. Also, letter to the author from Mrs. F. M. Hambly (daughter-in-law of the late W. H. Hambly), 27 May 1985.
19. See K. 5, pp. 19–20.
20. See K. 38, p. 16.
21. See K. 43, p. 37 and K. 50, pp. 1–2.
22. The demand for horses as draught animals in the western Transvaal started to decline in 1921, a trend that accelerated with the more widespread use of motor vehicles after the recession. See, for example, Union of South Africa, *Report of the Land and Agricultural Bank of South Africa for the years ended 31st December 1921, 1922, 1923 and 1924* (Pretoria U.G. 16-22; U.G. 10 1923; U.G. 17 1924; and U.G. 24 1925), pp. 7, 7,

7 and 27 respectively. The same reports note the transport riders' continuing demand for mules and, to a lesser extent, donkeys.

23. O.R.D.P., Farm Register Vol. 13, HO 232–251, Folio 234/1, Entry No. 8, 15 February 1922.
24. For a record of the 1922 and 1923 rainfall in the district, see C.A.D., JUS, Vol. 367, File 1/554/23, 'Annual Report for the year ended 31st December 1923, in respect of Bloemhof and Christiana wards, District of Bloemhof.'
25. The daily dust storms are recounted in C.A.D., JUS, Vol. 304, File 3/514/22, Magistrate Christiana to Secretary of Justice, 4 January 1922. The desertification of the far western Transvaal, which was the result of ecological abuse that started with the provision of firewood for the Kimberley diamond fields and was greatly exacerbated by the increase in acreage devoted to maize after the passage of the Marketing Act in late 1937, has yet to be documented. For a North American parallel see D. Worster, *Dust Bowl: The Southern Plains in the 1930s* (Oxford 1979). See also K. 10, pp. 18–19.
26. The 1922 locust invasion is recorded in Union of South Africa, *Report of the Land and Agricultural Bank of South Africa for the year ended 31st December* 1922, pp. 6–7. See also K. 10, p. 17; K. 22, p. 15 and K. 34, p. 7.
27. See C.A.D., NTS, Vol. 1243, File 12/214, Union of South Africa, 'Application for Appointment as Marriage Officer,' 21 June 1922 and Sub-Native Commissioner, Groot Spelonken, P. O. Doornboom to Secretary for Native Affairs, 16 September 1922. I am indebted to Jim Campbell for this material.
28. The role of the A.M.E.C. in the life of the Maine family is recalled in K. 7, pp. 34–35; K. 43, pp. 23–25; FAM. 26 and, less satisfactorily, FAM. 27. See also K.O.D. No. 1, 'African Methodist Episcopal Church, Certificate of Membership,' 18 March 1922. White farmers were of the opinion that these 'church meetings had more to do with beer drinking than God.' See C. van Onselen, interview with J. C. Reyneke (b. 27 July 1909), Olifantshoek, 21 May 1986.
29. Leetwane later said that they made these joint excursions only in the earlier years of their marriage, and that in later years, she was expected to walk to church; FAM. 1, p. 40. It would appear that by the mid-1940s Kas considered church attendance a women's matter; see K. 50, p. 24. The A.M.E.C.'s reservations about diviners are recalled in K. 43, pp. 23–25. For an introduction to the A.M.E.C. in South Africa see H. Hughes, 'Black Mission in South Africa, 1892–1953' (B.A. [Hons.] diss., University of the Witwatersrand [U.W.], 1976).
30. The recovery of the market in late 1922 is noted in Union of South Africa, *Report of the Land and Agricultural Bank* 1922, p. 7. The pressure on the land, especially grazing land at Hartsfontein, is recalled in K. 37, p. 12.
31. See K. 22, pp. 14–15.
32. K. 22, p. 16. For an official assessment of the extent of the locust in-

vasion see C.A.D., JUS, Vol. 367, File 1/554/23, 'Annual Report on the Wolmaransstad Magisterial District, 1923,' p. 3; 'Annual Report on the Sub-District of Schweizer-Reneke, 1923,' pp. 1–2; and 'Annual Report, 1923, in respect of Bloemhof and Christiana Wards.'

33. See K. 22, pp. 17–18.
34. See K. 18, pp. 14–22; K. 22, p. 16 and K. 44, pp. 12–14.
35. See K. 15, pp. 19 and 33–35, and K. 34, p. 10.
36. The Moshodi brothers' stay at Hartsfontein is outlined in K. 39, p. 1 and K. 48, pp. 18–24. For Kas's scorn, see K. 44, pp. 13–14.
37. See K. 53, pp. 30–34. Although Kas put the age of the boys at between twelve and fifteen this does not match the ages of the Nieman boys at the time, the most likely contender being Johannes Jakobus (Hans) Nieman, born 25 December 1914. See C. van Onselen, interview with R. J. Nieman.
38. The entire episode is recounted in K. 53, pp. 31–34.
39. The history of Mdebeniso Tabu, also known as 'Kleinbooi,' and his family of immigrant sharecroppers drawn to the Triangle from the Cape Colony, is in P. 9, pp. 1–10. Kas's recollection of Tabu ('Boy Tapa') and Tjalempe's partnership with J. J. ('Koos') Meyer is in K. 48, pp. 46–48. See also C. van Onselen, interview with J. J. Meyer, Schweizer-Reneke, 11 June 1984.
40. See K. 37, pp. 10–12; K. 40, p. 38 and FAM. 3, p. 17.
41. See C. van Onselen, interview with J. C. Reyneke. See also O.R.D.P., Farm Register HO, Folio 234/1/1, Entry No. 1, 15 February 1922.
42. To avoid the confusion that arises from the repeated use of the name 'Kommissierust' in this study, that portion of Kommissierust which P. A. Reyneke acquired in 1922 is referred to as 'Vlakfontein.' See especially K. 3, p. 4—'Thereafter I moved to Koppie Alleen, but it was still Reyneke's farm because he hired that land. I was looking after his sheep.'
43. See K. 20, p. 9; K. 40, p. 29 and K. 45, p. 6.
44. At this point all cattle were said to belong to Sekwala—see K. 2, pp. 4–5. The use of the donkeys is noted in K. 2, p. 9 and K. 34, p. 2.
45. See K. 21, pp. 2–5.
46. See K. 12, p. 14 and K. 47, p. 22.
47. See Union of South Africa, *Report of the Department of Justice for the year 1916* (Pretoria U.G. 39-16), p. 87 and C.A.D., JUS, Vol. 257, File 1/307/19, 'Annual Report for the year ending 31st December 1919 of the Special Justice of the Peace, Bloemhof,' p. 5.
48. William Henry Coleman, J. J. Edwards's father-in-law, was one of the earliest owners of the property; see O.R.D.P., Vol. 13, HO 232–251, Folio 240/1, Entry No. 2. Even though Coleman had died two years before his birth, Kas was aware of this line of descent and the changing ownership; see K. 12, pp. 14–15 and K. 63, pp. 12–13.
49. See C. van Onselen, interview with J. R. Grant, Bloemhof, 13 June 1984. Jack Grant, son of the late W. W. Grant who was employed as the production manager at Sewefontein from 1927 until the 1950s, replaced

the earlier 'foreman,' S. Erlank; see K. 63, p. 13. On transport riders see C. van Onselen, interview with A. C. G. Nieman.

50. See K. 12, pp. 16–17.
51. See K. 27, pp. 31–32.
52. See K. 34, p. 2.
53. See K. 20, pp. 8–9 and pp. 17–18. The same procedure was followed in dealing with a cow with a fractured leg; see K. 20, pp. 18–19.
54. See K. 20, pp. 13–16.
55. See K. 20, pp. 20–22. It was probably no coincidence that the composite fee for the treatment amounted to a guinea, the basic charge of many practising professionals between the wars.
56. See K. 53, pp. 29–30.
57. See C. van Onselen, interview with J. J. Meyer. See also K. 5, p. 24.
58. See K. 20, p. 11.
59. See K. 20, p. 10: 'The Boers told him to abandon me. But he refused, telling them that I had helped him over many years.' In effect Hendrik Swanepoel won the contract from Meyer, and then subcontracted a large part of his business to Kas. See K. 49, pp. 25–26 and K. 55, p. 21. It is possible, of course, that Swanepoel's effort to win the contract was helped by Piet van Schalkwyk, an occasional transport rider who, it was said, was also Koos Meyer's son-in-law. See K. 20, p. 13.
60. See K. 49, pp. 24–26. The expansion of Koos Meyer's sharecropping empire during the 1920s was noted in C. van Onselen, interview with R. S. Meyer.
61. See K. 20, pp. 16–17 and K. 49, pp. 24–26. What Kas may or may not have known is that, in 1925, Swanepoel also sold a 400-morgen portion of Koppie Alleen to F. J. Dauth, which helped him pay off the larger original portion of the farm acquired in 1906. See O.R.D.P., Farm Register Vol. 12, HO 208–231, Folio 221/2/1, Entry No. 1, 27 October 1925.
62. See K. 24, p. 6; K. 41, pp. 13–14 and K. 45, p. 6.
63. On Asvat see K. 20, pp. 22–24 and 25–27. For Essop Suliman, who had a business at Holloway's Rust, see fragment of an invoice in the Maine papers, K.O.D. No. 2, 1928.
64. See K. 20, pp. 24–26.
65. See K. 7, p. 17.
66. See K. 17, p. 18.
67. See K. 37, p. 12: 'Now with Meyer, you had to farm on-the-half and still work as well'—you were expected to be a sharecropper as well as to undertake the duties usually associated with a labour tenant. Meyer's takeover of the property is recorded in K. 3, p. 4; K. 15, p. 27 and K. 34, p. 47.
68. See C. van Onselen, interviews with J. C. Reyneke, R. S. Meyer, and J. J. Meyer.
69. The picture of Koos Meyer held by white farmers is built up from, among other sources, C. van Onselen's interviews with P. G. Labuschagne, Bloemhof, 25 February 1981 and R. J. Nieman. For black perceptions of the man, other than Kas Maine's, see FAM. 11, p. 54 and,

especially, FAM. 18, pp. 22–35, in which Hwai Maine (Kas's nephew) recalls three decades' work with Meyer.

70. Meyer's insistent demands encouraged a good deal of theft of crops and stock by his sharecroppers. See K. 63, p. 10.
71. See C. van Onselen, interview with P. G. Labuschagne.
72. K. 63, p. 9. The threat, uttered at a time when Meyer still needed the help of Nini Tjalempe and Mdebeniso Tabu to hire Pienaarsfontein (see K. 48, pp. 46–48), probably held greater meaning for Meyer than it would have for farmers who did not share his impoverished background.
73. See K. 7, pp. 29–30; K. 34, pp. 47–48 and K. 63, p. 9 where these events and their sequel are recounted with minor variations. It was said amongst local white farmers that Koos Meyer's father, Mr. R. S. C. L. Meyer, had committed suicide. In later years, Koos Meyer himself experienced what members of his family referred to as a 'nervous breakdown.'

## 5. Consolidation

1. C. van Onselen, interview with J. C. ('Koos') Reyneke, Olifantshoek, 21 May 1986. See also K. 44, p. 3.
2. Ibid. For Kas's concurring judgment see K. 15, p. 36; K. 32, p. 9; K. 34, p. 54; K. 36, pp. 16–17 and K. 57, p. 2.
3. See C. van Onselen, interview with J. C. Reyneke.
4. See K. 53, pp. 7–8 and 17.
5. See, for example, H. Bradford, A *Taste of Freedom: The ICU in Rural South Africa, 1924–1930* (London 1987), pp. 21–59.
6. See K. 2, p. 9; K. 3, pp. 5–6; K. 7, p. 24; K. 11, p. 8 and K. 34, p. 2. Kas alluded in passing to rivalry amongst black sharecroppers prompted by white landlords: K. 2, p. 10. Yet he did not set out to *market* as much grain in the mid-1920s as he had right after World War I, when he had more labour at his disposal and maize and sorghum prices had increased. See his comments in K. 2, p. 5 and K. 34, p. 1.
7. See K. 5, p. 24; K. 9, p. 4 and K. 11, p. 12. The fact that it was left to the sharecropper to determine what sort of grain he planted is noted in K. 53, pp. 1–3. On Tshabatsie, and on the popularity of Botman and other varieties of maize in the western Transvaal see also P. L. Breutz, *The Tribes of the Mafeking District* (Ethnological Publication No. 32), p. 53, para. 146.
8. See K. 9, pp. 7–8. See also K. 22, p. 18: 'Later on they were sprayed with insecticide. That is when the world started to go wrong.'
9. See K. 2, pp. 11–13 and K. 18, pp. 1–2. Above average rainfall in the western Transvaal is recorded in Union of South Africa, *Fourteenth Yearly Report of the Central Board of the Land and Agricultural Bank of South Africa for the year ended 31st December 1925* (Pretoria 1926), p. 34 and, more particularly, Union of South Africa, *Annual Report of the Agricultural and Pastoral Production of South Africa, 1925* (Pretoria 1926), pp. 6–13. This turned out to be the first year in one of the

alternating nine-year wet and dry cycles as proposed by P. D. Tyson, in 'The Great Drought,' *Leadership S.A.*, Vol. 2, No. 3, 1983, pp. 49–57.

10. See K. 2, p. 13; K. 9, pp. 8–9 and K. 34, pp. 2 and 9–10.
11. See K. 2, pp. 11–12; K. 12, pp. 11–12 and FAM. 3, p. 25.
12. For a brief history of the co-operative movement in the district and Makwassie, see Suid-Westelike Transvaalse Landboukoöperasie, *'n Half-eeu van Koöperasie, 1909–1959* (Leeudoringstad 1959), pp. 10, 33 and 51. In 1914, Makwassie boasted a hotel, eight general dealer stores and about 350 European residents. See Union of South Africa, *Hansard*, 4th Session of the 1st Parliament, Col. 2420, 12 May 1914. On problems confronting the co-operative movement in 1919–29, see, for example, Union of South Africa, *Reports of the Land and Agricultural Bank of South Africa for the year ended 31st December 1919*, p. 15; *for the year ended 31st December 1920* (Pretoria U.G. 9-21), p. 17 and *for the year ending 31st December, 1929* (Pretoria U.G. 16-30), p. 14.
13. See K. 30, p. 26 and K. 44, pp. 9–10.
14. See K. 15, pp. 2 and 33–35; K. 18, p. 1; K. 34, p. 9; K. 40, p. 26; K. 50, pp. 23–24; FAM. 1, p. 9; FAM. 9, p. 1 and FAM. 17, p. 10. In K. 15, p. 34, it is incorrectly suggested that Morwesi was born while Kas was working with J. J. ('Koos') Meyer. Likewise, in K. 34, p. 9 and FAM. 1, p. 30, it is suggested that the family was then based at Koppie Alleen. This confusion can be traced to the way in which the Maines often referred to the events that took place under P. A. Reyneke as having occurred at Koppie Alleen rather than at Vlakfontein. Only through detailed knowledge, the search for internal consistency and chronological clues will the casual reader be able to tell whether primary evidence relates to 'Koppie Alleen,' the brief period spent under Koos Meyer, or the far longer period spent under Piet Reyneke at Vlakfontein.
15. See K. 15, pp. 31–35.
16. See K. 38, pp. 6–9; K. 48, pp. 45–48 and P. 9, p. 5.
17. On the Moshodi brothers see K. 39, p. 1; K. 44, pp. 13–14 and K. 48, pp. 19–24. Mmereki Molohlanyi's life is outlined in P. 3, especially pp. 10–35 and P. 8, pp. 1–20. Lerata Masihu's career is sketched in P. 1, pp. 1–5. Kas's uncomplimentary assessment of Masihu can be found in K. 21, p. 26 and K. 38, p. 27. Thloriso Kadi, or RaKapari as he was more commonly known, was for many years based on the Griesels' farm at Leeubos; an outline of his life is to be found in P. 5, pp. 1–5. Kas's views on this distant relative whom he greatly admired are recorded in many places, including K. 44, p. 12; K. 47, p. 15 and K. 61, pp. 7–10.
18. On Miss Tjalempe's Cape education see M. T. Nkadimeng, Field Report, Western Transvaal, 22–25 April 1986. Lerata Masihu's formal education is sketched in P. 1, pp. 3–8; he later sought to make a living from teaching in black farm schools, and worried that sharecroppers as a group tended to neglect their children's education; see P. 1, p. 9.
19. The limited need for literacy at Vlakfontein is recorded in K. 15, pp. 36 and 69–70; K. 36, pp. 16–17 and K. 52, p. 16.
20. Dust storms on the diggings are recorded in C.A.D., Government Native

Labour Bureau (GNLB), Vol. 87, Monthly Report of the Bloemhof Pass Officer, H. A. S. Irvine, 6 October 1922. According to one school of thought 1925 marked both the end of one nine-year 'wet cycle' and, beginning with the storms, the start of a new and devastating era of drought. See Tyson, 'The Great Drought,' pp. 49–57. See also D. Worster, *Dust Bowl* (Oxford 1979).

21. The account of the storm and its aftermath is taken from two interviews with Sellwane Maine: see FAM. 30, pp. 1–5 and FAM. 31, pp. 5–8.
22. Narrow, usually dry, watercourse—Afrikaans.
23. See K. 33, p. 5; K. 46, pp. 6–7 and K. 50, p. 24.
24. See FAM. 32, p. 12.
25. See K. 15, pp. 26–27; FAM. 17, pp. 8–10 and FAM. 32, p. 12.
26. On the drought see, for example, Union of South Africa, *Fifteenth Yearly Report of the Central Board of the Land and Agricultural Bank of South Africa for the year ended 31st December 1926* (Pretoria U.G. 12), p. 10; and *Annual Report of the Agricultural and Pastoral Production of the Union of South Africa for 1926*, pp. 7–10.
27. See K. 57, pp. 28–29 and 32 where it is unclear as to whether this service took place in 1926 or 1927. Intuition rather than intellect has shaped my choice.
28. On harvests at Vlakfontein at this time see, for example, K. 2, pp. 12–13.
29. See K. 40, p. 30 and C. van Onselen, interview with J. J. Meyer, Schweizer-Reneke, 11 June 1984.
30. C. van Onselen, interview with Mrs. M. J. Nieman (b. Greisel, 14 November 1912), Bloemhof, 16 July 1985. Piet Reyneke's acquisition of 259 morgen at Leeubos is recorded in the Office of the Registrar of Deeds, Pretoria (O.R.D.P.), Farm Register HO, Folio 236/3/1/, Entry No. 1, 12 February 1926.
31. See FAM. 15, p. 4 and C. van Onselen, interview with Mrs. Nieman, who remembered the house being built when she was eleven years old, by three BaSotho 'kaffirs'—'Andries' (Mphaka), 'Willem' (Phitise) and Kas.
32. See K. 32, p. 10 and K. 34, pp. 51–52.
33. See K. 27, p. 19 and K. 40, pp. 43–45. See also C. van Onselen, interview with J. C. ('Koos') Reyneke (a nephew of W. Griesel's).
34. K. 34, p. 54.
35. See K. 23, pp. 18–20; K. 38, pp. 19–20; K. 39, pp. 12–13 and C. van Onselen, interview with J. C. Reyneke.
36. A. V. Lindbergh acquired a Holt tractor in the mid-1920s and Kas suggested that Koos Meyer got one in 1927; though Meyer himself remembered that he had begun ploughing by tractor only in 'the early 1930s.' See C. van Onselen, interviews with F. M. Lindbergh, Wolmaransstad, 12 June 1984 and J. J. Meyer; also K. 17, p. 5. This trend was also commented on by the local magistrate; see C.A.D., JUS, Vol. 429, File No. 1/667/27, 'Annual Report: 1927,' p. 5. On the broader picture of the Transvaal and South African countryside, see for example, Union of

South Africa, *Annual Report on the Agricultural and Pastoral Production of the Union of South Africa 1926*, p. 40; and on the national context, Union of South Africa, *Report of the Central Board of the Land and Agricultural Bank of South Africa for the year ending 31st December 1929* (Pretoria U.G. 16-30), p. 35, para. 410.

37. See K. 7, p. 28 and K. 17, p. 4. The estimate, by E. K. du Plessis of Ventersdorp, a leading figure in the South African Maize Breeders, Growers and Judges' Association, was reported in the *Rand Daily Mail*, 21 September 1927.
38. See K. 5, p. 25; K. 38, p. 26 and K. 53, p. 11.
39. See K. 14, pp. 16–17 and K. 36, pp. 12–14.
40. Kas said these traditional rain-making ceremonies had taken place at Mahemspanne, Vaalboschfontein and Vlakfontein; K. 50, pp. 10–11. These correspond to the 'dry cycles' of 1906–15 and 1926–34, as suggested in Tyson, 'The Great Drought.'
41. See H. P. Maree, *Die Geskiedenis en Ontwikkeling van die Dorp en Distrik van Schweizer-Reneke, 1870–1952* (roneo 1952), p. 21; and Union of South Africa, *Report of the Central Board of the Land and Agricultural Bank of South Africa for the year ending 31st December 1927* (Pretoria U.G. 18-28), p. 10.
42. See, for example, C.A.D., JUS, Vol. 429, File No. 1/667/279, Magistrate's Annual Report for the Wolmaransstad District, 1927, p. 1.
43. Standard Bank Archives (S.B.A.), Johannesburg, 442/Bloemhof (1928), Inspection Report on the Bloemhof Branch by P. A. Tunstall, 26 May 1928.
44. On wool prices see Union of South Africa, *Report of the Central Board of the Land and Agricultural Bank*, 1927, p. 38 and, more especially, Union of South Africa, *Report of the Central Board of the Land and Agricultural Bank of South Africa for the year ending 31st December 1928* (U.G. 18-29), p. 37, para. 372. Reyneke's shortlived experience on the Lichtenburg diggings is mentioned in K. 1, p. 22 and K. 10, pp. 1–2.
45. See FAM. 15, p. 13.
46. On the harvest, see Union of South Africa, *Report of the Central Board of the Land and Agricultural Bank*, 1928, p. 39. Reyneke's role as rural banker is evident from, amongst others, K. 27, p. 31 and K. 38, p. 18.
47. See K. 6, pp. 36 and 51; K. 7, p. 22 and K. 12, pp. 32–33.
48. See K. 12, p. 4; K. 50, pp. 5–6 and K. 57, pp. 9–10.
49. O.R.D.P., Vol. HO 80, Folio 80/1, Entry No. 5 records that Petrus Albertus Reyneke acquired 682 morgen of the farm Zorgvliet from W. P. Jelliman on 30 July 1927. Koos Reyneke married Anna Kleynhans on 28 August 1928 and settled at Zorgvliet. See C. van Onselen, interview with J. C. Reyneke.
50. The presence of Sempane Maine at Zorgvliet was noted by both Kas and Koos Reyneke. See C. van Onselen, interview with J. C. Reyneke.
51. See K. 9, p. 1 and K. 17, p. 1.
52. See K. 11, pp. 7–8 and K. 2, p. 6.
53. See K. 9, p. 1 and K. 17, p. 1.

54. The influence of Thloriso Kadi as role model is recorded in K. 44, p. 12.
55. See K. 54, p. 11 and K. 57, p. 24. At an earlier point during his stay at Vlakfontein Kas had acquired yet another plough from Gabbe's store in Bloemhof; a single-share Little Chief drawn by eight donkeys, see K. 38, p. 17; K. 50, p. 2 and K. 57, pp. 22–23.
56. See K.O.D. No. 4. On the rest of the goods involved in this transaction with G. W. de Beer, see K. 54, p. 11.
57. See K. 38, p. 18.
58. See FAM. 3, pp. 20–21. On the acquisition and use of the chain plough see K. 43, p. 37.
59. At Pienaarsfontein, Nini Tjalempe had four spans of oxen in the field at any one time; see K. 48, p. 46.
60. See K. 43, pp. 33–35; FAM. 3, pp. 23–24 and FAM. 15, p. 15.
61. The late summer rainfall in the 1927–28 season is recorded in the Union of South Africa, *Report of the Central Board of the Land and Agricultural Bank, 1928*, p. 43.
62. See K. 11, p. 9 and K. 38, p. 27. For southern highveld harvesting practices see T. J. Keegan, *Rural Transformations in Industrializing South Africa* (Johannesburg 1986), pp. 124–28.
63. For some generalised accounts see K. 30, pp. 8–10 and K. 38, pp. 22–29.
64. The rates paid to harvesting teams are drawn from C. van Onselen, interviews with L. T. Pienaar, formerly of Schoonsig, Bloemhof, 11 June 1984 and P. A. Goosen, formerly of Mooiplaas, Bapsfontein, 29 August 1985.
65. See K. 2, pp. 9–13; K. 9, p. 12 and K. 34, pp. 9 and 28.
66. The wider impact of the co-operative in the southwestern Transvaal during the late 1920s and early 1930s was noted in discussions with several black and white informants. See, for example, K. 1, p. 28 and C. van Onselen, interviews with P. G. Labuschagne, Bloemhof, 25 February 1981 and L. T. Pienaar. See also Keegan, *Rural Transformations*, p. 107.
67. See K. 15, p. 36.
68. See K. 11, p. 17 and K. 57, pp. 14–17.
69. Contrast Kas's description of Reyneke's behaviour on this occasion, K. 15, p. 36, with his account of Nieman's performance under similar circumstances, K. 15, pp. 20 and 29.
70. See K. 15, p. 36.
71. See FAM. 3, p. 24 and K. 46, p. 6.
72. See K. 29, p. 12 and FAM. 15, pp. 4 and 9.
73. Mphaka's wife voiced the complaint about bad debts in FAM. 33, p. 4. The charge against Abraham Nieman is levelled by Mphaka's son, in FAM. 28, p. 13. Nieman himself remembered moving into a house at Kommissierust in 'about 1928 or 1929' where building repairs and extensions were done by 'Andries' (Mphaka), 'Willem' (Phitise), and Kas. He suggested that these 'excellent' but 'very lazy' BaSotho builders were paid in grain, meat and sheep. See C. van Onselen, interview with

A. C. G. Nieman, Bloemhof, 16 July 1985. Kas could not recall his brothers ever having been defrauded by Nieman: K. 55, p. 8.

74. This account of the rise of the I.C.U. is drawn almost exclusively from Bradford, *A Taste of Freedom*, pp. 1–20.
75. Ibid., pp. 164–67.
76. See K. 10, p. 5, but more especially K. 11, pp. 21–23.
77. See Bradford, *A Taste of Freedom*, pp. 127–28 and J. and C. Perry (eds.), *A Chief is a Chief by the People: The Autobiography of Stimela Jason Jingoes* (London 1975), pp. 99–126.
78. 'Moate and Jingoes's comparative conservatism when in the southwestern Transvaal is commented on by Bradford, *A Taste of Freedom*, p. 163.
79. On Makatini's speech, see K. 17, pp. 12–17 and P. 3, p. 31.
80. Quoted in Bradford, *A Taste of Freedom*, p. 163.
81. See K. 10, p. 7 and K. 11, pp. 10–11. On the I.C.U. branch at Schweizer-Reneke, which lasted until 1931, see C.A.D., JUS, Vol. 517, File 6175/29, R. Makatini to [?], 19 April 1929. On the branch at Makwassie, which lasted until 1937, see Bradford, *A Taste of Freedom*, pp. 164 and 334, n. 21; Union of South Africa, *Report of the Native Economic Commission 1930–1932* (Pretoria U.G. 22-32); and Perry (eds.), *A Chief is a Chief*, p. 125.
82. For comments by Sellwane Maine, Motlagomang Maine, Baefesi Maine and Mmereki Molohlanyi, see FAM. 31, p. 8; FAM. 33, p. 7; FAM. 28, p. 2 and P. 3, p. 34.
83. Mphaka's involvement in the I.C.U. and his relationships with Makatini, 'Moate and Jason Jingoes are traced in FAM. 28, pp. 1–7; FAM. 31, pp. 7–12; FAM. 33, pp. 5–11 and FAM. 34, pp. 14–19.
84. The links between Zionist churches, the land question and African Nationalist aspirations have long been the subject of speculation. See, for example, B. Sundkler, *Bantu Prophets* (London 1948), pp. 33–35. Seabata Koaho's role is alluded to in Perry (eds.), *A Chief is a Chief*, pp. 111–12.
85. See FAM. 31, pp. 7–12. More generally, see Bradford, *A Taste of Freedom*, p. 76.
86. FAM. 31, pp. 10 and 12; of the Maine women only Motlagomang, Mphaka's second wife, joined the I.C.U.
87. See C. van Onselen and J. Campbell, interview with Kas Maine, Ledig, 22 August 1985 and FAM. 31, p. 11. See, for example, C.A.D., Resident Magistrate, Vryburg, to Secretary for Native Affairs (S.N.A.), Cape Town, 23 July 1906, affidavit of George Mashwa taken at Vryburg, 21 July 1906, in which it was suggested, '. . . the blacks were going to set up a new Kingdom and the white people would be driven out of Africa.' My thanks to Jim Campbell for drawing this to my attention.
88. Quotations from K. 11, pp. 20 and 21; P. 8, p. 14 and P. 10, p. 8.
89. FAM. 33, p. 7. The white farmers' propensity to liken their black employees to monkeys, which we have seen in relation to the birth of the Maine children, formed a central strand in the ideological counter-attack by I.C.U. activists in the southwestern Transvaal. See, for example, Mbe-

ki's speech at Christiana in 1928 as reported in Bradford, *A Taste of Freedom*, p. 163.

90. P. 8, p. 13.
91. See Perry (eds.), *A Chief is a Chief*, pp. 114–17.
92. See C.A.D., JUS, Vol. 245, File No. 3/1007/20, Acting Director of Native Labour to S.N.A., Pretoria, 14 October 1920.
93. Quotations from P. 8, p. 12 and FAM. 31, p. 8. See also P. 3, p. 33.
94. On changing black attitudes to white farmers during the height of the I.C.U. campaigns, see K. 2, pp. 33–34.
95. See FAM. 31, p. 11.
96. See FAM. 34, p. 18. Violent disagreement amongst Afrikaner farmers about the fate of black sharecroppers was not, of course, without historical precedent. See, for example, Keegan, *Rural Transformations*, p. 181.
97. See K. 2, p. 34. See also C. van Onselen, interviews with J. C. Reyneke and L. T. Pienaar.
98. University of the Witwatersrand (U.W.), Archives of the Church of the Province of South Africa, Saffery Papers, AD 1178, B3, 1929, C. Doyle Modiakgotla, 'Oppressive Laws and their Validity,' *The Workers' Herald Weekly News Bulletin*, 5 May 1929.
99. FAM. 33, p. 7. What Motlagomang said was, '*the magistrate* of Bloemhof.' Throughout her interviews she consistently referred to lawyers as 'magistrates.' She also recounted at least one incident in which a white farmer was said to have lost his life when he tried to enter a hut and assault a 'coloured' farm labourer who was an active I.C.U. member—FAM. 34, pp. 25–26.
100. P. 10, p. 7.
101. Perry (eds.), *A Chief is a Chief*, pp. 116–17.
102. See especially K. 2, pp. 29, 30 and 33.
103. Quotations from K. 11, p. 20 and K. 17, pp. 12–14.
104. Quotations from K. 2, pp. 29 and 33, and K. 10, p. 6.
105. Quotations from K. 10, p. 7 and K. 17, p. 15.
106. Quotations from K. 17, p. 16 and FAM. 31, p. 9.
107. See S.B.A., Johannesburg INSP. 1/1/290, Inspection Report on the Bloemhof Branch by C. G. Collier, 17 September 1929, p. 11; and Union of South Africa, *Report of the Central Board of the Land and Agricultural Bank of South Africa for the year ending 31st December 1929* (Pretoria U.G. 16-30), p. 28.
108. See FAM. 32, pp. 2–4.
109. For an overview of the effect of the co-operative on grain trading in 1928–31, see K. 1, p. 28, and S. Trapido and C. van Onselen, interview with P. G. Labuschagne, Sewefontein, 25 February 1981. The way this worked to the detriment of grain traders such as the Gabbes can also be detected in S.B.A., Jnb., INSP 1/1/299, Inspection Report on the Bloemhof Branch of the Standard Bank by F. Humby, 27 December 1930, pp. 11–13.

110. The conflict with Piet Reyneke is recounted in two passages: K. 1, pp. 21–22 and K. 7, pp. 31–32.

## 6. Struggle

1. On the decline of Bloemhof and rise of Wolmaransstad see, Standard Bank Archives (S.B.A.), Johannesburg, INSP. 1/1/299, Inspection Report on the Bloemhof Branch by F. Humby, 27 November 1930, pp. 13–14 and INSP 1/1/318, Inspection Report on the Bloemhof Branch by G. Collier, 30 April 1932, p. 12.
2. See Central Archives Depot (C.A.D.), Pretoria, JUS, Vol. 275, File 1/307/19, 'Annual Report of the Wolmaransstad Magistrate for the calendar year 1919,' p. 4 and Office of the Registrar of Deeds, Pretoria (O.R.D.P.), Vol. HO, Kareepoort, Folio 210/1, entries 1–5, covering the period 1870–1903.
3. See C.A.D., Government Native Labour Bureau (GNLB), Vol. 87, 3694/12/27, 'Monthly Reports of Pass Officer, Bloemhof Labour District,' 13 January and 6 February 1923. On the ethnic composition of the labour force at work in the district during the 1920s see, C.A.D., NTS, Vol. 2478, File 13/293, 'Christiana Labour District, Staff and Tax Collection 1917–1925' and Item 2/2/20, C. J. Jooste, Magistrate, Christiana to Secretary for Native Affairs, Re: 'Tax Collection Bloemhof,' 9 June 1922. I am indebted to Tim Clynick for these and other references.
4. See K. 18, p. 28; K. 52, pp. 7–8 and K. 61, pp. 22–24. See also P. 18, p. 10 and pp. 23–24. More generally, see Union of South Africa, *Report of the Native Economic Commission 1930–1932* (U.G. 22-1932), Paras. 853–861.
5. See K. 7, p. 35; K. 12, p. 9; K. 15, p. 37 and FAM. 15, pp. 5–6. Also, O.R.D.P., Vol. HO, Wolmaransstad, Folio 210/1/1, Entry No. 1, 21 November 1928 and Vol. HO, Wolmaransstad, Folio 210/2/1, Entry No. 1, 10 November 1936.
6. See K. 1, pp. 23–25 and K. 7, p. 33.
7. See K. 61, pp. 23–24; FAM. 9, p. 11 and FAM. 13, p. 4.
8. See FAM. 9, p. 19.
9. See K. 15, pp. 68–69 and FAM. 8, p. 33.
10. See FAM. 13, p. 47 and S.B.A., Johannesburg, INSP 1/1/299, Inspection Report on the Bloemhof Branch by F. Humby, 27 November 1930, p. 13.
11. See K. 34, pp. 12–13 and K. 38, pp. 23–24.
12. It is unclear exactly when this alleged prosecution took place, see K. 1, p. 25 and K. 44, p. 15; however, see also K. 52, p. 1. Whatever the date, there is no doubt about growing hostility towards sharecropping during the Depression years. See especially K. 1, pp. 24–25 and, more generally, K. 11, pp. 26–27.
13. This passage, slightly reformulated for the sake of clarity, is in K. 11, pp. 25–26. See also K. 2, p. 13 and K. 7, p. 33.

14. See Mphaka's experience in FAM. 15, p. 6 and Sebubudi's in FAM. 3, pp. 19–20.
15. See K. 40, p. 16 and K. 43, p. 9.
16. See K. 25, p. 17; K. 50, p. 3; K. 57, pp. 9–10; FAM. 4, p. 11; FAM. 7, p. 4 and C. van Onselen, Field Report, Ledig, 22 July 1985.
17. See K. 29, pp. 21–23; K. 41, pp. 20–21; K. 43, p. 37; FAM. 4, pp. 13–14 and FAM. 9, pp. 50–51.
18. *"Ga o utlwa batho ba Kareepoort ba ne ba re; ba mpitsa ba re ke ramalebaleba ke phela ka polane."* K. 50, p. 3.
19. See K. 18, pp. 21–22 and K. 50, pp. 3–18 and 21–22.
20. See K. 40, p. 16.
21. See K.O.D. No. 6, 30 March 1931. Also P. 3, pp. 17–19. On the I.C.U. at Makwassie see H. Bradford, 'The Industrial and Commercial Workers' Union of Africa in the South African Countryside, 1924–1930' (Ph.d. diss., University of the Witwatersrand [U.W.] 1985), p. 240, and J. and C. Perry (eds.), A *Chief is a Chief by the People: The Autobiography of Stimela Jason Jingoes* (London 1975), pp. 120–22.
22. See P. D. Tyson, 'The Great Drought,' *Leadership* S.A., Vol. 2, No. 3, 1983, p. 55, where it is suggested that the 'wet cycle' of 1916–25 was followed by the 'dry cycle' of 1926–34.
23. See S.B.A., Jnb., INSP 1/1/299, Inspection Report on the Bloemhof Branch by F. Humby, 27 November 1930, p. 13; K. 43, pp. 33–34 and FAM. 3, p. 23.
24. FAM. 2, pp. 9–13 and 18–19.
25. See K. 15, pp. 3–4 and 37.
26. See K. 1, p. 25; K. 34, p. 13 and K. 38, p. 24.
27. See K. 49, p. 8; K. 58, p. 29; P. 12, p. 58 and P. 41, p. 2.
28. C.A.D., Makwassie Periodic Court Criminal Register, 28 October 1930–7 January 1933; Case No. 186 records that 'Kas Tau,' a thirty-four-year-old 'labourer' from Kareepoort, was arrested in Police District No. 41 for contravening Section Two, Paragraph 1 of Act 23 of 1907. The dog, Bles, is described in K. 21, p. 7.
29. The exact year involved in this experiment is in doubt, but see, K. 22, p. 10. It is in less doubt that Kas later again reverted to his beloved Botman and also experimented with the notoriously hard-seeded Platpit.
30. FAM. 13, pp. 1–5 and FAM. 17, pp. 12–13.
31. See FAM. 1, pp. 39–40.
32. See K. 34, p. 13.
33. C.A.D., GNLB, Vol. 148, File 102/14/77, 'Representations made by native Ngogo on behalf of Transvaal Native Congress. Complaints re: Treatment of natives in Bloemhof Labour District,' 14 October 1920. Also, 'A.N.C. Calls for Passive Resistance' and 'Report of Special Emergency Convention of the A.N.C.,' *Umteteli wa Bantu*, 27 June 1931 and 23 July 1932.
34. FAM. 33, pp. 6 and 8 and Perry (eds.), A *Chief is a Chief*, pp. 118–19.
35. FAM. 33, pp. 6–8.
36. What happened next is described in FAM. 28, pp. 4 and 6. Also Perry

(eds.), *A Chief is a Chief*, pp. 118–19. See also C.A.D., NTS, Vol. 7645, File 48/331, 'Native Administration Act 30/27: Native Agitators, Returns of Cases—Jason Ka Jingoes'; and K. 49, pp. 35–38; P. 7, pp. 9–10 and P. 41, p. 2.

37. On the *African Leader* see M. L. Faison, 'Pixley Ka Isaka Seme, President-General of the African National Congress 1930–1937: A Study of the Impact of his Leadership and Ideology on the Congress' (M.A. diss., Columbia University, 1983), pp. 47–48. For Jingoes's correspondence in the *African Leader* see issues of 3 September, 15 October and 22 October 1932. On Khayi see K. 61, p. 12 and P. 13, p. 2.
38. See K. 19, p. 7.
39. See K. 18, pp. 5–8; K. 19, p. 39; K. 20, p. 28 and K. 39, pp. 3–7.
40. See FAM. 25, pp. 1–16; P. 3, pp. 29–30 and M. T. Nkadimeng, Field Report, Western Transvaal, 24–29 June 1985.
41. On this story and the Bull Nines in general, see K. 18, pp. 8–10 and K. 39, pp. 5–7.
42. Leetwane's last quarterly ticket, K.O.D. No. 6, African Methodist Episcopal Church, is dated March 1931.
43. See K. 43, p. 11 and FAM. 1, p. 43.
44. See K. 43, p. 11.
45. See K. 1, p. 25 and K. 34, p. 13.
46. See K. 15, p. 37.
47. On Kgosiemang see K. 26, p. 12; K. 34, p. 45 and K. 46, pp. 18–19. Also, P. 2, pp. 11–13 and M. T. Nkadimeng, Field Report, Western Transvaal, 15–22 April 1984, p. 3. On C. G. Smith see K. 1, p. 25; K. 7, p. 36; K. 11, p. 9; K. 12, p. 9; K. 26, p. 12 and K. 34, p. 13.
48. See T. J. Kriel, *The New SeSotho-English Dictionary* (Johannesburg 1950), p. 148 and T. Brown, *SeTswana Dictionary* (Gaborone 1980), p. 185.
49. See FAM. 13, pp. 1–6; FAM. 17, p. 12 and K. 50, p. 22.
50. On the dust as Morwesi remembered it see FAM. 35, p. 2.
51. See K. 25, pp. 9 and 14, and S.B.A., Jnb., INSP 1/1/336, Inspection Report on the Bloemhof Branch by C. G. Collier, 6 June 1933, p. 13.
52. See Tyson, 'The Great Drought.'
53. See FAM. 15, pp. 5–8 and FAM. 33, pp. 2–5.
54. See K. 25, p. 9.
55. See FAM. 30, pp. 1–5.
56. On Boskuil see O.R.D.P., HO, Folio 229/2, Entry No. 11 which records how the ubiquitous Lewis & Marks Co. disposed of part of the property to the African and European Investment Co. in 1904.
57. See K. 40, pp. 11–12.
58. See FAM. 34, p. 6.
59. K. 9, p. 10.
60. See K. 61, p. 13 and M. T. Nkadimeng, Field Report, Ledig, 9–10 July 1985.
61. See K. 28, p. 40.
62. See K. 40, pp. 24–25.

63. See K. 47, pp. 11–12.
64. K. 25, p. 10; see also, K. 28, p. 33.
65. See S.B.A., INSP. 1/1/351, Inspection Report on the Bloemhof Branch by R. J. Bradshaw, 30 November 1934, p. 13.
66. See K. 25, p. 11 and K. 40, pp. 11–15.
67. K. 44, p. 14. See also, P. 2, p. 12 and FAM. 4, p. 3.
68. See K. 34, pp. 45–46 and K. 46, p. 18.
69. K. 44, p. 14. It also meant that he circumvented the resistance of labour tenants who worked with less enthusiasm on his fields than on their own. See E. Kgomo, interview with A. Seipthelo, Makwassie, January 1988 (M. M. Molepo Oral History Collection, Institute for Advanced Social Research [I.A.S.R.], U.W.).
70. See S.B.A., Jnb., INSP. 1/1/351, Inspection Report on the Bloemhof Branch by R. J. Bradshaw, 30 November 1934, p. 13.
71. See K. 18, p. 28 and K. 19, p. 18.
72. See K. 61, p. 15; FAM. 9, pp. 16–17 and FAM. 15, p. 10.
73. For this and the consequences see K. 40, pp. 15–20 and K. 49, pp. 8–10.
74. See FAM. 34, p. 24 and K. 49, pp. 35–39.
75. On this and the aftermath see K. 49, pp. 7–10.
76. See K. 40, p. 7.
77. FAM. 9, pp. 40–48.
78. Kas and Leetwane differed significantly on how this animal was acquired. According to Leetwane, the hog was exchanged for a heifer which had previously belonged to a cousin, Slatile; FAM. 1, p. 11. Kas contended that the pig was first sold to 'some whites' and that he used the proceeds, which admittedly were Leetwane's, to buy the heifer himself; K. 43, p. 10.
79. The cause and treatment of 'stiff-sickness' and 'bloated-belly' are recalled in K. 2, p. 24. See also Ministry of Agriculture, *Handbook for Farmers in South Africa* (Pretoria 1929), p. 213.
80. See FAM. 1, p. 11; K. 15, p. 38 and K. 41, p. 5.
81. P. 6, p. 3.
82. See FAM. 33, pp. 3–5 and FAM. 34, pp. 19–24 for an excellent account of these events. On malt factories at Bloemhof see C.A.D., GES, Vol. 698, File No. 18/328/13, 'Report on the Systematic Inspection of Health and Sanitary Conditions in the Area of the Village Council of Bloemhof conducted by Dr. F. W. P. Cluver, on 24 July 1931,' p. 8. One factory was apparently owned by a man named Eides, and a second by the Levitas brothers, Sam and Mike. See also C. van Onselen, interview with P. A. Goosen, Bapsfontein, 29 August 1985.
83. See K. 43, p. 9. The opening of the low-level road bridge over the Vaal is recorded in S.B.A., Jnb., INSP. 1/1/361, Inspection Report on the Bloemhof Branch by C. M. B. Skottowe, 24 October 1936, p. 13.
84. See K. 43, p. 41.
85. See R. A. Paroz, *Southern Sotho-English Dictionary* (Morija 1983), p. 37. The history of the family is reconstructed from an account by Jacob

Lebone's daughter, Mrs. S. Kadi. See P. 6, pp. 1–7. Solomon Hyman Meyer's acquisition of a 400 morgen portion of Boskuil is recorded in O.R.D.P., HO 229, Folio 229/2/1, Entry No. 2, 14 March 1933.

86. On postwar Johannesburg see P. Bonner, 'The Transvaal Native Congress 1917–1920: The Radicalisation of the Black Petty Bourgeoisie on the Rand' in S. Marks and R. Rathbone (eds.), *Industrialisation and Social Change in South Africa* (London 1982), pp. 270–313. See also P. 6, pp. 6–11.
87. See K. 43, pp. 41–42 and K. 49, pp. 15–16.
88. See K. 46, pp. 8–9.
89. See K.O.D. No. 6, 19 June 1935, transaction with F. M. du Toit, and K.O.D. No. 12, 14 August 1935, recording the purchase of ten bags of lime from Patel & Co.
90. On Patel's store, see C. van Onselen, interviews with Mrs. M. Hafejee, Lenasia, 5 June 1985; and Mrs Z. Isane and Mrs. A. Nanabhay (daughters of the late G. M. H. Patel), Johannesburg, 11 November 1987.
91. See K. 2, p. 21 and K. 28, p. 6.
92. These transactions are recorded in documents K.O.D. Nos. 9, 10 and 11, dated 26 July, 8 August and 13 August 1935 respectively.
93. On wool sales see K. 3, p. 7.
94. These three deals, on the second, third and fourth days of November 1936, are recorded in K.O.D. Nos. 14, 15 and 16.
95. See K. 41, pp. 3–5.
96. See T. R. H. Davenport, *South Africa: A Modern History* (Johannesburg 1987), pp. 302–317.
97. See M. Basner, *Am I an African? The Political Memoirs of H. M. Basner* (Johannesburg 1993), p. 2.
98. See Basner, *Am I an African?*, chapters 2–7.
99. For the broad context see C. Booth, 'Liberalism in the Countryside: J. D. Rheinallt Jones, the Joint Council's and the Farm Labour Question in the 1930's and 1940's' (B.A. [Hons.] diss., U.W., 1987), pp. 7–11. On Moult, see U.W., Church of the Province of South Africa Collection, Rheinallt Jones Papers, Senatorial Correspondence, Makwassie, J. G. F. Moult to J. D. Rheinallt Jones, 9 August 1939 and K. Donaldson (ed.), *Who's Who of South Africa* (Johannesburg 1939), p. 178. On J. C. Kuhn, see O.R.D.P., HO Wolmaransstad, Folio 229/2/1, Entry No. 3, 22 February 1937. Mrs. Kuhn's role is outlined in P. 14, pp. 41–54.
100. On gathering anti-Semitism in the countryside during this period see D. O' Meara, *Volkskapitalisme: Class, Capital and Ideology in the Development of Afrikaner Nationalism, 1934–1948* (Johannesburg 1983), p. 168.
101. The biographical details relating to L. and S. Wolpe are drawn from C. van Onselen, interview with the Wolpe daughters, Mrs. R. Liebman and Mrs. P. Mandelstam, Johannesburg, 24 February 1988. On Basner's link with the Wolpes, see C. van Onselen, interview with Mrs. A. Gordon, Johannesburg, 5 March 1987.
102. Biographical data relating to Errol Behrman drawn from C. van Onselen,

interview with Mrs. A. Gordon (Behrman's aunt) and a letter from her to the author, 8 July 1988. Also, letter to the author from G. Grant Behrman, February 1988.

103. See K. 18, p. 6.
104. For the background to the 1937 electoral campaign see Basner, *Am I an African?*, chapter 11.
105. Kas made no mention of these elections despite repeated questioning about the impact of organised politics in the countryside in the post-I.C.U. period. Although he clearly knew of the African National Congress it remained, for him, simply one more urban-based movement. See, for example, K. 17, p. 19.
106. See K. 1, p. 25 and K. 28, pp. 34–35.
107. See K.O.D. No. 20, receipt for an ox purchased from J. J. Coetzee at Kareepoort, 3 June 1937.
108. See K.O.D. No. 19, 3 July 1937.
109. K. 41, p. 18.

## 7. Excursion

1. Office of the Registrar of Deeds, Pretoria (O.R.D.P.), Vol. 13, HO 232–251, Folio 233/1, entries 1–6. On Hill & Paddon see, for example, K. Shillington, *The Colonisation of the Southern Tswana, 1870–1900* (Johannesburg 1985), pp. 216 and 221.
2. C. van Onselen, interview with P. A. Goosen, Bapsfontein, 29 August 1985.
3. C. van Onselen, interview with P. A. Goosen and O.R.D.P., Vol. 3, HO 55–65, Folio 60/4/1, entries 1 and 2 dated 2 June 1916 and 16 May 1924.
4. C. van Onselen, interview with P. A. Goosen. Also, letter to the author from the late J. M. Klopper's daughter-in-law, Mrs. M. J. A. Klopper, 10 September 1985.
5. Kas's knowledge of Klopper dated back at least to the time of the latter's marriage at Schoonsig; see K. 53, p. 27.
6. See K. 1, p. 26; K. 7, pp. 37–38 and FAM. 9, pp. 14–15.
7. Based on C. van Onselen, interview with P. A. Goosen and K. 23, p. 3; K. 40, p. 31 and FAM. 18, p. 24.
8. See K. 20, pp. 42–43. On horse-tail grass (*Equistetum Ramosisissimum*), see C. A. Smith, *Common Names of South African Plants* (Pretoria 1966), p. 368.
9. See K. 41, p. 18.
10. Cornelius Fick, as his name suggests, came to the Triangle from the Ficksburg district, which bordered on Basutoland, and in the late 1920s, married Lettie Erlank, daughter of the then manager of the Sewefontein salt works; C. van Onselen, interview with P. A. Goosen.
11. See K. 3, p. 7 and K. 20, pp. 29 and 39.
12. See K. 20, pp. 28 and 38.
13. See K. 7, pp. 20–21 and FAM. 9, pp. 39–40.
14. See K. 15, p. 39 and FAM. 5, p. 1. Leetwane, somewhat misleadingly,

suggested that the birth of Pulane took place at Sewefontein; FAM. 1, p. 30.

15. See K. 54, pp. 2–4.
16. See K. 20, p. 30.
17. On the gradual improvement in Triangle agriculture after the Depression see, Standard Bank Archives (S.B.A.), Johannesburg, INSP. 1/1/377, Inspection Report on the Bloemhof Branch by S. H. Heald, 21 February 1939, p. 13.
18. See K. 38, p. 18. In K. 2, p. 22, Kas mistakenly suggested that this loan was obtained from 'Meyer.'
19. See K.O.D. No. 24, 'Pass to Bloemhof' signed by J. M. Klopper, 18 February 1938.
20. This scenario is reconstructed from slightly contradictory evidence in K. 5, p. 22; K. 46, p. 14; K. 38, pp. 18–19 and K. 57, p. 24. See also K.O.D. Nos. 26 and 28: receipt issued by Massey-Harris, 1 February 1939, signed by L. Marks, and letter from Wentzel & van der Westhuizen to Mr. Kas Teeu, Klippan, P.O. Lorena, 21 September 1939.
21. See K. 20, pp. 29–30. There is a minor discrepancy as regards the number of cattle involved; given the price of grain prevailing at the time I opted for the figure of thirteen rather than sixteen cattle.
22. See K. 34, p. 15.
23. The figure of three hundred bags as an average harvest at the time was offered by a former thresher: see C. van Onselen, interview with L. T. Pienaar, Bloemhof, 11 June 1984. On the origins and functioning of the 1937 Marketing Act and the Maize Board, see A. Minnaar, 'The South African Maize Industry's Response to the Great Depression and the Beginnings of Large-Scale State Intervention, 1929–1934,' *The South African Journal of Economic History*, Vol. 4, No. 1, 1989, pp. 68–78; D. O' Meara, *Volkskapitalisme: Class, Capital and Ideology in the Development of Afrikaner Nationalism, 1934–1949* (Johannesburg 1983), pp. 184–185 and Suid-Westelike Transvaalse Landboukoöperasie Beperk, *'n Halfeeu van Koöperasie, 1909–1959* (Leeudoringstad 1959), p. 42.
24. See K.O.D. No. 25, Certificate of Ownership, signed by J. J. Coetzee of Kareepoort, 17 June 1938. On the store at Rietpan, see C. van Onselen, interview with Mrs. M. Hafejee (daughter of the late E. S. Seedat), Lenasia, 5 June 1985.
25. See R. A. Paroz (ed.), *Southern Sotho-English Dictionary* (Morija 1983), p. 24.
26. See K. 17, pp. 7–8; K. 61, p. 10 and P. 5, p. 3.
27. See K. 39, p. 1 and P. 5, p. 1.
28. See K. 9, p. 12; K. 39, p. 1; K. 48, pp. 12 and 15 and K. 61, p. 9.
29. On farm labour requirements see, for example, O' Meara, *Volkskapitalisme*, pp. 190 and 231. The surge in diamond production in the south-western Transvaal—from 6,097 carats valued at £24,476 in 1937 to 11,470 carats valued at £51,225 in 1938—is recorded in S.B.A., INSP. Vol. 1/1/337, Inspection Report on the Bloemhof Branch by S. H. Heald, 21 February 1939, p. 13.

30. K. 39, p. 1 (emphasis added).
31. See K. 58, p. 31. Kas, slightly mistakenly, identified 'Billy' Potgieter as Hendrik Goosen's brother-in-law; K. 27, pp. 15–16. In fact, Sergeant 'Ross' Potgieter was Piet Goosen's brother-in-law—see C. van Onselen, interview with P. A. Goosen.
32. On the use of this repressive law see, for example, C.A.D., GES, Vol. 698, File No. 328/13, 'Report on Inspection of Bloemhof Location held on 17th and 18th October 1938,' p. 17 and, for more personal testimony, P. 15, pp. 1–3. Kas's meeting with Dladla is in K. 47, pp. 1–2.
33. On Dladla see K. 47, pp. 1–7 and the brief discussion with his widow reported on in M. T. Nkadimeng, Field Report, Ledig, 15–16 February 1985.
34. See K. 47, pp. 1–2 and 6.
35. See S.B.A., Jnb., INSP. Vol. 1/1/377, Inspection Report on the Bloemhof Branch by S. H. Heald, 21 February 1939, p. 13.
36. C. van Onselen, interviews with L. T. Pienaar and J. J. Meyer, Schweizer-Reneke, 11 June 1984. See also *Report of the Native Farm Labour Committee 1937–1939* (Pretoria 1939), Part IV, pp. 32–34, paras. 150, 152 and 160.
37. See K. 17, pp. 4–5.
38. This attitude, which went largely unchanged for twenty years or more, is recorded in K. 17, pp. 4–5.
39. According to Tyson the years 1935–43 were a 'wet cycle.' See P. D. Tyson, 'The Great Drought,' *Leadership* S.A., Vol. 2, No. 3, 1983, p. 55.
40. See K. 20, pp. 31–41, for the full account of this incident and its aftermath.
41. According to a register kept in the Magistrate's Office in Bloemhof, R. M. W. Hawes was the Resident Magistrate from 1 June 1938 until 31 October 1941.
42. For obvious reasons this issue was never explored directly with Kas. The information here is drawn from FAM. 31—an interview with Nthakwana Maine (about ten years old when the family was at Klippan).
43. See K. 2, p. 14; K. 17, pp. 8–9 and K. 55, p. 10.
44. See C. van Onselen, interview with L. T. Pienaar (son of the late J. P. Pienaar).
45. See K. 38, p. 29.
46. See K. 2, p. 23.
47. On Kas's rivalry with Xaba see K. 2, p. 7; K. 7, p. 48 and K. 9, pp. 17–18.
48. See K. 2, p. 23.
49. See K. 2, p. 23 and K. 27, p. 8. Abraham Johannes Marais's two decades of legal practice in Bloemhof is remarked on in S.B.A., Jnb., INSP. Vol. 1/1/336, Inspection Report on the Bloemhof Branch by C. G. Collier, 29 June 1933, p. 13.
50. See K. 2, p. 23. Marais's relationship to Marks, unknown to Kas at the time, was confirmed by C. van Onselen, interview with Mrs. N. Marks (A. J. Marais's daughter), Johannesburg, 12 September 1988.

51. The letter survives as K.O.D. No. 28, Wentzel & van der Westhuizen to Mr. Kas Teen (sic), P. O. Lorena via Bloemhof, 21 September 1939. It should be read in conjunction with K.O.D. No. 26, Promissory Note No. 26902, entered into with Massey-Harris Co. (South Africa), Sydney Road, Durban, 1 February 1939.
52. See K. 57, p. 7 and K.O.D. No. 29: 'I the undersigned promise that after the next harvest—the 1940 harvest—I shall hand over six bags of sorghum to Kas.' Signed at 'Schoonsicht' by J. R. Bloem, 2 November 1939.
53. A search through the Makwassie court register has failed to establish any record of these proceedings. Possibly there was a less formal proceeding than Kas was led to believe or than is evident from his account. See K. 52, pp. 1–4.
54. K. 43, p. 39.
55. K. 49, p. 40. The 'tickey pass,' so called because it cost thruppence to renew at each month end, was required of both black men and black women; see FAM. 34, pp. 22–23 and P. 15, p. 5. The broader context of the larger campaign to enforce the new pass laws in 'urban areas' is briefly alluded to in D. Hindson, *Pass Controls and the Urban African Proletariat* (Johannesburg 1987), p. 45.
56. See K. 43, pp. 39–40.
57. See P. 12, p. 59.
58. K. 43, p. 40, and see also FAM. 4, p. 11.
59. K. 20, p. 41 (emphasis added).
60. See K. 43, p. 42 and, for the Transvaal more generally, P. Bonner, 'Desirable or Undesirable Women? Liquor, Prostitution and the Migration of Sotho Women to the Rand, 1920–45' (Institute for Advanced Social Research [I.A.S.R.], University of the Witwatersrand [U.W.]), 8 May 1988, p. 15.
61. K. 43, pp. 40–43. On objections to the provisions of the Precious Stones Act, see P. 15, p. 6 and the Native Commissioner's [?] unsigned notes arising from J. D. Rheinallt Jones's visit to Wolmaransstad in 1940 in U.W., Church of the Province of South Africa Collection (C.P.S.A.), Rheinallt Jones Papers: Senatorial Correspondence, Makwassie. See also D. Hindson, *Pass Controls and the Urban African Proletariat*, p. 45 and, for a detailed case study, K. Eales, 'Gender Politics and the Administration of African Women in Johannesburg, 1903–39' (M.A. diss., U.W., 1991), pp. 143–95.
62. On the Israelites' presence at Kareepoort see the testimony of Ishmael Moeng, a lay preacher in the A.M.E.C., as recorded in P. 14, p. 53. For the broader context of the Bulhoek Massacre see, for example, H. J. and R. E. Simons, *Class and Colour in South Africa, 1850–1950* (Harmondsworth 1969), pp. 252–55. For a detailed study see R. Edgar, *Because They Chose The Plan of God: The Story of the Bulhoek Massacre* (Johannesburg 1988).
63. See P. 12, pp. 44–46; P. 15, p. 8 and FAM. 34, p. 13.
64. FAM. 34, p. 23.
65. See P. 14, pp. 9, 47, and 50.

66. See P. 12, p. 11; P. 15, pp. 12–15; FAM. 33, p. 4 and FAM. 34, p. 22.
67. See FAM. 33, p. 8. On the impact of Government Notice No. 1632 of 1938 on the Triangle see, for example, C.A.D., GES, Vol. 698, File No. 328/13, 'Report on Inspection of Bloemhof Location held on 17 and 18 October 1938,' p. 17. The right to brew beer for domestic consumption is recognised in the 1937 Native Laws Amendment Act. See also C. Rogerson, 'A Strange Case of Beer: The State and Sorghum Beer Manufacture in South Africa,' *Area*, Vol. 18, No. 1, 1986, p. 18.
68. FAM. 34, p. 24 and P. 14, p. 4.
69. See P. 15, p. 27: 'People were arrested for beer brewing for many years, but the harassment was intensified once it was decreed that no house should exceed the four-gallon limit.'
70. This and all subsequent letters in the exchange are in U.W., C.P.S.A. Collection, Rheinallt Jones Papers: Senatorial Correspondence, Makwassie.
71. P. 15, pp. 25–29.
72. See P. 15, pp. 2–5 and 19.
73. A copy of this manifesto is in the University of Cape Town, Archival Collection, W. G. Ballinger Papers.
74. On the links between the A.M.E.C. and the 'Communist Party' in and around Kareepoort see, for example, P. 12, p. 26; P. 13, pp. 20 and 23–24; P. 14, pp. 58–59 and P. 15, pp. 8–9. In a letter to the author, dated 16 February 1987, the late Senator H. M. Basner's wife recalled how 'some lively women from the A.M.E.C.' had helped in her husband's successful election campaign in 1942.
75. See P. 12, p. 43; P. 13, p. 20 and P. 14, p. 53.
76. There is a good account of the openness of the 'Communist Party' in P. 12, p. 41.
77. P. 12, pp. 50–51. On Moeng's work for the Communist Party around Boskuil see P. 12, p. 30; P. 13, pp. 30 and 43 and P. 14, p. 61.
78. See K. 15, pp. 4–5 and FAM. 1, p. 30.
79. Hersch Gabbe's retirement to an old age home in Doornfontein, Johannesburg, at about this time, is noted in C. van Onselen, interview with J. Kanareck (former employee of Gabbe & Isaacs, Makwassie), Johannesburg, 3 September 1985. See also K.O.D. No. 32, invoice from Wentzel & van der Westhuizen, 25 January 1940.
80. K. 47, p. 29. The South African Weather Bureau, Pretoria, recorded the monthly rainfall at Cupid's Valley in the Wolmaransstad district during these three midsummer months as: December 1939, 27.4 mm; January 1940, 85.7 mm; and February 1940, 25.1 mm.
81. K. 47, pp. 29–30. Records of the South African Weather Bureau, Pretoria, indicate that the heaviest rainfall recorded at the Wolmaransstad station occurred on 28, 29 and 30 January 1940. Compare this with K. 47, p. 29 for yet another confirmation of the remarkable accuracy of Kas's memory.
82. See K. 43, p. 38 and FAM. 4, p. 13.
83. See K.O.D. No. 33, Invoice No. 2682 issued by Slabbert & Campbell

(Pty) Ltd., two oxen at £7.13.9, dated 16 April 1940. Also, more generally, K. 18, p. 22; K. 28, pp. 18–19 and 24 and FAM. 11, p. 45.

84. The collection of cobs, but not the gendered competition to which it gave rise, is noted in K. 43, p. 42; FAM. 4, p. 32 and FAM. 7, pp. 1–2.
85. On Nomahuka, see FAM. 33, pp. 11–12 and FAM. 36, p. 1.
86. The whipping of the prostitute is recorded in K. 55, pp. 11–12.
87. See K.O.D. No. 34, receipt for the sum of thirteen pounds issued to 'Joubert,' 30 September 1940, signed by E. G. D. Elliot of Makwassie.
88. See K.O.D. No. 35, an acknowledgement of debt signed by W. M. Swanepoel and written in broken Afrikaans, dated 9 October 1940.
89. See K.O.D. No. 36, Transvaal Wheel Tax Receipt No: 33232, 17 February 1941.
90. K. 49, p. 40.
91. See K.O.D. No. 37, note issued and authorised by the Pass Officer, Wolmaransstad, 5 March 1941.
92. Quotation from P. 7, p. 9, an interview with P. T. Baraganyi, a peer of the Maine brothers and resident at Boskuil from 1934 until his death in the mid-1980s.
93. K. 49, pp. 38–39 and P. 7, p. 10.
94. This book is in the M. M. Molepo Oral History Collection, I.A.S.R., U.W.
95. See K.O.D. No. 38, Invoice from I. Gordon, Attorney and Notary, 21 May 1941.
96. FAM. 4, p. 21.

## 8. Re-entry

1. These three paragraphs draw on D. O' Meara, *Volkskapitalisme: Class, Capital and Ideology in the Development of Afrikaner Nationalism 1934–1948* (Johannesburg 1983), pp. 119–21.
2. Ibid. pp. 132–33. O' Meara's analysis of the relative importance of the Cape and highveld to Afrikaner Nationalism during this period has, however, been questioned. See H. Giliomee, 'Constructing Afrikaner Nationalism,' *Journal of Asian and African Studies*, Vol. XVIII, No. 1–2, 1983, pp. 89–94.
3. PRICE INDEX OF SELECTED AGRICULTURAL AND LIVESTOCK PRODUCTS, 1939–1944

(1936 = 100)

| YEAR | BEEF (per 100 lbs.) | MAIZE (per 200 lbs.) | WOOL (per lb.) |
|---|---|---|---|
| 1939 | 132 | 119 | 84 |
| 1940 | 131 | 118 | 119 |
| 1941 | 138 | 144 | 102 |
| 1942 | 170 | 147 | 104 |
| 1943 | 197 | 200 | 129 |
| 1944 | 226 | 213 | 131 |

Department of Agriculture, *Handbook of Agricultural Statistics 1904–1950* (Pretoria 1961), Table 6, p. 11.

4. See J. M. Tinley, *South African Food and Agriculture in World War II* (Stanford, Ca. 1954), pp. 23 and 96, and more generally, Die Oliebeheerraad, *Beknopte Historiese Oorsig van die Oliesaadbedryf in Suid-Afrika* (Pretoria 1968), p. 5.
5. From Department of Agricultural Economics and Marketing, *Handbook of Agricultural Statistics, 1904–1950* (Pretoria 1961), Table 8, 'Agricultural Machinery and Implements,' p. 13.
6. See Tinley, *South African Food*, pp. 23–25.
7. See K. 1, pp. 26 and 48; K. 9, pp. 17–18.
8. See FAM. 9, p. 19 and FAM. 16, p. 49.
9. On Thakane see FAM. 9, pp. 6–8 and 28–31. On Morwesi see FAM. 9, pp. 21–31 and K. 18, pp. 29–30.
10. On the family's views of Bodule's relationship with Kas see FAM. 9, pp. 39–40 and FAM. 16, p. 53; on Kas's view see K. 53, p. 13.
11. See FAM. 9, p. 35.
12. Matlakala's childhood is recounted in FAM. 13, pp. 2–26. Mosala's early childhood, including gambling, is recorded in FAM. 8, pp. 7–22.
13. See K. 7, p. 38; FAM. 4, p. 15; and FAM. 8, p. 29.
14. The contract with Union Congo is recorded in C. van Onselen, interview with P. A. Goosen, Bapsfontein, 29 August 1985. The introduction of sunflowers is recalled in FAM. 37, p. 2.
15. The price increase for beans during 1941 is noted in Tinley, *South African Food*, pp. 96–97. The Maine family's bean growing is recorded in FAM. 37, p. 2 and K. 43, pp. 43–44; Goosen's observations on wartime agriculture are in K. 14, p. 19.
16. K. 53, pp. 22–23.
17. See K.O.D. Nos. 39 and 40, 22 and 31 October 1941; K. 3, p. 8; K. 34, p. 18 and K. 48, pp. 12–15.
18. See FAM. 9, p. 19; FAM. 37, p. 2 and K. 43, p. 43. Leetwane later attributed much of her poor health in old age to that one season at Klippan; see FAM. 1, p. 29.
19. See FAM. 6, p. 26 and FAM. 9, p. 32.
20. On this arrangement and its consequences see K. 20, pp. 42–45 and K. 48, p. 38.
21. See FAM. 9, pp. 30–35 for this development and its aftermath.
22. See K. 53, p. 22. On *Xanthium Strumarium*, also known as the cockleburr, see Ministry of Agriculture, *Handbook for Farmers in South Africa* (Pretoria 1929), p. 565 and C. A. Smith, *Common Names of South African Plants* (Pretoria 1966), p. 276.
23. See K. 53, pp. 21–23.
24. According to their daughter, Abner and Amelia Teeu were of the Mantwana-Funyela clan while the Maine-Teeus were of the Mawana-Phutelana clan; FAM. 16, pp. 2–3. On Mokobo and her daughters, see K. 37, p. 22.
25. See fragments in K. 2, p. 14; K. 7, p. 48; K. 9, pp. 17–18; K. 17, p. 9;

K. 37, p. 22; K. 43, p. 43 and K. 48, pp. 7–9. The drought and maize shortfall in the 1942 season are noted in Tinley, *South African Food*, p. 47.

26. See FAM. 9, p. 29. Thakane's version of these events, in which her grandfather played a prominent part, is corroborated by a letter that Kgofu Maine wrote to his uncle later that year. See K.O.D. No. 41, Kgofu Johannes Maine, Mokgareng, P.O. Taung Station, to Malome Kasa, 11 September 1942.
27. See FAM. 9, pp. 30–31.
28. See K. 28, pp. 28–31.
29. See FAM. 4, p. 15; K. 43, pp. 13–14 and K. 61, pp. 15–16.
30. See FAM. 11, p. 28 and FAM. 9, p. 23.
31. See FAM. 9, p. 28.
32. See FAM. 9, pp. 24–25. This vision, amongst the first to be seen by the younger women in the tightly controlled Maine family, points toward hysteria. In later years Thakane saw several prophetic visions and eventually became a fully fledged faith healer: see FAM. 9, pp. 55–56. For the broader context of hysteria see L. Swartz, 'Transcultural Psychiatry in South Africa,' Part 1, *Transcultural Psychiatric Research Review*, Vol. 23, 1986, pp. 273–303.
33. See K. 15, p. 5 and FAM. 9, p. 4.
34. On the history of the store at Rietpan, see C. van Onselen, interview with Mrs. M. Hafejee (daughter of the late E. S. Seedat), Lenasia, 5 June 1985. The purchase of the harness is recalled in K. 57, pp. 20–21.
35. On Labuschagne, see K. 34, p. 19 and K. 44, p. 24.
36. See K. 2, p. 14 and K. 3, pp. 8–9.
37. See P. 12, p. 32; FAM. 33, p. 4 and FAM. 34, p. 13. The new magistrate at Wolmaransstad, who not only ensured that Basner enjoyed reasonable access to the electorate but helped to curb the police excesses in the district, was C. J. Humphreys, who was based there from 1 June 1942 to 31 July 1944. See C. van Onselen, interview with Mrs. G. E. Humphreys, Ficksburg, 12 April 1988. The A.M.E.C. women's contribution to Basner's electoral success is noted in a letter to the author from Mrs. M. Basner, 16 February 1987. Given the circumstances outlined in Chapter Six above, it seems difficult to accept the suggestion that '. . . Basner probably exaggerated the state of unrest in this constituency,' as argued by M. Roth, 'The Natives' Representative Council, 1937–1951' (Ph.d. diss., University of the Witwatersrand [U.W.], 1987), p. 231.
38. See FAM. 40, pp. 3 and 9.
39. For these flowers—*Sutera Atropurpurea*, *Sutera Caerulea* and *Helichrysum*—see C. A. Smith, *Common Names of South African Plants*, Department of Agricultural Technical Services, Botanical Survey Memoir No. 35 (Pretoria 1963), pp. 221, 306 and 404.
40. 'Blue Grass,' or *Andropogon Appendiculatus* Nees, was well suited to lower-lying or marshy areas. See, for example, T. K. Lowrey and S. Wright, *The Flora of the Witwatersrand* (Johannesburg 1987), p. 85.

It was probably this that was 'machine-cut,' 'mixed with salt' and placed in a special camp that Goosen set aside for feeding draught oxen; see K. 34, p. 18.

41. Kas suggested later that he disposed of the cow because after it had borne its first calf its teats had become blocked and it could not nourish its offspring; see K. 43, pp. 13–14. But he was acutely aware that ownership of the animal was in dispute. Rather untypically, he used the plural 'we' when recounting the history of the beast; see K. 43, pp. 13–14. Significantly, Leetwane gave the animal the name 'Disu'—'dried cow dung.' See FAM. 1, p. 12.
42. K.O.D. No. 42, receipt for cart, signed by P. J. van Schalkwyk of Klippan, 27 January 1943.
43. K.O.D. Nos. 46 and 47, 16 March 1943. On the nickname see FAM. 40, p. 7.
44. See K.O.D. No. 48, Receipt No. M407120, 18 March 1943. The letter of authorisation to 'Kaas Tou' was dated 3 August 1943; see K.O.D. No. 49. By then Kas was no longer resident on a farm with an absentee owner, and the school was never convened.
45. The onus for recruiting a harvesting team lay with the sharecropper rather than the landlord; see K. 44, p. 8. In this case it is unlikely that the Chief's name was 'Makgeta' (see K. 17, p. 9); in all probability it was merely a popular name derived from the SeSotho verb khetha—'to choose' or 'to elect'—hence *mokhethi*, 'one who chooses' members of the harvesting team. See R. A. Paroz (ed.), *Southern Sotho-English Dictionary* (Morija 1983), p. 153.
46. On jute bags, see Suid-Westelike Transvaalse Landboukoöperasie Beperk (S.W.T.L.), *'n Halfeeu van Koöperasie, 1909–1959* (Leeudoringstad 1959), p. 43 and Tinley, *South African Food*, p. 75. The hiring of Matthews's machine is recalled in K. 44, p. 16.
47. See K. 17, p. 9 and K. 34, p. 11.
48. See C. van Onselen, interview with P. A. Goosen. Also, Archives of the Johannesburg City Health Dept, Vol. 159, File No: 11/1, City of Johannesburg, Non-European Affairs Dept., 'Report on Questionnaire relating to the Kaffir Beer Profits Enquiry, 1945,' 'Particulars of the source of supply of ingredients and prices,' covering the period July 1944–June 1945. On the attitude of the co-operatives, who were often poorly disposed towards Jews and 'foreigners' (there was a sharp rise in anti-Semitism amongst rural Afrikaners at this time), see S.W.T.L., *'n Halfeeu van Koöperasie, 1909–1959*, p. 43. On the increase in the number of municipal beer halls see, *Report of the Native Affairs Commission Appointed to Enquire into the Workings of the provisions of the Natives (Urban Areas) Act relating to the Use and Supply of Kaffir Beer* (G.P.S. 4620 1942/43), Annexure 1939–1943. For the wider context see P. la Hausse, *Brewers, Beerhalls and Boycotts* (Johannesburg 1988).
49. For maize prices, see Department of Agriculture, *Handbook of Agricultural Statistics, 1904–1950* (Pretoria 1961), Table 6, p. 11.

50. See K.O.D. No. 43, S. Altshuler to 'Kas Dew' (n.d.). On Altshuler's estate see Central Archives Depot (C.A.D.), Pretoria, Estate No. 2648/43.
51. See FAM. 19, p. 5. 'Early in 1943 the situation with regard to stocks and imports of many types of agricultural requisites, including tractors, windmills, wire of all sorts, and binder twine, had become so acute that all stocks in the hands of dealers, importers, and manufacturers were frozen. Thereafter, sales to individual farmers could be made only by a permit issued by the Controller of Agricultural Implements, Machinery and Requisites.' Tinley, *South African Food,* pp. 74–75. On the involvement of the younger children in Kas's smithying see, for example, FAM. 19, p. 5.
52. Labuschagne's persistence is shown in the SeSotho verb which Kas used to describe his 'recruitment'—*rabella,* 'to strike repeatedly.' See K. 44, p. 25. 'He [Labuschagne] said I must come to work on his farm because at Goosen's farm there was too little grass to feed my cattle.' K. 34, p. 19.

## 9. Advance

1. C. van Onselen, interview with P. G. Labuschagne, Bloemhof, 25 February 1981.
2. The purchase of the property at Klippan is recorded in the Office of the Registrar of Deeds, Pretoria (O.R.D.P.), H.O. 233, Folio 233/1, Entry No. 8, 27 January 1940. See C. van Onselen, interview with P. G. Labuschagne.
3. The broad principles underlying the Vaalrand accord are outlined in K. 2, p. 15 and K. 34, p. 19. The children's labour is recalled in FAM. 8, p. 7 and FAM. 37, p. 1.
4. See FAM. 4, p. 19; FAM. 8, pp. 28–30; and FAM. 9, pp. 19–20.
5. See K. 2, p. 15; K. 7, p. 39 and FAM. 5, pp. 5–6.
6. This is a composite quotation drawn from K. 53, pp. 9–10.
7. See K. 7, p. 39 and K. 53, p. 10.
8. See K. 9, p. 12; K. 39, p. 12 and K. 47, pp. 6–7. The differing ideologies, labour practices, religious beliefs and social origins of highveld landlords in the late nineteenth and early twentieth century still await detailed examination. South Africa, unfortunately, lacks the equivalent of Eugene D. Genovese, *The World the Slaveholders Made* (New York 1971).
9. See K. 7, p. 41; K. 52, p. 10 and FAM. 8, p. 12. On the dangers posed to sheep farmers by blue-tongue and wireworm, see, for example, *Handbook for Farmers in South Africa* (Pretoria 1929), pp. 243 and 256.
10. On Kolbooi, see FAM. 9, p. 37; FAM. 11, pp. 23–24; FAM. 40, p. 8 and K. 49, p. 34. On Baby, see K. 21, p. 6.
11. See K. 45, pp. 3–4.
12. The record harvest of 1944, despite the rain damage in parts of the Triangle, is recorded in Suid-Westelike Transvaalse Landboukoöperasie

Beperk (S.W.T.L.), *'n Halfeeu van Koöperasie 1909–1959* (Leeudoringstad 1959), pp. 45 and 66.

13. There was a 25 percent increase in the price of wool over the 1942–43 season. See K. 3, p. 7; K. 40, pp. 1–2 and K. 61, p. 10.
14. For Kas's version of the events described below, see K. 19, pp. 11–14; K. 39, p. 10 and K. 50, p. 14. A briefer account by Mmereki Molohlanyi is in P. 3, pp. 21–23; and there is a gloss on Moilwa's involvement in M. T. Nkadimeng, Field Report, Western Transvaal, 15–22 April 1984, p. 3. In reconstructing the story I have relied largely on the version told to me by the late Mr. E. S. Seedat's daughter, Mrs. Miriam Hafejee. See C. van Onselen, interview with Mrs. Hafejee, Lenasia, 5 June 1985. For further details on Seedat see Central Archives Depot (C.A.D.), Pretoria, Estate No: 1676/1944.
15. For the broader context of anti-Asian prejudice in the Transvaal see J. Hyslop, 'The Representation of White Working Class Women in the Construction of a Reactionary Populist Movement: "Purified" Afrikaner Nationalist Agitation for Legislation against "Mixed" Marriages 1934–1939' (Institute for Advanced Social Research [I.A.S.R.], University of the Witwatersrand [U.W.], 1993), pp. 23–25.
16. See C. van Onselen, interview with Mrs. Hafejee.
17. Contrast the evidence elicited by C. van Onselen, interview with Mrs. Hafejee, with that provided by Kas Maine in K. 19, p. 13.
18. See K. 2, p. 15; K. 34, p. 15; K. 38, p. 27 and K. 64, p. 2.
19. See K. 38, p. 29 and S.W.T.L., *'n Halfeeu van Koöperasie, 1909–1959*, p. 44.

INDEX OF PRICES PAID FOR JUTE BAGS, 1938–47
(1938 = 100)

| *YEAR* | *PRICE* |
|---|---|
| 1938 | 100 |
| 1939 | 146 |
| 1940 | 171 |
| 1941 | 175 |
| 1942 | 206 |
| 1943 | 247 |
| 1944 | 307 |
| 1945 | 315 |
| 1946 | 308 |
| 1947 | 325 |

J. M. Tinley, *South African Food and Agriculture in World War II* (Stanford, Ca. 1954), p. 133.

20. See S.W.T.L., *'n Halfeeu van Koöperasie, 1909–1959*, p. 44.
21. See K. 22, pp. 2–3 and C. van Onselen, interview with Mrs. J. Labuschagne (daughter-in-law of the late P. G. Labuschagne), Bloemhof, 29 November 1988. Since Kas later complained that only in 1948 did he become aware of the agterskot—the outstanding difference between the

preliminary payment and the end price, usually determined in February of the following year—it must be assumed that Labuschagne was one of the landlords who 'robbed' him of his share of the agterskot.

22. See K. 21, pp. 27–33; K. 26, p. 8 and K. 28, p. 39. *Handbook for Farmers in South Africa*, p. 261, does not mention salt as a preventative of measly pork.
23. See K. 56, pp. 5–7.
24. See K. 56, pp. 3–4.
25.

INDEX OF PRICES PAID FOR AGRICULTURAL IMPLEMENTS, 1938–47
(1938 = 100)

| *YEAR* | *PRICE* |
|---|---|
| 1938 | 100 |
| 1939 | 105 |
| 1940 | 120 |
| 1941 | 124 |
| 1942 | 123 |
| 1943 | 144 |
| 1944 | 161 |
| 1945 | 159 |
| 1946 | 153 |
| 1947 | 164 |

J. M. Tinley, *South African Food*, p. 133. Quotations taken from P. 1, p. 13; FAM. 4, p. 19 and C. van Onselen, interview with P. G. Labuschagne.

26. See K. 38, p. 7 and K. 57, pp. 24–25.
27. See FAM. 8, pp. 15–19, FAM; 44, p. 9 and K. 61, pp. 20–21.
28. These events are recalled in considerable detail by Kas in K. 15, pp. 40–42. See also, however, FAM. 9, p. 5, in which it is suggested that Nneang died of 'sunstroke' (dehydration?) and FAM. 37, p. 1, in which Nthakwana recalls observing the funeral as a young girl from the Labuschagnes' kitchen window.
29. On the links between the A.M.E.C. and the Ethiopian Church see H. Hughes, 'Black Mission in South Africa: Religion and Education in the African Methodist Episcopal Church in South Africa, 1892–1953' (B.A. [Hons.] diss., U.W., 1976), pp. 23–57. Kas's uncharacteristic involvement in this decision, together with the painful recall of Nneang's death, can be seen in K. 50, p. 24. The Maine couple's church membership is recorded in K.O.D. Nos. 51, 52 and 54, covering the period September–December 1944. K.O.D. No. 52, 23 September 1944, is the first document in which Kas Maine is referred to as 'Phillip Maine.'
30. For Kas's mediating role in procuring Padimole Kadi a wife, as well as the part he played at the wedding itself, see K. 61, pp. 8–9.
31. See K. 15, p. 65.
32. See FAM. 8, p. 40. Kas himself was uncharacteristically reluctant to

discuss this incident with a relatively junior outsider like Malete Nkadimeng, arguing: 'Look man, this machine of yours [the tape-recorder] is busy recording. You are too young for me to be able to discuss this matter with you.' K. 15, p. 63. Lerata Masihu also recalled the patriarchs coming to blows; P. 1, p. 11.

33. See P. 1, p. 12 and K. 23, pp. 2–22.
34. See K. 2, p. 15; K. 3, p. 9; K. 7, pp. 39–40 and FAM. 8, p. 10.
35. See K. 40, p. 11; K. 44, p. 4 and K. 47, p. 27.
36. See K.O.D. No. 53, letter from Modise Tsubane to Kas Maine, 1 December 1944.
37. See K. 52, p. 28; the baptismal certificates of Leetwane and Mmusetsi Maine survive as K.O.D. Nos. 55 and 56, 4 February 1945.
38. Kas and Leetwane Maine met their financial commitments to the Ethiopian Church with great regularity throughout the year; see K.O.D. Nos. 58 and 60–64.
39. Elsewhere in the country the 1944–45 season was marked by below average rainfall and drought. See, for example, Tinley, *South African Food*, p. 56.
40. Reconstructed from K. 27, p. 14; K. 31, pp. 6–11 and K. 46, p. 18 and from various unrecorded discussions with Kas Maine.
41. See K. 3, p. 9; K. 7, pp. 41–42 and K. 53, pp. 7–8.
42. See K. 31, p. 10 and K.O.D. No. 59, Receipt No. 766 issued to 'boy Kas Teew' by S. E. Seedat, Kingswood, 28 April 1945.
43. See K. 21, p. 26 and K. 38, p. 27.
44. For changing jute prices, see above, note 19.
45. See K. 2, p. 15 and S.W.T.L., *'n Halfeeu van Koöperasie, 1909–1959*, p. 66.
46. C. van Onselen, interview with Mr. J. J. Wentzel, Bloemhof, 15 March 1989. Pretorius was elected to represent Christiana as a Member of the Provincial Council in 1914, 1918 and 1924—see University of the Orange Free State, Institute for Contemporary History, Archives of the National Party, Vol. IG 6/5/5. Wentzel (b. 27 September 1899) was elected to the same post in 1930; in 1933 he was elected Member of Parliament for the same constituency, which he represented for ten years and, after the war, for a further ten years. According to Wentzel, Commandant Lodewyk ('Ampie') Roos was a prominent member of the Schweizer-Reneke branch of the National Party during the 1940s. The desire for commercial success amongst Afrikaner Nationalists during this period is outlined in D. O' Meara, *Volkskapitalisme*, pp. 143–44.
47. Not least of all because the Eendrag Branch was presided over by Pretorius. See C. van Onselen, interview with J. M. du Plooy, Bloemhof, 14 March 1985; and Archives of the National Party, Vol. IG 6/5/5.
48. C. van Onselen, interview with J. M. du Plooy.
49. The comments from P. G. Labuschagne are taken from K. 7, p. 39 and K. 23, pp. 20–22.
50. See K. 23, p. 20.
51. See K. 12, p. 14; K. 21, p. 28 and K. 26, p. 8.

52. See K. 26, p. 8.
53. See FAM. 44, pp. 9–11.
54. See K. 56, pp. 19–20. (Emphasis added.) The question of social mobility and 'coloured' identity was not a remote one within the Maine family. Not only did Kas's sister Sellwane take a 'coloured' as her first husband, but a family myth centred on the American state of 'Maine,' conjured up images of 'coloured' American-Canadian Maines. See, for example, FAM. 12, pp. 11–12.
55. See K. 2, p. 15; K. 7, p. 39 and K. 44, pp. 24–26.
56. See *Annual Report of the Board of the Land and Agricultural Bank of South Africa for 1950* (Pretoria 1951), p. 5, para. 90, 'Percentage of Arrears against capital invested' and Department of Agricultural Economics and Marketing, *Handbook of Agricultural Statistics, 1904–1950* (Pretoria 1961), Table 8, p. 13, 'European-owned Agricultural Machinery in the Transvaal, 1911–1950.'
57. See K. 56, p. 15.
58. See especially K. 44, p. 25.
59. See K. 44, pp. 18–19.
60. K.O.D. No. 65, the reply from Malcolmess to W. H. Edwards, 13 December 1945, is written in Afrikaans. Edwards's repeated invitations to Kas at about this time are recorded in, amongst other places, K. 34, p. 19.
61. Hans Coetzee was a kinsman of the Coetzees of Kareepan, where Sekwala Maine and his family had been based during the closing years of the First World War.
62. For the whole story, see K. 23, p. 23.
63. See K. 15, pp. 48–49.
64. See K. 35, p. 18.
65. See K. 15, p. 49; K. 55, p. 23 and FAM. 37, p. 2. Sekwala was—almost certainly incorrectly—said to have lived to the age of one hundred and six; FAM. 11, p. 23. Kas Maine's tombstone, unveiled at Ledig in September 1995, incorrectly gives *his* age as being in excess of 140 years at the time of his death! The Maines expect their men to be long-lived.
66. See K. 46, p. 5; K. 55, pp. 22–23; FAM. 34, p. 22 and FAM. 37, p. 2.
67. K. 3, pp. 8–9 and K. 40, pp. 1–2.
68. K. 48, pp. 39–40.
69. K. 40, pp. 9–10 and K. 48, p. 42. Marais was the same Cambridge-trained lawyer who had earlier saved Kas's harvest from the 'agent' Belcher at Klippan.
70. The importance of Kas's identity as an *individual* amongst white farmers at the time was clearly recalled by members of the Maine family. See, for example, FAM. 40, p. 7.
71. See K. 2, p. 15 and K. 30, p. 18.
72. K. 15, pp. 45–46.
73. While Kas himself thought it had been 'a kidney that was damaged,' his injuries were more consistent with those of a hernia; see K. 15, pp. 44–47 and FAM. 44, p. 3. This misunderstanding can perhaps be attrib-

uted to the linguistic static bedevilling his original discussion with Hutchinson: the Afrikaans term *nier* (kidney), bears a passing phonetic similarity to the English word 'hernia.'

74. See FAM. 8, p. 37.
75. See K. 44, p. 26.
76. K. 7, pp. 42–43.
77. K. 15, pp. 43–44.
78. K. 7, p. 44.
79. Quotation from K. 44, p. 25 and C. van Onselen, interview with P. G. Labuschagne.

## 10. Turnaround

1. See Alice Blanche Clarkson (b. Coleman 1857), 'Grandmother Remembers,' p. 29 (unpublished manuscript, Institute for Advanced Social Research [I.A.S.R.], University of the Witwatersrand [U.W.]).
2. Office of the Registrar of Deeds, Pretoria (O.R.D.P.), Vol. 13, HO 232–251, Zevenfontein, Folios 240/1 and 240/2/1 entry No. 1, 23 July 1862. For Kas Maine's understanding of the history of the property which dated back to the time of W. H. Coleman, see K. 12, pp. 14–15; K. 36, pp. 1–2 and K. 37, p. 24.
3. See Clarkson, 'Grandmother Remembers,' pp. 24–44.
4. See C. van Onselen, interview with Mrs. E. Lecornu (daughter of Edwards), Johannesburg, 20 March 1987. Edwards's willingness to identify himself with the Boer cause won him respect among most Afrikaner Nationalist farmers. See, for example, C. van Onselen, interview with P. G. Labuschagne, Bloemhof, 25 February 1981.
5. Central Archives Depot (C.A.D.), Pretoria, W.L.D. Criminal Register, 1/2/1937–27/10/42, Case No. 95 of 1940, p. 166 and record of the court proceedings.
6. See K. 58, pp. 26–28.
7. See K. 48, p. 23 and K. 53, p. 3. On the postwar search for protein and fats, see C.A.D., LDB, Vols. 1378, 4042 and 4308.
8. See K. 53, p. 4 and K. 58, pp. 27–28.
9. See O.R.D.P., Vol. 13, HO 232-51, Folios 240/6/1, 240/7/1 and 240/8/1.
10. On the 'Pegging' and 'Ghetto' Acts, see, for example, T. R. H. Davenport, *South Africa: A Modern History* (Johannesburg 1987), pp. 350–51. On the boycott, see D. J. Millar, 'To Save the "Volk"! The 1947 Consumer Boycott of Retail Traders in the Transvaal' (B.A. [Hons.] seminar paper, Department of Geography, U.W., 1988).
11. Based on interviews by C. van Onselen with J. H. du Plooy, Bloemhof, 16 February 1989 (Mr. du Plooy [b. 1914] was secretary to the Eendrag Branch of the National Party for fifteen years after its formation in the early 1940s); J. J. Wentzel, Bloemhof, 15 March 1989; and G. Hatia, Bloemhof, 29 November 1988 (Goolam Hatia [b. 1931] worked at Lakhi & Co. at the time of the boycott).
12. On these and the following paragraphs, see C. van Onselen, interview

with M. M. Kathrada, Schweizer-Reneke, 15 March 1989. P. G. Labuschagne did eventually open a small store at Sewefontein, which was still functioning in the mid-1990s.

13. C. van Onselen, interview with G. Hatia.
14. C. van Onselen, interview with J. H. du Plooy. The exception was Ishmael Moeng, who once worked as an assistant in an Indian-owned store. Kas purchased four blankets from S. A. Nanabhai at the Bloemhof Cash Store on 18 November 1947 (K.O.D. No. 77), i.e. during the boycott.
15. See FAM. 18, p. 22.
16. See K. 48, pp. 21–24.
17. See FAM. 19, pp. 3–7 and FAM. 41, pp. 8–10.
18. FAM. 41, p. 2. See also the Kadi family's perception of the Maines as recorded in P. 5, p. 7.
19. The pass granted to Kas to sell a 'trolley-load' of wool in Bloemhof was signed by a neighbouring farmer, J. Terblanche; K.O.D. No. 71, 14 May 1947.
20. See K. 46, pp. 15–16; FAM. 5, p. 12 and FAM. 19, pp. 4 and 15.
21. See FAM. 4, p. 17 and FAM. 8, p. 21. Kas's perspective is evident from K. 2, p. 20. Membership fees paid to the Ethiopian Church, September 1947 to May 1948, are documented in K.O.D. Nos. 74, 82 and 84.
22. See FAM. 35, p. 1.
23. On this matter, see K. 8, p. 16.
24. FAM. 3, p. 29.
25. See K. 47, p. 21 and K. 53, pp. 24–25.
26. See K. 15, pp. 49–50.
27. See K. 14, p. 17.
28. See K. 39, p. 4; K. 41, pp. 19–20 and K. 47, pp. 3 and 6. See also K.O.D. Nos. 75 and 76, 20 September 1947 and 14 October 1947.
29. See K. 40, pp. 22–23 and K. 47, pp. 12–13.
30. See FAM. 35, pp. 1–2.
31. See K. 27, p. 17; K. 40, p. 35 and FAM. 15, p. 9.
32. In 1948 arrears on both interest and capital redemption owed by white farmers to the South African Land and Agricultural Bank reached their lowest levels since the Depression, dropping from an average of 4.92 percent in 1933 to an all-time low of 0.6 percent in 1948. See *Annual Report of the Board of the Land and Agricultural Bank of South Africa 1950*, p. 4, para. 72. Likewise, the number of tractors at work in the Transvaal countryside virtually trebled—from 5,782 in 1946 to 14,341 by 1950; see Department of Agricultural Economics and Marketing, *Handbook of Agricultural Statistics 1904–1950* (Pretoria 1961), Table 8, 'Agricultural Machinery and Implements,' p. 13. In the Triangle, the economic upsurge *preceded* rather than followed the 1948 election.
33. See K. 2, pp. 17–19 and K. 11, p. 9.
34. See K. 34, pp. 19–20 and K. 44, pp. 6–7. There is an estimate of the sunflower crop in K. 61, p. 4. For outside verification from Baefesi Maine, see FAM. 15, p. 9.
35. Looking back, however, Kas was acutely aware of this; K. 1, pp. 26–29.

36. See K. 1, pp. 29–30; K. 22, pp. 2–3; K. 61, pp. 4–6 and M. T. Nkadimeng, Field Report, Ledig, 9–10 July 1985.
37. In September 1947 Kas had had to pay a ten-shilling fine as an admission of guilt on some unspecified charge; K.O.D. No. 73. On 29 July 1948, he was summoned to appear in the Bloemhof magistrate's court for having failed to produce a dog licence; K.O.D. No. 87. The note with the fraudulent alteration is K.O.D. No. 80.
38. On Bodule's recreational activities see K. 21, pp. 7–11 and K. 40, p. 36. On Kas's dislike of Dikeledi, K. 55, pp. 12–17 and, more insidiously, K. 56, p. 21. On the movements of the Teeu family during the 1940s, FAM. 16, pp. 16–19.
39. See K. 21, pp. 23–24; K. 40, pp. 36–37 and K. 45, pp. 1–3. While these transcripts mention 'Abraham's' store, I have followed K.O.D. No. 72, a fragment of which refers to Ebrahim's Cash Store.
40. See K. 45, p. 3; K. 54, p. 13 and FAM. 9, pp. 39–40.
41. See K. 40, p. 6 and K. 53, p. 35. It would seem that K.O.D. No. 75, 20 September 1947, also had some bearing on Kas's business with Jacob Dladla (Hlahla) at this time.
42. K. 40, p. 21.
43. See K. 40, pp. 20–23; K. 47, pp. 13–14 and K. 49, p. 1.
44. This account of the funeral of Jacomina Griesel, who died on 10 October 1948 (not in the late 1920s as Kas suggested), is drawn from K. 23, p. 10; C. van Onselen, interview with Mrs. A. C. G. Nieman (b. Martha Jacomina Griesel), Bloemhof, 16 July 1985 and C. van Onselen, Field Report, Ledig, 22 July 1985.
45. See K. 7, pp. 24–25 and K. 57, p. 26.
46. K. 53, pp. 4–5. For the contemporary debates on the virtues of various postwar maize hybrids, see, for example, 'Revolution in the Corn Belt —Hybrid Vigour,' *Libertas*, Vol. 7, No. 6, 1947, pp. 42–47.
47. These documents, along with the 'pass' issued by Moormeister to Kas to cover his journey into Bloemhof, are K.O.D. Nos. 90–93, 29 October, 1 November, 2 November and 3 November 1948.
48. See K. 5, p. 22 and K. 34, p. 20.
49. Thus, for example, on 8 October 1948 the Reverend J. M. Mkwane had written to Kas from Makwassie asking him to organise a reception party to meet him at Kingswood as well as a meeting of congregants; K.O.D. No. 89. For membership fees during this period see K.O.D. Nos. 97–99, all dated December 1948.
50. See FAM. 44, p. 1 and K. 33, pp. 3–11.
51. PRICE INDEX OF WOOL, 1939–49

| *Year* | *Farm* | *Price per lb.* |
|---|---|---|
| 1939 | Kareepoort | 84 |
| 1940 | " | 119 |
| 1941 | Klippan | 102 |
| 1942 | " | 104 |
| 1943 | " | 129 |

| *Year* | *Farm* | *Price per lb.* |
|---|---|---|
| 1944 | Vaalrand | 131 |
| 1945 | " | 132 |
| 1946 | " | 122 |
| 1947 | Sewefontein | 169 |
| 1948 | " | 266 |
| 1949 | " | 342 |

*Handbook of Agricultural Statistics 1904–1950*, Table 6, p. 11. See also K. 1, p. 27; K. 3, p. 10 and K. 41, pp. 19–20. We know, too, that only a few months later Kas owned more than two hundred and twenty sheep since, once the family reached Jakkalsfontein, Nthakwana was responsible for counting them; see FAM. 4, pp. 22–23 and FAM. 7, p. 7.

52. C.A.D., Magistrate's Records, Bloemhof, N 1/10/3 (93), File 4/17/2. The years 1944–53 were a 'dry cycle' according to Tyson's model of the South African climate. See P. D. Tyson, 'The Great Drought,' *Leadership S.A.*, Vol. 2, No. 3, 1983, p. 55.
53. See K. 53, pp. 5–6.
54. FAM. 40, p. 10.
55. On the meeting, see K. 11, p. 7; K. 17, pp. 19–20; K. 40, pp. 4–5 and K. 58, p. 12. In two of these sources, Kas refers to the presence of an official described as a 'commissioner' or 'magistrate.' Since magistrates in rural areas were often called on to act as 'Native Commissioners,' there is no inherent contradiction. The 'new law' and a 'government ruling,' indicating official endorsement of a local initiative, are mentioned in K. 2, p. 29 and K. 11, p. 7. The pertinent archival sources fail to reveal any record of the meeting.
56. K. 2, p. 29. The gist of this message is also found in K. 11, p. 7; K. 17, pp. 19–20 and K. 38, p. 3.
57. See K. 38, pp. 2–3 and K. 48, pp. 32–33.
58. K. 40, p. 5.
59. See K. 2, p. 29.
60. On the sale of the oxen, see K. 40, p. 5 and K. 48, p. 33. On the Daggabult initiative and Jason Jingoes see K. 17, pp. 19–20; K. 48, p. 52; K. 52, p. 17 and K. 58, pp. 12–13.
61. On Kas's assessment of Moormeister's attitude at this time, see K. 38, p. 3; K. 40, p. 5 and K. 44, p. 26.
62. K. 44, p. 24 and K. 48, p. 34.
63. K. 44, p. 24.
64. See K. 34, p. 20.
65. K. 44, p. 26 and K. 1, p. 28.
66. Moormeister's accommodating attitude is alluded to in K. 38, p. 2. On the Kadis see K. 15, pp. 66–67 and K. 40, p. 5.
67. See K. 1, pp. 27 33; K. 3, p. 10; K. 26, p. 9; K. 32, pp. 15–16 and K. 45, p. 1.
68. See FAM. 13, pp. 36–39 and FAM. 16, p. 15.
69. K. 32, p. 15.

70. See K. 32, pp. 15–16, where Pieterse—mistakenly referred to as Peterson—is said to have been in the employ of 'a Jew' (Charles Hotz). See also C. van Onselen, interview with Mrs. I. M. Pieterse (b. Klue), Klerksdorp, 29 March 1984.
71. K. 32, p. 15.
72. See K. 46, pp. 20–21.
73. Ibid.
74. See K.O.D. No. 106, Suid-Westelike Transvaalse Landboukoöperasie Beperk, Receipt No. 3101, 13 July 1949 and K. 38, p. 2.
75. The pass is K.O.D. No. 107, 15 July 1949.
76. See K. 46, pp. 23–24.
77. See K. 46, p. 24.
78. See K.O.D. No. 118, Poll Tax Receipt No. J 316742 issued at Jakkalsfontein, 13 July 1949 and K.O.D. No. 108, Pass authorising Kas to travel to Harrisburg station over two days, signed L. J. Klue, 22 July 1949.
79. See K.O.D. No. 109, S.W.T.L. Invoice No. A 7716, 13 August 1949; K. 37, pp. 17–19 and K. 54, pp. 7–8. Also K.O.D. No. 110, note signed by L. J. Klue, 26 August 1949.
80. On the role and function of the *senohe* see, W. D. Hammond-Tooke (ed.), *The Bantu-speaking Peoples of Southern Africa* (London 1974), p. 349.
81. See K. 33, pp. 10–11. My interpretation has been shaped by I. M. Lewis, *Ecstatic Religion: An Anthropological Study of Spirit Possession and Shamanism* (London 1971), pp. 66–99.
82. See K.O.D. No. 102, 'Pass for Old Kas to Bloemhof and Schweizer-Reneke by cart and horses,' signed by L. J. Klue at Jakkalsfontein, 2 October 1949.
83. See K. 38, p. 3.
84. See K. 1, p. 30; K. 36, p. 22 and K. 46, p. 23.
85. See K. 33, pp. 8–9 and, tangentially, FAM. 16, p. 15.
86. The treatment that allowed women to work with cattle is discussed in K. 56, pp. 17–18, while Dikeledi's early life is recalled in FAM. 16, pp. 1–15.
87. See K. 33, p. 13 and K. 55, pp. 12–17. For Dikeledi's version, see FAM. 16, pp. 30 and 47–48.

## 11. Retreat

1. For a history of gold mining on Rhenosterspruit and the adjacent farms, see H. Guest, *Klerksdorp's Fifty Years of Mining* (Klerksdorp 1938), pp. 96–98.
2. See P. Venter, 'Die Staatsversorging van 'n Tipiese Boeregemeenskap in S.W. Transvaal' (M.A. diss., University of Potchefstroom, 1938), p. 73, also pp. 63–66.
3. See Venter, 'Tipiese Boeregemeenskap,' p. 70.
4. Office of the Registrar of Deeds, Pretoria (O.R.D.P.), Farm Register, Vol. 29, LG 655/62, Jakkalsfontein, Folios 323/1, 323/1/2 and 323/9/1. The use of the term 'Cape Malay' in a highveld land register is presum-

ably meant to distinguish the old and respectable (Abdurahman and Pietersen) families from supposedly less worthy Transvaal 'coloured' counterparts.

5. O.R.D.P., Farm Register, Vol. 29, LG 655/62, Folio 323/12/1, Entry No. 1, 23 March 1938. See also K. 44, p. 27.
6. See A. Grundlingh, 'God het ons Arm Mense die Houtjies Gegee: Poor White Woodcutters in the southern Cape Forest Area, c. 1900–1939' in Morrell (ed.), *White but Poor: Essays on the History of Poor Whites in Southern Africa, 1880–1940* (Pretoria 1992), pp. 40–56.
7. C. van Onselen, interview with Mrs. I. M. Pieterse (daughter of L. J. Klue), Klerksdorp, 29 March 1984. Kas, quick as ever to grasp the importance of Afrikaner kinship networks, alludes to the Klue-Meyer connections in K. 7, p. 45; K. 18, p. 35 and K. 34, pp. 27–28.
8. This skirmish, witnessed by one or more of the residents on the property and recounted later to Kas, earned 'Majolo' a reputation for bravery amongst SeSotho-, if not Nguni-speakers. See K. 24, p. 1 and K. 32, p. 17.
9. See K. 1, p. 30; K. 2, p. 16; K. 5, p. 23; K. 7, pp. 45–46; K. 10, p. 20 and K. 34, pp. 21–34.
10. On the use of commercial fertilisers see K. 53, pp. 16–17. The contract with van Vuuren is recalled in somewhat vague and contradictory terms in K. 34, p. 21 and K. 44, pp. 26–27. Said Tutu was probably part of the same family of Fingo immigrants into the greater Klerksdorp district from which Archbishop Desmond Tutu is descended. See S. du Boulay, *Tutu: Voice of the Voiceless* (London 1989), p. 22. On the 1944–53 dry cycle see P. D. Tyson, "The Great Drought,' *Leadership* S.A., Vol. 2, No. 3, 1983, p. 55.
11. See K.O.D. No. 112(a), 'to Mr. Kas Maine from Your Brother, Maine,' 25 October 1949. This was probably one of the several encounters with the Motsuenyanes during 1949–50 that are alluded to in K. 45, p. 6 and K. 52, p. 14.
12. Byl and Wildebeespan had been bought from the Nederduitse Hervormde Church and neighbouring Afrikaner farmers by Moses Motsuenyane, his five sons and two sons-in-law between 1908 and 1912. See O.R.D.P., N.R.D.A. (2)/L.G. 835/61, Entry 8444 of 1912. Kas's knowledge of the Motsuenyanes dated back to the 1943–44 season which he spent at Prairieflower; see K. 32, pp. 12–13 and K. 45, p. 3. The Motsuenyanes, descendants of distinguished Free State sharecroppers, were distantly related to the Maines—see K. 32, p. 13 and Ms. B. Motau, interview with A. Motsuenyane, Mohlakeng, 6 March 1984.
13. See FAM. 13, pp. 28 and 32.
14. See FAM. 13, pp. 27–32 and FAM. 21, p. 11.
15. Matlakala refers to having been 'circumcised at Motsitlane near Madibogo station' at the age of fifteen; see FAM. 13, pp. 21–22 and FAM. 21, p. 10. See also V. van der Vliet, 'Girls Initiation Schools,' in W. D. Hammond-Tooke (ed.), *The Bantu-speaking Peoples of Southern Africa* (London 1974), pp. 232–33, in which it is suggested that only notional circumcision is likely to take place.

16. See FAM. 13, pp. 9–35 and FAM. 21, pp. 8–12.
17. FAM. 9, pp. 50–56.
18. K. 56, p. 13.
19. See K. 52, pp. 11–13; K. 56, p. 14 and K. 58, pp. 19–20.
20. See FAM. 13, pp. 43–44.
21. See K. 52, p. 12. Not surprisingly only one document—a pass signed by A. J. du Toit (a neighbouring farmer), on 1 December 1949—survives from this period, K.O.D. No. 114.
22. See K. 50, p. 17; K. 52, pp. 11–13; K. 56, p. 14 and K. 58, pp. 19–20.
23. See K. 33, pp. 1–4 and 10–11.
24. See K. 12, p. 10, in which Klue is described as a '*vuilgat*,' a 'dirty arsehole.'
25. See K. 34, pp. 37–40 and K. 44, pp. 36–39. For the Maine children's observations see FAM. 10, pp. 18–19 and FAM. 19, p. 10.
26. See K. 34, p. 21; FAM. 5, p. 8 and FAM. 41, pp. 2–3.
27. Kas's admission that the children had to work too hard at Klue's is in K. 12, pp. 10–12. The sale of peaches at Dominionville is noted in K. 34, pp. 35–38 and FAM. 19, pp. 12–13.
28. See FAM. 21, pp. 27–28.
29. See K. 2, p. 16 and K. 34, p. 21.
30. See FAM. 13, p. 27.
31. This homecoming is recalled in FAM. 16, pp. 19–22.
32. See FAM. 16, p. 24. In 1985, the year of his death, Kas was still of the opinion that 'Dikeledi brought herself here [among the Maines] and was not married in the conventional way.' See K. 55, p. 13.
33. This version, pieced together largely from Dikeledi's slightly jaundiced account in FAM. 16, pp. 31–33, squares with later developments. For Kas's recollections see K. 3, p. 10 and K. 15, p. 59.
34. In a rather confused passage, Dikeledi later suggested that this process took a full month to complete, see FAM. 16, p. 35. For Lebitsa's role in Bodule's marriage, see FAM. 13, p. 38.
35. See K. 24, p. 9 and K. 32, p. 17.
36. See K. 50, p. 8. This, presumably, was an attempt to link up with BaKubung sharecroppers who had moved through the Potchefstroom district on their way south across the Vaal River between 1884 and 1894. This would have been an offshoot of the same faction to which Hwai Maine had made an unsuccessful appeal for land in the early 1890s, and the descendants of the same group that Seonya Maine had 'betrayed' nearly one hundred and twenty years earlier, when he had first sided with Lesele and then fled to the Malutis at the outbreak of the *difaqane*. See H. L. Dugmore, 'Land and the Struggle for Sekama: The Transformation of a Rural Community—the BaKubung, 1884–1937' (B.A. [Hons.] diss., University of the Witwatersrand [U.W.], 1985), p. 34.
37. See K. 24, p. 9 and K. 38, p. 5. The Maines were exactly the sort of highveld immigrant family, capable of reconstructing its ethnic identity at short notice, that was most feared by the resident BaFokeng chiefs. As early as 1930 one such chief, August Mokhatle, had noted that those

of his kinsmen who worked on white-owned farms in good times wanted very little to do with tribal levies and obligations but that '. . . when bad times come and they have to leave the farms they want to return home and say, "I am MoKwena" or whatever it is, and they say "we want our rights" and they claim them.' See U.W., Church of the Province of South Africa (C.P.S.A.) Collection, 'Evidence to the Native Economic Commission 1930–32,' evidence of Mutle Mokhatle given on behalf of Chief August Mokhatle of the BaFokeng, 28 August 1930, p. 1005.

38. See FAM. 16, pp. 24–25.
39. K. 56, p. 21.
40. FAM. 2, p. 21.
41. See K.O.D. No. 119, 'Rosinah' (Thakane) Maine to Mr. Kas Maine, 28 August 1950. One person whom he did treat was Matthews Tshwane of the farm Sendelingsfontein. See M. T. Nkadimeng, Field Report, Western Transvaal, 1 August 1989.
42. See K. 32, pp. 17–18.
43. See K. 7, p. 46; K. 23, p. 7 and K. 46, pp. 11–12. On the ruling of the local court, see Venter, ''n Tipiese Boeregemeenskap,' p. 72.
44. See K. 46, pp. 12–13 and C. van Onselen, interview with J. Adamson, Klerksdorp, 15 February 1984.
45. C. van Onselen, interview with J. Adamson.
46. See K. 7, p. 56; K. 46, p. 12 and FAM. 4, p. 23.
47. See K. 41, p. 9. See K.O.D. Nos. 120 and 122, 15 September 1950 and 26 January 1951.
48. See K. 32, p. 23 and K. 38, p. 5. The receipt is K.O.D. No. 126, 19 March 1951.
49. See FAM. 16, pp. 47–48.
50. See K. 14, p. 21; K. 30, pp. 4–6; K. 55, pp. 12–13 and FAM. 12, p. 8. For Dikeledi's version see FAM. 16, pp. 29–49.
51. See K. 23, p. 7; K. 40, pp. 32–33 and K. 50, p. 12.
52. See K. 34, p. 24.
53. See K. 21, p. 33; K. 34, p. 50 and FAM. 7, p. 7.
54. See K. 34, pp. 21, 39, pp. 35–36 and K. 52, pp. 6–7.
55. See K. 50, p. 12.
56. These discussions, as recalled by Dikeledi, are in FAM. 16, p. 54.
57. See K.O.D. No. 130, Klerksdorp, 28 August 1951.
58. Most unusually, there are two contradictory versions as to where the Rhenosterhoek harvest was marketed: one—K. 44, pp. 35–36—asserts that it was sold to the S.W.L. co-operative depot in Potchefstroom while the other—K. 52, pp. 6–7—makes the more plausible suggestion that it was sold to a miller in Klerksdorp. I opted for the latter.
59. See K.O.D. No. 121 (n.d.) and K. 54, p. 13.
60. K. 30, p. 5; see also FAM. 12, p. 2 and FAM. 16, p. 49.
61. See FAM. 16, p. 37.
62. See K. 32, p. 19; K. 44, p. 29 and K. 50, p. 12.
63. See K. 13, p. 2 and K. 44, pp. 30–31.

64. On the history of the Smit family see K. 44, pp. 31–32 and FAM. 19, p. 18.
65. See C. van Onselen, Field Trip, Ledig, 13 March 1985.
66. See K. 32, p. 19. 'West Rand,' Map No. 2626, in the 'Land Type Series' on the 1:250,000 scale (South African Soil and Irrigation Research Institute 1984) reveals that there are at least four different soil types at Muiskraal.
67. See K. 16, pp. 4–5; K. 44, p. 36; FAM. 4, pp. 23–24 and FAM. 5, p. 20.
68. See K. 52, p. 26 and FAM. 16, p. 55.
69. See K. 13, p. 3; K. 32, p. 20 and K. 47, p. 35.
70. See, for example, K. 36, p. 1 and K. 45, p. 5, where Kas noted that Willem (Phitise) had to plough for Dinta (Smit).
71. On repairs to the truck see K.O.D. Nos. 132–34, 20 October 1951. Bodule's problems in acquiring a driver's licence are noted in FAM. 4, pp. 23–24 and FAM. 5, p. 20. Permission to use the truck for passenger transport is recorded in K.O.D. No. 153, 8 March 1952.
72. See K. 36, p. 1 and K. 64, p. 4.
73. See K. 52, p. 6 and K.O.D. No. 144, 'Permission to Non-European to keep a dog for the year ending 31st December 1952.'
74. See K. 61, p. 5 and K.O.D. No. 146, 'Agreement of Sale and Purchase,' 14 February 1952, signed at Ventersdorp 30 March 1953. The BaKwena-ba-Molocoane had lived at Doornkop since at least February 1936, when the state, eager to give geographical expression to segregation, had authorised their purchase of Portion "C" of the farm Doornkop in terms of the Native Trust and Land Act of 1936; see O.R.D.P., Deed of Transfer from Gert Johannes Yssel to the BaKwena-ba-Molocoane, 11 November 1936.
75. Kas must have paid a deposit of at least two hundred pounds, since years later he claimed to have paid off at least *half* the altered purchase price of four hundred and fifty pounds when the deal was cancelled. See M. M. Molepo, Field Report, Rustenburg, 20 October 1984. This reading is consistent with K.O.D. No. 175, F. B. Gillett, Solicitor, to Messrs. H. Braude Ltd., 19 February 1953. Phitise's purchase is recorded in K. 61, p. 6.
76. As always, hindsight made for perfect vision. M. M. Molepo, Field Report, Rustenburg, 30 October 1984, mentions a discussion in which Kas regretted 'not having bought land while he was still working for himself. In his view, the current situation (of effective landlessness) could have been avoided had he done so.'
77. K. 32, p. 20.
78. See K. 13, p. 3 and K. 47, pp. 25–26. Beyers's role as a National Party organiser is discussed in C. van Onselen, telephonic interview with J. C. Greyling (M.P. for Carletonville, 1953–77), 4 July 1985. See also University of the Orange Free State, Institute for Contemporary History, Collection PV 2, National Party Transvaal, File No. 191, Ventersdorp Branch, Circular on Office-bearers, signed by A. C. van Wyk, 15 June 1953.

79. See K. 44, pp. 29–31 and K. 64, p. 4.
80. See K. 26, p. 12 and K. 34, pp. 30–31. For the daughters' views on Smit, see FAM. 5, p. 19 and FAM. 19, p. 18.
81. See K.O.D. No. 140, H. Braude, Director, to Mr. Kas Deeu, Rysmierbult, P.O. Box 22, Muiskraal.
82. See K.O.D. Nos. 151, 153 and 154, respectively 26 February, 8 March and 11 March 1952.
83. See K.O.D. Nos. 155 and 156, 7 and 30 April 1952.
84. See FAM. 4, pp. 23–24 and FAM. 5, p. 20.
85. See K.O.D. No. 157, 28 June 1952, signed by J. J. N. Smit.
86. See K. 31, pp. 23–26.
87. See FAM. 13, pp. 52–57 and FAM. 19, pp. 8–9.
88. FAM. 19, p. 9.
89. See K. 34, p. 22 and K. 44, p. 35.
90. See FAM. 6, pp. 14–15.
91. See FAM. 5, p. 14; FAM. 19, pp. 18–19 and FAM. 21, pp. 20–23.
92. See K. 31, p. 24.
93. See K. 34, p. 23 and K. 44, p. 32.
94. See K. 31, p. 25. The receipt for the legal expenses Kas incurred is K.O.D. No. 160, 30 August 1952.
95. K. 13, p. 3.

## 12. Defence

1. C. van Onselen, interview with Mrs. M. Becker (b. Hersch), Johannesburg, 6 July 1985.
2. This list compiled from E. Msimango, interview with Boas Ngakane (son of Shaodi 'Saul' Ngakane), Ikageng, Potchefstroom, 3 March 1985.
3. As late as 1952 sixteen spans of oxen were still drawing double-share ploughs at Rietvlei; see K. 1, p. 32.
4. See E. Msimango, interview with Mrs. M. Mehala (b. Ngakane), Mamelodi West, Pretoria, 15 March 1985.
5. See C. van Onselen, interview with Mrs. M. Becker and E. Msimango, Field Report, Potchefstroom, 3 March 1985.
6. Verwoerd citation and the quoted passage from T. R. H. Davenport, *South Africa: A Modern History* (Johannesburg 1987), pp. 372–73.
7. See K. 13, p. 4 and K. 30, pp. 46–47.
8. K. 30, p. 47.
9. K.O.D. No. 161, Harry Braude Ltd. to Mr. Kas Deeu, postmarked 11 September 1952.
10. On Letsebele, see FAM. 10, p. 15. The letter to Braude is K.O.D. No. 162.
11. K.O.D. No. 163, Director, H. Braude Ltd. to Mr. Kas Deeu, 18 September 1952. The Reverend Jwili's land is indirectly alluded to in K.O.D. No. 223, J. Jwili to Mr. Maine, 18 July 1956.
12. See K.O.D. No. 160, receipt for 'fees re: consultation and advice' issued by Nel & Klynveld, 29 August 1952.

13. On beans at Rietvlei, see FAM. 13, p. 48 and FAM. 21, pp. 23–24. On dagga cultivation see E. Msimango, Field Report, Mamelodi West, 15 March 1985.
14. Labour at Rietvlei is recalled in K. 5, p. 7; FAM. 10, p. 7 and FAM. 21, pp. 23–25.
15. See FAM. 7, pp. 5–6 and FAM. 21, pp. 14 and 24.
16. FAM. 5, pp. 15–16; see also FAM. 21, pp. 23–24.
17. FAM. 21, p. 24.
18. See FAM. 5, pp. 13–14 and FAM. 16, p. 36.
19. See K.O.D. Nos. 167 and 169: Director to Mr. Kas Deeu, Rysmierbult, 6 January 1953 and Director to Mr. Kas Deeu, 72 Broad St., Potchefstroom, 10 January 1953.
20. See K.O.D. No. 171, 'Fee for letter and deposit in case re: Self vs. J. J. N. Smit,' 19 January 1953.
21. See K.O.D. No. 175, F. Gillett to Messrs. Harry Braude Ltd., 19 February 1953 and K.O.D. No. 146, 'Agreement of Sale and Purchase' signed by Kas Deeu and witnessed by F. B. Gillett at Ventersdorp, 30 March 1953.
22. See K. 61, p. 6.
23. This routine, which involved getting up between three and four each morning, is recalled in FAM. 19, p. 19 and FAM. 21, pp. 20–23.
24. See FAM. 19, p. 19 and FAM. 21, pp. 20–23.
25. See P. 40, p. 19 and FAM. 21, p. 31.
26. See K. 52, p. 24 and FAM. 21, p. 13.
27. FAM. 5, p. 10; see also FAM. 7, pp. 5–6 and FAM. 21, pp. 13–16.
28. For a critical analysis of the success and failures of Verwoerd's policies during this period, see D. Posel, *The Making of Apartheid, 1948–1961: Conflict and Compromise* (Oxford 1991).
29. South Africa, *House of Assembly Debates*, 1 June 1954, col. 7135 and, for similar sentiments, 1 April 1957, col. 3961.
30. Details of Greyling's career are in *Who's Who of Southern Africa, 1966* (Johannesburg 1966), p. 377. See also Office of the Registrar of Deeds, Pretoria (O.R.D.P.), Vol. IQ, De Beerskraal, Folio 94/11/1, Entry No. 4, recording the sale of a portion of the farm by D. J. Crafford to C. J. Greyling, 16 June 1945.
31. See K. 13, pp. 1–2.
32. See FAM. 40, p. 12.
33. See K. 7, p. 47 and K. 37, pp. 20–21.
34. See K. 2, p. 16 and FAM. 40, p. 12.
35. K. 2, p. 16.
36. See FAM. 6, p. 3 and FAM. 10, p. 8.
37. See K.O.D. No. 178, receipt of fifteen shillings for the sale of two donkeys acknowledged by J. S. Moller of Rietvlei, 7 May 1953.
38. On the maintenance and use of the truck see K.O.D. Nos. 173, 179 and 180, receipts and invoices for insurance, tools and spares purchased in Potchefstroom, 6 February, 12 May and 15 May 1953; and FAM. 6, p. 24. On the changing pattern of livestock holdings compare K. 26, p. 9 (Sewefontein) with K. 3, pp. 10–11 (Rietvlei).

39. Dinta's disappearance from Muiskraal is recorded in K. 31, pp. 23–25. Quotation from South Africa, *House of Assembly Debates 1959*, Debate on Vote No. 33, 'Labour,' 21 May 1959, col. 6352.
40. See K.O.D. No. 181, Receipt for payment made by Kas Deeu into 'Trust Account' for 'holdings 31 and 34' at Doornkop, 20 June 1953 (emphasis added).
41. See K.O.D. No. 177, 'Credit Note issued to Kastou (native) by Wes-Transvaalse Vendu-Afslaers (Edms.) Bpk,' 6 July 1953.
42. Dikeledi's complaints about unpaid bohadi are in FAM. 16, pp. 48–51, and the construction of the hut is in K. 44, pp. 1–2. There was a more charitable reason for building the hut: Dikeledi was pregnant at the time; see FAM. 16, p. 37.
43. See K.O.D. No. 182, 'I give Kas this pass to search for his daughter who has run away,' signed by 'D. Jacobs, Rietfly' (sic), 12 August 1953.
44. See FAM. 21, pp. 14–15.
45. 'My mistress would sometimes ask why I did this [slept in the kitchen] and I would tell her that I was afraid of my father. I could not even tolerate the presence of a man's hat in my room. I would dispose of it very quickly.' FAM. 21, p. 17.
46. FAM. 21, p. 16.
47. FAM. 21, p. 15.
48. FAM. 21, p. 17.
49. See K. 52, p. 25 and FAM. 43, p. 1.
50. See FAM. 6, p. 14 and FAM. 10, p. 24.
51. The passes covering these journeys into Potchefstroom and the associated business documents are numbered K.O.D. Nos. 184–87, 1 September and 25 September.
52. See K. 49, pp. 22–23. Kas's capacity to mete out discipline was undiminished; see FAM. 10, pp. 6 and 21.
53. See K.O.D. No. 188, Pass 'for four days to Doringkop' (sic) signed by D. Jacobs, 16 October 1953; and K. 55, p. 8.
54. K. 61, p. 5 (emphasis added).
55. See FAM. 10, pp. 15–16.
56. See FAM. 10, pp. 21–23.
57. K. 49, p. 23; see also FAM. 10, p. 23.
58. On the truck see K.O.D. Nos. 191 and 192, 12 January and 1 February 1954; and FAM. 41, p. 5. Bodule's movements are recorded in K.O.D. Nos. 193 and 194, passes of 15 and 20 February 1954.
59. See K. 2, p. 16.
60. C. van Onselen, interview with Mrs. M. Becker. Greyling confirmed his wish to hire property from Mrs. Hersch in a telephone discussion with the author, 4 July 1985.
61. See K. 49, p. 23 and FAM. 10, p. 24.
62. See FAM. 5, pp. 13–14; FAM. 16, pp. 36–40; FAM. 19, p. 14; FAM. 21, pp. 2–23 and FAM. 41, p. 13.
63. See K. 53, p. 13.
64. The 1954–55 season was the end of a 'wet cycle' and the start of a new

'dry cycle.' See P. D. Tyson, 'The Great Drought,' *Leadership* S.A., Vol. 2, No. 3, 1983, p. 57.

65. See K. 37, pp. 20–21 and K. 43, p. 5.
66. See K. 31, p. 13 and K. 43, pp. 5–6.
67. See K. 54, pp. 5–6.
68. See K.O.D. No. 197 which outlines the deal, signed by J. C. Marx at 'Cyferbult,' Buckingham, on 12 November 1954. Kas maintained the vehicle was sold for 'spare parts,' which implies that not only Bodule's absence but also the cost of repairs prompted the sale; see K. 16, p. 17.
69. K. 44, p. 33.
70. Ibid.
71. See K. 46, p. 13.
72. See K. 1, pp. 33–34 and K. 12, p. 12.
73. See K.O.D. No. 202, acknowledgment of debt by 'Alinah Deeu' (Thakane) at Zwartkop, 20 June 1955, witnessed by Elias Rampou, D. E. Rampou and Willem (Phitise) Maine.
74. See K.O.D. No. 204, 'This is to certify that the boy Kaas (sic) has permission to sell two black oxen from this farm,' signed by O. Hersch, Rietvlei, 28 June 1955.
75. See K. 36, p. 10.
76. See K. 31, p. 13; K. 34, pp. 23–24 and K. 43, p. 5.
77. See K. 12, p. 12.
78. K. 38, p. 4.
79. K. 13, p. 7 and K. 31, p. 31.
80. See K. 1, p. 34; K. 13, p. 7; K. 21, p. 12 and K. 38, pp. 4–5.
81. See K.O.D. Nos. 205 and 206, S.W.K.M.B. 'Grain Settlement Statements,' 30 August 1955.
82. See K. 1, p. 4 and K. 38, pp. 4–5.
83. See K. 34, p. 24.
84. See FAM. 16, pp. 57–58.
85. See K. 53, p. 13 and FAM. 16, p. 57.
86. See K. 31, p. 4.
87. See FAM. 6, pp. 25–26.
88. See K. 3, pp. 10–11 and K. 26, p. 10.
89. See K.O.D. Nos. 211, 212, and 220—dog licences and wheel tax receipts, 1956; also K. 21, p. 12.
90. See FAM. 38, pp. 4–6; K. 34, p. 24 and K. 38, pp. 4–5.
91. See K.O.D. No. 218, 29 June 1956.
92. Building the hut is recalled in K. 44, pp. 1–2.
93. FAM. 21, pp. 33–34.
94. See K. 13, p. 6; for the wider context see also K. 1, p. 36; K. 38, p. 5 and K. 44, p. 34. On Geldenhuys, see also O.R.D.P., Vol. IQ 85-96, Varkenskraal, Folio 93/4/1.
95. Bodule's departure is noted in FAM. 38, p. 5.
96. See K.O.D. Nos. 224 and 225, 16 August and 21 August 1956.
97. K.O.D. No. 227, 4 September 1956 (translated from Afrikaans).

## 13. Defiance

1. See H. L. Dugmore, 'The Rise to Power of the Monnakgotla Family of the Bakubung, 1830–1896,' *Africa Perspective* (Johannesburg), New Series, Vol. 1, Nos. 3 & 4, 1987, pp. 93–116.
2. See H. L. Dugmore, 'Land and the Struggle for Sekama: The Transformation of a Rural Community—the BaKubung, 1884–1937' (B.A. [Hons.] diss., University of the Witwatersrand [U.W.], 1985). For Kas's portrayal of these differences see K. 13, pp. 16–24. See also, J. S. Mohlamme, *Forced Removals in the People's Memory: The BaKubung of Ledig* (Johannesburg 1989), p. 3.
3. See Office of the Registrar of Deeds, Pretoria (O.R.D.P.), Farm Register IQ, Elandsfontein No. 21, Folio 21/9/1, Entry No. 5, 19 February 1952, and Folio 21/22/1, Entry No. 1.
4. See P. 4, p. 2; K. 26, p. 11 and K. 41, p. 10.
5. See P. 4, pp. 1–12; K. 50, p. 12 and K. 62, p. 1. See also C. Murray, 'Land, Power and Class in the Thaba Nchu District, Orange Free State, 1884–1983,' *Review of African Political Economy*, No. 29, 1984, pp. 42–43 and *Black Mountain: Land, Class and Power in the Eastern Orange Free State, 1880s–1980s* (Johannesburg 1992), pp. 149, 182 and 199.
6. See O.R.D.P., Farm Register IQ, Elandsfontein No. 21 and P. 24, pp. 10–12.
7. See O.R.D.P., Farm Register IQ, Elandsfontein No. 21, Folio 21/21/1, and P. 24, p. 14.
8. The construction of Mosala's house is recalled in K. 44, pp. 1–2. Mosala and Nthakwana retraced the exact layout of the domestic complex during a field excursion to Elandsfontein with the author in October 1990.
9. See FAM. 6, p. 26 and FAM. 7, pp. 7–8.
10. See K. 26, p. 9 and K. 49, pp. 22–23.
11. See K. 49, p. 22 and FAM. 6, p. 26.
12. See FAM. 6, p. 18; FAM. 16, pp. 58–60; K. 21, p. 12 and K. 26, p. 9.
13. K. T. Nkaba recalls this purchase as having taken place in 1956; see P. 4, p. 22. According to records in O.R.D.P., however, the transfer took place on 27 September 1957. See Farm Register IQ, Elandsfontein No. 21, Folio 21/21/1, Entry No. 3.
14. For Kas's attitude to Masihu see K. 21, p. 26.
15. See K. 13, pp. 8–11; K. 16, p. 25; K. 32, p. 22 and, especially, P. 4, pp. 19–24.
16. See P. 4, p. 7.
17. See K. 41, p. 9; FAM. 6, p. 10 and FAM. 10, p. 21.
18. See FAM. 6, p. 18; FAM. 13, p. 51; P. 4, p. 4; K. 21, p. 12 and K. 26, p. 10.
19. See K. 3, p. 11 and K. 17, p. 3.
20. On Lebitsa see FAM. 47, pp. 1–7.
21. See K. 50, pp. 12–13.
22. See S. Mofokeng, M. T. Nkadimeng and C. van Onselen, Field Report, Western Transvaal, 11–12 October 1990.

23. See K. 9, p. 2; K. 26, p. 11 and K. 50, p. 18. Shortly before this Bodule had also purchased fifty new maize bags from the co-operative for Kas's use; see K.O.D. No. 231, Cash Slip No. 45953, 17 July 1957.
24. See K. 43, pp. 2–4.
25. In 1957–58 the Maine family's spiritual needs were attended to by the Reverend Matabane. By 1959, Matabane had been replaced by the Reverend J. M. Ndzeku. See K.O.D. No. 232, membership card issued to 'Phillip Maine' dated 13 October 1957, and K.O.D. No. 245B, card issued to Martha Deeu for the period March–December 1959.
26. See FAM. 16, pp. 4–41 and FAM. 19, p. 14.
27. See FAM. 16, pp. 50–52 and FAM. 21, pp. 18–19.
28. See FAM. 7, pp. 5–8.
29. See K. 56, p. 12.
30. See S. Mofokeng, M. T. Nkadimeng and C. van Onselen, Field Report, Western Transvaal, 11–12 October 1990.
31. See K. 32, p. 22 and, on the Motsuenyane holdings, Ms. B. Motau, interview with A. M. Motsuenyane, Mohlakeng, 6 March 1984 (M. M. Molepo Oral History Collection, Institute for Advanced Social Research [I.A.S.R.], U.W.).
32. See K. 26, p. 9 and P. 29, p. 4.
33. See K. 26, p. 10.
34. See K.O.D. No. 238, receipt for a cow sold to Kas Teeu by Petrus Sentlwadi, 27 March 1958.
35. See K. 1, p. 35 and P. 4, pp. 6–7. Kas's use of fertiliser at Molote is in K. 38, p. 30.
36. See P. 29, p. 5.
37. See P. 29, pp. 7–9.
38. See K. 53, p. 13 and K.O.D. No. 241, 14 September 1958.
39. See K. 57, p. 33.
40. See K. 36, p. 3; K. 41, p. 11 and K. 47, pp. 22–24.
41. See K. 36, p. 4.
42. K. 53, p. 42.
43. See K. 26, p. 10; K. 36, p. 9 and P. 4, pp. 4–6. Documentary evidence shows that Kas bought mules at about this time. See K.O.D. No. 245, receipt for a mule purchased at a cost of six pounds, 26 March 1959.
44. See K. 47, p. 24.
45. See K. 41, p. 11.
46. See K. 36, pp. 3–4.
47. See K. 53, pp. 43–44.
48. See FAM. 16, p. 41.
49. See FAM. 16, pp. 44–46.
50. See FAM. 47, pp. 6–7 and M. T. Nkadimeng, Field Report, Ledig, 17–19 July 1990.
51. The tablet stands on the hillside above Mathopestat at the time of writing. The inscription, in SeSotho, reads simply: 'Mrs. Miriam Maine died on 9-7-1959 at the age of fifty-three. Hymn 111.' (Hymn 111, *Lefu le Kahlolo*—'Death and Judgment.')

52. Jonas Maponyane, the secretary to Chieftainess Monnakgotla's kgotla at the time recalled, 'We were first informed of the impending removals in 1959.' P. 29, p. 4. This accords with documentary sources, which suggest that these issues were first raised, in the broadest possible terms, in the Department's 'General Circular No: 22 of 1959'; see '*Verskuiwing van Bantoe wat Oortollig of Onwettig in Blankegebiede is na Bantoegebiede*,' Aanhangsel 'B,' *Konferensie van Hoof-Bantoesake Kommissarisse te Pretoria, 16–18 November, 1961*. (I am indebted to Deborah Posel for this reference.)

## 14. THREAT

1. See K.O.D. No. 246, 'Boons Station Cash Store,' Invoice No. 17, half a ton of 'Super 19%' at £5-11-6, 12 November 1959.
2. See K. 15, p. 62; FAM. 7, p. 8 and FAM. 42, p. 3.
3. See K. 53, p. 14.
4. See, for example, K.O.D. Nos. 247 and 248, receipts for cattle impounded at Klerkskraal, 30 December 1959.
5. See K. 53, p. 14.
6. See K. 36, p. 22.
7. See K. 47, p. 3; FAM. 6, pp. 12–13 and FAM. 48, pp. 3–6.
8. See K. 29, p. 25.
9. See K. 48, p. 6.
10. See, for example, Herman Charles Bosman, 'Border Bad-Man' in *Jurie Steyn's Post Office* (Cape Town 1971), pp. 109–17. See also S. N. Milton, 'South African Economic Nationalism, Imperial Decline and Cattle "Smuggling" in the Inter-War-Years' (M.A. diss., University of London, 1990).
11. See K. 15, pp. 60–61.
12. K. 53, p. 41.
13. Smithying during this period is recalled in P. 4, pp. 4–6.
14. See K. 20, pp. 14–15.
15. See FAM. 16, pp. 51–61.
16. Passing acknowledgment of this matter is in K. 30, p. 6. Details of the 'settlement' are drawn from Dikeledi's account, FAM. 16, pp. 51–52.
17. See K. 42, p. 6 and P. 4, p. 5.
18. See K. 53, p. 42.
19. The harvest at Nkaba's is recalled in K. 26, p. 11.
20. See K. 53, p. 41.
21. See FAM. 48, pp. 3–6.
22. See K. 46, p. 9.
23. See FAM. 12, p. 6; FAM. 14, p. 16 and K. 46, p. 9.
24. See K. 52, pp. 25–26.
25. K.O.D. No. 258, 1 July 1961, Receipt for twelve pounds paid to S. Malan of Onverwacht; see also K.O.D. No. 260, 9 August 1961.
26. See K. 16, p. 24.
27. See K. 42, p. 6.

28. See K. 45, p. 7.
29. See FAM. 48, p. 4 and K. 26, p. 11.
30. See K. 36, p. 23.
31. See K. 1, p. 35; K. 53, pp. 14–15 and FAM. 6, p. 22.
32. K. 26, p. 11.
33. K.O.D. No. 263, 14 June 1962, gave 'Kas Teu' written authority to visit the Krugersdorp 'location.'
34. K. 36, p. 25.
35. This necessarily sketchy story is reconstructed from bits and pieces provided by Mosala in FAM. 39, pp. 22–23, and from evidence provided by his son, Ramoritshana, in FAM. 50, pp. 1–2.
36. An examination of the relevant documentation has thus far failed to yield any record of these proceedings.
37. See K. 24, p. 3.
38. See K. 26, p. 11; FAM. 6, p. 17 and FAM. 16, pp. 42–43. K.O.D. No. 264, 27 June 1962, records Nkaba's debt of three hundred and fifty-five pounds.
39. See K. 24, p. 3.
40. See P. 4, p. 12.
41. See P. 31, pp. 5–6 and P. 32, p. 8.
42. See FAM. 7, pp. 8–9.
43. Annual Rainfall Totals, Derby 1960–1975, drawn from the databank of the South African Weather Bureau, Department of Environment Affairs, Pretoria.
44. See FAM. 21, p. 5 and FAM. 49, p. 3.
45. See FAM. 6, p. 10.
46. See FAM. 48, pp. 2–3.
47. See FAM. 48, pp. 2–5.
48. See K. 36, pp. 2–3; M. T. Nkadimeng, Field Report, Ledig/Molote, 11–12 October 1990 and K.O.D. No. 265, undated membership card issued to 'Phillip Deeu.'
49. See K. 16, pp. 8–13; K. 24, p. 5 and K.O.D. No. 262, 25 October 1961, signed by A. J. C. Claasen.
50. See K. 49, p. 3. Conventional medical wisdom has it that the pulsating of the fontanelle is a sign of advanced dehydration, a symptom associated with cases of severe diarrhoea.
51. The funeral is described in FAM. 49, pp. 3–5.
52. See FAM. 49, p. 2.
53. For useful comment on the incidence and importance of lightning in the lives of rural people in the northern Transvaal see, for example, R. C. Anderson, 'Keeping the Myth Alive: Justice, Witches and the Law in the 1988 Sekhukune Killings' (B.A. [Hons.] diss., University of the Witwatersrand [U.W.], 1990).
54. See K. 41, pp. 12–14 and, on the ability of certain ngakas to deal with the problem of lightning, W. D. Hammond-Tooke (ed.), *The Bantu-speaking Peoples of Southern Africa* (London 1974), p. 342.
55. See P. 4, pp. 22–24.

56. See FAM. 43, pp. 2–5.
57. See K. 62, pp. 1–3 and K.O.D. No. 281, receipt for fees paid for grazing hired at Booyskraal, 25 January 1964.
58. See K.O.D. No. 280, 4 January 1964.
59. See K. 54, pp. 8–9.
60. See K. 62, p. 2.
61. Ibid. and K.O.D. No. 283, 19 June 1964.
62. On the tokoloshe, see, for example, W. D. Hammond-Tooke, 'Urbanization and the Interpretation of Misfortune,' *Africa*, Vol. 40, 1970, p. 427.
63. See K. 41, p. 7.
64. See K. 56, pp. 12–13.
65. Kas's attitude is recalled in K. 56, p. 12, Bodule's visit in FAM. 21, p. 36.
66. See FAM. 21, p. 34. The exact psychosocial dynamics at work in this situation can only be speculated on. 'Mampharane' is more usually taken to mean a 'zombie,' a corpse that has been resurrected to perform menial tasks and labour for a person enjoying mystical powers. Possibly this throws light on the true position which Matlakala held in the Pule household at the time. But, in the interview conducted with her many years after these events, Matlakala used the word to allude to a 'lizard' —a small reptile that is often produced by 'witchdoctors' when called upon by their patients to produce evidence of the 'thing' that had brought on their illness.
67. See FAM. 21, pp. 35–36.
68. See K. 16, pp. 20–21.
69. See K.O.D. No. 288, 17 August 1964.
70. See P. D. Tyson, 'The Great Drought,' *Leadership* S.A., Vol. 2, No. 3, 1983, p. 57.
71. See K. 20, pp. 20–21; K.O.D. No. 289, invoice of 14 October, 1964, 2 February 1965 and 5 May 1965. Also, K.O.D. No. 293, S.A.R. & H. Consignment Note, 24 October 1964.
72. See K.O.D. Nos. 294 and 295, 'African Dingaka Association Certificate' and 'Membership Card,' both 1 January 1965.
73. See K.O.D. No. 289, 14 October 1964.
74. See K. 13, p. 32. Also, Jo-Anne Collinge, interview with Arthur Monnakgotla, *Saturday Star*, 16 March 1991.
75. Livestock sales prompted by the fear of heartwater are recalled in K. 13, p. 38 and P. 4, p. 7. See also Department of Agriculture, *Handbook for Farmers in South Africa* (Pretoria 1929), p. 242.
76. See K.O.D. No. 301, 2 September 1965.
77. The exact price of this fourth tractor is not clear but, in K. 45, pp. 8–9, mention is made of 1,500 'pounds'—presumably rands. Even at this price, however, it was certainly the most expensive of the four tractors Kas bought at Syferbult; in K. 1, p. 35, Kas refers to it as having cost £800 (R1,600).
78. See K.O.D. Nos. 302 and 303, receipts for R140 and R100, 7 and 12

October 1965. The liquidation of almost all Kas's livestock holdings before he left Molote is recalled in K. 13, p. 38. It was, of course, evident to other members of the family; see, for example, FAM. 5, p. 17.

79. See K.O.D. No. 304, 24 November 1965 and FAM. 43, pp. 3–4.
80. See K.O.D. Nos. 305 and 306, 2 February and 25 March 1966.
81. See FAM. 7, pp. 8–9 and K. 1, p. 35.
82. See Jo-Anne Collinge, interview with Arthur Monnakgotla.
83. See K. 13, pp. 29–35 and FAM. 42, p. 4.
84. See K. 13, pp. 29–35. See Jo-Anne Collinge, interview with Arthur Monnakgotla and J. S. Mohlamme, *Forced Removals in the People's Memory: The BaKubung of Ledig* (Johannesburg 1989), p. 6.
85. See P. 33, pp. 1–12. See also Mohlamme, *Forced Removals*, p. 12 and A. Claasens, 'Who Owns South Africa?' (Occasional Paper No. 11, Centre for Applied Legal Studies, U.W., February 1991), p. 1.
86. See FAM. 7, p. 9; FAM. 42, p. 4; and Mohlamme, *Forced Removals*, p. 6.
87. See K. 13, p. 17 and FAM. 42, pp. 1–4.
88. See, especially, K. 62, pp. 2–3.
89. See K.O.D. Nos. 309 and 311, 1 June and 31 August 1966.
90. See K. 50, p. 22; K.O.D. No. 447 (n.d.); and K.O.D. No. 312, 10 September 1966.
91. See K. 50, p. 22 and M. M. Molepo, Field Report, Ledig, 19 April 1982.
92. Kas's view that it was Baloyi who was responsible for damage to the Austin is in K. 50, p. 22.
93. See K.O.D. No. 317, 31 January 1967.
94. See K. 26, p. 12 and K. 61, p. 6.
95. See K.O.D. No. 319, 20 February 1967, and K.O.D. No. 320, 28 February 1967.
96. K. 26, p. 12.
97. See K.O.D. No. 324, 25 May 1967.
98. K.O.D. No. 325.

## 15. Reflexes

1. See R. G. Cawthorn, *The Geology of the Pilanesberg* (National Parks Board of Bophuthatswana, Mafikeng), Information Series No. 1.
2. See FAM. 42, p. 8.
3. See FAM. 8, pp. 33–34.
4. See M. M. Molepo, Field Report, Ledig, 30 November 1981.
5. See K.O.D. No. 326, 21 June 1967, and K.O.D. No. 329, 29 September 1967.
6. See K.O.D. No. 327, 31 July 1967 and, on the BaKgatla, Office of the Registrar of Deeds, Mafikeng (O.R.D.M.), Vol. JP, Welgeval, Folio 171/1, Entry No. 3, 10 July 1933.
7. See O.R.D.M., Mafikeng, Vol. JQ, Folio 91/1, Entry Nos. 3 and 4, 7 April 1899 and 10 August 1911. Also K.O.D. No. 332, 4 November 1967.
8. See K. 26, p. 10 and FAM. 54, pp. 4–5.

9. See P. 37, pp. 4–5 and 9–11; FAM. 40, p. 12 and M. M. Molepo, Field Report, Ledig, 19 April 1982.
10. See K. 16, p. 13 and C. van Onselen, Field Report, Ledig, 13 March 1985.
11. See K. 56, p. 2 and FAM. 40, p. 13.
12. The early history of the Maseko and Mngomezulu families is touched on in P. 37, p. 2 and P. 38, pp. 2–3.
13. See P. 35, pp. 7–8 and P. 36, pp. 2–3.
14. See FAM. 40, p. 12 and M. T. Nkadimeng, Field Report, Ledig, 17–18 September 1991.
15. See FAM. 54, pp. 2–7.
16. See C. Desmond, *The Discarded People* (Johannesburg 1969), pp. 121–23 and J. S. Mohlamme, *Forced Removals in the Memory of the People: The BaKubung of Ledig* (Johannesburg 1989), pp. 6–12.
17. See K. 13, p. 30; P. 36, pp. 2–9 and C. Desmond, *The Discarded People*, p. 122.
18. As Scotch Mngomezulu commented later, 'Kas Maine, the Maseko brothers, my brother Jan and my uncle, Jacob Mngomezulu, owned a lot of cattle. *In fact, most Zulu families had cattle far in excess of those held by BaKubung families.*' P. 36, p. 7.
19. On the changing terms of trade in BaKubung-'Ndebele' economic relationships, see P. 37, p. 11 and P. 38, p. 4. On the political marginalisation of Nguni-speakers see, for example, P. 36, p. 8.
20. See M. R. Bret, *Pilanesberg—Jewel of Bophuthatswana* (Johannesburg 1989), pp. 109–115.
21. Mahlangu and his Thanda Bantu ('Love the People') Store are reported on in M. T. Nkadimeng, Field Report, Ledig, 18 August 1991.
22. See FAM. 6, pp. 30–32.
23. See M. T. Nkadimeng, Field Report, Ledig, 17–18 September 1991, p. 3.
24. See K.O.D. Nos. 339–43, 19 August 1969, 4 September 1969, 12 September 1969, 29 December 1969 and 27 January 1970. Also FAM. 54, p. 3.
25. See K. 12, p. 34.
26. This programme was heard by, amongst others, Kas's old friend Kgaje Malebje who had once witnessed the 'miracle' of the cakes at the wedding in Molote; see P. 31, p. 3.
27. See K.O.D. No. 346, 14 March 1970.
28. See K. 53, p. 34.
29. See FAM. 54, pp. 3–6.
30. Morwesi's return to Ledig in 1970 is noted in M. T. Nkadimeng, Field Report, Ledig, 17–18 September 1991.
31. See K. 16, pp. 23–24; K. 41, p. 2; K. 50, p. 26 and K. 53, pp. 15–16.
32. See K. 16, p. 24 and K. 41, p. 2.
33. See K. 16, p. 23; K. 41, p. 2 and K. 53, p. 15.
34. See K. 41, p. 3.
35. See K. 13, p. 36; also, Data Bank of the South African Weather Bureau,

Pretoria: the trend line for annual rainfall in 1967–77, as reported at the Saulspoort Hospital station, is well above 600 mm, while from 1977 it is well below.

36. See K. 13, p. 39 and K.O.D. Nos. 348–50, 352–54, 18 January, 26 January, 10 May, 19 July, 5 August and 11 October 1971.
37. See FAM. 40, p. 30.
38. See FAM. 40, p. 13. In practice, the primary economic function of the Board was to protect the privileged access of the South African Railways motor service to the local market.
39. The rainfall in the 1972–73 season was in excess of 800 mm, well above the 600 mm average for the Pilanesberg. See also P. 38, p. 4.
40. See K. 13, p. 40; K. 16, pp. 14–16; P. 39, p. 12 and K.O.D. No. 363, 10 April 1973.
41. See P. 39, p. 12; and K.O.D. Nos. 368 and 371, 29 July and 30 December 1973.
42. See K. 13, pp. 40–41.
43. See K.O.D. Nos. 373, 375, 382 and 383, dated respectively 22 February, 30 March, 4 November and 5 November 1974.
44. See K. 13, p. 35.
45. See FAM. 54, p. 9.
46. See FAM. 55, p. 5.
47. See FAM. 54, p. 8.
48. NET PRICE OF WHITE MEALIES PAID TO PRODUCERS BY THE SOUTH AFRICAN MAIZE BOARD, 1965–78
(1964/65 season = 100)

| Season | Index |
|---|---|
| 1964–65 | 100 |
| 1965–66 | 105 |
| 1966–67 | 115 |
| 1967–68 | 93 |
| 1968–69 | 99 |
| 1969–70 | 109 |
| 1970–71 | 97 |
| 1971–72 | 106 |
| 1972–73 | 93 |
| 1973–74 | 132 |
| 1974–75 | 110 |
| 1975–76 | 112 |
| 1976–77 | 116 |
| 1977–78 | 113 |

49. See FAM. 55, p. 7 and, more generally, P. 34, pp. 2–6.
50. See FAM. 55, pp. 4–5 and, more generally, D. Posel, *The Making of Apartheid 1948–1961: Conflict and Compromise* (Oxford 1991).
51. See FAM. 55, pp. 4–5.
52. See K. 13, p. 35.
53. The sense of alienation and drift is well conveyed by Scotch and Jan Mngomezulu—and by Jonas Maponyane, who was at one time a mem-

ber of the kgotla; see, for example, P. 36, pp. 2–3 and 8–9; P. 39, pp. 7–10 and P. 35, pp. 5–10.

54. See P. 35, pp. 2–7; P. 36, p. 2 and K. 13, pp. 18–25. The stabbing at Mohlakeng was witnessed by the late Solomon Ratheo's sister, Mrs. S. Ntsoelengoe, and recounted by her to S. Lebelo in a telephonic discussion, 7 November 1991.
55. For the perception of Catherine Monnakgotla as a creature of the white administration, see K. 13, pp. 17–20.
56. See K. 13, p. 20; P. 36, pp. 2–6 and P. 37, p. 16. For a different interpretation, however, see P. 35, p. 10.
57. See K. 16, pp. 14–16.
58. See P. 39, p. 12.
59. See K. 7, p. 48.
60. See P. 39, p. 15.
61. See K. 4, p. 4; K. 7, p. 49; P. 35, p. 10 and P. 37, pp. 5–8.
62. See P. 35, p. 10.
63. For a selection of such cuttings—as well as a few tracts on religion found amongst Kas's possessions after his death—see K.O.D. Nos. 485, 486, 490, 492, 493, 495, 501, 502 and 516.
64. See FAM. 43, pp. 6–8 and K. 4, pp. 2–5.
65. See K.O.D. Nos. 394 and 396, 4 April and 25 May 1978.
66. See K. 1, p. 39; K. 43, p. 6; FAM. 1, p. 10; and K.O.D. No. 399, 24 July 1978.
67. See K. 24, p. 7.
68. It seems likely that Kas, like many elderly folk, suffered from high blood pressure and that blood from the back of his nose was entering his mouth, thereby giving the appearance of bleeding from the 'tongue.' I am indebted to Bruce Sparks of the University of the Witwatersrand Medical School for the suggested explanation.
69. See K. 24, p. 8.
70. See K. 24, p. 6.
71. See K. 24, pp. 7–9.
72. See K. 43, p. 7.
73. See K. 13, p. 37.
74. K. 15, p. 51.
75. See FAM. 16, pp. 59–63.
76. See K. 15, p. 51.
77. K. 15, pp. 51–52.

## 16. Survival

1. See P. 35, p. 9 and P. 36, p. 7.
2. On 'Operation Genesis' see M. R. Bret, *Pilanesberg—Jewel of Bophuthatswana* (Johannesburg 1989), pp. 109–115. See also K. 7, p. 49; P. 37, p. 5; P. 38, p. 5 and P. 39, p. 11.
3. See P. 37, pp. 5–7.
4. See P. 35, p. 9 and K. 13, p. 36.

5. FAM. 1, pp. 10–13.
6. For his perception, see K. 15, p. 54—for hers, FAM. 1, p. 13.
7. See K. 13, p. 36.
8. See K.O.D. No. 401, 20 November 1979, and M. T. Nkadimeng, Field Report, Ledig, 21 November 1979.
9. See K. 4, p. 5 and K. 16, p. 29.
10. See K. 13, p. 37.
11. See K.O.D. No. 405, 6 January 1985, and M. T. Nkadimeng, Field Report, Ledig, 19 February 1980.
12. See K.O.D. No. 406, receipt for a cow bought for 'seventy pounds' from Stephen Sechele, 20 January 1980. For Kas's love of horses, which he traced back to the time he spent under A. V. Lindbergh, see K. 26, p. 8.
13. See K. 13, p. 36.
14. See K.O.D. No. 402 (n.d.). On forced removals from Tsetse see, for example, S. Lebelo, interview with Mrs. L. Monnye, Ramatlabama, 23 August 1990 (Tape No. 310, M. M. Molepo Oral History Collection, Institute for Advanced Social Research [I.A.S.R.], University of the Witwatersrand [U.W.]).
15. See K. 43, pp. 3–4; K. 45, p. 13; and C. van Onselen, Field Report, Ledig, 13 March 1985.
16. See M. M. Molepo, Field Report, Rustenburg/Groot Marico, 18 February 1982.
17. See M. T. Nkadimeng, Field Report, Ledig, 23–28 July 1980.
18. Ibid.
19. See M. M. Molepo and M. T. Nkadimeng, Field Reports, Ledig, 16 September 1980 and 1 October 1980.
20. See K.O.D. No. 408, 25 November 1980.
21. See K.O.D. No. 409, 25 November 1980 and K.O.D. No. 410(a), 5 December 1980.
22. See K. 9, pp. 18–19.
23. See K. 1, pp. 1–24; K. 9, pp. 19–20; M. M. Molepo, Field Report, Ledig/Slurry, 10 December 1980 and Field Report, Ledig, 3 February 1981.
24. See K. 9, p. 18 and K. 13, p. 41.
25. See K.O.D. No. 412, 29 May 1985.
26. See K. 15, p. 54.
27. See K. 15, pp. 54–56 and K.O.D. No. 415, 'Abridged Death Certificate,' 7 June 1981, on which the cause of Leetwane's death is listed as 'cerebrovascular.'
28. K. 15, p. 55.
29. See K. 15, pp. 56–57; also B. Bozzoli, *Women of Phokeng* (Johannesburg 1991), p. 28.
30. See K. 16, p. 29; K. 21, p. 12 and K. 42, pp. 3–4. On the size of the 1980–81 harvest, see, amongst others, K. 11, pp. 15 and 31; K. 50, p. 29; M. M. Molepo, Field Report, Ledig, 4 August 1981 and C. van Onselen, Field Report, Ledig, 2 November 1981.
31. See K. 9, p. 24; K. 15, pp. 56–57 and notes on a discussion that took

place between M. T. Nkadimeng and Nthakwana at Ledig, 6–8 March 1991.

32. See C. van Onselen, Field Report, Ledig, 2 November 1981.
33. See K.O.D. No. 416, 17 September 1981; K.O.D. No. 418, 30 October 1981, and K.O.D. No. 419, Invoice No. 542639, 27 November 1981. See C. van Onselen, Field Report, Ledig, 2 November 1981 and M. M. Molepo, Field Report, Ledig, 30 November 1981.
34. See, for example, K.O.D. Nos. 420 and 421, cash sale diesel purchases of R82.70 and R25.50, 14 December and 22 December 1981.
35. See K. 14, pp. 7–8 and M. M. Molepo, Field Report, Ledig, 11 January 1982. Rainfall data for Saulspoort station as supplied by the Weather Bureau, Department of Transport, Pretoria.
36. See K. 25, pp. 1 and 6; and FAM. 6, pp. 24–25.
37. See M. M. Molepo, Field Report, Rustenburg/Pretoria, 1 March 1982.
38. See M. M. Molepo, Field Report, Ledig, 17 May 1982 and C. van Onselen, Field Report, Ledig, 13 March 1985.
39. See M. T. Nkadimeng, Field Report, Ledig, 30 June 1982.
40. See M. T. Nkadimeng, Field Report, Ledig, 30 July 1982 and K. 15, pp. 25–26.
41. See M. T. Nkadimeng, Field Report, Ledig, 30 July 1982 and a discussion between the author and M. T. Nkadimeng on 28 January 1992.
42. See K. 49, p. 21.
43. K. 15, p. 25.
44. See K. 15, p. 25; K. 46, pp. 5–6; M. T. Nkadimeng, Field Report, Ledig, 30 July 1982 and a discussion between the author and M. T. Nkadimeng on 29 January 1992.
45. K. 15, p. 25.
46. See K. 14, pp. 20–22.
47. See K. 46, pp. 5–6 and FAM. 16, p. 64.
48. FAM. 15, p. 64.
49. See K. 53, pp. 13–14 and K. 55, pp. 12–17.
50. See K. 20, p. 5.
51. See FAM. 6, pp. 24–25.
52. See 'Obituary,' William Radikomo Maine, K.O.D. No. 428.
53. K. 49, p. 2.
54. This last-minute scramble is recorded in M. T. Nkadimeng, Field Report, Ledig/Gannalaagte, 26 October 1982; the permit is K.O.D. No. 430, 22 October 1982. On Nkamang's part-time employment see M. T. Nkadimeng, Field Report, Ledig/Gannalaagte/Lichtenburg, 3 September 1982.
55. See M. T. Nkadimeng, Field Report, Ledig/Gannalaagte, 26 October 1982.
56. See K. 46, pp. 5–6 and M. T. Nkadimeng, Field Report, Ledig/Gannalaagte, 26 October 1982.
57. See K.O.D. No. 431, 9 November 1982.
58. See M. T. Nkadimeng, Field Reports, Ledig, 12 January and 28 February 1983.

59. See M. T. Nkadimeng, Field Report, Ledig, 22 April 1983.
60. See K. 46, pp. 5–6.
61. See M. T. Nkadimeng, Field Report, Ledig, 22 April 1983.
62. See K. 50, pp. 27–30.
63. See M. T. Nkadimeng, Field Report, Ledig, 28–31 May 1983; K. 14, pp. 10 and 17 and K. 41, p. 17.
64. See M. T. Nkadimeng, Field Report, Ledig, 28–31 May 1983.
65. See K.O.D. Nos. 432–33, 5 August 1983.
66. See K.O.D. No. 435, 23 September 1983.
67. The women's complaints about lack of education are in K. 18, p. 31.
68. See H. Mohlabane, M. T. Nkadimeng and C. van Onselen, Field Reports, Ledig, 20 March, 22 April, and 2 August 1984.
69. See K. 50, p. 29.
70. See C. van Onselen, Field Report, Ledig, 7 June 1984.
71. See FAM. 8, p. 21 and H. Mohlabane, Field Report, Ledig, 2 March 1984.
72. See H. Mohlabane, Field Report, Ledig, 16 June 1984.
73. Ibid.
74. See M. T. Nkadimeng, Field Report, Ledig, 11 July 1984 and K. 42, pp. 16–17.
75. See M. T. Nkadimeng, Field Reports, Ledig, 11 and 15 July 1984.
76. K. 42, pp. 7–8.
77. See K. 42, p. 8 and C. van Onselen, Field Report, Ledig, 2 August 1984.
78. See K. 42, pp. 8–9 and 16–17.
79. See M. T. Nkadimeng, Field Report, Ledig, 7 November 1984.
80. See K. 50, pp. 29–30; K.O.D. No. 440; C. van Onselen, Field Report, Ledig, 8 February 1985; M. T. Nkadimeng, Field Report, Ledig, 3 March 1985, and C. van Onselen, Field Report, Ledig, 13 March 1985.
81. See M. T. Nkadimeng, Field Report, Ledig, 10 July 1985.
82. Ibid.
83. K. 42, p. 12.
84. See K. 48, p. 30.
85. K. 61, p. 3.
86. See K.O.D. No. 443, 23 July 1985.
87. See M. T. Nkadimeng, Field Report, Ledig, 14 September 1985.
88. Ibid.
89. See E. Msimango and M. T. Nkadimeng, Field Report, Ledig, 5 October 1985.
90. See K.O.D. No. 443, Medical Certificate of the Cause of Death of Phillip Maine, 25 September 1985. Cause of death ascribed to 'old age, enlarged prostate.'
91. See E. Msimango and M. T. Nkadimeng, Field Report, Ledig, 5 October 1985.

# INDEX